HIGH-MASS X-RAY BINARIES: ILLUMINATING THE PASSAGE FROM MASSIVE BINARIES TO MERGING COMPACT OBJECTS

IAU SYMPOSIUM 346

COVER ILLUSTRATION: HIGH-MASS X-RAY BINARY AND A MASSIVE STAR CLUSTER

This composite image shows a sky region in the Small Magellanic Cloud galaxy. On the right-hand side is a high-mass X-ray binary. This binary, known as SXP 1062, consists of a massive Be-type star and a neutron star. The diffuse X-rays and optical shell around SXP 1062 both evidence for a supernova remnant. On the left-hand side of the image is the spectacular HII region, LHA 115-N 90, that surrounds a group of young massive stars. X-rays from Chandra and XMM-Newton telescopes have been colored blue and optical data from the Cerro Tololo Inter-American Observatory in Chile are colored red and green. The image is taken from the Chandra X-ray observatory photo album: chandra.harvard.edu/photo/2011/sxp1062/ where further details may be found.

Source: This image illustrates the discussions on deep links between massive stars, X-ray binaries, and compact objects which we had in Vienna, Austria, 27-31 August 2018.

IAU SYMPOSIUM PROCEEDINGS SERIES

Chief Editor
PIERO BENVENUTI, IAU General Secretary
IAU-UAI Secretariat
98-bis Blvd Arago
F-75014 Paris
France
iau-general.secretary@iap.fr

Editor
MARIA TERESA LAGO, IAU Assistant General Secretary
Universidade do Porto
Centro de Astrofísica
Rua das Estrelas
4150-762 Porto
Portugal
mtlago@astro.up.pt

INTERNATIONAL ASTRONOMICAL UNION

UNION ASTRONOMIQUE INTERNATIONALE

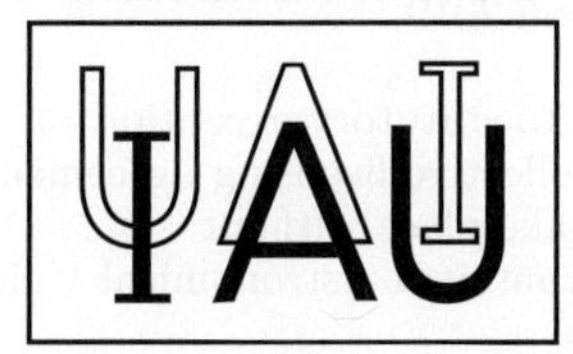

HIGH-MASS X-RAY BINARIES: ILLUMINATING THE PASSAGE FROM MASSIVE BINARIES TO MERGING COMPACT OBJECTS

PROCEEDINGS OF THE 346th SYMPOSIUM OF THE INTERNATIONAL ASTRONOMICAL UNION HELD IN VIENNA, AUSTRIA 27–31 AUGUST, 2018

Edited by

LIDIA M. OSKINOVA
University of Potsdam, Germany

ENRICO BOZZO
University of Geneva, Switzerland

TOMASZ BULIK
Warsaw University, Poland

and

DOUGLAS R. GIES
Georgia State University, USA

CAMBRIDGE UNIVERSITY PRESS
University Printing House, Cambridge CB2 8BS, United Kingdom
1 Liberty Plaza, Floor 20, New York, NY 10006, USA
10 Stamford Road, Oakleigh, Melbourne 3166, Australia

First published 2019

Printed in the UK by Bell & Bain, Glasgow, UK

Typeset in System LaTeX 2ε

A catalogue record for this book is available from the British Library Library of Congress Cataloguing in Publication data

ISBN 9781108471589 hardback
ISSN 1743-9213

Table of Contents

X-ray binaries with Be-type donors

Population in Galaxies and X-ray Luminosity Function

High Energy and Early Universe

HMXB and LMXB evolution and their links with gravitational wave sources

Summary

Preface

The idea of an IAU Symposium devoted to high-mass X-ray binaries (HMXBs) was born following the realization that a comprehensive understanding of these objects and their cosmic roles could be achieved only when diverse communities studying HMXBs come together. Consisting of a young massive donor star and an accreting degenerate object (a neutron star or a black hole), these objects are unique astrophysical laboratories for studies of stellar evolution, donor star winds and disks, and the compact objects they feed. Importantly, HMXBs are among our cosmic neighbors, and they provide us with deep and detailed insights into the physics of accretion and matter under extreme conditions. HMXBs are also present in the farther realms, and are, undoubtedly, important sources of stellar feedback across cosmic time. Their role in the early Universe was highly significant. The populations of HMXBs trace star formation, uncover their parental star cluster evolution, as well as ionize the interstellar medium of their host galaxies by hard radiation. Following the detection of gravitational wave events produced by merging stellar-mass black holes and neutron stars, HMXBs were put in the focus of current astrophysical research as a key transitional stage between young massive stars and degenerate binaries. Even in very dense star clusters, where degenerate binaries may form dynamically, a significant fraction of the compact object population must have passed through a HMXB stage. At present, there is a burst of theoretical work on the evolution of massive binary systems towards double degenerate mergers. Besides the standard population synthesis models that have predicted gravitational wave detection rates, new models and scenarios are being actively investigated. These models rely on our fundamental understanding of massive star physics. Hence, a scientific organizing committee (SOC) came together to organize an IAU Symposium aimed at developing a synergistic approach, comprising the physics of stars, compact objects, and their interactions.

HMXBs were discovered more than 50 years ago, and during this time an incredibly rich trove of facts, models, and theories has accumulated, calling for a broad interdisciplinary meeting to exchange ideas and share insights. The XXXth IAU General Assembly held in Vienna was uniquely suited for the broad scope meeting the SOC envisioned. Our goal was to build scientific bridges between the well advanced field of massive binary astrophysics and the newly emerging field of gravitational wave astronomy.

The IAUS 346 brought together, perhaps for the first time in the framework of an IAU Symposium, the communities focusing on the studies of compact objects and accretion processes, on the donor stars and binary evolution, as well as on gravitational wave event progenitors. The participants discussed different types of HMXBs – Be/X-Ray binaries (BeXRBs), supergiant HMXBs, Wolf-Rayet XRBs, microquasars, γ-ray binaries, and ultraluminous X-ray sources (ULXs) – trying to place them in one broad astrophysical picture. The SOC encouraged the participants to address and to seek the answers to astrophysical questions of general importance, such as: How do massive stars and binaries evolve, lose mass, and collapse? What are the answers to the enigmas of BeXRBs and Supergiant Fast X-ray Transients (SFXTs)? What is the physics of ULXs and their winds? Which multiwavelength transient phenomena are associated with HMXBs? Why are there different populations of HMXBs in different galaxies? What was the role of HMXBs in the early Universe? How are double degenerate binaries formed? Are properties of neutron stars and black holes in HMXBs similar to those observed in merger events by gravitational wave observatories?

Following endorsement by the IAU Commissions and the IAU Division G "Stars and Stellar Physics" and subsequent approval by the IAU Executive Committee, the IAUS 346 was included in the program of the XXXth IAU General Assembly. Thanks to the overwhelmingly positive response, a broad and diverse scientific program was formed. Fifteen speakers were invited to review their research fields, including three IAU GA plenary talks on massive stars, HMXBs, and gravitational wave astronomy. The science topics were addressed and clarified in 35 contributed talks, more than 200 posters, and through intense discussions. Altogether more than 400 participants from all around the world expressed their interest and attended the sessions of the IAUS 346.

The IAUS 346 adopted a comprehensive and unified approach to the problem of massive star lives and deaths, theory and observations of HMXBs, and gravitational wave astronomy. This led to many mutually enriching discussions, to building new contacts, and to developing advanced research strategies. For the younger generation of scientists, the Symposium's novel approach allowed them to see a broad astrophysical picture especially in the setting of the General Assembly of the IAU.

This scientific diversity is reflected in this volume. Each manuscript was reviewed prior to the acceptance for the publication. We are indebted to the invited speakers who provided reviews published in this volume that promote the exchange of knowledge among different fields. We are indebted as well as to the contributing authors, whose manuscripts reflect the current thinking and work in all areas related to the subject of the IAUS 346.

We warmly thank the organizing committee of the XXXth General Assembly of the IAU for making the symposium devoted to HMXBs possible. The advent of future facilities for astrophysical research from both space and ground, as well as the broadened on-going theoretical efforts across the different fields involving HMXBs, their progenitors and their descendants, will certainly lead to new discoveries and many unexpected challenges. We hope that this volume of proceedings, realized thanks to the significant effort of many esteemed colleagues coming from different research areas and backgrounds, will remain as a reference for future steps forward in our understanding of HMXBs.

Lidia M. Oskinova, Enrico Bozzo, Tomasz Bulik, Douglas R. Gies

Editors

Lidia M. Oskinova
University of Potsdam, Germany

Enrico Bozzo
University of Geneva, Switzerland

Tomasz Bulik
Warsaw University, Poland

Douglas R. Gies
Georgia State University, USA

Organizing Committees

Scientific Organizing Committee

SOC Co-Chairs

Lidia Oskinova	University of Potsdam, Germany
Enrico Bozzo	University of Geneva, Switzerland
Doug Gies	Georgia State University, USA
Daniel Holz	University of Chicago, USA

SOC Members

John Blondin	North Carolina State University, USA
Tomek Bulik	Warsaw University, Poland
Malcolm Coe	University of Southampton, UK
Gloria Koenigsberger	Universidad Nacional Autonoma de Mexico, Mexico
Xiang-Dong Li	Nanjing University, China
Atsuo Okazaki	Hokkai-Gakuen University, Japan
Biswajit Paul	Raman Research Institute, India
Konstantin Postnov	Moscow State University, Russia
Pablo Reig	Universtiy of Crete, Greece
Lara Sidoli	INAF-IASF Milano, Italy
John Tomsick	University of California, Berkeley, USA
Jose Miguel Torrejon	EPS Universidad de Alicante, Spain

Please note: As IAUS 346 was part of the IAU General Assembly there is no LOC.

CONFERENCE PHOTOGRAPH

Participants

Aftab, Nafisa
Allen, Christine
Anastasopoulou, Konstantina
Andrews, Jeff
Antoniou, Vallia
Arefiev, Vadim
Artale, Maria Celeste
Atamurotov, Farruh
Barlett, Elizabeth
Batta, Aldo
Belczynski, Krzysztof
Bisnovatyi-Kogan, Gennagy
Bogomazov, Alexey
Borvák, Levente
Bozzo, Enrico
Brown, Rory
Bulik, Tomasz
Carpano, Stefania
Cemeljic, Miljenko
Chaty, Sylvain
Chen, Li
Chen, Wang
Chruslinska, Martyna
Coe, Malcolm
Corbet, Robin
D'Aì, Antonino
de Freitas, Daniel Brito
Ding, Guoqiang
Dogan, Suzan
Driessen, Florian
Dwarkadas, Vikram
Echevarria, Juan
Egron, Elise
El Mellah, Ileyk
Erkut, M. Hakan
Fabrika, Sergei
Fornasini, Francesca
Fortin, Francis
Fragos, Tassos
Fuerst, Felix
García, Federico
Garofali, Kristen
Giacobbo, Nicola
Gies, Douglas
Goranskij, Vitaly
Grant, David
Grinberg, Victoria
Gvaramadze, Vasilii
Haberl, Frank
Hainich, Rainer
Hakkila, Jon
Harrison, Fiona
Heger, Alexander
Heida, Marianne
Hell, Natalie
Higgins, Erin
Hubrig, Swetlana
Icli, Tugce
Iliev, Lubomir
Inoue, Hajime
Ishii, Ayako
Islam, Nazma
Jaisawal, Gaurava K
Kadowaki, Luis H.S.
Kaper, Lex
Karitskaya, Eugenia
Kashi, Amit
Kavila, Indulekha
Kawai, Nobuyuki
Keszthelyi, Zsolt
Kim, Soo Hyun
Klencki, Jakub
Kobayashi, Chiaki
Kocak, Dolunay
Kouroumpatzakis, Konstantinos
Kovlakas, Konstantinos
Kritcka, Jrji
Kuiper, Rolf
Kurfürst, Petr
Kuznetsova, Ekaterina
Lages, José
Leahy, Denis
Lehmer, Bret
Lei, Weihua
Liu, Zhengwei
Li, Yan
Madura, Thomas
Maitra, Chandreyee
Malanchev, Konstantin
Mandel, Ilia
Mapelli, Michela
Maravelias, Grigoris
Marchant, Pablo
Martinez-Nunez, Silvia
Matsumoto, Ryoji
McBride, Vanessa
McSwain, M. Virginia
Meszaros, Attila
Miller, Nathan
Mirabel, Felix
Mönkkönen, Juhani
Montgomery, Michele
Nakajima, Motoki
Negoro, Hitoshi
Negueruela, Ignacio
Nichols, Joy
Nikolaeva, Svetlana
Okazaki, Atsuo
Ok, Samet
Oskinova, Lidia
Porter, Augustus
Postnov, Konstantin
Pradhan, Pragati
Qin, Ying

Qu, Jinlu
Ribó, Marc
Ricker, Paul
Rivinius, Thomas
Roberts, Tim
Rodriguez-Ramirez, Juan Carlos
Rojas Montes, Eliceth Y.
Rouco Escorial, Alicia
Samec, Ronald
Sander, Andreas
Schulz, Norbert
Sell, Paul
Shakura, Nikolay
Shao, Yong
Shenar, Tomer
Sholukhova, Olga
Shtykovsky, Andrey
Sidoli, Lara
Skulskyy, Mykhaylo
Soria, Roberto
Spurzem, Rainer
Sugawara, Yasuharu
Tagawa, Hiromichi
Takahashi, Koh
Torrejon, Jose Miguel
Townsend, Lee
Tsygankov, Sergey
van den Heuvel, Edward
van Jaarsveld, Johanna
Vink, Jorick
Vinokurov, Aleksandr
Wang, Bo
Wang, Ding-Xiong
Wei, Erhu
Weisberg, Idel
Wilms, Joern
Wilson-Hodge, Colleen
Wolter, Anna
Xiao, Lin
Xu, Xiao-Tian
Yakut, Kadri
Yang, Jun
Yan, Jingzhi
Yankova, Krasimira
Zakharov, Alexander
Zezas, Andreas
Zhang, Chengmin
Zhang, Shu
Zhang, Yue
Zuo, Zhao-Yu

High-mass X-ray Binaries: Illuminating the Passage from Massive Binaries to Merging Compact Objects
Proceedings IAU Symposium No. 346, 2019
L. M. Oskinova, E. Bozzo, T. Bulik & D. Gies, eds.

doi:10.1017/S1743921319001315

High-Mass X-ray Binaries: progenitors of double compact objects

Edward P. J. van den Heuvel

Anton Pannekoek Institute of Astronomy, University of Amsterdam,
Postbus 92429, NL-1090GE, Amsterdam, the Netherlands
email: E.P.J.vandenHeuvel@uva.nl

Abstract. A summary is given of the present state of our knowledge of High-Mass X-ray Binaries (HMXBs), their formation and expected future evolution. Among the HMXB-systems that contain neutron stars, only those that have orbital periods upwards of one year will survive the Common-Envelope (CE) evolution that follows the HMXB phase. These systems may produce close double neutron stars with eccentric orbits. The HMXBs that contain black holes do not necessarily evolve into a CE phase. Systems with relatively short orbital periods will evolve by stable Roche-lobe overflow to short-period Wolf-Rayet (WR) X-ray binaries containing a black hole. Two other ways for the formation of WR X-ray binaries with black holes are identified: CE-evolution of wide HMXBs and homogeneous evolution of very close systems. In all three cases, the final product of the WR X-ray binary will be a double black hole or a black hole neutron star binary.

Keywords. Common Envelope Evolution, neutron star, black hole, double neutron star, double black hole, Wolf-Rayet X-ray Binary, formation, evolution

---

1. Introduction

My emphasis in this review is on evolution: on what we think to know about how High Mass X-ray Binaries (HMXBs) were formed and how they may evolve further to form binaries consisting of two compact objects: double neutron stars, double black holes and neutron star-black hole binaries. After a brief description in section 2 of the different types of HMXBs, I summarize in section 3 what I consider to be the most important new developments in the field of HMXBs of the past one-and-a-half decades. In section 4, I describe the past evolution of HMXBs and give a rough outline of the expected future evolution of HMXBs that contain neutron stars, as we presently think we understand them.

In the later evolution of the HMXBs that contain neutron stars, Common-Envelope Evolution (CEE) is expected to play a crucial role. It is expected that only wide neutron-star HMXBs, with orbital periods upwards of about one year, will survive as binaries, with very short orbital periods and consisting of a helium star and a neutron star. Such systems later evolve into close eccentric-orbit double neutron stars, of which presently some twenty are known.

Neutron-star HMXBs with orbital periods shorter than about one year will merge into a single object, possibly resembling a Thorne-Zytkow star. In section 5 the predictions of the "standard model" for the formation and further evolution of neutron-star HMXBs are compared with the observations.

Section 6 summarizes our knowledge of the black-hole X-ray binaries and Wolf-Rayet X-ray Binaries (WRXBs), and focuses on the different ways in which BH-HMXBs may evolve into WRXBs, and on possible other channels for the formation of WRXBs. In HMXBs with short orbital periods that contain a black hole, stable Roche-lobe overflow is possible, such that CEE can be avoided, and the systems may survive as close binaries consisting of a helium star (WR star) and a black hole. It is argued that the compact stars in practically all WRXBs must be black holes, making these systems ideal progenitor systems of double black holes and Black hole-Neutron star (BH-NS) binaries.

2. The different types of High-Mass X-ray Binaries

There are three main types of HMXBs, with the following characteristics:

(i) The first type, discovered by Schreier *et al.* (1972) and Webster & Murdin (1972), is that of the supergiant HMXBs. In these systems the donor star is an O- or early B-type supergiant star that is close to filling its Roche lobe. The orbital periods of these systems, many of which are eclipsing, are mostly shorter than 15 days. These systems are persistent (permanent) X-ray sources, mostly powered by the capture of matter from the strong stellar wind of the supergiant companion. In a few cases, such as the 2.1 d orbit eclipsing and regularly pulsating X-ray source Centaurus X-3 (Schreier *et al.* 1972), the X-ray source is powered by beginning Roche-lobe overflow. The supergiant HMXBs are relatively rare, their total known number in the Galaxy being about 30, and, in practically all of them, the compact star is a neutron star: a X-ray pulsar. The blue supergiants have masses typically in the range 20 to $50 M_{\odot}$.

(ii) The second type of systems, practically all containing neutron stars, is that of the B-emission X-ray Binaries (short: BeXBs), discovered in 1975 with the Ariel V satellite, and first recognized and explained as a separate class by Maraschi *et al.* (1976). Most of these systems are recurrent transients, which can be quiet for many decades and then suddenly flare up as a strong pulsating X-ray source for weeks to months. The companion stars here are rapidly rotating B-stars that are in or very close to the main sequence and are deep inside their Roche lobes; they have a variable emission-line spectrum of hydrogen. These lines are formed in a rapidly rotating disk of gas that surrounds the star in its equatorial plane. The emission lines may be absent for years, then return, due to ejection of gas from the equatorial regions of the star (see Rivinius, this volume). If the Be-star has a compact companion, the motion of the latter through this ejected equatorial disk of gas will cause it to accrete matter and temporarily become a strong X-ray source. The BeXBs tend to have relatively long orbital periods, ranging from about 15 days to over 4 years. The Be stars in these systems typically have masses in the range 8 to $20 M_{\odot}$.

The BeXBs form the largest group of HMXBs, with a presently recognized number of around 220. Particularly the SMC is very rich in these systems, with a total number at least 120 (Haberl, this volume). In about half of the BeXBs regular X-ray pulsations have been observed, and the other half have similar X-ray spectra, suggesting they also harbor neutron stars. Only one Black hole BeXB is known (Casares *et al.* 2014, see also Ribo, this volume).

(iii) The third class of HMXBs as that of the Wolf-Rayet X-ray Binaries (WRXBs), of which presently only seven are known. Except for one, Cyg X-3, they are all located in external galaxies (Esposito *et al.* 2015, see also Carpano, this volume, and Soria, this volume). With the exception of the system M101 ULX-1, they all have very short orbital periods, of around one day or less (see Table 2). Wolf-Rayet (WR) stars are helium stars, and the very strong emission lines of He, C and N which are characteristic for these stars, are produced in an extremely strong radiation-driven stellar wind, with mass-loss rates around $10^{-5}\, M_{\odot}\,\mathrm{yr}^{-1}$, and velocities $2000 - 5000\,\mathrm{km\,s}^{-1}$ (e.g. Hamann *et al.* 2006; Crowther 2007; Conti, Crowther and Leitherer 2008). As the measured masses of WR-star in binaries are at least $8 M_{\odot}$ (Crowther 2007; Shenar *et al.* 2016), the large luminosity and strong radiation pressure required for driving WR winds apparently develop only in helium stars with masses above this lower mass limit. To produce a helium core larger than $8M_\odot$, the progenitor star of the WR star must have had a mass of at least $30M_\odot$. WRXBs must therefore have had HMXB progenitors with donor masses above $30M_\odot$. As we will show in section 6, based on arguments from binary evolution, the compact stars in WRXBs most likely are black holes (van den Heuvel *et al.* 2017).

3. Important developments in the HMXB field in the past one-and-a-half decades

I list here what in my personal view were the most important developments in the HMXB field in the past 15 years:

1. The discovery with the INTEGRAL satellite of two new classes of supergiant HMXBs, which increased the known galactic number of supergiant systems by a factor of four (see Sidoli, this volume): (a) the highly obscured supergiant systems, and (b) the Supergiant Fast X-ray Transients (SFXTs). These discoveries are an illustration of the fact that astronomy is a science that is heavily affected by observational selection effects. Earlier X-ray survey missions were sensitive for relatively soft X-rays, with energies below about 10 keV, where X-ray absorption by neutral hydrogen plays an important role. Objects that are highly obscured due to a high hydrogen column density towards the source, were missed by these missions. The IBIS/ISGRI soft gamma-ray telescope of INTEGRAL is in fact a hard X-ray telescope, working in a spectral region above 15 keV, where X-ray absorption by neutral hydrogen is much less important. IBIS discovered many new supergiant HMXBs with hydrogen column density in or around the system larger than 10^{23} H-atoms cm^{-2} such that X-rays below 10 keV are very heavily absorbed. A key example is the first such source discovered IGR J16318-4848 with $N_H = 2 \times 10^{24}\,cm^{-2}$ (Walter *et al.* 2003). Furthermore, with the large field of view ($30 \times 30\,degree^2$) of IBIS and the long stretches of observing times required for the gamma-ray spectroscopy with the SPI telescope, which looks in the same direction as the IBIS telescope, IBIS also turned out to be an excellent instrument to detect short-duration hard X-ray flares when these happen. These flares, lasting only a fraction of a day, are the special property of the Supergiant Fast X-ray Transients. Earlier instruments never stared for sufficiently long time intervals at the same field on the sky, and therefore missed these short-lasting very intense flares, which turned out to arise from this new class of supergiant HMXBs. In the Corbet diagram of High-Mass X-ray Binaries, in which the pulse periods of the supergiant systems and BeXBs are plotted against their orbital periods (Sidoli 2012), one finds that the new supergiant systems occupy the same region of the diagram as the earlier-discovered supergiant systems, the only difference being that now a few systems with quite long orbital periods have been added.

2. The discovery of extragalactic Wolf-Rayet X-ray binaries (see section 6 and Carpano, this volume, and Soria, this volume), and of the first Black Hole BeXB (Casares *et al.* 2014, Ribo, this volume);

3. The discovery of hundreds of extragalactic Ultra-Luminous X-ray sources (ULXs); I refer to the papers in this volume by Harrison, by Heida and by Walter, and to the many poster presentations of this conference;

4. The discovery of a class of BeXBs with nearly circular orbits (Pfahl *et al.* 2002), which demonstrates that there is a class of neutron stars that receive hardly any velocity kick in their birth events. The same class has now also been recognized among the double neutron stars, a considerable fraction of which have very low orbital eccentricities, indicating that the second-born neutron star received hardly any kick at birth (e.g. Tauris *et al.* 2017). An example is the double pulsar PSR J0737-3039, which has e=0.088. These low-kick neutron stars may either have been formed by electron-capture collapse, or by the collapse of ultra-stripped iron cores (Tauris *et al.* 2015). Electron-capture collapse may possibly occur only in binary systems (Podsiadlowski *et al.* 2004; Dessart *et al.*

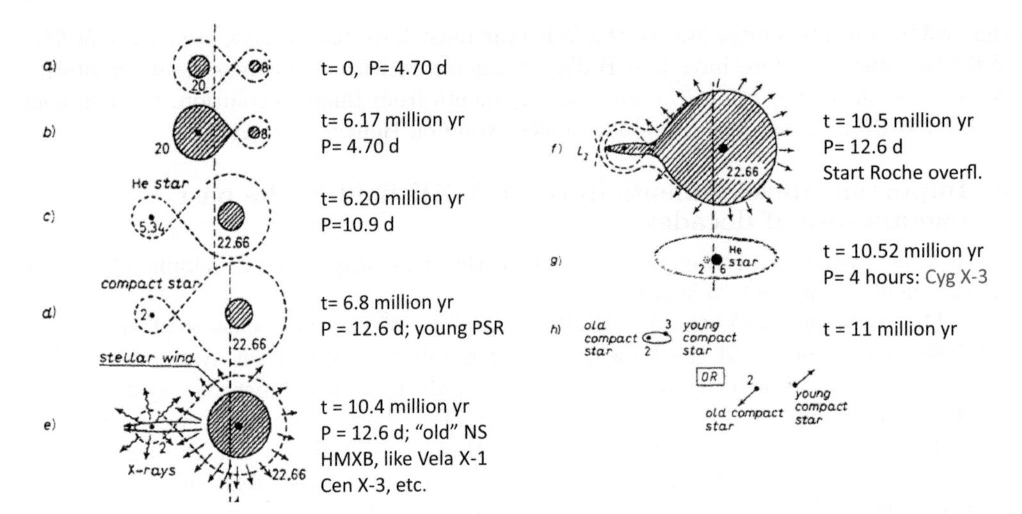

Figure 1. Model of the evolution of a close binary with initial components of 20 and $8M_\odot$ into a supergiant HMXB (left) and its further evolution into a close binary consisting of two compact stars (right) (after van den Heuvel 1976). Numbers near the stars indicate stellar masses. As explained in the text, in case the first-formed compact star is a neutron star, survival of stage f of the evolution requires that the orbital period of the system is much longer than indicated here; see also Fig. 2.

2006; Kitaura *et al.* 2006), and also the formation of ultra-stripped cores requires binary interaction.

5. The development of the "Settling Accretion Theory" for magnetized neutron stars (Shakura *et al.* 2012). This theory gives a consistent explanation of how neutron stars accreting from the stellar wind of a companion can be spun down to very long spin periods. It also can give an explanation for the SFXT outbursts of supergiant HMXBs (Postnov *et al.* 2014, see also Postnov, this volume).

4. Past and future evolution of HMXBs

The basic model for the formation and later evolution of HMXBs, as depicted in Fig. 1 (from van den Heuvel 1976)), was developed in the years 1972-1974 (van den Heuvel & Heise 1972; Tutukov & Yungelson 1973; van den Heuvel & De Loore 1973; Flannery & van den Heuvel 1975; De Loore *et al.* 1975).

The essential ingredient of the model for the formation of HMXBs is the occurrence of extensive mass transfer in the binary prior to the first supernova explosion in the system, such that by the time of this explosion the exploding star has become the less massive component of the binary. The supernova mass ejection (assumed to occur in a spherically symmetric way) will then not disrupt the binary, because if less than half of the total mass of the system is explosively ejected, the binary remains bound (Blaauw 1961). This is simply a consequence of the virial theorem. In the case that the less massive star of the binary explodes, the orbit will become eccentric and the system becomes a runaway star ("slingshot effect"); the orbital changes and runaway velocities for this case were calculated by van den Heuvel (1968). If the compact object formed in the supernova explosion receives a kick-velocity at birth, the system may still be disrupted, even if it is the less massive component that explodes. The first calculations of the effects of birth-kicks on the orbits were made by Flannery & van den Heuvel (1975).

As to the expected further evolution of a HMXB after the massive companion of the compact star begins to overflow its Roche lobe: van den Heuvel & De Loore (1973) had assumed, as depicted in Fig. 1 f-h, that a HMXBs with a donor star with a radiative envelope and a short orbital period could evolve with stable Roche-lobe overflow. In this case the entire H-rich envelope of the donor is lost, carrying off much orbital angular momentum, such that a very narrow system will remain, consisting of a helium star (the helium core of the donor) and the compact star. We suggested that the highly peculiar X-ray binary Cygnus X-3, with a 4.8 h orbital period, is such a system, which was confirmed 19 years later by IR spectroscopy which showed its companion to be a Wolf-Rayet star of type WN5 (van Kerkwijk *et al.* 1992). Such a close system consisting of a helium star and a compact star will, after the supernova explosion of the helium star - if not disrupted - produce a close eccentric binary consisting of two compact stars. When Hulse & Taylor (1975) discovered the first double neutron star, with a very short orbital period (P = 7h 45m) and high orbital eccentricity (e=0.615), it therefore was clear to us that this must be a later evolutionary product of a HMXB, after it went in Roche-lobe overflow and spiraled-in (Flannery & van den Heuvel 1975; De Loore *et al.* 1975).

Although this picture looked straightforward, it was shown somewhat later by Paczynski (1976) that, because of the extreme mass ratio of a system consisting of a blue supergiant and a neutron star, Roche-lobe overflow in such a system is unstable, and once Roche-lobe overflow starts, the mass transfer will run out of hand, and the neutron star will be engulfed by the envelope of the massive star, such that a Common Envelope will form, in which the neutron star and the compact helium core of the massive star spiral-in towards each other, due to the large friction on their orbital motion in the envelope. Numerical computations of the spiral-in process were pioneered by Taam, Bodenheimer & Ostriker (1978), and further developed by Taam and his collaborators over many years (e.g. Taam & Sandquist 2000) and others. For recent developments in the CEE field I refer to the paper by Ricker in this volume. The computations by Taam and collaborators showed that, for a neutron-star HMXB to survive Common Envelope Evolution (CEE), the system must start out with an orbital period longer than about one year (Taam 1996). For these reasons, the system of Fig. 1-f will not survive spiral in: the neutron star will spiral into the core of its companion, leading to a single massive star with a neutron star in its center, a so-called Thorne-Zytkow star (Thorne & Żytkow 1977). Computations of the structure of such stars showed that they are expected to look like red supergiants with peculiar element abundances, particularly of p-process elements (Podsiadlowski *et al.* 1995).

The only HMXBs with orbital periods longer than about one year are the BeXBs. Therefore, the progenitors of the double neutron stars are the long-period BeXBs. As the supergiant neutron-star HMXBs almost all have relatively short orbital periods, they will not survive CEE, and are expected to terminate as single Thorne-Zytkow stars. So far, such stars have never been identified with certainty. Still, as the galactic formation rate of HMXBs is of the order of $2 \times 10^{-4}\,\mathrm{yr}^{-1}$, one would expect Thorne-Zytkow stars to be formed at about the same rate, leading to some 20 to 200 such stars in the Galaxy (Podsiadlowski *et al.* 1995). The absence of evidence of their existence remains a great puzzle. Fig. 2 depicts the generally accepted model for formation of a close double neutron star as a later evolution product of a BeXB (Tauris *et al.* 2017). Presently, some 20 of such binaries are known (15 of them were listed in the above-mentioned paper by Tauris et al.). About half of them have orbital periods such that they will merge by GW losses within a Hubble time.

If the compact star in the evolutionary picture of Fig. 1f-h is not a neutron star, but a black hole with a mass above 20 to 30 per cent of the mass of the blue supergiant companion, the Roche-lobe overflow is stable (van den Heuvel *et al.* 2017; Pavlovski *et al.* 2017), and also systems with a short orbital period will survive as a close system consisting of a WR-star (helium star) and a black hole (van den Heuvel *et al.* 2017).

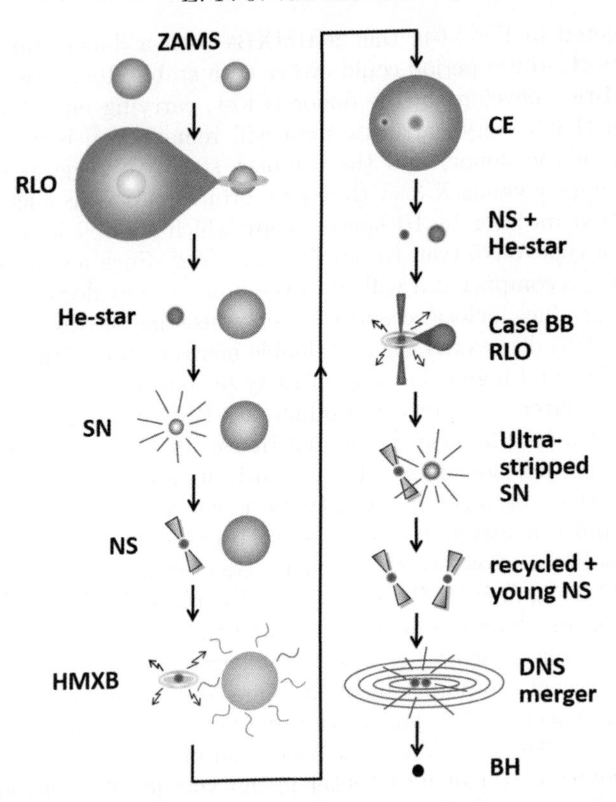

Figure 2. Evolution of a wide neutron-star HMXB, with orbital period longer than about one year, into a close double neutron star. The wide HMXB evolves through a Common Envelope phase into a close helium star plus neutron star binary, which produces a close double neutron star. This system, consisting of an old recycled neutron star and a newborn neutron star, may finally merge into a black hole (from Tauris *et al.* 2017).

So, for black-hole HMXBs the evolutionary picture of Fig 1f-h is valid. I deal with this type of evolution in more detail in section 6.

5. Predictions of the standard model for formation and evolution of HMXBs, compared to the observations

The model depicted in Fig 1 makes the following predictions:

(i) HMXBs must be runaway stars (van den Heuvel & Heise 1972; Tutukov & Yungelson 1973);

(ii) There must be massive stars with a young pulsar companion (van den Heuvel 1974);

(iii) Since many young neutron stars receive velocity kicks of several hundreds of $\mathrm{km\,s^{-1}}$ at birth, there must also be disrupted systems;

(iv) There must be double neutron stars in which one of the stars is a young strong-magnetic-field pulsar (Srinivasan & van den Heuvel 1982).

All of these predictions have in later years been confirmed by the observations. The first prediction was confirmed by the work of Kaper *et al.* (1997) who found the bow shock of the Vela X-1 (4U0900-40) supergiant HMXB, and measured its $45\,\mathrm{km\,s^{-1}}$ excess transverse velocity, indicating that the system originated in the association Vela OB1 some 2 to 3 million years ago. Later also a high excess transverse velocities were found for two other supergiant systems: $76\,\mathrm{km\,s^{-1}}$ for 4U1700-37 (its likely origin is in the association Sco OB1), and 85 km s^{-1} for 4U1538-52 (Kaper 2001). For the BeXBs the runaway velocities are much smaller, as expected because of their lower masses and wider orbits (van den Heuvel *et al.* 2001).

As to the second prediction: the discovery of gamma-ray emission from several OB binaries has shown that indeed there are OB stars that must have young pulsar companions (Dubus *et al.* 2017). Young Crab-like pulsars emit a highly relativistic electron-positron pulsar wind. Both positron annihilation and inverse Compton boosting of optical photons from the OB star by the relativistic electrons lead to the emission of gamma rays. Presently, there are 7 such systems known.

Two of the systems (LS 5039 and LMC P3) are short-period O-type binaries that are ideal progenitors of the supergiant HMXBs. The other ones have wide orbits and are ideal progenitors for BeXBs.

The third prediction was beautifully confirmed by the discovery by Dincel *et al.* (2016) that in the supernova remnant Semeis 147 there is a B0V runaway star with an excess transverse velocity of 74($\pm$7.5) km s^{-1}, plus a high-velocity pulsar, PSR J0538+2817, with a transverse velocity of 357 km s^{-1}. Both these velocities are directed away from the center of the supernova remnant. This leads to kinematic age of 30 000 yr, when both stars originated in the center of the supernova remnant. The high-velocity of the B0V star indicates that the pre-disruption system had a relatively short orbital period (less than 15 d), because only the disruption of a close system can have produced such a high velocity. The short orbital period then means that prior to the supernova there must have been extensive mass transfer in the system, and that at the time of the explosion the exploding star was the less massive star of the system. The fact that the system still was disrupted means that the disruption can only be due to the high kick velocity imparted to the neutron star at its birth. This is direct proof of birth kicks of neutron stars. Also, it is direct proof of the Blaauw (1961) mechanism for producing runaway early-type stars. Finally, also the fourth prediction was confirmed, by the discovery of the double radio pulsar PSR J0737-3039, in which the recycled (old) pulsar A with a pulse period of 22.7 ms and a weak magnetic field, has as companion J0737-3039B, which is a normal "garden variety" young pulsar with a period of 2.773 s and a strong magnetic field ($B \approx 5 \times 10^{11}$ G) at the age around 10^7 yr (Lyne *et al.* 2004). Clearly, this is the second-born neutron star in the system, whose birth event induced the orbital eccentricity of the system. The fact that the orbital eccentricity of this system (and of about half of the double neutron stars) is quite low ($\leqslant 0.20$) indicates that, like in the BeXBs with nearly circular orbits, the second-born neutron stars in these systems received hardly any kick velocity at birth. As mentioned above, this means that these neutron stars resulted either from electron-capture collapse, or from the collapses of ultra-stripped iron cores.

6. From Black-hole HMXBs to WRXBs and double black holes

The Black-hole X-ray binaries

The bulk of the about 60 known BH-XBs (Corral-Santana *et al.* 2016) consists of Low-Mass X-ray Binaries (LMXBs), with typical donor masses $\leqslant 2M_{\odot}$. These are the so-called "Soft X-ray Transients" or "X-ray Novae", which may be dormant for decades, and then go into a bright X-ray outburst. A spectacular example is the outburst of V404 Cygni (GS2023+338) in 2016. Although the black holes in these systems no doubt are remnants of massive stars, these systems are not HMXBs, and therefore I will not discuss them here. We know only five black-hole HMXBs, of which Cygnus X-1 is the best-known example. Table 1 lists these 5 systems. One of them is the recently discovered BeXB MWC 656 (Casares *et al.* 2014, Ribo, this volume). Two of the BH-XBs are in the Large Magellanic Cloud and one of them is in M33. The only system that probably is massive enough to produce a double black hole is the latter one: M33 X-7, a 15.7$M_{\odot}$ black hole

Table 1. The five known Black Hole High-Mass X-ray Binaries

Name	$P_{\rm orb}$ (d)	$M_{\rm don} M_\odot$	$M_{\rm BH} M_\odot$	Ref.
Cyg X-1	5.6	19.2(±1.9)	14.8(±1.0)	(1)
LMC X-1	3.9	31.8(±3.5)	10.9(±1.4)	(2)
LMC X-3	1.7	3.6(±0.6)	7.0(±0.6)	(3)
MCW 656	∼ 60	∼ 13	4.7(±0.9)	(4)
M33 X-7	3.45	70(±7)	15.7(±1.5)	(5)

References: (1) Orosz *et al.* (2011), (2) Orosz *et al.* (2009) (3) Orosz *et al.* (2014), (4) Casares *et al.* (2014), (5) Orosz *et al.* (2007)

with a $70 M_\odot$ companion in a 3.45-day orbit (Orosz *et al.* 2007). Its evolutionary origin has been described by Valsecchi *et al.* (2010). The rareness of BH-HMXBs with respect to the BH-LMXBs is, of course, a selection effect: the donor stars in the LMXB systems are very long-lived ($\geqslant 10^9$ yr), while those in the HMXB systems are very short-lived ($\leqslant 5 \times 10^6$ yr). The chances for observing a BH-LMXB, even though these turn on only occasionally, is therefore much larger than the chances of finding a BH-HMXB. The most massive BH-HMXBs are the ones which we expect to evolve into WRXBs, which we will consider now.

The WR X-ray Binaries

The first-discovered system of this type is Cygnus X-3 (see section 4). This is one of the most spectacular X-ray binaries known, and the only X-ray binary to which once an entire issue of Nature Physical Science was devoted (vol. 239, October 23, 1972). On September 23, 1972 its radio brightness increased by a factor of over 10^3 (Hjellming and Balick 1972), which was the start of several giant radio outbursts, making it temporarily the brightest radio source in the sky. The source is right in the galactic plane, and the three hydrogen 21-cm radio absorption lines at different doppler shifts visible during its radio outbursts showed that the source is behind three spiral arms, yielding a distance of about 10 kpc. The evolution of its radio spectrum during the outbursts was exactly as observed in quasar outbursts, indicating that it was synchrotron emission of an expanding cloud of relativistic electrons with magnetic fields. The adiabatic expansion of this cloud produces the characteristic quasar-like evolution of its radio spectrum (Hjellming and Balick 1972). So, Cyg X-3 in 1972 was already a *micro-quasar* long before this name was introduced by Mirabel *et al.* (1992). Around this time also its 4.8 h X-ray orbital period was discovered by Bert Brinkman (Parsignault *et al.* 1972), and following its radio outbursts it was discovered that it is a strong infra-red (IR) source with the same periodicity (Becklin *et al.* 1972). Because its radio behaviour and strong IR emission are totally different from what is observed for LMXBs, while its position right in the galactic plane strongly suggests that it is a Population I object, and its orbital period fitted exactly with the outcome of our calculations at that time of the later evolution of a HMXB, we suggested Cyg X-3 to be a close helium star plus compact star binary (van den Heuvel & De Loore 1973). This indeed was later proven to be correct (see section 4). During radio outbursts its IR spectrum is that of a WN7 star, during radio quiescence it is that of a WN5 star (Hanson *et al.* 2000). Its IR luminosity of 3×10^{39} erg s^{-1}, requires a helium star with a mass in the range $8 - 12\, M_\odot$ (e.g. Crowther 2007).

For a long time, Cyg X-3 was the only WRXB known, but in the past decade a half dozen such systems have been discovered in external galaxies, as summarized in Table 2. See for details Esposito *et al.* (2015) and the papers of Carpano and of Soria in this volume.

The orbital periods of all but one of the systems are of order one day or shorter, showing that the systems must be the result of drastic spiral in evolution. The masses of the WR stars in IC 10 X-1 and NGC 300 X-1 were derived from the optical brightness

Table 2. The Wolf-Rayet X-ray Binaries. Except for Cygnus X-3 the data are taken from the compilation by Esposito *et al.* (2015), where the original references to the data of the different systems can be found. The mass of the WR star in Cyg X-3 is estimated from its IR luminosity, as mentioned in the text. As explained in the text, the mass estimates of the compact stars in WRXBs are very uncertain, but on the basis of binary evolution, these compact stars are expected to be black holes.

Galaxy	Source	Orbital Period (h)	WR mass$M_\odot$	accretor mass
IC 10	X-1	34.9	35	33(?)
NGC 300	X-1	32.8	26	20(?)
NGC 4490	CXOUJ123030.3+413853	6.4	–	–
NGC 253	CXOUJ004732.0-251722.1	14.5	–	–
Circinus	CG X-1	7.2	–	–
M 101	ULX-1	196.8	19	20(?)
Milky Way	Cyg X-3	4.8	$8-12$	$\geqslant 3$(?)

of these stars. The masses of their compact companions are very uncertain, as discussed by Laycock *et al.* (2015) and Carpano and Soria (this volume). The reason is that these masses were derived from the observed radial velocity curves of the emission lines of the WR stars, which curves are almost 90 degrees out of phase with the radial velocity curves expected from the X-ray light curves of these binaries. Therefore, these radial velocity curves cannot be those of the centers of mass of the WR stars, but probably are due to shock-features in the WR wind.

Formation of WRXBs, reason why their compact stars likely are black holes

There are basically three ways in which a close WRXB can be formed: (i) by CE evolution of a wide BH-HMXB; (ii) by stable Roche-lobe overflow of a BH-HMXB with a relatively short orbital period, and (iii) through homogeneous evolution of a massive binary with a very short orbital period. We consider now each of these mechanisms separately.

(i) *Formation by CE evolution*

This model is similar to that for the NS-HMXBs, leading to the formation of double neutron stars, as depicted in Fig. 2, but scaled up to higher initial stellar masses. This is the model proposed by Bogomasov (2014) and Belczynski *et al.* (2016) for the formation of close double black holes. In this case one must, like in Fig. 2, start from a wide binary system. CE evolution then makes the orbit of the system shrink by a large factor, leading to a system like Cyg X-3.

(ii) *Formation through stable Roche-lobe overflow from a BH-HMXB with a blue supergiant donor star and a relatively normal (short) orbital period.*

When the HMXB consists of a blue supergiant and a black hole, Roche-lobe overflow from the supergiant to the black hole does not need to become unstable, like in the case of a neutron star companion, and CE evolution can be avoided, as was shown by van den Heuvel *et al.* (2017) and Pavlovski *et al.* (2017). The conditions for stable Roche-lobe overflow are: (i) the donor star should have a radiative envelope (King *et al.* 2000), and (ii) the mass ratio q of compact star and donor is q $\geqslant 0.3$ (for references to papers in which this condition was derived see van den Heuvel *et al.* 2017); Pavlovski *et al.* (2017) found that for very massive systems the latter condition can even be relaxed to q $\geqslant 0.2$.

A confirmation of the stability of mass transfer in the case of BH-HMXBs with $q \geqslant 0.3$ is provided by the system of SS 433, a 13-day binary consisting of an A-giant donor star and a compact star which is surrounded by a huge and very luminous accretion disk, which completely dominates the light of the system (e.g. King *et al.* 2000). In this system Roche-lobe overflow is going on, in which the bulk of the transferred matter is ejected by the compact star in the form of the famous relativistic jets (with $v = 0.265c$ and a mass-loss rate of $10^{-6}M_{\odot}\mathrm{yr}^{-1}$), and in the form of a very strong wind ($10^{-4}M_{\odot}\,\mathrm{yr}^{-1}$) from the huge accretion disk of the compact star (Begelman *et al.* 2006). The mass transfer in this system is stable, as it has been going on already for thousands of years, without the system going into CE evolution (King *et al.* 2000). This is demonstrated by the presence of the large W50 nebula which surrounds the system and the shape of which has been strongly influenced by the precessing relativistic beams together with the disk wind (e.g Begelman *et al.* 2006)). The mass ratio of the compact star and the donor in this system is indeed $\geqslant 0.3$, as the estimated component masses are $M_{\mathrm{d}} = 12.1(\pm 3.3)M_{\odot}$ and $M_{\mathrm{c}} = 4.3(\pm 0.8)M_{\odot}$ (Hillwig and Gies 2008). The latter values may be underestimated, as pointed out recently by Cherepaschuk *et al.* (2018), which authors do, however, agree with a mass ratio of the system $\geqslant 0.3$, which implies, like for the mass estimates of Hillwig and Gies (2008) that the compact star in this system is indeed a black hole.

Supergiant BH-HMXBs typically have orbital periods $\leqslant 15\,\mathrm{d}$ and are evolutionary products of normal WR+O-star close binaries, in which the core of the WR star has collapsed to a black hole. During their further evolution with stable Roche-lobe overflow, such systems spiral-in due to the SS433-like evolution in which the transferred matter is ejected from the surroundings of the compact star, carrying off the specific orbital angular momentum of the latter. Such systems terminate as WRXBs with orbital periods of the order of about one day (van den Heuvel *et al.* 2017). Since this type of evolution cannot occur for NS-HMXBs, this implies that in WRXBs formed in this way the compact stars always are black holes.

The same is true for the WRXBs formed through CE evolution, since the progenitor stars of the WR stars are more massive than $30M_{\odot}$; with such a massive donor star the orbital period required with a $1.4M_{\odot}$ neutron star to survive the CEE-phase is many years (much longer than for the case of a black hole companion, which typically has a mass $\geqslant 5M_{\odot}$). Systems with such extremely long periods are expected to be very rare, such that the surviving WRXBs also in this case will practically exclusively harbor black holes.

(iii) *Formation through homogeneous evolution*

This model, for the formation of close double black holes, was proposed by Marchant *et al.* (2016) and de Mink & Mandel (2016), based on the "homogeneous evolution" model of close very massive binaries with mass ratio close to unity, put forward by de Mink *et al.* (2008). This model is based on the fact that in massive binaries with orbital periods $\leqslant 2-3\,\mathrm{d}$ (such as are known in the Doradus region of the LMC) the strong tidal friction will cause the rotation period of the nearly equal-mass components to always be fully synchronized with the orbital period. This means that the two stars are kept in very rapid rotation. In such rapidly rotating massive stars, strong meridional circulation will develop, which keep the stellar material mixed throughout the star (Maeder 1987). Thus, the helium produced by the hydrogen burning in the core will be fully mixed through the star, and the star will keep a fully homogeneous composition throughout its entire hydrogen-burning phase, and will end as a pure helium star: a WR star. This implies that, contrary to the case of normal massive stars, the stellar radius never increases during its evolution, and the star never overflows its Roche lobe. The final helium star is smaller than the original H-rich star. As the two stars will never have a mass ratio exactly equal to unity, the more massive component will be the first to become a WR star, while its companion then is still burning hydrogen, and will look like an O-star. By the time the WR star collapses to a black hole, its lower-mass companion may now have itself become a WR star, such that for a while the system will be a short-period WRXB. After the core collapse of the second WR star the system terminates as a close double black hole. This evolutionary sequence for forming double black holes is depicted in Fig. 3 (after Marchant *et al.* 2017).

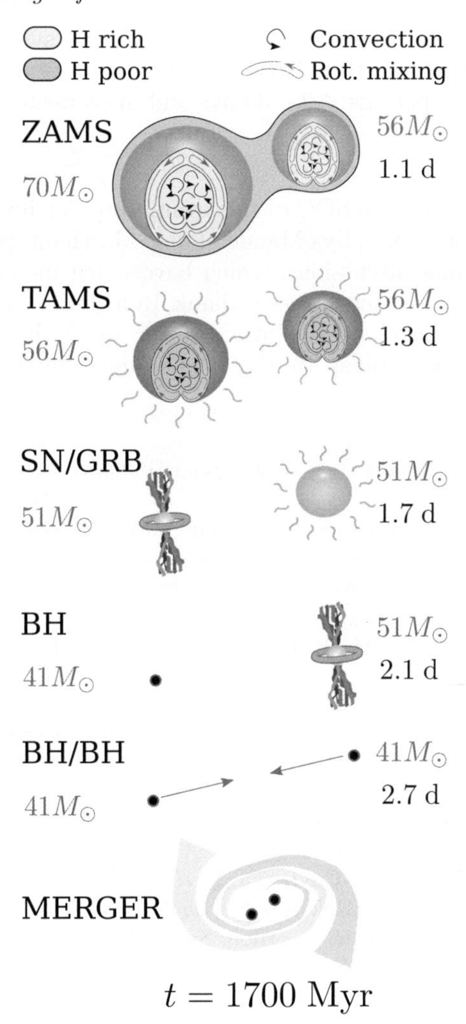

Figure 3. Homogeneous evolution of a very massive very close binary, orbital period $\leqslant 2 - 3$ d, into a close double black hole, which merges within the lifetime of the universe. Explanation in the text. Figure courtesy Pablo Marchant.

7. Summary and conclusions

We have seen that:

(1) The close double neutron stars are the later evolutionary products of wide neutron-star BeXBs, with orbital periods upwards of about one year. Neutron-star HMXBs with shorter orbital periods will merge and are expected to produce Thorne-Zytkow stars. Although this result has been known for over 40 years, and Thorne Zytkow stars should be quite common, so far never such an object has been identified with certainty.

(2) Close double black holes (and black-hole neutron star systems) that formed through binary evolution, are later evolutionary products of the short-period WRXBs.

For the formation the latter systems three models have been identified: (i) through CE evolution from wide BH-HMXBs, which is basically the same model as that for the formation of double neutron stars, scaled up to higher masses; (ii) by in-spiral due to stable Roche-lobe overflow from BH-HMXBs with "normal" supergiant HMXB orbital periods, upwards from a few days. Only systems with orbital periods less than about 10 days will probably be able to terminate as double black hole systems with orbital periods short enough to merge within a Hubble time; (iii) by homogeneous evolution of massive close binaries with orbital periods $\leqslant 2-3$ days and mass ratios $\geqslant 0.7-0.8$.

Acknowledgements

I thank Lida Oskinova and the SOC of this symposium for inviting me to present this overview. I thank Thomas Tauris, Ilya Mandel, Pablo Marchant, Selma de Mink and Chris Belczynski for enlightening discussions, which have much increased my understanding of the formation of double black holes. I thank Ron Taam and Natasha Ivanova for discussions on Common Envelope Evolution during more than 35 and more than 10 years, respectively. I thank Thomas Tauris for providing figures 2 and 3.

References

Becklin, E. F., Kristian, J., Neugebauer, G. & Wynn-Williams, C. G. 1972, *Nature Phys. Sci.*, 239, 130

Begelman, M. C., King, A. R., & Pringle, J. E. 2006, *MNRAS*, 370, 399

Belczynski, K., Holz, D. E., Bulik, T., & O'Shaughnessy, B. 2016, *Nature*, 534, 512

Blaauw, A., 1961, *Bull. Astr. Inst. Netherlands*, 15, 265

Bogomasov, A. I., 2014, *Astron. Rep.*, 58, 126

Casares, J., Negueruela, I., Ribo, M., Ribas, I., Paredes, J. M., Herrero, A., & Simon-Diaz, S., 2014, *Nature*, 505, 378

Cherepashchuk, A. M., Postnov, K. A., & Belinski, A. A. 2018, *MNRAS, in press*

Conti, P. S., Crowther, P. A., & Leitherer, C. 2008, *From Luminous Hot Stars to Starburst Galaxies* Cambridge University Press, Cambridge Astrophys. Series, 45

Corral-Santana, J. M., Casares, J., Munoz-Darias, T., Bauer, F. E., Martinez-Pais, I. G., & Russell, D. M. 2016, *A&A*, 587A, 61

Crowther, P. A. 2007, *ARAA*, 45, 177

De Loore, C., De Greve, J. P., & De Cuyper, J. P. 1975, *Ap&SS*, 36, 219

de Mink, S. E., & Mandel, I. 2016, *MNRAS*, 460, 3545

de Mink, S. E., Cantiello, M., Langer, N., Yoon, S.-C., Brott, I., Glebbeek, E., Verkoulen, M., & Pols, O. R. 2008, *Proc. IAU Symp. 252, L. Deng and K.L. Chang, eds.*, 365

Dessart, L., Burrows, A., Ott, C. D., Livne, E., Yoon, S.-C., & Langer, N. 2006, *ApJ*, 644, 1063

Dincel, B., Neuhauser, R., Yerli, S. K., Ankay, A., Pannicke, A., & Sasaki, M. 2016, *Sros.conf E36, online at: http://snr2016.astro.noa.gr, id.37*

Dubus, G., Guillard, N., Petrucci, P.-O., & Martin, P. 2017, *A&A*, 608, A59

Esposito, P., Israel, G. L., Milisavljevic, D., Mapelli, M., Zampieri, L., Sidoli, L., & Rodriguez-Castillo, G. A., 2015, *MNRAS*, 482, 1112

Flannery, B. P. & van den Heuvel, E. P. J., 1975, *A&A*, 39, 61

Hamann, W.-R., Gräfener, G., Liermann, A., 2006, *A&A*, 457, 1015

Hanson, M. M., Still, M. D. & Fender, R. P., 2000, *ApJ*, 541, 308

Hillwig, T. C. & Gies, D. R., 2008, *ApJ*, 676, L37

Hjellming, R. M. & Balick, B., 1972, *Nature Phys. Sci*, 239, 443

Hulse, R. A. & Taylor, J. H., 1975, *ApJ*, 195, L51

Kaper, L., van Loon, J. Th., Augusteijn, T., Goudfrooij, P., Patat, F., Waters, L. B. F. M. & Zijlstra, A. A., 1997, *ApJ*, 479, L153

Kaper, L., 2001, *ASSL*, 264, 125

Kiing, A. R., Taam, R. E., & Begelman, M. C., 2000, *ApJ*, 530, L25

Kitaura, F. S., Janka, H.-T. & Hillebrandt, W., 2006, *A&A*, 450, 345

Laycock, S. G. T., Maccarone, T. J. & Christodoulou, D. M., 2015, *MNRAS*, 452, L31

Lyne, A., Burgay, M., Kramer, M. *et al.* 2004, *Science*, 303, 1153

Maeder, A. 1987, *A&A*, 178, 159

Maraschi, L., Treves, A. & van den Heuvel, E. P. J. 1976, *Nature*, 259, 292

Marchant, P., Langer, N., Podsiadlowski, P., Tauris, T. M., & Moriya, T. J., 2016, *A&A*, 588, A50

Marchant, P., Langer, N., Podsiadlowski, P., Tauris, T. M., de Mink, S., Mandel, Y., & Moriya, T. J. 2017, *A&A*, 604, A55
Mirabel, I. F., Rodriguez, L. F., Cordier, B., Paul, J., & Lebrun, F. 1992, *Nature*, 358, 215
Orosz, J. A., McClintock, J. E., Narayan, R., Bailyn, C. D., Hartman, J. P., Macri, L., Liu, J., Pietsch, W., Remillard, R. A., Shporer, A., & Mazeh, T. 2007, *Nature*, 449, 872
Orosz, J. A., Steeghs, D., McClintock, J. E., Torres, M. A. P., & Bochkop, I., *et al.* 2009, *ApJ*, 697, 573
Orosz, J. A., McClintock, J. E., Aufdenberg, J. P., Remillard, R. A., Reid, M. J., Narayan, R., & Gou, L., 2011, *ApJ*, 742, 84
Orosz, J. A. Steiner, J. F., McClintock, J. E., Buxton, M. M., Bailyn, C. D., Steeghs, D., Guberman, A. & Torres, M. A. P., 2014, *ApJ*, 794, 154
Paczynski, B., 1976 *in: Structure and Evolution of Close Binary Systems, P. Eggleton, S., Mitton and J., Whelan, eds.* Dordrecht, Reidel Publ. Comp, 75
Parsignault, D. R., Gursky, H., Kellogg, E. M., Matilsky, T., Murray, S., Schreier, E., Tananbaum, H., Giacconi, R., & Brinkman, A. C., 1972, *Nature Phys. Sci.*, 239, 123
Pavlovski, K., Ivanova, N., Belczynski, K., & Van, K. X. 2017, *MNRAS*, 465, 2092
Pfahl, E., Rappaoprt, S., Podsiadlowski, P., & Spruit, H. 2002, *ApJ*, 574, 364
Podsiadlowski, P., Cannon, R. C., & Rees, M. J. 1995, *MNRAS*,274, 485
Podsiadlowski, P., Langer, N., Poelarends, A. J. T., Rappaport, S., Heger, A., & Pfahl, E., 2004, *ApJ*, 612, 1044
Postnov, K., Shakura, N., Sidoli, L., & Paizis, A. 2014, *Proceedings of 10th INTEGRAL Workshop: A Synergistic View of the High-Energy Sky http://pos.sissa.it/*
Schreier, E., Levinson, R., Gursky, H., Kellogg, E., Tananbaum, H., & Giacconi, R. 1972, *ApJ*, 172, L79
Shakura, N. I., Postnov, K. A., Kochetkova, A. Yu., & Hjalmarsdotter, L. 2012, *MNRAS*, 420, 216
Shenar, T., Hainich, R., Todt, H., *et al.* 2016, *A&A*, 591, 22
Sidoli, L., 2012 *Proc. 9th INTEGRAL Workshop: An INTEGRAL view of the high-energy sky(the first 10 years), online at: http://pos.sissa.it/cgi-bin/reader/conf.cgi?confid=176, id.11*
Srinivasan, G., & van den Heuvel, E. P. J. 1982, *A&A*, 108,143
Taam, R. E., 1996, *in: Compact Stars in Binaries, Proc. IAU Symp. 165, J. van Paradijs, E. P. J. van den Heuvel and E. Kuulkers, eds.*, 3
Taam, R. E., Bodenheimer, P., & Ostriker, J. P., 1978, *ApJ*, 222, 269
Taam, R. E. & Sandquist, E. L., 2000, *ARAA*, 38, 113
Tauris, T. M., Langer, N., & Podsiadlowski, P. 2015, *MNRAS*, 451, 2123
Tauris, T. M., Kramer, M., Freire, P. C. C., *et al.*, 2017, *ApJ*, 846, 170
Thorne, K. S. & Żytkow, A. N. 1977, *ApJ*, 212, 832
Tutukov, A. V. & Yungelson, L. R. 1973, *Nautsnie Informatsie*, 27, 58
Valsecchi, F., Glebbeek, E., Farr, W. M., Fragos, T., Willems, B., Orosz, J., Liu, J., & Kalogera, V. 2010, *Nature*, 468, 77
van den Heuvel, E. P. J. 1968, *BAN*,19, 432
van den Heuvel, E. P. J. 1974, *Proc. 16th Solvay Conf. on Physics, Univ. of Brussels Press*, 119
van den Heuvel, E. P. J. 1976, *in: Structure and Evolution of Close Binary Systems, P. Eggleton, S. Mitton and J. Whelan, eds.* Dordrecht, Reidel Publ. Comp., 35
van den Heuvel, E. P. J. & Heise, J., 1972, *Nature Phys. Sci.*, 239, 67
van den Heuvel, E. P. J. & De Loore, C., 1973, *A&A*, 25, 387
van den Heuvel, E. P. J., Portegies Zwart, S. F., Bhattacharya, D., & Kaper, L., 2001, *A&A*, 364, 563
van den Heuvel, E. P. J., Portegies Zwart, S. F. & de Mink, S. E. 2017, *MNRAS*, 471, 4256
van Kerkwijk, M. H., Charles, P. A., Geballe, T. R., King, D. L., Miley, G. K., Molnar, L. A., & van den Heuvel, E. P. J. 1992, *Nature*, 355, 703
Walter, R., Rodriguez, J., Foschini, L., de Plaa, J., Corbel, S., Courvoisier, T. J.-L., den Hartog, P. R., Lebrun, F., Parmar, A. N., Tomsick, J. A., & Ubertini, P., 2003, *A&A*, 411, L427
Webster, B. L. & Murdin, P., 1972, *Nature*, 235, 37

Massive stars

High-mass X-ray Binaries: Illuminating the Passage from Massive Binaries to Merging Compact Objects
Proceedings IAU Symposium No. 346, 2019
L. M. Oskinova, E. Bozzo, T. Bulik & D. Gies, eds.

doi:10.1017/S1743921318007652

Massive star winds and HMXB donors

Andreas A. C. Sander[1,2]

[1]Armagh Observatory and Planetarium,
College Hill, Armagh BT61 9DG, Northern Ireland, UK
email: Andreas.Sander@armagh.ac.uk

[2]Institut für Physik und Astronomie, Universität Potsdam,
Karl-Liebknecht-Str. 24/25, D-14476 Potsdam, Germany

Abstract. Understanding the complex behavior of High Mass X-ray binaries (HMXBs) is not possible without detailed information about their donor stars. While crucial, this turns out to be a challenge on multiple fronts. First, multi-wavelength spectroscopy is vital. As such systems can be highly absorbed, this is often already hard to accomplish. Secondly, even if the spectroscopic data is available, the determination of reliable stellar parameters requires sophisticated model atmospheres that accurately describe the outermost layers and the wind of the donor star.

For early-type donors, the stellar wind is radiatively driven and there is a smooth transition between the outermost layers of the star and the wind. The intricate non-LTE conditions in the winds of hot stars complicate the situation even further, as proper model atmospheres need to account for a multitude of physics to accurately provide stellar and wind parameters. The latter are especially crucial for the so-called "wind-fed" HXMBs, where the captured wind of the supergiant donor is the only source for the material accreted by the compact object.

In this review I will briefly address the different approaches for treating stellar winds in the analysis of HMXBs. The fundamentals of stellar atmosphere modeling will be discussed, also addressing the limitations of modern models. Examples from recent analysis results for particular HMXBs will be outlined. Furthermore, the path for the next generation of stellar atmosphere models will be outlined, where models can be used not only for measurement purposes, but also to make predictions and provide a laboratory for theoretical conclusions. Stellar atmospheres are a key tool in understanding HMXBs, e.g. by providing insights about the accretion of stellar winds onto the compact object, or by placing the studied systems in the correct evolutionary context in order to identify potential gravitational wave (GW) progenitors.

Keywords. accretion, hydrodynamics, methods: numerical, radiative transfer, stars: atmospheres, stars: early-type, stars: individual (4U 1700-37, 4U1907+09, 4U 1909+07, GX304-01, IGR J16465-4507, IGR J17544-2619, Vela X-1), stars: mass loss, stars: neutron, stars: winds, outflows, X-rays: binaries

---

1. Introduction

High-mass X-ray binaries (HMXBs) consist of a compact object - either a neutron star or a black hole - accreting material from a companion. This companion is a massive star with initially more than about 8 times the mass of our sun. HMXBs can be classified into three different types of systems:

(*a*) **Wind-fed** systems, where the compact object accretes material by capturing a fraction the donor star's wind

(*b*) **Roche-lobe overflow** (RLOF) systems, where the material from the donor is (mainly) accreted via the mass-loss through the inner Lagrangian point (L_1)

(*c*) **Be X-ray binaries** (BeXRBs) where the donor is a Be-star with a decretion disk and the compact object is (mainly) fed by accreting disk material

Although not inherent to their definition, the donor stars usually found in these systems are all hot and luminous stars, having $T_{\rm eff} > 10\,{\rm kK}$ and thus a strong UV flux. This makes HMXBs excellent tracers of star formation and young populations in galaxies. The spectral types of the massive donor stars are typically early B or late O-type stars, but also earlier O-subtypes and even some Wolf-Rayet (WR) stars are known as donors.

While BeXRB systems are quite distinct, the border between wind-fed and RLOF systems is not that obvious, especially if the radii of the donor stars are not well constrained and thus it can be unclear whether a donor really fills its Roche lobe or not. Moreover, Bondi-Hoyle-Lyttleton (BHL) accretion mechanism (Hoyle & Lyttleton 1939; Bondi & Hoyle 1944; Davidson & Ostriker 1973) yields that the possible X-ray luminosity due to accretion $L_{\rm X}$ crucially depends on the wind parameters of the donor star

$$L_{\rm X} \propto \frac{\dot{M}_{\rm donor}}{\left[v_{\rm orb}(d)^2 + v_{\rm wind}(d)^2\right]^{3/2} \cdot v_{\rm wind}(d)} \approx \frac{\dot{M}_{\rm donor}}{v^4_{\rm wind}(d)}, \tag{1.1}$$

in particular the wind speed $v_{\rm wind}$ at the orbital separation d of the compact object. Sophisticated spectral analyzes of the donor stars using state-of-the-art model atmospheres to determine the stellar and wind parameters of the donor stars in HMXBs are therefore a pivotal instrument for a better understanding of this important evolutionary stage towards compact object binaries.

2. Modeling the winds in HMXBs

To study the impact of winds in HMXB systems, two sorts of approaches have been applied, both having their advantages and disadvantages. On the one side, the situation can be studied with the help of hydrodynamics codes, allowing potentially for a multi-dimensional and time-dependent treatment in a rather complex geometry. However, the detailed treatment in terms of space and resolution comes at the cost of simplified wind physics. Two typical examples are the radiative transfer, which is usually approximated, and the use of a so-called "ionization parameter", which approximates the calculation of a detailed ionization balance as a single number entering the actual calculations. Seminal work in this field has been done by Blondin *et al.* (1990), followed up nowadays by the works of e.g. Manousakis & Walter (2015) or El Mellah *et al.* (2018a,b).

A second, rather different approach is the use of stellar atmospheres. Well established in the field of isolated massive stars, they provide a detailed physical treatment for the donor, including a full non-LTE calculation of the ionization structure and a radiative transfer performed in the co-moving frame. Here, the costs are a simplified geometry and usually also a stationary wind. Traditionally developed for spherically symmetric situations, the compact object in this approach is rather treated as a "disturbance" instead of being fundamentally taken into account from first principles. Applications to HMXBs are quite limited in terms of the number of systems studied, but have been done for quite a while, ranging from classic spectroscopic stellar analysis (e.g. Clark *et al.* 2002) to studies of the effect of X-ray irradiation on the radiative driving of the donor star (e.g. Krtička *et al.* 2012, Sander *et al.* 2018). Essentially, the two approaches complement each other. To fully understand the role of the donor stars in HMXBs, a proper atmosphere analysis is essential and thus the second approach will be the focus of this article.

Hot, massive stars have inherent mass outflows, called "stellar winds". These winds can reach up to several thousand ${\rm km\,s^{-1}}$ with typical mass-loss rates on the order of $10^{-6}\,M_\odot\,{\rm yr}^{-1}$ for O supergiants at solar metallicity. The wind is accelerated by momentum transfer from stellar continuum photons to metal ions, which happens mainly due to absorption in spectral lines. Due to the fact that the re-emission is isotropic, while the absorbed radiation is radially coming from the star, there is a net momentum away from the stellar surface. For O and B stars, the most prominent spectral signatures of stellar winds are the P Cygni profiles found in the ultraviolet resonance lines of e.g. C IV. For more dense winds, some optical line profiles such as Hα will turn from an absorption into an emission. For sufficiently dense winds, such as found in Galactic WR stars, the whole spectrum is formed in the wind and all optical lines are in emission. However, the lack of optical emission lines on the other hand does not imply the absence of a considerable wind.

The theoretical description of these radiation-driven winds was pioneered by Lucy & Solomon (1970) and Castor, Abbott, & Klein (1975). The initials of the latter provided the name for the so-called "CAK theory". While considerably improved in the 1980s (e.g. Friend & Abbott 1986, Pauldrach *et al.* 1986), the fundamental premises of CAK such as the assumption that all acceleration is provided either by free electron or line scattering remain until today. A major outcome of the theory is the so-called "β-law" describing the radial behavior of the wind velocity

$$v(r) = v_\infty \left(1 - \frac{R_\star}{r}\right)^\beta \tag{2.1}$$

for a massive star with the help of a parameter β. This essentially analytic description of $v(r)$ is widely used, allowing an easy determination of the velocity or density of the stellar wind in more complex calculations or simulations. Nonetheless, there are various cases where the CAK formalism fails to correctly describe the observed stellar winds, especially but not limited to the situations where winds get more dense and the average photon leaving the star is scattered more than once ("multiple scattering"). It is thus favorable to go beyond these limitations in a sophisticated modeling of a stellar atmosphere. Multiple challenges have to be accounted for in this task, including

- the intricate non-LTE situation,
- a radiative transfer in an expanding atmosphere accounting for multiple scattering,
- sufficient model atoms accounting for all elements that are either detectable in the wind or influence the stratification.

The non-LTE situation implies that we cannot rely on the Saha-Boltzmann statistics, but instead assume a statistical equilibrium, i.e. a balance of total gain and loss rates for all considered atomic levels. Unlike in cool stars, it is not sufficient to determine just the ionization balance. Instead we need to obtain the population numbers for a considerable amount of levels in each relevant ionization stage of each element taken into account. Even when considering only a handful of elements, this quickly leads to a system of a few hundreds of rate equations that needs to be solved.

Obtaining a proper model atom is especially tricky for the elements of the iron group which have thousands of levels and millions of line transitions. This introduces a so-called "blanketing" effect, significantly altering the atmosphere stratification. An explicit treatment of all these levels is impossible. Thus, basically all modern atmosphere codes use a concept going back to Anderson (1989) and Dreizler & Werner (1993) where levels in a certain energy range are grouped into a "superlevel". While the superlevels are then treated in full non-LTE, the levels inside a superlevel are assumed to follow the Boltzmann statistics.

In an expanding, non-LTE environment, the determination of the electron temperature stratification is also not trivial. Going back to the ideas of Unsöld (1951, 1955) and Lucy (1964), temperature corrections can be obtained from the equation of radiative equilibrium and it's integral describing the conservation of the total flux. Alternatively, one can also obtain the corrections from calculating the electron thermal balance going back to Hummer & Seaton (1963). Since both of these two – or three, depending on how one counts – methods have their strengths and weaknesses, some stellar atmosphere codes can also use a combination of them.

The radiative transfer is performed in the co-moving frame (CMF). While this is computationally considerably more expensive than the CAK formalism, this rigorous approach considers all contributions to the radiative acceleration and implicitly accounts for various effects such as multiple scattering. Combining the previously sketched techniques is a task of its own, as the problems are highly coupled. The CMF radiative transfer introduces a coupling in space, and the solution of the statistical equations comes with an intrinsic coupling in frequency, thus we are faced with a problem coupled in both, space and frequency. The solution is an accelerated lambda iteration (ALI, e.g. Hamann 1985) until a consistent solution of the calculations (i.e. solution of the statistical equations, radiative transfer, temperature corrections) is reached. Beside this, one more layer of complexity is added by the requirement to account for both the quasi-hydrostatic layers of the star as well as the supersonic wind on top. While for an analysis purpose it is often sufficient to assume a β-type velocity law in the wind, the proper quasi-hydrostatic treatment requires additional updates of the density stratification (see, e.g. Sander *et al.* 2015).

When a converged atmosphere stratification is reached, the emergent spectrum is calculated in a final radiative transfer calculation, now performed in the observer's frame. The resulting synthetic spectrum can then be compared to observations in order to check whether a model with certain assumed parameters is a sufficient representation of the star. If matching, the complexity of the model atmospheres then becomes a huge advantage: Assuming sufficient observed spectra are available, preferably in multiple wavelength regimes, stellar atmosphere models provide a multitude of information about the donor star, including the stellar and wind parameters (e.g. T_{eff}, $\log g$, L, v_∞, $\dot{M}$, ...), the chemical abundances, and also the mechanical and ionizing feedback. This level of knowledge is essential for a proper understanding of the donor and thus the HMXB system in general.

3. The parameters of HMXB donors

Given the progress in analyzing single massive stars, it is tempting to apply the insights gained there to an HMXB donor star of the same spectral type. However, donor stars are not necessarily identical to massive single stars. Firstly, the compact object next to them might be small in size, but not in influence. The breaking of the spherical symmetry and the influence of the X-rays irradiating the donor star can make an important difference. Secondly, the donor star in an HMXB system has a very distinctive history. The existence of the compact object points to the fact that the current donor has been the secondary in a former binary system and thus was likely a mass gainer during an earlier stage of mass transfer (e.g. Vanbeveren & De Loore 1994). From the observational and analytical point of view, various studies have confirmed that donor stars are more luminous than single stars of the same mass (e.g. Conti 1978, Vanbeveren *et al.* 1993). Kaper (2001) summarized that for their luminosity, the donor stars have lower masses and radii compared to normal OB supergiants. While the total number of systems studied in detail so far is too low to give final answers, these results nonetheless outline the peculiarity of HMXB donor stars, thus making it problematic to simply adopt parameters from isolated stars. Still this is and sometimes even has to be done as HMXB systems are often well studied in X-rays, but poorly at other wavelength regimes. This discrepancy of knowledge has two major reasons of very different origin: The observational reason is the high interstellar absorption for many of the systems which prevents their observation in the UV and often even in the optical. With the UV being the most important wind diagnostic regime for O and B-type stars, this leads to a significant limitation of statistics as only a handful of systems can actually be studied in this regime. The second, more sociological reason is the interdisciplinary nature of HMXB systems, combining stellar winds with

Table 1. Examples of wind-fed system analyses

System	sp. Type	$T_{\rm eff}$ [kK]	$\log g$ [cgs]	$X^{\rm a}_{\rm He}$	$X^{\rm a}_{\rm N}$ $[10^{-3}]$	$\log \dot{M}$ $[M_\odot\,{\rm yr}^{-1}]$	v_∞ $[{\rm km}^{-1}]$	Reference
GX 301-2	B1I	19	2.38	0.52	5.25	−5.0	305	Kaper *et al.* (2006)
4U 1909+07	B0I..B3I	23	3.2	$0.26^{\rm b}$	$0.6^{\rm b}$	−6.5	500	Martínez-Núñez *et al.* (2015)
Vela X-1	B0.5I	26	2.86	0.34	1.8	−6.2	700	Giménez-García *et al.* (2016)
IGR J16465-4507	B0.5I	26	3.10	0.5	5.5	n/a	n/a	Chaty *et al.* (2016)
IGR J17544-2619	O9.5I	29	3.25	0.37	2.2	−5.8	1500	Giménez-García *et al.* (2016)
4U 1907+09	O8I..O9I	30	3.1	$0.26^{\rm b}$	$0.6^{\rm b}$	−5.1	1700	Cox *et al.* (2005)
4U 1700-37	O6.5I	35	3.5	0.44	10	−5.0	1750	Clark *et al.* (2002)

Notes: [a]mass fraction; [b]assumed solar abundance

accretion mechanisms and high-energy physics. A proper understanding requires experts from various fields and when new systems are discovered in X-rays, there is a delay until observations in other wavelength regimes have been made and follow-up science is underway. Thus, it is more important than ever to bring experts from different fields together early-on in the process.

The analysis of the donor star with different modern stellar atmospheres has been done for about a handful of systems. While differing in their computational details, sometimes quite significantly, three of them, namely PoWR (Hamann & Schmutz 1987, Hamann & Koesterke 1998, Gräfener *et al.* 2002), CMFGEN (Hillier 1990, Hillier & Miller 1998), and FASTWIND (Santolaya-Rey *et al.* 1997, Puls *et al.* 2005), are state-of-the-art tools for analyzing hot stars with expanding atmospheres. PoWR and CMFGEN exactly reflect the scheme sketched in Sect. 2, while FASTWIND uses an approximate blanketing approach, getting lower computation times for their models at the cost of a smaller range of applicable stars, e.g. excluding Wolf-Rayet stars. For purely hydrostatic atmospheres, also the TLUSTY code (Hubeny 1988, Hubeny & Lanz 1995) is an option. However, since in most HMXB systems – maybe aside from the BeXRBs – the wind is not negligible, a model atmosphere inherently accounting for the wind is highly recommended as otherwise the deduced stellar parameters could be wrong due to wind emission filling up the absorption profiles.

The METUJE code by Krtička *et al.* (2012, 2015) takes a slightly different approach by focussing on the calculation of mass-loss rates instead of reproducing observed spectra. This concept will be discussed further in Sect. 4. There are more atmosphere codes on a comparable level of sophistication, such as WM-basic (Pauldrach 1987, Pauldrach *et al.* 1994, 2001) or PHOENIX (Hauschildt 1992; Hauschildt & Baron 1999, 2004), but both have shifted their focus to other areas (supernovae for WM-basic, cooler stars for PHOENIX) and thus to our knowledge they have so far not been applied to HMXB donors.

A selection of wind-fed HMXB systems analyzed with current state-of-the-art atmosphere codes is compiled in Table 1. Essentially, all of these systems have late O- or early B-type donor stars. In systems where abundances could be determined, the helium and nitrogen fractions are enhanced, clearly indicating that they are evolved stars. The wind parameters of the various donors differ quite significantly, even if their spectral type is rather similar. The terminal velocities show a trend with increasing $T_{\rm eff}$, but the mass-loss rates do not reflect this. Instead we see a signficantly higher mass-loss rate for the coolest star in the sample compared to those with slightly higher temperatures. While there are effects due to the compact objects and its accretion as we will discuss below, this significant change is most likely not a result of this, but of the so-called bi-stability jump. This phenomenon is not exclusive to HMXB donors, but a general effect seen in the regime of early B supergiants in a temperature regime of $T_{\rm eff} \approx 20...25\,{\rm kK}$. Found theoretically by Pauldrach & Puls (1990) and confirmed observationally by Lamers *et al.* (1995), a drastic increase in the mass-loss rate $\dot{M}$ is seen compared to stars with higher $T_{\rm eff}$, accompanied by a considerable decrease of the terminal wind velocity v_∞. Studied since its discovery (e.g. Vink *et al.* 1999, Petrov *et al.* 2016, Keszthelyi *et al.* 2017), the bi-stability jump is attributed to a change in the ionization balance - mainly of iron - but a lot of details still remain unclear, including the precise jump conditions and the metallicity dependence.

In HMXBs, X-ray irradiation might further influence the ionization structure, thus further complicating the situation. Hatchett & McCray (1977) noticed that since the irradiation is coming from the orbiting compact object, the effects should be seen in the phase-dependent profiles of UV resonance lines, but in fact this so-called "Hatchett-McCray effect" is not observed for all systems. Approximated radiative transfer models by van Loon *et al.* (2001) revealed that the precise orbital configuration can have a significant influence on the appearance of this effect and demonstrated that for systems like 4U 1700−37 indeed no significant phase-dependent profile change should occur.

A significant change of the ionization balance also affects the stellar wind. A considerable X-ray irradiation effectively harms the wind in the direction towards the compact object. The depopulation of important ionization stages leads to the effect that the wind is no longer accelerated and might even slow down (Krtička *et al.* 2012). However, for the mass-loss rate itself to be changed, almost the whole donor atmosphere down to the stellar surface would need to be affected, which is probably not the case in most systems (Sander *et al.* 2018).

The X-ray flux variabilities seen in HMXBs further suggest that the winds of the donor stars are not homogeneous (e.g. int Zand 2005, van der Meer 2005, Negueruela *et al.* 2008, Oskinova *et al.* 2012). Indeed, smooth winds are not expected even from single stars, where small scale structures, often referred to as "clumping" are a part of the standard picture. However, the theoretical concepts for clumping in single star winds predict rather small clumps, which do not really match the larger clump sized inferred from the X-ray observations so far. The origin of clumping itself is still a matter of debate, with the classic idea of the line-driven or line deshadowing instability (LDI, e.g. Owocki *et al.* 1988, Feldmeier *et al.* 1997, Dessart & Owocki 2003, Sundqvist *et al.* 2018) facing certain challenges, e.g. regarding the predicted onset of the clumps. Envelope instability or sub-surface convection (e.g. Glatzel 2008, Cantiello *et al.* 2009, Jiang *et al.* 2015) has been suggested as an alternative that would allow for a deeper clumping onset around opacity peaks, e.g. of iron in main sequence stars. Small-scale clumping also likely coexists with large scale structures, attributed to co-rorating interaction regions (CIRs) spectroscopically manifesting in so-called discrete absorption components (DACs). A detailed discussion about clumping and winds in HMXBs was recently published in an extensive review by Martínez-Núñez *et al.* (2017).

4. From measurements to laboratories

Traditionally, stellar atmosphere models have been used to measure the stellar and wind parameters. However, their complexity also allows us to use them as virtual laboratories, where we can test what for example the imprint of a certain change of parameters is on the stellar spectrum. Moreover, we can predict the wind parameters ($v_{\rm wind}(r)$, $\dot{M}_{\rm donor}$) instead of measuring them by including the solution of the hydrodynamic (HD) equation of motion

$$v\left(1-\frac{a_{\rm s}^2}{v^2}\right)\frac{{\rm d}v}{{\rm d}r} = a_{\rm rad}(r) - g(r) + 2\frac{a_{\rm s}^2}{r} - \frac{{\rm d}a_{\rm s}^2}{{\rm d}r}, \tag{4.1}$$

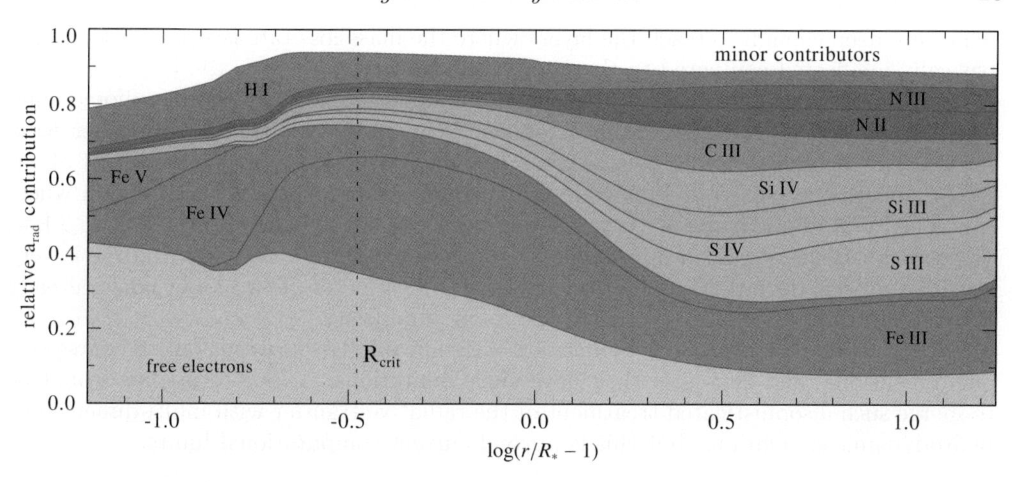

Figure 1. Relative contributions to the radiative acceleration in the donor star of Vela X-1.

which has to be fulfilled at all depths in a self-consistent atmosphere model. Due to the fact that the hydrodynamic Eq. (4.1) has a critical point, the additional constraint of requiring a smooth transition of $v(r) \equiv v_{\mathrm{wind}}(r)$ through this point is introduced. This can be translated into a condition for the mass-loss rate $\dot{M}_{\mathrm{donor}}$. Thus, a converged model that also fulfills Eq. (4.1), essentially predicts the fundamental wind parameters from a given set of stellar parameters.

While this concept might sound rather simple and goes back to Lucy & Solomon (1970), its actual implementation is not. Early efforts were made by Pauldrach *et al.* (1986), using a pure CMF line force implementation. WM-basic Pauldrach *et al.* (2001) later used the concept together with a Sobolev-based approach. Predicting mass-loss rates in this way is also the heart of the more theory-focussed METUJE code. In Krtička & Kubát 2004, a Sobolev approach was used, but since Krtička & Kubát (2010) also a CMF radiative transfer is standard. Gräfener & Hamann (2005) performed the first complete implementation into a CMF-based analysis code, namely PoWR. Successfull applications for a WC and later also a grid of WN models (Gräfener & Hamann 2008) were performed using an implementation based on a generalized force multiplier concept. To extend this method to further regimes, a new implementation with a different technique was added to PoWR by Sander *et al.* (2017), yielding also hydrodynamically consistent models for O and B stars. Recently, Sundqvist & Puls (2018) announced the solution of the hydrodynamic equation of motion also as one of their goals for a future FASTWIND update.

Utilizing the new PoWR implementation, Sander *et al.* (2018) published a self-consistent model for the donor star of Vela X-1. This included a standard model as well as additional models with different levels of X-ray irradiation in order to study the irradition effect on the radiative driving and the resulting wind velocity field. The study revealed that in the donor of Vela X-1, apart from free electrons, Fe III is the leading ion driving the wind, followed by S III (see Fig. 1). However, just because Fe III is the leading driver, this does not imply the Fe III is the main Fe ionization stage in the wind. In fact, most of the material is in Fe IV, but its driving contribution is just not important in the outer wind. The irradiation of X-rays from the accretion onto the neutron star can be large enough to shut down the further acceleration in the affected cone and thus reduce the terminal wind velocity there compared to an isolated massive star, but they do not penetrate deep enough to reach the layers where the mass-loss rate is set, thus leaving it basically unaffected compared to the unperturbed situation.

An important detail revealed by the model in Sander *et al.* (2018) is the tailored wind velocity law and its deviation from a typical β-law in the wind onset and inner wind region. This is crucial as the neutron star in Vela X-1 is sitting at a distance of about $1.8\,R_*$, a rather typical value for wind-fed HMXBs, and it is exactly this region where the velocity is about a factor of two lower than one would infer from using a β-law. Since Bondi-Hoyle accretion approximately yields $L_\mathrm{X} \propto \dot{M}/v^4$, the potential L_X is very sensitive to such deviations and a velocity that is by a factor of two lower could already power almost an order of magnitude more X-ray luminosity.

Hydrodynamically consistent models are a significant step forward, but of course still one-dimensional and thus also they have their limitations. A long-term thus would be to merge such a sophisticated treatment of the radiative transfer with multi-dimensional hydrodynamic simulations, but this is beyond current computational limits.

5. Summary and conclusions

The donor stars of high-mass X-ray binary systems are – with the exception of the BeXRBs – typically late O- or early B-type supergiants. The details of their wind parameters can differ significantly due to the bi-stability regime and further the different orbital configurations. In many cases, the donor properties are rather poorly constrained. Photometry is usually not sufficient to give a valid characterization of the donor. Spectra are required to avoid inferring wrong quantities or evolutionary scenarios. To derive the stellar wind parameters of the donor stars is a difficult task, especially as donor stars are peculiar in their properties when compared to isolated massive stars of the same spectral type. An atmosphere analysis with a state-of-the-art model atmosphere code is strongly recommended to get a reliable handle on the wind parameters. This is even more important if no UV spectra of the donor are available, despite the fact that also the atmosphere analysis might be limited with regards to some parameters. The non-homogeneous nature (or "clumpyness") of stellar winds are supported both from the single star observations as well as the accretion X-ray variability. Clumps might be able to explain a variety of observed phenomena, but also add a considerable layer of complexity to the models. The presence of a strong X-ray source near the donor star has a noticeable effect on its (outer) atmosphere and can change the wind ionization and acceleration. As the causing X-ray illumination is phase-dependent, some systems also show phase-dependent changes of the spectrum, in particular in the UV resonance lines. Beside the detailed, but one-dimensional stellar atmosphere models, another important tool to tackle the complex behaviour of donor winds in HMXB systems as multi-dimensional, hydrodynamic models, having a kind of complementary approach with their complex geometry, but simplified wind physics. So far, we are not in a stage to combine both approaches, but first steps, where results from one approach are incorporated in the other one in a parametrized form are on the way.

References

Anderson, L. S. 1989, *ApJ*, 339, 558
Blondin, J. M., Kallman, T. R., Fryxell, B. A., & Taam, R. E. 1990, *ApJ*, 356, 591
Bondi, H. & Hoyle, F. 1944, *MNRAS*, 104, 273
Cantiello, M., Langer, N., Brott, I., *et al.* 2009, *A&A*, 499, 279
Castor, J. I., Abbott, D. C., & Klein, R. I. 1975, *ApJ*, 195, 157
Chaty, S., LeReun, A., Negueruela, I., *et al.* 2016, *A&A*, 591, A87
Clark, J. S., Goodwin, S. P., Crowther, P. A., *et al.* 2002, *A&A*, 392, 909
Conti, P. S. 1978, *A&A*, 63, 225

Cox, N. L. J., Kaper, L., & Mokiem, M. R. 2005, *A&A*, 436, 661
Davidson, K. & Ostriker, J. P. 1973, *ApJ*, 179, 585
Dessart, L. & Owocki, S. P. 2003, *A&A*, 406, L1
Dreizler, S. & Werner, K. 1993, *A&A*, 278, 199
El Mellah, I., Sundqvist, J. O., & Keppens, R. 2018, *MNRAS*, 475, 3240
El Mellah, I., Sander, A. A. C., Sundqvist, J. O., & Keppens, R. 2018, ArXiv e-prints, 1810.12933
Feldmeier, A., Puls, J., & Pauldrach, A. W. A. 1997, *A&A*, 322, 878
Friend, D. B. & Abbott, D. C. 1986, *ApJ*, 311, 701
Giménez-García, A., Shenar, T., Torrejón, J. M., *et al.* 2016, *A&A*, 591, A26
Glatzel, W. 2008, in Astronomical Society of the Pacific Conference Series, Vol. 391, Hydrogen-Deficient Stars, ed. A. Werner & T. Rauch, 307
Gräfener, G. & Hamann, W.-R. 2005, *A&A*, 432, 633
Gräfener, G. & Hamann, W.-R. 2008, *A&A*, 482, 945
Gräfener, G., Koesterke, L., & Hamann, W. 2002, *A&A*, 387, 244
Hamann, W.-R. 1985, *A&A*, 148, 364
Hamann, W.-R. & Koesterke, L. 1998, *A&A*, 335, 1003
Hamann, W.-R. & Schmutz, W. 1987, *A&A*, 174, 173
Hatchett, S. & McCray, R. 1977, *ApJ*, 211, 552
Hauschildt, P. H. 1992, J. Quant. Spec. Radiat. Transf., 47, 433
Hauschildt, P. H. & Baron, E. 1999, Journal of Computational and Applied Mathematics, 109, 41
Hauschildt, P. H. & Baron, E. 2004, *A&A*, 417, 317
Hillier, D. J. 1990, *A&A*, 231, 116
Hillier, D. J. & Miller, D. L. 1998, *ApJ*, 496, 407
Hoyle, F. & Lyttleton, R. A. 1939, Proceedings of the Cambridge Philosophical Society, 35, 405
Hubeny, I. 1988, Computer Physics Communications, 52, 103
Hubeny, I. & Lanz, T. 1995, *ApJ*, 439, 875
Hummer, D. G. & Seaton, M. J. 1963, *MNRAS*, 125, 437
in't Zand, J. J. M. 2005, *A&A*, 441, L1
Jiang, Y.-F., Cantiello, M., Bildsten, L., Quataert, E., & Blaes, O. 2015, *ApJ*, 813, 74
Kaper, L. 2001, in Astrophysics and Space Science Library, Vol. 264, The Influence of Binaries on Stellar Population Studies, ed. D. Vanbeveren, 125
Kaper, L., van der Meer, A., & Najarro, F. 2006, *A&A*, 457, 595
Keszthelyi, Z., Puls, J., & Wade, G. A. 2017, *A&A*, 598, A4
Krtička, J. & Kubát, J. 2004, *A&A*, 417, 1003
Krtička, J. & Kubát, J. 2010, *A&A*, 519, A50
Krtička, J., Kubát, J., & Krtičková, I. 2015, *A&A*, 579, A111
Krtička, J., Kubát, J., & Skalický, J. 2012, *ApJ*, 757, 162
Lamers, H. J. G. L. M., Snow, T. P., & Lindholm, D. M. 1995, *ApJ*, 455, 269
Lucy, L. B. 1964, SAO Special Report, 167, 93
Lucy, L. B. & Solomon, P. M. 1970, *ApJ*, 159, 879
Manousakis, A. & Walter, R. 2015, *A&A*, 584, A2
Martínez-Núñez, S., Kretschmar, P., Bozzo, E., *et al.* 2017, Space Sci. Rev., 212, 59
Martínez-Núñez, S., Sander, A., Gímenez-García, A., *et al.* 2015, *A&A*, 578, A107
Negueruela, I., Torrejón, J. M., Reig, P., *et al.*, 2008, in A Population Explosion: The Nature and Evolution of X-ray Binaries in Diverse Environments, eds. R. M. Bandyopadhyay, S. Wachter, D. Gelino, & C. R. Gelino, AIP Conf. Ser., 1010, 252
Oskinova, L. M. and Feldmeier, A. and Kretschmar, P. 2012, *MNRAS*, 421, 2820
Owocki, S. P., Castor, J. I., & Rybicki, G. B. 1988, *ApJ*, 335, 914
Pauldrach, A. 1987, *A&A*, 183, 295
Pauldrach, A., Puls, J., & Kudritzki, R. P. 1986, *A&A*, 164, 86
Pauldrach, A. W. A., Hoffmann, T. L., & Lennon, M. 2001, *A&A*, 375, 161
Pauldrach, A. W. A., Kudritzki, R. P., Puls, J., Butler, K., & Hunsinger, J. 1994, *A&A*, 283, 525
Pauldrach, A. W. A. & Puls, J. 1990, *A&A*, 237, 409
Petrov, B., Vink, J. S., & Gräfener, G. 2016, *MNRAS*, 458, 1999

Puls, J., Urbaneja, M. A., Venero, R., *et al.* 2005, *A&A*, 435, 669
Sander, A., Shenar, T., Hainich, R., *et al.* 2015, *A&A*, 577, A13
Sander, A. A. C., Fürst, F., Kretschmar, P., *et al.* 2018, *A&A*, 610, A60
Sander, A. A. C., Hamann, W.-R., Todt, H., Hainich, R., & Shenar, T. 2017, *A&A*, 603, A86
Santolaya-Rey, A. E., Puls, J., & Herrero, A. 1997, *A&A*, 323, 488f
Sundqvist, J. O., Owocki, S. P., & Puls, J. 2018, *A&A*, 611, A17
Sundqvist, J. O. & Puls, J. 2018, ArXiv e-prints, 1805.11010
Unsöld, A. 1951, Naturwissenschaften, 38, 525
Unsöld, A. 1955, Physik der Sternatmosphären, mit besonderer Berücksichtigung der Sonne. (Berlin, Springer, 1955. 2. Aufl.)
Šurlan, B., Hamann, W.-R., Aret, A., *et al.* 2013, *A&A*, 559, A130
Šurlan, B., Hamann, W.-R., Kubát, J., Oskinova, L. M., & Feldmeier, A. 2012, *A&A*, 541, A37
van der Meer, A. and Kaper, L. and di Salvo, T., *et al.* 2005, *A&A*, 432, 999
van Loon, J. T., Kaper, L., & Hammerschlag-Hensberge, G. 2001, *A&A*, 375, 498
Vanbeveren, D. & De Loore, C. 1994, *A&A*, 290, 129
Vanbeveren, D., Herrero, A., Kunze, D., & van Kerkwijk, M. 1993, Space Sci. Rev., 66, 395
Vink, J. S., de Koter, A., & Lamers, H. J. G. L. M. 1999, *A&A*, 350, 181

Discussion

L. KAPER: The X-ray eclipse presents a way to measure the radius of the OB supergiant with high accuracy. This parameter is often poorly constrained by stellar atmosphere models. Do you use these radius determination?

A.A.C. SANDER: This is a good point. We do not explicitly use this in the model for the Vela X-1 donor, since the stellar parameters are taken from Giménez-García *et al.* (2016), but as far as I am aware their analysis makes use of this. Moreover, we did use radius constraints in the analysis of the X1908+075 donor where we went with the rather unconventional way to fix T_* and R_* to constrain the luminosity and distance.

E.P.J. VAN DEN HEUVEL: In some of the highly obscured Integral supergiant HMXBs, the donor is a later O supergiant, but from the high obscuration you would expect the wind to be much slower and denser than in a single O supergiant. Has any work been done on how such winds are slowed down by the X-ray emission?

A.A.C. SANDER: As far as I am aware there is no explicit work with respect to these highly obscured Integral sources. However, depending on the orbital situation, a significant shut down of the wind acceleration and thus low velocities due to X-ray irradiation seems quite reasonable. This effect has been studied in Krtička *et al.* (2012) and also Sander *et al.* (2018), but both times based on the Vela X-1 system. J. Krtička will talk much more about the effect of X-ray irradiation (see these proceedings).

A. KASHI: Do you assume clumping in your simulations or do you get the clumping factor as a result?

A.A.C. SANDER: A depth-dependent clumping stratification is assumed in the Vela X-1 donor model. We account for the so-called "microclumping", i.e. optically thin clumps, which is the standard approach in current stellar atmosphere models. For a detailed optically thick clumping a two-component calculation would be required, which is computationally expensive and requires a considerable extension of a model code. Using a Monte Carlo approach, detailed calculations have been performed by Šurlan *et al.* (2012, 2013). For standard model atmospheres, an approximate treatment of optically thick clumping that does not rely on a two-component calculation has recently be introduced to FASTWIND by Sundqvist & Puls (2018).

D. Gies: Are the ionization calculations for the on-axis position in the binary?

A.A.C. Sander: Yes. Since the atmosphere model is essentially 1D, the performed calculations would best represent the on-axis situation.

F. Mirabel: Feedback from compact HMXBs is in the form of X-rays and jets (e.g. Cyg X-1, SS 433), and we know that the mechanical energy from the jets may be larger than that by X-rays. This feedback from jets complicate further the modeling of accretion winds.

A.A.C. Sander: Fair point. While the jets usually should not directly hit the surface of the donor star and modify the atmosphere, their existence would complicate the proper interpretation of the observations and measurements. Fortunately the Vela X-1 system is not known for having jets.

High-mass X-ray Binaries: Illuminating the Passage from Massive Binaries to Merging Compact Objects
Proceedings IAU Symposium No. 346, 2019
L. M. Oskinova, E. Bozzo, T. Bulik & D. Gies, eds.

doi:10.1017/S1743921318007573

Wind inhibition in HMXBs: the effect of clumping and implications for X-ray luminosity

Jiří Krtička[1], Jiří Kubát[2] and Iva Krtičková[1]

[1]Ústav teoretické fyziky a astrofyziky, Masarykova univerzita, Brno, Czech Republic
email: krticka@physics.muni.cz

[2]Astronomický ústav, Akademie věd České republiky, Ondřejov, Czech Republic
email: kubat@sunstel.asu.cas.cz

Abstract. Winds of hot stars are driven by the radiative force due to absorption of light in the lines of heavier elements. Consequently, the mass-loss rate and the wind velocity depend on the ionization state of the wind. As a result of this, there is a feedback between the ionizing X-ray source and the stellar wind in HMXBs powered by wind accretion. We study the influence of the small-scale wind structure (clumping) on this feedback using our NLTE hydrodynamical wind models. We find that clumping weakens the effect of X-ray irradiation. Moreover, we show that the observed X-ray luminosities of HMXBs can not be explained by wind accretion scenario without introducing the X-ray feedback. Taking into account the feedback, the observed and estimated X-ray luminosities nicely agree. We identify two cases of X-ray feedback with low and high X-ray luminosities that can explain the dichotomy between SFXTs and sgXBs.

Keywords. Stars: winds, outflows, stars: mass-loss, stars: early-type, X-rays: binaries, hydrodynamics

1. Introduction

A class of high-mass X-ray binaries (HMXBs) is powered by accretion of stellar wind on a compact component – either a neutron star or a black hole (see Martínez-Núñez *et al.* 2017, for a review). Stellar winds of massive hot stars are accelerated by the absorption of radiation mainly in resonance lines of heavier elements (e.g., C, N, O, and Fe). Therefore, the radiative force is sensitive to the ionization state of the stellar wind. As a result, while the wind powers X-ray emission, there is a feedback effect, because the emitted X-rays influence the radiative force. We will study the feedback effect of X-ray irradiation using our NLTE wind models.

2. Adopted model

We study the effect of X-ray irradiation using our METUJE code (Krtička & Kubát 2017) for calculation of spherically symmetric stationary hot star wind models. The radiative force is calculated in the comoving frame using solution of statistical equilibrium equations. Wind density, velocity and temperature are derived from hydrodynamical equations. The models enable us to predict the wind velocity, density, and temperature structure including wind mass-loss rate $\dot{M}$ and terminal velocity v_∞.

The flow in accreting high mass X-ray binary has a complex 3D structure. Calculation of detailed radiative force in such complex environment is beyond the possibilities of current computers. Therefore, to simplify the problem, we solve the wind equations only

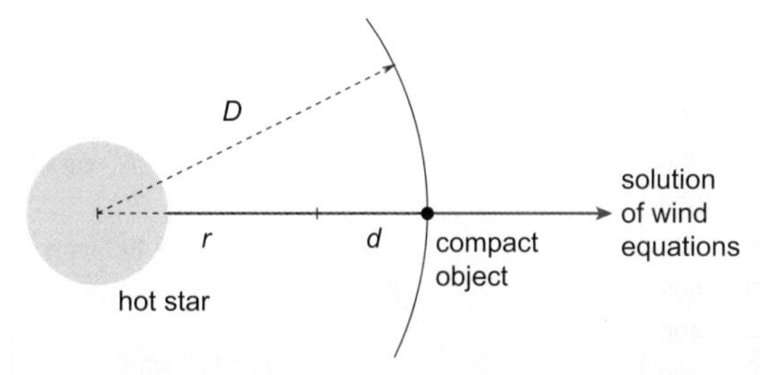

Figure 1. Assumed geometry of the solution. Here D is the binary separation, and r and d are distances of a given point from the donor and compact star, respectively.

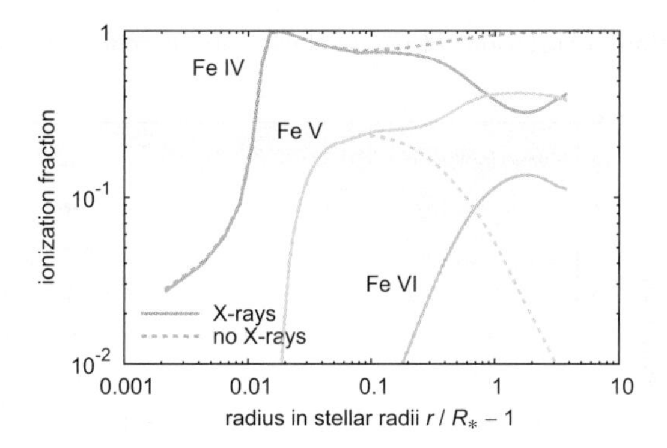

Figure 2. Radial variations of ionization equilibrium of iron in a model corresponding to Vela X-1. Dashed lines denote model without X-ray irradiation and solid lines model with X-ray irradiation.

in the direction of the X-ray source (see Fig. 1). This enables us to understand the effect of X-rays on the radiative force while keeping the whole problem tractable.

The effect of the compact companion is included only by the X-ray irradiation of the donor star due to the wind accretion on the compact star. We include an additional term to the mean intensity

$$J_\nu^{\mathrm{X}} = \frac{L_\nu^{\mathrm{X}}}{16\pi^2 d^2} \mathrm{e}^{-\tau_{\nu(r,d)}},$$

where L_ν^{X} is luminosity per unit of frequency, d is the distance from the neutron star, and the optical depth along a given ray is

$$\tau_{\nu(r,d)} = \int_0^d \kappa_\nu(z)\rho(z)\,\mathrm{d}z.$$

3. Influence of X-rays and the effect of clumping

The presence of strong X-ray radiation leads to a change of the ionization structure due to X-ray ionization (Fig. 2). Higher ionization states become more populated, while the population of lower ionization states decreases. Because the higher ions are less effective in line driving, the change of the wind ionization implies decrease of the radiative force (Hatchett & McCray 1977, Krtička *et al.* 2012, 2015, Sander *et al.* 2018). This is seen in the radial velocity plot in Fig. 3 as decrease of the wind velocity.

For a too high X-ray luminosity or too low binary separation, the X-rays penetrate deeply the wind base. This can lead to the wind disruption. Consequently, there is a

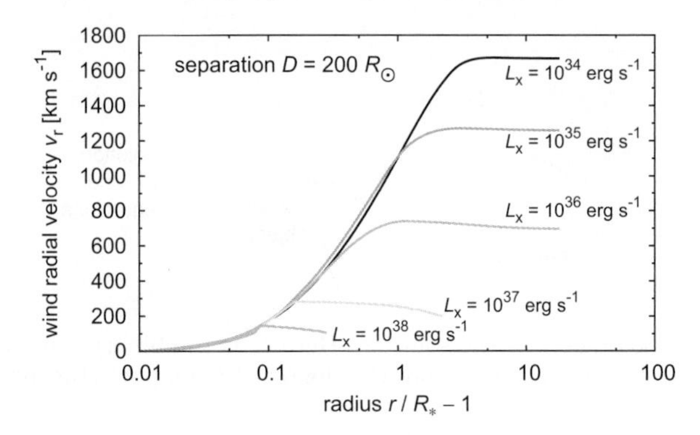

Figure 3. Radial variations of wind velocity of different X-ray luminosities.

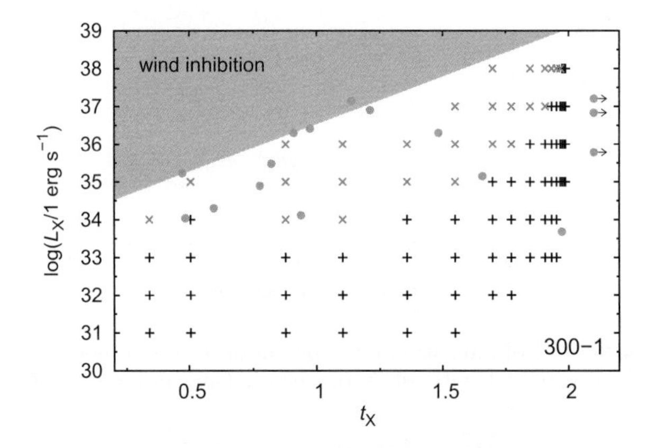

Figure 4. Models and massive binaries in $L_X - t_X$ diagram. Black plus symbols denote models where the X-ray irradiation does not significantly affect the radiative force. Red crosses denote models where the X-ray irradiation leads to decrease of the radiative force. The filled red area corresponds to a region where the X-ray irradiation is so strong that it leads to wind disruption. Real systems (blue circles) lie mostly outside the area of wind inhibition.

limiting X-ray irradiation given mostly by the wind optical depth (D is the binary separation)

$$\tau_X = \int_{R_*}^{D} \kappa_\nu \rho \, \mathrm{d}r \sim \frac{\kappa_\nu \dot{M}}{4\pi v_\infty} \left(\frac{1}{R_*} - \frac{1}{D} \right). \tag{3.1}$$

Thus the effect of wind disruption can be described by the optical depth parameter

$$t_X = \frac{\dot{M}}{v_\infty} \left(\frac{1}{R_*} - \frac{1}{D} \right) \left(\frac{10^3 \,\mathrm{km\,s^{-1}}\, 1\, R_\odot}{10^{-8}\, M_\odot\, \mathrm{yr}^{-1}} \right). \tag{3.2}$$

These effects can be manifested in the $L_X - t_X$ diagram (see Fig. 4 and Krtička *et al.* 2015). Here models with negligible influence of X-ray irradiation (black plus symbols) appear either for high optical depth parameters or for low X-ray luminosities. With lower optical depths or higher X-ray luminosities, the influence of X-rays becomes stronger decreasing with terminal velocity (red crosses). Winds are inhibited in filled red region. Real systems (blue dots) appear mostly outside the region of wind inhibition.

A good agreement between the position of real systems and the location of region of wind inhibition in Fig. 4 was achieved by inclusion of clumping (small scale wind inhomogeneities). Clumping increases the local wind density; therefore, it favours recombination

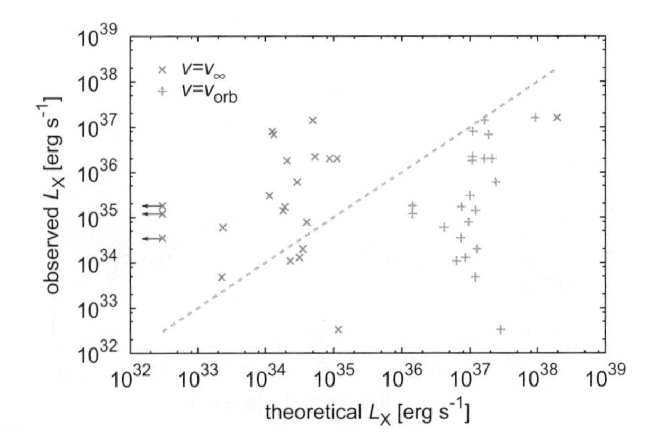

Figure 5. Predicted X-ray luminosity for sample of X-ray binaries powered by wind accretion. Derived using Eq. (4.1) assuming two limiting values of v.

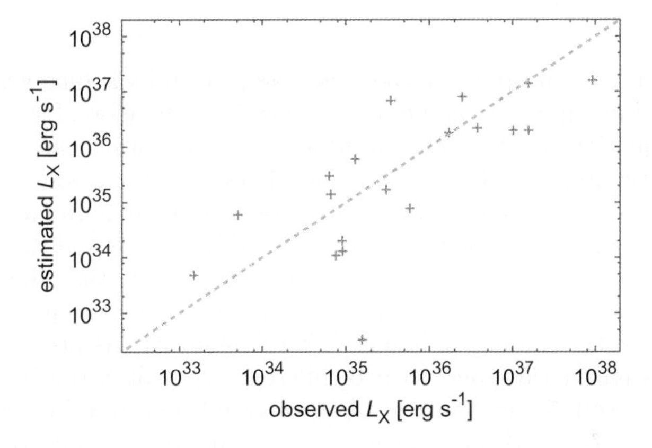

Figure 6. Predicted X-ray luminosity for sample of X-ray binaries powered by wind accretion. Derived using Eq. (4.1) assuming the dependence of v_∞ on L_X predicted by our wind models.

leading to weaker influence of X-rays (Oskinova *et al.* 2012). Furthermore, clumping increases the mass-loss rate (Muijres *et al.* 2011), causing larger X-ray optical depth and larger X-ray absorption.

4. Prediction of X-ray luminosity

The X-ray luminosity of wind accreting systems can be estimated within Bondi-Hoyle-Lyttleton model (Hoyle & Lyttleton 1941; Bondi & Hoyle 1944) as

$$L_X = \frac{G^3 M_X^3}{R_X D^2 v^4} \dot{M}, \tag{4.1}$$

where R_X is radius of the compact companion, $v^2 = v_\infty^2 + v_{orb}^2$, and v_{orb} is the orbital velocity. Assuming $v = v_\infty$ for real wind-fed systems, Eq. 4.1 predicts too low luminosity in comparison with observations (red crosses in Fig. 5). However, inserting $v = v_{orb}$ gives theoretical predictions that are always above the experimental values (blue plus symbols in Fig. 5). This demonstrates the importance of accounting for realistic wind velocity when predicting the X-ray luminosity (Ho & Arons 1987, Sander *et al.* 2018).

We used fits of the results of our models to test if the predicted X-ray luminosity is consistent with observed X-ray luminosity. Fig. 6 shows that these values nicely agree.

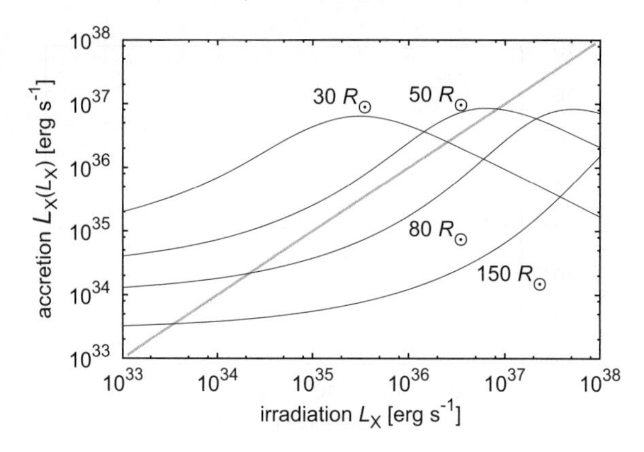

Figure 7. Predicted accretion X-ray luminosity Eq. (4.1) as a function of X-ray irradiation (black lines) for different binary separations D denoted in the plot. Red line corresponds to one-to-one relation. The intersection of the black and red lines gives solution of implicit equation Eq. (4.1) for L_X.

In a presence of X-ray irradiation, the mass-loss rate and terminal velocity in Eq. (4.1) are function of X-ray luminosity and binary separation. Therefore, for a fixed D, Eq. (4.1) is an implicit equation for L_X. The solution of this equation can be either obtained numerically or estimated from Fig. 7. Here black lines give the X-ray luminosity generated by the accretion as a function of X-ray irradiation. For a low X-ray irradiation the wind mass-loss rate and terminal velocity are not affected by X-rays and therefore the amount of accreted matter and also the X-ray luminosity are low. On the other hand, for higher X-ray irradiation the wind terminal velocity decreases, and consequently the amount of accreted mass is higher leading to higher X-ray luminosity (as predicted by Eq. (4.1)). These two states may correspond to two different types of X-ray binaries (supergiant X-ray binaries and soft X-ray transients, respectively). For even higher luminosities the mass-loss rate decreases with increasing X-ray irradiation, leading to a decrease of the accretion X-ray luminosity in Fig. 7. This shows that supergiant X-ray binaries are in self-regulated state.

5. Conclusions

We studied the effect of X-ray irradiation and small-scale wind structures (clumping) on the wind in HMXBs. As a result of X-ray irradiation, the radiative force decreases. This causes the decrease of the wind terminal velocity and also of the mass-loss rate for extreme X-ray irradiation. We found that clumping weakens the effect of X-ray irradiation. Moreover, we show that the observed X-ray luminosities of HMXBs can not be explained by wind accretion scenario without introducing the X-ray feedback. Taking into account the feedback, the observed and estimated X-ray luminosities nicely agree. We identify two cases of X-ray feedback with low and high X-ray luminosities that can explain the dichotomy between SFXTs and sgXBs.

This research was supported by grant GA ČR 18-05665S.

References

Bondi, H., & Hoyle, F. 1944, *MNRAS*, 104, 273
Hatchett, S., & McCray, R. 1977, *ApJ*, 211, 552
Ho, C., & Arons, J. 1987, *ApJ*, 316, 283
Hoyle, F., & Lyttleton, R. A. 1941, *MNRAS*, 101, 227
Krtička, J., & Kubát, J. 2017, *A&A*, 606, A31

Krtička, J., Kubát, J., & Skalický, J. 2012, *ApJ*, 757, 162
Krtička, J., Kubát, J., & Krtičková, I. 2015, *A&A*, 579, A111
Martínez-Núñez, S., Kretschmar, P., Bozzo, E., *et al.* 2017, Space Sci. Rev; 212, 59
Muijres L., de Koter A., Vink J., *et al.* 2011, *A&A*, 526, A32
Oskinova, L. M., Feldmeier, A., & Kretschmar, P. 2012, *MNRAS*, 421, 2820
Sander, A. A. C., Fürst, F., Kretschmar, P., *et al.* 2018, *A&A*, 610, A60

High-mass X-ray Binaries: Illuminating the Passage from Massive Binaries to Merging Compact Objects
Proceedings IAU Symposium No. 346, 2019
L. M. Oskinova, E. Bozzo, T. Bulik & D. Gies, eds.

doi:10.1017/S1743921318008414

Clumpy wind accretion in Supergiant X-ray Binaries

Ileyk El Mellah[1], Andreas A. C. Sander[2,3], Jon O. Sundqvist[4] and Rony Keppens[1]

[1]Centre for mathematical Plasma Astrophysics, Department of Mathematics, KU Leuven, Celestijnenlaan 200B, B-3001 Leuven, Belgium
email: ileyk.elmellah@kuleuven.be

[2]Armagh Observatory and Planetarium, College Hill, Armagh, BT61 9DG, Northern Ireland

[3]Institut für Physik und Astronomie, Universität Potsdam, Karl-Liebknecht-Str. 24/25, 14476 Potsdam, Germany

[4]KU Leuven, Instituut voor Sterrenkunde, Celestijnenlaan 200D, B-3001 Leuven, Belgium

Abstract. Supergiant X-ray Binaries host a compact object, generally a neutron star, orbiting an evolved O/B star. Mass transfer proceeds through the intense radiatively-driven wind of the stellar donor, a fraction of which is captured by the gravitational field of the neutron star. The subsequent accretion process onto the neutron star is responsible for the abundant X-ray emission from those systems. They also display variations in time of the X-ray flux by a factor of a few 10, along with changes in the hardness ratios believed to be due to varying absorption along the line-of-sight. We used the most recent results on the inhomogeneities (aka clumps) in the non-stationary wind of massive hot stars to evaluate their impact on the time-variable accretion process. We ran three-dimensional simulations of the wind in the vicinity of the accretor to witness the formation of the bow shock and follow the inhomogeneous flow over several spatial orders of magnitude, down to the neutron star magnetosphere. In particular, we show that the impact of the clumps on the time-variability of the intrinsic mass accretion rate is severely damped by the crossing of the shock, compared to the purely ballistic Bondi-Hoyle-Lyttleton estimation. We also account for the variable absorption due to clumps passing by the line-of-sight and estimate the final effective variability of the mass accretion rate for different orbital separations. These results are confronted to recent analysis of Vela X-1 observations with Chandra by Grinberg *et al.* (2017). It shows that clumps account well for time-variability at low luminosity but can not generate, per se, the high luminosity activity observed.

Keywords. accretion, accretion disks, methods: numerical, hydrodynamics, stars: neutron, X-rays: binaries, stars: winds, outflows, stars: supergiants, plasmas, stars: early-type.

---

1. Introduction

Supergiant X-ray binaries (SgXB) are thought to be the ideal stage to witness wind accretion. A SgXB hosts a Supergiant star which looses mass via a dense and fast line-driven wind, with a mass loss rate of the order of $10^{-6}M_{\odot}\cdot yr^{-1}$, while it is orbited by a compact companion, generally a neutron star (NS). The NS is deeply embedded in the wind, standing at approximately one stellar radius above the stellar photosphere. It captures a fraction of the wind and as it falls onto the compact object, the accreted flow emits a plethora of X-rays which account for the observed X-ray luminosity, ranging from 10^{35} to $10^{37}erg\cdot s^{-1}$ in SgXB. In this proceedings, I want to provide new insights about the possibility to study these systems through the coupling of multiple numerical simulations.

In SgXB, the main challenge we face when carrying out a numerical investigation is to bridge the scale gap between, at the lower end, where most of the X-rays we observe are emitted, the size of the compact object, and, at the upper end, the orbital separation, 6 orders of magnitude larger. Fortunately, the dominant physics at stake at each scale is different: if the immediate vicinity of the accretor requires a relativistic treatment, the NS magnetosphere needs to be handled in a magneto-hydrodynamical framework. While at the orbital scale, the bulk motion of the wind is essentially ballistic and a hydrodynamical (HD) bow shock forms within the Roche lobe of the compact object.

In this proceedings, I want to focus on what happens in-between: how is the flow carried from the bow shock down to the outer rim of the NS magnetosphere? And in particular :

• what is the impact of the overdense regions in the wind on the time variability of the mass accretion rate?

• does the flow gain enough angular momentum at the orbital scale and does it retain enough of it downstream the shock to form a disc-like structure before being truncated by the NS magnetosphere? Is there enough room for a disk between the shock and the NS magnetosphere?

2. Clumps in the wind

Wind launching for hot massive stars relies on the resonant line absorption of stellar UV photons by partly ionized metal ions in the outer layers of the star. As the flow accelerates, it keeps tapping Doppler-shifted previously untouched photons (Lucy & Solomon 1970, Castor, Abbott & Klein 1975). Due to the line-deshadowing instability Owocki & Rybicki (1984), we expect strong internal shocks to develop in the wind and lead to the formation of overdense regions a.k.a. "clumps". The shape, dimension and amount of mass contained in these clumps has been for long a matter of debate: the radiative-HD computation required to tackle this question are computationally expensive and only uni-dimensional radial simulations could be performed Feldmeier *et al.* (1997) but they would not tell us about the transverse extension of the clumps.

But last year, Sundqvist, Owocki & Puls (2017) performed two-dimensional simulations of the wind launching, resolved the micro-structure and derived a density contrast of the order of 100 in the wind. To evaluate the impact of this micro-structure on the wind accretion process, we plunge the orbiting compact object in the wind. We define a zone of gravitational and radiative influence around the accretor: within this zone, the X-ray ionizing feedback from the accreted flow onto the wind inhibit the line-driven acceleration (Hatchett & McCray 1977, Blondin *et al.* 1990, Stevens 1991, Manousakis & Walter 2015). The outer boundary conditions in the upstream hemisphere of the 3D spherical simulation space are entirely determined by the wind simulation of Sundqvist, Owocki & Puls (2017): we directly inject the clumps within the simulation space and monitor their HD evolution, using the new version of the `MPI-AMRVAC` code described in Xia *et al.* (2017).

The initial state of the simulations is a 3D extension of the axisymmetric planar uniform Bondi-Hoyle-Lyttleton problem (Hoyle & Lyttleton 1939; Bondi & Hoyle 1944) computed in El Mellah & Casse (2015). Upstream the shock, the clumps are ballistically advected in the supersonic flow. However, they do not preserve their structure when they cross the shock: the clumps are not zero-temperature bullets and do experience HD effects such as the sudden increase in entropy at the shock. Contrary to the uniform planar case, the instantaneous net amount of angular momentum is not zero since the captured clumps arriving with a non-zero impact parameter carry their own angular momentum and do not have a counterpart with opposite impact parameter to cancel out with. Consequently, the accretion of a clump can be delayed if it enhances the absolute value of the angular

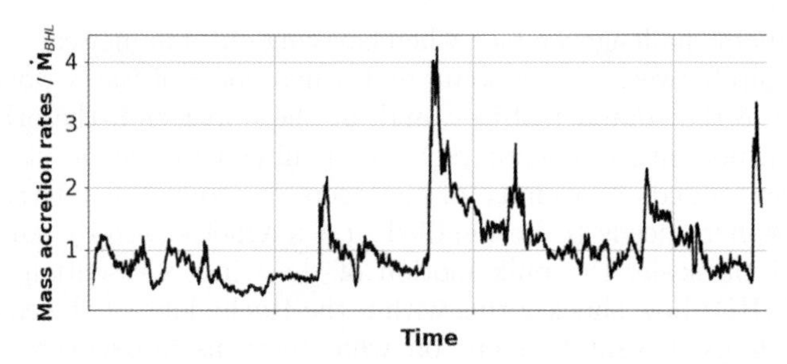

Figure 1. Mass accretion rate compared to the Bondi-Hoyle-Lyttleton proxy as a function of time (code units).

momentum already available in the shocked region while it can trigger the flush of a larger amount of matter if it lowers the absolute value of the angular momentum in the shocked region.

Because of this mixing of the clumps with the material already present downstream the shock, this region acts as a buffer where the clumps are restructured. It also means that the peak observed in mass accretion rate through the inner border of the simulation space (whose radius is a few times the NS magnetosphere radius) in Figure 1 do not correspond to one clump in particular but to the cumulative and contingent contribution of several of them: deriving the mass of the clump responsible for a given X-ray flare always leads to an overestimation compared to the actual mass of the clumps which participated in producing this flare.

The overall peak-to-peak variability in mass accretion rate reaches 20 for an orbital separation of 2 stellar radii and 10 for an orbital separation of 1.6 stellar radii. It is an order of magnitude lower than the observed time variability. The origin of this discrepancy might be numerical (e.g. an integration time too low compared to the characteristic time of occurrence of the extreme cases) but more likely, additional instabilities might occur within and at the edge of the NS magnetosphere which might amplify this time variability (see e.g. the propeller effect, Bozzo, Falanga & Stella, 2008) and account for the lack of high mass accretion rate regimes in our simulations.

The work presented in this section has been reported in more detail in El Mellah, Sundqvist & Keppens (2017).

3. Orbital bending and disc formation

The reader might also wonder about the contribution to time-variability of serendipitous absorption by unaccreted clumps passing by the line-of-sight. This question has been addressed in more detail in Grinberg *et al.* (2017) but the main conclusion is that with such a high wind speed, the duration of the coherent absorption events observed can not be reproduced.

But is the wind really that fast? In this preliminary model, we just cared about the clumps and assumed that outside of this limited simulation space the whole structure of the wind did not depart from the one of an isolated massive star, but it is only correct if the wind is too fast to see the Roche potential i.e. if the wind speed is large compared to the orbital speed. In a system such as Vela X-1, recent observations by Gimenez-Garcia *et al.* (2016) confirmed that the terminal wind speed was only twice the orbital speed and Sander *et al.* (2017) showed that the wind might accelerate slower than expected and consequently, upstream the accreting NS, the wind speed is not large compared to the orbital speed. It means that the dynamics of the wind will be dominated by orbital effects. From now on, I set aside the micro-structure of the wind, no more clumps, but I address the question of the systematic deviation of the wind from a purely radial wind

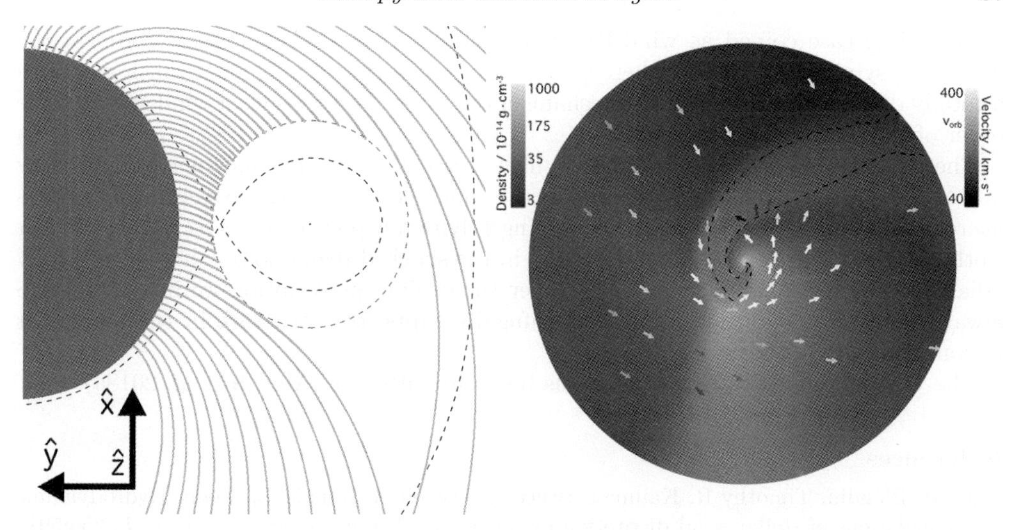

Figure 2. (left) Wind streamlines in orange in the orbital plane of the co-rotating frame, for the case Heavy Slow (HS). The black dashed line stands for the critical Roche potential curve passing by the first Lagrangian point while the green dashed line is the HD simulation space where we inject the wind. (right) Logarithmic colormap of the mass density in the orbital plane. The arrows stand for the velocity field and the black dashed line indicates a Mach-1 locus.

by the Roche potential and the Coriolis force. How much angular momentum does the accreted flow carry?

To evaluate the motion of the wind, we assimilate it to test-masses and compute in 3D in the co-rotating frame the steady streamlines using an integrator developed and validated in El Mellah & Casse (2016). The test-masses which reach an extended Roche lobe centered provide natural outer boundary conditions at the outer edge of the HD simulation space. The dynamics is set by the Roche potential, the Coriolis force and the line-driven acceleration taken from Sander *et al.* (2017). We consider two cases :

• Heavy slow (HS) : the line-driven acceleration is not altered and leads to wind speed of the order or smaller than the orbital speed upstream the accretor (of mass $2.5M_{\odot}$).

• Light fast (LF) : the line-driven acceleration is enhanced by 50%, leading to wind speeds 20% larger than the orbital speed. The accretor has now a mass of $1.5M_{\odot}$.

The result of this computation is illustrated for the HS case in Figure 2 (left) where the bending of the streamlines is significantly more important than in the LF case. The latter displays streamlines more radial while the former shows that a non negligible amount of angular momentum flows into the Roche potential of the accretor.

The consequences are dramatic. While the LF configuration leads to structures qualitatively similar to the classic fast wind Bondi-Hoyle-Lyttleton picture, the HS configuration produces a totally different geometry. In the LF case, the flow is still essentially planar and axisymmetric around the mean direction of arrival, which deviates from the line joining the 2 bodies by only ∼20 degrees. The bow shock is still present, along with the inner sonic surface. In the HS case (Figure 2, right), where the wind speed entering the simulation space is only a few 10% lower but enough to reach the orbital speed, the flow is highly compressed in the equatorial plane which leads to a more important density enhancement. It is also more beamed along a channel of matter reminiscent of the one observed in Roche lobe overflowing systems, although the star does not fill its Roche lobe. As explained in El Mellah, Sundqvist & Keppens (2018), it leads to a significant enhancement of the mass transfer rate and can lead to levels suitable for ultra-luminous X-ray sources. The shock is now totally misaligned and takes a spiral shape. It is the intermediate case coined as wind-RLOF by Mohamed & Podsiadlowski (2007) in the context of symbiotic binaries, where the donor is an Asymptotic Giant Branch star with a totally different wind-launching mechanism but where you retrieve a wind speed of the order of the orbital speed.

These HD simulations are adiabatic which means that matter does not radiate away any of the entropy it has been granted at the shock, leading to excessively high temperatures. We empirically represent the cooling relying on polytropic prescriptions: either isothermal or assuming a constant entropy in the shocked region lower than what a fully adiabatic simulation would yield. Whatever the cooling prescription we invoke, the HS always leads to the formation of a centrifugally supported structure in the innermost region while the LF never.

The work presented in this section has been submitted (El Mellah *et al.* 2018a)

References

John M. Blondin, Timothy R. Kallman, Bruce A. Fryxell, & Ronald E. Taam. Hydrodynamic simulations of stellar wind disruption by a compact X-ray source. *Astrophys. J.*, 356:591–608, jun 1990

H. Bondi & F. Hoyle. On the mechanism of accretion by stars. *Mon. Not. R. Astron. Soc.*, 104:273, 1944.

E. Bozzo, M. Falanga, & L. Stella. Are There Magnetars in HighMass XRay Binaries? The Case of Supergiant Fast XRay Transients. *Astrophys. J.*, 683(2):1031–1044, aug 2008.

J. I. Castor, D. C. Abbott, & R. I. Klein. Radiation-driven winds in Of stars. *Astrophys. J.*, 195:157, jan 1975.

L. Ducci, L. Sidoli, & A. Paizis. INTEGRAL results on supergiant fast X-ray transients and accretion mechanism interpretation: ionization effect and formation of transient accretion discs. *Mon. Not. R. Astron. Soc.*, 408(3):1540–1550, nov 2010.

I. El Mellah & F. Casse. Numerical simulations of axisymmetric hydrodynamical BondiHoyle accretion on to a compact object. *Mon. Not. R. Astron. Soc.*, 454(3):2657–2667, oct 2015.

I. El Mellah & F. Casse. A numerical investigation of wind accretion in persistent Supergiant X-ray Binaries I - Structure of the flow at the orbital scale. *Mon. Not. R. Astron. Soc.*, 467(3):2585–2593, sep 2016.

I. El Mellah, J. O. Sundqvist, & R. Keppens. Accretion from a clumpy massive-star wind in Supergiant X-ray binaries. *Mon. Not. R. Astron. Soc. Vol. 475, Issue 3, p.3240-3252*, 475:3240–3252, nov 2017.

I. El Mellah, Andreas A. C. Sander, J. O. Sundqvist, & R. Keppens. Formation of wind-captured discs in Supergiant X-ray binaries Consequences for Vela X-1 and Cygnus X-1. submitted, arXiv : 1810.12933

I. El Mellah, J. O. Sundqvist, & R. Keppens. Wind Roche lobe overflow in high mass X-ray binaries - A possible mass transfer mechanism for Ultraluminous X-ray sources. submitted, arXiv : 1810.12937

A. Feldmeier, R.-P. Kudritzki, R. Palsa, A. W. A. Pauldrach, & J. Puls. The X-ray emission from shock cooling zones in O star winds. *Astron. Astrophys.*, 320:899–912, 1997.

Fürst, F., Kreykenbohm, I., Pottschmidt, K., Wilms, J., Hanke, M., Rothschild, R. E., Kretschmar, P., Schulz, N. S., Huenemoerder, D. P., Klochkov, D. & Staubert, R. X-ray variation statistics and wind clumping in Vela X-1 *Astron. Astrophys.*, 519, 2010.

A. Gimenez-Garcia, T. Shenar, J. M. Torrejon, L. Oskinova, S. Martinez-Nunez, W.-R. Hamann, J. J. Rodes-Roca, A. Gonzalez-Galan, J. Alonso-Santiago, C. Gonzalez-Fernandez, G. Bernabeu, & A. Sander. Measuring the stellar wind parameters in IGR J17544-2619 and Vela X-1 constrains the accretion physics in Supergiant Fast X-ray Transient and classical Supergiant X-ray Binaries. *Astron. Astrophys.*, 591(A26):25, mar 2016.

V. Grinberg, N. Hell, I. El Mellah, J. Neilsen, A. A. C. Sander, M. Leutenegger, F. Fürst, D. P. Huenemoerder, P. Kretschmar, M. Kühnel, S. Martínez-Núñez, S. Niu, K. Pottschmidt,

N. S. Schulz, J. Wilms, & M. A. Nowak. The clumpy absorber in the high-mass X-ray binary Vela X-1. *Astron. Astrophys. Vol. 608, id.A143, 18 pp.*, 608, nov 2017.

S. Hatchett & R. McCray. X-ray sources in stellar winds. *Astrophys. J.*, 211:552, jan 1977.

F. Hoyle & R. A. Lyttleton. The effect of interstellar matter on climatic variation. *Math. Proc. Cambridge Philos. Soc.*, 35(03):405–415, oct 1939.

Shigeyuki Karino. Bimodality of Wind-fed Accretion in High Mass X-ray Binaries. *Publ. Astron. Soc. Japan*, 66(2):2–3, mar 2014.

Jiri Krticka, & Jiri Kubat. Influence of X-ray radiation on the hot star wind ionization state and on the radiative force. *Adv. Sp. Res.*, 58(5):710–718, feb 2016.

L. B. Lucy & P. M. Solomon. Mass Loss by Hot Stars. *Astrophys. J.*, 159:879, mar 1970.

A. Manousakis & R. Walter. The stellar wind velocity field of HD 77581. *Astron. Astrophys. Vol. 584, id.A25, 5 pp.*, 584, jul 2015.

S. Mohamed & Philipp Podsiadlowski. Wind Roche-Lobe Overflow: a New Mass-Transfer Mode for Wide Binaries, 2007, ASPC, 372, 397.

S. P. Owocki & G. B. Rybicki. Instabilities in line-driven stellar winds. I - Dependence on perturbation wavelength. *Astrophys. J.*, 284:337, sep 1984.

Andreas A. C. Sander, Felix Fürst, Peter Kretschmar, Lidia M. Oskinova, Helge Todt, Rainer Hainich, Tomer Shenar & Wolf-Rainer Hamann. Coupling hydrodynamics with comoving frame radiative transfer: II. Stellar wind stratification in the high-mass X-ray binary Vela X-1. *Astron. Astrophys.*, 610:A60, feb 2017.

Ian R. Stevens. X-ray-illuminated stellar winds - Optically thick wind models for massive X-ray binaries. *Astrophys. J.*, 379:310, sep 1991.

J. O. Sundqvist, S. P. Owocki, & J. Puls. 2D wind clumping in hot, massive stars from hydrodynamical line-driven instability simulations using a pseudo-planar approach. *Astron. Astrophys. Vol. 611, id.A17, 10 pp.*, 611, oct 2017.

Walter, Roland, Lutovinov, Alexander A., Bozzo, Enrico and Tsygankov, & Sergey S. High-Mass X-ray Binaries in the Milky Way: A closer look with INTEGRAL. *Astron. Astrophys. Rev.*, 23, 2015

C. Xia, J. Teunissen, I. El Mellah, E. Chané, & R. Keppens. MPI-AMRVAC 2.0 for Solar and Astrophysical Applications. *Astrophys. J. Suppl. Ser.*, 234(2):30, oct 2018.

Discussion

KARINO: What about the subsonic quasi-spherical shell settling model by Shakura *et al.* (2013)? Is it compatible with the wind-capture disc you observe?

EL MELLAH: In the presence of a disc, we should rather resort on the Gosh and Lamb coupling with the NS magnetosphere. But keep in mind that for slightly (20%) faster winds, you go over the orbital speed and a disc does not form. Also, if the cooling is inefficient, a disc does not form neither. In these 2 cases, the quasi-spherical subsonic model remains valid.

POSTNOV: Did you characterize the statistical information/behavior of the mass accretion rate in your simulations and compared it to the observed ones?

EL MELLAH: In El Mellah, Sundqvist & Keppens (2017), I plotted the activity diagrams (see Figure 10), for different orbital separations and absorption, and compared it to the observed ones by Fürst *et al.* (2010) and Walter *et al.* (2015). The peak-to-peak variability is an order of magnitude lower but I retrieve approximately log-normal distributions.

High-mass X-ray Binaries: Illuminating the Passage from Massive Binaries to Merging Compact Objects
Proceedings IAU Symposium No. 346, 2019
L. M. Oskinova, E. Bozzo, T. Bulik & D. Gies, eds.

doi:10.1017/S1743921318007585

Studying the presence of magnetic fields in a sample of high-mass X-ray binaries

Swetlana Hubrig[1], Alexander F. Kholtygin[2], Lara Sidoli[3], Markus Schöller[4] and Silva P. Järvinen[1]

[1]Leibniz-Institut für Astrophysik Potsdam (AIP), An der Sternwarte 16, 14482 Potsdam, Germany
email: shubrig@aip.de

[2]Saint-Petersburg State University, Universitetskij pr. 28, 198504 Saint-Petersburg, Russia

[3]INAF, Istituto di Astrofisica Spaziale e Fisica Cosmica, Via E. Bassini 15, 20133 Milano, Italy

[4]European Southern Observatory, Karl-Schwarzschild-Str. 2, 85748 Garching, Germany

Abstract. Previous circular polarization observations obtained with the ESO FOcal Reducer low dispersion spectrograpgh at the VLT in 2007–2008 revealed the presence of a weak longitudinal magnetic field on the surface of the optical component of the X-ray binary Cyg X-1, which contains a black hole and an O9.7Iab supergiant on a 5.6 d orbit. In this contribution we report on recently acquired FORS 2 spectropolarimetric observations of Cyg X-1 along with measurements of a few additional high-mass X-ray binaries.

Keywords. stars: magnetic fields, stars: individual (BP Cru, Cyg X-1, Vela X-1, LS 5039), (stars:) supergiants, (stars:) binaries: general, X-rays: stars

---

1. Introduction

High-mass X-ray binaries are fundamental for studying stellar evolution, nucleosynthesis, structure and evolution of galaxies, and accretion processes. The classical high-mass X-ray binaries (HMXBs) with supergiant companions (SgHMXBs) are known since the birth of X-ray astronomy and are persistent X-ray emitters, with a limited range of intensity variability (around a factor of 10). These targets are among the brightest X-ray sources in the sky. For them, the observed spectral and time variability is best explained by assuming that accretion onto the compact object is taking place from a highly structured stellar wind, where cool dense clumps are embedded in a rarefied photoionized gas.

While the first spectropolarimetric observations using the FOcal Reducer low dispersion Spectrograph (FORS 1/2; Appenzeller *et al.* 1998) mounted on the 8 m Antu telescope of the Very Large Telescope indicated the presence of a rather strong longitudinal magnetic field of a kG order in the supergiant fast X-ray transient (SFXT) IGR J11215-5952 (Hubrig *et al.* 2018), no systematical search for magnetic fields was carried out in SgHMXBs. SFXTs are a subclass of HMXBs associated with early-type supergiant companions, and characterized by sporadic, short and bright Xray flares reaching peak luminosities of 10^{36}–10^{37} erg s^{-1} and typical energies released in bright flares of about 10^{38}–10^{40} erg (see the review by Sidoli in 2017 for more details). Different accretion regimes - transient in the settling accretion mode versus persistent in the free-fall Bondi mode - were suggested in the last years for SFXTs and SgHMXBs, respectively.

The magnetic field of the eclipsing binary Vela X-1 with a B0.5Ia component and an orbital period of 8.96 d was previously studied by Hubrig *et al.* (2013). However, no magnetic field detection at a significance level of 3σ has been achieved in this system. The spectral behaviour of Vela X-1 is known to be very complex due to the presence of bumps and wiggles in the line profiles, and an impact of tidal effects producing orbital phase-dependent variations in the line profiles leading to asymmetries such as extended blue or red wings (Koenigsberger *et al.* 2012). Furthermore, Kreykenbohm *et al.* (2008) detected flaring activity and temporary quasi-periodic oscillations in INTEGRAL X-ray observations. Magnetic field measurements of the X-ray binary Cyg X-1, with the historically first black-hole candidate, using FORS 2 low-resolution spectropolarimetric observations were reported by Karitskaya *et al.* (2010). The authors detected a relatively weak mean longitudinal magnetic fields of the order of 100 G with a few measurements at a significance level in the range between 3.5 and 6.2σ.

In this contribution we discuss the most recent FORS 2 spectropolarimetric observations of the Cyg X-1 system and two other SgHMXB systems, BP Cru, and LS 5039. For completeness we also present the older results for Vela X-1, as this system can be considered as the prototype of persistent HMXBs.

2. Magnetic field measurements

The FORS 2 multi-mode instrument is equipped with polarisation analysing optics comprising super-achromatic half-wave and quarter-wave phase retarder plates, and a Wollaston prism with a beam divergence of $22''$ in standard resolution mode. We used the GRISM 600B and the narrowest available slit width of $0.4''$ to obtain a spectral resolving power of $R \approx 2000$. The observed spectral range from 3250 to 6215 Å includes all Balmer lines, apart from Hα, and numerous helium lines. For the observations, we used a non-standard readout mode with low gain (200kHz,1×1,low), which provides a broader dynamic range, hence allowed us to reach a higher signal-to-noise ratio in the individual spectra. The spectral appearance of all targets in the FORS 2 spectra is presented in Fig. 1.

The longitudinal magnetic field was measured in two ways: using the entire spectrum including all available lines or using exclusively the hydrogen lines. Furthermore, we carried out Monte Carlo bootstrapping tests. These are most often applied with the purpose of deriving robust estimates of standard errors. The measurement uncertainties obtained before and after the Monte Carlo bootstrapping tests were found to be in close agreement, indicating the absence of reduction flaws. The results of our magnetic field measurements, those for the entire spectrum or only for the hydrogen lines, are presented in Table 1.

Our measurements do not reveal the presence of significant mean longitudinal magnetic fields in any of the four studied binaries. We observe changes of the field polarity in the measurements of all targets, but the measurement uncertainties are rather large, leading to significance levels of only 2.3–2.5σ and less. On the other hand, as we show in Fig. 2, typical Zeeman features are detected in the FORS 2 Stokes V spectra of BP Cru and Cyg X-1.

3. Discussion

The recent spectropolarimetric observations of four SgHMXBs with FORS 2 showed changes in the field polarities, but the measurement uncertainties are too large to allow us to conclude on the presence of magnetic fields. Apart from the search for magnetic fields, the acquired spectra of the optical components allowed us to detect significant spectral variability: spectral lines belonging to hydrogen and other elements show changes of the line intensities and radial velocities over different observing nights. We present a few examples in Figs. 3–6 showing individual Stokes I helium and hydrogen line profiles.

Table 1. Longitudinal magnetic field values obtained using FORS 2 observations of four high-mass X-ray binaries. In the first two columns we show the name of the binary and the modified Julian date of mid-exposure, followed by the mean longitudinal magnetic field using the Monte Carlo bootstrapping test, for all lines and for the hydrogen lines. In the last column, we present the significance of the measurements of $\langle B_z \rangle_{all}$ using the set of all lines. All quoted errors are 1σ uncertainties.

Name	MJD	$\langle B_z \rangle_{all}$ [G]	$\langle B_z \rangle_{hyd}$ [G]	Significance σ
BP Cru	57528.0781	344 ± 149	441 ± 173	2.3
BP Cru	57533.1965	−144 ± 162	−319 ± 193	0.9
BP Cru	57591.0529	254 ± 162	82 ± 214	1.6
Cyg X-1	57585.1653	159 ± 63	447 ± 221	2.5
Cyg X-1	57593.2230	−61 ± 82	− 274 ± 237	0.8
Cyg X-1	57644.0254	−88 ± 44	−111 ± 157	2.0
Cyg X-1	57645.0189	−79 ± 45	−52 ± 114	1.7
Vela X-1	55686.0962	−15 ± 30	−11 ± 45	0.5
Vela X-1	55687.0584	−80 ± 32	−114± 50	2.5
Vela X-1	55688.0479	57 ± 37	88 ± 69	1.5
LS 5039	57585.1260	794 ± 277	634 ± 317	2.5
LS 5039	57591.0970	−517 ± 291	−118 ± 383	1.8
LS 5039	57646.1463	534 ± 268	199 ± 340	2.0
LS 5039	57611.2456	525 ± 450	953 ± 617	1.2

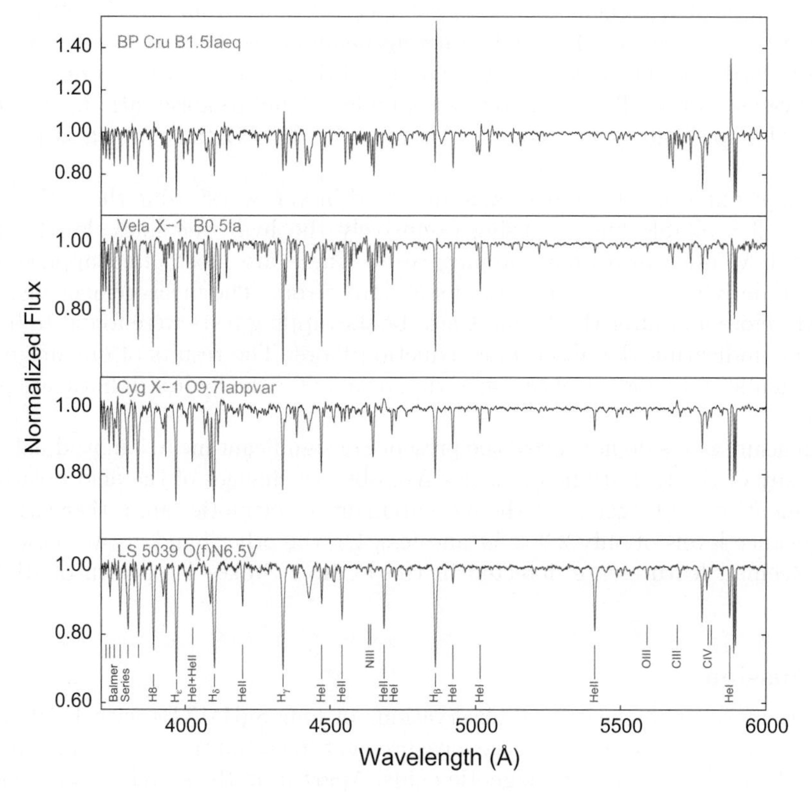

Figure 1. Normalised FORS 2 spectra of LS 5039, Cyg X-1, Vela X-1, and BP Cru. Well known spectral lines are indicated.

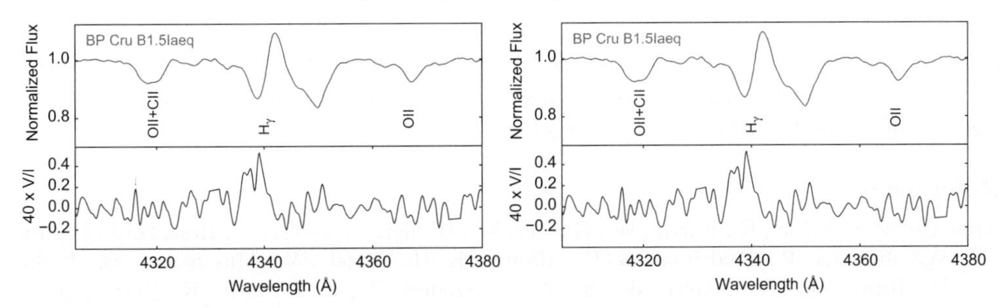

Figure 2. Examples of Stokes I and Stokes V spectra of BP Cru (left panel) and Cyg X-1 (right panel) in the vicinity of the $H\gamma$ line and the He I 5876 line, respectively.

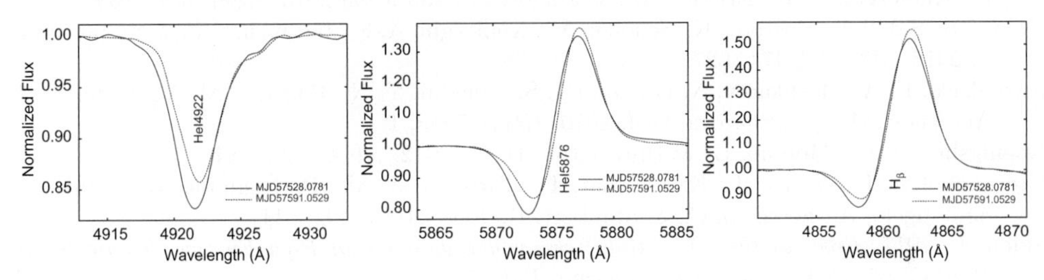

Figure 3. Variability of spectral lines in the FORS 2 spectra of BP Cru recorded on two different nights.

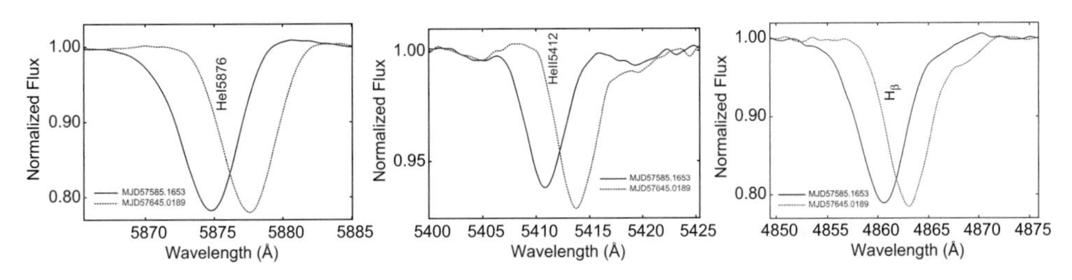

Figure 4. Same as in Fig. 3, but for Cyg X-1.

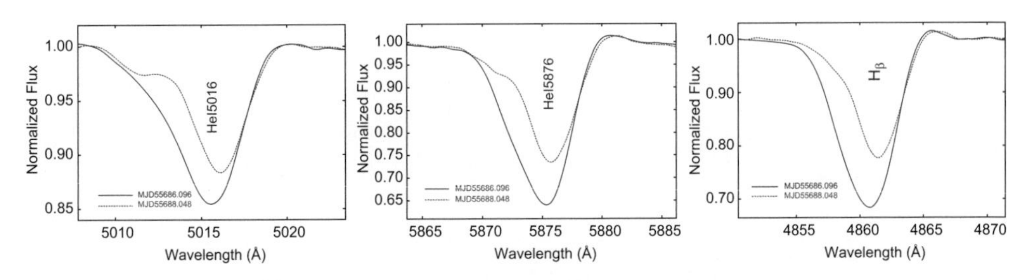

Figure 5. Same as in Fig. 3, but for Vela X-1.

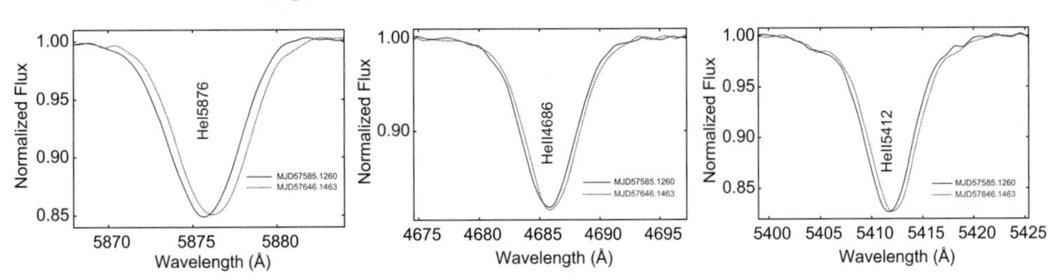

Figure 6. Same as in Fig. 3, but for LS 5039.

It is not clear yet whether the detected spectral variability is caused by the presence of magnetospheres or by pulsational variability, frequently detected in massive OB-type stars. Future observations are urgently needed to be able to draw solid conclusions about the role of magnetic fields in these targets.

References

Appenzeller, I., Fricke, K., Fürtig, W., Gässler, W., Häfner, R., Harke, R., Hess, H.-J., Hummel, W., Jürgens, P., Kudritzki, R.-P., Mantel, K.-H., Meisl, W., Muschielok, B., Nicklas, H., Rupprecht, G., Seifert, W., Stahl, O., Szeifert, T., & Tarantik, K. 1998, *The ESO Messenger*, 94, 1

Hubrig, S., Schöller, M., Ilyin, I., Kharchenko, N. V., Oskinova, L. M., Langer, N., González, J. F., Kholtygin, A. F., Briquet, M., & Magori Collaboration 2013, *A&A*, 551, A33

Hubrig, S., Sidoli, L., Postnov, K., Schöller, M., Kholtygin, A. F., Järvinen, S. P., & Steinbrunner, P. 2018, *MNRAS*, 474, L27

Karitskaya, E. A., Bochkarev, N. G., Hubrig, S., Gnedin, Y. N., Pogodin, M. A., Yudin, R. V., Agafonov, M. I., & Sharova, O. I. 2010, *IBVS*, 5950, 1

Koenigsberger, G., Moreno, E., & Harrington, D. M. 2012, *A&A*, 539, A84

Kreykenbohm, I., Wilms, J., Kretschmar, P., Torrejón, J. M., Pottschmidt, K., Hanke, M., Santangelo, A., Ferrigno, C., & Staubert, R. 2008, *A&A*, 492, 511

Sidoli, L. 2017, *Proc. of the "XII Multifrequency Behaviour of High Energy Cosmic Sources Workshop", 12-17 June, 2017 Palermo, Italy*, 52

High-mass X-ray Binaries: Illuminating the Passage from Massive Binaries to Merging Compact Objects
Proceedings IAU Symposium No. 346, 2019
L. M. Oskinova, E. Bozzo, T. Bulik & D. Gies, eds.

doi:10.1017/S1743921318007597

First empirical constraints on the low Hα mass-loss rates of magnetic O-stars

Florian A. Driessen[1], **Jon O. Sundqvist**[1] **and Gregg A. Wade**[2]

[1]Institute of Astronomy, KU Leuven,
Celestijnenlaan 200D box 2401, BE-3001, Leuven, Belgium
email: florian.driessen@kuleuven.be

[2]Dept. of Physics & Space Science, Royal Military College of Canada,
PO Box 17000, Station Forces, Kingston, Ontario, Canada

Abstract. A small subset of Galactic O-stars possess surface magnetic fields that alter the outflowing stellar wind by magnetically confining it. Key to the magnetic confinement is that it induces rotational modulation of spectral lines over the full EM domain; this allows us to infer basic quantities, e.g., mass-loss rate and magnetic geometry. Here, we present an empirical study of the Hα line in Galactic magnetic O-stars to constrain the mass fed from the stellar base into the magnetosphere, using realistic multi-dimensional magnetized wind models, and compare with theoretical predictions. Our results suggest that it may be reasonable to use mass-feeding rates from non-magnetic wind theory if the *absolute* mass-loss rate is scaled down according to the amount of wind material falling back upon the stellar surface. This provides then some empirical support to the proposal that such magnetic O-stars might evolve into heavy stellar-mass black holes (Petit *et al.* 2017).

Keywords. stars: massive, stars: magnetic field, stars: mass-loss, stars: winds, outflows

1. Introduction

Hot, luminous, massive OB stars are known to have high-speed, radiation driven winds (Castor *et al.* 1975). Spectropolarimetric surveys over the past decade have shown that about 7% of massive stars harvest a strong, often dipolar, surface magnetic field (Wade *et al.* 2016). The wind-magnetic field interaction around these stars leads to magnetic confinement of the wind and the formation of a circumstellar magnetosphere (ud-Doula & Owocki 2002). For the slowly rotating magnetic O-stars studied here, trapped material falls back onto the star on a dynamical timescale and the formation of a Dynamical Magnetosphere (DM) occurs (Sundqvist *et al.* 2012; Petit *et al.* 2013). The DM structure is the origin of a range of spectral line diagnostics which we study here by means of high-quality, high-resolution spectra acquired by the Magnetism in Massive Stars survey (MiMeS; Wade *et al.* 2016).

The predicted fall-back of wind material upon the star leads naturally to the question if these magnetic stars lose less mass than their non-magnetic counterparts, and so might evolve into Galactic heavy mass black holes when ending their lives (Petit *et al.* 2017). However, so far models studying this have assumed that the wind mass fed into the magnetosphere was unaltered by the presence of the magnetic field. Hence, the mass launched from the magnetic star's base (the feeding rate) is assumed to be equal to that of a non-magnetic star (Vink *et al.* 2000). To examine this assumption empirically, here we perform a systematic study of Hα diagnostics of a small sample of confirmed Galactic magnetic O-stars.

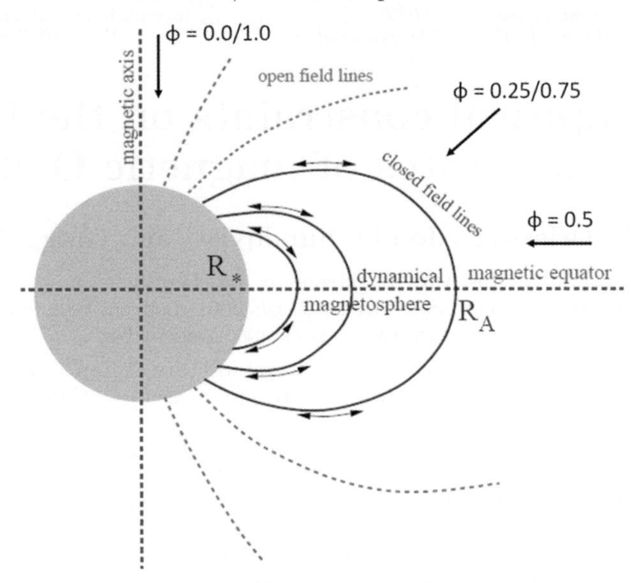

Figure 1. Cartoon of a massive star with oblique dipolar magnetic field. Because of magnetic obliquity the stellar rotation results in a change of projected surface area of the magnetosphere, i.e., the angle between observer and magnetic field axis changes in time. This leads to modulated line-profile emission with rotation phase ϕ.

2. Hα line-formation and mass-loss rates

To compute theoretical Hα line-profiles we follow the procedure of Sundqvist *et al.* (2012) to solve the formal solution of radiative transfer in 3D cylindrical space for an observer viewing under angle α w.r.t. the magnetic axis. We complement the 3D radiative transfer with a description of the DM as provided by the Analytical Dynamical Magnetosphere (ADM) formalism (Owocki *et al.* 2016) to describe the velocity and density inside the DM.

Due to magnetic confinement the mass that can effectively escape the star $\dot{M}_B$ is much smaller than the mass that gets launched from the base of the star $\dot{M}_{B=0}$. The rates are related via $\dot{M}_B = f_B \dot{M}_{B=0}$, with f_B essentially the fraction the magnetosphere covers (ud-Doula *et al.* 2008; their Eq. 23).

3. Sample study and empirical Hα mass-loss rate constraints

The rotationally phase-modulated emission of the Hα line (Figure 1) can be used to infer and constrain both the magnetic geometry (i, β), setting the shape of the rotationally modulated emission, and the mass-feeding rate $\dot{M}_{B=0}$ that primarily affects the absolute amount of Hα emission.

Figure 2 displays the fitted phased Hα equivalent widths (EW) of our sample stars, showing overall good agreement between the ADM model and observations. We note that θ^1 Ori C exhibits an asymmetry in its lightcurve; the origin of this feature is unknown, but might be due to star spots or non-dipolar field contributions (ud-Doula *et al.* 2013), both which cannot currently be modelled with the ADM. Because we are primarily interested in constraining mass-feeding rates we leave the asymmetry out (effectively by fitting only the region $0 \leq \phi \leq 0.5$) as the absolute amount of emission is only marginally affected by this (Figure 2, panel 5 & 6).

The ADM is a steady-state model that predicts a smooth wind outflow. However, more realistically a transient infall of matter occurs onto the star which leads to statistical density variations. These are here modelled by a clumping factor $f_{cl} = \langle\rho^2\rangle/\langle\rho\rangle^2$,

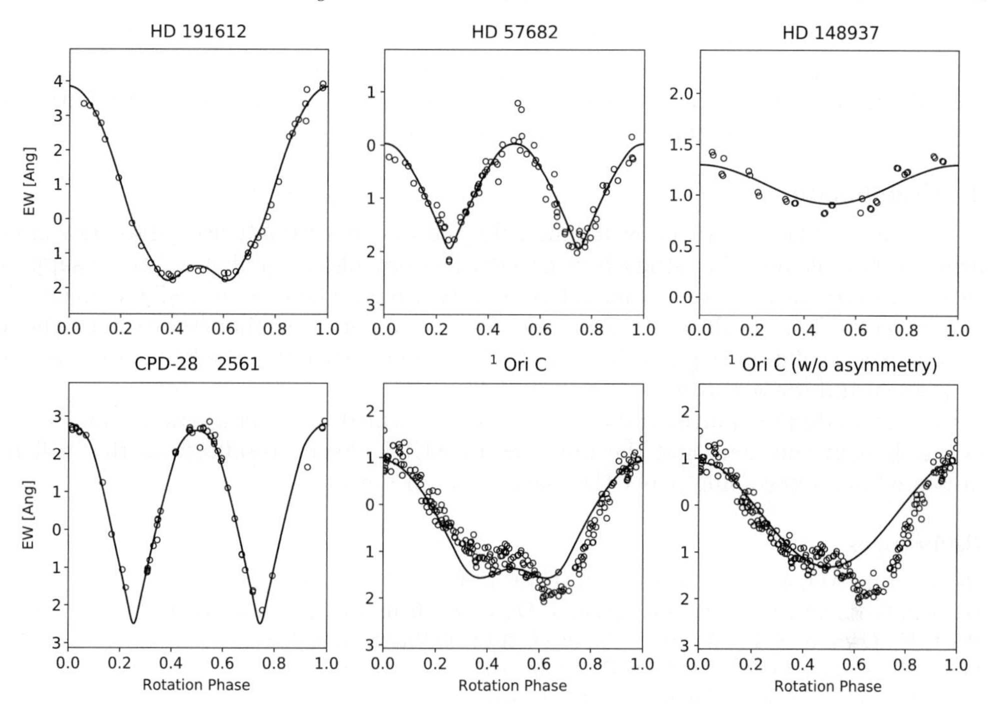

Figure 2. Fits of phased Hα EW variations of the investigated sample of confirmed Galactic magnetic O-stars by the MiMeS survey. Open black dots show observations and the black solid line is the ADM best-fit theoretical profile.

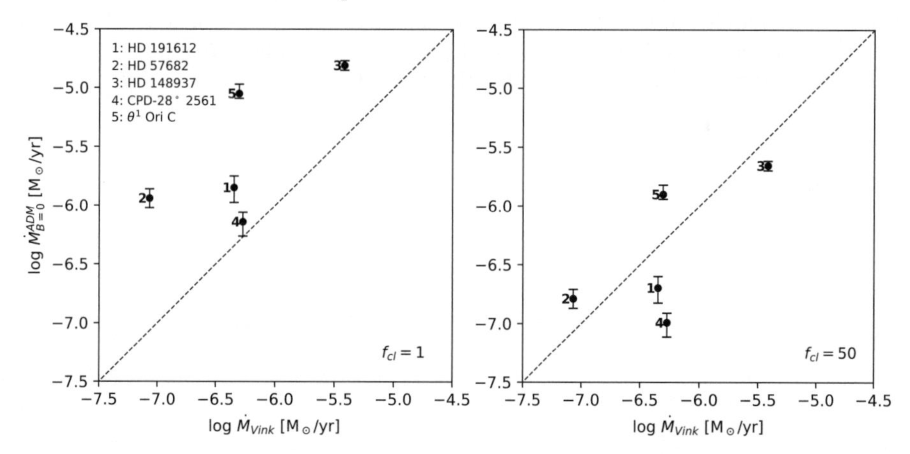

Figure 3. Hα ADM non-magnetic mass-feeding rates as a function of the non-magnetic mass-feeding rates predicted by Vink *et al.* (2000). The black dashed line is a one-to-one correspondence. Error bars are 3σ confidences.

which scales down the mass-loss rate by $\sqrt{f_{cl}}$ appropriate for ρ^2-diagnostics like Hα (e.g., Puls *et al.* 2008).

Time-dependent magnetohydrodynamic simulations predict $f_{cl} \approx 50$ for a magnetic O-star (Owocki *et al.* 2016). We apply the above correction to our constrained $\dot{M}_{B=0}$ (from Figure 2) and compare with the Vink *et al.* (2000) prescription (Figure 3). The latter are computed according to the stellar parameter compilation of Petit *et al.* (2013). Though the scatter is large, the indication is that such a clumping factor indeed gives a rather small overall off-set, thereby lending some support to previous studies that have used the Vink *et al.* prescription for $\dot{M}_{B=0}$ in their studies of magnetic O-stars (e.g., Petit *et al.* 2017). We emphasize, however, that the absolute mass-loss rate is much lower than the mass-feeding rate; e.g., for HD 191612 one gets $\dot{M}_B = f_B \dot{M}_{B=0} \approx 10^{-8}\ M_{\odot}/\mathrm{yr}$ when using an Alfvén radius $R_A = 3.5R_{\star}$ (Owocki *et al.* 2016).

4. Conclusions

Although further observations and modelling efforts are certainly needed to draw more firm conclusions, our pilot-study here nevertheless provides some first empirical support that non-magnetic massive star mass-feeding rates can also be used in studies of magnetic massive stars. This result may then also lend some support to similar studies in magnetic massive star evolution (e.g., Petit *et al.* 2017), provided that the *absolute* mass-loss rate $\dot{M}_B$ is scaled down accordingly (see Section 2).

Currently additional investigations are being performed to better assess the uniqueness of our best-fit solutions and the influence of NLTE effects; results from this will be presented in an upcoming paper (Driessen *et al.*, in prep.).

References

Castor, J. I., Abbott, D. C. & Klein, R. I. 1975, *ApJ*, 195, 157
Owocki, S. P., ud-Doula, A., Sundqvist, J. O., *et al.* 2016, *MNRAS*, 462, 3830
Petit, V., Owocki, S. P., Wade, G. A., *et al.* 2013, *MNRAS*, 429, 398
Petit, V., Keszthelyi, Z., MacInnis R., *et al.* 2017, *MNRAS*, 466, 1052
Puls, J., Vink, J. S. & Najarro, F. 2008, *A&AR*, 16, 209
Sundqvist, J. O., ud-Doula A., Owocki, S. P., *et al.* 2012, *MNRAS*, 423, L21
ud-Doula, A. & Owocki, S. P. 2002, *ApJ*, 576, 413
ud-Doula, A., Owocki, S. P., & Townsend R. H. D. 2008, *MNRAS*, 385, 97
ud-Doula, A., Sundqvist, J. O., Owocki, S. P., *et al.* 2013, *MNRAS*, 428, 2723
Vink, J. S., de Koter, A., & Lamers, H. J. G. L. M. 2000, *A&A*, 362, 295
Wade, G. A., Neiner, C., Alecian, E., *et al.* 2016, *MNRAS*, 456, 2

High-mass X-ray Binaries: Illuminating the Passage from Massive Binaries to Merging Compact Objects
Proceedings IAU Symposium No. 346, 2019
L. M. Oskinova, E. Bozzo, T. Bulik & D. Gies, eds.

doi:10.1017/S1743921318007743

Spectroscopic identication of INTEGRAL high-energy sources with VLT/ISAAC†

F. Fortin[1], S. Chaty[1], A. Coleiro[2], J. A. Tomsick[3] and C. H. R. Nitschelm[4]

[1]Laboratoire AIM (UMR 7158 CEA/DRF - CNRS - Université Paris Diderot), Irfu / Département dAstrophysique, CEA-Saclay, 91191 Gif-sur-Yvette Cedex, France

[2]APC, Université Paris Diderot, CNRS/IN2P3, CEA/Irfu, Observatoire de Paris, 10 rue Alice Domon et Léonie Duquet, 75205 Paris Cedex 13, France

[3]Space Science Laboratory, 7 Gauss Way, University of California, Berkeley, CA 94720-7450, USA

[4]Unidad de Astronomía, Universidad de Antofagasta, Avenida Angamos 601, Antofagasta 1270300, Chile

Abstract. *INTEGRAL* has been observing the γ-ray sky for 15 years and has discovered many high-energy sources of various nature. Among them, active galactic nuclei (AGN), low or high-mass X-ray binaries (LMXB and HMXB) and cataclysmic variables (CV) are rather difficult to differentiate from one another at high energies and require further optical or near-infrared observations to constrain their exact nature. Using near-infrared photometric and spectroscopic data from ESO VLT/ISAAC, we aim to reveal the nature of 14 high-energy INTEGRAL sources and improve the census of X-ray binaries. By comparing their spectral features to stellar spectra atlases, we identified 5 new CVs, 2 low or intermediate mass X-ray binaries, 2 HMXBs and 5 AGNs.

Keywords. infrared: stars, X-rays: binaries, binaries: general

---

1. Introduction

The *INTEGRAL* satellite has been looking at the high-energy sky between 15 keV and 10 MeV for 15 years. The nature of the high-energy sources is often ambiguous and further observations in optical and near-infrared (nIR) are required to constrain it. This is why a significant fraction ($\sim$20%) of the *INTEGRAL* sources (IGR) need lower energy followups. According to the catalogue of *INTEGRAL* sources (Bird *et al.* 2016), we expect that a majority of the unknown IGR sources are accreting binaries or active galactic nuclei (AGN); the former are binary stars in which one component is an accreting compact object.

We distinguish three main categories of accreting binaries depending on the compact object and the mass of the companion star: cataclysmic variables (CV), low-mass X-ray binaries (LMXB), and high-mass X-ray binaries (HMXB). LMXBs host either a neutron star (NS) or a black hole (BH); CVs host a white dwarf. Both have a low-mass companion star ($M \leqslant 1\,M_{\odot}$). Roche lobe overflow allows the accretion of matter from the companion star and releases high-energy photons. Compact objects can be surrounded by an accretion disc, which may lead to transient behaviours. Intermediate-mass X-ray binaries (IMXBs) are less common, and have a companion of mass between 1 and $10\,M_{\odot}$.

† Based on observations made with ESO Telescopes at the La Silla Paranal Observatory under programme ID 089.D-0181(A).

The accretion mechanisms are similar to that of LMXBs. For the sake of consistency with the literature and especially Bird *et al.* (2016), we group IMXBs and LMXBs in the same class of binaries.

High-mass X-ray binaries host a massive star ($M \geqslant 10\,M_{\odot}$) with a NS or a BH as primary component. Among HMXBs are two sub-categories, based on the evolutionary state of the companion star. In Be binaries (BeHMXB), the secondary is a fast-rotating main-sequence O/B star that loses matter as a consequence of high centrifugal force, forming a decretion disc around itself. Accretion occurs when the compact object crosses through the decretion disc. Supergiant binaries (sgHMXB) host an evolved O/B supergiant star with an intense stellar wind driven by its luminosity. The compact object thus feeds off that wind. *INTEGRAL* made it possible to differentiate two new subclasses of sgHMXBs thanks to its increased sensitivity at high energies, as reviewed in Chaty (2013). Obscured HMXBs have intrinsic absorption ($N_H > 10^{23}\,\mathrm{cm}^{-2}$), while supergiant fast X-ray transients (SFXTs) have short bursts of high-energy radiation with low quiescent states.

To accurately identify high-energy sources, further observations are required, for which nIR is well adapted. Firstly, many *INTEGRAL* sources (IGR) are located near the Galactic plane, where optical photons are absorbed by dust while infrareds are not. Secondly, most of the nIR emission of a binary comes from the companion star or the accretion disc, which is ideal to pinpoint their nature by deriving their spectral type.

We worked on a sample of 14 IGR sources for which nIR photometry and/or spectroscopy was performed. We aim to confirm unambiguous nIR counterparts for each of these IGR sources and provide constraints on their nature, such as the spectral type of companion stars in X-ray binaries.

2. Observations

The observations were carried out in 2012 (P. I. S. Chaty) on 14 *INTEGRAL* sources (programme ID 089.D-0181). Near-infrared photometry and spectroscopy were performed at ESO in Chile on the 8 m Very Large Telescope Unit 3 Nasmyth A (VLT/UT3) using the near-infrared spectro-imager ISAAC.

2.1. *Photometry - Finding nIR counterparts*

2.1.1. Data reduction

Near-infrared images were taken with a K_s filter (1.98 – 2.35 μm) with a $2'5 \times 2'5$ field of view. For each source, five frames were taken with a random spatial offset following the jitter procedure standardly used in nIR ESO acquisitions.

We performed the data reduction with standard Image Reduction and Analysis Facility (IRAF†) routines. After subtracting dark and correcting the flat-field, the sky background was subtracted through the median of five jittered images for each source. Images were then aligned based on precise astrometry and averaged.

We performed aperture photometry with the *IRAF.apphot.qphot* tool to derive the apparent K_s magnitudes. The radii of the integration circle and background annulus were chosen to minimize sky background and pollution from nearby bright stars. We used eight photometric standard stars to derive the average zero-point of photometry ($Z_p = 0.972 \pm 0.056$). Extinction was corrected by using the value‡ given by ESO for Paranal: $\kappa_{K_s} = 0.07\,\mathrm{mag\,airmass}^{-1}$.

† IRAF is distributed by the National Optical Astronomy Observatories, which are operated by the Association of Universities for Research in Astronomy, Inc., under cooperative agreement with the National Science Foundation.

‡ http://www.eso.org/sci/facilities/paranal/decommissioned/isaac/tools/imaging_standards.html#Extinction

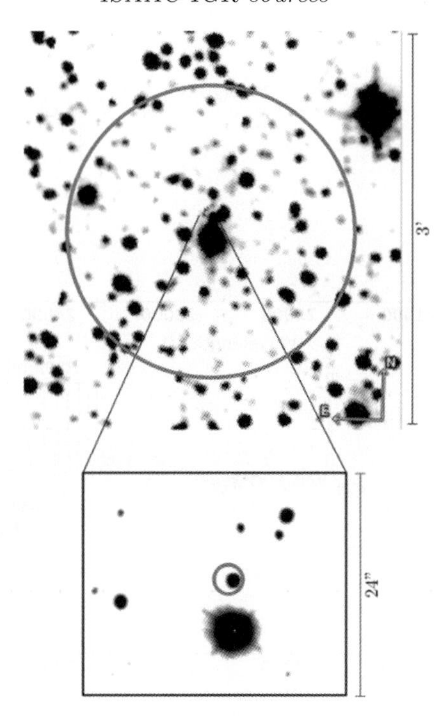

Figure 1. K-band field of view of IGR J13020-6359 from 2MASS (Skrutskie *et al.* 2006) along with the *INTEGRAL* error circle in red. The insert is the ISAAC K_s field of view around the Swift position (Rodriguez *et al.* 2009).

2.1.2. Finding nIR counterparts of IGR sources

Finding nIR counterparts to high-energy detections requires good astrometry. We refined the astrometric solution of ISAAC images using GAIA (Graphical Astronomic Image Analysis) by matching the positions of the stars in each field of view with 2MASS sources from the 2MASS Point Source Catalogue (PSC) and/or Gaia DR1.

The IBIS instrument on board *INTEGRAL* has a wide field of view, but does not have enough spatial resolution to associate accurately an optical/nIR counterpart to the high-energy detections. Precise X-ray localization is thus given by either *Chandra*, *XMM-Newton,* or *Swift* telescopes. We used available positions from these facilities in the litterature to associate an unambiguous nIR counterpart to the *INTEGRAL* sources (see example in Fig. 1).

2.2. *Spectroscopy*

2.2.1. Data reduction

Spectroscopy was performed with ISAAC with a long slit in short wavelength spectroscopy low resolution mode (SWS-LR). The 0.6" slit allowed us to obtain a spectral resolution of $R = 750$ in the K band (1.8–2.5 μm). We measured the width of narrow OH lines from sky emission to be 26±1 Å at 22 000 Å, which is compatible with the theoretical instrumental resolution of $R = 750$. For each source, eight spectral frames were taken. A slight spatial offset ($\sim$30") along the slit was added between each spectral acquisition, following the standard ESO nodding procedure.

We performed data reduction with standard IRAF tools. Each spectrum was corrected by dark and flat frames, and the overall sky value was estimated with the median of the

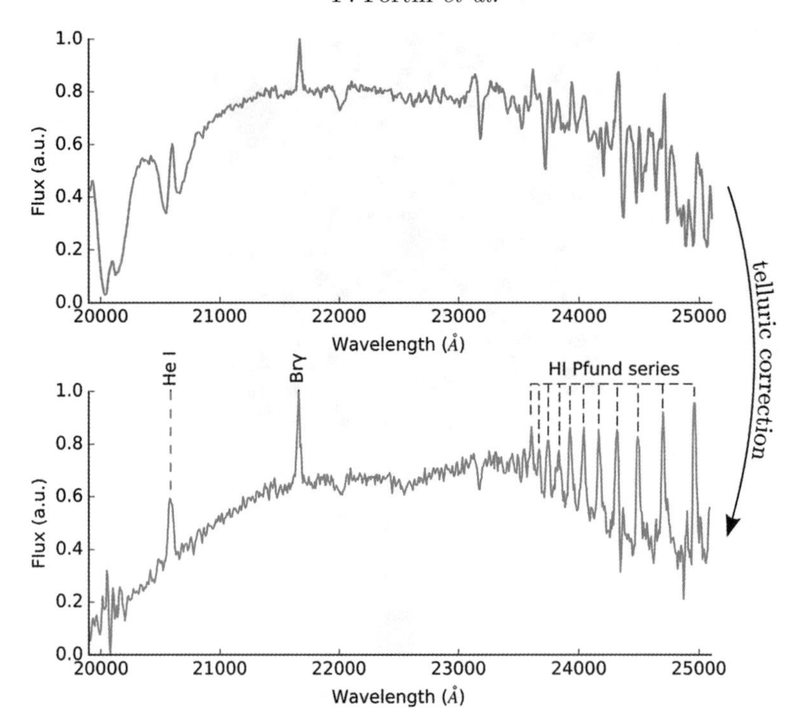

Figure 2. IGR J13020-6359 K-band ISAAC spectrum before (*top*) and after (*bottom*) telluric correction. Molecfit allows us to exploit deeply absorbed parts of the spectrum, revealing in this case HI Pfund emission lines, a feature that is distinctive of early-type stars.

eight spectra. We extracted 1D spectra via *IRAF.apall* package. The extracted spectra were stacked using a median to remove cosmic rays.

Telluric absorption was corrected on each reduced spectrum using *Molecfit* (Kausch *et al.* 2015, Smette *et al.* 2015). This software fits atmospheric features based on meteorological conditions on the date acquisitions were performed. This method was used instead of the classical telluric standard star correction, since it might have introduced artifacts because of the difference in spectral types between the target and the standard star.

The wavelength solution was derived using the argon and xenon lamp spectra provided by the standard calibration procedure in ESO for ISAAC. We derived a solution with a RMS of 0.15Å, and then derived individual wavelength zero-point correction for each source using eight OH lines from the sky spectra.

2.2.2. Deriving spectral types

We compared the features in the spectrum of our 14 IGR sources to spectral atlases (Kleinmann & Hall 1986, Hanson *et al.* 1996, 2005, Ramirez *et al.* 1997, Lenorzer *et al.* 2002, Harrison *et al.* 2004) to derive their spectral type (see example Fig. 2).

3. Results

Among our 14 sources, we identified 5 AGNs, 5 CVs, 2 I/LMXBs and 2 HMXBs (see Tab. 1). All but one are situated within the Galactic plane. Two AGNs (IGR J18457+0244 and IGR J18532+0416) were identified through imaging (extended sources) since their spectrum did not show any feature, which could be associated to synchrotron emission from flaring AGNs. Two CVs (IGR J12489-6243 and IGR J174004-3655) lack the typical

Table 1. Summary of the identifications derived in this study with VLT/ISAAC nIR data.

RAJ2000 (nIR)	DEJ2000 (nIR)	Unc. (")	Previous identification (comment)	Our identification (spectral type / comment)
IGR J00465	**−4005**			
00:46:20.681	−40:05:49.26	0.060	AGN Sey 2 (z=0.201)	AGN (Sey 2, z=0.202±0.002)
IGR J10447	**−6027**			
10:44:51.925	−60:25:11.78	0.080	?	AGN (Sey 2, z=0.047±0.001)
IGR J12489	**−6243**			
12:48:46.422	−62:37:42.53	0.053	CV / HMXB ?	CV (K/M companion)
IGR J13020	**−6359**			
13:01:58.723	−63:58:08.88	0.164	HMXB (NS)	BeHMXB (B0–6Ve companion)
IGR J13186	**−6257**			
13:18:25.041	−62:58:15.66	0.072	HMXB ?	BeHMXB (B0–6Ve companion)
IGR J15293	**−5609**			
15:29:29.394	−56:12:13.42	0.136	CV (K-type giant ?)	CV (K5V–III companion)
IGR J17200	**−3116**			
17:20:05.920	−31:16:59.62	0.056	HMXB	Symbiotic CV (KIII companion)
IGR J17404	**−3655**			
17:40:26.862	−36:55:37.39	0.125	HMXB (NS) ?	CV (K3–5V companion)
IGR J17586	**−2129**			
17:58:34.558	−21:23:21.55	0.092	HMXB ?	Symbiotic CV (KIII companion)
IGR J17597	**−2201**			
17:59:45.518	−22:01:39.48	0.110	LMXB (NS) ?	LMXB (G8–K0III companion)
IGR J18457	**+0244**			
18:45:40.388	+02:42:08.88	0.043	Pulsar / AGN ?	AGN
IGR J18532	**+0416**			
18:53:16.028	+04:17:48.24	0.037	HMXB / AGN ?	AGN ($z = 0.051$)
IGR J19308	**+0530**			
19:30:50.756	+05:30:58.12	0.252	IMXB (F4V companion)	IMXB (F8-G0V-III companion)
IGR J19378	**−0617**			
19:37:33.029	−06:13:04.76	0.204	Sey1.5 (z=0.011)	AGN (Sey1.5, z=0.011±0.001)

CO bandheads that usually are the signature of cooler stellar atmospheres from the companion star. This may be due to the dominant emission from an accretion disc, or to the possibility of the secondary to have depleted part of its atmosphere were CO banheads arise.

4. Conclusion

The photometric and spectroscopic data allowed us to find unambiguous nIR counterparts to the high-energy detections and identify or better constrain the nature of these sources. Among them, we find 5 AGNs, 5 CVs, 2 BeHMXBs, and 2 I/LMXBs. Even though the proportions between the different types of sources are not in full agreement with those published in Bird *et al.* (2016), we still expect that the remaining unidentified *INTEGRAL* sources are mainly AGNs, X-ray binaries, and CVs. The current census of binaries would benefit from having more candidates with a well-constrained nature, hence the need to identify the rest of the IGR sources. This will help population studies and answer general questions on binary evolution in the context of gravitational wave astronomy and compact mergers.

References

Bird, A. J., Bazzano, A., Malizia, A., *et al.* 2016, *ApJS*, 223, 15
Chaty, S. 2013, Advances in Space Research, 52, 2132
Hanson, M. M., Conti, P. S., & Rieke, M. J. 1996, *ApJS*, 107, 281
Hanson, M. M., Kudritzki, R.-P., Kenworthy *et al.*2005, *ApJS*, 161, 154
Harrison, T. E., Osborne, H. L., & Howell, S. B. 2004, *AJ*, 127, 3493
Kausch, W., Smette, S. N. A., Kimeswenger, S., *et al.* 2015, *A&A*, 576, A78
Kleinmann, S. G. & Hall, D. N. B. 1986, *ApJS*, 62, 501
Lenorzer, A., Vandenbussche, B., Morris, P., *et al.* 2002, *A&A*, 384, 473
Ramirez, S. V., Depoy, D. L., Frogel, J. A. *et al.* 1997, *AJ*, 113, 1411
Rodriguez, J., Tomsick, J. A., & Chaty, S. 2009, *A&A*, 494, 417
Skrutskie, M. F., Cutri, R. M., Stiening, R., *et al.* 2006, *AJ*, 131, 1163
Smette, A., Sana, H., Noll, S., *et al.* 2015, *A&A*, 576, A77

High-mass X-ray Binaries: Illuminating the Passage from Massive Binaries to Merging Compact Objects
Proceedings IAU Symposium No. 346, 2019
L. M. Oskinova, E. Bozzo, T. Bulik & D. Gies, eds.

doi:10.1017/S1743921318007913

The interaction of core-collapse supernova ejecta with a stellar companion

Zheng-Wei Liu[1,2], T. M. Tauris[3], F. K. Röpke[4,5], T. J. Moriya[6], M. Kruckow[1,2], R. J. Stancliffe[3] and R. G. Izzard[7]

[1]Yunnan Observatories, Key Laboratory for the Structure and Evolution of Celestial Objects, CAS, Kunming 650216, China,

[2]Center for Astronomical Mega-Science, CAS, Beijing, China
email: zwliu@ynao.ac.cn

[3]Argelander-Institut für Astronomie, Auf dem Hügel 71, D-53121 Bonn,

[4]Heidelberger Institut für Theoretische Studien, Schloss-Wolfsbrunnenweg 35, D-69118 Heidelberg, Germany,

[5]Zentrum für Astronomie der Universität Heidelberg, Institut für Theoretische Astrophysik, Philosophenweg 12, D-69120 Heidelberg, Germany,

[6]National Astronomical Observatory of Japan,

[7]University of Surrey, Guildford, Surrey GU2 7XH, United Kingdom.

Abstract. The progenitors of many core-collapse supernovae (CCSNe) are expected to be in binary systems. By performing a series of three-dimensional hydrodynamical simulations, we investigate how CCSN explosions affect their binary companion. We find that the amount of removed stellar mass, the resulting impact velocity, and the chemical contamination of the companion that results from the impact of the SN ejecta, strongly increases with decreasing binary separation and increasing explosion energy. Also, it is foud that the impact effects of CCSN ejecta on the structure of main-sequence (MS) companions, and thus their long term post-explosion evolution, are in general not dramatic.

Keywords. stars: supernovae: general, stars: kinematics, binaries: close

1. Introduction

The discovery of many low-mass X-ray binaries and millisecond pulsars in tight orbits, i.e. binary neutron stars with orbital periods of less than a few hours, provides evidence for supernova (SN) explosions in close binaries with low-mass companions. The nature of the SN explosion determines whether any given binary system remains bound or is disrupted (Hills 1983). An additional consequence of the SN explosion is that the companion star is affected by the impact of the shell debris ejected from the exploding star (Wheeler *et al.* 1975). Besides chemical enrichment, such an impact has kinematic effects and may induce significant mass loss and heating of the companion star. Core-collapse supernovae (CCSNe) arise from massive stars. There is growing observational evidence that the fraction of massive stars in close binary systems is large. Sana *et al.* (2012) found that more than 70% of massive stars are in close binary systems, which supports the idea that binary progenitors contribute significantly to the observed CCSNe.

After a SN explosion occurs in a binary system, the ejected debris is expected to expand freely and eventually impact the companion star. The companion star may be significantly heated and shocked by the SN impact, causing the envelope of the companion star to be partially removed due to the stripping and ablation mechanism (e.g., Wheeler *et al.* 1975, Marietta *et al.* 2000, Liu *et al.* 2012, Liu *et al.* 2013, Pan *et al.* 2012,

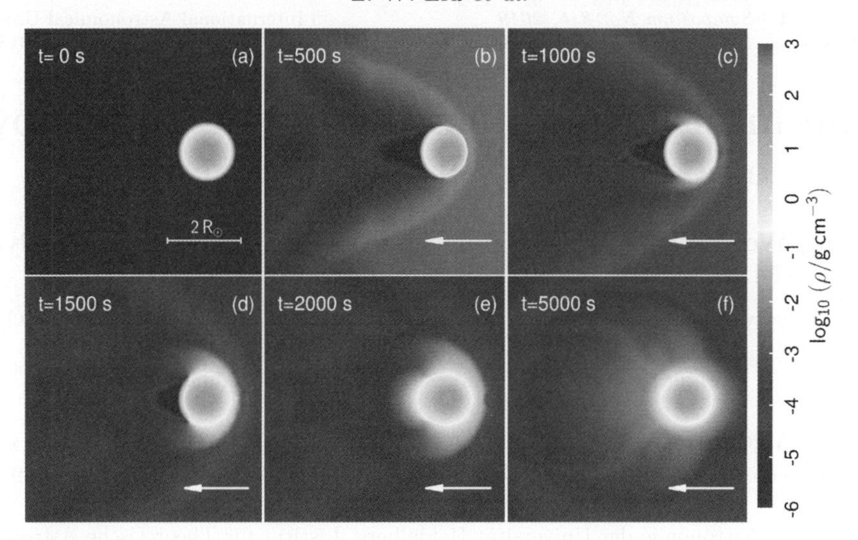

Figure 1. Density distributions of all gas material as a function of the time in our impact simulations for a G/K-dwarf companion model with a binary separation of $5.48\,R_{\odot}$. The direction of motion of the incoming SN shell front is from right to left (see arrow symbols). The color scale shows the logarithm of the mass density in $\mathrm{g\,cm^{-3}}$.

Hirai *et al.* 2018). In this work, we perform impact simulations using a three-dimensional (3D) smoothed particle hydrodynamics (SPH) method to systematically study, for the first time, the impact of CCSN ejecta on MS companion stars.

2. Results and conclusions

We use the BEC stellar evolution code to construct the detailed companion structure at the moment of SN explosion. The impact of the SN blast wave on the companion star is followed by means of 3D SPH simulations using the STELLAR GADGET code (Pakmor *et al.* 2012). Figure 1 illustrates the temporal density evolution of the SN ejecta and companion material of our hydrodynamics simulations for a G/K-dwarf companion model. Figure 2 shows the effects of varying the orbital separation parameter a/R_2, by a factor of about 6, on the total amount of removed companion mass (ΔM_2), the resulting impact velocity ($v_{\rm im}$), and total accumulated ejecta mass ($\Delta M_{\rm acc}$), for the $0.9\,M_{\odot}$ and $3.5\,M_{\odot}$ companion star models.

To discuss the effects of an explosion on the companion star in CCSNe of Type Ib/c, we perform populations synthesis calculations for SNe with the `binary_c/nucsyn` code (Izzard *et al.* 2004, Izzard *et al.* 2009). Under an assumption of a Galactic star-formation rate of $0.68 - 1.45\,M_{\odot}\,\mathrm{yr}^{-1}$ and the average stellar mass of the Kroupa initial mass function ($0.83\,M_{\odot}$), the total CCSN rate of the Galaxy predicted from their population synthesis is $0.93–1.99 \times 10^{-2}\,\mathrm{yr}^{-1}$ (36% SNe Ib/c, 10% SNe IIb, 54% SNe II), consistent with the estimated Galactic CCSN rate of $2.30 \pm 0.48 \times 10^{-2}\,\mathrm{yr}^{-1}$ from recent surveys (Li *et al.* 2011).

In our populations synthesis calculations, most SNe Ib/c have an orbital separation of $\gtrsim 5.0\,R_2$, which is about a fraction of $\gtrsim 95\%$ in our binary population synthesis calculations. Furthermore, with the distributions of a/R_2 in Fig. 3, we can simply estimate ΔM_2, $v_{\rm im}$ and $\Delta M_{\rm acc}$ by applying our power-law relationships stated in Fig. 2. We caution that there are large uncertainties in population synthesis studies, which may influence the results. In this case, these mainly relate to the input physics of common-envelope evolution and the subsequent Case BB roche-lobe overflow from the naked helium star prior to its explosion.

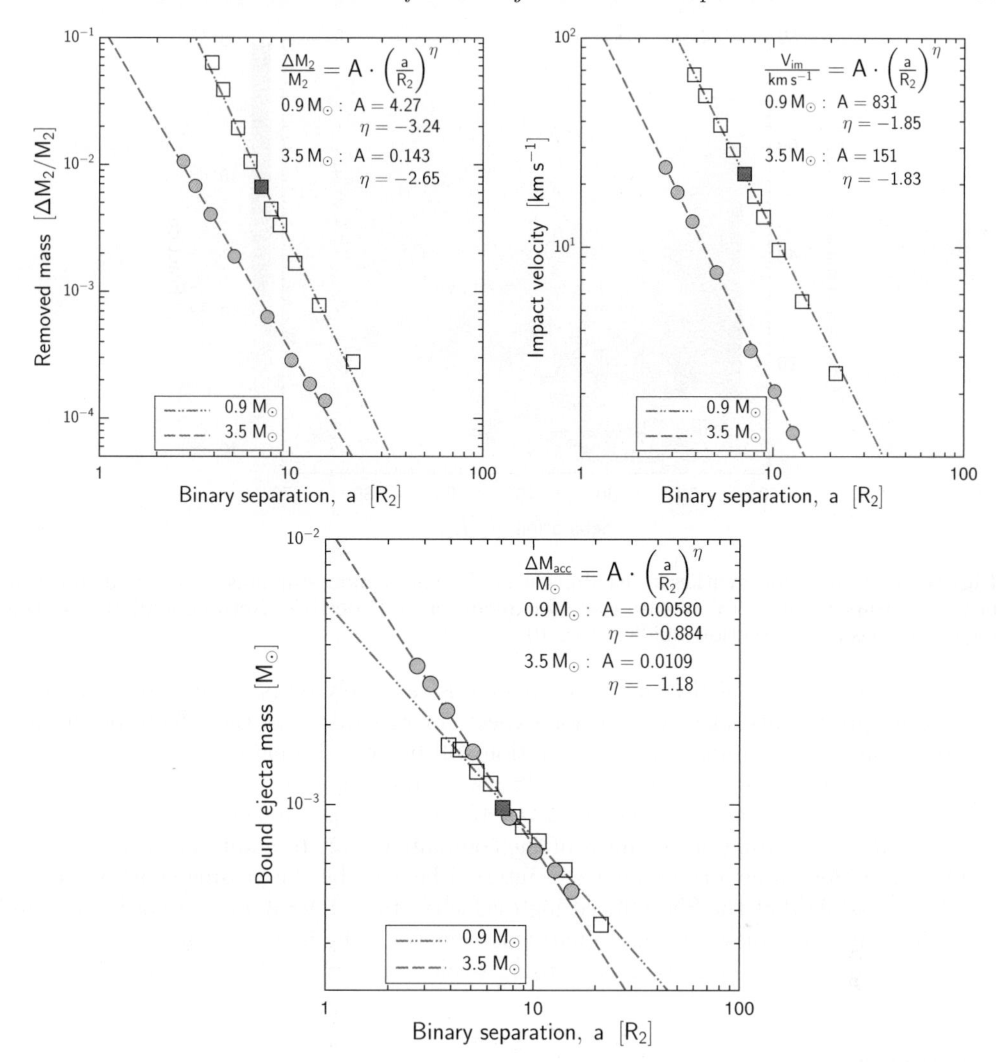

Figure 2. Total removed companion mass (top-left panel), resulting impact velocity of the companion star (top-right panel) and the amount of accreted contamination from the SN ejecta (bottom panel), as a function of initial binary separations for a G/K-dwarf (square symbols, $M_2 = 0.9\, M_\odot$ and $R_2 \approx 0.77\, R_\odot$) and a late-type B-star (filled circle symbols, $M_2 = 3.5\, M_\odot$ and $R_2 \approx 2.18\, R_\odot$) companion model. Power-law fits are also presented in each panel.

We have investigated the impact of SN ejecta on the companion stars in CCSNe of Type Ib/c using the SPH code STELLAR GADGET. Our main results can be summarized as follows (see also Liu *et al.* 2015):

i) The dependence of total removed mass (ΔM_2), impact velocity ($v_{\rm im}$) and the amount of accreted SN ejecta mass ($\Delta M_{\rm acc}$) on the pre-SN binary separation (a) can be fitted with power-law functions. All three quantities are shown to decrease significantly with increasing a, as expected (see Fig. 2).

ii) If our population synthesis is correct, we predict that in most CCSNe less than 5% of the MS companion mass can be removed by the SN impact (i.e. $\Delta M_2/M_2 < 0.05$). In addition, the companion star typically receives an impact velocity, $v_{\rm im}$, of a few 10 km s^{-1}, and the amount of SN ejecta captured by the companion star after the explosion, $\Delta M_{\rm acc}$, is most often less than $10^{-3}\, M_\odot$.

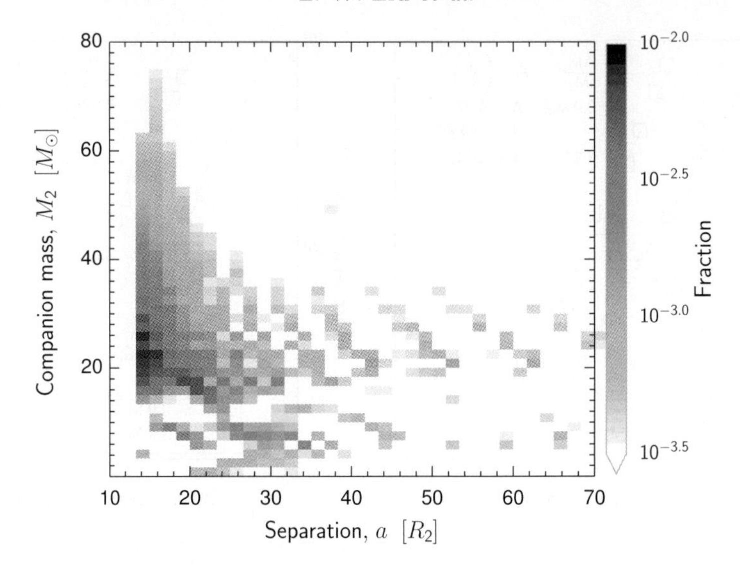

Figure 3. Population synthesis distribution of the companion star mass (M_2) as a function of the binary separation (a) in case the SN explodes as a Type Ib/c. Nothing is plotted in the regions with number fraction smaller than $10^{-3.5}$.

iii) Because a typical CCSN binary companion is relatively massive and can be located at a large pre-SN distance, we do not expect, in general, that the effects of the SN explosion on the post-impact stellar evolution will be very dramatic.

iv) In the closest pre-SN systems, the MS companion stars are affected more strongly by the SN ejecta impact, leading to $\Delta M_2/M_2 \simeq 0.10$, $v_{\rm im} \simeq 100\,{\rm km\,s^{-1}}$ and $\Delta M_{\rm acc} \simeq 4 \times 10^{-3}\,M_\odot$, depending on the mass of the companion star. In addition, these stars are significantly bloated as a consequence of internal heating by the passing shock wave.

v) It is possible that the SN-induced high velocity stars (HVSs), or more ordinary, less fast, runaway stars, may be contaminated sufficiently to be identified by their chemical peculiarity as former companion stars to an exploding star if mixing processes are not efficient on a long timescale.

References

Hills, J. G. 1983, *ApJ*, 267, 322

Hirai, Ryosuke, Podsiadlowski, Ph., & Yamada, S. 2018, *ApJ*, 864, 119

Izzard, R. G., Tout, C. A., Karakas, A. I., & Pols, O. R. 2004, *MNRAS*, 350, 407

Izzard, R. G., Glebbeek, E., Stancliffe, R. J., & Pols, O. R. 2009 2009, *A&A*, 508, 1359

Li, Weidong, Chornock, R., Leaman, J., Filippenko, A. V., Poznanski, D., Wang, Xiaofeng, Ganeshalingam, M., & Mannucci, F. 2011, *MNRAS*, 412, 1473

Liu, Zheng-Wei, Pakmor, R., Röpke, F. K., Edelmann, P., Wang, B., Kromer, M., Hillebrandt, W., & Han, Z. W. 2012, *A&A*, 548, A2

Liu, Zheng-Wei, Pakmor, R., Seitenzahl, I. R., Hillebrandt, W., Kromer, M., Röpke, F. K., Edelmann, P., Taubenberger, S., Maeda, K., Wang, B. & Han, Z. W. 2013, *ApJ*, 774, 37

Liu, Zheng-Wei, Tauris, T. M., Röpke, F. K., Moriya, T. J., Kruckow, M., Stancliffe, R. J., Izzard, R. G. 2015, *A&A*, 584, A11

Marietta, E., Burrows, A., & Fryxell, B. 2000, *ApJS*, 128, 615

Pakmor, R., Edelmann, P., Röpke, F. K., & Hillebrandt, W. 2012, *MNRAS*, 424, 2222

Pan, Kuo-Chuan, Ricker, P. M., & Taam, R. E. 2012, *ApJ*, 750, 151

Sana, H., de Mink, S. E., de Koter, A., Langer, N., Evans, C. J., Gieles, M., Gosset, E., Izzard, R. G., Le Bouquin, J.-B., & Schneider, F. R. N. 2012, *Science*, 337, 444

Wheeler, J. C., Lecar, M., & McKee, C. F. 1975, *ApJ*, 200, 145

High-mass X-ray Binaries: Illuminating the Passage from Massive Binaries to Merging Compact Objects
Proceedings IAU Symposium No. 346, 2019
L. M. Oskinova, E. Bozzo, T. Bulik & D. Gies, eds.

doi:10.1017/S1743921318007925

Orbital resolved spectroscopy of GX 301–2: wind diagnostics

Nazma Islam

Harvard-Smithsonian Center for Astrophysics, 60 Garden Street, Cambridge, MA 02138, USA
email: nazma.syeda@cfa.harvard.edu

Abstract. GX 301–2, a bright high-mass X-ray binary with an orbital period of 41.5 days, exhibits stable periodic orbital intensity modulations with a strong pre-periastron X-ray flare. Several models have been proposed to explain the accretion at different orbital phases. In Islam & Paul (2014), we presented results from an orbital resolved spectroscopic study of GX 301–2 using data from MAXI Gas Slit Camera. We have found a strong orbital dependence of the absorption column density and equivalent width of the iron emission line. A very large equivalent width of the iron line along with a small value of the column density in the orbital phase range 0.1–0.3 after the periastron passage indicates the presence of high density accretion stream. We aim to further investigate the characteristics of the accretion stream with an *AstroSat* observation of the system.

Keywords. stars: individual: GX 301–2, stars: neutron

1. Introduction

GX 301–2 is a bright High Mass X-ray binary pulsar, with an orbital period $\sim$ 41.5 days of the binary system and spin period of the neutron star $\sim$ 685 sec (Koh *et al.* 1997). It exhibits periodically varying intensity modulations: a bright phase during X-ray flare (pre-periastron passage around orbital phase 0.95), dim or low intensity phase (after periastron passage around orbital phase 0.15–0.3) and intermediate intensity phase (during the apastron passage around orbital phase 0.5). A strong X-ray flare occurs before the periastron passage as well as a medium intensity peak is observed at the apastron passage, indicating accretion onto the neutron star due to both spherical stellar wind along with a possible equatorial disk or accretion stream (Pravdo & Ghosh 2001, Leahy & Kostka 2008).

GX 301–2 has a highly absorbed X-ray spectrum with a partial covering high energy cutoff power-law component and several emission lines. It has a very high line of sight photoelectric absorption ($\sim 10^{23}$ cm^{-2}), which is attributed to the dense circumstellar environment in which the neutron star moves. A prominent Fe Kα line is found to exist in almost all orbital phases. This fluorescence line is produced due to reprocessing of X-ray photons from the pulsar by the surrounding circumstellar matter. The equivalent width of the Fe Kα line depends on the distribution (geometry and column density) of the surrounding matter (Kallman *et al.* 2004). Therefore, by comparing the equivalent width of Fe Kα line with N_H, we can study the distribution of circumstellar matter around the neutron star at different orbital phases and can be further used to examine various accretion models.

Monitor of All sky X-ray Image (MAXI) is all sky X-ray monitor, operating on the International Space Station since 2009 (Matsuoka *et al.* 2009). The main instrument on MAXI, Gas Slit Camera, are proportional counters, operating in energy range 2–20 keV (Mihara *et al.* 2011). Its uniform orbital coverage of GX 301-2 for multiple orbital

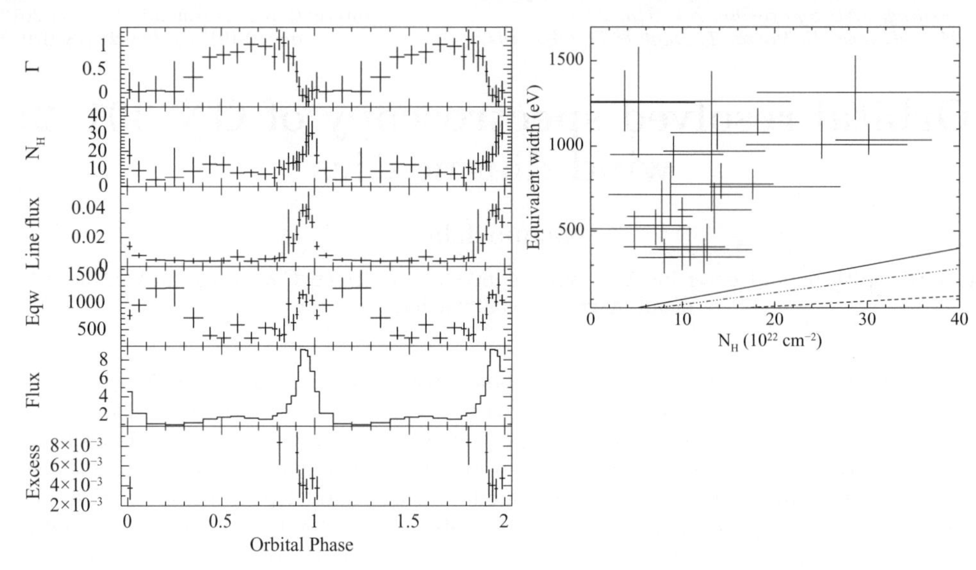

Figure 1. Left panel (a): Orbital variation of Photon index (Γ), column density (N_H in 10^{22} cm^{-2}), Line flux of Fe Kα (photons cm^{-2} s^{-1}), Equivalent width of Fe Kα line (Eqw in eV) and Flux of source (F in 10^{-9} ergs s^{-1} cm^{-2}) for power-law model with high energy cut-off model. Right panel (b): Plot of equivalent width of Fe Kα versus N_H. Solid line and dashed lines represents the relation between equivalent width and column density of absorbing matter for isotropically distributed matter at different Γ (Inoue 1985, Kallman *et al.* 2004).

cycles smears out of short time scale variations and long-term accretion characteristics are brought forth. In Islam, & Paul (2014), we carried out orbital phase resolved spectroscopic study of GX 301–2 using long term data from MAXI–Gas Slit Camera. We studied the orbital phase dependence of the column density and the line equivalent width, which are then used to examine the various models about the distribution of circumstellar matter and the mode of accretion in GX 301–2.

2. Data and analysis

Using MAXI on demand data† we extracted orbital resolved spectra in 21 independent orbital bins. These orbital resolved spectra are fitted with two models: an absorbed power-law continuum, with and without a high energy cut-off. A Fe fluorescence line was found in all the orbital phases, which was modelled by a single Gaussian line. For some orbital phases near the X-ray peak, a low energy excess is found to be present in the spectra. To only estimate the flux in the soft excess, we have modelled the low energy excess with an unabsorbed blackbody component.

3. Discussions and conclusions

Figure 1(a) shows the orbital variation of Γ, N_H, flux and equivalent width of Fe fluoresence line, total flux of the system and ratio of flux included in the low excess to the total flux, for the absorbed power-law with a high energy cutoff model. The column density and flux of the Fe fluorescence line has a large value around the pre-periastron passage, suggesting the possible origin of X-ray flare due to enhanced mass accretion. The column density N_H is found to vary with a pattern similar to the flux of the system, indicating a possible origin of flare due to increased mass accretion. The orbital variation of equivalent width of Fe Kα line shows a different trend as compared

† http://maxi.riken.jp/mxondem/

to the orbital variation of column density. The highest equivalent width occurs at the dim phase of 0.1–0.3 which also has lowest N_H along the line of sight. Figure 1(b) is the plot of equivalent width of the iron line and the absorbing column density N_H in different orbital bins. These observations highly deviate from the relation expected for an isotropically distributed gas (Inoue 1985, Kallman *et al.* 2004). Instead, there seems to exist high anisotropicity in the distribution of circumstellar matter around the X-ray pulsar, especially in some orbital phases. The optical studies of GX 301–2 done by Kaper *et al.* (2006) confirms the presence of gas stream trailing the X-ray pulsar around the orbital phases 0.18–0.34. These results strongly favour a high density gas stream plus a stellar wind model for mode of accretion on to the neutron star in GX 301–2 and provide stronger constraints to the model (Leahy & Kostka 2008).

4. *AstroSat* observations

Due to limited statistics with MAXI/GSC, we cannot further investigate the characteristics of the inferred high density accretion stream. *AstroSat* is an Indian astronomical observatory (Agrawal 2006, Singh *et al.* 2014), with five payloads to carry out simultaneous multi-wavelength observations. With the objective to further investigate the accretion stream characteristics, we carried out a 40 kilosec observation of GX 301–2 with Soft X-ray Telescope (SXT) and Large Area Xenon Proportional Counters (LAXPC) of *AstroSat*. This observation was carried out at the dim phase (orbital phase 0.1–0.3) where maximum anisotropicity in the distribution of the circumstellar matter is found. Further work in analysing these observations and interpreting the results are in progress.

References

Agrawal, P. C. 2006, *Advances in Space Research*, 38, 2989
Inoue, H. 1985, *SSRv*, 40, 317
Islam, N., & Paul, B. 2014, *MNRAS*, 441, 2539
Kallman, T. R., Palmeri, P., Bautista, M. A., Mendoza, C., & Krolik, J. H. 2004, *ApJS*, 155, 675
Kaper, L., van der Meer, A., & Najarro, F. 2006, *A&A*, 457, 595
Koh, D. T., Bildsten, L., Chakrabarty, D., *et al.* 1997, *ApJ*, 479, 933
Leahy, D. A., & Kostka, M. 2008, *MNRAS*, 384, 747
Matsuoka, M., Kawasaki, K., Ueno, S., *et al.* 2009, *PASJ*, 61, 999
Mihara, T., Nakajima, M., Sugizaki, M., *et al.* 2011, *PASJ*, 63, S623
Pravdo, S. H., & Ghosh, P. 2001, *ApJ*, 554, 383
Singh, K. P., Tandon, S. N., Agrawal, P. C., *et al.* 2014, *Space Telescopes and Instrumentation 2014: Ultraviolet to Gamma Ray*, 91441S

High-mass X-ray Binaries: Illuminating the Passage from Massive Binaries to Merging Compact Objects
Proceedings IAU Symposium No. 346, 2019
L. M. Oskinova, E. Bozzo, T. Bulik & D. Gies, eds.

doi:10.1017/S1743921318007949

3D time-dependent hydrodynamical and radiative transfer modeling of Eta Carinae's innermost fossil colliding wind structures

Thomas Madura[1], T. R. Gull[2], N. Clementel[3], M. Corcoran[2,4], A. Damineli[5], K. Hamaguchi[2,6], D. J. Hillier[7], A. F. J. Moffat[8], N. Richardson[9] and G. Weigelt[10]

[1]San José State University,
One Washington Square, San José, CA 95192-0106, USA
email: thomas.madura@sjsu.edu

[2]NASA Goddard Space Flight Center, Greenbelt, MD 20771, USA

[3]Katholieke Universiteit Leuven, Celestijnenlaan 200D, 3001 Leuven, Belgium

[4]The Catholic University of America, Washington, DC 20064, USA

[5]IAG–USP, Rua do Matao 1226, Cidade Universitaria, Sao Paulo 05508-900, Brazil

[6]University of Maryland, Baltimore County, 1000 Hilltop Circle, Baltimore, MD 21250, USA

[7]University of Pittsburgh, 3941 OHara Street, Pittsburgh, PA 15260, USA

[8]Universite de Montreal, CP 6128 Succ. A., Centre-Ville, Montreal, Quebec H3C 3J7, Canada

[9]University of Toledo, Toledo, OH 43606-3390, USA

[10]Max-Planck-Institut fur Radioastronomie, Auf dem Hugel 69, D-53121 Bonn, Germany

Abstract. Eta Carinae is the most massive active binary within 10,000 light-years. While famous for the largest non-terminal stellar explosion ever recorded, observations reveal a supermassive ($\sim$120 $M_{\odot}$) binary consisting of an LBV and either a WR or extreme O star in a very eccentric orbit ($e = 0.9$) with a 5.54-year period. Dramatic changes across multiple wavelengths are routinely observed as the stars move about in their highly elliptical orbits, especially around periastron when the hot ($\sim$40 kK) companion star delves deep into the denser and much cooler ($\sim$15 kK) extended wind photosphere of the LBV primary. Many of these changes are due to a dynamic wind-wind collision region (WWCR) that forms between the stars, plus expanding radiation-illuminated fossil WWCRs formed one, two, and three 5.54-year orbital cycles ago. These fossil WWCRs have been spatially and spectrally resolved by the *Hubble Space Telescope*/Space Telescope Imaging Spectrograph (*HST*/STIS) at multiple epochs, resulting in data cubes that spatially map Eta Carinae's innermost WWCRs and follow temporal changes in several forbidden emission lines (e.g. [Fe III] 4659 Å, [Fe II] 4815 Å) across the 5.54-year cycle. We present initial results of 3D time-dependent hydrodynamical and radiative-transfer simulations of the Eta Carinae binary and its WWCRs with the goal of producing synthetic data cubes of forbidden emission lines for comparison to the available *HST*/STIS observations. Comparison of the theoretical models to the observations reveals important details about the binary's orbital motion, photoionization properties, and recent (5 – 15 year) mass loss history. Such an analysis also provides a baseline for following future changes in Eta Carinae, essential for understanding the late-stage evolution of a nearby supernova progenitor. Our modeling methods can also be adapted to a number of other colliding wind binary systems (e.g. WR 140) that are scheduled to be studied with future observatories (e.g. the James Webb Space Telescope).

Keywords. hydrodynamics, radiative transfer, line: formation, methods: numerical, stars: individual (Eta Carinae), stars: mass loss, stars: winds, outflows

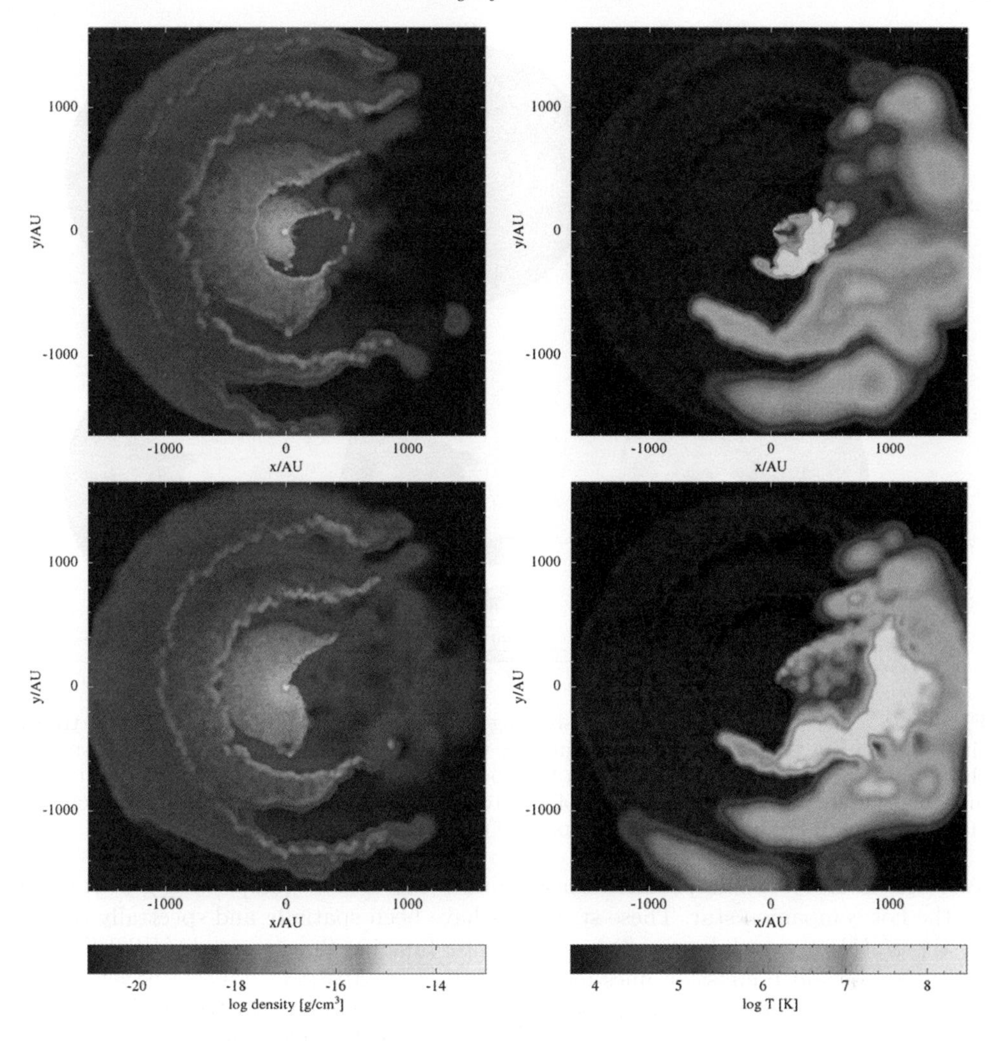

Figure 1. Slices showing log density (left) and log temperature (right) at apastron (top row) and periastron (bottom row) in the orbital xy plane from a 3D SPH simulation of Eta Carinae's colliding stellar winds, assuming primary and secondary mass loss rates of $8.5 \times 10^{-4}\ M_{\odot}/\text{yr}$ and $1.4 \times 10^{-5}\ M_{\odot}/\text{yr}$, and wind terminal speeds of 420 km/s and 3000 km/s, respectively. The orbital semimajor axis length $a = 15.45$ au and the eccentricity $e = 0.9$. The computational domain radius $r \approx 1545$ au $\approx 0.67''$. Axis tick marks correspond to an increment of 155 au.

At 2.3 kpc (Smith 2006), Eta Carinae is the closest and most luminous evolved massive star and supernova progenitor that we can study in great detail, making it an ideal astrophysical laboratory for studying massive binary interactions and stellar wind-wind collisions. Due to their intense luminosities, the stars in Eta Carinae have strong radiation-driven stellar winds. The LBV primary has an incredibly dense wind ($\approx 8.5 \times 10^{-4}\ M_{\odot}/\text{yr}$, $v_{\infty} \approx 420$ km/s; Hillier *et al.* 2001, Groh *et al.* 2012), while the less luminous companion has a much lower density, but faster wind ($\approx 1.4 \times 10^{-5}\ M_{\odot}/\text{yr}$, $v_{\infty} \approx 3000$ km/s; Pittard & Corcoran 2002, Parkin *et al.* 2009). These winds violently collide and generate a series of shocks and wind-wind collision regions (WWCRs) that give rise to numerous forms of time-variable emission and absorption seen across a wide range of wavelengths (Damineli *et al.* 2008). Due to the high orbital eccentricity, the

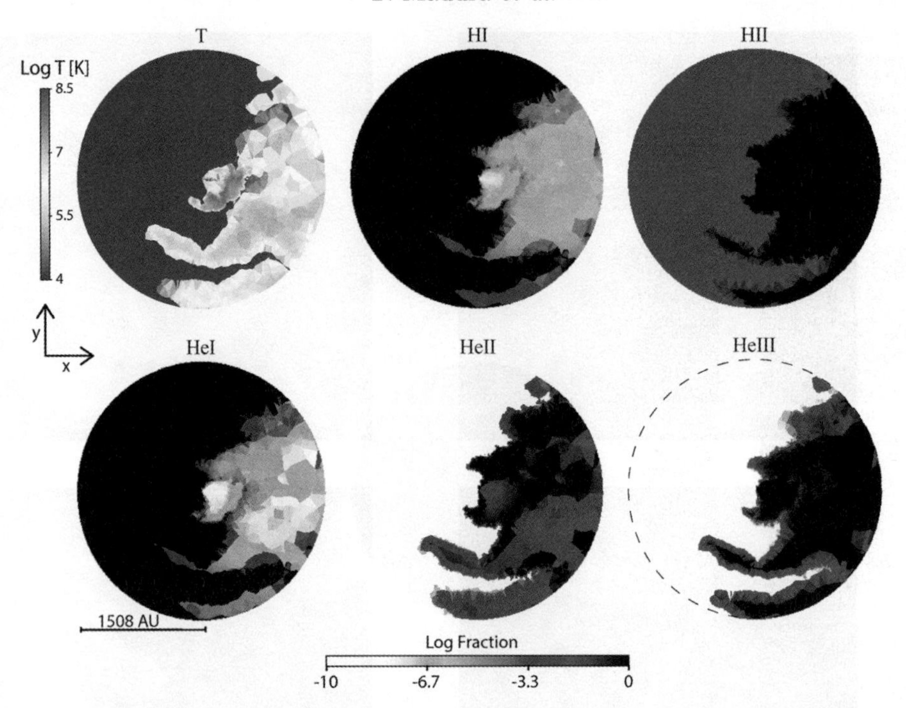

Figure 2. Slices in the orbital plane showing results from a 3D SimpleX radiative transfer simulation of Eta Carinae at apastron. SimpleX was applied to the 3D SPH simulation snapshot shown in the top row of Fig. 1. Color in the above panels shows log temperature (K, top left) and the computed fractions (log scale) of H I (top middle), H II (top right), He I (bottom left), He II (bottom middle), and He III (bottom right).

WWCRs in Eta Carinae produce dense spiral structures of compressed gas irradiated by the hot companion star. These structures have been spatially and spectrally resolved by the *Hubble Space Telescope*/Space Telescope Imaging Spectrograph (*HST*/STIS) in numerous forbidden emission lines (see e.g. Gull *et al.* 2016).

Three-dimensional (3D) Smoothed Particle Hydrodynamics (SPH) simulations have helped to greatly increase our understanding of the Eta Carinae system's WWCRs and how they affect numerous observational diagnostics (see e.g. Okazaki *et al.* 2008, Madura *et al.* 2012, 2013, Richardson *et al.* 2016). When coupled with 3D radiative transfer simulations, such SPH simulations can be used to obtain detailed 3D maps of ionization fractions of hydrogen, helium, and other elements (Clementel *et al.* 2014, 2015a,b). These model ionization maps help constrain the regions where observed forbidden emission lines can form. From the ionization maps, synthetic data cubes for various forbidden emission lines can be generated and directly compared to available and upcoming *HST*/STIS observations. By comparing the model data cubes to the observations, we hope to place tighter constraints on the binary's orbital, stellar, wind, and ionization parameters, as well as the system's recent $(5-15$ year) mass loss history.

Fig. 1 shows density and temperature slices in the orbital plane from a large-scale (computational domain radius $r \approx 1545$ au $\approx 0.67''$) 3D SPH simulation of Eta Carinae's colliding stellar winds. Visible in the density slices in the left column are the 'shells' of compressed primary wind formed after each periastron passage (to the left in the panels) and the extended, open WWCRs that can be illuminated by the companion (to the right in the panels). The temperature panels in the right column of Fig. 1 show the presence of extremely hot ($\gtrsim 10^7$ K) gas that is mostly companion star wind material. Such hot gas is responsible for Eta Carinae's observed X-ray emission.

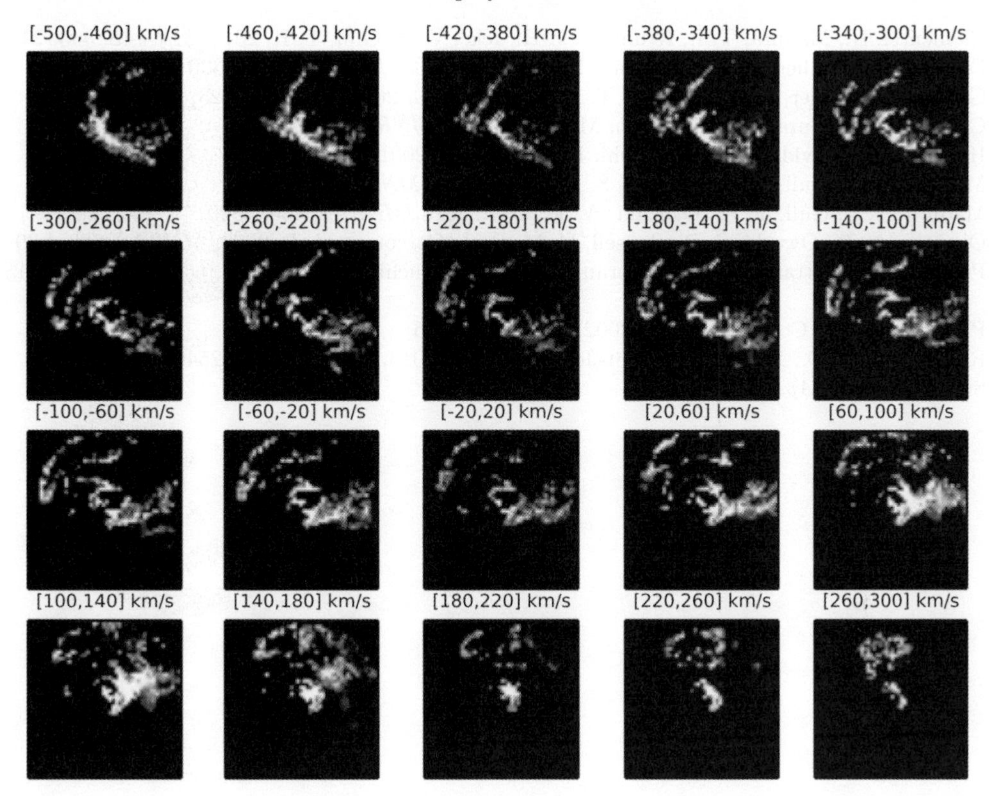

Figure 3. Slices from a synthetic data cube showing the predicted spatial distribution of [Fe III] emission on the sky (North is up) from Eta Carinae's collding winds in 40 km/s velocity bins ranging from −500 km/s (upper left) to +300 km/s (bottom right). The synthetic [Fe III] emission was computed using the SimpleX results in Fig. 2 and the equations/methods outlined in Madura *et al.* (2012). The width of each panel is $\approx 1.34''$. The color bar is on a square root scale.

Fig. 2 shows the results of applying the SimpleX 3D radiative transfer algorithm to the 3D SPH simulation in Fig. 1 (see Clementel *et al.* 2014, 2015a,b for details on SimpleX and our methods). The equations and methods in Madura *et al.* (2012) can then be applied to the SimpleX results in order to compute synthetic data cubes showing the predicted spatial distribution on the sky of emission from the observed forbidden lines. Example slices from a synthetic data cube of [Fe III] emission are in Fig. 3. Results like those in Fig. 3 can then be compared directly to the observed data cubes presented in e.g. Gull *et al.* (2016) and used to help constrain Eta Carinae's stellar, wind, and orbital parameters. Work is currently underway to compute synthetic data cubes at multiple orbital phases for direct comparison to the observed [Fe II] and [Fe III] data cubes obtained at multiple epochs and published in Gull *et al.* (2016). Any future observed changes in Eta Carinae's extended forbidden line emission can also be modeled and used to better understand any changes in the primary LBV's mass loss rate or the system's other stellar, wind, or orbital parameters. Other colliding wind binaries (e.g. WR 140) can be similarly modeled.

References

Clementel, N., Madura, T. I., Kruip, C. J. H., & Paardekooper, J.-P. 2015, *MNRAS*, 450, 1388
Clementel, N., Madura, T. I., Kruip, C. J. H., Paardekooper, J.-P., & Gull, T. R. 2015, *MNRAS*, 447, 2445

Clementel, N., Madura, T. I., Kruip, C. J. H., Icke, V., & Gull, T. R. 2014, *MNRAS*, 443, 2475
Damineli, A., Hillier, D. J., Corcoran, M. F., *et al.* 2008, *MNRAS*, 386, 2330
Groh, J. H., Hillier, D. J., Madura, T. I., & Weigelt, G. 2012, *MNRAS*, 423, 1623
Gull, T. R., Madura, T. I., Teodoro, M., *et al.* 2016, *MNRAS*, 462, 3196
Hillier, D. J., Davidson, K., Ishibashi, K., & Gull, T. 2001, *ApJ*, 553, 837
Madura, T. I., Gull, T. R., Owocki, S. P., *et al.* 2012, *MNRAS*, 420, 2064
Madura, T. I., Gull, T. R., Okazaki, A. T., *et al.* 2013, *MNRAS*, 436, 3820
Okazaki, A. T., Owocki, S. P., Russell, C. M. P., & Corcoran, M. F. 2008, *MNRAS*, 388, L39
Parkin, E. R., Pittard, J. M., Corcoran, M. F., Hamaguchi, K., & Stevens, I. R. 2009, *MNRAS*, 394, 1758
Pittard, J. M. & Corcoran, M. F. 2002, *A&A*, 383, 636
Richardson, N. D., Madura, T. I., St-Jean, L., *et al.* 2016, *MNRAS*, 461, 2540
Smith, N. 2006, *ApJ*, 644, 1151

High-mass X-ray Binaries: Illuminating the Passage from Massive Binaries to Merging Compact Objects
Proceedings IAU Symposium No. 346, 2019
L. M. Oskinova, E. Bozzo, T. Bulik & D. Gies, eds.

doi:10.1017/S1743921318008013

Circumstellar structures around high-mass X-ray binaries

Vasilii V. Gvaramadze[1,2]

[1]Sternberg Astronomical Institute, Lomonosov Moscow State University, Universitetskij Pr. 13, Moscow 119992, Russia
email: vgvaram@mx.iki.rssi.ru

[2]Space Research Institute, Russian Academy of Sciences, Profsoyuznaya 84/32, 117997 Moscow, Russia

Abstract. Many high-mass X-ray binaries (HMXBs) are runaways. Stellar wind and radiation of donor stars in HMXBs along with outflows and jets from accretors interact with the local interstellar medium and produce curious circumstellar structures. Several such structures are presented and discussed in this contribution.

Keywords. Circumstellar matter, stars: individual (4U 1907+09, EXO 1722-363, HD 34921, GX 304-01, Vela X-1, IGR J16327−4940), ISM: bubbles, X-rays: binaries.

1. Introduction

The high space velocities of HMXBs could be revealed via measurement of proper motions and/or radial velocities of these systems, or through the detection of bow shocks – the secondary attributes of runaway systems. The first detection of a bow shock produced by a HMXB was reported by Kaper *et al.* (1997), who discovered an Hα arc around Vela X-1. Later, Huthoff & Kaper (2002) searched for bow shocks around eleven high-velocity HMXBs using *IRAS* maps, but did not find new ones.

Our search for bow shocks around HMXBs from the sample of Huthoff & Kaper (2002) using data from the *Spitzer Space Telescope* led to the discovery of a bow shock associated with 4U 1907+09 (Fig. 1; cf. Gvaramadze *et al.* 2011). An asymmetric shape of the bow shock could be caused by density inhomogeneities in the local interstellar medium (ISM), as evidenced by the *Herschel* images of the field around 4U 1907+09 (see Fig. 1). The runaway nature of this HMXB is supported by proper motion measurements. Particularly, the *Gaia* DR2 (Gaia Collaboration 2018) proper motion and distance (≈ 4 kpc; Bailer-Jones *et al.* 2018) of 4U 1907+09 indicate that this system has a peculiar (transverse) velocity of $\approx 200 \, \mathrm{km \, s^{-1}}$, which is the highest peculiar velocity measured for HMXBs.

2. EXO 1722-363, HD 34921 & GX 304-01

We also searched for bow shocks around other HMXBs covered by *Spitzer* but, surprisingly, did not find any. Instead, we detected curious infrared nebulae around several HMXBs, two of which are presented below (both were independently discovered by Prisegen 2018). Fig. 2 shows a tau-shaped nebula associated with EXO 1722-363. The shape of the nebula and position of EXO 1722-363 within it exclude the bow shock interpretation for this nebula. Although one cannot exclude the possibility that the nebula is produced by collimated outflows (jets) from this HMXB (cf. Gallo *et al.* 2005; Heinz *et al.* 2008), the more plausible explanation is that we deal with a local ISM heated by radiation from the B0–1 Ia (Mason *et al.* 2009) companion star in EXO 1722-363.

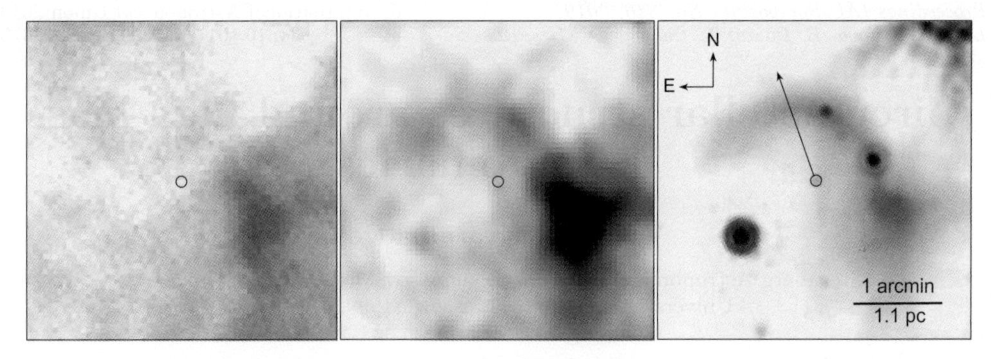

Figure 1. From left to right: *Herschel* 160 and 70 μm, and *Spitzer* 24 μm images of the field containing 4U 1907+09 (indicated by a circle). The arrow shows the direction of motion of 4U 1907+09, as follows from the *Gaia* proper motion measurement.

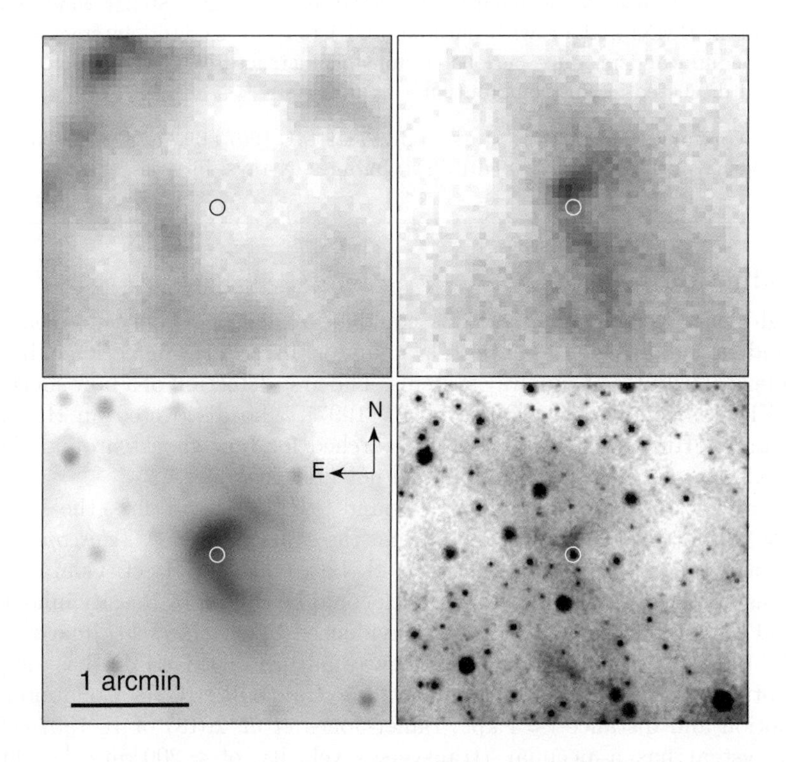

Figure 2. From left to right and from top to bottom: *Herschel* 160 and 70 μm, and *Spitzer* 24 and 8 μm images of the field containing EXO 1722-363 (indicated by a circle).

Fig. 3 shows the *Spitzer* 24 μm image of a barrel-like nebula around HD 34921 in two intensity scales to highlight some details of its filamentary structure. From the *Gaia* data and the heliocentric radial velocity of HD 34921 of $-20.5\,\mathrm{km\,s^{-1}}$ (Gontcharov 2008), we derive the peculiar velocity of this star of $\approx 30\,\mathrm{km\,s^{-1}}$. Although this velocity is typical of runaway stars, the complex shape of the nebula excludes its interpretation as a pure bow shock.

With the advent of the *Wide-field Infrared Survey Explorer* (*WISE*), it became possible to search for bow shocks around all ($\sim$100; e.g. Liu *et al.* 2006) known HMXBs, and what is absolutely amazing is that we did not find new bow shocks at all! The only interesting discovery is a bow-like structure attached to GX 304-1 (independently detected by Prisegen 2018). The geometry of this structure (see Fig. 4) and the weak

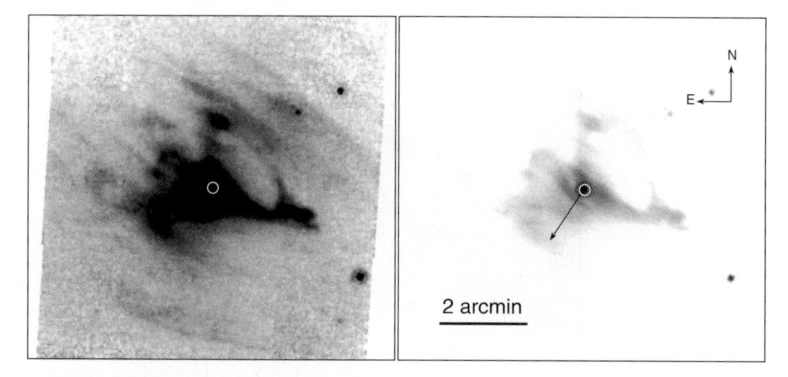

Figure 3. *Spitzer* 24 μm image of the barrel-like nebula around HD 34921 (indicated by a circle) in two intensity scales. The arrow shows the direction of motion of HD 34921.

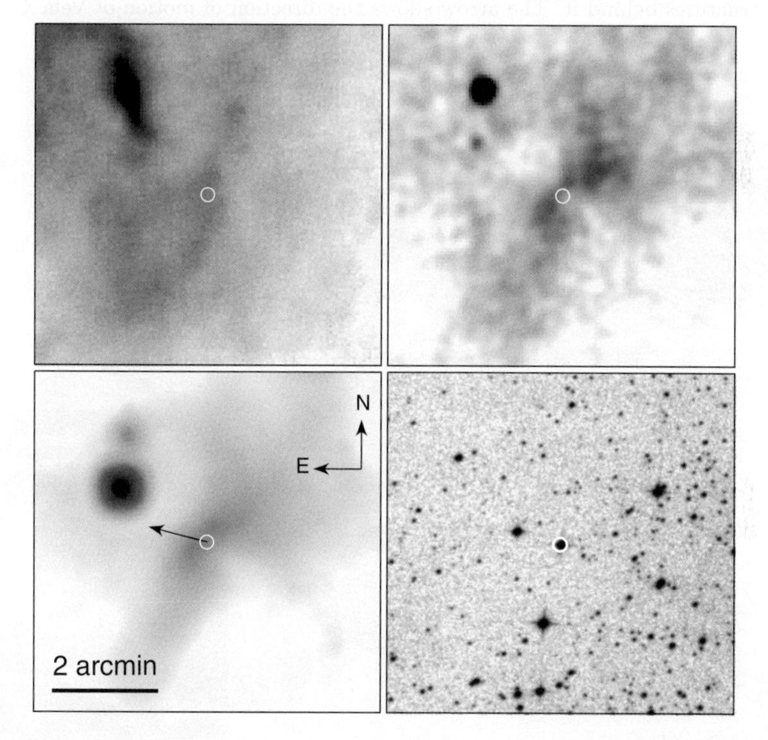

Figure 4. From left to right and from top to bottom: *Herschel* 160 and 70 μm, *WISE* 22 μm and DSS-II red-band images of the field containing GX 304-01 (indicated by a circle). The arrow shows the direction of motion of GX 304-01.

wind of the B2 Vne (Parkes *et al.* 1980) donor star in GX 304-1 suggest that here we deal with a radiation-pressure-driven bow wave (cf. van Buren & McCray 1988; Ochsendorf *et al.* 2014), although other explanations (involving jets or illumination of the local ISM) cannot be excluded as well.

3. Vela X-1

We also discovered a filamentary structure stretched behind the high-velocity ($\approx$ 50 km s^{-1}) HMXB Vela X-1. Fig. 5 shows the SuperCOSMOS H-alpha Survey (SHS; Parker *et al.* 2005) Hα image of this structure along with the already known bow shock

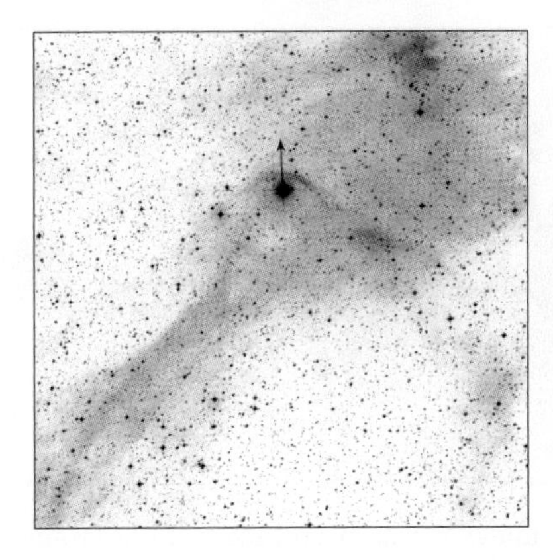

Figure 5. SHS Hα image of a 30 arcmin$\times$30 arcmin field containing Vela X-1 and filamentary structures behind it. The arrow shows the direction of motion of Vela X-1.

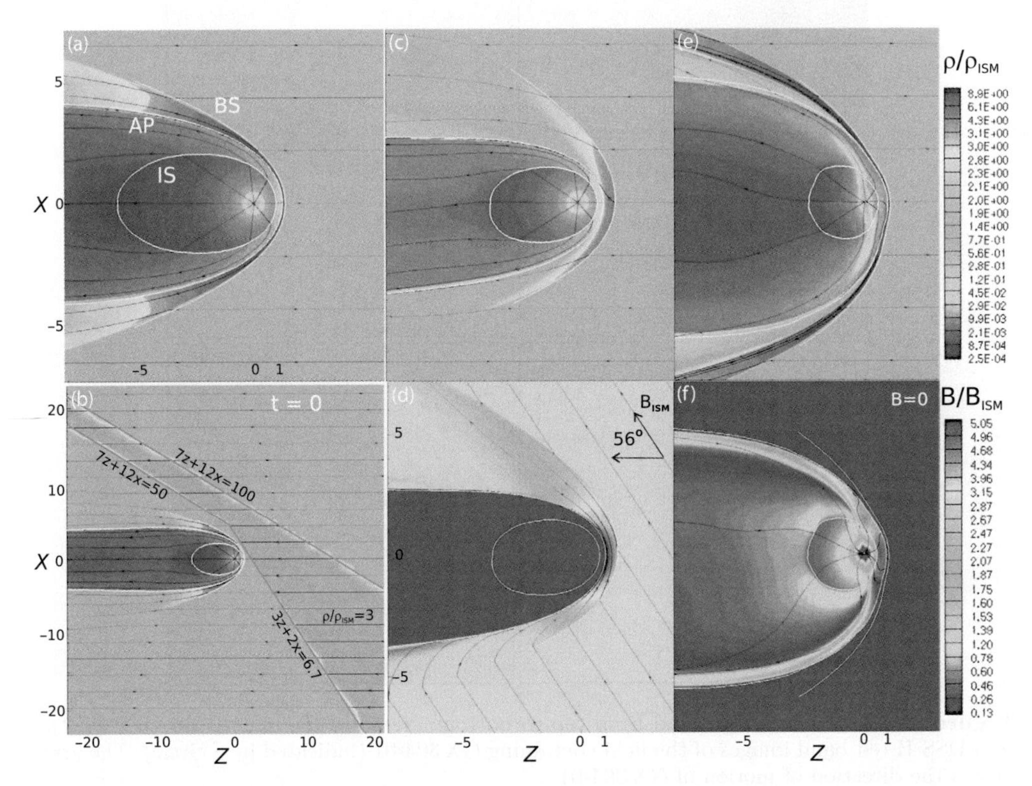

Figure 6. 2D distributions of the plasma density and streamlines (panels A, C, and E) and the magnetic field with the field lines (panels D and F) in the steady-state models. Panel A corresponds to model 1, panels C and D to model 2, and panels E and F to model 3. Panel B shows the initial condition in the ISM for the non-stationary models (shown in Fig. 7). The inner shock (IS), the astropause (AP) and the bow shock (BS) are plotted with white lines. Adopted from Gvaramadze *et al.* (2018a).

ahead of Vela X-1. The geometry of the filaments suggests that Vela X-1 has met a wedge-like layer of enhanced density on its way and that the shocked material of this layer outlines a wake downstream of Vela X-1.

To substantiate this suggestion, we carried out 3D MHD simulations of interaction between Vela X-1 and the layer for three limiting cases (see Fig. 6): the stellar wind and

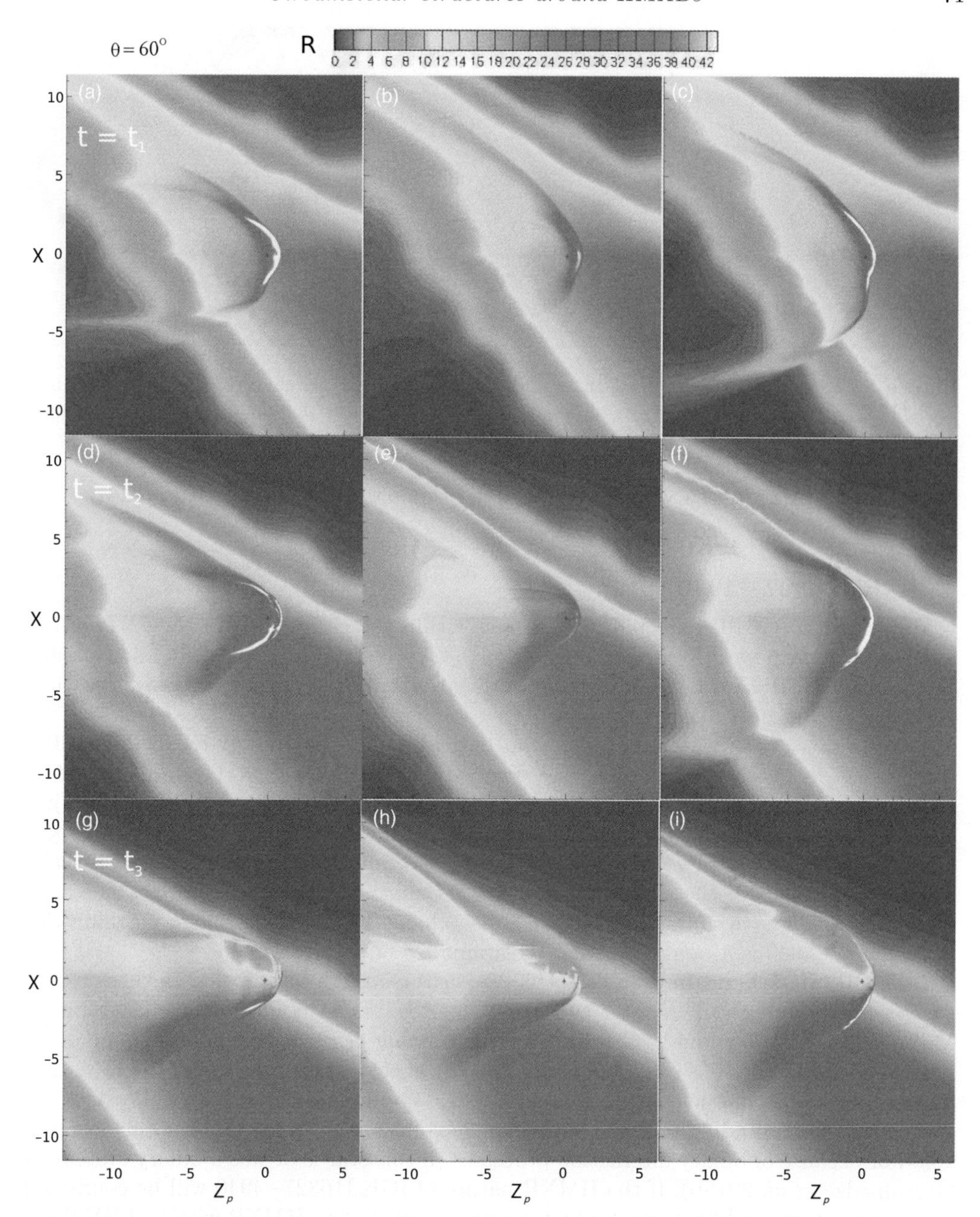

Figure 7. Projection of synthetic Hα intensity maps with a line of sight at an angle of $\theta = 60°$ to the symmetry axes of the non-stationary models 1, 2, and 3 (left to right) at three times (top to bottom). Adopted from Gvaramadze *et al.* (2018a).

the ISM were treated as pure hydrodynamic flows (model 1); a homogeneous magnetic field was added to the ISM, while the stellar wind was assumed to be unmagnetized (model 2); the stellar wind was assumed to possess a helical magnetic field (described by the Parker solution; Parker 1958), while there was no magnetic field in the ISM (model 3). We found that although the first two models can provide a rough agreement with the observations (cf. Fig. 7 with Fig. 5), only the third one allowed us to reproduce not only the wake behind Vela X-1, but also the opening angle of the bow shock and the apparent

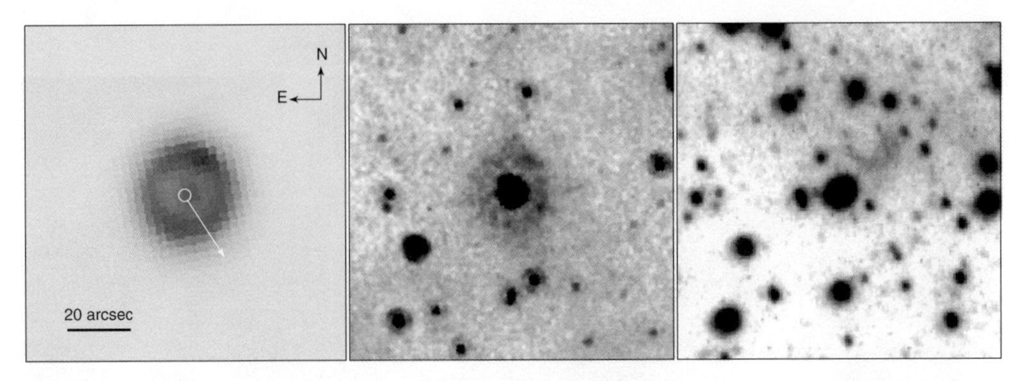

Figure 8. From left to right: *Spitzer* 24 and 8 μm, and SHS Hα images of the field containing IGR J16327−4940 (indicated by a circle) and its circumstellar nebula. The arrow shows the direction of motion of IGR J16327−4940.

detachment of its eastern wing from the wake (for more details see Gvaramadze *et al.* 2018a).

4. IGR J16327−4940

Masetti *et al.* (2010) detected an OB star within the error circle of the *INTEGRAL* transient source of hard X-ray emission IGR J16327−4940 (Bird *et al.* 2010) and classified this source as a HMXB because of its 'overall early-type star spectral appearance, which is typical of this class of objects'.

Using *Spitzer* data, we found that the optical counterpart to IGR J16327−4940 is surrounded by a circular nebula (see Fig. 8), named MN44 in Gvaramadze *et al.* (2010). A spectrum of the central star of MN44 taken in 2009 shows hydrogen and iron lines in emission, which is typical of luminous blue variables (LBVs) near the visual maximum (Gvaramadze *et al.* 2015). New observations carried out in 2015 revealed significant changes in the spectrum, indicating that the star became hotter. The spectral variability was accompanied by ≈ 1.6 mag changes in the brightness, meaning that IGR J16327−4940 is a bona fide LBV (Gvaramadze *et al.* 2015).

Gaia DR2 data indicate that IGR J16327−4940 is a high-velocity ($\approx 75\,\mathrm{km\,s^{-1}}$) runaway system ejected $\approx 4-5$ Myr ago from one of the most massive star clusters in the Milky Way — Westerlund 1 (Gvaramadze 2018), while a perfectly circular shape of the associated nebula implies that it does not feel the effect of ram pressure of the ISM. This means that the stellar wind still interacts with a co-moving dense material lost by the star during the preceding (e.g. red supergiant) evolutionary stage (cf. Gvaramadze *et al.* 2009) or because of binary interaction processes (if this star is or was a binary system; cf. Gvaramadze *et al.* 2018b). If the HMXB nature of IGR J16327−4940 will be confirmed, then this system would represent a first known example of an HMXB with an LBV donor star.

5. Conclusions

A possible explanation of the non-detection of bow shocks around HMXBs is that most of these systems are moving through a low-density, hot medium, so that the emission measure of their bow shocks is below the detection limit or the bow shocks do not form at all because the sound speed in the local ISM is higher than the stellar peculiar velocity (Huthoff & Kaper 2002). This provides a reasonable explanation of why only one-fifth of runaway OB stars produce (observable) bow shocks (van Buren *et al.* 1995). The detection rate of bow shocks around HMXBs is, however, a factor of ten less than that for OB stars (cf. Prisegen 2018). This difference could be understood if the HMXBs have systematically lower space velocities compared to the ordinary runaway stars (ejected in the field mostly because of dynamical few-body interactions), which could be connected to the formation mechanism of HMXBs (e.g. the supernova explosions in the HMXB progenitors should not be too energetic to unbind them). Moreover, the large proportion of HMXBs with (weak-wind) Be donor stars (Coleiro *et al.* 2013) could also contribute to this difference.

This work was supported by the Russian Science Foundation grant No. 14-12-01096.

References

Bailer-Jones, C. A. L., Rybizki J., Fouesneau, M., Mantelet, G., & Andrae, R. 2018, *AJ*, 156, 58
Bird A. J., *et al.* 2010, *ApJS*, 186, 1
Coleiro, A., Chaty, S., Zurita Heras, J. A., Rahoui, F., & Tomsick, J. A. 2013, *A&A*, 560, A108
Gaia Collaboration Brown, A. G. A., Vallenari, A., Prusti, T., de Bruijne, J. H. J., Babusiaux, C., & Bailer-Jones, C. A. L. 2018, *A&A*, 616, A1
Gallo, E., Fender, R., Kaiser, C., Russell, D., Morganti, R., Oosterloo, T., & Heinz S. 2005, *Nature*, 436, 819
Gontcharov, G. A. 2006, *Astron. Lett.*, 32, 759
Gvaramadze, V. V. 2018, *RNAAS*, in press; preprint arXiv:1811.07899
Gvaramadze, V. V., *et al.* 2009, *MNRAS*, 400, 524
Gvaramadze, V. V., Kniazev, A. Y., & Fabrika, S. 2010, *MNRAS*, 405, 1047
Gvaramadze, V. V., Röser, S., Scholz, R.-D., & Schilbach, E. 2011, *A&A*, 529, A14
Gvaramadze, V. V., Kniazev, A. Y., & Berdnikov, L. N. 2015, *MNRAS*, 454, 3710
Gvaramadze, V. V., Alexashov, D. B., Katushkina, O. A., & Kniazev, A. Y. 2018a, *MNRAS*, 474, 4421
Gvaramadze, V. V., Maryeva, O. V., Kniazev, A. Y., Alexashov, D. B., Castro, N., Langer, N., & Katkov, I. Y. 2018b, *MNRAS*, in press; arXiv:1810.12916
Heinz, S., Grimm, H. J., Sunyaev, R. A., & Fender, R.P. 2008, *ApJ*, 686, 1145
Huthoff, F., & Kaper, L. 2002, *A&A*, 383, 999
Kaper, L., van Loon, J. Th., Augusteijn, T., Goudfrooij, P., Patat, F., Waters, L. B. F. M., & Zijlstra, A. A. 1997, *ApJ* (Letters), 5475, L37
Liu, Q. Z., van Paradijs, J., & van den Heuvel, E. P. J. 2006, *A&A*, 455, 1165
Masetti, N., *et al.* 2010, *A&A*, 519, A96
Mason, A. B., Clark, J. S., Norton, A. J., Negueruela, I., & Roche, P. 2009, *A&A*, 505, 281
Ochsendorf, B. B., *et al.* 2014, *A&A*, 563, A65
Parker, E. N. 1958, *ApJ*, 128, 664
Parker, Q., *et al.* 2005, *MNRAS*, 362, 689
Parkes, G. E., Murdin, P. G., & Mason, K. O. 1980, *MNRAS*, 190, 537
Prišegen, M. 2018, *A&A*, in press; arXiv:1811.06781
van Buren, D., & McCray, R. 1988, *ApJ*, 329, L93
van Buren, D., Noriega-Crespo, A., & Dgani, R. 1995, *AJ*, 110, 2914

High-mass X-ray Binaries: Illuminating the Passage from Massive Binaries to Merging Compact Objects
Proceedings IAU Symposium No. 346, 2019
L. M. Oskinova, E. Bozzo, T. Bulik & D. Gies, eds.

doi:10.1017/S1743921318008074

The formation of massive binaries as a result of the dynamical decay of trapezium systems

Christine Allen, Alejandro Ruelas-Mayorga, Leonardo J. Sánchez and Rafael Costero

Instituto de Astronomía, Universidad Nacional Autónoma de México,
Cd. Universitaria, Ciudad de México. 04510, México,
email: chris@astro.unam.mx

Abstract. We propose that a significant fraction of the wide massive binaries in the field are formed as a result of the disintegration of multiple systems of trapezium type. As examples we discuss here the binaries formed from the evolution of the mini-cluster associated with the B component of the Orion Trapezium, from that of the Orion Trapezium itself, and from 10 additional massive trapezia for which we found reliable data in the literature.

Keywords. binaries: general — stars: early-type — stars: kinematics and dynamics — stars: formation

1. Introduction

The formation of massive wide binaries, which are quite abundant among early-type field stars, is still an open problem. One plausible mechanism is the dynamical disintegration of small star clusters (Reipurth 2000; Kouwenhoven *et al.* 2010, Reipurth & Mikkola *et al.* 2012, Reipurth *et al.* 2014). Indeed, as a by-product of our studies of the dynamical evolution of massive trapezium-type multiple systems we have found that at least one binary remains at the end of the evolution (Allen *et al.* 2015, 2017, 2018). These binaries will end up as field binaries, and it is interesting to compare their properties with the observational data.

We follow the definition of Ambartsumian (1954) for a trapezium. Let a multiple star system (of 3 or more stars) have components a, b, c,... If three or more distances among the components are of the same order of magnitude, then the system is a trapezium; otherwise, the system is of ordinary (or hierarchical) type. Two distances are of the same order of magnitude in this context if their ratio is greater than 1/3 but less than 3. Since only the angular separations are available, the sample of observed trapezia will contain also optical trapezia and pseudotrapezia. An optical trapezium is an apparent multiple system whose components are not physically connected. A pseudo-trapezium is a hierarchical system that appears as a trapezium due to projection. Ambartsumian and later Abt & Corbally (2000) showed that the number of pseudotrapezia is small only among multiple systems of spectral types B3 and earlier.

To model the dynamical evolution of trapezia we performed Monte Carlo N-body simulations for each system. As initial conditions we used the planar positions, transverse velocities, distances and masses from the best observations found in the literature. Radial velocities (when unavailable) and z-positions were modeled by Monte Carlo simulations. In the following sections we discuss the results we obtained for the mini-cluster associated with the B component of the Orion Trapezium, for the Orion Trapezium itself, and for 10 additional massive trapezia with data from the literature, with particular emphasis on the properties of the binaries formed.

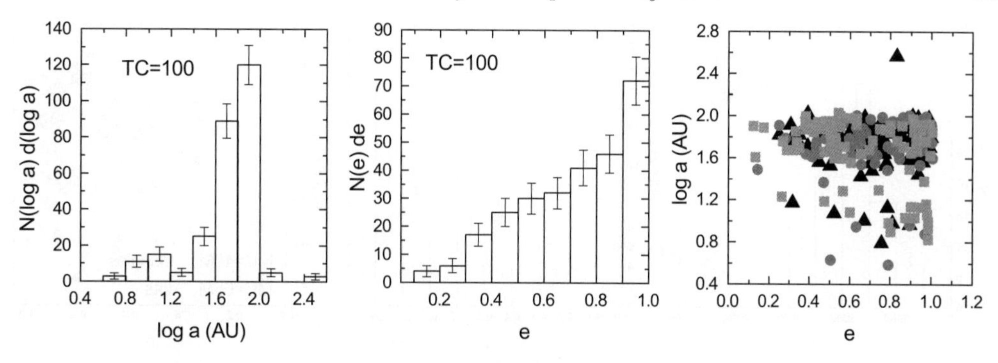

Figure 1. Properties of the binaries formed during the dynamical evolution of θ^1 Ori B at TC=100 (100 crossing times or about 30,000 years). Left panel: Frequency distribution of the major semi-axes. Middle panel: Frequency distribution of eccentricities. Right panel: Semi-axes as a function of eccentricities. Triangles, dots, and squares correspond to simulations with slightly different initial conditions.

2. N-body simulations of the mini-cluster θ^1 Ori B

This mini-cluster was studied in detail by Close *et al.* (2013), providing us with a suitable set of initial conditions to simulate its dynamical evolution (Allen *et al.* 2015). We obtained 300 Monte Carlo realizations for this mini-trapezium, by randomly perturbing each quantity according to its associated observational uncertainty. The N-body integrations showed that the systems evolved and disintegrated rapidly, with lifetimes of about 30 thousand years. Interestingly, the most massive stars, designated as (B1+B5) and B2 by Close *et al.* (2013) formed a close stable binary in 60 % of the cases. Binaries with major semi-axes as small as 6 AU were formed. Most of the escaping single stars were of low mass, but in 4 % of the cases the most massive binary (B1+B5) escaped. Most of the escapers were ejected with low velocities, but in 7 % of the cases runaway stars (stars with velocities larger than 3 times the escape velocity) were ejected. The statistical properties of the resulting binaries are shown in Figure 1. These properties are similar to those of field binaries (Duquennoy & Mayor 1991; Raghavan *et al.* 2010; Sana *et al.* 2012, 2014; Duchêne & Kraus 2013; Moe & Di Stefano 2017).

3. N-body simulations of the Orion Trapezium

The initial conditions for this system (distance, planar positions, radial velocities and masses) were taken from the literature (Allen *et al.* 2017). The separation velocities were computed from a combination of historical and modern measures of the separations among the components as a function of time. This provided us with a time span of over 180 years. For the other quantities, we tried to select the most reliable observational values. Since the z-positions are not available, they were randomly assigned, with a dispersion equal to the radius of the system. Random perturbations to each quantity were applied to obtain 100 realizations of the Orion Trapezium. We found that the systems disintegrated in less than 10 thousand years when we adopted the recently determined value for the mass of component C, $M_C = 45\ M_\odot$. We obtained more plausible lifetimes adopting $M_C = 65\ M_\odot$. With this value,the systems survived for about 30 to 40 thousand years, a lifetime compatible with that we found for θ^1 Ori B. The result of the integrations was usually a wide massive binary, always with star C as the primary. In 66% of the cases Star C was accompanied by the second most massive star. Hence, the dynamical decay of such systems is able to populate the field with wide massive binaries. The properties of the resulting binaries are shown in Figure 2. As mentioned before, these properties are similar to those observed for field binaries.

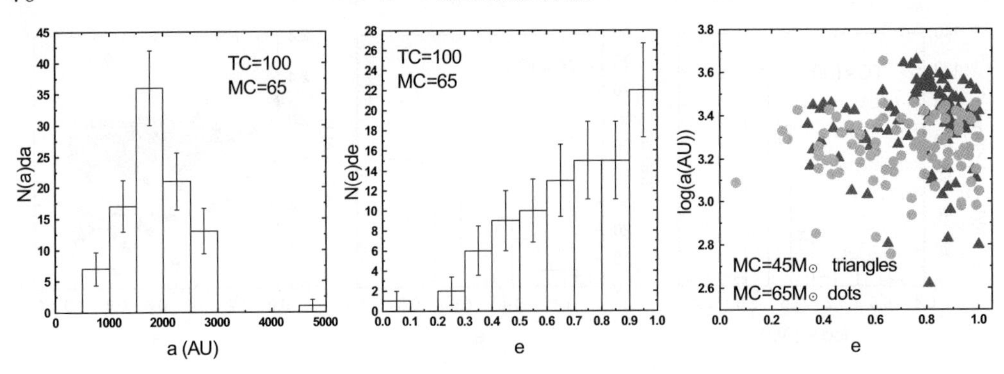

Figure 2. Properties of the binaries formed during the dynamical evolution of the Orion Trapezium at TC=100 (100 crossing times or about one million years). Left panel: Frequency distribution of the major semi-axes. Middle panel: Frequency distribution of the eccentricities. Right panel: Semi-axes as a function of eccentricities.

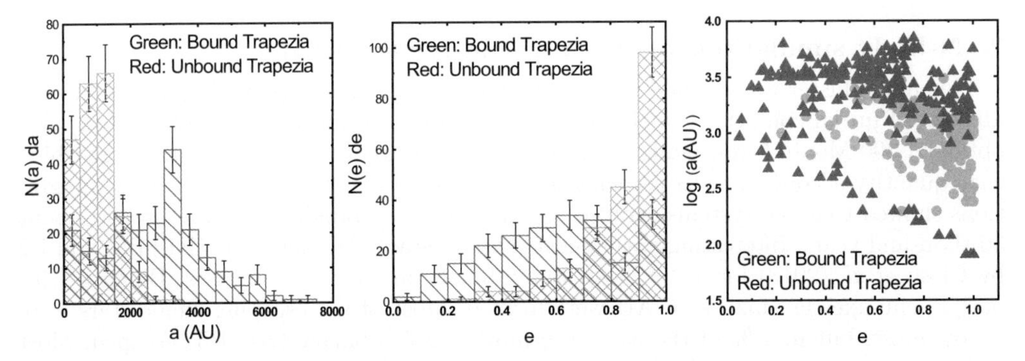

Figure 3. Properties of the binaries formed during the dynamical evolution of 10 massive trapezia. Binaries from bound systems are depicted as green-cross hatched bins, those from unbound systems as red-single hatched bins. Left panel: Frequency distribution of the major semi-axes. Middle panel: Frequency distribution of the eccentricities. Right panel: Semi-axes as a function of the eccentricities. Triangles and dots correspond to unbound and bound systems respectively.

4. N-body simulations of ten massive trapezia

To complete our study of the dynamical evolution of trapezia we conducted an extensive search of the literature. This search produced only 10 early-type systems likely to be true trapezia and with sufficient data to compute separation velocities among their components.

Here again, we obtained some initial conditions (planar positions, distances and masses) from the literature. Separation velocities among the components were calculated combining historical and modern measures of the separations among the components, as they change over time. Since neither z-positions nor radial velocities are available, they were randomly assigned, with dispersions compatible with the positions and transverse velocities of each trapezium. Random perturbations representing the observational uncertainties were applied to the observed data. In this way we obtained 100 Monte Carlo realizations for each of the 10 trapezia studied.

The numerical integrations showed that most systems resulted unbound; they remained unbound even doubling the values of the component masses to account for probable undiscovered binaries in them. The unbound systems rapidly dispersed, but in most of them energy redistribution among the components took place, and binaries were formed.

The bound systems evolved dynamically and disintegrated in about 10 thousand years. The result was usually a binary, sometimes a triple system of either hierarchical or non-hierarchical type. Hierarchical triples remained stable over the whole integration time of 1 million years. Non-hierarchical triples could survive for up to 300 thousand years.

Among the realizations for the bound trapezium ADS 719 we found that in all but 2 cases the most massive star formed a binary, and in 63 % of the cases it was accompanied by the second most massive star.

The properties of the resulting binaries are shown in the three panels of Figure 3. The binaries formed in bound systems have properties similar to those of field binaries, those stemming from unbound systems do not accord well with observations.

5. Conclusions

Our results show that the dynamical lifetimes of trapezium-type systems are extremely short, much shorter than the evolutionary lifetimes of their massive components. Only assuming much larger masses for the components could the lifetimes become longer. The end result of the simulations is usually a massive binary, sometimes a triple system. The dynamical disintegration of trapezium systems thus produces wide binaries that end up as field massive binaries.

Most semiaxes are of a few thousands AU, but binaries as close as a few AU are formed in some trapezia. Finally, the distributions of major semi-axes and eccentricities of binaries formed during the integrations are similar to those observed for wide massive field binaries.

Acknowledgments

A. R.-M and L. J. S. thank DGAPA-UNAM for financial support under PAPIIT projects IN102617 and IN102517.

References

Abt, H. A. & Corbally, C. J., 2000, *ApJ*, 541, 841
Allen, C., Costero, R. & Hernández, M. 2015, *AJ*, 150, 167
Allen, C., Costero, R., Ruelas-Mayorga, A. & Sánchez. L. J. 2017, *MNRAS*, 466, 4937
Allen, C., Ruelas-Mayorga, A. & Sánchez. L. J. & Costero, R. 2018, *MNRAS*, 481, 3953
Ambartsumian, V. 1954, *LIACo*, 5, 293
Close, L. M. *et al.*2013, *ApJ*, 774, A94
Duchêne, G. & Kraus, A., 2013, *ARAA*, 51, 269
Kouwenhoven, M. B. N., Goodwin, S. P., Parker, R. J. *et al.* 2010, *MNRAS*, 404, 1835
Moe, M. & Di Stefano, R. 2017, *ApJS*, 230, 15
Raghavan, D. McArthur, H. A., Henry, T. J. *et al.*, 2010, *ApJS*, 190, 1
Reipurth, B. 2000, *AJ*, 120, 3177
Reipurth, B. & Mikkola, S. 2012, *Nature*, 492, 221
Reipurth, B., Clarke, C. J., Boss, A. P. *et al.* 2014, in Protostars and Planets VI, ed. H Beuther *et al.*(U. of Arizona Press), p267
Sana, H., de Mink, S. E., de Koter, A., Langer, N., *et al.* 2012, *Sci*, 337, 444
Sana, H., Le Bouquin, J. B., Lacour, S., *et al.* 2014, *ApJS*, 215, 15

High-mass X-ray Binaries: Illuminating the Passage from Massive Binaries to Merging Compact Objects
Proceedings IAU Symposium No. 346, 2019
L. M. Oskinova, E. Bozzo, T. Bulik & D. Gies, eds.

doi:10.1017/S1743921318008128

Formation of the SMC WO+O binary AB8

Chen Wang[1], Norbert Langer[1], Götz Gräfener[1] and Pablo Marchant[2]

[1]Argelander-Institut fur Astronomie, Universität Bonn,
Auf dem Hügel 71, 53121 Bonn, Germany
email: cwang@astro.uni-bonn.de

[2]Dept. of Physics & Astronomy, Northwestern University,
2145 Sheridan Road, Evanston, IL 60208, USA

Abstract. Wolf-Rayet (WR) stars are stripped stellar cores that form through strong stellar wind or binary mass transfer. It is proposed that binary evolution plays a vital role in the formation of WR stars in low metallicity environments due to the metallicity dependance of stellar winds. However observations indicate a similar binary fraction of WR stars in the Small Magellanic Cloud (SMC) compared to the Milky Way. There are twelve WR stars in the SMC and five of them are members of binary systems. One of them (SMC AB8) harbors a WO type star. In this work we explore possible formation channels of this binary. We use the MESA code to compute large grids of binary evolution models, and then use least square fitting to compare our models with the observations. In order to reproduce the key properties of SMC AB8, we require efficient semiconvection to produce a sufficiently large convective core, as well as a longer He-burning lifetime. We also need a high mass loss rate during the WN stage to assist the removal of the outer envelope. In this way, we can reproduce the observed properties of AB8, except for the surface carbon to oxygen ratio, which requires further investigation.

Keywords. Stars, Wolf-Rayet, stellar evolution, binary

1. Introduction

Wolf-Rayet (WR) stars are hot and luminous stars characterised by broad emission lines. They are believed to be in a late evolutionary phase of massive stars. Due to their strong stellar wind, they provide an important contribution to the chemical enrichment of galaxies (for a review of WR stars, see Crowther 2007). They may be progenitors of long gamma-ray bursts (GRBs, Woosley & Bloom 2006), and explode as type Ib/c supernova. Thus it is of great importance to study the formation and evolution of WR stars.

Based on the intensity of spectra lines, WR stars are divided into three subtypes: WN type with strong He and N lines; WC type with strong He and C lines; and WO type with strong He, C and O lines. This difference is believed to be the result of different nuclear processed material being exposed at stellar surface (Lamers *et al.* 1991; Crowther & Hadfield 2006). Generally a star firstly exposes the H-burning products at its surface due to stellar wind or binary interaction and is seen as a WN type star, then exposes the He-burning products and is observed as a WC/WO type star (Crowther 2007 and references there in).

It is important for us to study massive star evolution in low metallicity environment, since the strong stellar wind can be avoided, making it possible for us to check the influence of other processes such as binary interaction. In the SMC, binary interaction should play a more important role in the formation of WR stars than in the Galaxy. However Foellmi *et al.* (2003) found that the binarity in the SMC is similar to that in the Galaxy as well as in the LMC. Apparently single WR stars have been studied by

Schootemeijer & Langer (2018), who concluded that they may in fact be undetected binaries. In this work we investigate the formation channel of the binary system AB8 that consists the only observed WO star in the SMC.

2. Method

We use version 8845 of MESA code (Paxton *et al.* 2011; Paxton *et al.* 2013; Paxton *et al.* 2015) to establish binary models, taking into account the physics of mass-loss, rotation and binary interaction.

We use Ledoux criterion for convection and the mixing length parameter $\alpha = l/H_P$ is set to be 1.5. We use step-overshooting to extend the convective core by 0.335 H_P, where H_P is the pressure scale height at the boundary of the convective core. We consider both inefficient and efficient semiconvection with parameter $\alpha_{\rm sc} = 0.01$ and $\alpha_{\rm sc} = 1$ respectively. The efficiency parameter for thermohaline mixing is $\alpha_{th} = 1$.

For stellar wind mass loss, we follow Brott *et al.* (2011), with wind for main sequence hydrogen rich stars computed following the recipe by Vink *et al.* (2001). For temperatures below that of the bi-stability jump, we take the maximum of the rate of Vink *et al.* (2001) and Nieuwenhuijzen & de Jager (1990). When surface hydrogen abundance is lower than 0.4, we use prescription from Hamann *et al.* (1995). For stars that have surface hydrogen abundances in range of $0.4 - 0.7$, we interpolate between the mass loss rate of Vink *et al.* (2001) and Hamann *et al.* (1995). Stellar mass loss rate scales as $\dot{M} \propto Z^m$. For Hamann *et al.* (1995) prescription we check $m = 0.85, 0.75, 0.65$ and 0.55 to account for different mass loss rate during WR phase, while for other prescriptions we assume $m = 0.85$ (Vink *et al.* 2001). During Roche lobe overflow the accretor spins up. If the star reaches $\Omega/\Omega_{crit} = 0.99$, we implicitly increase the mass loss rate such that the star could rotate just below the critical value.

Rotational mixing is modeled as a diffusive process, including the effects of dynamical and secular shear instabilities, the Goldreich-Schubert-Fricke instability, and Eddington-Sweet circulations, as described in Heger *et al.* (2000). We also include the transport of angular momentum due to magnetic fields from the Tayler-Spruit dynamo (Spruit 2002). Mass transfer is modeled using a contact scheme as described in Marchant *et al.* (2016).

We use a standard χ^2 minimization algorithm to find the best-fitting model in our dataset. We take into account six observables: $\log T_1$, $\log T_2$, $\log L_1$, $\log L_2$, $X_{H,1}$, and P corresponding to the temperature of the two stars, the luminosity of the two stars, the surface H abundance of the primary star and the orbital period in units of days. Then the standard χ^2 value is expressed as:

$$\chi^2(\log M_{i,1}, q_i, \log P_i, t) = \sum_{n=1}^{6} \left(\frac{O_n - E_n(\log M_{i,1}, q_i, \log P_i, t)}{\sigma_n} \right)^2 ,$$

where O_n are the six observables and $E_n(\log M_{i,1}, q_i, \log P_i, t)$ is corresponding theoretical values at time t defined by the three initial parameters. σ_n is the corresponding observational error. The observational data are listed in Table 1.

3. Results

We establish binary systems with the initial primary mass $\log M_{i,1}$ ranging from 1.65 to 2.10 in intervals of 0.05, the mass ratio $q = M_{i,2}/M_{i,1}$ from 0.3 to 0.95 in intervals of 0.05 and the initial orbital period $\log P_i$ from 0.0 to 2.5 in intervals of 0.025. All of the binary systems are assumed to be initially synchronised by tidal interaction and are in circular orbits.

Table 1. Different properties of our best fitting models, and observed properties of AB8 from Shenar *et al.* (2016)

	m=0.55	m=0.65	m=0.75	m=0.85	AB8
$\log T_{\rm eff,1}$	5.154	5.182	5.121	5.106	$5.149^{+0.154}_{-0.066}$
$\log T_{\rm eff,2}$	4.632	4.636	4.616	4.616	$4.653^{+0.046}_{-0.051}$
$\log L_1/{\rm L}_\odot$	6.15	6.17	6.16	6.20	$6.15^{+0.1}_{-0.1}$
$\log L_2/{\rm L}_\odot$	5.81	5.82	5.87	5.87	$5.85^{+0.1}_{-0.1}$
$M_1/{\rm M}_\odot$	35	35	35	37	19^{+3}_{-8}
$M_2/{\rm M}_\odot$	51	51	53	53	61^{+14}_{25}
$X_{\rm C,s}$	0.46	0.43	0.00	0.00	$0.30^{+0.05}_{-0.05}$
$X_{\rm O,s}$	0.26	0.37	0.00	0.00	$0.30^{+0.1}_{-0.1}$
$R_1/{\rm R}_\odot$	2.0	1.8	2.3	2.6	2^{+1}_{-1}
$R_2/{\rm R}_\odot$	14.6	14.5	16.9	16.8	14^{+6}_{-4}
$P_{\rm orb}/{\rm days}$	16.58	16.21	16.34	16.14	16.6
$\log \dot{M}_{\rm WO}/{\rm M}_\odot{\rm yr}^{-1}$	-4.20	-4.27	-4.38	-4.40	$-4.8^{+0.1}_{-0.1}$

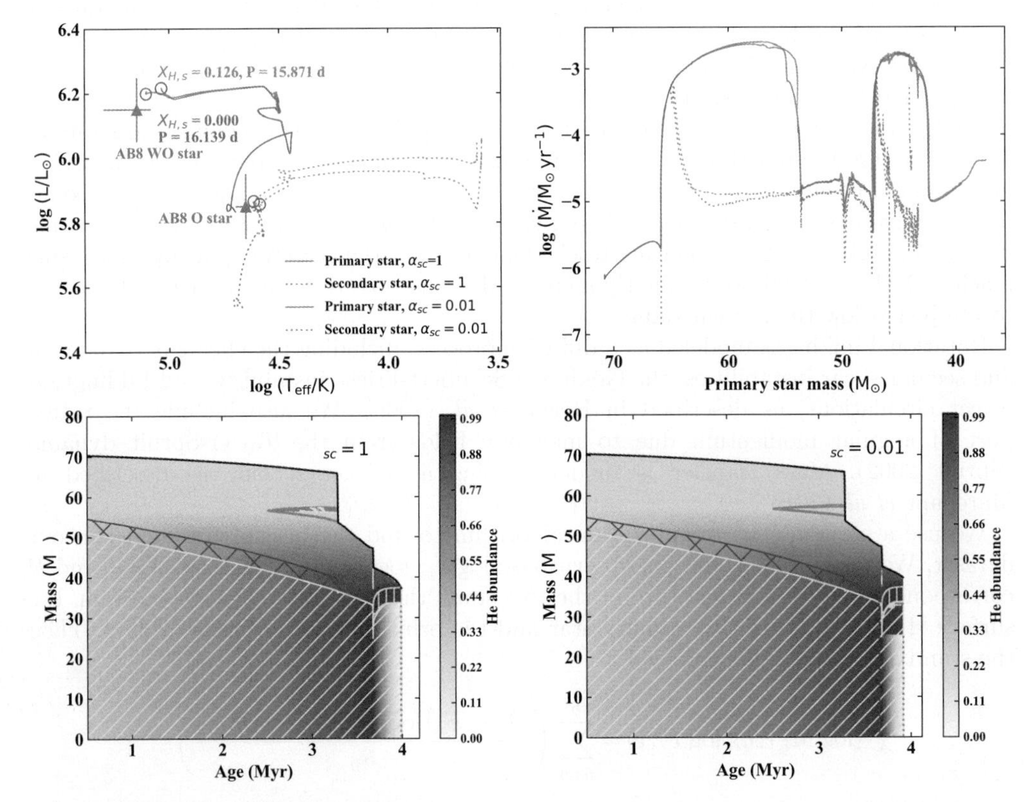

Figure 1. Evolutionary history of the model that can best reproduce the observed properties of AB8 by adopting $\alpha_{\rm sc} = 1$ and $m = 0.85$. This model has initially $M_{i,1} = 70\,M_\odot$, $M_{i,2} = 50\,M_\odot$ and $P_i = 12.5$ days. The system with the same initial parameters but computed with $\alpha_{\rm sc} = 0.01$ is also shown. The first panel shows their evolutionary paths in the HR diagram. Solid and dotted lines correspond to the evolution of the primary and secondary stars, respectively. Green triangles represent the observations. Circles indicate the best fitting positions. The surface H abundances as well as the orbital periods at the best fitting time are listed. The second panel shows the mass transfer history of the two systems. We use the same labels as in the first panel. The third and fourth panels are Kippenhahn diagrams for the primary stars computed from different $\alpha_{\rm sc}$ values. We use green, purple, red and yellow lines to delineate the convective, overshooting, semiconvective and thermohaline mixing regions, respectively.

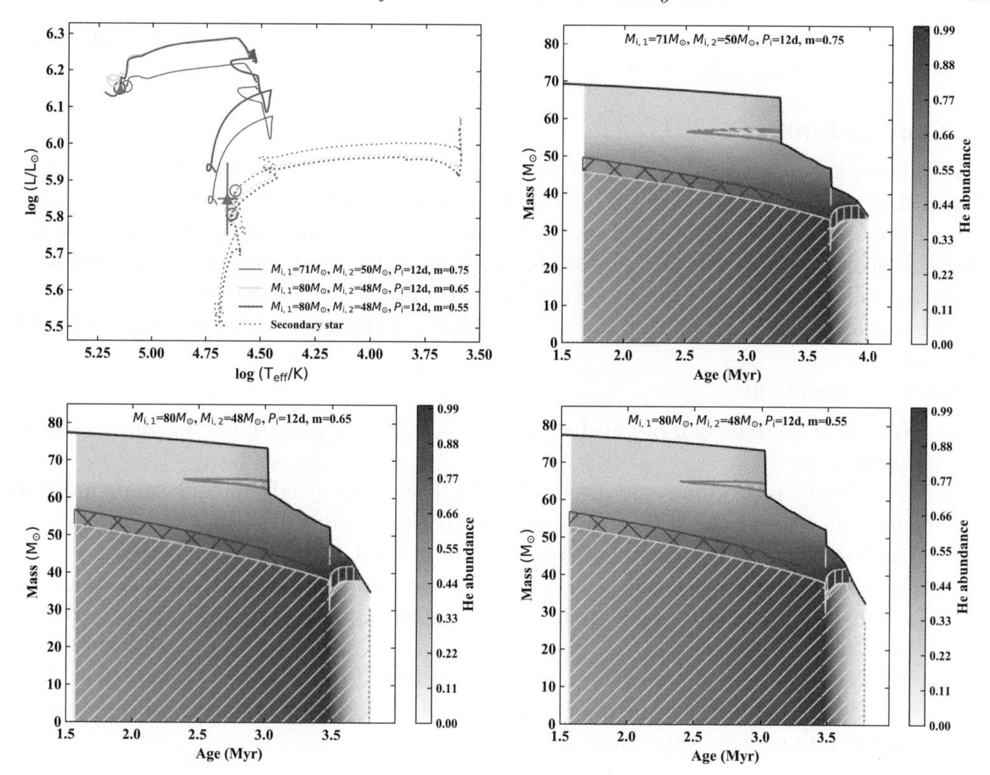

Figure 2. Models that can best reproduce the properties of AB8 adopting $\alpha_{\rm sc} = 1$ and different m values. In the first panel, solid and dotted lines with different colors correspond to the evolution of the primary and secondary stars in binaries with different intial parameters and different m values, respectively. The second to fourth panels are the Kippenhahn diagrams for the primary stars. The inital parameters and adopted m values are listed in each panel.

Fig. 1 shows the evolutionary history of the model with $\log M_{i,1}/M_\odot = 1.850$, $q_i = 0.700$ and $\log P_i/d = 1.100$ ($M_{i,1} = 70\,M_\odot$, $M_{i,2} = 50\,M_\odot$ and $P_i = 12.5\,$d). This model best reproduces the observations, based on the assumption of $\alpha_{\rm sc} = 1$ and $m = 0.85$. We also show the system with the same initial parameters but computed with inefficient semi-convection. With inefficient semiconvection $\alpha_{\rm sc} = 0.01$, there is a thin H layer left at the best fitting time. With efficient semiconvection, the CO core mass is $\sim 8\,M_\odot$ larger. The He core mass is nearly the same in the two situations. However with efficient semiconvection, due to the injection of fresh nuclear fuel, the primary star has a longer He burning lifetime ($\sim 0.07\,$Myr longer), making it possible to expose its He core. Still this WR star model has a thin pure He layer left, meaning that it is a WN type star rather than a WC/WO type star.

The highly uncertain WR stellar wind mass loss rate may play an important role in the formation of WC/WO stars in low metallicity environments. The most uncertain part in the prescription of the WR mass loss is its dependence on metallicity. Therefore we increase the stellar wind during the WR phase by changing the power of the metallicity dependance m to 0.75, 0.65 and 0.55 which will increase the stellar wind mass loss rate by a factor of 1.23, 1.51 and 1.86, respectively.

Fig. 2 depicts the models that can best reproduce the observations, computed from assuming $\alpha_{\rm sc} = 1$ and different m values. The initial parameters are listed. We can see from the third and fourth panels that with the help of an enhanced stellar wind, the pure He layer of the primary stars can be totally removed. In this way, we can obtain progenitors that can fit the observed properties of AB8. The best fitting model has initially $M_{i,1} \simeq 80\, M_\odot$, $M_{i,2} \simeq 48\, M_\odot$, $P_i \sim 12$ d.

4. Conclusions

In this work we have evolved tens of thousands of binary systems with SMC metallicity. The initial parameter space is $1.65 \leqslant \log M_{i,1} \leqslant 2.10$, which securely covers the progenitors of the WO star in AB8 based on current assumptions, the mass ratio $0.3 \leqslant q_i \leqslant 0.95$ and the orbital period $0.0 \leqslant \log P_i \leqslant 2.5$. We have calculated models using both inefficient and efficient semiconvection with $\alpha_{\rm sc} = 0.01$ and $\alpha_{\rm sc} = 1$, respectively. We have also varied the metallicity scaling of the WR wind mass loss rate with, $m = 0.85, 0.75, 0.65$ and 0.55.

Our main results are:

(a) WC/WO systems can be produced in low metallicity environments through binary interaction.

(b) In order to reproduce a hydrogen-free WR star, we need efficient semiconvection that can extend the stellar life during core He burning such that the primary star can lose more of its envelope through stellar wind.

(c) In order to reproduce SMC AB8, we need to adopt a high stellar wind mass loss rates. In our models we can find progenitor systems with both, $m = 0.65$ and $m = 0.55$.

(d) The SMC WO+O binary AB8 formed most likely from a ZAMS binary with $M_{i,1} \simeq 80\, M_\odot$, $M_{i,2} \simeq 48\, M_\odot$ and an initial orbital period of $\sim$ 12 days, through stable highly non-conservative mass transfer.

References

Brott, I., de Mink, S. E., Cantiello, M., *et al.* 2011, *A&A*, 530, A115
Crowther, P. A. 2007, *ARAA*, 45, 177
Crowther, P. A. & Hadfield, L. J. 2006, *A&A*, 449, 711
Foellmi, C., Moffat, A. F. J., & Guerrero, M. A. 2003, *MNRAS*, 338, 360
Hamann, W.-R., Koesterke, L., & Wessolowski, U. 1995, *A&A*, 299, 151
Heger, A., Langer N., & Woosley, S. E. 2000, *ApJ*, 528, 368
Lamers, H. J. G. L. M., Maeder, A., Schmutz, W., & Cassinelli, J. P. 1991, *ApJ*, 368, 538
Marchant, P., Langer, N., Podsiadlowski, P., *et al.* 2016, *A&A*, 588, A50
Nieuwenhuijzen, H. & de Jager, C. 1990, *A&A*, 231, 134
Paxton, B., Bildsten, L., Dotter, A., *et al.* 2011, *ApJS*, 192, 3
Paxton, B., Cantiello, M., Arras, P., *et al.* 2013, *ApJS*, 208, 4
Paxton, B., Marchant, P., Schwab, J., *et al.* 2015, *ApJS*, 220, 15
Schootemeijer, A. & Langer, N. 2018, *A&A*, 611, 75
Shenar, T., Hainich, R., Todt, H., *et al.* 2016, *A&A*, 591, A22
Spruit, H. C. 2002, *A&A*, 381, 923
Vink, J. S., de Koter, A., & Lamers, H. J. G. L. M. 2011, *A&A*, 369, 574
Woosley, S. E. & Bloom, J. S. 2006, *ARAA*, 44, 507

High-mass X-ray Binaries: Illuminating the Passage from Massive Binaries to Merging Compact Objects
Proceedings IAU Symposium No. 346, 2019
L. M. Oskinova, E. Bozzo, T. Bulik & D. Gies, eds.

doi:10.1017/S1743921318008438

Massive star mass-loss revealed by X-ray observations of young supernovae

Vikram V. Dwarkadas

Dept. of Astronomy and Astrophysics, Univ of Chicago
5640 S Ellis Ave, Chicago, IL 60637
email: vikram@astro.uchicago.edu

Abstract. Massive stars lose a considerable amount of mass during their lifetime. When the star explodes as a supernova (SN), the resulting shock wave expands in the medium created by the stellar mass-loss. Thermal X-ray emission from the SN depends on the square of the density of the ambient medium, which in turn depends on the mass-loss rate (and velocity) of the progenitor wind. The emission can therefore be used to probe the stellar mass-loss in the decades or centuries before the star's death.

We have aggregated together data available in the literature, or analysed by us, to compute the X-ray lightcurves of almost all young supernovae detectable in X-rays. We use this database to explore the mass-loss rates of massive stars that collapse to form supernovae. Mass-loss rates are lowest for the common Type IIP supernovae, but increase by several orders of magnitude for the highest luminosity X-ray SNe.

Keywords. radiation mechanisms: thermal; shock waves; astronomical data bases: miscellaneous; circumstellar matter; stars: mass loss; supernovae: general; stars: winds, outflows; stars: Wolf-Rayet; X-rays: general

1. Introduction

Core-collapse Supernovae (SNe) arise from massive stars. These stars have strong mass-loss in the form of stellar winds, with mass-loss rates ranging from 10^{-7} to 10^{-4} $M_\odot$ yr^{-1}, and therefore lose a considerable amount of mass over their evolution. When the SN explodes, the resulting shock wave expands in this wind-blown medium. The evolution of the SN shock wave, and the resulting emission, depends on the structure and density profile of this medium. The radiation signatures from the expansion of the SN shock wave can be used to trace the density profile of this medium, which is formed by pre-supernova wind mass-loss. Since the shock wave heats the gas to high temperatures, X-ray emission is one of the signposts of the shock wave interaction with the circumstellar medium (Chevalier & Fransson 2003)

2. X-Ray Emission

X-ray emission from young SNe has been found to be both thermal and non-thermal. If the X-ray emission from the SN is thermal, due to thermal bremsstrahlung with line emission, the X-ray luminosity of the SN shock wave can be written as (Chevalier & Fransson 2003):

$$L_i = 3 \times 10^{39} g_{ff} C_n \left(\frac{\dot{M}_{-5}}{v_{w10}}\right)^2 \left(\frac{t}{10\,\mathrm{d}}\right)^{-1} \mathrm{ergs\ s^{-1}} \tag{2.1}$$

where i can refer to either the forward or reverse shock, g_{ff} is the Gaunt factor, $C_n = 1$ for the circumstellar shock and $(n-3)(n-4)^2/4(n-2)$ for the reverse shock. $\dot{M}_{-5}$ is the mass-loss rate of the wind into which the SN is expanding, in units of 10^{-5} $M_{\odot}$ yr^{-1}, v_{w10} is the wind velocity in units of 10 km s^{-1}, and t is the time in days. The mass-loss rate and wind velocity are assumed constant, leading to a density in the wind medium that decreases as r^{-2}, and an X-ray luminosity decreasing as t^{-1}. Note that the X-ray luminosity implied by this equation is technically the luminosity over the entire X-ray range, while current X-ray satellites such as *Chandra*, *XMM-Newton* and *Swift* are only effective in a limited X-ray range of around 0.5-10 keV. Electron-ion equilibration behind the shock front is assumed, which is probably not true for young SNe except when the density is very high. However the equation can be easily modified to account for any specified ratio of electron to ion temperature.

2.1. *X-Ray Spectra of SNe*

Inspection of the X-ray spectra can reveal whether the emission is thermal or non-thermal. A non-thermal spectrum would resemble a power-law. Although it appears simple to distinguish between them in principle, in practice it is much more difficult, due to low counts, and the fact that it is possible some may have both a thermal and non-thermal emission component. Nevertheless, some general points can be made, keeping in mind that due to the small statistics, exceptions are always possible.

- *Type IIn SNe* Observed Type IIn spectra are clearly thermal, showing distinct lines of various elements, especially α elements. All IIns that have been observed in X-rays thus far (Bauer *et al.* 2008, Chandra *et al.* 2009, Chandra *et al.* 2012, Chandra *et al.* 2015, Dwarkadas *et al.* 2016) display thermal spectra. Chandra *et al.* (2015) tried to invoke a non-thermal component for SN 2010jl but did not get it to fit.
- *Type Ib/c SNe* Chevalier & Fransson (2006) have suggested that Type Ib/c SNe have X-ray emission that is non-thermal, either inverse Compton or synchrotron.
- *Type IIb SNe* The prototype Type IIb SN, SN 1993J, shows a clearly thermal X-ray spectrum, with distinct lines (Chandra *et al.* 2009, Dwarkadas *et al.* 2014). However Chevalier & Soderberg (2010) have suggested that there is another class of Type IIb SNe with compact progenitors, whose X-ray emission may be non-thermal. The counts for SNe observed in the latter category are generally low, and it has not been easy to categorize the emission mechanism.
- *Type IIL SNe* There are not too many observations of Type IIL SNe in X-rays. SN 1979C was a IIL whose spectrum was fitted with two components. One component was definitely a thermal plasma, while the other could be thermal (Immler *et al.* 2005) or nonthermal (Patnaude *et al.* 2011).
- *Type IIP SNe* The emission from Type IIP was proposed by Chevalier, Fransson & Nymark (2006) to be non-thermal. This was found to be the case for SN 2011ja (Chakraborti *et al.* 2013) and SN 2013ej (Chakraborti *et al.* 2016).

3. X-Ray Lightcurves

Analysis of an X-ray observation of a SN provides a source flux and luminosity at a given epoch. Plotting the luminosity over a series of epochs gives rise to the X-ray lightcurve of the SN. Over the years we have compiled together a library of X-ray lightcurves of young SNe, both those that have appeared in the literature and those analysed by us. This process started with the publication of a first compilation of lightcurves in 2012 (Dwarkadas & Gruszko 2012), to which we have gradually added more over the years (Dwarkadas 2014 and Dwarkadas *et al.* 2016.) Many of these lightcurves are available in an online database that we have created for this purpose (Ross & Dwarkadas 2017), which can be viewed at kronos.uchicago.edu/snax.

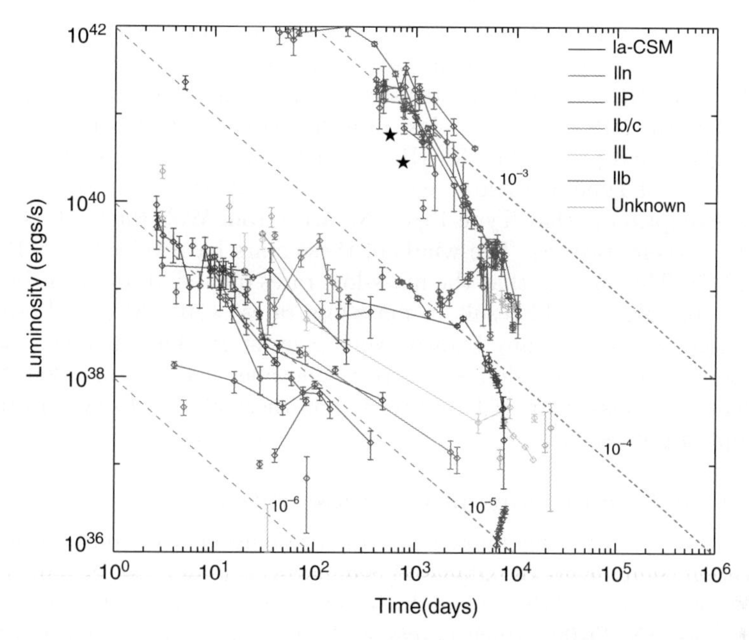

Figure 1. X-Ray Lightcurves of most observed X-Ray SNe. Grouped by type. Dashed lines are lines of constant mass-loss, with t^{-1} slope. Mass-loss rates in $M_\odot$ yr^{-1} assuming wind with constant parameters (density $\propto r^{-2}$), $v_{wind} = 10$ km s^{-1}, electron temperature assumed 10% of the ion temperature behind the shock [Chevalier & Fransson (2003)].

Fig. 1 shows the currently available set of lightcurves. We have grouped them according to the type of SN. Almost all the SNe, barring one, are of the core-collapse variety. The single exception (black stars) is a Type Ia-CSM SN (Bochenek *et al.* 2018). These are a rare and unusual class of SNe (Silverman *et al.* 2013) that show a spectrum very similar to those of a Type Ia SN, but including hydrogen lines (the general characteristic of a Ia is that it has no hydrogen).

We then invert Equation 2.1 to overplot lines of X-ray luminosity at a constant mass-loss rate. We use an electron-to-proton temperature ratio of 0.1 which is generally true for many young SNe, although we caution that at the highest densities, typical of the Type IIn SNe for example, Coulomb equilibration will bring the electron and proton temperatures closer to equilibrium.

The mass-loss rate for any SN can be directly read off the plot. Although these rates are approximate, and sometimes different from what more detailed analysis reveals, the plot reveals several interesting characteristic of the mass-loss rates of various SN types

- *Type IIP SNe* These have the lowest mass-loss rates, in general lower than about 10^{-5} $M_\odot$ yr^{-1}. These mass-loss rates led Dwarkadas (2014) to conclude that the progenitors of Type IIP SNe were red supergiants with initial mass $< 19\,M_\odot$. This result is in agreement with results derived via direct optical measurements (Smartt 2009, Smartt 2015), as well as theoretical calculations (Sukhbold *et al.* 2016).
- *Type IIn SNe* These have the highest mass-loss rates, generally higher than $10^{-4}\,M_\odot$ yr^{-1}, and even exceeding $10^{-3}\,M_\odot$ yr^{-1}. Note however that they show the maximum deviation from the t^{-1} slope of the lightcurves, and consequently from an r^{-2} density profile for the wind medium. This suggests that the mass-loss rate, or wind velocity, or both, are functions of time, and hence radius. The slopes are generally much steeper than t^{-1}. If the wind velocity is assumed constant, as is frequently taken to be the case, then this implies that the mass-loss rates are much higher closer to the onset of core-collapse,

and could significantly exceed even the rates suggested here (see below). If the wind velocity is larger than 10 km s^{-1}, the mass-loss rate will also be correspondingly higher.

• *Type IIL, Type IIb* Although the statistics are low, these have mass-loss rates somewhat higher than the IIPs. If the mass-loss rates increase with stellar mass, these suggest marginally higher mass progenitors for IILs and IIbs as compared to IIPs.

• *Type Ib/c* SNe of Type Ib/c show mass-loss rates that are $> 10^{-5}$ $M_\odot$ yr^{-1}. However, the general assumption is that Type Ib/c SNe arise from Wolf-Rayet stars (due to the lack of H/He in their spectra). The winds of these stars have velocities > 1000 km s^{-1} (Crowther 2007). This implies that the mass-loss rates quoted above must be multiplied by a factor $\geqslant 100$. This would result in mass-loss rates $> 10^{-3}$ $M_\odot$ yr^{-1}, substantially exceeding mass-loss rates for any known Wolf-Rayet stars. The inference may be that the assumption of thermal emission is incorrect. Chevalier & Fransson (2006) have in fact suggested that the emission from Ib/c SNe is non-thermal, due to synchrotron and/or Inverse Compton processes.

4. Accurate determinations of mass-loss rates

Using well-sampled X-ray lightcurves and high resolution X-ray data, it is possible to determine the mass-loss rates of individual SNe much more accurately. This was demonstrated in Fransson *et al.* (1996). Chandra *et al.* (2015) determined a mass-loss rate of 0.06 $M_\odot$ yr^{-1} for SN 2010jl, which agrees well with the optically determined value of 0.1 $M_\odot$ yr^{-1} by Fransson *et al.* (2014). For SN 2005kd, Dwarkadas *et al.* (2016) find a mass-loss rate of 4.3 $\times 10^{-4}$ $M_\odot$ yr^{-1} for a wind velocity of 10 km s^{-1} at 10^{16} cm, and infer that it could have been $> 10^{-3}$ $M_\odot$ yr^{-1} closer to core-collapse. If the wind velocity is higher, as is likely for Type IIn SNe, this mass-loss rate would be even higher.

For the other SN types, the mass-loss rates may be more similar to those deduced from the plot above. Figure 1 suggests a mass-loss rate somewhat greater than 10^{-5} $M_\odot$ yr^{-1} for SN 1993J. This is consistent with the mass-loss rate of 4 $\times 10^{-5}$ $M_\odot$ yr^{-1} deduced by Tatischeff (2009) via fitting of the radio lightcurves of the SN.

5. Conclusions

The X-ray emission from young SNe can be used to deduce the density of the medium into which the SN is expanding, and thus the mass-loss rate of the progenitor star for a medium formed by stellar mass-loss. This can provide insight into the mass-loss rates of massive stars in the years, decades and centuries before core-collapse, which are otherwise difficult to determine. The values so derived agree with those obtained from observations at other wavelengths. They show that Type IIPs are expanding in winds with the lowest mass-loss rates, while IIns have the highest mass-loss rates, which may exceed 10^{-3} $M_\odot$ yr^{-1}, and perhaps higher depending on their wind velocities. It is unclear which progenitor stars could sustain such high mass-loss rates, although luminous blue variable stars have been suggested (Smith 2014).

Acknowledgements

VVDs work on X-ray SNe is supported by NASA Astrophysics Data Analysis program grant # NNX14AR63G awarded to PI V. Dwarkadas at the University of Chicago. We also acknowledge several Chandra grants over the years, including GO1-12095A, GO2-13092B, GO4-15075X and GO7-18066X. VVD would like to thank the IAU for a travel grant that enabled him to attend the 2018 IAU General Assembly.

References

Bauer, F. E., Dwarkadas, V. V., Brandt, W. N., Immler, S., Smartt, S., Bartel, N. & Bietenholz, M. F. 2008, *ApJ*, 688, 1210

Bochenek, C. D., Dwarkadas, V. V., Silverman, J. M., Fox, O. D., Chevalier, R. A.; Smith, N, & Filippenko, A.V. 2018, *MNRAS*, 473, 336
Chakraborti, S., Ray, A., Smith, R., Ryder, S., Yadav, N., Sutaria, F., Dwarkadas, V. V., Chandra, P., Pooley, D., & Roy, R. 2013, *ApJ*, 774, 30
Chakraborti, S., Ray, A., Smith, R., Margutti, R., Pooley, D., Bose, S., Sutaria, F., Chandra, P., Dwarkadas, V.V., Ryder, S., & Maeda, K. 2016, *ApJ*, 817, 22
Chandra, P., Dwarkadas, V. V., Ray, A., Immler, S., & Pooley, D. 2009, *ApJ*, 699, 388
Chandra, P., Chevalier, R. A., Chugai, N., Fransson, C., Irwin, C. M., Soderberg, A. M., Chakraborti, S., & Immler, S. 2012, *ApJ*, 755, 110
Chandra, P., Chevalier, R. A., Chugai, N., Fransson, C., Soderberg, A. M., Chakraborti, S., & Immler, S. 2012, *ApJ*, 810, 32
Crowther, P. A. 2007, *ARAA*, 45, 177
Chevalier, R. A., & Fransson, C. 2003, in: K. Weiler (ed), *Supernovae and Gamma-Ray Bursters, Lecture Notes in Physics*, (Berlin: Springer Verlag), 598, 17
Chevalier, R. A., & Fransson, C. 2006, *ApJ*, 651, 381
Chevalier, R. A., Fransson, C., & Nymark, T. 2006, *ApJ*, 641, 1029
Chevalier, R. A., & Soderberg, A. M. 2010, *ApJ*, 711, L40
Dwarkadas, V. V, & Gruszko, J. 2012, *MNRAS*, 419, 1515
Dwarkadas, V. V., Bauer, F. E., Bietenholz, M., & Bartel, N. 2014, in J-U. Ness (ed), *The X-ray Universe 2014*, 248
Dwarkadas, V. V, 2014, *MNRAS*, 440, 1917
Dwarkadas, V. V., Romero-Caizales, C., Reddy, R., & Bauer, F. E. 2016, *MNRAS*, 462, 1101
Fransson, C., Lundqvist, P., & Chvelier, R. A. 1996, *ApJ*, 461, 993
Fransson, C., Ergon, M., Challis, P. J., Chevalier, R. A., France, K., Kirshner, R. P., Marion, G. H., *et al.* 2014, *ApJ*, 797, 118
Immler, S., Fesen, R. A., Van Dyk, S. D., Weiler, K. W., Petre, R. Lewin, W. H. G., Pooley, D., Pietsch, W., Aschenbach, B., Hammell, M. C., & Rudie, G. C. 2005, *ApJ*, 632, 2381
Patnaude, D. J., Loeb, A., & Jones, C. 2011, *NewA*, 16, 187
Silverman, J. M., Nugent, P. E., Gal-Yam, A., Sullivan, M., Howell, D. A., Filippenko, A. V., *et al.* 2013, *ApJS*, 207, 3
Smartt, S. J. 2009, *ARAA*, 47, 63
Smartt, S. J. 2015, *PASA*, 32, 16
Smith, N. 2014, *ARAA*, 52, 487
Sukhbold, T., Ertl, T., Woosley, S. E., Brown, J. M., & Janka, H.-T. 2016, *ApJ*, 821, 38
Tatischeff, V. 2009, *A&A*, 499, 191

High-mass X-ray Binaries: Illuminating the Passage
from Massive Binaries to Merging Compact Objects
Proceedings IAU Symposium No. 346, 2019
L. M. Oskinova, E. Bozzo, T. Bulik & D. Gies, eds.

doi:10.1017/S1743921319001340

X-ray spectroscopy of massive stellar winds: previous and ongoing observations of the hot star ζ Pup

N. Miller[1], W. Waldron[2], J. Nichols[3], D. Huenemoerder[4], M. Dahmer[5], R. Ignace[6], J. Lauer[7], A. Moffat[8], Y. Nazé[9], L. Oskinova[10], N. Richardson[11], T. Ramiaramanantsoa[12], T. Shenar[13] and K. Gayley[14]

[1]Department of Physics & Astronomy, University of Wisconsin-Eau Claire, 105 Garfield Avenue, Eau Claire, WI 54701 USA

[2]Eureka Scientific, Inc., USA

[3]Harvard-Smithsonian Center for Astrophysics, USA

[4]Massachusetts Institute of Technology, USA

[5]Northrop Grumman, USA

[6]East Tennessee State University, USA

[7]Harvard-Smithsonian Center for Astrophysics, USA

[8]University of Montreal, Canada

[9]Fund for Scientific Research-FNRS, Belgium and University of Liège, Belgium

[10]University of Potsdam, Germany

[11]University of Toledo, USA

[12]Arizona State University, USA

[13]KU Leuven, Belgium

[14]University of Iowa, USA

Abstract. The stellar winds of hot stars have an important impact on both stellar and galactic evolution, yet their structure and internal processes are not fully understood in detail. One of the best nearby laboratories for studying such massive stellar winds is the O4I(n)fp star ζ Pup. After briefly discussing existing X-ray observations from Chandra and XMM, we present a simulation of X-ray emission line profile measurements for the upcoming 840 kilosecond Chandra HETGS observation. This simulation indicates that the increased S/N of this new observation will allow several major steps forward in the understanding of massive stellar winds. By measuring X-ray emission line strengths and profiles, we should be able to differentiate between various stellar wind models and map the entire wind structure in temperature and density. This legacy X-ray spectrum of ζ Pup will be a useful benchmark for future X-ray missions.

Keywords. stars: early-type, stars: mass loss, stars: winds, outflows, X-rays: stars

---

1. Introduction: Deep HETG spectrum of ζ Pup and multi-wavelength campaign 2018-2019

Massive stars have an important impact on the evolution and energy budget of a galaxy, producing comparable energy to a supernova over their ≈4 Myr lifetimes. The structure (including mass loss rates) and mechanisms of massive stellar winds have an important impact on galactic evolution, yet remain only partially understood. Our target here (ζ Pup (O4I(n)fp)) is one of the closest X-ray bright massive stars. XMM has long used ζ Pup as a reference target, so 18 years of observations are available from this spacecraft. These observations have been extensively analyzed in Nazé *et al.* (2012), Nazé *et al.* (2013), Hervé *et al.* (2013), and Nazé *et al.* (2018). One early Chandra HETG spectrum with a 67 kilosecond exposure is also currently available (Cassinelli *et al.* 2001).

The existing HETG spectrum hints at the interesting complexity of X-ray emission line profiles for this star. Even cursory inspection reveals that the observed line profiles greatly differ from the instrumental line spread function. However, insufficient photon counts per bin limited the utility of that spectrum to fully exploit the high spectral resolution of the HETG. We have therefore initiated a program to obtain unprecedentedly high S/N X-ray data for this star. With the HETG's high resolution and improved coverage of short wavelength lines (especially lines shortward of $\approx$8 Å), the spectra being obtained in this project will complement and extend current high S/N measurements performed with the RGS on XMM. The core of our program is 840 ks of Chandra HETG spectral data to be taken over an 8-9 month period from July 2018 through early 2019. Some of the later X-ray spectra will be complemented by a contemporaneous multiwavelength campaign. Ground-based optical observations will occur in locations including Australia, New Zealand, SAAO, and CTIO. Space-based observations will take place using the BRITE constellation of satellites, providing us with a dual-band precision optical photometry time series during the relevant time period.

2. Line-Shape Models and Wind Properties

Winds of hot stars are believed to be propelled by an inherently unstable, line-driven process (Lucy & White 1980, Feldmeier 1995). It is thought that this instability allows small-scale perturbations in the wind to steepen into shocks. Interestingly, a recent investigation by Ramiaramanantsoa *et al.* (2018) found that these wind perturbations can be triggered by perturbations in the photosphere. After the wind perturbations form shocks, the resultant shock heating raises wind material to X-ray emitting temperatures. Due to Doppler broadening and wind absorption, we would expect asymmetric X-ray emission lines (Owocki *et al.* 1988, MacFarlane *et al.* 1991). Any detailed emission line model will need to include these effects, as discussed in Owocki & Cohen (2001), Cohen (2009), and Ignace (2016).

However, on the observational side, blue-shifted, asymmetric emission lines initially predicted by this scenario are in fact not commonly seen in hot stars, a situation referred to as the "line asymmetry problem" (Waldron & Cassinelli 2007). A number of investigations in recent years have made progress towards understanding the nature of this discrepancy for hot stars by exploring the ramifications of lower estimated mass loss rates (Oskinova *et al.* 2007, Cohen *et al.* 2014, Leutenegger *et al.* 2013, Oskinova 2016). Nonetheless, in some ways this paucity of profiles showing the initially expected widths and shifts gives those O stars which have asymmetric line profiles (such as ζ Pup) even more particular interest. The XMM observations of ζ Pup's X-ray emission lines have been analyzed by Hervé *et al.* (2013).

One important outstanding issue for understanding hot star X-ray emission line profiles is quantifying the presence and importance of inhomogeneities in the wind. Line profile models for different clumping properties can be seen in Oskinova *et al.* (2006). In that study, stellar wind properties were kept constant except for giving the winds different clumping properties. The large effects of clumping they find there illustrate the sensitivity of the X-ray line profiles as a probe for the structure of the wind. Another model including clumps is described in Leutenegger *et al.* (2013). For comparison, a typical smooth-wind model is described in Cohen *et al.* (2014).

The high resolution and long exposure time for the spectra in this investigation will allow us to perform detailed modeling of the X-ray emission line shapes. Analysis of mass

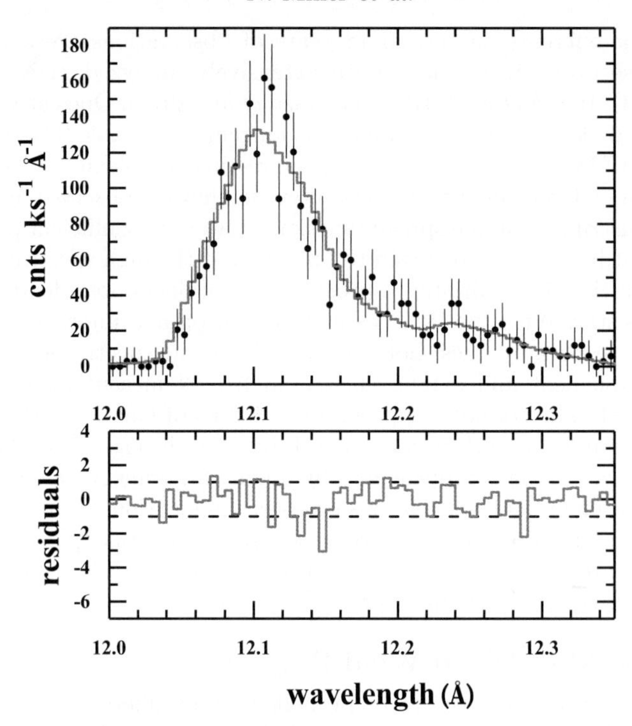

Figure 1. Comparison of the MEG Ne X region clump model (solid line) to a simulated smooth wind dataset with errors appropriate to the existing 2000 HETG observation. The large errors per bin make it difficult to make precise determinations of the model parameters, and there is no clear pattern in the residuals. The residuals are in the sense of (model) minus (simulated model) divided by the simulated error for each bin. This diagram is constructed using the standard bin size for the MEG detector (0.005 Å).

loss rates, porosity (caused by clumps in the wind), and wind opacity will require very high S/N in each bin across each emission line. To explore the ability of observations to diagnose the importance of clumps, we simulated data for a no-clump case, and then fit it using a model which includes clumping. Because the simulated dataset is constructed with one model and the fit with an entirely different model, an examination of the residuals will indicate if an observation allows discrimination between the two cases. Figure 1 shows typical error bars and residuals for the existing 67 ks Chandra observation, while Figure 2 shows a simulation of the improvement expected when the full 840 ks Chandra observation is in hand. These figures illustrate how the increase in the S/N ratio of the observation will allow greater discrimination between models.

3. X-ray Temperature Distribution and Analysis of Weak Lines

The analysis of the overall spectral properties of hot stars is complicated by the line shape effects described in the previous sections. For many ordinary coronal X-ray sources X-ray emission lines are not broadened by bulk motions of the X-ray emitting plasma, but for hot stars the emission line shape must be understood to disentangle the contribution of each spectral line to each spectral bin. This is especially true in regions with many blends. After detailed line modeling is complete, the next step will be to determine line fluxes. An examination of the ratios of H-like to He-like lines of prominent elements can be used to obtain temperature information (Miller *et al.* 2002). This can be used with the f to i ratios of He-like ions to probe the temperature structure in the wind, as described in Waldron & Cassinelli (2007).

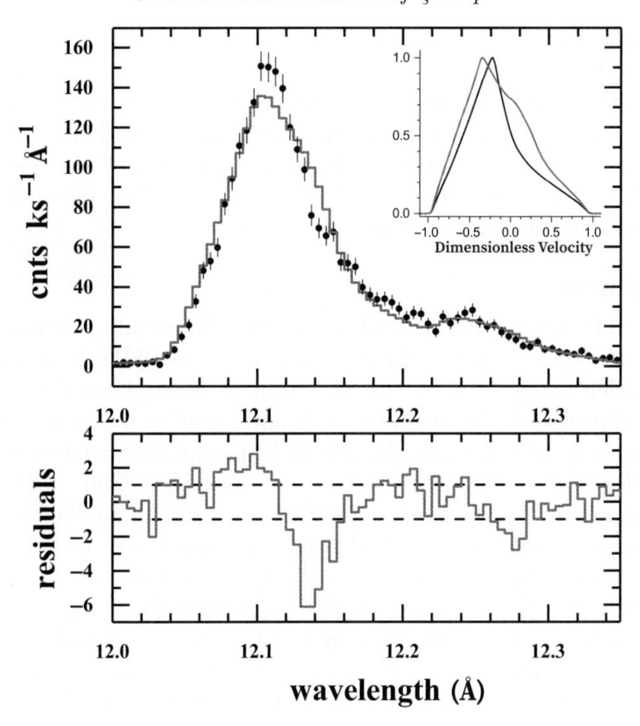

Figure 2. Same as Figure 1 except this panel illustrates a simulation of the precision we expect for the spectrum using the full approved Cycle 19 Chandra exposure time. This will greatly enhance our ability to differentiate models as evident in the clear trends which are apparent in the simulated residuals. The inset shows the theoretical line shapes for the smooth wind (black) and clumped (red) models before the inclusion of any instrumental effects. The normalized theoretical line profile models in the inset are displayed as a function of the dimensionless velocity (i.e. the line-of-sight wind velocity divided by the terminal velocity, with 0 representing the rest wavelength for the line). This calculation assumes a value of 2250 km s^{-1} for the terminal velociy (Puls *et al.* 2006).

In addition to previously-measured strong lines, there are many weak emission lines in ζ Pup's spectrum. These have been difficult to study in previous spectra due line blending in observations with insufficient exposure time to allow detailed modeling. For example, there are 50 lines in the 10-12 Å region that have T_x greater than 15 MK (mostly Fe lines). Also, He-like and H-like β lines for Ne, Mg, and Si, and weaker He-like Ar lines, and Ca lines may possibly be present. Additional Fe L shell lines may be measurable in this spectrum and would provide density diagnostic ratios. Measuring all the line strengths in this deep Chandra exposure will also allow accurate modeling and better determination of the overall temperature distribution of the X-ray emitting plasma.

4. X-ray Variability

XMM shows variability for ζ Pup on a timescale of days in the total band, which may reflect the presence of CIRs (Nazé *et al.* 2018, Nazé *et al.* 2013). Temporal variation in the X-ray emission from hot stars might also be expected due to the growth and fading of shocks. The XMM data for this star was examined for this kind of variation (Nazé *et al.* 2013). We will be exploring this Chandra dataset for line strength and profile variations over time. This Chandra dataset will especially allow us to analyze the temporal behavior of shorter-wavelength emission lines and will allow us to probe all emission lines at higher resolution. In 2019 we will be observing ζ Pup nearly contemporaneously with both earth- and space-based optical telescopes to probe for correlated temporal variation between the

X-ray and optical bands. These parallel observations will allow us to see if the X-rays are modulated on the same 1.78 d timescale found in the optical by Howarth & Stevens (2014). In a recent analysis Ramiaramanantsoa *et al.* (2018) were able to map bright surface features and their associated CIRs which had periodicity on this time scale, firmly connecting this 1.78 day period with stellar rotation. The exploration of related variability on this timescale could provide an important link between features on the photosphere and structures in the wind.

Acknowledgements

Support for this work was provided by the National Aeronautics and Space Administration through Chandra Award Number GO8-19011 issued by the Chandra X-ray Center, which is operated by the Smithsonian Astrophysical Observatory for and on behalf of the National Aeronautics Space Administration under contract NAS8-03060. NAM acknowledges support from an American Astronomical Society International Travel Grant and an SREU grant through the Univ. of Wisc.-Eau Claire Office of Research & Sponsored Programs. JSN acknowledges NASA contract NAS 8-03060 to the Smithsonian Astrophysical Observatory. Support of DPH for this work was provided by NASA through the SAO contract SV3-73016 to MIT for Support of the CXC and Science Instruments. AFJM thanks NSERC (Canada) and FQRNT (Quebec) for financial aid. YN (research asociate FNRS) acknowledges support from the FNRS and CFWB (Belgium) as well as the PRODEX Xmas contract. TS acknowledges funding from the European Research Council (ERC) under the European Unions DLV-772225-MULTIPLES Horizon 2020 research and innovation programme.

References

Cassinelli, J. P., Miller, N. A., Waldron, W. L., MacFarlane, J. J. & Cohen, D. H. 2001 *ApJ* 554, 55L

Cohen, D. H. 2009, AIP Conference Proceedings 1161

Cohen, D., Wollman, E., Leutenegger, M., Sundquist, J., Fullerton, A., Zsargo, J. & Owocki, S. 2014, *ApJ* 586, 495

Feldmeier, A. 1995 *A&A* 299, 523

Hervé, A., Rauw, G. & Nazé, Y. 2013, *A&A*, 551, A83

Howarth, I. D. & Stevens, I. R. 2014 *MNRAS* 445, 2878

Ignace, R. 2016, Adv. Sp. Res. 58,694

Leutenegger, M., Cohen, D., Sundquist, J. & Owocki, S. 2013, *ApJ* 770,80

Lucy, L. B. & White, R. L. 1980 *ApJ* 241, 300

MacFarlane, J. J., Cassinelli, J. P., Walsh, B. Y., Vedder, P. W., Vallerga, J. V. & Waldron, W. L. 1991 *ApJ* 380, 564

Miller, N. A., Cassinelli, J. P., Waldron, W. L., MacFarlane, J. J. & Cohen, D. H. 2002 *ApJ* 577, 951

Nazé, Flores, C. A. & Rauw, G. *A&A* 538, A22

Nazé, Y., Oskinova, L. & Gosset, E. 2013 *ApJ* 763,143

Nazé, Y., Ramiaramanantsoa, T., Stevens, I. R., Howarth, I. D. & Moffat, A. F. J. 2018, *A&A*, 609, A81

Oskinova, L. M., Hamann, W.-R. & Feldmeier, A. 2006, *MNRAS* 372,313

Oskinova, L. M., Hamann, W.-R. & Feldmeier, A. 2007 *A&A* 476, 1331

Oskinova, L. M., 2016 Adv. Sp. Res. 58, 679

Owocki, S. P., Castor, J. L. & Rybicki, G. B. 1988 *ApJ* 335, 914

Owocki, S. P. & Cohen D. H. 2001 *ApJ* 559, 1108

Puls, J., Markova, N., Scuderi, S., Stanghellini, C., Taranova, O. G., Burnley, A. W. & Howarth, I. D. 2006, *A&A*, 454, 625

Ramiaramanantsoa, T., Moffat, A. F. J., Harmon, R., *et al.* 2018, *MNRAS*, 473, 5532

Waldron, W. L. & Cassinelli, J. P., 2007 *ApJ* 668, 456

High-mass X-ray Binaries: Illuminating the Passage from Massive Binaries to Merging Compact Objects
Proceedings IAU Symposium No. 346, 2019
L. M. Oskinova, E. Bozzo, T. Bulik & D. Gies, eds.

doi:10.1017/S1743921319002059

Accretion simulations of Eta Carinae and implications to massive binaries

Amit Kashi

Department of Physics, Ariel University, Ariel, POB 3, 40700, Israel
email: kashi@ariel.ac.il

Abstract. Using high resolution 3D hydrodynamical simulations we quantify the amount of mass accreted onto the secondary star of the binary system η Carinae during periastron passage on its highly eccentric orbit. The accreted mass is responsible for the spectroscopic event occurring every orbit close to periastron passage, during which many lines vary and the x-ray emission associated with the destruction wind collision structure declines. The system is mainly known for its giant eruptions that occurred in the nineteenth century. The high mass model of the system, $M_1 = 170 M_{\odot}$ and $M_2 = 80 M_{\odot}$, gives $M_{\rm acc} \approx 3 \times 10^{-6} M_{\odot}$ compatible with the amount required for explaining the reduction in secondary ionization photons during the spectroscopic event, and also matches its observed duration. As accretion occurs now, it surely occurred during the giant eruptions. This implies that mass transfer can have a huge influence on the evolution of massive stars.

Keywords. accretion, accretion disks — stars: mass loss — stars: variables: other — stars: winds, outflows — methods: numerical — (stars:) binaries: general

---

1. Introduction

Simulating accretion in a binary stellar system entangles many physical processes and requires special treatment. Most accretion simulations are performed in the context of accretion into compact objects that do not have winds at all. There are challenges in this regime of accretion as well, especially if the accretor is a NS with strong magnetic fields. In this case the accretion flows to the magnetosphere and later tunneled along the dipole field onto the magnetic poles, a process that functions as a bottleneck and significantly reduces the accretion rate (compared to conventional disk accretion or wind accretion). If stars are involved they are usually the donors and not the accretors, and even if they accrete they have negligible wind. Massive interacting binary systems, with accretors that have their own winds, are different. Most obviously, since in those systems both stars eject winds, the winds collide.

First 3D simulations of colliding winds were performed for systems such as γ^2 Vel and WR 140 (Folini & Walder 2000, 2002; Walder & Folini 2000, 2002, 2003). These simulations showed the formation of the pinwheel structure of the colliding winds, explained the variability obtained in x-ray, and demonstrated the formation of instability and clumping. The resolution in those early simulations was not high enough to obtain the small-scale structure close to the stars. These simulations and their follow-ups dealt with colliding winds of comparable momentum. However, cases in which one of the stellar winds is much stronger than the other, to the extent that the star with the weaker wind accretes, have not been explored.

An extreme example for a colliding winds system is η Carinae (η Car). The system is composed of a very massive star at late stages of its evolution, the primary, and a hotter and less luminous star, the secondary (Davidson & Humphreys 1997, 2012).

The binary system has a highly eccentric orbit (e.g., Davidson *et al.* 2017), and strong winds (Akashi *et al.* 2006) resulting in unique period of strong interaction every 5.54 years during periastron passage known as the spectroscopic event. During the event many spectral lines and emission in basically all wavelengths show rapid variability (e.g., Davidson & Humphreys 2012; Mehner *et al.* 2015 and many references therein). The x-ray intensity, which also serves as an indicator to the intensity of wind interaction, drops for a duration of a few weeks, changing from one spectroscopic event to the other (Corcoran *et al.* 2015). The last three spectroscopic events showed different spectral features, and reflected a trend in the intensities of various lines (Mehner *et al.* 2015; Davidson *et al.* 2018a). Observations of spectral lines across the 2014.6 were interpreted as weaker accretion onto the secondary close to periastron passage compared to previous events, indicating a decrease in the mass-loss rate from the primary star. This 'change of state' of the primary was already identified by Davidson *et al.* (2005), and theoretically explained by Kashi *et al.* (2016).

Soker (2005b) interpreted the line variations during spectroscopic events as a result of accreting clumps of gas onto the secondary near periastron passages, disabling its wind. The suggestion was later developed to a detailed model accounting for different observations in the accretion model framework (Akashi *et al.* 2006; Kashi & Soker 2009a). An estimate of the amount of accreted mass during the spectroscopic event was first obtained by Kashi & Soker (2009b), who found that accretion should take place close to periastron and the secondary should accrete $\sim$ few $\times$ 10^{-6} $M_{\odot}$ each cycle.

Previous grid-based simulations (Parkin *et al.* 2009, 2011) and SPH simulations (Okazaki *et al.* 2008; Madura *et al.* 2013) of the colliding winds did not obtain accretion onto the secondary. Teodoro *et al.* (2012) and Madura *et al.* (2013) advocated against the need of accretion in explaining the spectroscopic event. The resultion of their simulations was too low to capture the important phsyics of clumping and fragmentation as a result of instabilities and thus they did not obtain accretion.

Higher resolution 3D hydrodynamical numerical simulations (Akashi *et al.* 2013) found that a few days before periastron passage clumps of gas are formed due to instabilities in the colliding winds structure, and some of them flow towards the secondary. The clumps reached the secondary wind injection zone, implying accretion. However, the resolution of their simulation was still not good enough to see the accretion itself.

The final evidence for accretion came from the simulations of Kashi (2017), that showed the destruction of the colliding winds structure into filaments and clumps that later were accreted onto the secondary. Kashi (2017) demonstrated that dense clumps are crucial to the onset of the accretion process. The clumps were formed by the smooth colliding stellar winds that developed instabilities that later grew into clumps (no artificial clumps were seeded). This confirmed the preceding theoretical arguments by Soker (2005a,b) who suggested accretion of clumps. Furthermore, as the simulations in Kashi (2017) included a radiation transfer unit which treats the photon-gas interaction, so the momentum of the accreted gas is being changed appropriately along its trajectory. It thus quantitatively showed that radiative braking cannot prevent the accretion, by that confirming theoretical arguments given by Kashi & Soker (2009b).

2. Simulations results

We tested four approaches to understand the way the secondary wind would respond to the high accretion rate: (1) Approaching gas removal: removing dense gas that reaches the secondary wind injection region, and replacing it by fresh secondary wind with its regular mass loss and velocity. Namely, we do not make any changes to secondary wind and let it continue to blow as if the accreted gas did not cause any disturbance. (2) Exponentially reduced mass loss: reducing the mass loss rate of the secondary as it approaches periastron passage. This is an artificial approach that does not relate to the actual accretion situation in the simulation, and does not allow recovery of the secondary wind and was used only for comparison. (3) Accretion dependent mass loss: dynamically changing the mass loss rate of the secondary wind in response to the mass that has been accreted. We lowered $\dot{M}_2$ by changing the density of the ejected wind by the *extra* density of the accreted gas, namely

$$\frac{d\dot{M}_{2,\rm eff}}{d\Omega} = \frac{d\dot{M}_2}{d\Omega}\frac{\rho_u(\Omega) - [\rho_a(\Omega) - \rho_u(\Omega)]}{\rho_u(\Omega)} = \frac{d\dot{M}_2}{d\Omega}\left(2 - \frac{\rho_a(\Omega)}{\rho_u(\Omega)}\right), \tag{2.1}$$

where Ω is a solid angle, $d\dot{M}_2/d\Omega$ is the differential mass loss of the secondary, $\rho_u(\Omega)$ is the undisturbed density of the secondary wind as if it blows without the interruption of accreted gas, and $\rho_a(\Omega)$ is the density of the incoming accreted gas. This approach gives non-isotropic mass loss rate which depends on the direction from where accreted gas parcels arrived. (4) No intervention: not changing the mass loss rate of the secondary and not removing any accreted gas from the simulation. Cells in the secondary wind injection zone where dense blobs arrive are not replaced by fresh secondary wind.

For each approach we measured the duration of accretion and calculates the accreted mass according to the following scheme. With no accretion, the density in the injection zone of the wind should be $\rho_u(r) = \dot{M}_2/4\pi r^2 v_2$. As the simulations ran, high density clumps and filaments approached the injection zone of the secondary wind and even the cells of the secondary itself. Whenever the actual density $\rho_{a,\rm cell}$ in a cell in the injection zone increased above the expected undisturbed value of $\rho_{u,\rm cell}$, we counted the extra mass as accreted, so that the contribution to the accreted mass is

$$\Delta M_{\rm acc} = (\rho_{a,\rm cell} - \rho_{u,\rm cell})V_{\rm cell}, \tag{2.2}$$

where $V_{\rm cell}$ is the volume of the cell. We then summed all the contributions from all cells in the injection zone to obtain the total mass accreted for that time step.

We found that accretion is obtained for both the conventional mass model ($M_1 = 120M_\odot, M_2 = 30M_\odot$) and the high mass model ($M_1 = 170M_\odot, M_2 = 80M_\odot$; Kashi & Soker 2016). For the high mass model the stronger secondary gravity attracts the clumps and we get higher accretted mass of $M_{\rm acc} \simeq$ few$\times 10^{-6} M_\odot$ yr^{-1} and longer accretion periods, in the order of a month, which much better match the observed ones. We also calculated the increase in optical depth in any line of sight and the reduction in the effective temperature ($T_{\rm eff}$) as a result. Observations of lines during the spectroscopic event indicate ionizing radiation from the secondary equivalent to that of a star with $T_{\rm eff} \lesssim 25\,000$K. We therefore concluded that the approach that best fits observation of the spectroscopic event is approach (4). Namely, our simulations are able to start accretion and shut it down without needing a prescription code to intervene with the natural process. For reasonable and even high mass loss rates, only the high-mass model matched the observed decline in $T_{\rm eff}$ (Fig. 1).

An important parameter we studied is the mass loss rate of the primary, for which we used values within the range explored in the literature (see Kashi 2017). We demonstrated that the mass loss rate of the primary affects the accretion rate of the secondary in non-linear way, and found strong dependency between the accreted mass and the mass loss rate of the primary. The simulations showed that if the mass loss rate of the primary is lowered by a factor of a few the accretion can stop, and by that supported the claim of Mehner *et al.* (2015), who suggested that following the observed 'change of state', it may well happen that the mass loss rate of the primary will continue to decrease and therefore future spectroscopic events will be very weak or might not occur at all. Our findings show that a close binary companion can significantly influence on the evolution of a massive star, especially at later stages where it may undergo giant episodes of mass loss.

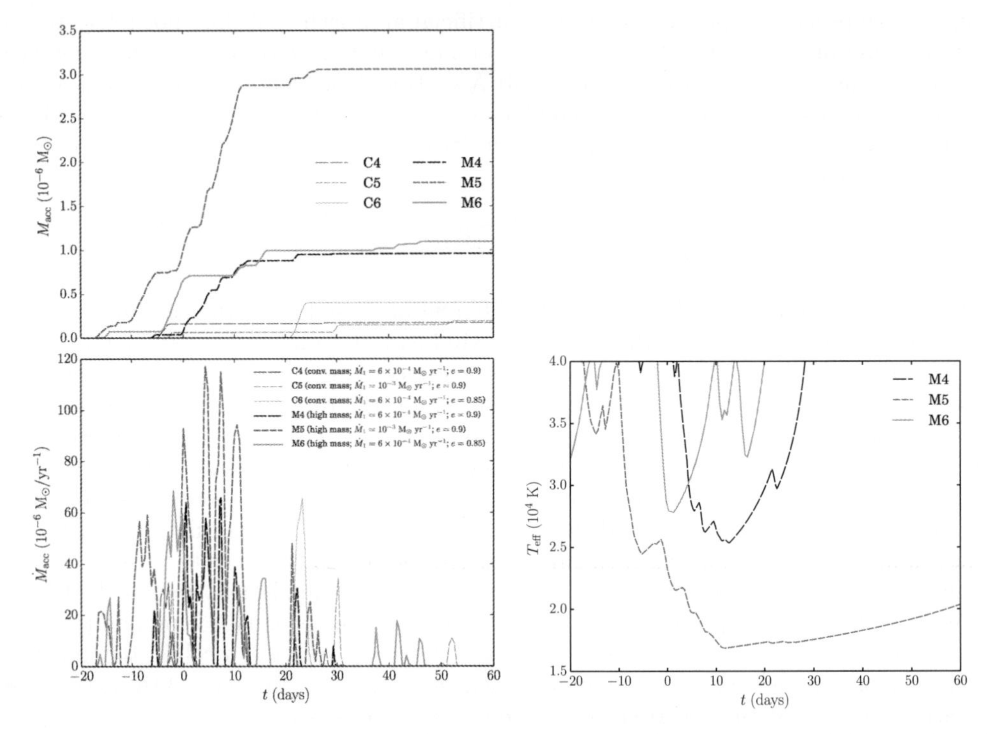

Figure 1. The accreted mass (upper left panel), the accretion rate (lower left panel), and the obtained effective temperature of the secondary, computed for six of our simulations (see parameters in legend). Time is given in days relative to periastron passage ($t = 0$). It can be seen that for the high mass model (M-runs) much more mass is accreted onto the secondary than for the conventional mass model (C-runs). The main reason is the stronger gravity of the secondary. It is also very clear that stronger mass loss rate of the primary (runs C5 and M5) causes a large increase in the accreted mass. The dependence on eccentricity is more complicates as lower eccentricity (runs C6 and M6) means larger periastron distance but also longer periastron passage. These two effects can combine in different ways, making the results difficult to predict. For calculating the effective temperature, first we determine the density $\rho(t)$ by the accreted mass and the mass loss of the secondary, and we then obtain the effective temperature assuming a grey photosphere, averaged over all directions $T_{\rm eff}^4(t) = \frac{4}{3}\left(\tau(t) + \frac{2}{3}\right)^{-1} T^4(\tau = 2/3)$, where we take $T(\tau = 2/3) = 40\,000$ K to be the isotropic effective temperature of the undisturbed secondary,

Another parameter we varied is the eccentricity, for which we also tested $e = 0.85$. This value was favored by Davidson *et al.* (2017) because it gives the smallest possible separation distance at the time when the spectroscopic event begins. It was therefore expected that $e = 0.85$ would produce earlier accretion compared to $e = 0.9$, even though the periastron distance is 50% larger for the smaller eccentricity. Fig. 1 shows our results, confirming that for $e = 0.85$ the accretion duration is indeed longer, and more mass was accreted. Run M6 also showed early accretion exactly as expected by Davidson *et al.* (2017). For the conventional mass model (run C6) we did not see this behavior, because the larger periastron distance and smaller secondary mass combined to reduce the gravitational attraction of the secondary and therefore early accretion could not occur.

3. Conclusions

We conclude that the high mass model for η Car better explains the observations of the decline of lines during the spectroscopic event. Furthermore, we confirm our previous finding that the radiation of the secondary cannot prevent the accretion of primary gas by radiation braking.

In a future work we intend to explore in more details directional effects of the accreted gas and quantitatively study the angular momentum of the accreted gas and how it effects the binary system at times of spectroscopic events. More effects will be included, especially the acceleration of the winds and the mutual effect of the ratdiations from the stars on the winds.

Acknowledgements

This work used the Extreme Science and Engineering Discovery Environment (XSEDE) TACC/Stampede2 at the service-provider through allocation TG-AST150018. This work was supported by the Cy-Tera Project, which is co-funded by the European Regional Development Fund and the Republic of Cyprus through the Research Promotion Foundation.

References

Akashi, M., Soker, N., & Behar, E. 2006, *ApJ*, 644, 451
Akashi, M. S., Kashi, A., & Soker, N. 2013, *New Astron.*, 18, 23
Corcoran, M. F., Hamaguchi, K., Liburd, J. K., *et al.* 2015, arXiv:1507.07961
Davidson, K., Helmel, G., & Humphreys, R. M. 2018, RNAAS, 2, 133
Davidson, K., & Humphreys, R. M. 1997, *ARA&A*, 35, 1
Davidson, K., & Humphreys, R. M. 2012, *Astrophysics and Space Science Library, Eta Carinae and the Supernova Impostors*, 384
Davidson, K., Ishibashi, K., Martin, J. C., & Humphreys, R. M. 2018, *ApJ*, 858, 109
Davidson, K., Ishibashi, K., Gull, T. R., Humphreys, R. M., & Smith, N. 2000, *ApJL*, 530, L107
Davidson, K., Ishibashi, K., & Martin, J. C. 2017, RNAAS, 1, 6
Davidson, K., Martin, J., Humphreys, R. M., *et al.* 2005, *AJ*, 129, 900
Folini, D., & Walder, R. 2000, *Ap&SS*, 274, 189
Folini, D., & Walder, R. 2002, *Interacting Winds from Massive Stars*, 260, 605
Kashi, A. 2017, *MNRAS*, 464, 775
Kashi, A., Davidson, K., & Humphreys, R. M. 2016, *ApJ*, 817, 66
Kashi, A., & Soker, N. 2009a, *MNRAS*, 397, 1426
Kashi, A., & Soker, N. 2009b, *New Astron.*, 14, 11
Kashi, A., & Soker, N. 2016, *ApJ*, 825, 105
Madura, T. I., Gull, T. R., Okazaki, A. T., *et al.* 2013, *MNRAS*, 436, 3820
Mehner, A., Davidson, K., Humphreys, R. M., *et al.* 2015, *A&A*, 578, A122
Okazaki, A. T., Owocki, S. P., Russell, C. M. P., & Corcoran, M. F. 2008, *MNRAS*, 388, L39
Parkin, E. R., Pittard, J. M., Corcoran, M. F., Hamaguchi, K., & Stevens, I. R. 2009, *MNRAS*, 394, 1758
Parkin, E. R., Pittard, J. M., Corcoran, M. F., & Hamaguchi, K. 2011, *ApJ*, 726, 105
Soker, N. 2005a, *ApJ*, 619, 1064
Soker, N. 2005b, *ApJ*, 635, 540
Teodoro, M., Damineli, A., Arias, J. I., *et al.* 2012, *ApJ*, 746, 73
Walder, R., & Folini, D. 2000, *Ap&SS*, 274, 343
Walder, R., & Folini, D. 2002, *Interacting Winds from Massive Stars*, 260, 595
Walder, R., & Folini, D. 2003, *A Massive Star Odyssey: From Main Sequence to Supernova*, 212, 139

High-mass X-ray Binaries: Illuminating the Passage from Massive Binaries to Merging Compact Objects
Proceedings IAU Symposium No. 346, 2019
L. M. Oskinova, E. Bozzo, T. Bulik & D. Gies, eds.

doi:10.1017/S1743921319001297

Testing how massive stars evolve, lose mass, and collapse at low metal content

Eliceth Y. Rojas Montes[1,2] **and Jorick Vink**[2]

[1]School of Mathematics and Physics, Queen's University Belfast, Belfast, BT7 1NN, Northern Ireland
email: eliceth.rojasmontes@armagh.ac.uk

[2]Armagh Observatory and Planetarium, Armagh, BT61 9DG, Northern Ireland
email: jorick.vink@armagh.ac.uk

Abstract. In order to test massive star evolution above 25 M⊙, we perform spectral analysis on a sample of massive stars in the Small Magellanic Cloud that includes both O stars as well as more evolved Wolf-Rayet stars. We present a grid of non-LTE stellar atmospheres that has been calculated using the CMFGEN code, in order to have a systematic and homogeneous approach. We obtain stellar and wind parameters for O stars, spectral types ranging from O2 to O6, and the complete sample of known Wolf-Rayet stars. We discuss the evolutionary status of both the O and WR stars and the links between them, as well as the most likely evolutionary path towards black hole formation in a low metallicity environment, including testing theoretical predictions for mass-loss rates at low metallicities.

Keywords. stars: early-type, stars:Wolf-Rayet, Small Magellanic Cloud, stars: fundamental parameters

1. Introduction

The Small Magellanic Cloud is a galaxy of low metallicity, and an excellent environment to constrain stellar evolution at metallicities resembling the early Universe. Stellar evolution of massive stars at different metallicities is not identical. Metallicity plays a role in the evolutionary path that each star follows until reaching its final stage. Quantitative spectroscopy provides a useful tool to obtain several stellar parameters to understand how massive stars evolve in several environments when comparing to evolutionary models. Progenitors of Wolf-Rayets (WR) have not been defined in the Small Magellanic Cloud (SMC), and there are multiple channels for their origin.

One possible channel considers an O star, rapidly rotating and going through chemically homogeneous evolution (CHE) to reach the WR stage. However, constraints obtained via spectropolarimetry of WR stars view this as unlikely, and a more classical post-LBV mass loss scenario is more probable (Vink & Harries, 2017). This channel is partially supported by quantitative spectral analysis of single WR stars in the SMC (Hainich *et al.* 2015); a possible exception might be the binary system AB5 (Shenar *et al.* 2016).

Massive stars have been reported to be in binary systems with a ~50% likelihood, where they can interact with their companion by merging, mass transfer, or they are expected to be spun up (Sana *et al.* 2013). Bouret *et al.* (2013) hinted at binary systems for some putatively single O dwarfs in the SMC to explain their high luminosities.

Then, a possible channel for the formation of WR stars in the SMC is binary interaction via mass transfer, to account for the surface abundances and luminosities obtained through quantitative analysis of these single WR stars (Hainich *et al.* 2015). Though, this scenario has been challenged by contemporary results for WR binaries, where the primaries would have gone through a WR phase independent of binary interaction, questioning mass transfer as the dominant mechanism (Shenar *et al.* 2016).

The origin of WR stars in the SMC is unclear, and future research is needed in this matter to solve the discrepancy between empirical results and the ones predicted by evolutionary models. Shedding some light on the origin of WR stars in the SMC might be expected by homogeneous analysis of O dwarfs, O supergiants and WR stars in this galaxy, covering a wide range of parameters for O star evolution in the Hertzprung-Rusell diagram.

2. Spectral fitting

Previous work on O and WR stars have been done with different radiative transfer codes (CMFGEN, PoWR), and there is a lack of analyses of O supergiants in the SMC. In order to have realistic error bars a homogeneous and systematic approach has been taken to perform the quantitative analysis of the O and WR stars spectra, using the radiative transfer code CMFGEN to create a grid of models in the parameter range of O dwarfs, O supergiants and WR stars in the SMC.

A fine multidimensional grid will be ideal to model systematically the spectra of these stars, however it would require a large computational time. It is important to choose parameters that have a large influence on stellar lives to reduce the computational time developing the grid. Therefore our grid has as main parameters the effective temperature ($T_{\rm eff}$ at $\tau = 2/3$) and the transformed mass-loss rate ($\dot{M}_{\rm t}$) (Gräfener & Vink, 2013). The abundances used are scaled solar (1/5 $Z_\odot$) by Asplund *et al.* (2009), and including ions of H, He, C, N, O, Si, P, S and Fe.

The use of the transformed mass loss rate will allow us to preserve the spectral shape for different values of mass loss rate if the luminosity, wind terminal velocity and clumping factor are kept constant. The basic grid has a fixed: $\log(g) = 4.0$, luminosity $L = 10^6\ L_\odot$, terminal velocity $V_\infty = 2800\,{\rm km\,s^{-1}}$, clumping factor $f_v = 0.1$ and a β value of the wind acceleration of $\beta = 1$.

The main parameters of the two-dimensional grid, effective temperature ($T_{\rm eff}$) and transformed mass loss rate $\dot{M}_{\rm t}$, range from 35.5 to 56.2 kK and $\log(\dot{M}_{\rm t})$ [$M_\odot\,{\rm yr}^{-1}$] from -6.28 to -4.78 respectively.

To cover the parameter range by WR stars in the SMC models with higher temperatures and higher $\log(g)$ values are needed, for transformed mass loss rates values ranging from $\log(\dot{M}_{\rm t})$ [$M_\odot\,{\rm yr}^{-1}$] of -5.16 to -4.78. These models are currently running, and future work is pending.

The quality of the fitting between models and data is measure through a χ^2_{fit} method similar to that of Bestenlehner *et al.* (2014), where only the sum of equivalent widths of diagnostic lines are taken into account to constrain the transformed mass loss rate and the effective temperature.

To obtain the helium abundance, crucial to determine the evolutionary status of these stars, an extra dimension is needed for the grid. This ranges from 25 % to 47 % on helium content in the basic grid, included in the χ^2_{fit} analysis. To determine luminosities and extinction values SED fitting is performed using photometric data obtained from Bonanos *et al.* (2010) and the Cardeli extinction law (Cardelli *et al.* 1989).

3. Preliminary results

Fitted spectra of an O dwarf and a Wolf-Rayet star are shown in Figures 1 and 2. The best fit CMFGEN model for the O dwarf AzV 243 classified as an O6 V, reproduced the

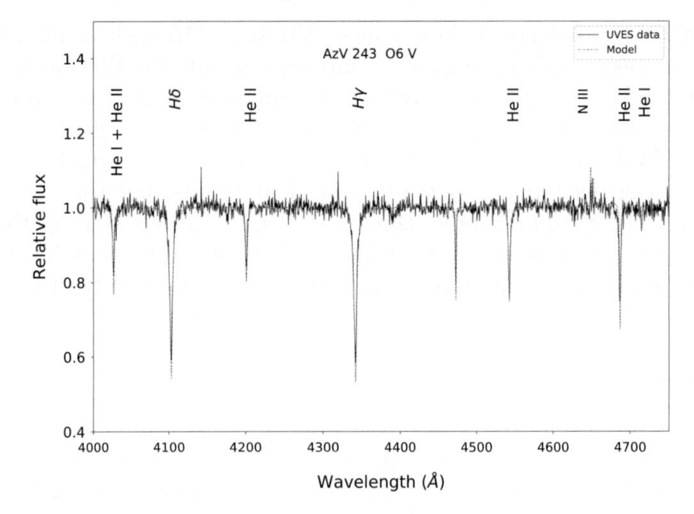

Figure 1. Normalised spectra of AzV 243 compared with the best fitting CMFGEN model from the grid. The normalised spectra for AzV243 is shown as a black solid line, while the CMFGEN model is shown as a red dashed line.

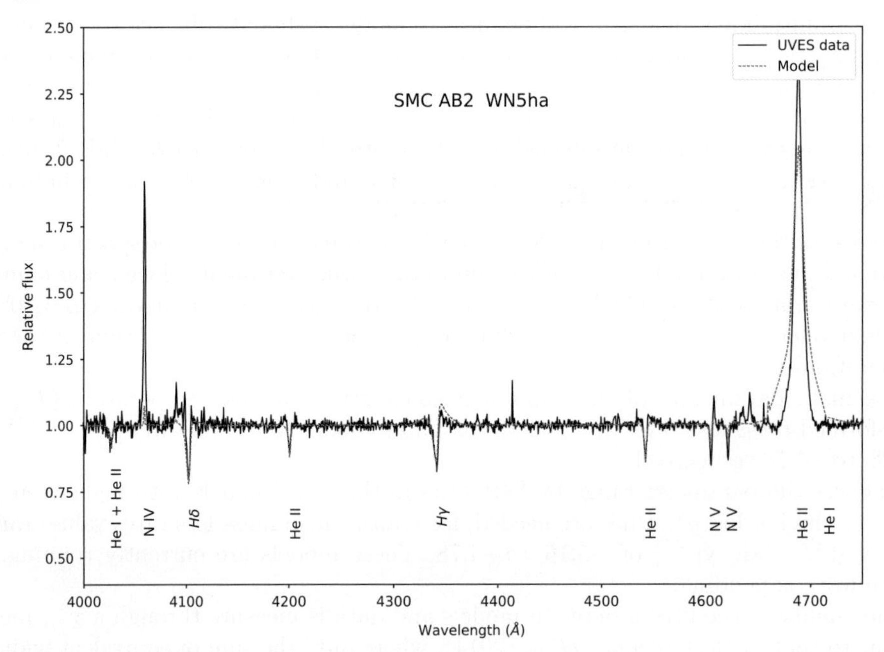

Figure 2. Normalised spectra of SMC AB2 compared with the best fitting CMFGEN model from the grid. The normalised spectra for SMC AB2 is shown as a black solid line while the CMFGEN model is shown as a red dashed line.

line profile for all absorption lines present in its UVES spectrum. While the strength of the lines is not fully reproduced, effective temperature and mass loss rate values obtained are similar to the ones reported by Bouret *et al.* (2013).

In the case of the Wolf-Rayet SMC AB2 classified as a WN5ha the line profile of HeII λ 4686 Å is not reproduced by the model. This is caused by our fixed value for the wind terminal velocity (V_∞), indicating a lower value for V_∞ in agreement with results by Hainich *et al.* (2015). Also the strength of NIV λ 4058 Å is not reproduced, pointing to a higher abundance for this element than $1/5\,Z_\odot$ corresponding with the WN nature of the star. Regardless of the fitting the obtained values for the effective temperature and mass loss rate are comparable to the ones obtained by Hainich *et al.* (2015).

In the current analysis of the O dwarfs and cooler WR stars in the SMC, the best fitted models from the grid (see Figs. 1 and 2) do not reproduce entirely the strength or line profiles displayed by the data. However the effective temperature and mass loss rate values, obtained through the method described in the previous section, are comparable to the ones found by Hainich *et al.* (2015) and Bouret *et al.* (2013); ergo validating our systematic and homogeneous approach. It is expected then that our objective method once applied to the O supergiants sample will obtain reliable values for their effective temperature and mass loss rate. This can be compared without a subjective component to the rest of the sample, eliminating systematic errors, and most probably unveiling links between these evolutionary phases.

4. Future Work

Further model computation to extend the parameter range of the grid to cover different $\log(g)$ values and higher temperatures is needed: to be able to model O supergiants; hotter Wolf-Rayets in the sample; and also to obtain He, C, and N abundances. Once the sample has been fully analysed a comparison with stellar evolutionary models is essential to unveil links between evolutionary stages and possible WR progenitors.

References

Asplund M., Grevesse N., Sauval A. J., & Scott P. 2009, *ARAA* 47, 481
Bestenlehner, J. M., Gräfener, G., Vink, J. S., *et al.* 2014, *A&A* 570, A38
Bonanos, A. Z., Lennon, D. J., Kölinger, F., *et al.* 2010, *AJ* 140, 416
Bouret, J. C., Lanz, T., Martins, F., *et al.* 2013, *A&A* 555, A1
Cardelli, J. A., Clayton, G. C., & Mathis, J. S. 1989, *ApJ* 345, 245
Gräfener, G., & Vink, J. S. 2013, *A&A* 560, A6
Hainich, R., Pasemann, D., Todt, H., *et al.* 2015, *A&A* 581, A21
Sana, H., de Koter, A., de Mink, S. E., *et al.* 2013, *A&A* 550, A107
Shenar, T., Hainich, R.,Todt, H., *et al.* 2016, *A&A* 591, A22
Vink, J. S., & Harries, T. J. 2017, *A&A* 603, A120

X-ray binaries with Be-type donors

High-mass X-ray Binaries: Illuminating the Passage from Massive Binaries to Merging Compact Objects
Proceedings IAU Symposium No. 346, 2019
L. M. Oskinova, E. Bozzo, T. Bulik & D. Gies, eds.

doi:10.1017/S1743921318008207

Be stars in the X-ray binary context

Thomas Rivinius

ESO — European Organisation for Astronomical Research in the Southern Hemisphere,
Casilla 19001, Santiago 19, Chile
email: triviniu@eso.org

Abstract. Rapidly rotating B-type stars with gaseous mass-loss disks in Keplerian rotation are common central objects in X-Ray binaries. These disks are physically well understood in the framework of the viscous decretion disk, and their typical parameters have been established for a large number of single Be stars in the recent years. According to the current observational evidence, the Be stars and disks found in BeXRBs are well within the boundaries known from single Be stars, i.e., they are normal Be stars. New results have also been obtained on the orbital disk truncation and other tidal effects of the companion objects on the disk.

Keywords. circumstellar matter, stars: emission-line, Be

1. Introduction

The primary stars in the binaries that form the class of the Be X-Ray binaries (BeXRBs) are classical Be stars. Since the first Be star was identified by Secchi (1867) by virtue of its Balmer line emission, it has become clear that many objects of different physical properties can be summarized under a not further distinguished Be star label. As a consequence, a more specific sub-classification has been introduced over the years, and Rivinius *et al.* (2013) give a general overview of the classical Be stars and their distinction from taxonomically similar objects. In the following, the article will deal only with classical Be stars, since only those are found in BeXRBs. Figure 1 illustrates the current schematic view of a classical Be star and how it relates to the spectroscopically observed appearances of Be stars.

In brief, classical Be stars are those Be stars that possess a circumstellar, typically equatorial disk. This disk is supported by Keplerian rotation, and the material of the disk originates from the Be star itself. In the past two decades, numerous interferometric and polarimetric studies have confirmed the basic geometry of Be star disks as geometrically thin disks in Keplerian motion (see, for instance Wood *et al.* 1997; Meilland *et al.* 2007), and the historical record of Be stars growing and losing disks excludes an external source of the disk material (Dalla Vedova *et al.* 2017, give a recent example).

The detailed mechanism by which the central star ejects the material and provides angular momentum for the Keplerian disk is still debated. Nevertheless, a major contributor could be identified to be common to all Be stars, namely rapid stellar rotation. At above 80% of the critical rotation, Be stars as a class are the most rapidly rotating non-degenerate objects (Rivinius *et al.* 2013).

The source of the remaining up to 20% of velocity to make the material escape remains elusive, but in a conceptual sense the reason for the existence of Be stars is much clearer. As main sequence stars evolve, their cores contract and their envelopes expand. As a consequence, the core rotates faster than the envelope and the stellar angular momentum is being re-distributed within the star. According to the current stellar evolution

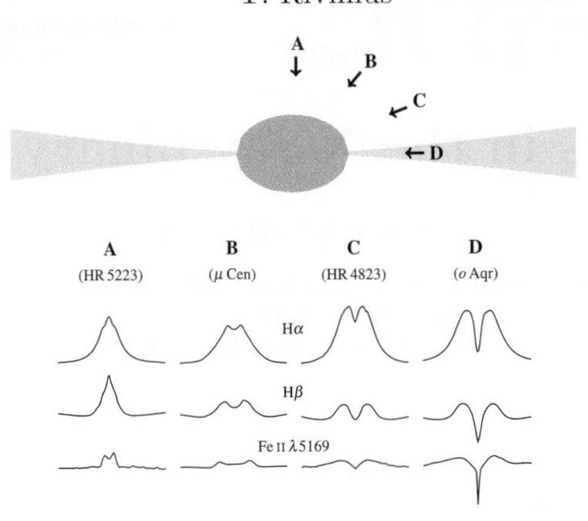

Figure 1. The upper part shows a schematic equator-on view of classical Be star, the lower part example spectral profiles from pole-on to shell Be stars for Hα, a very optically thick line, Hβ, and the mostly optically thin FeII λ 5169 line. Reproduced with permission from Rivinius *et al.* (2013).

paradigms, the overall effect is a gentle acceleration of the stellar surface towards critical rotation, and for sufficiently high initial rotation the critical limit is reached within the MS-lifetime of a B star (Granada *et al.* 2013). The ejection of a circumstellar disk is simply the means by which a star getting close to the limit sheds angular momentum to avoid becoming a supercritical rotator. Whether and which additional mechanism supports this shedding already at 80% upwards, instead of at 100% critical rotation, is irrelevant in this context.

Originally, it was not clear how the quite distinct appearance of the shell stars puts them apart from the Be stars, though an inclination effect was suspected. A close relation between Be stars and shell stars was recognized early, not at least because the enigmatic object γ Cas was observed to oscillate between both states in the 1940s. But exactly this sort of transition indeed cannot be understood in the frame of an equatorial disk combined with an inclination effect, since the stellar inclination does not change. Instead, it was taken as indication that Be- and shell-phases might rather form a temporal sequence than a geometrical one (Underhill & Doazan 1982, part II). However, the proof that shell stars really are simply equatorially seen Be stars was finally delivered with the above mentioned results with modern instrumentation. Hence for γ Cas, and a few other objects showing such transitional behavior, an explanation beyond the stable equatorial disk needed to be found (see Sect. 4).

2. Disk formation and evolution

For the understanding of BeXRBs, the most important aspect of the Be star is its circumstellar environment. Once the material has been ejected with sufficient angular momentum to remain in orbit, it settles into a disk. While the ejection process might be violent and is probably localized on the stellar surface, within several orbits the viscosity acts to circularize the material into Keplerian motion and to settle it into vertical hydrostatic equilibrium. Such a Keplerian disk in vertical hydrostatic equilibrium is geometrically thin, close to the star, with the scale height being governed by the thermal speed, and hence temperature. Under the not overly unrealistic assumption of isothermality throughout the disk, that disk will flare, i.e., increase its opening angle, at larger radii (see, e.g., Sigut *et al.* 2009; Halonen & Jones 2013).

Once the freshly ejected material has settled into this Keplerian, hydrostatic disk state, its further evolution is fully governed by viscous processes. One can say that the disk, once formed, has completely lost the memory of how it was formed, which is one of the problems in tracking down the exact formation mechanism. Such a disk has been named viscous decretion disk, VDD in short.

The physics governing the disk evolution in Be stars is the same as the physics in gaseous accretion disks, it is rather the boundary conditions that are responsible for any differences. In a Be star disk, the inner boundary is a net source of both matter and angular momentum, while the outer boundary is a sink for both.

2.1. *Disk life cycle*

One important property of the Be stars is that, on time scales from weeks to many decades, these disks can newly form, grow, decay, and vanish again. Even ignoring the details of the ejection process, this makes clear that the disks must form inward-out. For simplicity assume the Be star ejection is a simple process, either active or inactive, and can be parameterized as keeping the inner boundary at a constant density. From here on, the disk is let to evolve purely viscously. Because the orbital speeds and their differentials in the inner disk regions are much faster than further out, so are the time scales there. The inner disk, within the first few stellar radii, will quickly evolve into a steady state configuration, to arrive at this state at larger distances from the star may take decades (Haubois *et al.* 2012).

The steady state is characterized by a constant density slope vs. radius of the form $\Sigma(r) \propto r^2$, where Σ is the integrated surface density in the equatorial plane. The actual volume density has an additional factor of $r^{-1.5}$ (in the isothermal case) due to the flaring height of the disk. When the density slope is measured, it is important to keep in mind that this is not global density slope of the entire disk, but always the slope of the region of the disk that is probed by the observables used. With that disclaimer, a steeper density slope than the steady state one is the signature of a disk being built-up.

An important question is the outer radius of the disk, and the answer is that an outer radius, strictly speaking, does not exist. The most meaningful approximation for a non-binary Be star is the radius at which the orbital velocity becomes comparable to the local velocity of sound. At this radius, the viscous coupling is lost and the disk makes a transition from Keplerian motion to angular momentum conserving outflow (Krtička *et al.* 2011). However, this only happens very far from the star, at hundreds to thousands of stellar radii, and no currently accessible observable is formed in that region of a Be star disk.

For Be stars in binaries, the outer radius of the disk is typically assumed to be close to the companion's orbital radius, but this is a choice driven by numerical complexity, rising steeply with outer radius when it goes beyond the orbital one, since the tidal effects would become more and more dominant for the circumbinary structure. What happens to the disk structure at such radii is discussed in the next Section.

A system that has reached the steady state throughout the disk is not static, though. There will be a constant outflow of both matter and angular momentum, and not all matter that is originally ejected from the star will move out and be lost from the system. In fact, due to the viscous process, material that loses angular momentum in an interaction will go to a lower orbit, and eventually be re-accreted. Material that gains angular momentum will be lifted away from the star, and eventually be lost from the system.

For a star just beginning to build a disk, most newly ejected material will begin to move outwards. In a Be star that has a fully developed steady state disk, a lot more material returns to the star, and the re-accretion ratio might be as high as 99% (Panoglou *et al.* 2016; Ghoreyshi *et al.* 2018). This means that the actual *mass ejection* rate from the star is a factor of up to 100 higher than the *mass loss* ratio through the outer boundary. Typical values, from the two works cited above, are mass ejection rates of about $10^{-7}\,\mathrm{M}_\odot\,\mathrm{yr}^{-1}$, but mass loss rates of only $10^{-9}\,\mathrm{M}_\odot\,\mathrm{yr}^{-1}$. Typical disk masses in those scenarios are between 10^{-9} and $10^{-8}\,\mathrm{M}_\odot$. These values are observationally well supported for both the Milky Way and the Magellanic Clouds (Vieira *et al.* 2017; Rímulo *et al.* 2018).

Finally, let's consider a decaying disk. In a model computationally this can be achieved by allowing the density at the inner radius to change freely, and removing material as re-accreted when its angular momentum is lower than that needed to support that radius. Again, the innermost part of the disk will react quickest to this change of the inner boundary. Governed by viscosity, and not provided with fresh angular momentum from the star, most of this material will re-accrete, meaning the density slope of a disk in decay will be shallower than that of a disk in steady state (see Fig. 1 of Haubois *et al.* 2012).

Be stars, in particular early type Be stars, are typically not stable in either constant mass-ejection or quiescence for timescales longer than a few weeks, or at most years for later-type Be stars. Instead, they alternate between states of quiescence and ejections with various strength and length. Because the timescale is much faster close to the star, even if the inner part is highly dynamic, from a certain radius on outwards, the disk can still be considered in a quasi steady state condition, and inwards as a disk cycling through growth and decay states (see Fig. 7 of Haubois *et al.* 2012).

2.2. *Observables*

The above description creates a distinct observable signature. The most commonly used observables, like photometry or spectroscopy with optical techniques are formed within a few stellar radii of the central star. The only potential exception to this rule is the Hα line emission. In other words, all those observables are formed well within any potential binary companion's orbit (see Fig. 2 and 7 of Rivinius *et al.* 2013).

Photometry in the optical regime, for instance, is formed within a few stellar radii of the primary. It is hence highly dynamic, and to observe it in a steady state is quite unusual. This volatility decreases with wavelength, so that the near-infrared photometry is more stable, and the thermal infrared even more so.

This is in agreement with the picture described above: the time scales get longer further out, with the precise values depending on the properties of the individual feeding events. The higher the duty cycle, i.e., percentage of active time, of the disk feeding is, the closer to the star the stable region begins, and the more dense the disk will be, compared to a lower duty cycle.

For the purpose of understanding the variability originating from accretion onto a secondary, this means taking the variability observed close to the star and simply projecting this variability to larger radii will over predict the local variations at larger radii.

Unfortunately, a single Be star can undergo widely differing phases of Be activity in relatively short time. Over the 20th century, the prototypical γ Cas went from a strong emission with pole-on disk appearance through a phase oscillating between pole-on and equatorial appearance in the 1940s, then the disk dissipated almost completely, slowly recovering between 1950 and 1970, and since then having a similar appearance as it had around 1900 (Doazan *et al.* 1983). This means that to base a hypothesis on the current state of the circumstellar envelope on any single snapshot observation, instead of monitoring, is likely misleading.

3. Be stars as binary stars

What has been written so far is true for classical Be stars in general. Considering binary stars, a few statements can be made that set Be binaries apart from both single Be stars, as well as from non-Be binaries.

Other than for non-Be binaries, no short period Be binaries are known. For orbital periods below about a week this well explained by tidal forces, that will not allow any disk to settle into a VDD (Panoglou, priv. comm. The shortest orbital period for which a model is published is 10 d, by Panoglou *et al.* 2018). Observationally, the Be star binary with the shortest known period is SAX J2103.5+4545, with $P_{\rm orb} = 12.68$ d, as listed in the online catalog by Raguzova & Popov (2005)†. The period space between 12 and 28 d is not well populated, only a few BeXRBs are known. The shortest period Be star binaries with a companion other than a neutron star are 59 Cyg (Rivinius & Štefl 2000) and *o* Pup (Rivinius *et al.* 2012; Koubský *et al.* 2012), with orbital periods of about 28 d. They both have a subdwarf O-star companion, from which the envelope was stripped during binary evolution. So there is certainly a not yet explained discrepancy between the shortest observed orbital period and the shortest predicted ones.

Another surprising result, for both Be+NS and Be+sdO, is the spectral type distribution. While the incidence of Be stars with spectral subtype is somewhat debated, it is probably flatter than previously thought, i.e., the percentage of Be stars does not drop strongly for later B subtypes (Shokry *et al.* 2018, and references therein). In stark contrast, neither Be+NS (Reig 2011; Haberl & Sturm 2016), nor Be+sdO binaries with primaries later than about B3 are known. For Be+sdO stars, a recent search by Wang *et al.* (2018) gave a highly significant result: Even though their search method was quite in favor of detecting such hot companions around later type primaries, they did not find a single candidate. It is worth noting that Wang *et al.* explain this with a flat mass ratio distribution of the progenitor systems. Another class of Be potential binaries also adheres to this subtype limit, the γ Cas analogies (Smith *et al.* 2016). It is, however, not firmly established whether all of them are binaries (some certainly are), and what it would have to do with their properties as a class.

The tidal effects of a binary companion have also been modeled. Recent results have mostly confirmed and detailed earlier models (see, e.g., Panoglou *et al.* 2016, 2018, and references therein). In a binary Be star one finds a truncation radius, typically at a strong orbital resonance radius, close to maybe half or a bit more of the semi-major axis, at which the Be star disk does not vanish, but strongly changes its density structure: The density slope with radius becomes a lot steeper. As demonstrated above, for most observables shortward of the radio regime, this is effectively a truncation. For the few stars for which radio data at sufficiently long wavelengths are available, however, this region is observable (Klement *et al.* 2017). The observations are generally in good agreement with the model as far as it concerns the change of the density structure at the the mentioned "truncation" radius. In terms of the SED, this produces a turndown, i.e., a reduction of the SED slope at a wavelength that corresponds to the truncation (see the pseudo-photosphere concept by Vieira *et al.* 2015, for the transformation of

† http://xray.sai.msu.ru/~raguzova/BeXcat/

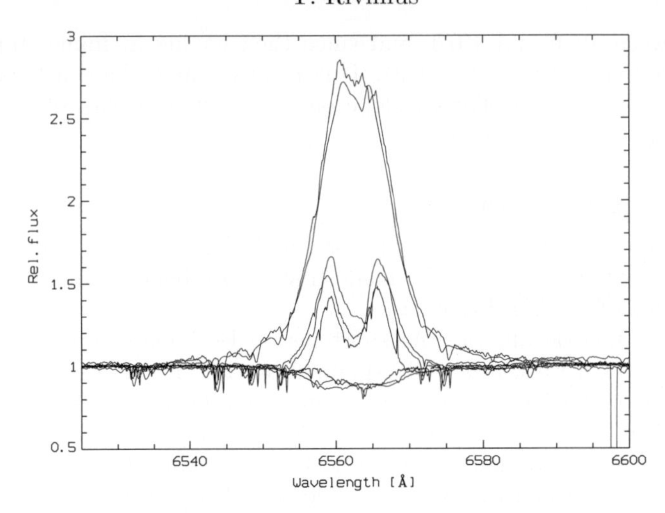

Figure 2. Selected Hα emission profiles of π Aqr from 2000 to 2018. Data from the BeSS database (http://basebe.obspm.fr/) and own observations.

radius to wavelength). At even longer wavelengths, however, there is a discrepancy. The current models do not predict another change of structure at even larger radii. However, the observations suggest that, outside the companions orbit, the disk may recover and return to the original slope, but on a lower density level. Plainly put, a very tenuous circumbinary Keplerian disk might be a possible explanation for the radio observations.

The second observable, that is potentially formed at radii far enough from the central star to show the presence of a companion, is the Hα line emission. As shown by Panoglou *et al.* (2016), a phase-locked spiral structure might arise in the disk, causing phase-locked cycling of the relative peak strength of the violet and red emission peaks, the V/R-ratio. If the disk is dense enough, one might as well observe the effects of the truncation on the emission profile directly. In Fig. 2 the Hα spectra of π Aqr show the growing emission from 2000 to 2018. π Aqr is a Be binary with an orbital period of about 84 d (Bjorkman *et al.* 2002). In the lowermost set, the star has almost no disk at all, while in the middle set the disk is present, but the morphology is fully consistent with an undisturbed, i.e., single star disk. Only when the disk has grown for a time and probably reached the steady state density at the truncation radius, this is no longer the case. Instead of the traditional double peaked profile, one now observes a flat-topped or multiply peaked profile. This morphology of the emission is very typical for Be stars in binaries, in particular when the emission is strong. In might be used as a strong candidate indicator for binarity of a Be star.

4. Disturbed Be star disks

The previous sections dealt with largely steady state disks, if at all only slightly disturbed by the tidal forces due to a companion. There are, though, much stronger instabilities that can act in a Be star disk. These can fully dominate the disk emission morphology and temporal behavior, and completely veil the presence of binary effects.

The first is that a strong, global density wave pattern can grow in a disk, and then precess around the central star on the time scale of a few years (see Štefl *et al.* 2009, for a comprehensive observational description). Because of the behavior of the emission lines, this is called violet-to-red, or V/R, cyclic variation. The density wave itself and its precession behavior is theoretically well understood (Carciofi *et al.* 2009, and references therein), but what triggers the disturbance, and what lets the density wave decay after a number of precession cycles, is not really known.

Finally, there is the drastic changes that have been mentioned in the Introduction. These are called "Spectacular Variations" (SV) and are best described as one and the same Be star changing from a pole-on or low inclination appearance to a high-inclination or shell star appearance within a few years and back again. Among the brighter classical Be stars, only three objects are known to have shown that behavior, γ Cas, 59 Cyg, and Pleione. However, there might be more of them among the BeXRBs, like Reig *et al.* (2000) have shown for LS I $+61^\circ$ 235. With the geometry discussion of the Be star circumstellar environment being settled, it became clear that a disk with stable orientation would not explain this behavior. Instead, an inclined and precessing disk was considered by Hummel (1998). Unfortunately, the SV episodes of γ Cas and 59 Cyg took place before modern instrumentation and detectors were available. In Pleione, however, the SV is ongoing. Hirata (2007) presented a series of spectroscopic and polarimetric observations, in which they showed that the rotation of the polarization angle is in very good agreement with the misaligned, precessing disk hypothesis proposed by Hummel.

Counting in LS I $+61^\circ$ 235 as fourth case of SV, it is striking that all the four stars are known binaries, with orbital periods from one month to a bit less than one year. Unfortunately, no model could yet reproduce this behavior.

5. Conclusions

In the context of BeXRBs, only the conditions as found in early type Be stars are of relevance. This is because no BeXRBs are found with primaries later than about B3. Compared to late type Be stars, early ones have shorter variability time scales and higher disk densities. The variability of Be star disks, though, is typically assessed by observables that form very close to the central star, like photometry or spectroscopy (other than Hα). At the position of a binary companion, this variability has typically been averaged out by viscous processes acting in the disk. The companion perceives a much more stable, steady state disk, from which it accretes, than one would expect from the observables mentioned above alone. Disk observables sensitive to the orbital region in the vicinity of the secondary might rather be the Hα emission line, at least for dense, well developed disks, and photometry and bolometry from the thermal infrared long-wards to cm-radio observation, depending on orbital dimension.

Yet, even if the disk has settled into a steady state configuration at the companion position, the historical context must be known for any reliable analysis, since Be stars may change also their long-term activity level. For this, fortunately no high cadence data are needed, but in turn a long time base. In the past decades many monitoring projects were started that are capable of providing these data, such as the OGLE projects for the Magellanic Clouds (Udalski *et al.* 2015), or the ASAS (Pojmanski 1997), KELT (Pepper *et al.* 2007), and similar surveys for the Milky Way.

There is a characteristic radius for the disk in a binary, commonly called the truncation radius. It is typically at some resonance radius at about half or a bit more of the semi-major axis of the binary. To imagine the truncation radius as an outer edge of the disk would be misleading, however. What really happens is that the density slope inside the truncation radius is slightly shallower than in the single star steady state case, but considerable steeper outside. For a dense disk, this becomes apparent when the Hα emission morphology becomes flat-topped or multiply peaked, instead of having a clear double peak behavior.

Finally, in case one wants to study the binary evolution in such a system, it is important to remember that the disks are not very massive. The disk is primarily the means by which the Be star gets rid of angular momentum *without* having to lift more than the bare minimum of mass needed.

References

Bjorkman, K. S., Miroshnichenko, A. S., McDavid, D., & Pogrosheva, T. M. 2002, *ApJ*, 573, 812

Carciofi, A. C., Okazaki, A. T., Le Bouquin, J.-B., *et al.* 2009, *A&A*, 504, 915

Dalla Vedova, G., Millour, F., Domiciano de Souza, A., *et al.* 2017, *A&A*, 601, A118

Doazan, V., Franco, M., Rusconi, L., Sedmak, G., & Stalio, R. 1983, *A&A*, 128, 171

Ghoreyshi, M. R., Carciofi, A. C., Rímulo, L. R., *et al.* 2018, *MNRAS*, 479, 2214

Granada, A., Ekström, S., Georgy, C., *et al.* 2013, *A&A*, 553, A25

Haberl, F. & Sturm, R. 2016, *A&A*, 586, A81

Halonen, R. J. & Jones, C. E. 2013, *ApJ*, 765, 17

Haubois, X., Carciofi, A. C., Rivinius, T., Okazaki, A. T., & Bjorkman, J. E. 2012, *ApJ*, 756, 156

Hirata, R. 2007, in Astronomical Society of the Pacific Conference Series, Vol. 361, Active OB-Stars: Laboratories for Stellare and Circumstellar Physics, ed. S. Štefl, S. P. Owocki, & A. T. Okazaki, 267

Hummel, W. 1998, *A&A*, 330, 243

Klement, R., Carciofi, A. C., Rivinius, T., *et al.* 2017, *A&A*, 601, A74

Koubský, P., Kotková, L., Votruba, V., Šlechta, M., & Dvořáková, Š. 2012, *A&A*, 545, A121

Krtička, J., Owocki, S. P., & Meynet, G. 2011, *A&A*, 527, A84

Meilland, A., Stee, P., Vannier, M., *et al.* 2007, *A&A*, 464, 59

Panoglou, D., Carciofi, A. C., Vieira, R. G., *et al.* 2016, *MNRAS*, 461, 2616

Panoglou, D., Faes, D. M., Carciofi, A. C., *et al.* 2018, *MNRAS*, 473, 3039

Pepper, J., Pogge, R. W., DePoy, D. L., *et al.* 2007, *PASP*, 119, 923

Pojmanski, G. 1997, *Acta Astron.*, 47, 467

Raguzova, N. V. & Popov, S. B. 2005, *Astronomical and Astrophysical Transactions*, 24, 151

Reig, P. 2011, *Ap&SS*, 332, 1

Reig, P., Negueruela, I., Coe, M. J., *et al.* 2000, *MNRAS*, 317, 205

Rímulo, L. R., Carciofi, A. C., Vieira, R. G., *et al.* 2018, *MNRAS*, 476, 3555

Rivinius, T., Carciofi, A. C., & Martayan, C. 2013, *A&A* Rev., 21, 69

Rivinius, T. & Štefl, S. 2000, in Astronomical Society of the Pacific Conference Series, Vol. 214, IAU Colloq. 175: The Be Phenomenon in Early-Type Stars, ed. M. A. Smith, H. F. Henrichs, & J. Fabregat, 581

Rivinius, T., Vanzi, L., Chacon, J., *et al.* 2012, in Astronomical Society of the Pacific Conference Series, Vol. 464, *Circumstellar Dynamics at High Resolution*, ed. A. C. Carciofi & T. Rivinius, 75

Secchi, A. 1867, *Astronomische Nachrichten*, 68, 63

Shokry, A., Rivinius, T., Mehner, A., *et al.* 2018, *A&A*, 609, A108

Sigut, T. A. A., McGill, M. A., & Jones, C. E. 2009, *ApJ*, 699, 1973

Smith, M. A., Lopes de Oliveira, R., & Motch, C. 2016, *Advances in Space Research*, 58, 782

Udalski, A., Szymański, M. K., & Szymański, G. 2015, *Acta Astron.*, 65, 1

Underhill, A. & Doazan, V. 1982, B Stars with and without emission lines (NASA)

Štefl, S., Rivinius, T., Carciofi, A. C., *et al.* 2009, *A&A*, 504, 929
Vieira, R. G., Carciofi, A. C., & Bjorkman, J. E. 2015, *MNRAS*, 454, 2107
Vieira, R. G., Carciofi, A. C., Bjorkman, J. E., *et al.* 2017, *MNRAS*, 464, 3071
Wang, L., Gies, D. R., & Peters, G. J. 2018, *ApJ*, 853, 156
Wood, K., Bjorkman, K. S., & Bjorkman, J. E. 1997, *ApJ*, 477, 926

High-mass X-ray Binaries: Illuminating the Passage from Massive Binaries to Merging Compact Objects
Proceedings IAU Symposium No. 346, 2019
L. M. Oskinova, E. Bozzo, T. Bulik & D. Gies, eds.

doi:10.1017/S1743921319001212

Optical interferometry of High-Mass X-ray Binaries: Resolving wind, disk and jet outflows at sub-milliarcsecond scale

Idel Waisberg[1], Jason Dexter[1], P.-O. Petrucci[2], Guillaume Dubus[2], Karine Perraut[2] and GRAVITY Collaboration†

[1]Max Planck Institute for extraterrestrial Physics,
Giessenbachstr., 85748 Garching, Germany
email: idelw@mpe.mpg.de

[2]Univ. Grenoble Alpes, CNRS, IPAG
Box 515, F-38000 Grenoble, France

Abstract. Because of their small angular size < 1 mas, spatial information on High-mass X-ray binaries (HMXB) has typically been inferred from photometry or spectroscopy. Optical interferometry offers the possibility to spatially resolve such systems, but has been traditionally limited to bright targets or low spectral resolution. The VLTI instrument GRAVITY, working in the near-infrared K band, achieves unprecedented precision in differential interferometric quantities at high spectral resolution, allowing to study HMXBs through the lens of optical interferometry for the first time. We present GRAVITY observations on two X-ray binaries: the microquasar SS 433 and the supergiant HMXB BP Cru. The former is the only known steady super-Eddington accretor in the Galaxy and is in a unique stage of binary evolution, with probable ties to at least part of the ULX population. With GRAVITY, we resolve its massive winds and optical baryonic jets for the first time, finding evidence for powerful equatorial outflows and photoionization as the main heating process along the jets. BP Cru harbors an X-ray pulsar accreting from the wind of its early-blue hypergiant companion Wray 977. The GRAVITY observations resolve the inner parts of the stellar wind and allow probing the influence of the orbiting pulsar on the circumstellar environment.

Keywords. techniques: interferometric — stars: binaries: close — stars: circumstellar matter — stars: winds, outflows — infrared: stars — X-rays: binaries — stars: individual: BP Cru — stars: individual: SS 433

---

1. Introduction

Direct spatial information on X-ray binaries has traditionally been limited to radio wavelengths, with Very Long Baseline Interferometry enabling $\sim$ mas spatial resolution. The movement of individual radio jet blobs, for example, has been imaged in many microquasars (e.g. Vermeulen *et al.* 1993, Mirabel & Rodriguez 1994). Optical interferometry, on the other hand, can achieve the necessary sub-mas resolution to spatially resolve such systems at scales comparable to the binary orbit $a_{orb} \lesssim 1$ mas, but has been limited to bright targets or low spectral resolution. This has changed with the GRAVITY instrument at the Very Large Telescope Interferometer (VLTI), which has achieved unprecedented precision in differential interferometric observables at high

† GRAVITY is developed in a collaboration by the Max Planck Institute for Extraterrestrial Physics, LESIA of Paris Observatory and IPAG of Universit Grenoble Alpes / CNRS, the Max Planck Institute for Astronomy, the University of Cologne, the Centro Multidisciplinar de Astrofísica Lisbon and Porto, and the European Southern Observatory.

spectral resolution in the near-infrared K band (GRAVITY Collaboration *et al.* 2017a). Even though the canonical imaging resolution of GRAVITY is $\approx \frac{\lambda}{B} \approx 3$ mas for baseline lengths ~ 100 m, spectral differential interferometry allows resolving sub-mas structures through differential visibilities across strong emission lines.

Here we present GRAVITY observations of two X-ray binaries: the canonical wind-accreting BP Cru / GX 301-2, where we resolve the deep layers of the donor star wind to probe the effects of the gravitational and radiation fields of the pulsar companion, and the microquasar SS 433, where we resolve the super-Eddington outflows in the form of baryonic jets, wind and equatorial outflows.

2. Resolving the Stellar Wind in BP Cru / GX 301-2

BP Cru / GX 301-2 is one the canonical wind-accreting High-Mass X-ray Binaries, consisting of a slowly rotating X-ray pulsar accreting from the wind of its hypergiant (B1Ia+; Kaper, van der Meer & Najarro 2006) donor star, called Wray 977. The latter is very massive ($\gtrsim 40 M_\odot$) and luminous ($5 \times 10^5 L_\odot$), with a very high mass-loss rate $\sim 10^{-5} M_\odot$/year, leading to particularly strong emission lines formed in an extended wind. It is relatively nearby (3 kpc), making it an ideal target for high spectral resolution optical interferometry, capable of resolving the inner parts of its wind, through which the pulsar moves in its unusually eccentric ($e = 0.46$) orbit. The gravitational field of the pulsar is expected to lead to the formation of a wind stream with enhanced density, and in fact the very modulated X-ray light curve has been shown to be consistent with the presence of an accretion stream, which the pulsar crosses twice per orbit (Leahy & Kostka 2008). The X-ray flux from the pulsar is also expected to affect the wind of Wray 977, leading to further distortions in its structure (e.g. Blondin 1994, Cechura & Hadrava 2015).

The K band spectrum of BP Cru contains two strong emission lines of Brγ and He I 2.06 μm. The GRAVITY observations of this system (Waisberg *et al.* 2017; Figure 1) show differential visibility amplitudes and phases across the emission lines, allowing to probe the spatial structure of the inner portions of the wind. They imply a very extended wind (FWHM $\sim 3 - 7 R_*$) that is also asymmetric across the line (being more extended on the blue/approaching part), as well as differential visibility phases that imply an asymmetry relative to the continuum and which are also stronger on the approaching part of the line. One possibility is that such asymmetries are caused by X-ray ionization of the stellar wind or by the putative accretion stream in this system. We can also compare the interferometric data with predictions from stellar wind codes. Figure 2 shows the intensity profiles of the near-infrared continuum and emission lines as a function of impact parameter as calculated with the stellar wind atmosphere code CMFGEN (Hillier & Miller 1998) with parameters derived from optical spectroscopy of BP Cru (Kaper, van der Meer & Najarro 2006). They can be directly compared to the interferometric data (black lines in Figure 1). The overall agreement on the spatial extent of the Brγ line is good, but spherically-symmetric models naturally cannot explain the asymmetries detected (e.g. they produce zero differential visibility phases).

Further observations are needed to disentangle natural variability from a wind from distortions caused by the gravitational and radiation fields of the pulsar. In particular, because the pulsar is on an eccentric orbit ($P_{orb} = 41.5$ days), comparing the spatial structure of the wind between apastron and periastron should be particularly promising. Another viable approach is to compare the spatial structure of the wind in BP Cru to isolated blue hypergiants of similar spectral type. This will help to constrain and characterize the influence of the pulsar on its environment, which is important for understanding the complex wind accretion process in HMXBs.

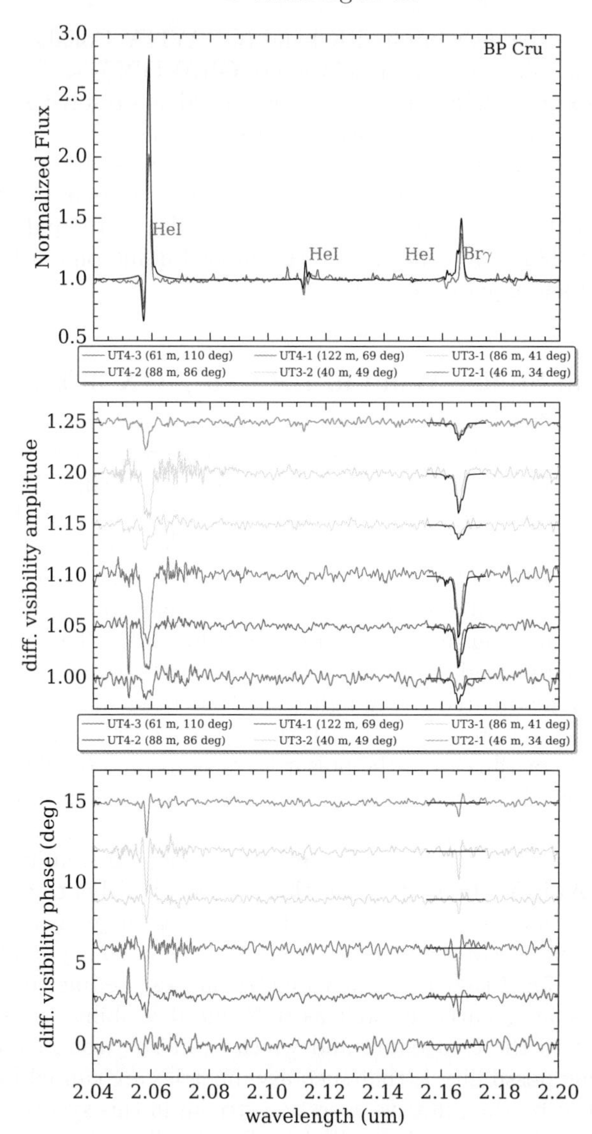

Figure 1. K band spectrum and differential visibility amplitude and phases in the GRAVITY observations of BP Cru. A CMFGEN stellar atmosphere model with parameters derived from optical spectroscopy is shown in black (visibilities are only computed for Brγ where the strength of the line is better matched). The strength of the differential visibility amplitudes are in good agreement with the model but spherically symmetric models cannot explain the nonzero differential visibility phases across the line.

3. Resolving the Super-Eddington Outflows in SS 433

SS 433 was the first microquasar discovered through the broad emission lines of hydrogen and helium moving across its optical spectrum spanning large redshifts (Margon *et al.* 1979), which arise from precessing, relativistic (0.26c) baryonic jets (Fabian & Rees 1979). The supercritical accretion disk in SS 433 outshines its donor star at all wavelengths and drives powerful outflows, not only in the form of jets but also in strong and broad "stationary" (as opposed to the jets) emission lines (Fabrika 2004). The estimated mass-loss rate $\dot{M} \sim 10^{-4} M_{\odot}/\mathrm{yr}$ (Fuchs, Miramond & Ábrahám 2006) clearly

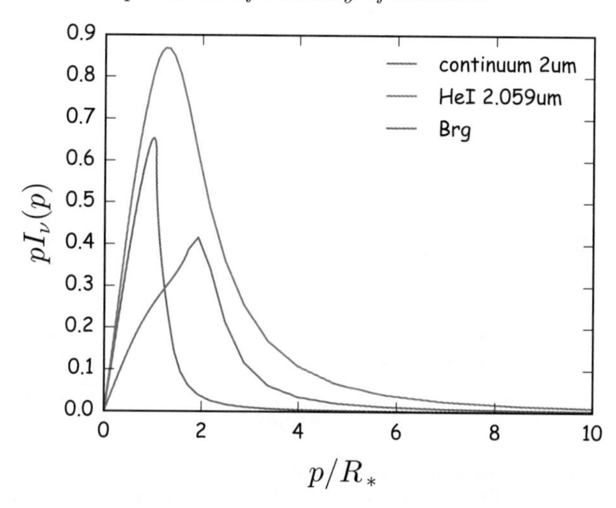

Figure 2. Intensity profile of the near-infrared continuum and K band emission lines as a function of impact parameter (p), as calculated with a CMFGEN stellar wind model for Wray 977. A 2d Fourier Transform of the profiles can be directly compared to the optical interferometry data as shown in Figure 1.

establishes SS 433 as an outflow-regulated supercritically accreting system. With a binary orbit size $a_{orb} = \left(\frac{M}{40M_\odot}\right)^{1/3} \times 0.07$ mas (where M is the total binary mass) resolving the inner parts of the outflow is not possible with current single telescopes, but is within the grasp of high precision spectro-differential optical interferometry.

The first GRAVITY observations of SS 433 happened in July 2016 and the results were presented in GRAVITY Collaboration *et al.* 2017b (Paper I). The optical jets were resolved for the first time, and their emission profile was shown to follow an exponential distribution which peaks at the center of the binary and decays on a scale ≈ 2 mas, in contradiction with previous indirect estimates from spectroscopic monitoring of the moving Hα lines which suggested that the optical bullets had a typical emission peak ≈ 5 mas and extend up to $\gtrsim 10$ mas from the central binary (Borisov & Fabrika 1987). The observations also spatially resolved the Brγ stationary line, revealing an extended structure dominated by emission following the jet direction (suggestive of bipolar outflow). Here we present preliminary results on the second set of GRAVITY observations of SS 433. These consist of three observations taken over a period of four nights (2017-07-07, 2017-07-09 and 2017-07-10).

3.1. *The Equatorial Outflows*

Figure 3 shows the model-independent centroid displacements across the Brγ stationary line derived from the differential visibility phases in one of the 2017 observations. The line is clearly dominated by an equatorial structure perpendicular to the jets. The presence of Gaussian components from a rotating circumbinary ring in stationary emission lines such as Hα and Brγ has been inferred from previous spectroscopic observations (e.g. Blundell, Bowler & Schmidtobreick 2008, Perez & Blundell 2009, Bowler 2010), where the assumption of Keplerian rotation at the innermost stable circumbinary radius has been used to derive a total binary enclosed mass $\gtrsim 40M_\odot$ suggestive of a massive black hole as the compact object.

Figure 4 shows the spectrum centered on the Brγ line, as well as the differential visibility amplitudes and phases across two representative baselines. For baselines aligned with the equatorial direction, the "S-shape" signature typical of rotation is very clear, in contrast with baselines more aligned with the jet direction. The decay in visibility

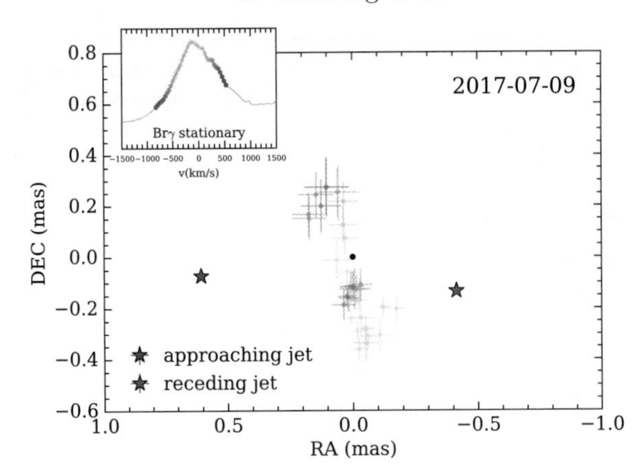

Figure 3. Model-independent centroid shifts across the Brγ stationary line in SS 433. The centroids for the jet emission lines are also shown. The stationary emission has a clear equatorial direction in 2017. The insets show the Brγ line spectrum. An angular size of 1 mas corresponds to 5.5 AU for the assumed distance 5.5 kpc.

amplitude (indicating structure more extended than the continuum, which is partially resolved) requires the presence of a second extended component which we interpret as a spherical wind. This component is also responsible for the high-velocity $\gtrsim 1000$ km/s wings of the emission line, and with a FWHM ~ 6 mas it engulfs the entire binary system and must therefore be optically thin.

We fit the data with a Keplerian disk model, which results in a too high enclosed mass $\gtrsim 400 M_{\odot}$, following from a deprojected rotational velocity ≈ 260 km/s at an outer disk radius ≈ 1.0 mas. This would suggest an intermediate mass black hole as the compact object, which would be in severe tension with all the known properties of the object (Fabrika 2004.) Furthermore, the resulting inner radius in the model is $\lesssim 0.1$ mas, which would be smaller than the semi-major axis for such a high mass and the disk would therefore not be circumbinary. We rather interpret the disk as an equatorial rotating outflow (which is also the model shown in Figure 4), with a resulting outflow radial velocity ~ 240 km/s. Equatorial outflows in SS 433 are typically seen in radio images either as outflowing knots (e.g. Paragi *et al.* 1999) or as a smooth structure, which has been called the "radio ruff" (Blundell *et al.* 2001). We postulate that the equatorial structure we interpret traces the inner portions of these outflows, which are clearly rotating. We note that in Paper I the equatorial structure was not detected; rather, the stationary Brγ line emission is dominantly along the jet direction, suggestive of a bipolar outflow. This suggests that the equatorial structure in SS 433 is unstable.

An interesting aspect of the equatorial outflows is that they carry a very large amount of specific angular momentum. The rotational velocity is ~ 220 km/s for an outer radius ~ 0.7 mas, which corresponds to a specific angular momentum $\gtrsim 10\times$ the specific orbital angular momentum of the compact object, assuming a radial velocity semi-amplitude ~ 200 km/s (Fabrika & Bychkova 1990) and a total binary mass $\lesssim 40 M_{\odot}$. The orbital angular momentum of the donor star is even lower since the mass ratio $q < 1$ (Cherepashchuk, Postnov & Belinski 2018). We propose that the equatorial outflows may be driven by the supercritial accretion disk and carry angular momentum either from the inner portions of the disk (Blandford & Payne 1982) or from the compact object itself through a magnetic propeller effect (Illiarionov & Sunyaev 1975). We note that our first spatially resolved observations of a stationary line in SS 433 cast severe doubt on previous mass estimates that assumed that their typical double-peaked structure arises in an accretion disk (Fillipenko *et al.* 1988; Robinson *et al.* 2017) or in a circumbinary ring in Keplerian rotation (Blundell, Bowler & Schmidtobreick 2008; Bowler 2010),

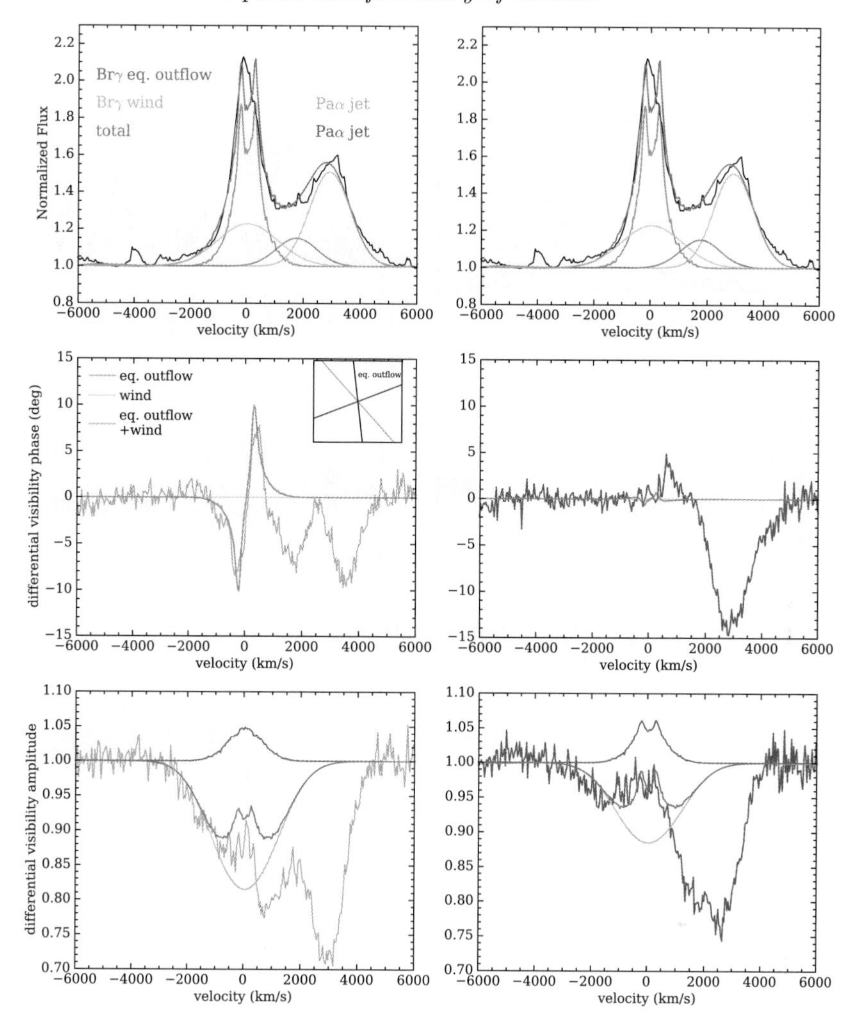

Figure 4. The **top** row shows the spectrum centered on the Brγ stationary line. The latter is decomposed into an equatorial outflow and a spherical wind. The former is responsible for the S-shape signatures in the differential visibility phases (**middle** row) for baselines which are close to perpendicular to the jet (left), and show almost no signature on baselines more aligned with the jet (right). The inset shows the position angle of the outflow from the fit as well as the baseline directions on the sky plane. The **bottom** row shows the differential visibility amplitudes. The equatorial outflow alone would lead to an increase in visibility amplitude across the Brγ line. The extended wind component can explain both the high velocity wings $\gtrsim 1000$ km/s in the spectrum as well as the net decrease in visibility amplitude across the line. Note that there are two Paα emission lines from the receding jet which are blended with the Brγ stationary line on its red side, and which also create strong visibility signatures. The model fits were done for all the components simultaneously, but here we show only the visibility model for the stationary line for clarity.

which concluded that the compact object must be a neutron star or massive black hole, respectively. We consider that the nature of the compact object is still unknown.

3.2. *The Optical Jets*

In the 2017 observations, the precessional phase of the jets was ≈ 0.9 (where phase 0.0 corresponds to lowest inclination/maximum visibility of the accretion disk). The strongest

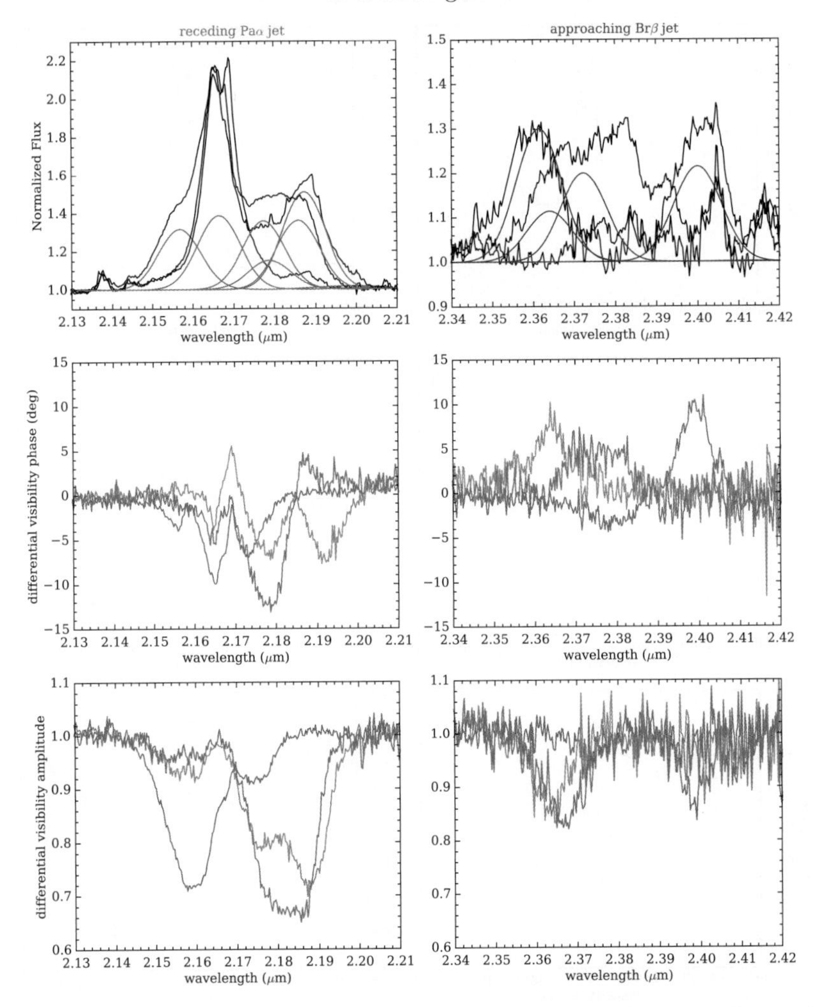

Figure 5. Same as Figure 4 but for the jets emission lines in the 2017 observations. The spectrum, differential visibility phase and differential visibility amplitudes for a representative baseline are shown in the **top**, **middle** and **bottom** panels, respectively. The left panels are for the Paα lines from the receding jet (which are blended with the Brγ stationary line), whereas the right panels correspond to the Brβ lines from the approaching jet (which are blended with stationary high-order Pfund lines). The red/blue components in the spectrum show the different jet knots in the observations throughout four nights. There are strong differential visibility signatures across the lines (shown in different colors for the three different nights), which allow to fit for the spatial profile of the optical jets.

jet emission lines in the K band spectrum at this epoch are the Paα line from the receding jet and Brβ line from the approaching jet (in contrast, in Paper I jet lines from Brγ, Brδ and He I 2.056 μm fell in the K band spectrum resulting from a precessional phase ≈ 0.7). Another difference relative to Paper I is that the jet lines often have at least two different components, one of them likely resulting from a previous jet knot ejection.

The three observations over four days allowed to follow the evolution of the jet components. In particular, the two observations over consecutive nights have jet components at similar redshifts that could correspond to the same jet knots that have brightened or faded over one day. Figure 5 shows the jet emission lines and corresponding differential visibility amplitudes and phases for a given baseline for illustration.

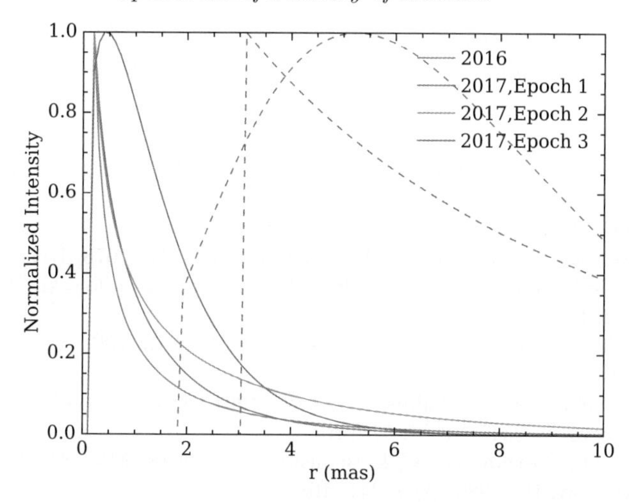

Figure 6. Collection of spatial emission profiles of the optical jets across several observations. The profiles can be divided into two classes: compact (solid lines), which peak very close to the binary and have a typical exponential profile with decay constant ~ 2 mas, and extended (dashed lines), which peak further away. The latter correspond to older jet ejections which have had more time to travel. The smooth and continuous profiles are strongly suggestive of a continuous heating mechanism along the entire jet, such as photoionization by the beamed radiation. An angular size of 1 mas corresponds to 5.5 AU for the assumed distance 5.5 kpc.

Figure 6 shows a collection of resolved spatial profiles of the optical jets over all GRAVITY observations. They can be broadly divided into two groups: compact profiles, which peak close to the accretion disk and have an exponential-like profile with decay constant ≈ 2 mas, and more extended profiles with broader shapes and which peak farther away from the binary. The latter correspond to older jet ejections which have had more time to travel along the beam.

The smooth, exponential-like spatial profiles for the optical jets suggest that the heating mechanism for the optical bullets must act smoothly along a long portion of the jet. We suggest that this is better realized through photoionization of the bullets rather than external heating processes such as interaction with a surrounding wind outflow or shocks. Although the latter had been preferred in past work (e.g. Begelman *et al.* 1980, Davidson & McCray 1980, Brown, Casinelli & Collins 1991), photoionization has also been suggested as a possible heating mechanism (Fabrika & Borisov 1987). This opens the possibility of constraining properties of the beamed radiation through spectral and spatial properties of the optical bullets.

4. Conclusion

Spectro-differential optical interferometry allows to resolve emission line structures in compact and X-ray binaries at sub-mas scale, comparable to the size of the binary orbits. GRAVITY at VLTI has significantly improved sensitivity and precision in optical interferometry, allowing observations of fainter targets at high spectral resolution with unprecedented precision in differential visibility quantities across spectral lines. We have shown two examples of X-ray binaries successfully resolved by GRAVITY. In the canonical wind-accreting BP Cru, we have resolved the inner parts of the stellar wind and found asymmetries that could probe the influence of the gravitational and radiation fields of the orbiting pulsar on the surrounding stellar environment. On the microquasar SS 433, we have found evidence for an equatorial, rotating outflow in the stationary Brγ line that carries very high specific angular momentum, and have resolved for the first time the spatial profile of the optical jets, which is strongly suggestive of photoionization by beamed radiation as the main heating mechanism of the optical bullets.

References

Begelman, M. C., Sarazin, C. L., Hatchett, S. P., McKee, C. F., & Arons, J. 1980, *ApJ*, 238, 722

Blandford, R. D. & Payne, D. G., 1982, *MNRAS*, 199, 883

Blondin, J. M. 1997, *ApJ*, 435, 756

Blundell, K. M., Bowler, M. G., & Schmidtobreick, L. 2008, *ApJ*, 678, L47

Blundell, K. M., Mioduszewski, A. J., Muxlow, T. W. B., Podsiadlowski, P. & Rupen, M. P., 2001, *ApJ*, 562, L79

Borisov, N. V. & Fabrika, S. N. 1987, *Soviet Astron. Letters*, 13, 200

Bowler, M. G. 2010, *A&A*, 521, A81

Brown, J. C., Cassinelli, J. P., & Collins, II, G. W., 1991, *ApJ*, 378, 307

Cechura, J. & Hadrava, P. 2015, *A&A*, 575, A5

Cherepashchuk, A. M., Postnov, K. A., & Belinski, A. A., 2018, *MNRAS*, 479, 4844

Davidson, K. & McCray, R., 1980, *ApJ*, 241, 1082

Fabian, A. C. & Rees, M. J. 1979, *MNRAS*, 187, 13P

Fabrika, S. 2004 *Space Sci. Revs*, 12, 1

Fabrika, S. N. & Borisov, N. V., 1987, *Soviet Astron. Letters*, 13, 279

Fabrika, S. N. & Bychkova, L. V., 1990, *A&A*, 240, L5

Filippenko, A. V., Romani, R. W., Sargent, W. L. W., & Blandford, R. D., 1988, *AJ*, 96, 242

Fuchs, Y., Koch Miramond, L., & Ábrahám, P. 2006, *A&A*, 445, 1041

Gravity Collaboration, Abuter, R., Accardo, M., *et al.* 2017a, *A&A*, 602, A94

Gravity Collaboration, Petrucci, P.-O.,Waisberg, I., *et al.*, 2017b, *A&A*, 602, L11

Hillier, D. J. & Miller, D. L. 1998, *ApJ*, 496, 407

Illarionov, A. F. & Sunyaev, R. A. 1975, *A&A*, 39, 185

Kaper, L., van der Meer, A. & Najarro, F. 2006, *A&A*, 457, 595

Leahy, D. & Kostka, M. 2008, *MNRAS*, 384, 747

Margon, B., Ford, H. C., Grandi, S. A. & Sone, R. P. S. 1979, *ApJ*, 233, L63

Mirabel, I. F. & Rodriguez, L. F. 1994, *Nature*, 371, 46

Paragi, Z., Vermeulen, R. C., Fejes, I., *et al.*, 1999, *A&A*, 348, 910

Perez M., S. & Blundell, K. M. 2009, *MNRAS*, 397, 849

Robinson, E. L., Froning, C. S., Jae, D. T. *et al.*, 2017, *ApJ*, 841, 79

Vermeulen, R. C., Schilizzi, R. T., Spencer, R. E. *et al.* 1993, *A&A*, 270, 177

Waisberg, I., Dexter, J., Pfuhl, O. *et al.* 2017, *ApJ*, 844, 72

Discussion

KRETSCHMAR It is fascinating to see X-ray binaries resolved at these scales. For which other sources can we expect similar results in the future?

WAISBERG The current limiting magnitude in GRAVITY for fringe tracking with the Unit Telescopes is $K \lesssim 10$ (it is possible to integrate on fainter targets if there is such a bright target within 2" but that is very rare in non-crowded fields). Unfortunately that excludes nearly all Low-Mass X-ray Binaries, but there are quite a few HMXBs besides the two presented here that are possible (e.g. some BeHMXBs). Very obscured systems that are bright in the K band but very faint in optical (such as the ones detected by INTEGRAL) should also be possible in the future once the near-infrared AO system CIAO is offered in on-axis mode.

High-mass X-ray Binaries: Illuminating the Passage from Massive Binaries to Merging Compact Objects
Proceedings IAU Symposium No. 346, 2019
L. M. Oskinova, E. Bozzo, T. Bulik & D. Gies, eds.

doi:10.1017/S1743921319001182

Studying the H-alpha line of the B[e] supergiant binary GG Carinae using high-cadence optical spectroscopy

Augustus Porter[1], **Katherine Blundell**[1] **and Steven Lee**[2]

[1]Department of Physics, University of Oxford, Oxford, United Kingdom
email: augustus.porter@physics.ox.ac.uk

[2]Anglo-Australian Telescope, Coonabarabran NSW 2357, Australia

Abstract. We present a case study of GG Carinae (GG Car), a Galactic B[e] supergiant binary having significant eccentricity (0.28), based on Global Jet Watch spectroscopy data which has been collecting high-time-sampled optical spectra since early 2015. GG Car has so far not been observed in the X-ray band, however it is of similar phenomenology to known X-ray binaries and may therefore be an obscured X-ray source. We have discovered that the absorption component of the H-alpha line displays a ~462-478-day period in both equivalent width and wavelength centroid indicating cycles in the dynamics of the circumstellar environment, such as precession of the circumbinary or circumprimary disk. Circumbinary disk precession is an as-of-yet under-explored origin of super-orbital variations in the X-ray flux of X-ray binaries, since the rate of precession is generally much longer than the orbital period of the inner binary.

Keywords. stars: binaries - stars: emission-line, Be - stars: supergiants - stars: individual: GG Car

1. Introduction

GG Car is a Galactic eclipsing binary comprising a B[e] supergiant (sgB[e]) in the post main sequence phase of its evolution and a secondary of an unknown type. GG Car has been known for a few decades to display both spectroscopic and photometric variability with a period of ~31 days (Hernandez *et al.* 1981; Gosset *et al.* 1984; Brandi *et al.* 1987). Lamers *et al.* (1998) classified GG Car as an sgB[e] based on its effective temperature and luminosity, and its exhibiting the B[e] characteristics. The orbital solution was revised by Marchiano *et al.* (2012) to a period of 31.033 days, eccentricity of 0.28, and mass ratio of 2.2. Kraus *et al.* (2013) discovered GG Car's circumbinary disk from CO emission, and Maravelias *et al.* (2018) found that the forbidden emission must originate from circumbinary regions. Doolin & Blundell (2011) found that the orbits of test particles in circumbinary orbits are expected to precess in inclination and longitude of ascending node, with the rate determined by the binary eccentricity and mass ratio. Martin & Lubow (2018) confirmed this behaviour applies to gaseous circumbinary disks.

2. Results

GG Car's H-alpha line can be adequately fit throughout all epochs using a model comprising five gaussians. The time-dependence of the centroid and the equivalent width of the absorption component was studied with Fourier analysis utilizing the CLEAN algorithm to extract the spectrum from unevenly sampled data (Roberts *et al.* 1987). Figure 1, left, displays the variation of the centroid of the absorption component and

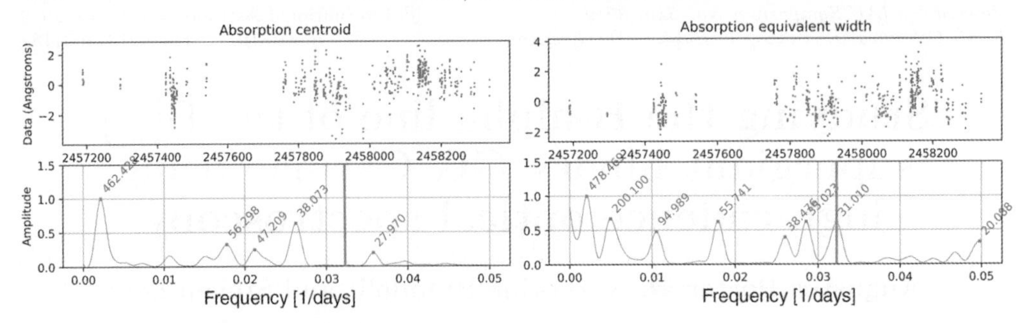

Figure 1. The variation of the centroid of the absorption component of the H-alpha line in GG Carinae (top figure, the data is mean subtracted) and the corresponding Fourier power spectrum (normalised to the strongest peak). Significant peaks have been labeled with their corresponding periods in days. A thin red line denotes a 31.033 day period.

its corresponding Fourier spectrum, while right displays the same but for the equivalent width of the absorption component.

The Fourier spectrum of the centroid has its strongest peak around 462 days, a period which has not been noted before for GG Car, and it does not have any variation with the ~31 day period of the inner binary. The Fourier spectrum of the equivalent width has its strongest peak at 478 days, which is consistent with the 462-day peak in the centroid power spectrum. The two long periodicities possibly indicate super-periodic variations in GG Car's circumstellar environment. The equivalent width of the absorption component, but not the centroid, has a dependency on the ~31 day period of the inner binary.

3. Discussion

A ~470-day candidate period is apparent for the centroid position and strength of the absorption component of the H-alpha line, indicating that there are varying amounts of absorbing atomic Hydrogen along the line of sight with this period. This long-term variation may be indicative of precession of the circumbinary disk as the orbits changes their inclination and longitude of ascending node, or due to changing conditions in the stellar wind of the sgB[e] primary. A similar study will be carried out on the Helium lines in GG Car, which also display strong P-Cygni profiles to disentangle the scenarios. This work explores the possibility that periodic precession of a circumbinary disk may obscure the inner binary with a super-periodic timescale. Circumbinary precession could therefore explain some of the super-periodic variations seen in many X-ray binary sources.

References

Brandi, E., Gosset, E. & Swings, J.-P. 1987, *A&A*, 175, 151
Doolin, S. & Blundell, K. 2011, *MNRAS*, 418, 2656
Farago, F. & Laskar, J. 2010, *MNRAS*, 401, 1189
Gosset, E., Surdej, J. & Swings, J. P. 1984, *A&A Supplement Series*, 55, 411
Hernandez, C. A., Sahade, J., Lopez, L. & Thackeray, A. D. 1981, *PASP*, 93, 747
Kraus, M., Oksala, M., Nickeler, D., *et al.* 2013, *A&A*, 549, A28
Lamers, H., Zickgraf, F.-J., de Winter, D., *et al.* 1998, *A&A*, 340, 117
Maravelias, G., Kraus, M., Cidale, L. S., *et al.* 2018, *MNRAS*, 480, 320
Marchiano, P., Brandi, E., Muratore, M. F., *et al.* 2012, *A&A*, 540, 9
Martin, R. G. & Lubow, S. H. 2018, *MNRAS*, 479, 1297
Roberts, D. H., Lehar, J., & Dreher, J. W. 1987, *ApJ*, 93, 968

High-mass X-ray Binaries: Illuminating the Passage from Massive Binaries to Merging Compact Objects
Proceedings IAU Symposium No. 346, 2019
L. M. Oskinova, E. Bozzo, T. Bulik & D. Gies, eds.

doi:10.1017/S1743921319002497

The young Be-star binary Circinus X-1

Norbert S. Schulz[1], Timothy E. Kallman[2], Sebastian Heinz[3], Paul Sell[4], Peter Jonker[5] and William N. Brandt[6]

[1]Kavli Institute for Astrophysics and Space Research,
Massachusetts Institute of Technology, Cambridge, MA 02139, USA
email: nss@space.mit.edu

[2]Goddard Space Flight Center, NASA, Green Belt, MD, USA
email: tim@milkyway.gsfc.nasa.gov

[3]Department of Astronomy, University of Wisconsin, Madison, WI, USA
email: heinzs@astro.wisc.edu

[4]Department of Physics, University of Crete, Heraklion, Greece
email: psell@physics.uoc.gr

[5]Department of Astrophysics, Radboud University, Nijmegen, The Netherlands
email: peterj@sron.nl

[6]Department of Astronomy & Astrophysics, 525 Davey Laboratory,
The Pennsylvenia State Univerity, University Park, PA, USA
email: niel@astro.psu.edu

Abstract. Cir X-1 is a young X-ray binary exhibiting X-ray flux changes of four orders of magnitude over several decades. It has been observed many times since the launch of the Chandra X-ray Observatory with high energy transmission grating spectrometer and each time the source gave us a vastly different look. At its very lowest X-ray flux we found a single 1.7 keV blackbody spectrum with an emission radius of 0.5 km. Since the neutron star in Cir X-1 is only few thousand years old we identify this as emission from an accretion column since at this youth the neutron star is assumed to be highly magnetized. At an X-ray flux of 1.8×10^{-11} erg cm^{-2} s^{-1} this implies a moderate magnetic field of a few times of 10^{11} G. The photoionized X-ray emission line properties at this low flux are consistent with B5-type companion wind. We suggest that Cir X-1 is a very young Be-star binary.

Keywords. stars: neutron, X-rays: binaries, techniques: spectroscopic

1. Introduction

Cir X-1 has shown a large range of brightness levels, variability patterns, and spectral changes in its X-ray emissions since its discovery a few decades ago (Margon *et al.* 1971). The true nature of this X-ray binary, one orbit lasts about 16.5 days (Kaluzienski *et al.* 1976), always was somewhat mysterious as the identification of its companion remained exceedingly unclear. We do know that the compact object is a neutron star because of direct observations of type I X-ray bursts (Tennant *et al.* 1986, Linares *et al.* 2010, Papitto *et al.* 2010). Whelan *et al.* (1977) suggested the companion star to be an early-type emission line or symbiotic star, while Moneti (1992) found three heavily reddened objects as possible counterparts. Photometric variability of a suggested optical counterpart, better determination of its orbital parameters, as well as X-ray spectral and timing patterns seemed to point to low-mass X-ray binary (LMXB) nature (Brandt & Podsiadlowski 1995, Tauris *et al.* 1999, Tennant 1987, Shirey *et al.* 1999).

Jonker *et al.* (2007) determined that the companion is very likely a massive supergiant of A0 to B5 type, which leads to an orbital eccentricity (e $\sim$ 0.45) and makes Cir X-1 a high-mass X-ray binary (HMXB). Such a companion would be consistent with an earlier tentative identification by Whelan *et al.* (1977). However, the Jonker *et al.* study could not entirely rule out effects caused by absorption in the accretion disk with respect to the supergiant nature.

Perhaps the most striking recent result is the discovery of the X-ray supernova remnant associated with Cir X-1 (Heinz *et al.* 2013). This allowed to place an upper limit of 4600 yr on its age making it the youngest known X-ray binary. Such a young age is also quite consistent with an earlier assessment of X-ray dip periodicity that Cir X-1 is a state of dynamical evolution as in a very young post-supernova system (Clarkson *et al.* 2004). This has striking consequences on the nature of Cir X-1. The observation of type I X-ray bursts on the surface of the neutron star indicates that the magnetic field is not very high suggesting either the possibility that accretion can rapidly de-magnetize a neutron star or there is the possibility that neutron stars can be born with low magnetic fields (Heinz *et al.* 2013). However, neither to date is further supported by theory and observations. In fact, all we know about young neutron stars is that they have magnetic fields or the order of 10^{12} Gauss (see Kaspi 2010, Reig 2011 and references therein). In Schulz *et al.* (2018) we argue that under the assumption that young neutron stars have high magnetic fields, Cir X-1 should as well. Under the further assumption that then the most likely emission site for a 1.7 keV blackbody with an emission radius of 0.5 km is the accretion column, that study estimates a magnetic field strength of the order of 10^{11} G. Furthermore, from blueshifted X-ray lines consistent with a B5 stellar wind, it was concluded that Cir X-1 is a HMXB, maybe a Be-star X-ray binary. In the following we discuss how such a binary nature holds up with what we know about X-ray binaries.

2. Properties of X-ray Binaries

Generally X-ray binaries containing a neutron star can be divided into low- and high-mass systems depending whether the mass of the donor star is below about 2 $M_{\odot}$ or above about 8 $M_{\odot}$, respectively. LMXBs are old systems with ages beyond 10^9 yr (Cowley *et al.* 1988). With their magnetic fields to be decayed down to about 10^8 G, mass accretion onto the neutron star is hardly affected by this field. Accretion from low-mass companion stars happens effectively via Roche-lobe overflow resulting in various X-ray spectral variation patterns depending on mass accretion and luminosity (Schulz, Hasinger & Truemper 1989) accompanied by distinct quasi-periodic timing patterns (Hasinger & van der Klis 1989). The brightest X-ray sources radiate close to the Eddington limit and due to their variation pattern in the X-ray color-color diagram are called Z-sources. Sources with X-ray luminosities more than an order of magnitude lower are called atoll sources due to their different and more disjunct pattern consisting of island and banana states. LMXBs also show a zoo of distinct features in their lightcurves and spectra such as type I and II X-ray bursts, accretion disk coronae, dips and in more rare cases eclipses. Type I X-ray bursts are thermonuclear explosions on the surface of the neutron star, type II X-ray bursts occur due to instabilities in the accretion disk (see Lewin, van Paradijs & van den Heuvel 1995 for a review). X-ray spectra can usually be modelled by multi-component models involving blackbody functions, multi-temperature disk blackbody functions, power laws, bremsstrahlung, reflection and Comptonisation.

In contrast, HMXB are comparatively young due to the fact that massive stars have much shorter life times, i.e. less than a few 10^7 yr. They divide further into supergiant X-ray binaries (SGXB) and Be X-ray Binaries (BeXB) depending on the evolutionary status of the optical companion (see Reig 2011 for a more recent review). SGXBs usually contain companions with a luminosity class I - II, BeXBs luminosity classes III - V. Most SGXBs are not known to have significant accretion disks and the bulk of accretion happens through Bondi Hoyle wind accretion. Prime examples are Vela X-1 and 4U 1700-37, rare exceptions are Cen X-3, LMC X-4, and SMC X-1. In BeXBs the massive

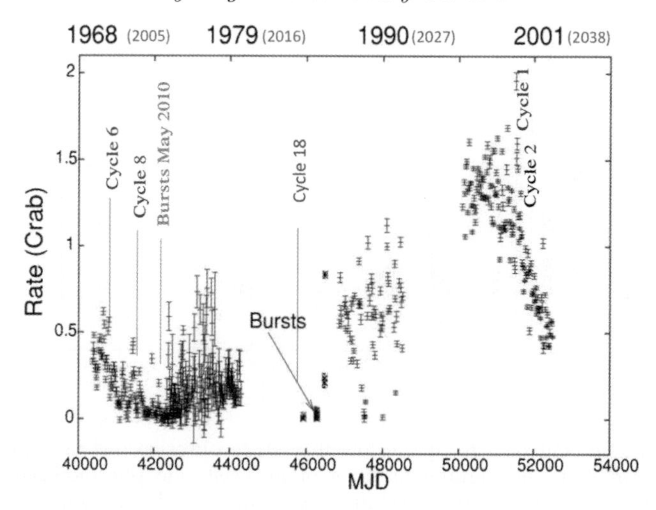

Figure 1. Long-term lightcurve from Parkinson *et al.* (2003). We marked the locations of Chandra observations so far as well as the locations of type I X-ray bursts.

companion star is a fast-rotating B-type star, optically identified through spectral line emission, which have their origin in a circumstellar disk formed from wind material expelled from a rapidly spinning B-star.

3. Where does Cir X-1 fit in?

Cir X-1 has given us many appearances throughout the decades. Figure 1 shows the long-term lightcurve from Parkinson *et al.* (2003). During its brighter X-ray phases in the late 1990s it behaved like a Z-source (Shirey *et al.* 1999), during its rapid decay in the early 2000s it looked more like an atoll source (Schulz *et al.* 2008), usually accretion trademarks of LMXBs. However, Homan *et al.* (2010) showed in recent studies of transient sources that sources do morph through Z- and atoll stages during their rise and decline in source brightness making these patterns more an imprint of Roche-lobe overflow accretion rather than defining a distinct X-ray binary type. If Cir X-1 is a LMXB, its neutron star was formed in an accretion induced collapse (AIC, Bhattacharya & van den Heuvel 1991) of a white dwarf. The analysis of the X-ray remnant could neither rule out nor confirm an AIC scenario (Heinz *et al.* 2013). However, the high eccentricity of the binary orbit together with the fast orbital evolution of the system are at odds with such an event. Neutron stars in AIC events like the ones from electron capture supernova events hardly receive an kick during these events and their binary orbits are not significantly affected (Tauris *et al.* 2013).

The most likely scenario is that Cir X-1 was born in a core collapse supernova event of a massive star. In that case the companion cannot be a low mass star for simple evolutionary reasons. Here it has to be a massive star of similar or somewhat later type than the progenitor type of the neutron star. The study by Jonker *et al.* (2007) suggests that the companion is a massive star of A0 to B5 type, which does fit into that paradigm. In order to produce a neutron star in a core collapse explosion, that progenitor star has to have had a mass of higher than 8 to 10 $M_\odot$ demanding at least an early B-type nature, i.e. B3 or earlier (Behrend & Meader 2001). The X-ray line centroid shifts measured in Schulz *et al.* (2018) narrows possibilities down to a companion being a B5 star (Fig. 2). Later type B-star winds have much lower velocities. The lines are attributed to the companion wind as they are observed at periastron, but not at apastron. If the identification by Jonker *et al.* (2007) for the companion to be of supergiant nature is correct, then this is

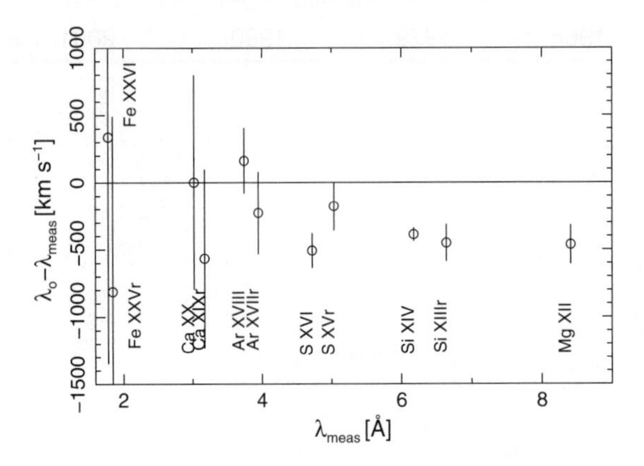

Figure 2. X-ray line centroids from the low flux state observations in Chandra cycle 8 and 18. The most prominent and significant lines of at least one photoionized plasma spectral components in the spectral fits are blueshifted by about 400 km s^{-1}(data from Schulz *et al.* 2018).

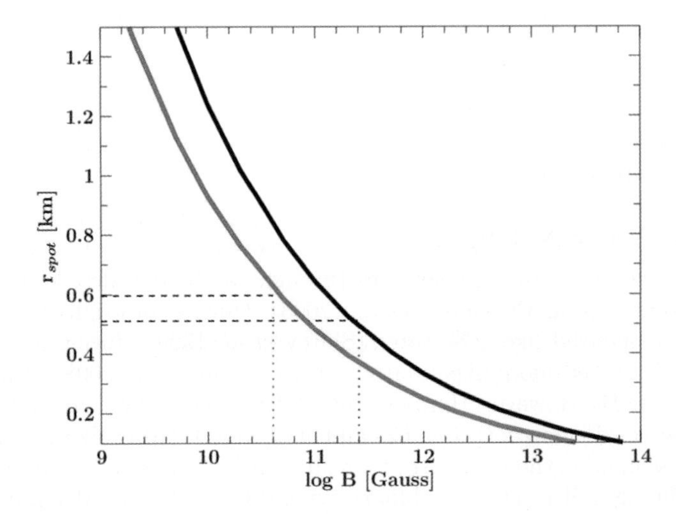

Figure 3. The magnetic field strength of Cir X-1 in dependence of the measured blackbody emission radii (from Schulz *et al.* 2018). From a range of radii between 0.5 and 0.6 km follows a range of magnetic field strength between 4×10^{10} G and 2.5×10^{10} G assuming that very young neutron stars have high magnetic fields.

almost the only choice left because any later type would not have had enough time to evolve that far. Under all these circumstances the most likely nature of Cir X-1 is that of an HMXB.

At this point we also want to put some attention onto the fact that the neutron star in Cir X-1 is extremely young and by that fact should have a high magnetic field. Heinz *et al.* (2013) argued that maybe neutron stars can be born with very low ($< 10^{10}$ G) magnetic fields or that accretion could de-magnetize a neutron star on very short time scales. However, this has never been observed before nor is there any theoretical backup for such a scenario. All we know is that neutron stars are born with high magnetic fields. How high is relative, most young neutron stars have magnetic fields as high as 10^{12} G and beyond. Halpern & Gotthelf (2010) presented an example of a young neutron star in Kes 79 of much less than 10^{11} G proposing the existence of 'anti-magnetars', i.e. neutron stars born with moderately high magnetic fields. Figure 3 shows the magnetic field range determined for Cir X-1 from the blackbody emission radii measured in Schulz *et al.* (2018) from very low X-ray flux data. This would indicate that the neutron star in Cir X-1 has a more moderate magnetic filed. This is an interesting possibility because it would allow the scenario for a very young accreting neutron star to exhibit type I X-ray bursts, an process only known to be effective in low magnetic fields of old LMXBs.

4. Cir X-1 as a BeXB

There are some inconsistencies in the identification whether Cir X-1 is a young SGXB or BeXB. One is the fact that in none of the known SGXB the massive star is of Be-type. The other one is that none of the companions in known BeXBs are later than of B2 type. While the latter might be an observational bias, the former needs more evolutionary understanding of Be stars. The orbital period of 16.5 days as well as the eccentricity of 0.45 are now consistent with kick velocities of several hundred km s^{-1}as observed in other post supernova HMXBs, specifically BeXBs (Reig 2011). BeXBs typically show low persistent X-ray luminosities of $\sim 10^{34-35}$ erg s^{-1} (Reig & Roche 1999). However, BeXBs with higher eccentricities also show two patterns of outbursts. One pattern consists of regular and periodic outbursts near periastron passage of the neutron star, another more longterm pattern involves X-ray flux increases of 10^3 - 10^4 times the quiescence flux. Both types of patterns as well as the range of luminosities are observed in Cir X-1, Fig. 1 shows the second pattern spanning over 30 yr.

In the case of Cir X-1 we invoke a longterm Be-star disk precession. Precessing Be-star disks may be rare, but not unheard of. Cir X-1 as a BeXB has the potential to explain the ~30 yr transient flux behavior as shown in Fig. 1. The requires to propose a precession period for the Be-star disk. Similar but physically different scenarios have been suggested by Brandt & Podsiadlowski (1995) and Heinz *et al.* (2013). The former study suggested an accretion disk precession, while the latter discussed spin-orbit coupling effects between the neutron star spin and the binary orbit. This is not unrealistic as super-orbital periods in accretion disks are not unusual in X-ray binaries. Examples of such super-orbital disk precession periods are the ones in Her X-1, LMC X-4, and SMC X-1. Another X-ray binary microquasar which has been compared to Cir X-1 many times before and resides in the a young remnant (W50) is SS 433. Even though we do not have direct knowledge of its compact object mass, much evidence points to a black hole accreting from a highly evolved supergiant primary (Blundell *et al.* 2008). This system also has high dynamical features in form of a helical precessing jet. Seward *et al.* (2012) identified a likely HMXB in the LMC remnant DEM L241, which consists of a O5III(f) star and an undetermined compact object. Another example in this context is SXP 1062, a BeXB in the SMC, which appears to be embedded in a shell-like structure, likely a SNR (Hénault-Brunet *et al.* 2012). More recently Lau *et al.* (2016) presented evidence of a precessing helical outflow from the massive star WR102c. Even though it is a single star, it shows that dynamic outflows can happen from a precessing massive star. All this shows that we now have plenty of observational evidence of young high mass systems in young remnants exhibiting precession action.

5. Conclusions

We now have a surmounting amount of evidence that the neutron star in Cir X-1 is not only very young but the system itself is a HMXB. Even though the companion has been identified as a supergiant, it phenomenologically shows many features and traits linked to BeXBs. More observations and modeling are needed, but a precessing Be-star disk provides an intriguing mechanism to explain the vast longterm X-ray flux variations.

References

Behrend, R. & Maeder, A. 2001 *Astron. & Astrophys.*, 373, 190
Bhattacharya, D., & van den Heuvel, E. P. J. 1991, *Phys. Rep.*, 203, 1
Blundell, K. M., Bowler, M. G. & Schmidtobreick, L. 2008 *ApJ*, 678, L47
Brandt, N. & Podsiadlowski, P. 1995 *MNRAS*, 274, 461
Clarkson, W. L., Charles, P. A. & Onyett, W. 2004 *MNRAS*, 348, 458
Cowley, A. P., Hutchings, J. B. & Crampton, D. 1988 *ApJ*, 333, 906
Halpern, J. P. & Gotthelf, E. V. 2010 *ApJ*, 709, 436
Hasinger, G. & van der Klis, M. 1989 *Astron. & Astrophys.*, 225, 79
Heinz, S., Sell, P., Fender, R. P., Jonker, P. G., Brandt, W. N., Calvelo-Santos, D. E., Tzioumis, A. K., Nowak, M. A., Schulz, N. S., Wijnands, R., & van der Klis, M. 2013 *ApJ*, 779, 171
Hénault-Brunet, V., Oskinova, L. M., Guerrero, M. A. *et al.* 2012 *ApJ*, 619, 503
Homan, J., van der Klis, M., Fridriksson, J. K., *et al.* 2010, *ApJ*, 719, 201
Jonker, P. G., Nelemans, G. & Bassa, C. G. 2007 *MNRAS*, 374, 999
Kaluzienski, L. J., Holt, S. S., Boldt, E. A., & Serlemitsos, P. J. 1976 *ApJ Letters*, 208, 71L
Kaspi, V. 2010 *Proceeding of the National Academy of Science*, 107, 7147
Lau, R. M., Hankins, M. J., Herter, T. L. *et al.* 2016 *ApJ*, 818, 117
Lewin, W. H. G., van Paradijs, J. & van den Heuvel, E. P. J. 1995 *Cambridge Astrophysics Series*, Vol. 26
Linares, M., *et al.* 2010 *ApJ Letters*, 719, L84
Margon, B., Lampton, M., Bowyer, S., & Cruddace, R. 1971 *ApJ Letters*, 169, L23
Moneti, A. 1992 *Astron. & Astrophys.*, 260, 7
Papitto, A., Riggio, A., di Salvo, T., Burderi, L., D'Aì, A., Iaria, R., Bozzo, E., & Menna, M. T. 2010 *MNRAS*, 407, 2575
Parkinson, P. M. S., Tournear, D. M., Bloom, E. D. *et al.* 2003 *ApJ*, 595, 333
Reig, P. & Roche, P. 1999 *MNRAS*, 306, 100
Reig, P. 2011 *Ap& SS*, 332, 1
Schulz, N. S., Hasinger, G. & Trumper, J. 1989 *Astron. & Astrophys.*, 225, 48
Schulz, N. S., Kallman, T. E, Galloway, D. K., & Brandt, W. N. 2008 *ApJ*, 572, 171
Schulz, N. S., Kallman, T. E, Heinz, S., Sell, P., Jonker, P. G., & Brandt, W. N. 2018 *ApJ*, submitted
Shirey, R. E., Bradt, H. V. & Levine, A. M. 1999*ApJ*, 517, 472
Steward, F. ., Charles, P. A., Foster, D. L. *et al.* 2012 *ApJ*, 759, 123
Tauris, T. M., Fender, R. P., van den Heuvel, E. P. J., Johnston, H. M., & Wu, K. 1999 *MNRAS*, 310, 1165
Tauris, T. M., Sanyal, D., Yoon, S.-C., & Langer, N. 2013 *Astron. & Astrophys.*, 558, A39
Tennant, A. F., Fabian, A. C., & Shafer, R. A. 1986 *MNRAS*, 221, 27
Tennant, A. F. 1987 *MNRAS*, 226, 971
Whelan, J. A. J., *et al.* 1977 *MNRAS*, 181, 259

Discussion

K. Postnov What is the spin of the neutron star?

N. S. Schulz This is a very important point as for a young pulsar one should expect a spin period of a few tens of millisecond. To date no spin period has been found. However, we also point out that at these moderate fields there is quite a range of angles between the magnetic axis with repect to the rotation axis, where the detection of such a period is very difficult if not impossible. There are a few Be-star binaries where no period has been detected so far.

S. Chaty If a low magnetic field is consistent with young neutron stars, how low do you reconcile this low magnetic field with young neutron stars?

N. S. Schulz There is not much theory tells us at this point. From observations we know the lowest field to be a few times 10^{10} G. The suggested field strength for Cir X-1 is consistent with that.

High-mass X-ray Binaries: Illuminating the Passage from Massive Binaries to Merging Compact Objects
Proceedings IAU Symposium No. 346, 2019
L. M. Oskinova, E. Bozzo, T. Bulik & D. Gies, eds.

doi:10.1017/S1743921319001820

Firm detection of 7-year X-ray periodicity from X Persei

Motoki Nakajima[1], Hitoshi Negoro[2], Tatehiro Mihara[3], Mutsumi Sugizaki[3,4], Fumiaki Yatabe[3] and Kazuo Makishima[3]

[1]School of Dentistry at Matsudo, Nihon University,
2-870-1, Sakaecho-nishi, Matsudo, Chiba 271-8587, JAPAN
email: nakajima.motoki@nihon-u.ac.jp

[2]Department of Physics, Nihon University,
1-8 Kanda-Surugadai, Chiyoda-ku, Tokyo 101-8308, Japan

[3]MAXI team, Institute of Physical and Chemical Research (RIKEN),
2-1 Hirosawa, Wako, Saitama 351-0198, Japan

[4]Department of Physics, Tokyo Institute of Technology,
2-12-1 Ookayama, Meguro-ku, Tokyo 152-8551, Japan

Abstract. We report on the detection of long-term X-ray periodicity from the Be/X-ray binary pulsar X Persei. Based on over 23 years of X-ray data observed using RXTE/ASM, Swift/BAT and MAXI/GSC, we confirmed that X Persei exhibits quasi-periodic X-ray flares with a period of ∼7 years. The recurrence timescale corresponds to approximately 10 times its binary orbital period of 250 days. Spectral and hardness ratio changes were not detected along with long-term periodic activity. If we interpret the observed 7 year periodicity of X-ray band flux as a superorbital modulation, then this would be the first observation among the Be/X-ray binaries.

Keywords. pulsars:individual(X Persei), X-rays:stars

1. Long-term X-ray and optical variations

X Persei (X Per or 4U 0352+309) is a classical persistent Be/X-ray binary composed of a slowly rotating X-ray pulsar ($P_{spin} \sim 835$ second; Yatabe *et al.* 2018) and (09.5III-B0Ve) optical companion (HD 24534; Lyubimkov *et al.* 1997). This source was independently discovered by Ariel 5 and Copernicus (White *et al.* 1976). Delgado-Marti *et al.* (2001) reported that the binary system has an orbital period of ∼ 250 days, a projected semi-major axis of the neutron star of $a_x \sin i = 454$ lt-s, and a moderate eccentricity of $e = 0.11$.

Unlike general Be/X-ray binaries, X Per does not show a normal X-ray outburst at periastron due to mass accretion from the circumstellar disc of the Be star (e.g. Reig 2011); however, it has been observed that its X-ray flux has increased at a rate of approximately twice every 7-years since 2003. This "superorbital" modulation with a 7-year period is clearly evident in Figure 1(a). Interestingly, the time interval between the 2003 X-ray flare and the 1975 event also exhibited a 7-year repetition. Furthermore, it was confirmed that all of the decrease timescales in three X-ray flares have same time duration (∼250 days). In contrast to the X-ray flux modulation, the hardness ratios did not exhibit clear variations as shown in Figure 1(b).

Figure 1(c) and 1(d) show the long-term fluctuations of the optical brightness and Hα equivalent width (EW). As already reported in previous studies on Be/X-ray binaries (e.g. A0535+26; Yan *et al.* 2012), there is a correlation between the Hα EW and the X-ray flux. Based on this figure, we can confirm that a clear X-ray flux and Hα EW

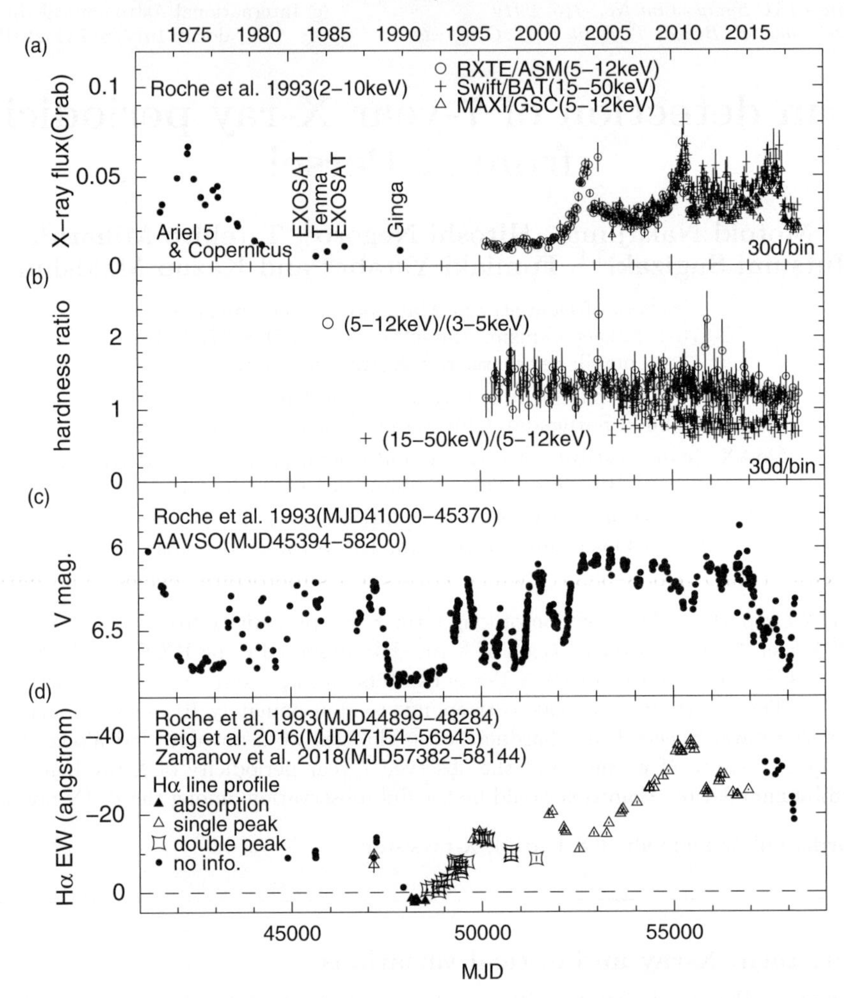

Figure 1. Long-term X-ray flux, hardness ratio, V-band magnitude and Hα equivalent width (EW) history.

correlation, and an anti-correlation between the optical brightness and X-ray flux was evident during the 2010 and 2017 X-ray flare event. As to the X-ray flux and Hα EW variation, Zamanov *et al.* (2018) reported that the correlation is probably due to wind Roche lobe overflow. However, an optical - X-ray correlation was not observed for the 2003 flare.

2. Period Search

In order to determine the superorbital periodicity, an epoch folding χ^2 search was performed on combined 5-12 keV RXTE/ASM and MAXI/GSC data. As a result, the superorbital modulation period was determined to be 2441 ± 83 d (6.7 ± 0.2 yr). This period corresponds to 10 times the binary orbital period. The 5-12 keV light curves were folded over 2441 d, with phase 0 set at the flux maximum at MJD 52800 as shown in the left panel of Figure 2.

3. X-ray spectral analysis

Since the pulse-phase resolved spectroscopy results and the orbital modulation of X Per have already been discussed in several papers (e.g. Lutovinov *et al.* 2012;

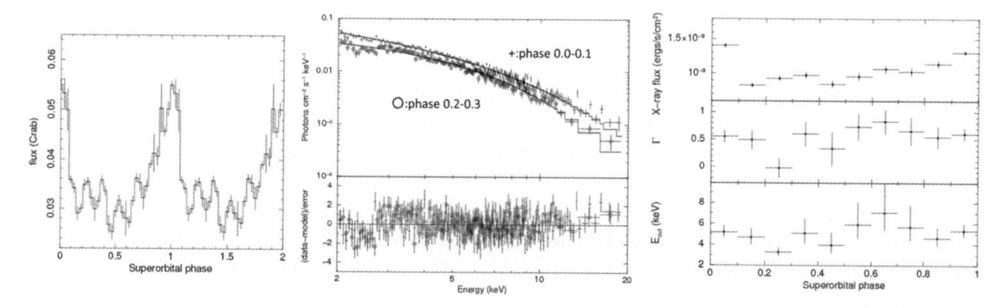

Figure 2. (left) The 5–12 keV light curve folded over the 2441 d superorbital period. (center) The unfolded X-ray spectra of X Per showing the different superorbital phase data. (right) Valiability of 2–20 keV flux, photon index of power-law with exponential cutoff, and cutoff energy over the superorbital period. The errors quoted in this panel are 90% confidence level.

Maitra *et al.* 2017; Yatabe *et al.* 2018), our analysis is focused on X-ray spectrum variability in superorbital modulation. We utilized the 2-20 keV data obtained by the MAXI/GSC (MJD 55058–58331) which covers 1.3 superorbital cycles. The data were divided into 10 phase resolved groups separated by 0.1 phase bins.

Firstly, we attempted to fit each data using a power-law model with an exponential cutoff. This model can be successfully applied to all the superorbital phase resolved data. The center panel of Figure 2 shows the representative spectra extracted from the 0.0-0.1 and 0.2-0.3 superorbital phase. The right panel of Figure 2 shows the dependence of the spectral parameters on the superorbital phase. Systematic parameter changes over the superorbital period were not observed.

In addition, we also examined the results extracted from other models (e.g. black-body plus power-law). However, we did not identify any phase dependent changes of the spectral parameters, such as black-body temperature or column density.

4. Conclusions

We investigated the superorbital variabilities with 6.7 years periodicity observed from X Per. This represents the first detection of superorbital modulation in X-rays among the Be/X-ray binaries. The modulation in the prototypical system (e.g. Her X-1; Kotze & Charles 2012) is caused by the precessing of the warped Be disc. Thus, a large variation of column density is expected. However, no such modulations could be confirmed based on our analyses. Therefore, we consider that the long-term periodicity seen in X Per is related to the variation of the mass accretion rate caused by an unknown mechanism.

In general, there are two accretion schemes in X-ray binaries; wind capture and disk accretion. In the case of wind-capture accretion in supergiant high-mass X-ray binaries, Corbet & Krimm (2013) discussed the relation between their modulation and the geometry of binary systems and determined that the superorbital period is proportional to the binary orbital period. In contrast, as claimed by Yatabe *et al.* (2018), X Per has a disk accretion scheme similar to that of the other Be/X-ray binaries. Thus, the variation of the mass-transfer rate from a circumstellar disc to a neutron star is the cause of the periodic superorbital modulation. Laplace *et al.* (2017) predicted that Kozai-Lidov oscillation might produce a periodic giant X-ray outburst. If this oscillation is present in X Per, then the oscillation would be ≥ 20 years, which is significantly longer than the observed modulation period. Consequently, another scenario is needed to explain the superorbital modulation in Be/X-ray binaries. Further discussions will be presented in a forthcoming paper.

References

Corbet, R. H., & Krimm, H. A. 2013, *ApJ*, 778, 45
Delgado-Marti, H., Levine, A. L., Pfahl, E., & Rappaport, S. A. 2001, *ApJ*, 546, 455
Kotze, M. M., & Charles, P. A. 2012, *MNRAS*, 420, 1575
Laplace, E., Mihara, T., Moritani, Y., Nakajima, M., Takagi, T., Makishima, K., & Santangelo, A. 2017, *A&A*, 597, 124
Lutovinov, A., Tsygankov, S., & Chernyakova, M. 2012, *MNRAS*, 423, 1978
Lyubimkov, L. S., Rostopchin, S. I., Roche, P., & Tarasov, A. E. 1997, *MNRAS*, 286, 549
Maitra, C., Raichur, H., Paradhan, P., & Paul, B. 2017, *MNRAS*, 470, 713
Reig, P. 2011, *Ap&SS*, 332, 1
Reig, P., Nersesian, A., Zezas, A., Gkouvelis, L., & Coe, M. J. 2016, *A&A*, 590, 122
Roche, P., Coe, M. J., Fabregat, J., McHardy, I. M., Norton, A. J., Percy, J. R., Reglero, V., Reynolds, A., & Unger, S. J. 1993, *A&A*, 270, 122
White, N. E., Mason, K. O., Sanford, P. W., Murdin, P. 1976, *MNRAS*, 176, 201
Yan, J., Li, H., & Liu, Q. 2012, *ApJ*, 744, 37
Yatabe, F., Makishima, K., Mihara, T., Nakajima, M., Sugizaki, M., Kitamoto, S., Yoshida, Y., & Takagi, T. 2018, *PASJ*, 99Y
Zamanov, R., Stoyanov, K. A., Wolter, U., Marchev, D. 2018, *arXiv:1811.09162*

High-mass X-ray Binaries: Illuminating the Passage from Massive Binaries to Merging Compact Objects
Proceedings IAU Symposium No. 346, 2019
L. M. Oskinova, E. Bozzo, T. Bulik & D. Gies, eds.

doi:10.1017/S1743921319001844

Superstrong magnetic fields of neutron stars in Be X-ray binaries

ChangSheng Shi[1,2], ShuangNan Zhang[3,5,6] and XiangDong Li[2,4]

[1]College of Material Science and Chemical Engineering, Hainan University, Hainan 570228, China
email: shics@hainu.edu.cn

[2]Key Laboratory of Modern Astronomy and Astrophysics (Nanjing University), Ministry of Education, Nanjing 210046, China

[3]Key Laboratory of Particle Astrophysics, Institute of High Energy Physics, Chinese Academy of Sciences, Beijing 100049, China

[4]Department of Astronomy, Nanjing University, Nanjing 210046, China

[5]National Astronomical Observatories, Chinese Academy of Sciences, Beijing 100012, China

[6]Physics Department, University of Alabama in Huntsville, Huntsville, AL 35899, USA

Abstract. A few Be X-ray binaries might constitute a group of special sources because the neutron stars in them may have superstrong magnetic fields. Generally, the neutron stars have long spin periods and some emission lines are shown from the B type star, which is attributed to an equatorial disc. We re-build new dimensionless torque models and obtain the superstrong magnetic fields of the neutron stars in the Be X-ray binaries in Large Magellanic Cloud, Small Magellanic Cloud and Milky Way when the compressed magnetosphere is considered. Although our conclusions are obtained when the disk accretion mode is considered, the results may be applied the Be X-ray binaries with wind accretion mode. SXP1323 and 4U 2206+54, in which the magnetic fields of the NSs may be close to the maximum 'virial' value, are the best objects to explore superstrong magnetic field.

Keywords. accretion, accretion disks, stars: neutron, X-rays: binaries.

---

1. Introduction

Be X-ray binaries (BeXBs) are the most numerous sub-class of the high mass X-ray binaries and have attracted many researchers due to the possible superstrong magnetic fields of the neutron stars (NS) in them (Shi, Zhang & Li 2015). There is a $8M_{\odot} - 18M_{\odot}$ B type star with emission lines and a rotating neutron star (NS) in every BeXB, where $M_{\odot}$ is the mass of the Sun. Generally, BeXBs are transient due to their eccentric orbits ($e > 0.3$) and the Be stars in these systems are often surrounded by a circumstellar disk, which may originate from the ejected photospheric matter with sufficient angular momentum and energy due to the fast rotating Be stars (Knigge *et al.* 2011; Reig 2011). When the neutron star passes through the disk, the matter in the disk can be accreted through the magnetosphere. Then the variable angular momentum produced from accretion or outflow leads to a change of the NSs spin period (P). Therefore the changed torque can be obtained from the parameters such as the derivative of the changing spin period ($\dot{P}$).

Klus *et al.* (2014) obtained two groups of magnetic field solutions for the NSs in 42 Be X-ray binaries in the Small Magellanic Cloud (SMC). Ho *et al.* (2014) argued that higher solutions of the magnetic fields are more correct. Shi, Zhang & Li (2015) recalculated the surface magnetic fields of these NSs and expanded this recalculation to the NSs of Be X-ray binaries (BeXBs) in the Large Magellanic Cloud (LMC) and Milky Way (MW).

2. New dimensionless torque

The accreting matter in the accretion disk rotates around the NS and it is funneled to the two poles of the NS with a dipolar magnetic field, which is aligned to the magnetic axis and perpendicular to the Keplerian accretion disk (Ghosh & Lamb 1979). The accretion disk interacts with the neutron star by the transportation of the accretion matter in the disk. Wang (1995) obtained different results for the relation between the azimuthal magnetic field and the toroidal magnetic field in the accretion disk when reconnection takes place outside the disk and the magnetosphere is force-free. Bozzo *et al.* (2018) reviewed the implications of the case with a non-orthogonal rotator in the magnetically threaded disk model and compared the magnetospheric radius from Ghosh & Lamb (1979) with that from Wang (1995). Generally, it is believed that the spin-up state for NSs in XBs can be differentiated from the spin-down state by the corotation radius ($r_{\rm co}$). The spin-up state originates from a positive torque that is imposed on a NS and the spin-down state from a negative one.

The total torque (N) acting on the NS system is made of the material torque from the accretion matter and the magnetic torque produced by the magnetic coupling between the NS and the accretion disk. It was considered that all the angular momentum of the accreting matter is transferred to the NS and the corresponding torque is N_0, i.e. $\dot{M}\sqrt{GMr_{\tt i}}$, where G is gravitational constant, M the mass of the NS, $\dot{M}$ the accretion rate, $r_{\tt i}$ the inner radius, which is often considered as the magnetospheric radius ($r_{\tt m}$). However, the NS and the dipolar magnetic field around the NS should be one system and the accretion disk is the other system. As a part of the accretion disk, the accreting plasma should keep the last angular momentum at the magnetospheric radius when it is controlled by the magnetosphere and corotates with the NS at the magnetospheric radius. Therefore, the torque acting on the NS system that corresponds to the lost angular moment of the plasma is $\dot{M}\sqrt{GMr_{\rm m}} - \dot{M}\Omega_{\rm s} r_{\rm m}^2$, where $\Omega_{\rm s}$ is the angular frequency of the spin. According to this hypothesis, we obtain the new dimensionless torque as follows,

$$n = \frac{N}{N_0} = \begin{cases} (1-\omega) + \frac{1}{3}\frac{\mu^2 r_{\rm m}^{-3}}{\dot{M}\sqrt{GMr_{\rm m}}}(\frac{2}{3} - 2\omega + \omega^2), & \omega \leqslant 1, \\ (1-\omega) + \frac{1}{3}\frac{\mu^2 r_{\rm m}^{-3}}{\dot{M}\sqrt{GMr_{\rm m}}}(\frac{2}{3}\omega^{-1} - 1), & \omega > 1. \end{cases} \tag{2.1}$$

where $\omega = (r_{\rm i}/r_{\rm co})^{3/2}$ is the fast parameter.

3. Superstrong magnetic field

As a rotating system, the NS is sped up due to the total interactional torque and thus we can obtain the relation $-\dot{P} = NP^2/2\pi I_{\rm eff}$, where $I_{\rm eff}$ is the effective moment of inertia of the NS. If equation (2.1) is substituted into the above equation and the magnetospheric radius is obtained, we can obtain an estimate for the surface magnetic field of the NS in BeXB. Finally, the surface magnetic field can be expressed as a relation, $B \propto PL^{\alpha}$, where α is a constant that reflects the amount of compression. As discussed by Shi, Zhang & Li (2014, 2015), the calculated surface magnetic field of a NS in BeXB for the compressed magnetosphere is much higher than the one for the uncompressed magnetosphere.

Besides the results for the magnetic field in the non-equilibrium state of a BeXB ($r_m < r_{\rm co}$), we can also estimate the magnetic field for the equilibrium state ($r_{\rm co} = r_m$). In fact, most BeXBs are close to spin equilibrium. The effect of a compression of the dipolar magnetic field for the spin equilibrium can be found in Figure 1. Three types of magnetospheric radii are considered in Figure 1, i.e. $r_{\rm m1}$ for the uncompressed magnetosphere, $r_{\rm m2}$ for the compressed magnetosphere in the plane of an accretion disk, $r_{\rm m3}$ for the compressed magnetosphere that comes from Kulkarni & Romanova (2013).

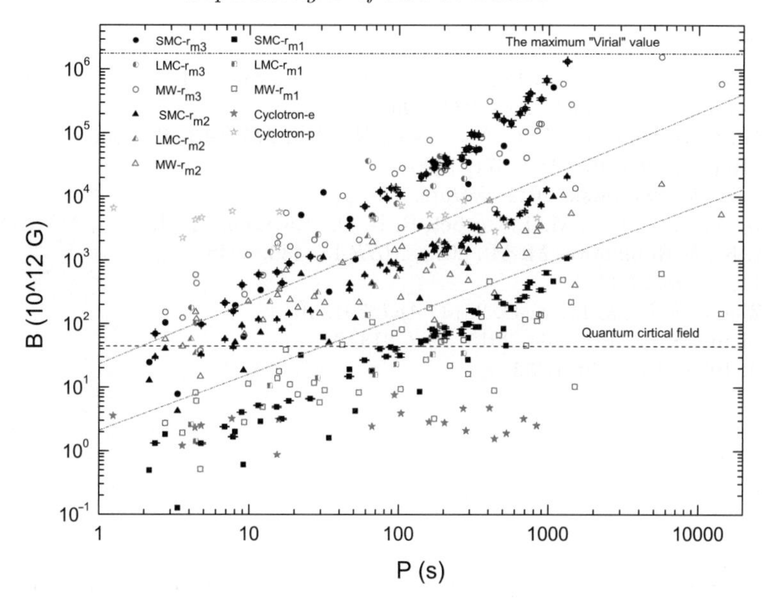

Figure 1. Relation between the spin period of NSs and the surface magnetic fields of NSs in BeXBs. The square points indicate for rm1, the triangle points for rm2, the circle points the solutions for rm3, the open points indicate sources for the MW, the half-filled points for LMC, and the solid points the solutions for the SMC as well. The stars show the magnetic field inferred from the observed cyclotron line sources (the solid points for electrons and the hollow for protons).

The whole compression from Kulkarni & Romanova (2013) leads to the calculated maximum magnetic field, which is close to the maximum 'virial' value, especially for the sources SXP1323 and 4U 2206+54. Most magnetic fields of NSs in BeXBs are shown to be higher than the quantum critical field (44.14 TG), which means that BeXBs are a kind of accreting magnetars. The surface magnetic field obtained from the observed cyclotron lines can be expressed as $B_{12} = 0.863 * E_{10\text{kev}}$ for the electrons in the ground state, but $B_{12} = 1585 * E_{10\text{kev}}$ for the protons, where $E_{10\text{kev}}$ expresses the energy of the cyclotron absorption line in units of 10 keV, B_{12} the surface magnetic field of NSs in BeXBs in units of 10^{12} G. The data for protons are in the range of the upper solutions discussed above. It seems as if the discussed cyclotron lines may be indeed produced by protons.

4. Discussion and conclusion

BeXBs with a high luminosity always include a long period NS, which may have a superstrong magnetic field. However, some ultraluminous X-ray sources (ULXs) whose distance is still unclear are found to be BeXBs. When the real surface magnetic field of the NS in a ULX is equally strong to the value estimated according to our calculations for the compressed magnetosphere, the derived distance from the Earth would be smaller and then the ULXs may not be found to be ultraluminous, i.e. some of these ULXs may be normal BeXBs. The relation between ULXs and BeXBs need to be explored further.

The magnetic field of a NS in some BeXB is found to be superstrong. However, supergiant X-ray binaries (SGXRs) might also be characterized by a similar accretion disk. Expanding our calculation to SGXRs, it might also be concluded that the magnetic fields of NSs in some SGXRs could be superstrong. A NS being formed from the terminal contraction of a dying massive star may keep the original magnetic flux and then end up obtaining a superstrong magnetic field. Many physical mechanisms in extreme conditions in BeXBs need to be explored.

References

Bozzo, E., Ascenzi, S., Ducci, L., Papitto, A., Burderi, L., Stella, L. 2018, *A&A*,617, A126
Ghosh, P., & Lamb, F. K. 1979, *ApJ*, 234, 296
Ho, W. C. G., Klus, H., Coe, M. J., Andersson, N. 2014, *MNRAS*, 437, 3664
Kluźniak & Rappaport 2007, *ApJ*, 671, 1990
Knigge, Coe, & Podsiadlowski 2011, *Nature*, 479, 372
Klus, H., Ho, W. C. G., Coe, M. J., Corbet, R. H. D., Townsend, L. J. 2014, *MNRAS*, 437, 3863
Kulkarni, A. K., & Romanova, M. M., 2013, *MNRAS*, 433, 3048
Reig, P. 2011, *Ap&SS*, 332, 1
Shi, C.-S., Zhang, S.-N., & Li, X.-D. 2014, *ApJ*, 791, 16
Shi, C.-S., Zhang, S.-N., Li, X.-D. 2015, *ApJ*, 813, 91
Wang, Y.-M. 1995, *ApJ*, 449, L153

High-mass X-ray Binaries: Illuminating the Passage from Massive Binaries to Merging Compact Objects
Proceedings IAU Symposium No. 346, 2019
L. M. Oskinova, E. Bozzo, T. Bulik & D. Gies, eds.

doi:10.1017/S174392131900111X

X-ray binary Beta Lyrae and its donor component structure

M. Yu. Skulskyy[1], M. V. Vavrukh[2] and S. V. Smerechynskyi[2]

[1]Department of Physics, Lviv Polytechnic National University,
Bandera str. 12, 79013, Lviv, Ukraine
email: mysky@polynet.lviv.ua

[2]Department of Astrophysics, Ivan Franko National University,
Kyryla & Methodiya str. 8, 79005 Lviv, Ukraine
emails: mvavrukh@gmail.com, svjatt@gmail.com

Abstract. We show that the structure of magnetized accretion gas flows between the components of the Beta Lyrae system can cause a scattering gas shell that masks completely these components in soft X-ray region. Also we have calculated the inner structure of the donor that is filling a Roche lobe and is preceding a forming of the degenerate dwarf. We show that mass of the degenerate core of the donor is in region $0.3 - 0.5 M_{\odot}$.

Keywords. X-rays: binaries, stars: Beta Lyrae, interiors

First of all, we propose an explanation of the steady state of X-rays of the massive interacting Beta Lyrae system. It was based mainly on our spectral observations that were conducted on the 2.6-m and 6-m telescopes and that led, in particular, to the first determination of the masses of both components ($2.9 M_{\odot}$ for a donor and $13 M_{\odot}$ for an accretor); to the discovery and research of the donor's magnetic field (that varies with the orbital phase within 1kGs); to the study of the dynamics and energetics of developed circumstellar structures of specific configuration, including the accretion disk of a complex structure (Skulskyy 2015). It was necessary to take into account the following factors: the presence of radio nebula surrounding of this system, and soft X-ray radiation, which does not change with the phase of the orbital period, and is associated with the Thomson scattering over the orbital plane. It is also based on an analysis of spectrophotometric studies of the Beta Lyrae system, including the progress in the latest modelings of the light curves from the far ultraviolet to the far infrared region (Mourard *et al.* 2018). The main problem is the harmonization of the results of these simulations with real light curves in the far ultraviolet beyond the Lyman limit, which does not resemble the light curves of the close binary systems (Skulskyy 2015). There is the significant contribution of the accretion disk radiation in the light curve of this system. In particular, two hot regions on the disk rim, that are located in the phases 0.80P and 0.40P with covering respectively 30% and 10% of the disk rim and with the temperatures that are 10% and 20% higher than the average on this disk, are detected (Mennickent *et al.* 2013). The hotter region of the rim disk at the phase 0.40P of the first quadrature is naturally explained by the Coriolis force deflection of the main gas flow that is directed from a donor through the inner Lagrangian point to the accretor's Roche lobe and with further collision of this flow with the accretion disk which is confirmed by own spectrophotometrical observations (Skulskyy 2015). However, this hydrodynamic picture can't explain a wide hot area of the disk, which is observed in the region of 0.80P phases of the second quadrature. This fact as well as a number of other observable facts should have a completely different explanation, connected with our consideration of Beta Lyrae system.

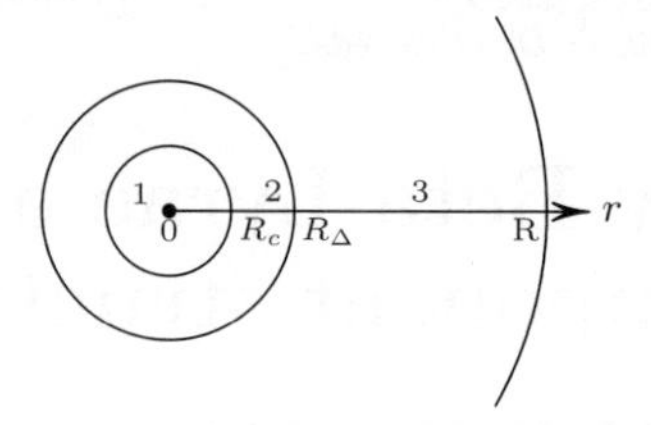

Figure 1. Model of the donor component in Beta Lyrae system.

We show that the structure of gas flows between the components of the Beta Lyrae system is above all due to the presence of the magnetic field of the donor that resembles a dipole with an axis directed along the orbital phases of 0.35 - 0.85P. In the phase range around 0.8-0.9P the magnetic field is maximal and its pole on the donor surface which reaches its Roche lobe is the closest to the accretion disk. This resumes the presence of more effective shock collisions of the magnetized plasma in the phases of the second quadrature 0.6-0.9P. The study of the curves of changes in the magnetic field and the curves of changes in the intensity and radial velocities of the spectral lines with the orbital phase, as well as the data of absolute spectrophotometry, confirm that the spatial structure of mass transfer in the Beta Lyrae system is due to the specific configuration of the donor magnetic field. The energy effect of collision between the disk and magnetized plasma, which is channeled by the donor's magnetic field towards a massive accretor at a speed of 200-700 km/s, is amplified in these phases due to the counter rotation of the disk edge with a speed of 250 km/s in front of the incident gas. This leads to the observation of the hot arc on the accretion disk edge facing to the donor, and this arc includes a wide hot region in the phases near 0.80P (Skulskyy 2015). As a result of high-energy collision of ionized plasma, channeled by donor's magnetic field with the accretion disk, a scattering gas shell is generated. This hot gas shell partially masks the components of this binary system outside the Lyman limit and completely in the soft X-ray region that is associated with Thomson scattering of X-rays in stellar wind and in jet-like structures.

Secondly, the massive binary system Beta Lyrae has donor component in the phase of active mass transfer and its core's structure is similar to that of hot white dwarf with helium degenerate core. Therefore simplified description of donor's structure can be obtained with generalization of white dwarf models (Vavrukh *et al.* 2018). We have used spherically symmetric three-phase model, depicted on Fig. 1, without taking into account rotation and orbital motion. Region 1 corresponds to degenerate core; region 2 is the transition layer, where the electron subsystem becomes non-degenerate; region 3 corresponds to surface layer, which fills Roche lobe.

We consider degenerate core with radius which consists of two regions 1 and 2, the former one is isothermal with temperature T_c. Also, we have modeled average molecular weight per electron in order to take into account radial dependence of chemical composition

$$\mu_e(r) = \mu_e\, t(r/R_\Delta), \quad 0 \leqslant r \leqslant R_\Delta,$$
$$t(r/R_\Delta) = \left\{1 + \alpha\,(r/R_\Delta)^2\right\}^{-1}, \tag{0.1}$$

where $\mu_e \equiv \mu_e(0) = 2.0$ and parameter α is close to one. In the transition layer ($R_c < r < R_\Delta$) we have assumed radial temperature distribution as follows

$$T(r) = T_c \left\{1 + \gamma \left(\frac{r - R_c}{R_\Delta}\right)^2\right\}^{-1}, \tag{0.2}$$

where γ is unknown parameter.

Mechanical equilibrium equation inside regions 1 and 2 can be rewritten as equation for local chemical potential of electrons $\mu(r) = m_0 c^2 \left\{ \left[1 + x^2(r)\right]^{1/2} - 1 \right\}$, where m_0 is electron mass, $x(r) = \hbar\,(m_0 c)^{-1} \left(3\pi^2 n(r)\right)^{1/3}$ - local relativistic parameter, $n(r)$ - electron number density on a sphere with radius r.

Mentioned above equation has the next form:

$$\frac{1}{r^2}\frac{d}{dr}\left(t^{-1}(r/R_\Delta)\, r^2 \frac{d\mu}{dr}\right) = -\frac{32\pi^2 G (m_u^2 \mu_e)^2 t(r/R_\Delta)}{3h^3} \int_0^\infty dp\, p^2 n_p(r),$$

$$n_p(r) = \{1 + \exp\left[\beta(r)(E_p - \mu(r))\right]\}^{-1}, \tag{0.3}$$

where $E_p = ((m_0 c)^2 + c^2 p^2)^{1/2} - m_0 c^2$, m_u - atomic mass unit (inside the core $T(r) \equiv T_c$).

Equation (0.3) satisfies initial conditions

$$\mu(0) = m_0 c^2 \left\{ \left(1 + x_0^2\right)^{1/2} - 1 \right\}, \quad \frac{d\mu}{dr} = 0 \text{ at } r = 0, \tag{0.4}$$

where $x_0 \equiv x(r)$ at $r = 0$. Core radius can be found from condition $\mu(R_c) = 0$ and outer radius of the transition layer R_Δ, in turn, from $\exp\{-\beta(R_\Delta)\ \mu(R_\Delta)\} \gg 1$ i.e. $\mu(R_\Delta) = -Ck_B T(R_\Delta)$, which means that degeneration is negligible (in our calculation we took $C = 2.7183$).

Equation of state in the surface layer ($R_\Delta \leqslant r \leqslant R$) we have taken in the polytropic form

$$P(r) = DT^{1+n_*}(r), \quad \rho(r) = DT^{n_*}(r)\,\frac{\mu_* m_u}{k_B} \tag{0.5}$$

with polytrope index $4.0 < n < 5.0$, where μ - average molecular weight in atomic mass units m_u.

In this region equilibrium equation in approximation (0.5) transforms to equation for temperature and is similar to Lane-Emden equation

$$(1 + n_*)\frac{1}{r^2}\frac{d}{dr}\left(r^2 \frac{dT}{dr}\right) = -4\pi G D \left(\frac{\mu_* m_u}{k_B}\right)^2 T^{n_*}(r). \tag{0.6}$$

Condition $T(r) = 0$ determines stellar radius and initial conditions at $r = R_\Delta$ can be found assuming equilibrium between matter in regions 2 and 3:

$$\begin{aligned} P_e(R_\Delta) &= DT^{1+n_*}(R_\Delta), \\ \left.\frac{dP_e}{dr}\right|_{r=R_\Delta-\delta} &= (1+n)DT^n(R_\Delta)\left.\frac{dT}{dr}\right|_{r=R_\Delta+\delta}, \quad \Delta \to +0, \end{aligned} \tag{0.7}$$

where $P_e(r)$ - pressure of electron gas at r.

Model we described here has seven parameters: core parameters ($x_0, T_c, \mu_e(0), \alpha$), transition layer parameters (α, γ) and ones of surface layer (μ_*, n). In our work we have assumed $\mu_e(0) = 2.0$, $\mu_* = 0.6$ (similar to solar chemical composition).

Grid of models was calculated for values of parameters $x_0, T_c, \alpha, \gamma, n$ ranging:

$$\begin{aligned} &4.1 \leqslant n \leqslant 4.9, && \Delta n = 0.1; \\ &0.1 \leqslant \alpha \leqslant 0.7, && \Delta\alpha = 0.3; \\ &0.1 \leqslant \gamma \leqslant 0.7, && \Delta\gamma = 0.3; \\ &0.3 \leqslant x_0 \leqslant 3.0, && \Delta x_0 = 0.1; \\ &5\cdot 10^6 K \leqslant T_c \leqslant 10^8 K, && \Delta T_c = 5\cdot 10^6 K. \end{aligned} \tag{0.8}$$

Table 1. Calculated characteristics of the donor component with polytrope index $n = 4.5$: model parameters α and γ; degenerate core parameters (x_0, T_c, R_c); total mass M and radius R of the donor; mass M_Δ and radius R_Δ of the core and transition layer.

α	γ	x_0	T_c, K	R_c/R_0	M/M_0	R/R_0	M_Δ/M_0	R_Δ/R_0
0.4	0.4	0.5	10^7	2.076	1.0091	1389.4	0.086	2.246
0.4	0.4	0.6	10^7	1.896	1.0254	1344.9	0.109	2.016
0.1	0.4	0.6	10^7	1.715	1.0138	1318.5	0.101	1.805
0.1	0.7	0.7	10^7	1.586	1.0467	1389.7	0.122	1.654

Basing on solutions of equations (0.3) and (0.6) we have chosen from the grid models with

$$0.95 \leqslant \frac{M}{M_0} \leqslant 1.05; \qquad 1300 \leqslant \frac{R}{R_0} \leqslant 1400, \tag{0.9}$$

which are in agreement with values of the donor in Beta Lyrae binary system obtained from observations. Here $M_0 \approx 2.887 M_\odot$, $R_0 \approx 1.116 \cdot 10^{-2} R_\odot$ are mass and radius scales, respectively (Vavrukh *et al.* 2018).

There were four models satisfying (0.9) with parameters given in Tab. 1.

As can be seen from our modeling results, within donor's mass and radius errors for mass of its core we have yielded values in range $0.25 - 0.35 M_\odot$. These ones are typical for hot low-mass field white dwarfs and agree with the assumption about white dwarfs as the endpoint of evolution of stars with masses below $8 M_\odot$. Our model doesn't account for two substantial factors - donor's axial rotation and magnetic field. The former one can increase of white dwarf mass up to 5% (Vavrukh *et al.* 2018) and the last one can cause partial spin polarization of the core electron subsystem resulting also in mass increase of the white dwarf (Vavrukh *et al.* 2018). In the case of full polarization mass increases in $\sqrt{2}$ times comparing with core's mass with paramagnetic electron subsystem (Vavrukh *et al.* 2018). Thus, the influence of axial rotation and magnetic field can lead to mass increase up to $0.5 M_\odot$. This is the upper limit for the core mass of the donor in Beta Lyrae interacting system.

References

Skulskyy, M. Yu. 2015, *Science and Education a New Dimension. Natural and Technical Sciences*, III(6), 54, 6

Mourard, D., Broz, M., Nemravova, J. A. *et al.* 2018, *Astron. and Astrophys.*, 618, A112

Mennickent, R. E., Djurasevic, G. 2013, *MNRAS*, 432, I(1), 799

Vavrukh, M. V., Skulskyy, M. Yu., Smerechynskyi, S. V. Models of massive degenerate dwarfs. Lviv: Rastr-7, 2018 (292 p.)

High-mass X-ray Binaries: Illuminating the Passage from Massive Binaries to Merging Compact Objects
Proceedings IAU Symposium No. 346, 2019
L. M. Oskinova, E. Bozzo, T. Bulik & D. Gies, eds.

doi:10.1017/S1743921319002217

Detection of the progenitors of Be X-ray Binaries

Douglas Gies[1], Luqian Wang[1] and Geraldine Peters[2]

[1]CHARA, Dept. of Physics & Astronomy, Georgia State University, P.O. Box 5060, Atlanta, GA 30302-5060, USA
emails: gies@chara.gsu.edu, lwang@chara.gsu.edu

[2]Space Sciences Center, University of Southern California, Los Angeles, CA 90089-1341, USA
email: gpeters@usc.edu

Abstract. A recent survey of the far-ultraviolet spectra of 264 B-emission line stars has revealed 16 systems with hot companions that are the stripped down remains of a former mass donor star. Some of these will probably become Be + neutron star X-ray binaries in the future. The actual numbers of such systems may be large, because the detected systems have companions that occupy the brief and bright, He-shell burning stage of evolution.

Keywords. binaries: spectroscopic, stars: emission-line Be, stars: evolution

1. Binary Origin of Be Stars

Be stars are rapidly rotating, B-type stars that eject equatorial disks, which are detected through their line and continuum emission (Rivinius *et al.* 2013). The origin of their rapid rotation may be related to mass and angular momentum accretion in an interacting binary (Pols *et al.* 1991). Depending on the original masses of the two component stars, Roche lobe overflow can strip the outer envelope of the mass donor star, while the gainer star increases in both mass and spin. The outcomes include Be + neutron star systems (observed as Be X-ray Binaries = BeXRBs; Reig 2011), Be + helium burning cores, and Be + white dwarf remnants. Some of the immediate progenitors of BeXRBs may consist of a Be star and He star remnant with a mass greater than the Chandrasekhar limit, and these remnants will explode as H-deficient supernovae.

It is important to search for such BeXRB progenitors among Be stars with stripped He star companions. However, it is difficult to detect such faint companions because they are lost in the glare of the much brighter Be stars. However, the hot companions do contribute relatively more flux at shorter wavelengths, and their line spectra are particularly rich in the far-ultraviolet, so the few detections thus far have resulted from analysis of their ultraviolet spectra (although some are detected through line emission from hot gas in the vicinity the He star: see the case of HD 55606 by Chojnowski *et al.* 2018). The detections from UV spectroscopy include ϕ Per (Mourard *et al.* 2015), FY CMa (Peters *et al.* 2008), 59 Cyg (Peters *et al.* 2013), 60 Cyg (Wang *et al.* 2017), and HR 2142 (Peters *et al.* 2016). All of these systems were known to be spectroscopic binaries in advance of their companion detection in the ultraviolet. However, successful detection of hot companions should be possible in favorable cases even if the system is not a known binary. This report summarizes a survey made by Wang *et al.* (2018) to find new cases of Be + He star binaries.

2. IUE Survey of Be stars

Wang *et al.* (2018) gathered all the available ultraviolet spectra from the archive of the International Ultraviolet Explorer Satellite for a sample of 264 Be stars from the catalog of Yudin (2001). These are Short Wavelength Prime camera observations made with the high dispersion grating (yielding a spectral resolving power of ≈ 10000). Each of the spectra were cross-correlated with a model spectrum appropriate for a He star with an effective temperature of $T_{\rm eff} = 45000$ K. All of the known He star companions have spectra with very sharp lines indicative of a small projected rotational velocity, so the presence of a hot companion is revealed by a sharp peak in the cross-correlation function (CCF). All of the CCFs were tested for the presence of a narrow peak that attained a CCF maximum greater than three times the standard deviation of the scatter in the extreme velocity parts of the CCF. The analysis led to the confirmation of detections for ϕ Per, FY CMa, 59 Cyg, and 60 Cyg, plus twelve new candidate Be + He star systems, effectively increasing the known sample by a factor of three.

Although none of the twelve candidate systems are known binaries, there was a sufficient number of archival spectra for eight of them to demonstrate the Doppler shifts and orbital motion of the He star companion. New spectroscopic observations are now underway at Apache Point Observatory and Cerro Tololo Interamerican Observatory to measure the small orbital motion of the Be star and to search for any optical band spectral features from the He star. The He star companions are relatively faint and generally contribute less than 5% of the combined monochromatic flux in the ultraviolet spectrum, but a new program of Hubble Space Telescope observations in Cycle 26 will provide the ultraviolet spectroscopy needed to characterize the orbital and physical parameters for these binaries.

3. Are there other stripped-down companions?

Mass estimates are only known for six systems with full orbital solutions, and of these only the He star companion of 60 Cygni has a mass greater than the Chandrasekhar limit ($M(\text{He star}) = 1.7 M_{\odot}$; Wang *et al.* 2017). Thus, the He star in 60 Cygni may be a future supernova candidate, and the binary may be a progenitor BeXRB system. All the Be + He systems discovered to date have relatively hot and massive Be primaries, and 12% of B0-B3 types in the survey have detected companions (Fig. 1). Schootemeijer *et al.* (2018) studied the evolutionary state of the He stars in ϕ Per and several other systems, and they argue that these stars are in a luminous and short-lived stage of He-shell burning that occurs after the longer duration He-core burning stage. If so, then there probably exists a much larger population of systems in this He-core burning stage with He stars that are too faint to detect by current means. For He stars like that in ϕ Per, Schootemeijer *et al.* (2018) estimate that the He-shell burning phase lasts only about 3% of the total lifetime, so that the total number of Be + He star systems should be about 30 times greater than the detected number. With a 12% detection rate among the early Be stars, this would suggest that all Be stars have stripped down companions.

Wang *et al.* (2018) found no Be + He star binaries among the cooler, low mass Be stars of types B4 - B9, even though their sample of targets was more than adequate to detect such systems if they occur with a fraction similar to that of the hotter Be stars (Fig. 1). The lack of Be + He star systems among the low mass Be stars was anticipated by Pols *et al.* (1991; see their Fig. 4b), who suggested that low mass systems probably host white dwarf companions with masses too small to ignite He burning. Such Be + white dwarf binaries may be detected in cases where the inclination permits mutual occultations of the components. One tell-tale case is KOI-81 (Matson *et al.* 2015), an eclipsing binary discovered in the NASA Kepler field-of-view. The primary of KOI-81 is a rapidly rotating

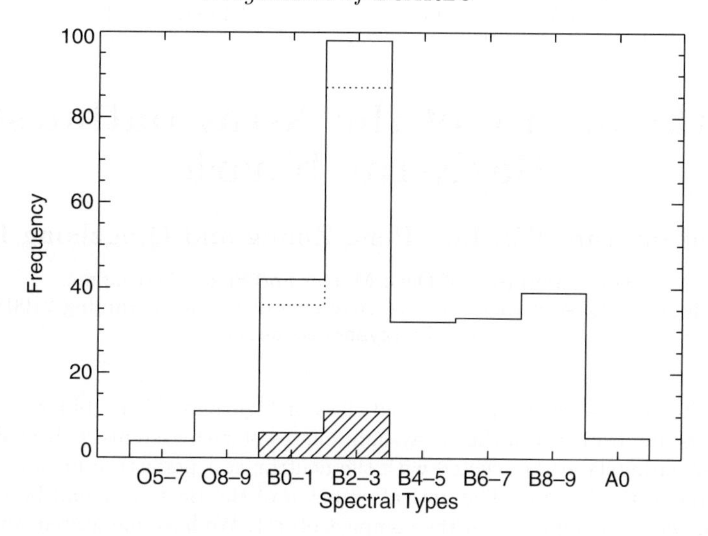

Figure 1. Histograms of the spectral type distributions of the full Be star sample (solid line), those with no detections (dotted line), and those that are known or candidate Be + He star systems (line filled) (from Wang *et al.* 2018).

B8 V star with a low mass $0.19M_{\odot}$, hot subdwarf in a 23.9 day orbit. The cooler Be stars may be the hosts of similar kinds of remnants that are destined to become white dwarfs.

Acknowledgments

This work was supported in part by NASA grant NNX10AD60G (GJP) and by the National Science Foundation under grant AST-1411654 (DRG).

References

Chojnowski, S.D., Labadie-Bartz, J., Rivinius, T., Gies, D., Panoglou, D., Borges Fernandes, M., Wisniewski, J. P., Whelan, D. G., Mennickent, R. E., McMillan, R., Dembicky, J. M., Gray, C., Rudyk, T., Stringfellow, G. S., Lester, K., Hasselquist, S., Zharikov, S., Levenhagen, R., Souza, T., Leister, N., Stassun, K., Siverd, R. J., & Majewski, S. R. 2018 *ApJ*, 865, 76

Matson, R. A., Gies, D. R., Guo, Z., Quinn, S. N., Buchhave, L. A., Latham, D. W., Howell, S. B., & Rowe, J. F. 2015, *ApJ*, 806, 155

Mourard, D., Monnier, J. D., Meilland, A., Gies, D., Millour, F., Benisty, M., Che, X., Grundstrom, E. D., Ligi, R., Schaefer, G., Baron, F., Kraus, S., Zhao, M., Pedretti, E., Berio, P., Clausse, J. M., Nardetto, N., Perraut, K., Spang, A., Stee, P., Tallon-Bosc, I., McAlister, H., ten Brummelaar, T., Ridgway, S. T., Sturmann, J., Sturmann, L., Turner, N., & Farrington, C. 2015, *A&A*, 577, A51

Peters, G. J., Gies, D. R., Grundstrom, E. D., & McSwain, M. V. 2008, *ApJ*, 686, 1280

Peters, G. J., Pewett, T. D., Gies, D. R., Touhami, Y. N., & Grundstrom, E. D. 2013, *ApJ*, 765, 2

Peters, G. J., Wang, L., Gies, D. R., & Grundstrom, E. D. 2016, *ApJ*, 828, 47

Pols, O. R., Cote, J., Waters, L. B. F. M., & Heise, J. 1991, *A&A*, 241, 419

Reig, P. 2011, *Ap&SS*, 332, 1

Rivinius, T., Carciofi, A. C., & Martayan, C. 2013, *A&AR*, 21, 69

Schootemeijer, A., Götberg, Y., de Mink, S. E., Gies, D., & Zapartas, E. 2018, *A&A*, 615, A30

Wang, L., Gies, D. R., & Peters, G. J. 2017, *ApJ*, 843, 60

Wang, L., Gies, D. R., & Peters, G. J. 2018, *ApJ*, 853, 156

Yudin, R. V. 2001, *A&A*, 368, 912

High-mass X-ray Binaries: Illuminating the Passage from Massive Binaries to Merging Compact Objects
Proceedings IAU Symposium No. 346, 2019
L. M. Oskinova, E. Bozzo, T. Bulik & D. Gies, eds.

doi:10.1017/S1743921319002357

On the nature of the X-ray outbursts in Be/X-ray binaries

Jingzhi Yan, Wei Liu, Peng Zhang and Qingzhong Liu

Key Laboratory of Dark Matter and Space Astronomy,
Purple Mountain Observatory, Chinese Academy of Sciences, Nanjing 210034, China
email: jzyan@pmo.ac.cn

Abstract. Be/X-ray binaries are a major subclass of high mass X-ray binaries. Two different X-ray outbursts are displayed in the X-ray light curves of such systems. It is generally believed that the X-ray outbursts are connected with the neutron star periastron passage of the circumstellar disk around the Be star. The optical emission of the Be star should be very important to understand the X-ray emission of the compact object. We have monitored several Be/X-ray binaries photometrically and spectroscopically in the optical band. The relationship between the optical emission and X-ray activity is described, which is very useful to explain the X-ray outbursts in Be/X-ray binaries.

Keywords. X-rays: binaries, stars: emission-line, Be, Be/X-ray binaries

---

1. Introduction

Be/X-ray binaries are composed of a Be star and a compact object, usually a nutron star (Reig 2011). The Be star is a non-supergiant B star whose spectrum has, or had at some time, one or more Balmer lines in emisison (Rivinius *et al.* 2013). The emission lines are attributed to a circumstellar disk around Be star. Neutron star moves around the Be star in an eccentric orbit. When neutron star approaches its periastron point, the material in the circumstellar disk will be accreted onto the surface of the neutron star and a kind of Type I or Type II X-ray outburst may result.

The X-ray activities of the Be/X-ray binaries should be connected with the physical states and evolution of the circumstellar disk. Only the inner part of the circumstellar disk has the contribution to the V-band continuum emission, while the Hα emission comes from the whole disk (Carciofi 2011). Therefore, we can do the optical photometric and spectroscopic observations at the same time to monitor the variability and evolution of the circumstellar disk around the Be star. Okazaki *et al.* (2013) and Reig & Blinov (2018) suggested that the Type II X-ray outbursts in Be/X-ray binaries might be connected with a warped disk around the Be star. Here we summarise our spectroscopic and photometric observational results on several Be/X-ray binaries.

2. Observations

We have been annually monitoring the visual spectra of a number of Be/X-ray binaries since the 1990s. Optical spectroscopic observations were carried out with the 2.16 m telescope at Xinglong Station of the National Astronomical Observatories of China. After 2012, we also carried out spectroscopic observations with Lijiang 2.4m telescope at Yunnan Astronomical Observatory. We do the photometric observations using the small telescopes at Xinglong Station, including 60cm and 80cm telescopes. Data reduction and analysis were introduced in Yan *et al.* (2012a).

After RXTE/ASM was ceased its science operations in January 2012, X-ray activities of the Be/X-ray binaries are monitored by Swift/BAT and MAXI. Combined with our optical data, we can study the relationship between the X-ray variability and its optical emission.

3. Results

We have analysed the optical observational data of five Be/X-ray binaries, including MXB 0656-072, A 0535+26, and BSD 24-491. We summarise the major results for each Be/X-ray binary.

MXB 0656-072 underwent a series of X-ray outbursts between 2007 November and 2008 November. A 101.2-day orbital period was reported for the first time for MXB 0656-072 (Yan *et al.* 2012a). Our optical observations indicate that the strength of the Hα line in our 2006 observations became stronger than that of 2005 and it had an extraordinary strength during our 2007 observations, which were taken just before the first X-ray outburst of 2007 November (see Fig. 8 in Yan *et al.* 2012a). In 2007, our simultaneous optical photometry and spectroscopy on MXB 0656-072 showed an interesting behaviour: while the Hα emission line strongly increased, the source brightness in UBV decreased by 0.2 mag in 2007 compared to the 2008-2009 observations.

We analysed the photometric and spectroscopic data of A 0535+26 from 1992 to 2010. Results indicated that each giant X-ray outburst of A 0535+26 usually occurred in a fading phase of the optical brightness. An anti-correlation between the optical brightness and the Hα intensity was observed during our 2009 observations (see Fig. 2 in Yan *et al.* 2012b): when the brightness of the system showed an obvious decline, the intensity of Hα line kept increasing. It reached an unprecedented maximum, with an EW of $\sim$ -25 Å, during our 2009 October observations, which were obtained just before the 2009 giant X-ray outbursts.

An X-ray outburst from RX J0440.9+4431 was observed between 2010 March 26 and 2010 April 15 by MAXI and the following two small X- ray flares were also detected by Swift/BAT. Several positive and negative correlations between the V-band brightness and the Hα intensity were found from the long-term photometric and spectroscopic observations (see Fig. 1 in Yan *et al.* 2016). When the optical brightness of the system began to decline, the intensity of Hα emission line was still in an increasing phase. The strongest Hα emission line during the last 20 years, with an EW of $\sim$ -12.9Å, was observed during our 2010 observations, which trigged three consecutive X-ray outbursts between 2010 March and 2011 February.

4. Conclusions

V-band photometry and Hα spectroscopy are useful tools to monitor the physical changes in the circumstellar disk. Rivinius *et al.* (2001) suggested that a low density region would be formed in the inner part of Be star disk after a strong mass ejection event, and the circumstellar envelope could change from a disk to rings. The low density region will move outward at a time scale of viscosity. With the movement of the ejected material through the low density region, gas may encounter the compact object. Therefore, the neutron star can be a probe to constrain the structure of the Be star disk.

We have been monitoring a sample of Be/X-ray binaries for a long time. Several anti-correlations between the V-band brightness and the Hα emission have been found in MXB 0656-072, A0535+26, and LS V+44 17. Such similar phenomena are also observed in other Be/X-ray binary systems. X-ray outbursts were usually triggered during a decline phase of the continuum emission. There should be a time lag between the mass ejection events and the X-ray outbursts. We can use this time lag to constrain the mass transportation in the circumstellar disk.

Acknowledgements

This work was partially supported by the National Natural Science of China under grants 11433009, 11573071, 11630367, and 11733009. This work was partially Supported by the Open Project Program of the Key Laboratory of Optical Astronomy, National Astronomical Observatories, Chinese Academy of Sciences. We also acknowledge the support of the staff of the Lijiang 2.4m telescope. Funding for the telescope has been provided by CAS and the People's Government of Yunnan Province.

References

Carciofi, A. C. 2011, *IAU Symposium*, 272, 325
Okazaki, A. T., Hayasaki, K., & Moritani, Y. 2013, *PASJ*, 65, 41
Reig, P. 2011, *Astrophysics and Space Science*, 332, 1
Reig, P., Blinov, D. 2018, *A&A*, 619, 19
Rivinius, T., Baade, D., Stefl, S., & Maintz, M. 2001, *A&A*, 379, 257
Rivinius, T., Carciofi, A. C., & Martayan, C. 2013, *A&ARv*, 21, 69
Yan, J. Z., Zurita Heras, J. A., Chaty, S., Li, H., & Liu, Q. Z. 2012, *ApJ*, 753, 73
Yan, J. Z., Li, H., & Liu, Q. Z. 2012, *ApJ*, 744, 37
Yan, J. Z., Zhang, P., Liu, W., & Liu, Q. Z. 2016, *AJ*, 151, 104

High-mass X-ray Binaries: Illuminating the Passage from Massive Binaries to Merging Compact Objects
Proceedings IAU Symposium No. 346, 2019
L. M. Oskinova, E. Bozzo, T. Bulik & D. Gies, eds.

doi:10.1017/S1743921319002345

Binarity of Pleione and its influence on the circumstellar disk

Lubomir Iliev

Institute of Astronomy and National Astronomical Observatory,
Bulgarian Academy of Sciences, 72 Tsarigradsko Shosse blvd., BG-1784, Sofia, Bulgaria
email: liliev@astro.bas.bg

Abstract. Pleione is a classical Be star well known for its cyclic transitions between Be-, shell- and normal B spectral phases. Its nature as a binary system was discussed by McAlister *et al.* (1989), Gies *et al.* (1990), Luthardt & Menchenkova (1994) and Nemravova *et al.* (2010). We present the results that trace the evolution of the dimensions of the circumstellar disk of Pleione that are related to the binary system.

Keywords. stars: binaries: spectroscopic, stars: emission-line: Be, stars: individual, Pleione

---

1. Introduction

Harmanec (1982) was the first to propose that Pleione's cyclic variations could be the result of interaction in a binary system with about 13000 days period. McAlister *et al.* (1989) found that Pleione is a visual binary system based on speckle interferometry observations. They reported about a component at 0.217 arcseconds. More detailed discussion of the Pleione system was provided by Gies *et al.* (1990). They supposed that shell episodes of the star are caused by tidal forces during periastron passages. Gies *et al.* (1990) estimated the total magnitudes and conclude that the secondary is a A5V star. They also calculated that the mass of the system is approximately 6 sol. masses. The semimajor axis was estimated to 19.1 AU which is in agreement with the results from speckle- interferometry. Luthardt & Menchenkova (1994) found an approximately 35 year period of radial velocity variations and assumed that they are due to orbital movement in wide binary system. Katahira *et al.* (1996) performed an analysis of photographical spectra of Pleione and determined period of 218 days. They also discussed that this period could reflect orbital movement in a system with secondary component evolved He star. Cases of a neutron star or white dwarf as secondary component were also discussed. Hirata (2007) interpreted his observations of polarization angle and of Hα profile as a result of the circumstellar disk precession caused by a secondary companion in a binary with a period of 218^d. Nemravova *et al.* (2010) analyzed a large set of CCD spectral observations of Pleione. They confirmed a 218^d period and, on the basis of an elliptical-orbit solution analyzed possibilities that the second component of this short-period system is an M-type dwarf or hot subdwarf.

2. Observations

All observations were carried out with the coudé-spectrograph of Rozhen National Observatory 2m RCC telescope. Bausch & Lomb grating with 632 lines/mm was used in combination with Photometrics AT200 (SITe S1003AB 1024x1024 24μm pixels) CCD camera. The spectrograph was set in configuration to provide a resolving power of about 34000 at wavelength of Hα line. Balmer lines of the Hydrogen and O I lines in the optical

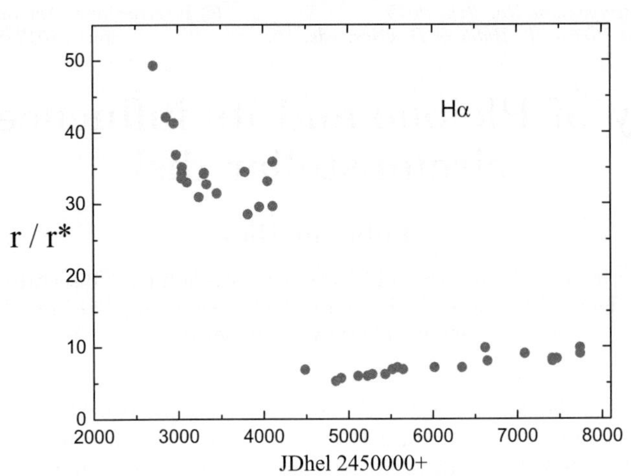

Figure 1. Variations of the dimension of Hα emitting region of Pleione during the 2003–2016 period. Estimations were made according to Huang (1972) conclusions. The complex shape of Hα profile makes it difficult to estimate emission peak velocities.

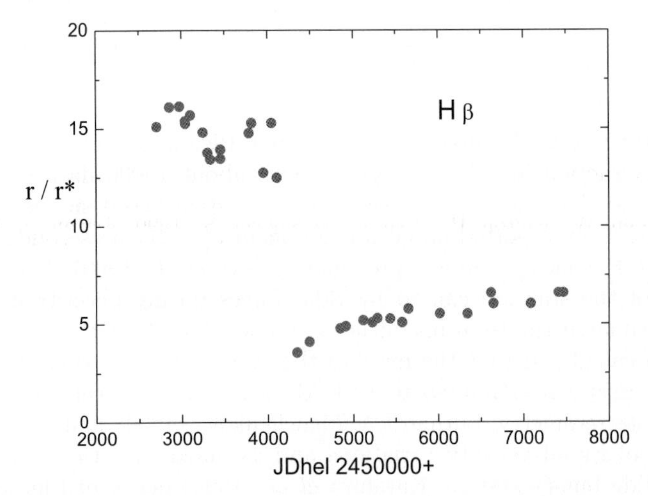

Figure 2. Variations of the dimension of Hβ emitting region around Pleione in the 2003–2016 period.

near IR region were chosen as the main targets of our observations as it was found that they are quite sensitive to the processes of Be star activity (Iliev 2015). The mean S/N ratio of the observations was usually better than 200 and in some cases exceeded 400.

3. Results and discussion

During spectral phase transitions, as well as during the development of the spectral phases, Pleione demonstrates changes that are typical for Be star spectral features. Fig. 1 and Fig. 2 show variations of the estimated dimensions of the emitting region in Hα and Hβ lines of Hydrogen. As can be seen from the figures, the start of the new "shell" phase is clearly distinguished by the corresponding abrupt change in the emitting circumstellar envelope. In general, the start of the new spectral phase follows the 34 years cycle with emission-shell phase, the transitions already observed for Pleione. This period of Pleione cyclic variations is suggested by Hirata (1995) and Katahira *et al.* (1996) as connected with the orbital movements in a wide binary.

The period of our observations cover the medium and concluding parts of the current shell-phase of Pleione. In that time span, the central absorption core of Balmer lines became deeper. Small V/R variations can be noted as well in the profiles of Hα and Hβ lines. At the same time, the shell lines of the metal ions also strengthened.

4. Conclusions

The variation in the dimension of the emitting regions in the circumstellar envelope of Pleione in general appeared to be synchronized with an orbital period of 34(35) years. This period is connected with a suspected wide component of the stellar system found with speckle-interferometric observations and radial velocity measurements. Such a synchronization could be the result of tidal forces from the wide component, that act to distort the disk around the Be primary. We are approaching the end of the current spectral phase of Pleione which provides a timely opportunity for more detailed studies of this fascinating star.

Acknowledgements

The research has made intensive use of the SIMBAD database, operated at CDS, Strasbourg, France.

The research used facilities and tools developed within the framework of the International Virtual Observatory Alliance (IVOA). The research has also made use of ESO-MIDAS image processing and data reduction software, created and maintained by European Southern Observatory.

References

Gies, D., McKibben, W., Kelton, P., Opal, C., & Sawyer, S., 1990, *Astron. J.*, 103, 1601

Harmanec, P., 1982, in *IAU Symp. 98*, 279

Hirata, R., 2007, in *ASP Conference Series*, 361, 267

Hirata, R., 1995, *Publ. Astron. Soc. Japan.*, 47, 195

Huang, S.-S., 1972, *Astrophys. J.*, 171, 549

Iliev. L., 2015, *Bul. Astron. J.*, 22, 37

Katahira, J.-I., Hirata, R., Ito, M., Katoh, M., Ballereau, D., & Chauville, J., 1996, *Publ. Astron. Soc..Japan*, 48, 317

Luthardt, R., & Menchenkova, E. V., 1994, *Astron. & Astrophys.*, 284, 118

McAlister, H., Hartkopf, W., Sowell, J., Dombrowski, E., & Franz, O., 1989, *Astron. J.*, 97, 510

Nemravová, J., Harmanec, P., Kubát, J., Koubský, P., Iliev, L., Yang, S., Ribeiro, J., Šlechta, M., Kotková, L., Wolf, M., & Škoda, P., 2010, *Astron. & Astrophys.*, 516, 80

High-mass X-ray Binaries: Illuminating the Passage from Massive Binaries to Merging Compact Objects
Proceedings IAU Symposium No. 346, 2019
L. M. Oskinova, E. Bozzo, T. Bulik & D. Gies, eds.

doi:10.1017/S1743921319001248

IGR J16318-4848: optical and near-infrared spectroscopy of the most absorbed B[e] supergiant X-ray binary with VLT/X-Shooter†

F. Fortin[1], S. Chaty[1], P. Goldoni[2] and A. Goldwurm[2]

[1]Laboratoire AIM (UMR 7158 CEA/DRF - CNRS - Université Paris Diderot), Irfu / Département dAstrophysique, CEA-Saclay, 91191 Gif-sur-Yvette Cedex, France

[2]APC, Université Paris Diderot, CNRS/IN2P3, CEA/Irfu, Observatoire de Paris, 10 rue Alice Domon et Léonie Duquet, 75205 Paris Cedex 13, France

Abstract. The supergiant high-mass X-ray binary IGR J16318-4848 was detected by *INTEGRAL* in 2003 and distinguishes itself by its high intrinsic absorption and B[e] phenomenon. It is the perfect candidate to study both binary interaction and the environment of supergiant B[e] stars. We report on VLT/X-Shooter observations from July 2012 in both optical and near-infrared, which provide unprecedented wide-range, well-resolved spectra of IGR J16318-4848 from 0.5 to 2.5 μm. Adding VLT/VISIR and Herschel data, the spectral energy distribution fitting allows us to further constrain the contribution of each emission region (central star, irradiated rim, dusty disc). We derive geometrical parameters using the numerous emitting and absorbing elements in each different sites in the binary. Various line shapes are detected, such as P-Cygni profiles and flat-topped lines, which are the signature of outflowing material. Preliminary results confirm the edge-on line of sight and the equatorial configuration of expanding material, along with the detection of a potentially very collimated polar outflow. These are evidence that the extreme environment of IGR J16318-4848 is ideal to have a better grasp of highly obscured high-mass X-ray binaries.

Keywords. infrared: stars - optical: stars, X-rays: binaries, X-rays: IGR J16318-4848, stars: binaries: general

1. Introduction

Since 2002, *INTEGRAL* (INTErnational Gamma-Ray Astrophysics Laboratory) has been observing the sky looking for gamma-ray sources of various nature. On top of significantly increasing the number of known X-ray binaries, *INTEGRAL* was able to discover a new type of highly obscured supergiant high-mass X-ray binaries (sgHMXB), as reviewed in Walter *et al.* (2015). These peculiar binaries host either a neutron star (NS) or a black hole (BH) in orbit around an early type supergiant star. Depending on the configuration of the binary, the compact object can be fed through the intense stellar wind of its giant companion, or by Roche Lobe overflow. The study of such extreme objects is crucial for understanding both the environment of supergiant stars and the products of binary interaction.

IGR J16318-4848 is the first source detected by *INTEGRAL*, and is the most absorbed sgHMXB known to this day. Discovered on January 29, 2003 (Courvoisier *et al.* 2003)

†Based on observations made with ESO Telescopes at the La Silla Paranal Observatory under program ID 089.D-0056(A).

with *INTEGRAL/IBIS* in the 15–40 keV band, the X-ray column density is so high ($N_H \simeq 2 \times 10^{24}\,cm^{-2}$, Matt & Guainazzi 2003, Walter *et al.* 2003) that it becomes nearly invisible below 2 keV. It is known to be a galactic persistent X-ray source with recurrent outbursts that last up to ~20 days.

Filliatre & Chaty (2004) use optical and nIR spectra to derive an absorption of A_V=17.4, which is far greater than the neighbouring value of 11.4, while still a hundred times lower than in X-rays. This leads the authors to suggest a concentration of X-ray absorbing material local to the compact object, and the presence of a shell around the whole binary absorbing optical/nIR wavelengths. The nIR spectrum in Filliatre & Chaty (2004) shows many prominent emission features in the same way CI Cam does, the first HMXB to be detected with an sgB[e] companion. P-Cygni profiles and forbidden [FeII] lines, also present in the nIR spectrum, are the evidence of a complex and rich environment, local to the binary.

Later, Kaplan *et al.* (2006) use photometry to show evidence of mid-infrared excess in IGR J16318-4848. The authors find that a ~1000 K blackbody can be fit to the mid-IR excess in the spectral energy distribution (SED) and they associate it to the presence of warm dust around the central star. Moon *et al.* (2007) provide *Spitzer* spectra from 5 to 40 μm that reveal a rich environment composed of an ionised stellar wind, a lower density region giving birth to forbidden lines, a photodissociated region and a two-component circumstellar dust (T >700 K and T~180 K). Rahoui *et al.* (2008) performed photometry on IGR J16318-4848 with VISIR and reach a similar conclusion, i.e. that warm circumstellar dust is responsible for the MIR excess. However, Ibarra *et al.* (2007) suggests that the column density is inhomogeneous and that the circumstellar matter could very well be concentrated in the equatorial plane, seen almost edge-on, hence the very high N_H. The outflow might then be bimodal, with a fast polar wind and a slow, dense equatorial outflow. Chaty & Rahoui (2012) use VLT/VISIR mid-IR along with NTT/SofI and Spitzer spectra to fit the SED of IGR J16318-4848. They report the presence of an irradiated rim around the star at T_{rim} = 3500–5500 K and a warm dust component at T_{dust} = 767 K in the outer regions of the binary using models of Herbig AeBe stars, as the authors suggest that IGR J16318-4848 has circumstellar material analogous to this class of objects.

Jain *et al.* (2009) suggest a possible 80 d period based on *Swift*-BAT and *INTEGRAL* data. Recently, Iyer & Paul (2017) provide the results of a long-term observation campaign on IGR J16318-4848 with *Swift/BAT*. An orbital period of 80.09± 0.01 d is derived.

2. Observations

The observations of target IGR J16318-4848 and standard star HD145412 were performed in July 2012 (P. I. S. Chaty) at the European Southern Observatory (ESO, Chile) under program ID 089.D-0056(A). Spectra from 300 to 2480 nm were acquired on the 8-meter Very Large Telescope Unit 2 Cassegrain (VLT, UT2) on three different arms (UVB, VIS and NIR) of the X-Shooter instrument. Because of the high intrinsic absorption of the source, the UBV spectra yielded no signal and we only present VIS and NIR data.

Optical echelle spectra were obtained through a 0.7"×11" slit giving a spectral resolution of R = 10640 (28 km s^{-1}) over a spectral range of 533–1020 nm. Four exposures of 300 s were taken, for a total integration time of 1200 s. Near-infrared echelle spectra were obtained through a 0.6"×11" slit giving a spectral resolution of R = 8040 (37 km s^{-1}) over a spectral range of 994–2480 nm. Twenty exposures of 10 s were taken, for a total integration time of 200 s. All the acquisitions followed the standard ESO nodding pattern. The data reduction was performed with ESOReflex, using the dedicated X-Shooter pipeline. It consists of an automated echelle spectrum extraction along with standard bias, dark and sky subtraction along with airmass correction. Median stacking was used to add individual spectra in order to correct for cosmic rays. Telluric absorption features were corrected using Molecfit (Kausch *et al.* 2015, Smette *et al.* 2015), a software that fits atmospheric features using a radiation transfer code and various parameters from the local weather.

3. Spectral features analysis

3.1. *P-Cygni lines*

We extracted, normalized and combined each hydrogen lines of the spectrum into an average profile, which is characteristic of expanding material (P-Cygni, see Fig. 1). The difference in velocity between the emitting and absorbing regions is $265 \pm 4\,\mathrm{km\,s^{-1}}$. The shape of the emission line is asymetric and we reckon that a careful modeling of the profile may provide further information about the emitting medium.

3.2. *Flat-topped lines*

[FeII] lines show a peculiar flat-topped profile (Fig. 2), which are supposed to arise from a low-density medium that undergoes a spherical expansion. The width of the line yields the terminal velocity of the corresponding outflow. We averaged every [FeII] line into a single profile, and fit it with a convolution of a Gaussian and a rectangle function. The half-width of the rectangle provides a terminal velocity of $262 \pm 3\,\mathrm{km\,s^{-1}}$. The Gaussian half-width of $53 \pm 2\,\mathrm{km\,s^{-1}}$ may correspond to the orbital motion of the medium.

3.3. *Narrow lines*

The Hα line and its surroundings shows abnormally narrow emission lines (Fig 3) compared to the rest of the spectrum. In Fig. 3, the dashed line is the average HI P-Cygni profile, and it highlights the presence of another narrow component on top. Other similar lines from [OI], [NII] and [SII] are also present in this region of the spectrum. The widths of those narrow lines are up to 15 times smaller than the rest of the lines. Such lines are known to arise from the polar wind of T-Tauri stars as suggested in Edwards *et al.* (1987). We note that these lines are not resolved in our spectrum, so that we do not detect any peculiar profile aside from a single Gaussian.

4. Spectral energy distribution

The fit of the SED reveals that no stellar component is necessary to reproduce the data (Fig. 4). We thus suggest that the P-Cygni HI and HeI lines are emitted from the irradiated rim and absorbed further away in the disc in outflowing material at $265\,\mathrm{km\,s^{-1}}$. Then their large width (HWHM=$170\,\mathrm{km\,s^{-1}}$) can be explained by the orbital velocity of the rim.

The narrow lines of Hα, [NII], [OI] and [SII] all point towards the presence of a fast polar wind. The fact that we neither detect significant broadening nor detect evidence of a double peaked profile suggests that the polar wind, and thus the whole system, is seen almost edge-on ($i = 90 \pm 2^\circ$ for a fiducial wind velocity of $1\,000\,\mathrm{km\,s^{-1}}$). Moreover, their slight blueshift of $-32 \pm 2\,\mathrm{km\,s^{-1}}$ should be indicative of the intrinsic velocity of the system. We found in Russeil (2003) that a star forming region (SFR) with the same velocity is present in the line of sight, located at 2.4 ± 0.3 kpc. We suggest IGR J16318-4848 can be associated to this SFR and is thus located at 2.4 ± 0.3 kpc.

While the expansion velocity derived from the flat-topped lines is consistent with the one of P-Cygni lines, they are likely coming from different regions: the iron lines may come from a spherical medium much further away than the rim where the hydrogen and helium lines form.

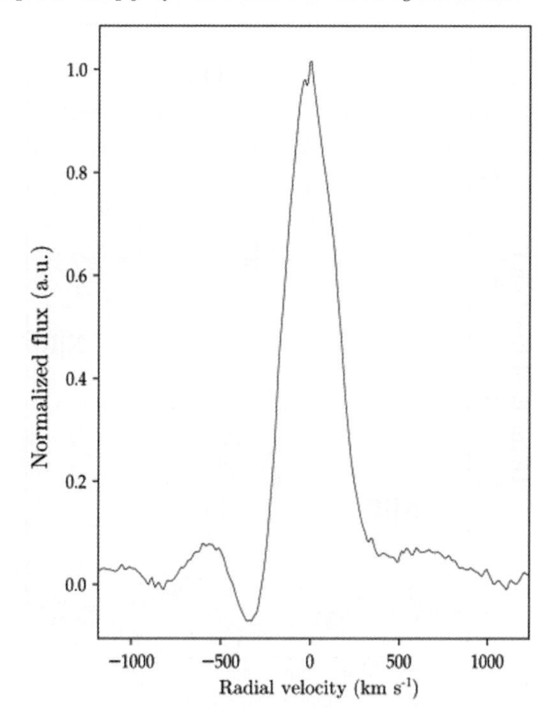

Figure 1. Average HI P-Cygni profile.

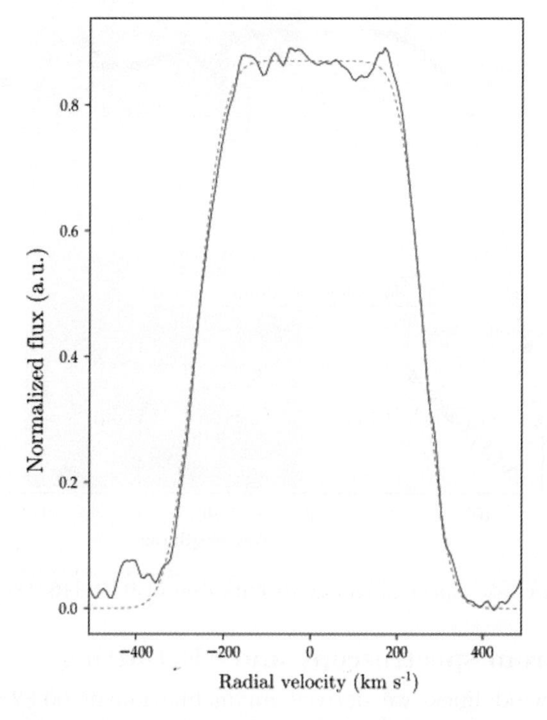

Figure 2. Average [FeII] flat-topped profile (*black*), fitted model (*red*).

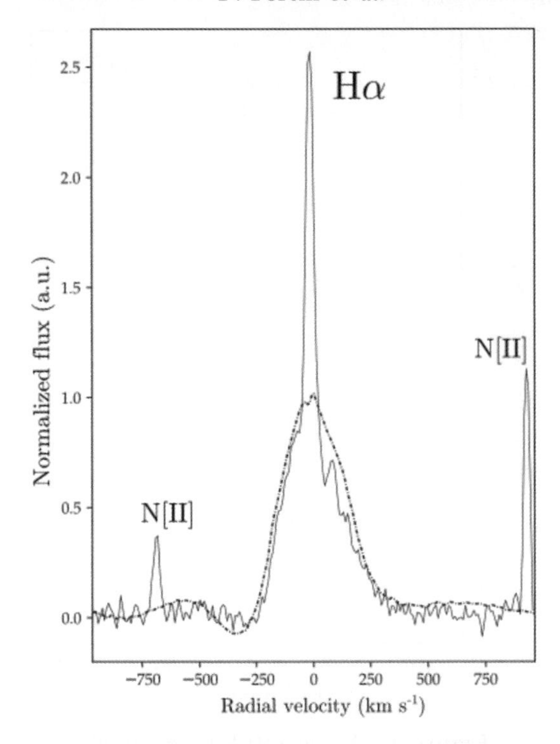

Figure 3. Hα region of the spectrum (*plain*), average P-Cygni profile (*dotted*).

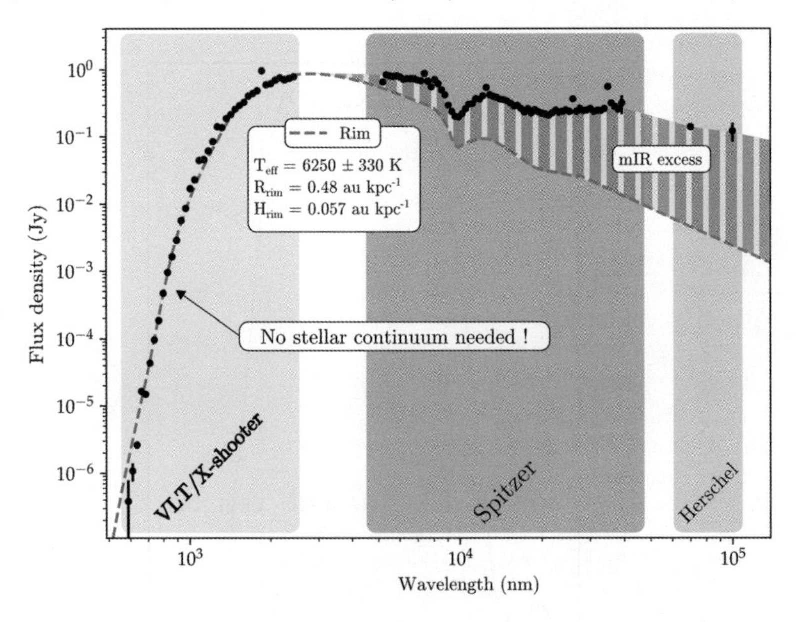

Figure 4. Spectral energy distribution of IGR J16318-4848.

5. Deductions from spectroscopy and SED fitting

From the polar wind lines, we derived an inclination of 90±2° and estimated the distance to IGR J16319-4848 to be 2.4±0.3 kpc. Using the orbital period from Iyer & Paul (2017), we fit the absolute irradiated rim scales: its radius is 1.15±0.2 au, and its half-height is 0.14±0.03 au. Since we neither detect any stellar feature nor need stellar

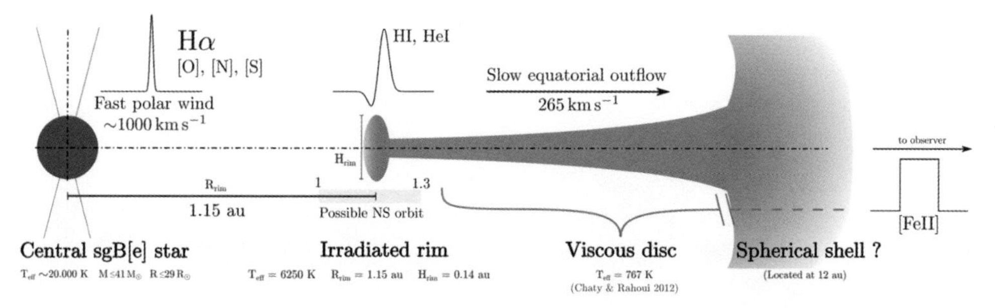

Figure 5. Schematics of IGR J16318-4848's geometry derived in this study (edge-on view).

continuum to fit the SED, we suggest that the rim might be obscuring the star from us. Given its size, this means the star cannot be bigger than $R_{sgBe} \leqslant 29\,R_{\odot}$. If we also assume that the rim is in keplerian orbit, the width of the emission lines in HI P-Cygni profiles provides us an orbital motion of $170\,\mathrm{km\,s^{-1}}$ at 1.15 au, thus meaning that the mass of the central star should be $M_{sgBe} \leqslant 41\,M_{\odot}$. Finally, if we associate a keplerian velocity to the gaussian broadening of the [FeII] flat-topped iron lines, we find that the corresponding medium is located at $R=12^{+2}_{-4}$ au from the central star. All this information is compiled under a graphical representation of IGR J16318-4848 in Fig. 5.

6. Conclusion

We acquired a broadband medium resolution spectrum of IGR J16318-4848 with VLT/X-Shooter, and it revealed many lines with peculiar profiles. We detected an equatorial outflow of $265\,\mathrm{km\,s^{-1}}$ from hydrogen and helium lines, which is likely correlated to the velocity measured on [FeII] flat-topped lines arising in the outer regions of the binary. We detected narrow lines from a polar wind, with no evidence of deviation from a perfect edge-on configuration. The flat-topped lines may indicate that on top of having a flat disc, IGR J16318-4848 might have an extra outer region undergoing spherical expansion. We reckon there is still much to discover about both the inner and outer parts of the binary, in particular the properties of the polar wind and the external region giving rise to flat-topped lines.

References

Chaty, S. & Rahoui, F. 2012, *ApJ*, 751, 150
Courvoisier, T. J.-L., Walter, R., Rodriguez *et al.* *A&A*. 2003, IAU Circular, 8063, 3
Edwards, S., Cabrit, S., Strom, S. E., *et al.* 1987, *ApJ*, 321, 473
Filliatre, P. & Chaty, S. 2004, *ApJ*, 616, 469
Ibarra, A., Matt, G., Guainazzi, M., *et al.* 2007, *A&A*, 465, 501
Iyer, N. & Paul, B. 2017, *MNRAS*, 471, 355
Jain, C., Paul, B., & Dutta, A. 2009, Research in *A&A* (RAA), 9, 1303
Kaplan, D. L., Moon, D.-S., & Reach, W. T. 2006, *ApJ*, 649, L107
Kausch, W., Smette, S. N. A., Kimeswenger, S., *et al.* 2015, *A&A*, 576, A78
Matt, G. & Guainazzi, M. 2003, *MNRAS*, 341, L13
Moon, D.-S., Kaplan, D. L., Reach, W. T., *et al.* 2007, *ApJ*, 671, L53
Rahoui, F., Chaty, S., Lagage, P.-O., & Pantin, E. 2008, *A&A*, 484, 801
Russeil, D. 2003, *A&A*, 397, 133
Smette, A., Sana, H., Noll, S., *et al.* 2015, *A&A*, 576, A77
Walter, R., Lutovinov, A. A., Bozzo, E., & Tsygankov, S. S. 2015, *A&A* Review, 23, 2
Walter, R., Rodriguez, J., Foschini, L., *et al.* 2003, *A&A*, 411, L427

Supergiant HMXBs

High-mass X-ray Binaries: Illuminating the Passage from Massive Binaries to Merging Compact Objects
Proceedings IAU Symposium No. 346, 2019
L. M. Oskinova, E. Bozzo, T. Bulik & D. Gies, eds.

doi:10.1017/S1743921319001273

The dark side of supergiant High-Mass X-ray Binaries

Sylvain Chaty[1], **Francis Fortin**[1], **Federico García**[1] **and Federico Fogantini**[2]

[1]AIM, CEA, CNRS, Université Paris-Saclay, Université Paris Diderot, Sorbonne Paris Cité, F-91191 Gif-sur-Yvette, France
email: chaty@cea.fr

[2]Instituto Argentino de Radioastronomía (CCT-La Plata, CONICET, CICPBA), C.C. No. 5, 1894 Villa Elisa, and Facultad de Ciencias Astronómicas y Geofisicas, Universidad Nacional de La Plata, Paseo del Bosque s/n, 1900 La Plata, Argentina

Abstract. High Mass X-ray Binaries (HMXB) have been revealed by a wealth of multi-wavelength observations, from X-ray to optical and infrared domain. After describing the 3 different kinds of HMXB, we focus on 3 HMXB hosting supergiant stars: IGR J16320-4751, IGR J16465-4507 and IGR J16318-4848, respectively called "The Good", "The Bad" and "The Ugly". We review in these proceedings what the observations of these sources have brought to light concerning our knowledge of HMXB, and what part still remains in the dark side. Many questions are still pending, related to accretion processes, stellar wind properties in these massive and active stars, and the overall evolution due to transfer of mass and angular momentum between the companion star and the compact object. Future observations should be able to answer these questions, which constitute the dark side of HMXB.

Keywords. (stars:) binaries (including multiple): close, circumstellar matter, early-type, emission-line, Be, stars: neutron, supergiants, winds, outflows, (ISM:) dust, extinction, infrared: stars, X-rays: binaries: IGR J16320-4751, IGR J16465-4507 and IGR J16318-4848

1. Introduction

Intensive programs, including imaging, photometry, low and high resolution spectroscopy, stellar spectra modeling, spectral energy distribution (SED) fitting, timing and interferometry, have shown that properties of HMXB are mainly dictated by the nature of their massive host stars. Imaging and photometry allow us to identify various types of HMXB; low and high resolution spectroscopy, combined to stellar spectra modeling, lead us to derive accurate parameters of the companion star (interstellar absorption, metallicity, rotation, gravity, etc); SED fitting gives us information on intrinsic absorption and characteristics of circumstellar enveloppe; mid-infrared imaging allows us to explore the impact of these active stars on their environment; timing brings us orbital and spin periods; and finally interferometry opens the way to directly imaging the dust cocoon surrounding HMXB. The *INTEGRAL* satellite has triggered the revival of HMXB studies, extending the population of supergiant HMXB –from only 5 in 1986 to 35 today–, revealing previously unknown highly obscured and transient HMXB (so-called supergiant Fast X-ray Transients, SFXT). The first detections of gravitational waves has confirmed the interest of studying compact binaries hosting massive stars, the obscured HMXB being the precursors of common enveloppe systems.

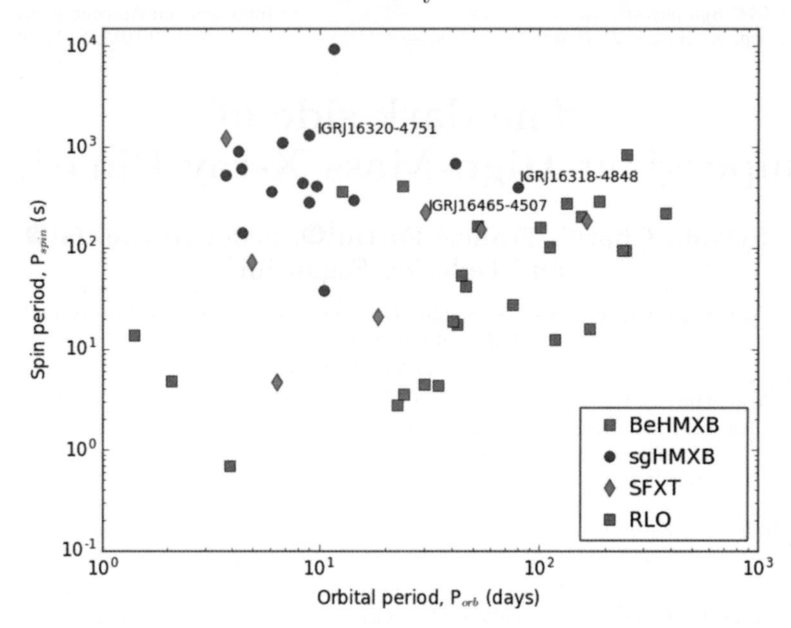

Figure 1. Corbet diagram showing the different populations of HMXB. We indicated the positions of IGR J16320-4751 and IGR J16465-4507. Concerning IGR J16318-4848, since we only know the orbital period (80d), we indicate the position of the source, assuming an average spin period for sgHMXB.

2. Introduction on HMXB systems

"High-Mass X-ray Binaries" (HMXB) are composed of a compact object (neutron star –NS– or black hole –BH–) orbiting a luminous and massive early OB spectral type companion star ($\geqslant 10\ M_\odot$). We can distinguish 3 different types of HMXB, according to the process of accretion, that we order here by decreasing number of systems of each class: (i) Be X-ray binaries (BeHMXB), (ii) supergiant X-ray binaries (sgHMXB), and (iii) Roche Lobe Overflow systems (RLO) (see more extensive discussion in Tauris *et al.* 2017).

(i) Be X-ray binaries (BeHMXB) BeHMXB host a main sequence donor star of spectral type B0-B2e III/IV/V, a rapid rotator surrounded by a circumstellar (so-called "decretion") disc of gas, as seen by the presence of a prominent Hα emission line. This disc is created by a low velocity/high density stellar wind of $\sim 10^{-7}\ M_\odot\ yr^{-1}$. Transient and bright ("Type I") X-ray outbursts periodically occur, each time the compact object (usually a NS on a wide and eccentric orbit) approaches periastron and accretes matter from the decretion disc (see Charles & Coe 2006; Tauris & van den Heuvel 2006 and references therein). These systems exhibit a correlation between the spin and orbital period, as shown by their location from the lower left to the upper right of the Corbet diagram (initial diagram shown in Corbet 1986, updated diagram in Fig. 1), due to efficient transfer of angular momentum at each periastron passage: rapidly spinning NS correspond to short orbital period systems, and slowly spinning NS to long orbit systems. Apart from MWC 656 hosting a BH (Casares *et al.* 2014), most BeHMXB seem to host a NS.

(ii) Supergiant X-ray binaries (sgHMXB) sgHMXB host a supergiant star of spectral type O8-B1 I/II, characterized by an intense, slow and dense, radiatively steady and highly supersonic stellar wind, radially outflowing from the equator. There are ~ 16

so-called "classical" sgHMXB, most of them being close systems, with a compact object orbiting on a short and circular orbit, directly accreting from the stellar wind, through e.g. Bondy-Hoyle-Littleton process. Such wind-fed systems exhibit a luminous and persistent X-ray emission ($L_X = 10^{36-38}$ erg s^{-1}), with superimposed large variations on short timescales, and a cut-off (10-30 keV) power-law X-ray spectrum. Located in the upper left part of the Corbet diagram, with small orbital period $P_{\rm orb} \sim 3-10$ days and long spin periods $P_{\rm spin} \sim 100-10000$ s (see Fig. 1), they do not show any correlation, due to absence of net transfer of angular momentum. Nearly half of sgHMXB (~ 8) exhibit a substantial intrinsic and local extinction $N_{\rm H} \geqslant 10^{23}$ cm^{-2}, with a compact object deeply embedded in the dense stellar wind (such as the highly obscured IGR J16318-4848; Chaty & Rahoui 2012). Likely in transition to Roche Lobe Overflow, these systems are characterized by slow winds causing a deep spiral-in of the compact object, and leading to Common Envelope Phase. Detection of long pulsations imply that they host young NS with $B \sim 10^{11-12}$ G. There exists also the possibility, for sgHMXB such as Cyg X-1, to accrete both through Roche lobe overflow and stellar wind accretion.

A significant subclass of sgHMXB is constituted of 17 (+5 candidate) Supergiant Fast X-ray Transients (SFXT, see Negueruela *et al.* 2006). These systems, characterized by a compact object orbiting with $P_{orb} \sim 3.3-100$ days on a circular or excentric orbit, and by $100-1000$ s spin periods, span on a vast location in the Corbet diagram, mostly inbetween BeHMXB and sgHMXB (see Fig. 1). They exhibit short and intense X-ray outbursts, an unusual characteristic among HMXB, rising in tens of minutes up to a peak luminosity $L_X \sim 10^{35-37}$ erg s^{-1}, lasting for a few hours, and alternating with long (~ 70 days) quiescence at $L_X \sim 10^{32-34}$ erg s^{-1}, with an impressive variability factor $\frac{L_{max}}{L_{min}}$ going up to $10^2 - 10^5$. Various processes have been invoked to explain these flares, such as wind inhomogeneities, magneto/centrifugal accretion barrier, transitory accretion disc, etc (see the reviews Chaty 2013 and Walter 2015, and references therein).

(iii) (Beginning Atmospheric) Roche Lobe Overflow systems (RLO) RLO host a massive star filling its Roche lobe, where accreted matter flows via inner Lagrangian point to form an accretion disc (similarly to LMXB). These systems are also called beginning atmospheric Roche lobe overflow. They constitute the classical bright HMXB (such as Cen X-3, SMC X-1 and LMC X-4), with accretion of matter occuring through the formation of an accretion disc, leading to a high X-ray luminosity ($L_X \sim 10^{38}$ erg s^{-1}) during outbursts. There are only a few sources, located in the lower left of the Corbet diagram (Fig. 1), characterized by short orbital and spin periods.

A number of 114 HMXB are reported in Liu *et al.* (2006), and 117 in Bird *et al.* (2016). By cross-correlating both catalogues, we find that they share 79 sources in common. Among these common sources, 6 sources identified as HMXB in Liu *et al.* (2006) are now assigned to other types by Bird *et al.* (2016). We therefore find that the total number of HMXB currently known in our Galaxy amounts to 152 (Fortin *et al.* 2018a), see also this volume (Fortin *et al.* 2018b). HMXB thus represent $\sim 40\%$ of the total number of high energy binary systems (i.e. adding all known LMXB and HMXB reported in Liu *et al.* 2007; Bird *et al.* 2016). Among HMXB, there are 63 BeHMXB (51 firmly identified and 12 candidates), 33 sgHMXB (30 firmly identified and 3 candidates) and 56 HMXB of unidentified nature. sgHMXB can be further divised in 16 "classical" sources and 17 SFXT. Thus, HMXB can be divided respectively in 41% of BeHMXB, 22% of sgHMXB and 37% of unidentified HMXB.

Contrary to LMXB located towards the Galactic center, HMXB are concentrated in the Galactic plane, towards tangential directions of Galactic arms, rich in star forming regions and stellar formation complexes (SFC, Coleiro & Chaty 2013 and references therein). The distribution of HMXB in our Galaxy is thus a good indicator of starburst activity, and their collective X-ray luminosity has been used to compute the star-formation rate of the host galaxy (Grimm *et al.* 2003). Coleiro & Chaty (2013) have correlated the position of HMXB (including BeHMXB and sgHMXB) with the position of SFC, and showed that HMXB are clustered within 0.3 kpc of the closest SFC, with an inter-cluster distance of 1.7 kpc, thus showing that HMXB remain close to their birthplace. Coleiro & Chaty (2013) have also shown that the HMXB distribution was offset by $\sim 10^7$ years with respect to the spiral arms, corresponding to the delay between star birth and HMXB formation. By taking into account the galactic arm rotation, they managed to derive parameters such as age, migration distance and kick for some BeHMXB and sgHMXB.

3. "The Good": IGR J16320-4751

IGR J16320-4751 is a highly obscured sgHMXB hosting an O8 I star, with column density $N_{\rm H} = 10^{23}$ cm^{-2}. We named it "The Good" source, because it is located well inside the sgHMXB domain in the Corbet diagram (Fig. 1), with $P_{spin} = 1300$ s and $P_{orb} = 9$ days. We studied this source aiming at constraining both its geometrical and physical properties, following the evolution of the NS orbiting the supergiant star. We give here a summary of the study described in detail in (García *et al.* 2018a), and also in this volume (García *et al.* 2018b).

To study this source, we retrieved i.) the *Swift*/BAT folded hard X-ray lightcurve from 2004 to 2017, and ii.) the *XMM*-Newton/PN lightcurves (soft 0.5-6.0 keV and hard 6.0-12.0 keV bands), from 11 observations spanning from 2003 to 2008. The source exhibits a high variability and flaring activity on several timescales. For instance, in one of the XMM observations, we found two flares with an increase of a factor ~ 10 in only 300 s. Using the *Swift*/BAT lightcurve, we refined the orbital period to 8.99 ± 0.01 days.

The *XMM*-Newton PN spectra of the source exhibit a highly absorbed continuum, with an Fe absorption edge at ~ 7 keV, along with Fe Kα, Fe Kβ and Fe XXV lines. We performed a spectral fitting, with thermally comptonized COMPTT model, adding three narrow Gaussian functions with 2 TBABS absorption components. We then extracted the line characteristics, the total continuum flux and the intrinsic $N_{\rm H}$.

We first found a clear correlation between Fe Kα line and the total continuum flux, showing that the fluorescence emission emanates from a small region close to the accreting pulsar. We then found another correlation between the Fe Kα line flux and the intrinsic $N_{\rm H}$, suggesting that the fluorescent matter is linked to the absorbing matter. Both results taken together show that absorbing matter is located within a small dense region surrounding the NS, likely due to accreted stellar wind.

We show in Fig. 2 the orbital evolution of the intrinsic absorption $N_{\rm H}$ derived from fitting *XMM*-Newton spectra (top left), and the orbital evolution of the folded hard X-ray *Swift*/BAT lightcurve (bottom left). In order to fit $N_{\rm H}$ and *Swift*/BAT lightcurves, we built a toy model of a NS orbiting a supergiant star, and accreting from its stellar wind (Castor *et al.* 1975 model, red lines): $M_\star = 25\ M_\odot$, $R_\star = 20\ R_\odot$, $\beta = 0.85$, $v_{inf} = 1300$ km s^{-1}, $\frac{dM}{dt} = 3 \times 10^{-6}\ M_\odot$ yr^{-1}. The best fit gives the solution $a = 0.25$ au $= 2.7\ R_\odot$, $e = 0.20$, $i = 62°$ and $A = +146°$ (Fig. 2, central and right panels).

Thus, we find that the model reproduces well the orbital evolution of a NS surrounded by absorbing matter, and modulated by the stellar wind density profile, as viewed by the observer. In particular, it reproduces: i.) the orbital modulation of $N_{\rm H}$ with the sharp peak at phase = 0.47 (Fig. 2, top left panel); ii.) the smooth evolution of the hard X-ray lightcurve (Fig. 2, bottom left panel); and iii.) the phase shift of their maxima, with the maximum $N_{\rm H}$ at phase = 0.47 as seen by the observer, and the maximum *Swift*/BAT counts at phase = 0.53 at the time when the NS crosses the periastron.

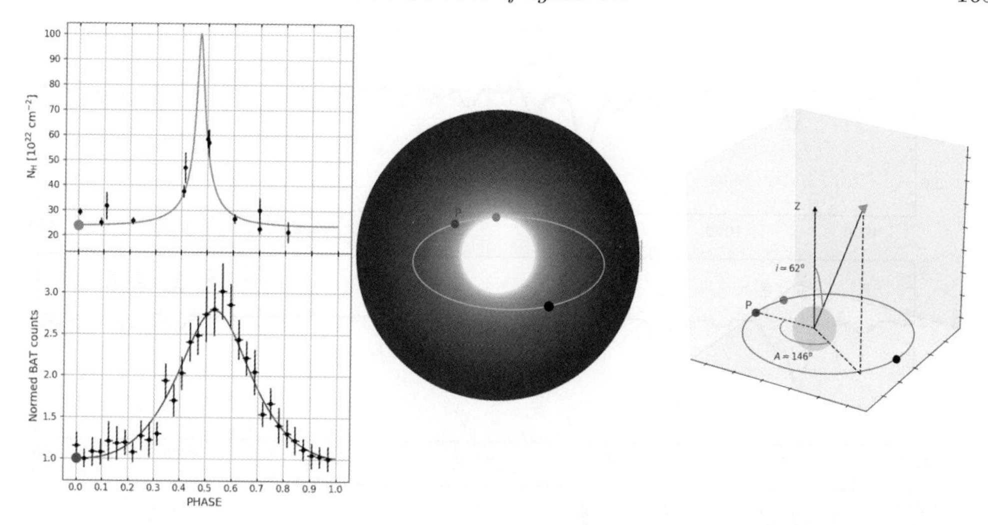

Figure 2. Top left: In black, data points showing the variation of $N_{\rm H}$ along orbital phase, in red: model fit. Bottom left: in black, data points showing the variation of normalised *Swift*/BAT counts along orbital phase, in blue: model fit. Middle: supergiant star surrounded by NS as seen by observer, Periastron point (P) indicated, the red point is the furthest point on the line-of-sight. Right: our simple model, with NS (in black) orbiting the supergiant star (in green), periastron (P) indicated, the red point is the furthest point on the line-of-sight, the pink arrow indicates the direction of the observer. i is the inclination angle, and A the azimuth.

4. "The Bad": IGR J16465-4507

IGR J16465-4507 is an obscured sgHMXB hosting a B0.5-1 Ib star, with $N_{\rm H} = 10^{22}$ cm^{-2}, that we call "The Bad" source, because it is located inbetween the sgHMXB domain and the BeHMXB domain in the Corbet diagram (Fig. 1), with $P_{spin} = 228$ s and $P_{orb} = 30$ days.

We analysed all optical and infrared (OIR) photometric and spectroscopic observations obtained at ESO with EMMI, SUSI2, SOFI, FORS1 and X-shooter instruments, from 2006 to 2012. This analysis, reported in detail in Chaty *et al.* (2016), first allowed us to accurately establish the spectral type of the companion star to B0.5-1 Ib. Observations with X-shooter allowed us to get large-band spectra (UVB, VIS and NIR, reported in Fig. 3), to build a whole photometric and spectroscopic spectral energy distribution (SED), and to detect an IR excess, likely due to circumstellar cold material.

We then fitted absorption and emission lines using the stellar spectral model FASTWIND -Fast Analysis of STellar atmospheres with WINDs–, a spherical, non-LTE model atmosphere code with mass loss and line-blanketing (Santolaya-Rey *et al.* 1997; Puls *et al.* 2005; Castro *et al.* 2012). The result is reported in Fig. 3. We point out that it is one of the few sgHMXB for which the spectral type of the companion star has been accurately fitted with realistic stellar atmospheric spectral model.

From this fit, we derive that the supergiant star is characterized by a high rotation velocity: $v \times sin(i) = 320 \pm 8$ km s^{-1}, implying that this star must have a small radius (smaller than 15 $R_\odot$), in order not to break up. We determine that $T_{eff} = 26000$ K, with an envelope expanding at $v \sim 170$ km s^{-1}.

Looking at the Corbet diagram (Fig. 1), we see that IGR J16465-4507 is located inbetween sgHMXB and BeHMXB domains, like two other supergiant fast X-ray transients (SFXT): IGR J18483-0311 ($P_{spin} = 21$ s and $P_{orb} = 18.5$ days) and IGR J11215-5952 ($P_{spin} = 187$ s and $P_{orb} = 165$ days), see Liu *et al.* (2011) and references therein. The question that arises is then: which process does make these SFXT look like BeHMXB?

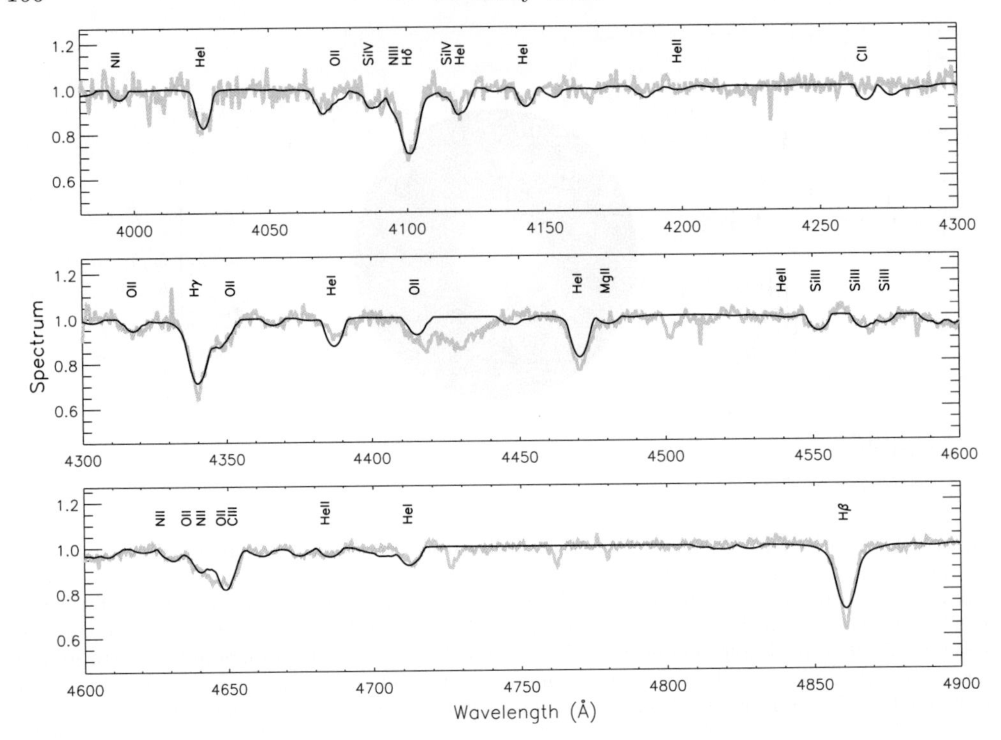

Figure 3. X-shooter spectrum of IGR J16465-4507 fitted with FASTWIND stellar spectral model. The spectrum is shown in grey, and the fit obtained is reported in black.

To answer to this question, there are 2 possibilities:

i.) The high rotation velocity of these SFXT originated from the rapid rotator nature of O-type emission line stars (main sequence Oe-X-ray binary). In this case, these systems exhibit 2 periods of accretion during their evolution, first during the main sequence, and then during the supergiant phase (the NS would spin at its equilibrium period for a sgHMXB hosting a B1 Ia star, with $B = 3 \times 10^{12}$ G, Liu *et al.* 2011).

ii.) Most SFXT tend to have an eccentric orbit, which would reduce the equilibrium period of the NS at the settling accretion stage, by a factor between 10 and 100 (Postnov *et al.* 2018a), and explain the position of some sgHMXB in the BeHMXB domain. This is also visible in the eccentricity vs orbital period plot of BeHMXB, sgHMXB and SFXT reported in the extensive review of HMXB seen by INTEGRAL (Sidoli & Paizis 2018).

5. "The Ugly": IGR J16318-4848

IGR J16318-4848 is a highly obscured sgHMXB hosting a sgB[e] star, with $N_{\rm H} = 6 \times 10^{24}$ cm^{-2}, that we call "The Ugly" source, because with an average spin period typical of supergiant systems, it would be located inside the BeHMXB domain in the Corbet diagram (Fig. 1), with $P_{orb} = 80$ days. Its companion star is a luminous sgB[e] star with a stratified circumstellar enveloppe, characterized by the following parameters: $L = 10^6\ L_\odot$, $M = 30\ M_\odot$, $T = 22000$ K, $R = 20\ R_\odot = 0.1$ au.

We studied this source thanks to a full set of photometric and spectroscopic observations, obtained at ESO NTT/SOFI, VLT/VISIR, *Spitzer*, and *Herschel*. It exhibits strong emission lines, a MIR excess, an ionised wind at a velocity of 400 km s^{-1}, shocked [FeII] lines, some dense regions of circumstellar matter ($> 10^{5-6}$ cm^{-3}), and other lense dense regions, with NaI, Ne, Si, characteristic of a stratified envelope (Filliatre & Chaty 2004; Chaty & Rahoui 2012).

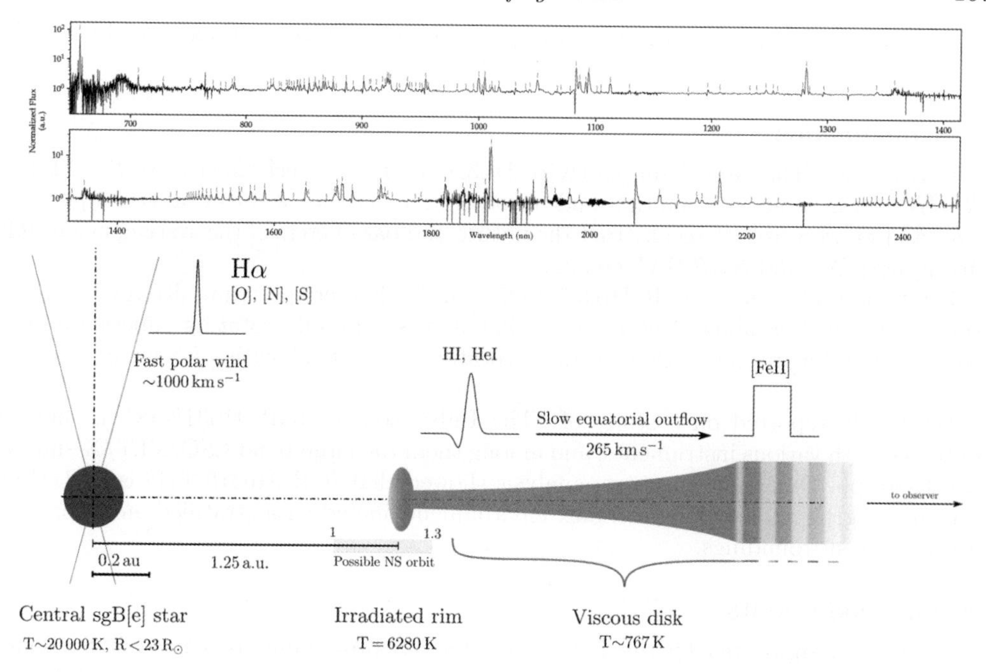

Figure 4. Top: ESO VLT/X-shooter UVB-VIS-NIR spectrum of IGR J16318-4848. Bottom: Artist view of the scenario explaining the rare properties of IGR J16318-4848.

We obtained in 2012 an ESO VLT/X-shooter UVB-VIS-NIR spectroscopy, shown in Fig. 4 (top). Preliminary results of the study of this spectrum are described in this volume (Fortin *et al.* 2018c). This spectrum shows prominent HI Balmer, Paschen, Bracket, and Pfund series, with a mean P-Cygni profile at 264 km s^{-1}. The Balmer Hα line exhibits some evidence for a double component, which, when subtracting a mean P-Cygni profile obtained by stacking all HI lines, shows a narrow Balmer Hα coming from a diluted and fast polar wind, suggesting that we see the binary system as edge-on. He lines also display P-Cygni profile at 164 km s^{-1}, with an extra broad emission at 840 km s^{-1}. Both H and He lines seem to come from the medium inbetween the supergiant star and the internal part of circumstellar material. The spectrum also shows forbidden [NII], [OI] and [SII] lines, likely coming from external part of the circumstellar material, characterized by a low density. In addition, we detect flat-topped profile of [Fe] lines, at a velocity of 262 km s^{-1}. These flat-topped forbidden lines are characteristic of lines formed inside a spherical shell of outflow and/or expanding material of low density: they are probably coming from a slow equatorial outflow, relatively far away from the star (see an artistic view of the scenario in Fig. 4 bottom).

The scenario that we built to explain the rare properties of this system is based on Herbig Ae/Be model, with a torus geometry, an irradiated rim and a viscous disk extending up to 12 $R_\star$ (Chaty & Rahoui 2012). By taking the orbital period of $P_{orb} = 80$ days (Iyer & Paul 2017), we can derive that the compact object orbits within the dense disk rim.

The binary system is currently transiting to Roche-lobe overflow, with a deep spiral-in, entering the Common Envelope Phase. Since the nature of the compact object is unknown, we have to consider two possibilities: i.) In case it is a NS, this short P_{orb} HMXB will not survive the Common Envelope Phase, and becomes an ideal Thorne-Zytkow candidate (Tauris *et al.* 2017). ii.) In case the compact object is a BH with a mass ratio $q < 3.5$, the HMXB should survive the spiral-in phase, and then become a

Wolf-Rayet X-ray binary with an orbital period of a few days, eventually leading to the merging of both compact objects (van den Heuvel *et al.* 2017).

6. Conclusions

Concerning "The Good" source (IGR J16320-4751), we performed a *Swift*-BAT and *XMM*-Newton study of accretion and absorption along the NS orbiting the supergiant star, and we were able to reconstruct the nature and parameters of the system, accurately fitting both $N_{\rm H}$ and *Swift*-BAT counts.

For "The Bad" source (IGR J16465-4507): our X-shooter spectrum, fitted on a stellar spectrum modeling allowed us to show that it was a rapidly rotating supergiant star, either descending from Oe (Be-like) rapid rotator, or an SFXT with a NS on an eccentric orbit.

We finally reported on a study of "The Ugly" source (IGR J16318-4848), that we observed with various instruments, and among them the large-band ESO/VLT/X-shooter spectrograph. Our multi-spectrum analysis showed that IGR J16318-4848 is an HMXB hosting a sgB[e] star, with a complex environment, including a stratified envelope, and a rim+disk surroundings.

Acknowledgments

This work, supported by the Labex UnivEarthS programme of Universit Sorbonne Paris Cité (Interface project I10 –From binaries to gravitational waves–, and by the Centre National d'Etudes Spatiales (CNES), was based on observations obtained with MINE –Multi-wavelength *INTEGRAL* NEtwork–.

References

Bird, A. J., Bazzano, A., Malizia, A., *et al.* 2016, *ApJS*, 223, 15
Casares, J., Negueruela, I., Ribó, M., *et al.* 2014, *Nature*, 505, 378
Castor, J. I., Abbott, D. C., & Klein, R. I. 1975, *ApJ*, 195, 157
Castro, N., Urbaneja, M. A., Herrero, A., *et al.* 2012, *A&A*, 542, A79
Charles, P. A. & Coe, M. J. 2006, *Compact stellar X-ray sources.*, ed. Lewin, W. H. G. & van der Klis, M., Vol. 39 (Cambridge Astrophysics Series, Cambridge University Press), 215–265
Chaty, S. 2013, *Advances in Space Research*, 52, 2132
Chaty, S., LeReun, A., Negueruela, I., *et al.* 2016, *A&A*, 591, A87
Chaty, S. & Rahoui, F. 2012, *ApJ*, 751, 150
Coleiro, A. & Chaty, S. 2013, *ApJ*, 764, 185
Corbet, R. H. D. 1986, *MNRAS*, 220, 1047
Filliatre, P. & Chaty, S. 2004, *ApJ*, 616, 469
Fortin, F., Chaty, S., Coleiro, A., Tomsick, J., & Nitschelm, C. 2018a, in IAU Symposium 346: *High-mass X-ray binaries: illuminating the passage from massive binaries to merging compact objects*, ed. L. M. Oskinova (IAU)
Fortin, F., Chaty, S., Coleiro, A., Tomsick, J. A., & Nitschelm, C. H. R. 2018b, A&A, 618, A150
Fortin, F., Chaty, S., Goldoni, P., & Goldwurm, A. 2018c, *in IAU Symposium 346: High-mass X-ray binaries: illuminating the passage from massive binaries to merging compact objects*, ed. L. M. Oskinova (IAU)
García, F., Fogantini, F. A., Chaty, S., & Combi, J. A. 2018a, *A&A*, 618, A61
García, F., Fogantini, F. A., Chaty, S., & Combi, J. A. 2018b, *in IAU Symposium 346: High-mass X-ray binaries: illuminating the passage from massive binaries to merging compact objects*, ed. L. M. Oskinova (IAU)
Grimm, H., Gilfanov, M., & Sunyaev, R. 2003, *MNRAS*, 339, 793
Iyer, N. & Paul, B. 2017, *MNRAS*, 471, 355
Liu, Q. Z., Chaty, S., & Yan, J. Z. 2011, *MNRAS*, 415, 3349
Liu, Q. Z., van Paradijs, J., & van den Heuvel, E. P. J. 2006, *A&A*, 455, 1165

Liu, Q. Z., van Paradijs, J., & van den Heuvel, E. P. J. 2007, *A&A*, 469, 807

Negueruela, I., Smith, D. M., Reig, P., Chaty, S., & Torrejón, J. M. 2006, *in ESA Special Publication*, ed. A. Wilson, Vol. 604, 165–170

Postnov, K., Kuranov, A.G., Yungelson, L.R., 2018a, *in IAU Symposium 346: High-mass X-ray binaries: illuminating the passage from massive binaries to merging compact objects*, ed. L. M. Oskinova (IAU)

Puls, J., Urbaneja, M. A., Venero, R., *et al.* 2005, *A&A*, 435, 669

Santolaya-Rey, A. E., Puls, J., & Herrero, A. 1997, *A&A*, 323, 488

Sidoli, L. & Paizis, A. 2018, *MNRAS*, 481, 2779

Tauris, T. M., Kramer, M., Freire, P. C. C., *et al.* 2017, *ApJ*, 846, 170

Tauris, T. M. & van den Heuvel, E. P. J. 2006, *Compact stellar X-ray sources*, ed. W. H. G. Lewin & M. van der Klis, Vol. 39 (Cambridge Astrophysics Series, Cambridge University Press), 623–665

van den Heuvel, E. P. J., Portegies Zwart, S. F., & de Mink, S. E. 2017, *MNRAS*, 471, 4256

Walter, R., Lutovinov, A. A., Bozzo, E. & Tsygankov, S. S. 2015, *A&A Review*, 23, 2

Discussion

ILEYK EL MELLAH: These results are very interesting! Are you able to detect the accretion wake in the observations of the Good source?

SYLVAIN CHATY: No, we can only derive the overall variation along the full orbit, but do not have the S/N high enough to get some additional details from the *Swift*-BAT lightcurve.

High-mass X-ray Binaries: Illuminating the Passage from Massive Binaries to Merging Compact Objects
Proceedings IAU Symposium No. 346, 2019
L. M. Oskinova, E. Bozzo, T. Bulik & D. Gies, eds.

doi:10.1017/S1743921319001285

On the nature of Supergiant Fast X-ray Transients

Ignacio Negueruela

Departamento de Física Aplicada, Facultad de Ciencias, Universidad de Alicante
Carretera de San Vicente del Raspeig s/n, E03690, San Vicente del Raspeig, Alicante, Spain
email: ignacio.negueruela@ua.es

Abstract. More than a decade after fast x-ray transients with an OB supergiant counterpart were identified as a distinct class of wind-accreting sources, we still have not reached a consensus on the physical origin of their similarities and differences with persistent sources. Both kinds seem to extend over the same range of every relevant parameter. Here I argue that, despite this overall overlap, persistent sources have – on average – later-type, more evolved counterparts, and discuss the hypothesis that SFXTs are – on average – a younger population, as well as some of its possible implications.

Keywords. accretion, stars: atmospheres, stars: early-type, stars: mass loss, stars: neutron, stars: winds, outflows, X-rays: binaries

---

1. Introduction

It is now more than a decade since Supergiant Fast X-ray Transients (SFXTs) were identified as a distinct class of x-ray sources with blue supergiant donors (Negueruela *et al.* 2006a; Smith *et al.* 2006). These objects spend most of the time at very low x-ray luminosity (or simply undetected), and occasionally present flares lasting a few hours, which normally consist of a few short (several hundred seconds) peaks, with large flux changes on a timescale of minutes (Sguera *et al.* 2006; Blay *et al.* 2008). The nature of their counterparts, together with their x-ray spectral properties, suggested that, just like the majority of Supergiant X-ray Binaries (SGXBs), they are fed by the radiation-driven wind of the companion. SGXBs are persistent x-ray sources, always detected by pointing instruments with adequate sensitiveness, typically displaying $L_X \approx 10^{36}$ erg s^{-1} with moderate short-time variability. Several ideas were put forward to provide an explanation for the obvious differences (together with many similarities) between SGXBs and SFXTs: accretion from a clumpy wind (Walter & Zurita Heras 2007), clumpy winds combined with orbital geometry (Negueruela *et al.* 2008a), inhibition of accretion by very strong magnetic fields in the neutron stars (Bozzo *et al.* 2008). However, it soon became clear that, although these hypotheses could explain the behaviour of a given source, none of them on its own can account for the wide variety of behaviours observed (see, e.g., discussion in González-Galán *et al.* 2014). A combination of several factors needs to be invoked.

Progress towards a global solution has been slow. One key difficulty is inherent to the very nature of SFXTs: they are only detectable as x-ray sources for very short time spans. As a consequence, it is very hard to obtain accurate information about them, because data tend to have low signal to noise ratio. For example, spectral evolution with luminosity is poorly constrained, as the sources rarely show high luminosity. Likewise, pulsation periods have been claimed for many SFXTs, but they have proved very difficult to confirm, as long intervals of good data cannot be obtained. Careful analysis of several

Table 1. Persistent SGXBs with known counterparts, ordered by increasing P_{orb}. Spectral types in boldface are consistent with results of quantitative spectral analysis, while those in italics are less secure than the others, as they are based on K-band spectra only. The reference for the spectral type is given.

System	Spectral Type	P_{orb} (d)	e	Ref.
4U 1700−37	**O6 Iafcp**	3.4	≈ 0	(1)
4U 1538−52	B0 I	3.7	$\lesssim 0.2$	(2)
4U 1909+07	**B1 − 3 I**	4.4	~ 0	(3)
SAX J1802.7−2017	B1 Ib	4.6	$\lesssim 0.2$	(4)
Cyg X-1	**O9.7 Iab**	5.6	≈ 0	(5)
XTE J1855−026	BN0.2 Ia	6.1	< 0.04	(6)
IGR J16493−4348	*∼B0.5 Ib*	6.8	~ 0	(7)
4U 1907+097	**O8-9 I**	8.4	~ 0.3	(8)
Vela X-1	**B0 Iab**	8.9	0.09	(9)
IGR J16320−4751	*BN0.5 Ia*	9.0	~ 0.2	(10)
EXO 1722−363	B0 − 1 Ia	9.7	~ 0.2	(11)
OAO 1657−415	Ofpe/WN9	10.4	0.11	(12)
2S 0114+65	B1 Ia	11.6	≈ 0.18	(13)
IGR J19140+0951	B0.5 Ia	13.6	?	(4)
1E 1145.1−6141	B2 Ia	14.4	0.2	(14)
GX 301−2	**B1 Ia$^+$**	41.5	0.46	(15)

References: (1) Clark *et al.* (2002); (2) Reynolds *et al.* (1992); (3) Martínez-Núñez *et al.* (2015); (4) Torrejón *et al.* (2010); (5) Herrero *et al.* (1995); (6) Own data; (7) Pearlman *et al.* (2018); (8) Cox *et al.* (2005); (9) Giménez-García *et al.* (2016); (10) Coleiro *et al.* (2013); (11) Mason *et al.* (2010); (12) Mason *et al.* (2012); (13) Reig *et al.* (1996); (14) Densham & Charles (1982); (15) Kaper *et al.* (2006).

years of monitoring with *INTEGRAL* (Sidoli & Paizis 2018) and *Swift* (Romano 2015) have finally allowed a good characterisation of SFXTs as a class. Following Sidoli (2017), SFXTs are hard x-ray sources with OB supergiant donors presenting:

- A low duty cycle ($< 5\%$) in bright x-ray flares (where bright means $L_X \gtrsim 10^{36}$ erg s^{-1}).
- A high dynamical range ($L_{max}/L_{min} \gtrsim 100$).
- A low time-averaged luminosity ($L_X \lesssim 10^{36}$ erg s^{-1}, below the typical time-averaged luminosity of SGXBs.)

Moreover, comparison of these large datasets with homogeneous observations of SGXBs shows that the behaviour of SFXTs as x-ray sources is different from that of classical SGXBs at a statistically significant level (Bozzo *et al.* 2015; Romano 2015) in terms of the properties listed above.

2. The optical counterparts

What are then the reasons for this difference? Both SFXTs and SGXBs are binaries consisting of a compact object (in fact, with the exception of Cyg X-1, all confirmed systems contain, or are believed to contain, a neutron star) and an OB supergiant. Table 1 lists (most of) the SGXBs with well-characterised counterparts, together with some of their orbital properties. The earliest spectral type is seen in 4U 1700−37, with a luminous O6 supergiant companion. The latest type counterparts are around B2. Most counterparts are moderate-luminosity B0 − 1 supergiants. Table 2 lists the same parameters for SFXTs and related objects. The counterparts to SFXTs do not show significantly different spectral types (see Sidoli 2017; fig. 2). The spin periods of the neutron stars in SGXBs range from a few hundred seconds to about one thousand, with only two exceptions: OAO 1657−415 has a shorter period of only 38 s, while 2S 0114+65 has a very long 2.6 h period. The spin periods of neutron stars in SFXTs, as noted, are not known. The neutron stars in SGXBs typically have surface magnetic fields between 10^{12} and 10^{13} G (Revnivtsev & Mereghetti 2015; although the long spin period of 2S 0114+65

Table 2. Objects that have been classified as SFXTs with known counterparts. The top panel lists systems that likely are high-eccentricity SGXBs. The second panel lists intermediate systems (though those in italics have also been classified within other categories). The bottom panel includes objects with typical SFXT behaviour. Spectral types in boldface are consistent with results of quantitative spectral analysis, while those in italics are less secure than the others, as they are based on K-band spectra only. The reference for the spectral type is given.

System	Spectral type	$P_{\rm orb}$ (d)	e	Ref
IGR J11215−5952	**B0.5 Ia**	165	High	(1)
IGR J00370+6122	**BN0.7 Ib**	15.7	0.6	(2)
IGR J16465−4507	**B0.5 Ibn**	30.2	Unknown	(3)
IGR J18483−0311	B0 − 1 Iab	18.5	High?	(4)
IGR J17354−3255	*O9 Iab*	8.4	Unknown	(5)
SAX J1818.6−1703	B0.5 Iab	30	0.3 − 0.4	(4)
IGR J16418−4532	*BN0.5 Ia*	3.7	Unknown	(5)
IGR J16328−4726	*O8 Iaf*	10.1	Unknown	(5)
IGR J16479−4514	**O8.5 Ib**	3.3	Moderate?	(6)
AX J1841.0−0536	B0.2 Ibp	6.5?	Low?	(7)
IGR J08408−4503	O8.5 Ib-II(f)p	9.5	0.63	(8)
XTE J1739−302	O8 Iab(f)	51.5??	Unknown	(9)
IGR J17544−2619	**O9 Ib**	4.9	Moderate?	(10)
AX J1845.0−0433	O9 Ia	5.7?	Low to moderate	(7)

References: (1) Lorenzo *et al.* (2014); (2) González-Galán *et al.* (2014); (3) Chaty *et al.* (2016); (4) Torrejón *et al.* (2010); (5) Coleiro *et al.* (2013); (6) Negueruela *et al. in prep.*; (7) Negueruela *et al.* (2008b); (8) Gamen *et al.* (2015); (9) Negueruela *et al.* (2006b); (10) Giménez-García *et al.* (2016)

has sometimes be interpreted in terms of a higher magnetic field). Spectral properties of SFXTs suggest similar values, with the detection of a cyclotron line in the prototypical IGR J17544−2619 (Bhalerao *et al.* 2015) providing strong evidence in this sense. The orbital periods of SGXBs range from 3.4 d to 14.4 d and almost all have low eccentricity (GX 301−2 is a very peculiar case: with a longer orbital period and higher eccentricity, and a very massive and luminous hypergiant companion, it cannot be considered typical of SGXBs). The orbital periods of SFXTs cover approximately the same range, although there is a possible 51.5 d period in XTE J1739−302 (Drave *et al.* 2010). In all, the average properties of both kinds of system seem very similar. The only possibility of a systematic difference lies in the spin periods, but there is no *a priori* strong reason to expect it. In fact, the theory of quasi-spherical accretion on to magnetised neutron stars (Shakura *et al.* 2012), the most widely accepted model for the production of x-rays in wind-accreting systems, assumes that the neutron stars rotate slowly.

Since the global x-ray behaviour must be determined by the interaction of the stellar wind and the neutron star magnetosphere (see Sander in this proceedings and references therein), the properties of mass donors should play a role in setting the differences. The possibility that the donors in SFXTs are not true supergiants has recently been raised. This is not a straightforward question, as O-type supergiants are still H-core burning objects and thus not fundamentally different from O-type dwarfs. Their morphological differences are mostly due to higher mass loss rates at higher luminosities (see, e.g., Holgado *et al.* 2018), and so a lower luminosity would imply weaker winds. To test this possibility, we collected high-quality VLT/ISAAC spectra of the IR counterpart to IGR J16479−4514, the SFXT with the shortest orbital period. In fact, this heavily-reddened eclipsing transient presents the shortest (by little) orbital period for any wind-accreting system, $P_{\rm orb} = 3.32$ d (Sidoli *et al.* 2013). Rahoui *et al.* (2008) estimated an early spectral time around O8.5 I. According to calibrations (e.g. Martins *et al.* 2005), such a star has a radius of $\approx 22\,R_\odot$, while a dwarf of the same spectral type only has $R_* \approx 8\,R_\odot$, allowing for a wider orbit. Using tailored CMFGEN models, we find stellar parameters typical of an O8.5 Ib star (Negueruela *et al.* in prep.), confirming that even in this extreme case the counterpart is a supergiant. The neutron star cannot be further away from the surface of the donor than in most classical SGXBs. Therefore the environmental conditions for the neutron star must be quite similar to those in some classical SGXBs. This is strong evidence for the existence of gating mechanisms close to the surface of the neutron star, and joins similarly strong evidence coming from analysis of x-ray data (Romano 2015; Bozzo *et al.* 2015; Pradhan *et al.* 2018).

Of course, the key to this discussion lies on the accuracy of the spectral types and the reliability of the stellar parameters derived from them. The counterparts to many x-ray binaries are distant and highly obscured by intervening material along the line of sight and thus obtaining high-quality optical spectra is not always feasible. Moreover, different groups use different techniques for spectral classification. The MK system is based on features lying in the blue side of the spectrum, which is much more heavily reddened than the red side and thus not always accessible. A good spectral classification of OB stars is possible with near-IR spectra if a wide spectral range is observed, so that many features can be used for the classification. On the other hand, classifications based on a small spectral range (e.g. K-band spectra only, as in Nespoli *et al.* 2008) have a much higher uncertainty, because there are very few features in the range and most are sensitive to more than one physical parameter, including the mass loss rate. Even when spectral types are accurate, their calibration against stellar parameters is necessarily loose, because of physical reasons (see Simón-Díaz *et al.* 2014; Holgado *et al.* 2018). Therefore stellar parameters based on quantitative spectral fitting with suitable model atmospheres (see the contribution by Sander) are always more reliable. To take these difficulties into account, the spectral types in Tables 1 and 2 have been coded: spectral types in boldface are supported by quantitative spectral analysis. Spectral types in roman type are based on blue spectra or a combination of several red and near-IR bands, while spectral types in italics are derived from single-band IR spectra or indirect methods.

3. A working hypothesis

When the reliability of spectral types is taken into account, we can see some interesting trends emerging. Given the size of existing samples, such trends cannot be considered statistically significant†, but are still highly suggestive. When we look at the SGXBs, two systems have donors with very strong winds, 4U 1700−37 with an O6 Iaf supergiant (pressumably a very massive star; see Clark *et al.* 2002) and OAO 1657−415, which likely has followed a different evolutionary path from most other systems (Mason *et al.* 2012). A third one, 4U 1907+097, has on O-type supergiant as companion. All the other ones have companions in a very narrow spectral range, from O9.7 to B2, with the vast majority concentrated between B0 and B1. Objects with orbital periods below 8 d have essentially circular orbits, while longer periods imply moderately eccentric orbits. The exception is again 4U 1907+097, with a higher eccentricity, only surpassed by the peculiar system GX 301−2.

If we look now at Table 2, we find in the top panel three objects that have been associated with SFXTs, but seem more closely related to SGXBs. They all have companions in the B0 − 1 range. The difference with the main SGXB group lies in their wide (and

† Indeed, if we take into account the many difficulties in obtaining reliable spectral types, it is quite possible that the whole Galactic population of wind-fed systems is insufficient to give a statistically significant sample (see Tabernero *et al.* 2018, for a robust estimation of the sample sizes needed to ascertain a difference in average spectral type between two populations).

eccentric) orbits. These subset most likely consists of the SGXBs with the widest orbits, for which orbital geometry alone likely leads to the transient-like behaviour, as has also been proposed by Walter *et al.* (2015) for a number of objects. The second panel contains objects that have sometimes been classified as intermediate between SGXBs and SFXTs. Their counterparts are again in the same spectral range and there are reasons to at least suspect that their orbits are eccentric. The third panel contains those objects that have been confirmed as SFXTs, following Romano (2015). The spectral types are decidedly earlier. With the exception of the peculiar counterpart to AX J1841.0−0536, all fall within O8 and O9.

Although the difference in spectral type is small and likely lacks statistical significance, it seems too well defined to be due to random sampling. If the counterparts of SFXTs are consistently earlier, this implies somewhat smaller stars and – crucially – faster, less dense winds. While the counterparts to SGXBs straddle the bi-stability jump – a sudden change in wind conditions happening at temperatures cooler than 25 000 K that results in higher mass loss rates and slower winds (e.g. Vink 2018), two conditions that favour accretion – the counterparts to SFXTs lie well to the hot side, with temperatures $> 30\,000$ K. These faster, less dense winds imply that – on average – conditions will be less favourable for accretion over a wide range of orbital parameters and neutron star properties. Interestingly, this small difference in spectral types also implies that, according to standard evolutionary tracks, the average donor in an SFXT will evolve into the average donor in an SGXB. This does not necessarily mean that all SGXBs must have had an earlier phase as SFXTs†, but is strongly suggestive of the idea that SFXTs, as a population, are younger than SGXBs.

What would this hypothesis of SFXTs as a younger population than SGXBs mean? In fact, there are two interpretations – not at all mutually exclusive – to such a statement. On the one hand, this youth may refer to the evolutionary status of the mass donor, as discussed in the previous paragraph. But it can also mean that the neutron star is younger, i.e. that the supernova explosion took place more recently. If so, the binary system has had less time to evolve. For example, assuming that all O star + NS systems form with some eccentricity due to mass loss and a kick during the explosion, the fact that all SGXBs with short ($\lesssim 10$ d) orbital period have (almost) circular orbits suggests an efficient mechanism for circularisation (see González-Galán *et al.* 2014). The very high eccentricity of a system like IGR J08408−4503, on the other hand, indicates that there has not been time for circularisation. Even the short-period systems IGR J16479−4514 and IGR J17544−2619 seem to require some eccentricity to explain their lightcurves (Ducci *et al.* 2010; Bozzo *et al.* 2016), again pointing to a relatively recent formation‡. If this second sense of youth also applies to SFXTs, then the properties of

† This idea of late-O supergiants evolving into B-type supergiants must be interpreted in a broad, general sense. According to the models in Martins & Palacios (2017), late-O supergiants are spread between the $30\,M_\odot$ and $40\,M_\odot$ tracks, with observations showing some objects at slightly lower masses. In the absence of dynamical mass determinations, we assume that counterparts to SFXTs lie in this range – those of Ib luminosity class not very far above $30\,M_\odot$, and perhaps even less massive, given the tendency of counterparts in HMXBs to be undermassive. Such objects evolve into B1 – 2 Ia supergiants. On the other hand, objects with classifications B0 – 1 Ib probably come from stars with masses $\approx 25\,M_\odot$, which have luminosity class II-III when late-O stars. IGR J00370+6122 has a B0.7 Ib counterpart (verging on luminosity class II) with a moderately low mass $\approx 15\,M_\odot$, a bit lower than expected for a star of its spectral type. This object cannot have been an O-type supergiant earlier in its life, but probably had a spectral type close to O9.5 III.

‡ This scenario is further reinforced by the high eccentricity of 4U 1907+097, the only SGXB whose counterpart is similar to those of SFXTs. It could be argued that these systems with short orbital periods and moderate eccentricity are the descendents of binaries that formed with such a high eccentricity that they required more time than the others to circularise. Again, we

the neutron stars in these systems may also show some differences with respect to those in classical SGXBs, having had less life time for spin down and magnetic field decay – with fast rotation and high fields again contributing to make accretion less effective. In this respect, it is worth remembering that the spin periods of neutron stars in wind-accreting systems are thought to be determined by evolution during the propeller phase, i.e. before accretion begins (see Li *et al.* 2016 and references therein). If the system formed when the mass donor was close to the O-supergiant phase – which, we should not forget, is still H-core burning – the equilibrium period may be noticeably different from that in a system formed when the mass donor was still a dwarf. In any case, in order to understand the effect of system age on its x-ray properties, we still need a much better knowledge of the different evolutionary pathways leading to HMXB formation and the consequences of rejuvenation on O-type stars that accrete substantial amounts of mass from their binary companions (cf. Dray & Tout 2007, and references therein.)

4. Conclusions

The main ideas discussed in this paper are:

• Gating mechanisms must be at work to explain the existence of SFXTs as a separate class. These mechanisms are seen to operate very differently in systems with similar orbital and wind parameters overall. The theory of quasi-spherical accretion on to magnetised neutron stars (Shakura *et al.* 2012) provides a firm base for such mechanisms, either through magnetic-field interaction (Shakura *et al.* 2014), or the accumulation mechanism proposed by Drave *et al.* (2014).

• In consequence, differences in behaviour must be due to specific parameter combinations, which are hard to identify and test. We are limited by small sample size in a very large parameter space.

• The idea that SFXTs represent an earlier stage for (some) SGXBs is probably worth exploring.

Acknowledgements

I would like to thank all my collaborators in binary work, especially David Smith, Sylvain Chaty and J. Simon Clark, for many fruitful discussions. This research is partially supported by MinECO/FEDER under grant AYA2015-68012-C2-2-P and Ministerio de Educación y Ciencia under grant PRX14-00169.

References

Bhalerao, V., Romano, P., Tomsick, J., *et al.* 2015, *MNRAS*, 447, 2274
Blay, P., Martínez-Núñez, Negueruela, I., *et al.* 2008, *A&A*, 489, 669
Bozzo, E., Falanga, M., & Stella, L. 2008, *ApJ*, 683, 1031
Bozzo, E., Romano, P., Ducci, L. *et al.* 2015, *AdSpR*, 55, 1255
Bozzo, E., Bhalerao, V., Pradhan, P., *et al.* 2016, *A&A*, 596, A16
Chaty, S., LeReun, A., Negueruela, I., *et al.* 2016, *A&A*, 591, A87
Clark, J.S., Goodwin, S.P., Crowther, P.A., *et al.* 2002, *A&A*, 392, 909
Coleiro, A., Chaty, S., Zurita Heras, J. A., *et al.* 2013, *A&A*, 560, A108
Cox, N.L.J., Kaper, L., & Mokiem, M.R. 2005, *A&A*, 436, 661
Densham, R. H., & Charles, P. A. 1982, *MNRAS*, 201, 171
Drave S. P., Clark, D. J., Bird, A. J., *et al.* 2010, *MNRAS*, 409, 1220
Drave, S. P., Bird, A. J., Sidoli, L., *et al.* 2014, *MNRAS*, 439, 2175

lack the numbers to show a statistically significant effect, but the sample available is suggestive of evolution, with 4U 1907+097 qualifying as a SGXB once its eccentricity has decreased below a threshold for which its stellar wind is strong enough to favour accretion.

Dray, L. M., & Tout, C. A. 2007, *MNRAS*, 376, 61
Ducci, L., Sidoli, L., & Paizis, A. 2010, *MNRAS*, 408, 1540
Gamen, R., Barbà, R. H., Walborn, N. R., *et al.* 2015, *A&A*, 583, L4
Giménez-García, A., Shenar, T., Torrejón, J. M., *et al.* 2016, *A&A*, 591, A26
González-Galán, A., Negueruela, I., Castro, N., *et al.* 2014, *A&A*, 566, A131
Herrero, A., Kudritzki, R.P., Gabler, R., *et al.* 1995, *A&A*, 297, 556
Holgado, G., Simón-Díaz, S., Barbá, R. H., *et al.* 2018, *A&A*, 613, A65
Kaper, L., van der Meer, A., & Najarro, F. 2006, *A&A*, 457, 595
Li, T., Shao, Y., & Li, X.-D. 2016, *ApJ*, 824, 143
Lorenzo, J., Negueruela, I., Castro, N., *et al.* 2014, *A&A*, 562, A18
Martínez-Núñez, S., Sander, A., Gímenez-García, A., *et al.* 2015, *A&A*, 578, A107
Martins, F., & Palacios, A. 2017, *A&A*, 598, A56
Mason, A. B., Norton, A. J., Clark, J. S., *et al.* 2010, *A&A*, 509, A79
Mason, A.B., Norton, A.J., Clark, J.S., *et al.* 2011, *A&A*, 532, A124
Mason, A. B., Clark, J. S., Norton, A. J., *et al.* 2012, *MNRAS*, 422, 199
Negueruela, I., Smith, D.M., Reig, P., *et al.* 2006a, *ESA SP*-604, 165
Negueruela, I., Smith, D.M., Harrison, T.E., & Torrejón, J.M. 2006b, *ApJ*, 638, 982
Negueruela, I., Smith, D.M., Torrejón, J.M., & Reig, P. 2008, *ESA SP*-622, 255
Negueruela, I., Torrejón, J. M., & Reig, P. 2008, *Proceedings of the 7th INTEGRAL Workshop*, 72
Nespoli, E., Fabregat, J., & Mennickent, R. E. 2008, *A&A*, 486, 911
Pearlman, A. B., Coley, J. B., Corbet, R. H. D., & Pottschmidt, K. 2018, *ApJ*, in press (arXiv:1811.06543)
Pradhan, P., Bozzo, E., & Paul, B. 2018, *A&A*, 610, A50
Ray, P.S., & Chakrabarty, D. 2002, *ApJ*, 581, 1293
Reig, P., Chakrabarty, D., Coe, M.J., *et al.* 1996, *A&A*, 311, 879
Revnivtsev, M., & Mereghetti, S. 2015, *SSRv*, 191, 293
Reynolds, A.P., Bell, S.A., & Hilditch, R.W. 1992, *MNRAS*, 256, 631
Romano, P. 2015, *Journal of High Energy Astrophysics*, 7, 126
Sguera, V., Bazzano, A., Bird, A.J., *et al.* 2006, *ApJ*, 646, 452
Shakura, N., Postnov, K., Kochetkova, A., & Hjalmarsdotter, L. 2012, *MNRAS*, 420, 216
Shakura, N., Postnov, K., Sidoli, L., & Paizis, A. 2014, *MNRAS*, 442, 2325
Sidoli, L., Esposito, P., Sguera, V., *et al.* 2013, *MNRAS*, 429, 2763
Sidoli, L. 2017, *Proceedings of the XII Multifrequency Behaviour of High Energy Cosmic Sources Workshop (MULTIF2017)*, A52
Sidoli, L., & Paizis, A. 2018, *MNRAS*, 481, 2779
Simón-Díaz, S., Herrero, A., Sabín-Sanjulián, C., *et al.* 2014, *A&A*, 570, L6
Smith, D. M., Heindl, W.A., Markwardt, C.A., *et al.* 2006, *ApJ*, 638, 974
Tabernero, H. M., Dorda, R., Negueruela, I., & González-Fernández, C. 2018, *MNRAS*, 476, 3106
Torrejón, J. M., Negueruela, I., Smith, D. M., & Harrison, T. E. 2010, *A&A*, 510, A61
Vink, J. S. 2018, *A&A*, 619, A54
Walter, R., & Zurita Heras, J.A. 2007, *A&A*, 476, 335
Walter, R., Lutovinov, A. A., Bozzo, E., & Tsygankov, S. S. 2015, *A&ARv*, 23, 2

Discussion

SANDER: Knowing that the amount of existing spectroscopic analyses for mass donors is very limited, did you take a look at the abundances of the donors from what you would define as the "true" SFXTs? Do they differ from the other ones?

NEGUERUELA: So far there are analyses for only two SFXTs. In the case of IGR J16479−4514, the data are only good enough for stellar parameter determination. Giménez-García *et al.* (2016) calculated CNO abundances for IGR J17544−2619 and the prototypical SGXB Vela X-1, finding almost identical values in the two objects. If the connection between the two types of object is indeed evolutionary, differences are likely to be marginal, in the sense of the B-type supergiants showing a slightly more advanced stage, with a higher He fraction, more N and less C. In most of the cases analysed, the CNO abundances clearly display the effects of evolution, but at values that are compatible with the effect of fast rotation on an isolated O-type star. The most obvious signs of recent binary interaction are the very high rotational velocities of some systems, most notably IGR J16465−4507 (Chaty *et al.* 2016).

PRADHAN: Would it be possible to distinguish the age of SGXBs and SFXTs from the nature of the companion stars?

NEGUERUELA Not really. The difference pointed out here is in terms of average spectral type, not the types of individual stars. In fact, this difference is quite small and it can be quite difficult to show its statistical significance even if larger samples of wind-accreting systems are discovered.

KARINO: Even though donors in SFXTs are young and emit fast winds, the wind condition should have a large variety, since orbital periods show large differences. So, how does the fact that donors are young affect the accretion properties?

NEGUERUELA: I have to stress again that I am using the term "younger" in a broad sense and that I am talking about the bulk properties of the population. The accretion conditions depend on the interaction between the wind and the neutron star. My hypothesis is that fast, low density winds allow gating mechanisms to work more effectively than dense, slow winds. But the efficiency of gating mechanisms must depend on orbital parameters and very likely also on neutron star parameters. Whether a system behaves like an SGXB, an SFXT or something intermediate will depend on the specific combination of all these variables in that system.

PRADHAN: So can we say that the difference between SFXTs and HMXBs lies in the wind velocity?

NEGUERUELA: Not quite. I think that your recent paper (Pradhan *et al.* 2018) provides strong evidence for the existence of gating mechanisms. But wind conditions very likely determine whether these mechanisms act effectively. Unfortunately, directly measuring wind velocities is in practice impossible for most wind-accreting system, as it can only be done with UV spectra, which cannot be obtained for highly obscured sources.

High-mass X-ray Binaries: Illuminating the Passage from Massive Binaries to Merging Compact Objects
Proceedings IAU Symposium No. 346, 2019
L. M. Oskinova, E. Bozzo, T. Bulik & D. Gies, eds.

doi:10.1017/S1743921319001145

Investigating High Mass X-ray Binaries at hard X-rays with INTEGRAL

Lara Sidoli and Adamantia Paizis

INAF/IASF Milano, Istituto di Astrofisica e Fisica Cosmica di Milano
via A. Corti 12, I-20133, Milano, Italy
emails: lara.sidoli@inaf.it, adamantia.paizis@inaf.it

Abstract. The *INTEGRAL* archive developed at INAF-IASF Milano with the available public observations from late 2002 to 2016 is investigated to extract the X-ray properties of 58 High Mass X-ray Binaries (HMXBs). This sample consists of sources hosting either a Be star (Be/XRBs) or an early-type supergiant companion (SgHMXBs), including the Supergiant Fast X-ray Transients (SFXTs). *INTEGRAL* light curves (sampled at 2 ks) are used to build their hard X-ray luminosity distributions, returning the source duty cycles, the range of variability of the X-ray luminosity and the time spent in each luminosity state. The phenomenology observed with *INTEGRAL*, together with the source variability at soft X-rays taken from the literature, allows us to obtain a quantitative overview of the main sub-classes of massive binaries in accretion (Be/XRBs, SgHMXBs and SFXTs). Although some criteria can be derived to distinguish them, some SgHMXBs exist with intermediate properties, bridging together persistent SgHMXBs and SFXTs.

Keywords. X–rays: binaries, accretion

1. Introduction

High Mass X-ray Binaries (HMXBs) host a compact object (most frequently a neutron star [hereafter, NS]) accreting matter from an O or B-type massive star. In the great majority of these systems the mass transfer to the accretor occurs by means of the stellar wind, while in a limited number of HMXBs (SMC X-1, LMC X-4, Cen X-3) it happens through Roche Lobe overflow (RLO). Before the launch of the *INTEGRAL* satellite (Winkler *et al.* 2003, Winkler *et al.* 2011), two types of HMXBs were known, depending on the kind of companion, either an early type supergiant star (SgHMXBs) or a Be star (Be/XRBs).

Nowadays the scenario has significantly changed, with a number of Galactic HMXBs tripled and new sub-classes of massive binaries (the "highly obscured sources" and the "Supergiant Fast X–ray Transients", SFXTs) discovered thanks to the observations of the Galactic plane performed by the *INTEGRAL* satellite. The first type includes HMXBs where the absorbing column density due to the local matter is more than one order of magnitude larger than the average in HMXBs (reaching 10^{24} cm^{-2} in IGR J16318–4848).

The SFXTs are HMXBs that undergo short (usually less than a few days) outbursts made of brief (typical duration of ~2 ks) and bright X–ray flares (peak $L_X \sim 10^{36}$ erg s^{-1}), while most of their time is spent below $L_X \sim 10^{34}$ erg s^{-1}. The physical mechanism producing this behavior is debated: the main models involve different ways to prevent accretion onto the NS (invoking opposite assumptions on the NS magnetic field and spin period), coupled with different assumptions on the donor (clumpy and/or magnetized) wind parameters (see Martínez-Núñez *et al.* 2017, Walter *et al.* 2015 and Sidoli 2017 for recent reviews, and references therein).

While SgHMXBs and Be/XRBs differ in the type of their companion star, the boundaries between SgHMXBs and SFXTs are based *only* on their X–ray phenomenology (persistent vs transient X–ray emission), since they both harbour an early-type supergiant donor. In fact, unlike SgHMXBs, SFXTs display a large dynamic range that can reach six orders of magnitude in X–ray luminosity, between quiescence and the outburst peak (as in IGR J17544-2619; in't Zand (2005), Romano *et al.* (2015)). For other SFXTs the range of flux variability is typically comprised between 10^2 and 10^4.

The present *paper* summarizes our systematic analysis of the *INTEGRAL* observations of HMXBs, spanning fourteen years of operations, from 2002 to 2016. The main aim of this work is to obtain an overall, quantitative, characterization of the different sub-classes of HMXBs at hard X–rays (above 18 keV) and to put this phenomenology into context of other known properties (like pulsar spin period, orbital geometry) and soft X–ray behavior (1–10 keV), as described in the literature. We refer the reader to Sidoli & Paizis (2018) for more details on this work.

2. The *INTEGRAL* archive and the selection of the sample

Our investigation is based on observations performed by IBIS/ISGRI on-board the *INTEGRAL* satellite and it is focussed on the energy range 18–50 keV. We built an *INTEGRAL* local archive of all public observations (see Paizis *et al.* 2013 and Paizis *et al.* 2016 for the technical details). For all known HMXBs we extracted the long-term light curves of the sources (bin time of ~2 ks, the typical duration of an *INTEGRAL* observation, called "Science Window") spanning fourteen years (from late 2002 to 2016). We retained in our final sample only the sources which were detected (above 5 sigma) in at least one *INTEGRAL* observation (i.e. one single Science Window), within 12° from the centre of the field-of-view. These selection criteria translated into a sensitivity threshold of a few 10^{-10} erg cm^{-2} s^{-1} (18–50 keV) for our survey, and into a total exposure time of ~200 Ms for the final HMXB sample.

The final list of sources includes 58 HMXBs, classified in the literature as SgHMXBs (18 sources), SFXTs (13 sources) and Be/XRBs (20 sources); the remaining 9 massive binaries are two pulsars accreting from early type giant stars (LMC X–4 and Cen X–3), three black hole (candidate) systems (Cyg X-1, Cyg X-3, SS 433) plus two peculiar massive binaries (IGR J16318-4848 and 3A 2206+543). Then, we also included a symbiotic binary XTE J1743-363, that is a different kind of wind-fed system, to compare it with massive binaries. The complete list of sources is reported in Table 1, together with their sub-class, as reported in the literature.

3. *INTEGRAL* results

Duty Cycles (18-50 keV). The long-term light curves for our sample of HMXBs were used to calculate the source duty cycle in the energy range 18-50 keV ($DC_{18-50\ keV}$), defined as the percentage of detections (at ~2 ks time bin) or, in other words, the ratio between the exposure time when the source is detected and the total exposure time at the source position. Table 1 (third column) lists the values obtained. Even in case of a persistent SgHMXB, the duty cycle can be lower than 100%, because of source variability leading the source flux below the IBIS/ISGRI threshold of detectability on the adopted time bin. Eclipses or off-states also reduce the source duty cycle in persistent sources (e.g. in Vela X-1, Kreykenbohm *et al.* 2008; Sidoli *et al.* 2015). The advantage of the adoption of a long-term archive, analysed here in a systematic way, is that we are confident that the source duty cycles are close to the real source activity, above the *INTEGRAL* sensitivity. We refer the reader to Sidoli & Paizis (2018) for a detailed discussion of the possible observational biases.

Table 1. Results of our survey of a sample of HMXBs.

Source	**sub-class**	$\mathbf{DC_{18-50keV}}$ [%]	**Av. Luminosity (18-50 keV)**[1] [erg s^{-1}]	$\mathbf{DR_{1-10keV}}$ $(\mathbf{F_{max}/F_{min}})$
SMC X-1	SgHMXB	49.05	1.7E+38	7.7
3A 0114+650	SgHMXB	14.63	2.1E+36	–
Vela X-1	SgHMXB	79.22	1.3E+36	1.7
1E 1145.1-6141	SgHMXB	31.95	3.0E+36	–
GX 301-2	SgHMXB	94.47	2.8E+36	2.6
H 1538-522	SgHMXB	30.15	9.2E+35	32.9
IGR J16207-5129	SgHMXB	0.39	1.1E+36	9.3
IGR J16320-4751	SgHMXB	21.32	5.9E+35	14.7
IGR J16393-4643	SgHMXB	0.40	3.4E+36	3.2
OAO 1657-415	SgHMXB	59.78	5.8E+36	10.0
4U 1700-377	SgHMXB	73.09	1.1E+36	12.0
IGR J17252-3616	SgHMXB	4.65	2.9E+36	17.3
IGR J18027-2016	SgHMXB	0.54	5.2E+36	375
IGR J18214-1318	SgHMXB	0.06	3.4E+36	–
XTE J1855-026	SgHMXB	9.64	4.2E+36	–
H 1907+097	SgHMXB	20.13	8.1E+35	546
4U 1909+07	SgHMXB	24.84	7.1E+35	11.5
IGR J19140+0951	SgHMXB	14.18	5.2E+35	769
LMC X-4	giant HMXB	47.23	1.2E+38	3.4
Cen X-3	giant HMXB	62.79	4.0E+36	5.0
IGR J08408-4503	SFXT	0.09	3.0E+35	6750
IGR J11215-5952	SFXT	0.64	1.6E+36	>480
IGR J16328-4726	SFXT	0.28	1.7E+36	300
IGR J16418-4532	SFXT	1.22	6.1E+36	308
IGR J16465-4507	SFXT	0.18	2.9E+36	37.5
IGR J16479-4514	SFXT	3.33	3.6E+35	1667
IGR J17354-3255	SFXT	0.01	3.0E+36	>929
XTE J1739-302	SFXT	0.89	4.8E+35	>2040
IGR J17544-2619	SFXT	0.54	5.6E+35	1.67×10^6
SAX J1818.6-1703	SFXT	0.81	2.9E+35	>1364
IGR J18410-0535	SFXT	0.53	3.8E+35	1.1×10^4
IGR J18450-0435	SFXT	0.35	1.5E+36	513
IGR J18483-0311	SFXT	4.63	5.2E+35	899
H 0115+634	Be/XRB	9.55	1.5E+37	1.4×10^5
RX J0146.9+6121	Be/XRB	0.11	1.1E+35	–
EXO 0331+530	Be/XRB	25.10	2.4E+37	1.07×10^6
X Per	Be/XRB	76.96	2.5E+34	10
1A 0535+262	Be/XRB	12.34	4.4E+36	2.7×10^4
GRO J1008-57	Be/XRB	8.87	2.4E+36	181
4U 1036-56	Be/XRB	0.35	7.5E+35	60
IGR J11305-6256	Be/XRB	0.41	1.9E+35	–
IGR J11435-6109	Be/XRB	2.68	1.4E+36	–
H 1145-619	Be/XRB	1.07	1.2E+35	250
XTE J1543-568	Be/XRB	0.14	2.7E+36	8
AX J1749.1-2733	Be/XRB	0.17	8.1E+36	–
GRO J1750-27	Be/XRB	4.88	2.9E+37	>10
AX J1820.5-1434	Be/XRB	0.15	2.1E+36	–
Ginga 1843+009	Be/XRB	3.39	5.8E+36	5660
XTE J1858+034	Be/XRB	5.34	8.8E+36	–
4U 1901+03	Be/XRB	10.44	1.2E+37	1000
KS 1947+300	Be/XRB	9.41	6.8E+36	800
EXO 2030+375	Be/XRB	28.99	7.8E+36	>2784
SAX J2103.5+4545	Be/XRB	11.14	2.0E+36	6364
IGR J16318-4848	other HMXB	35.17	7.4E+35	3.3
3A 2206+543	other HMXB	6.41	2.5E+35	250
Cyg X-1	other HMXB	99.88	2.5E+36	3.7
Cyg X-3	other HMXB	93.49	1.0E+37	4.9
SS 433	other HMXB	14.97	8.5E+35	5.0
XTE J1743-363	symbiotic	0.13	1.1E+36	6.2

Notes: [1]This 18-50 keV luminosity is an average over *INTEGRAL* detections only. This implies that, for transients, it is an average luminosity in outburst. See Sidoli & Paizis (2018) for the source distances adopted here.

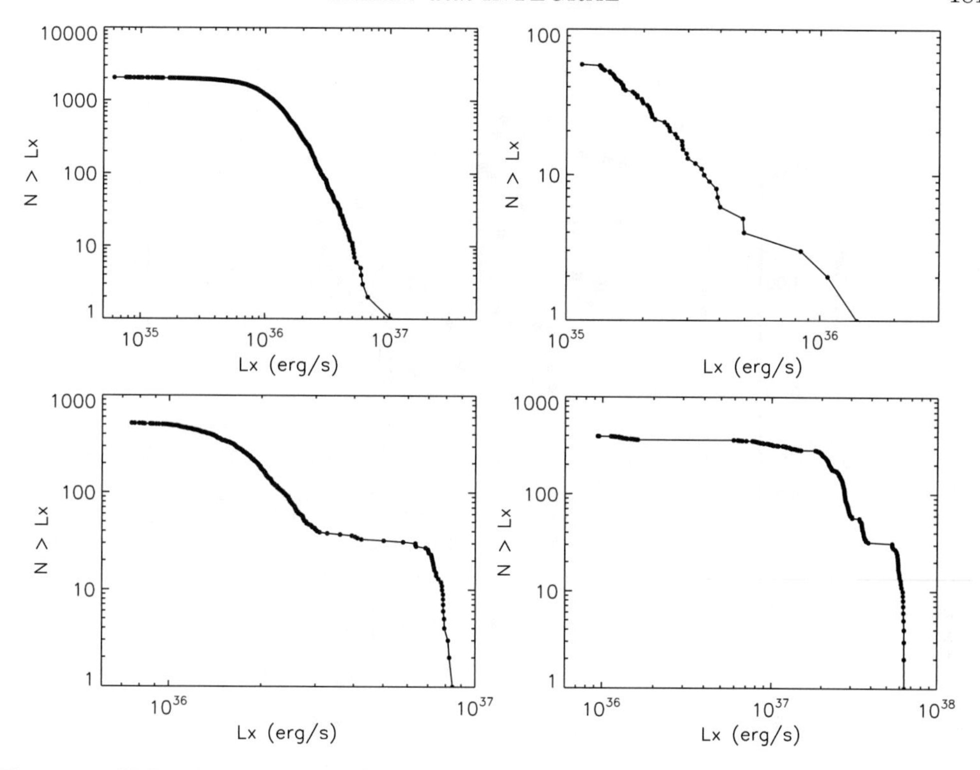

Figure 1. CLDs of sources representative of the three sub-classes of HMXBs: from left to right, from top to bottom, the persistent SgHMXB Vela X–1, the SFXT SAX J1818.6-1703 and the two Be transients SAX J2103.5+4545 and EXO 0331+530.

Cumulative Luminosity Distributions. The hard X-ray light curves were used to extract the Cumulative Luminosity Distributions (CLDs). We adopted a single average conversion factor of 4.5×10^{-11} erg cm^{-2} count^{-1} from IBIS/ISGRI count-rates to X-ray fluxes (18–50 keV) and assumed the source distances reported by Sidoli & Paizis (2018).

The CLDs of four sources are shown in Fig. 1, representative of the behavior of a persistent SgHMXB (Vela X-1), a SFXT (SAX J1818.6-1703) and of two transient Be/XRBs (SAX J2103.5+4545 and EXO 0331+530). Their shape appears different: a lognormal-like distribution is evident in Vela X-1, a powerlaw CLD in the SFXT, while a more complex behavior is present in the Be/X-ray transients.

Since the timescale of the SFXT flare duration is similar to the bin time of the *INTEGRAL* light curves, the SFXT CLDs are distributions of the SFXT flare luminosities (Paizis & Sidoli 2014). The difference among supergiant systems (SgHMXBs vs SFXTs), between lognormal and powerlaw-like luminosity distributions were already found by Paizis & Sidoli (2014) from the analysis of the first nine years of *INTEGRAL* observations of a sample of SFXTs, compared with three SgHMXBs.

This behavior might be ascribed to a separate physical mechanism producing the bright X–ray flares in SFXTs: in the framework of the quasi-spherical settling accretion regime (Shakura *et al.* 2012), hot wind matter, captured within the Bondi radius, accumulates above the NS magnetosphere; magnetic reconnection at the base of this shell (between the magnetized, captured, wind matter and the NS magnetosphere) has been suggested to enhance the plasma entry through the magnetosphere, opening the NS gate. This allows the sudden accretion of the shell material onto the NS and the emission of the SFXT flares (Shakura *et al.* 2014). The detection of a ~kG magnetic field from the companion of the SFXT IGR J11215–5952 supports this scenario (Hubrig *et al.* 2018).

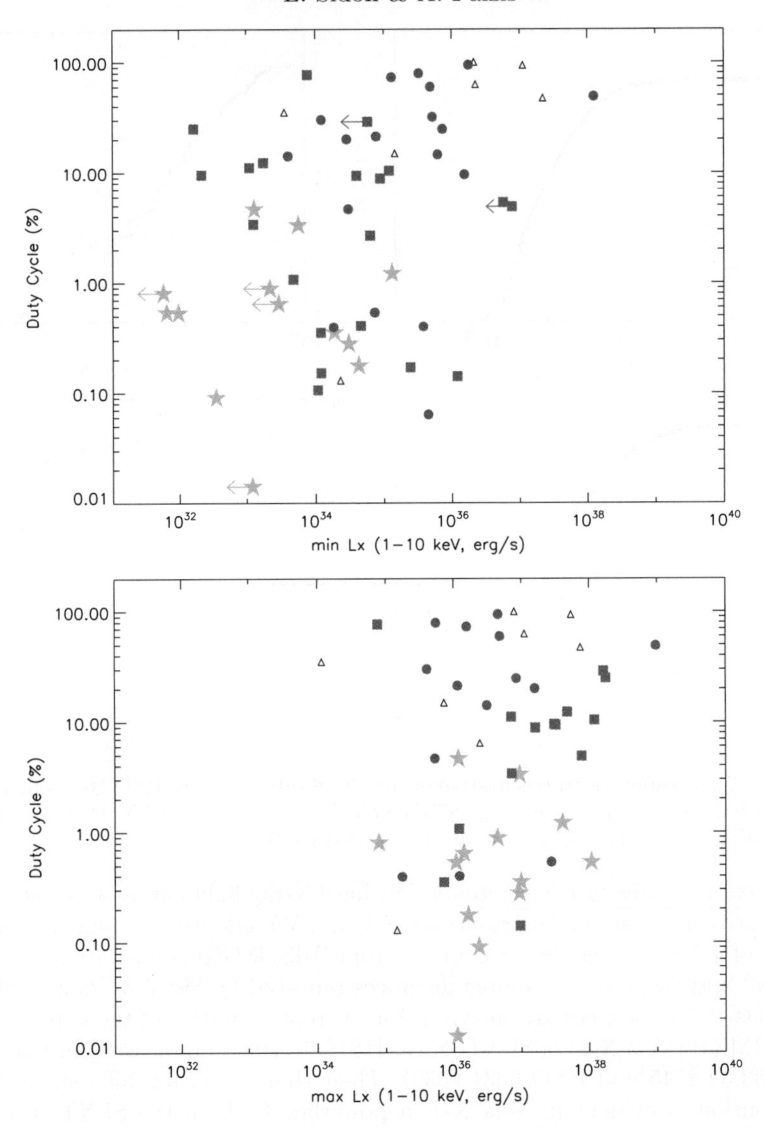

Figure 2. *INTEGRAL* source duty cycle ($DC_{18-50\ keV}$) versus the soft X–ray luminosities reported in the literature (minimum and maximum ones are in the upper and lower panels, respectively). Symbols indicate four sub-classes: blue circles for SgHMXBs, green stars for SFXTs, red squares for Be/XRBs and empty thin diamonds for the remaining types of sources, as reported in Table 1 (the "other" HMXBs, together with the two giant HMXBs and the symbiotic system). Arrows indicate upper limits on the minimum flux. Note that the different number of sources reported in the two panels are because for some HMXBs we have found in the literature only a single value for the soft X-ray flux (and we ascribed it to the "minimum 1-10 keV flux").

Transient Be/XRBs can show two types of outbursts, the "normal" and the "giant" ones (Stella *et al.* 1986; Negueruela *et al.* 1998, 2001,b; Okazaki *et al.* 2001; Reig 2011; Kuhnel *et al.* 2015). The first type happens periodically and is produced by the higher accretion rate when the NS approaches the decretion disc of the Be star, at each passage near periastron. The second type of outburst can occur at any orbital phase, is more luminous than the normal one and is thought to be produced by major changes in the Be decretion disc structure. We ascribe the bimodal behaviour evident in the CLD of

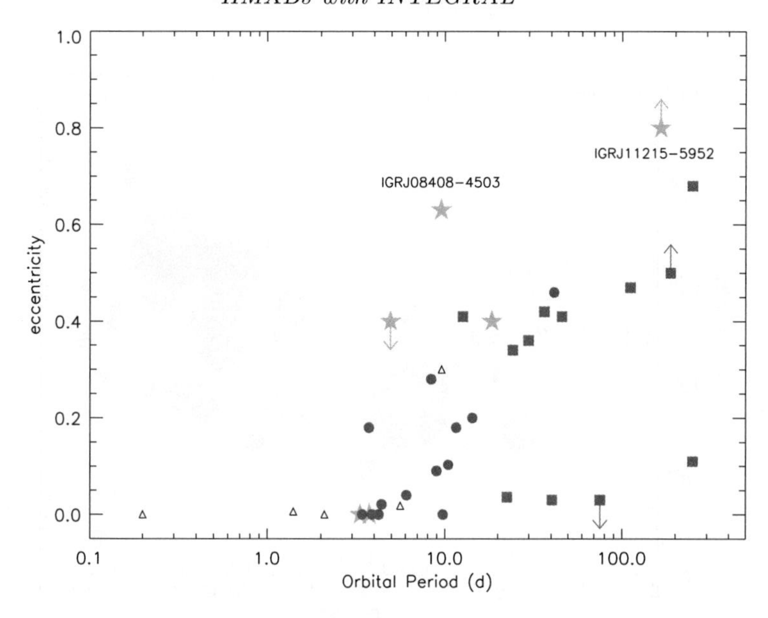

Figure 3. Orbital eccentricity versus orbital period for our sample of HMXBs. The symbols have the same meaning as in Fig. 2.

SAX J2103.5+4545 shown in Fig. 1 to the two different luminosities reached during the two types of outbursts: low (high) luminosity in normal (giant) one, respectively. Other Be/XRBs show more complex shapes, multi-modal distributions (like in the case of EXO 0331+530 shown in Fig. 1), indicative of multi peaks within the same outburst, or outbursts reaching different peak luminosities.

The CLDs of all HMXBs of our sample are reported by Sidoli & Paizis (2018; their Fig. 1-4). In their normalized version, these functions allow the reader to obtain in one go, not only an easy comparison between all kind of HMXBs, but also to quantify the time spent by each HMXB in any given luminosity state, above the instrumental sensitivity.

Average Luminosity (18-50 keV). An average luminosity (18-50 keV) was calculated for each source, over the *INTEGRAL* detections (at 2 ks timescale; see Table 1, forth column). Note that this definition implies that, for transient sources, this is an average luminosity *in outburst.*

4. Other HMXB properties from the literature

Dynamic Ranges (1-10 keV). Other source properties were collected from the literature, in order to put the *INTEGRAL* behavior into a wider context: distance, pulsar spin and orbital period, eccentricity of the orbit, maximum and minimum fluxes in soft X–rays (1–10 keV, corrected for the absorption). These latter were investigated since the instruments observing the sky at soft X–rays are much more sensitive than *INTEGRAL* and can probe the true quiescent state in transient sources, together with their variability range between quiescence and outburst peak.

When the published soft X–ray fluxes were not available in the 1–10 keV range, we extrapolated them using WEBPIMMS and the appropriate model found in the literature. Then, we calculated their ratio (the dynamic range "$DR_{1-10\ keV}$" = F_{max} / F_{min}, reported in Table 1, last column). When only a single value for the soft X-ray flux was found, the dynamic range was not calculated ("–" in Table 1) and the flux was ascribed to the "minimum flux". Note that we considered only spin-phase-averaged fluxes for X-ray pulsars, and out-of-eclipse minimum fluxes for eclipsing systems, to obtain the intrinsic range of X–ray variability.

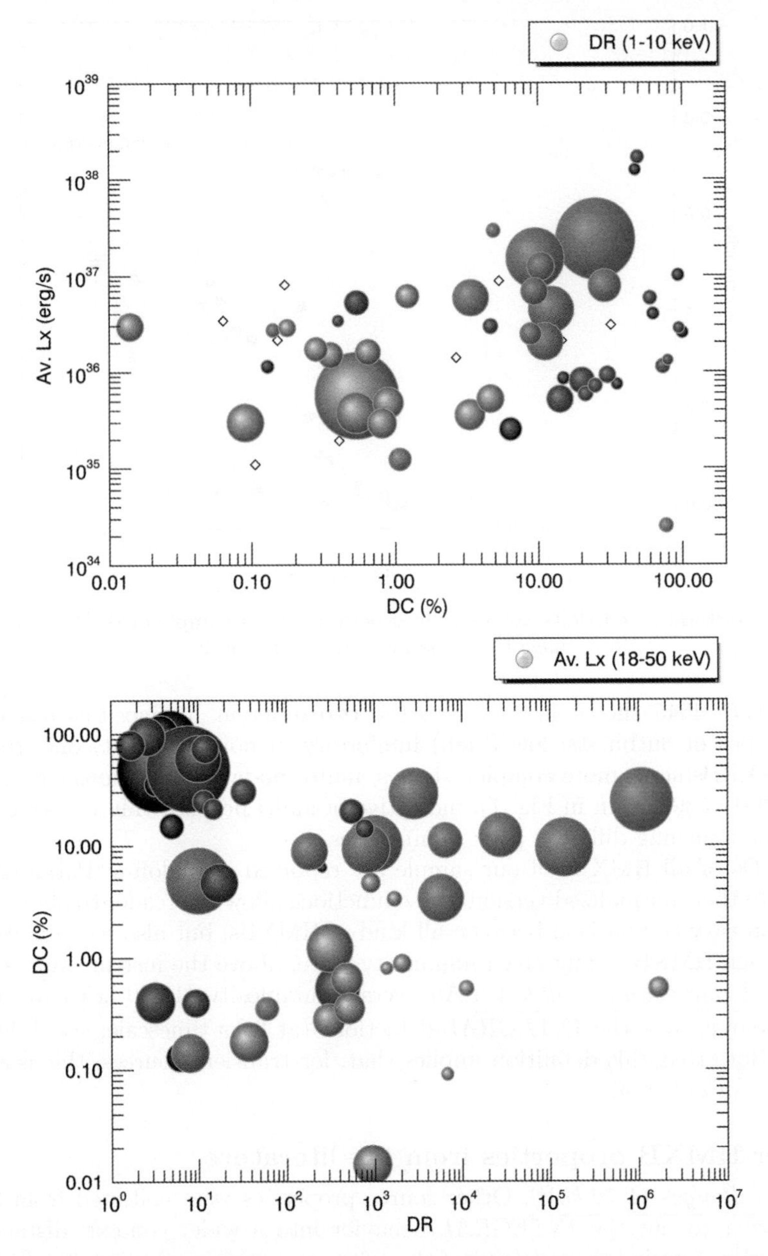

Figure 4. Bubbleplots summarizing the results obtained, focussing on three main quantities: $DC_{18-50\,keV}$, $DR_{1-10\,keV}$ and the average luminosity in the energy range 18–50 keV (note that, for transient sources, this is a luminosity during outburst, as observed by *INTEGRAL*). *Upper panel*: average $L_{18-50\,keV}$ vs $DC_{18-50\,keV}$; each bubble indicates a single source, with the bubble size correlating with the $DR_{1-10\,keV}$. The diamonds mark the position of sources for which the $DR_{1-10\,keV}$ could not be calculated (only a single flux in the 1-10 keV energy band was found in the literature). In blue we mark the SgHMXBs, in red the Be/XRBs and in green the SFXTs. In black, all other systems are indicated (the "other" HMXBs reported in Table 1, together with the giant HMXBs and the symbiotic binary). *Lower panel*: $DC_{18-50\,keV}$ vs $DR_{1-10\,keV}$, with bubble sizes correlating with the average $L_{18-50\,keV}$.

In Fig. 2 we show the source $DC_{18-50\ keV}$ plotted against the minimum and maximum luminosities (1–10 keV), for different HMXB sub-classes: the scatter is huge in the upper panel where the duty cycle is plotted versus the minimum soft X–ray luminosity. The SFXTs are located in the lower left part of the plot, at low $DC_{18-50\ keV}$ and X–ray luminosity in quiescence, while the persistent SgHMXB mostly lie in the upper right part, at both high luminosities and large duty cycles. Be/XRBs appear located in-between them. In the lower panel, where the maximum soft X–ray luminosity is considered, the sub-classes regroup to the right, at more similar luminosities (in outbursts for SFXTs and Be/X-ray transients). A few sources, classified in the literature as SgHMXB (blue circles in Fig. 2), display a very low $DC_{18-50\ keV}$, similar to SFXTs. They might be either mis-classified transients or persistent sources emitting X–rays at a level just below the instrumental sensitivity, that are detected only during sporadic flaring. Note that the HMXBs almost reaching the Eddington luminosity are the RLO systems SMC X–1 and LMC X–4.

Orbital geometry. Among the many trends of source properties we have investigated for our sample (see Sidoli & Paizis, 2018), we report here on the plot showing the system eccentricity versus the orbital period (Fig. 3). Two trends are evident, above P_{orb}~10 d: low eccentricity Be/XRBs with no correlation with the orbital period (X Per is the prototype) and a group of binaries (mostly Be/XRBs) where the eccentricity correlates with the orbital period. SgHMXBs are located at lower eccentricities and orbital periods. This plot has already been investigated in the literature (Townsend *et al.* 2011). The novelty here is the inclusion of SFXTs (not considered by Townsend *et al.* 2011): some of them display circular orbits, while others very eccentric geometries, like IGR J08408-4503 (e=0.63 and P_{orb}=9.54 d) and IGR J11215-5952 (e> 0.8 and P_{orb}=165 d). These SFXTs enable the HMXBs hosting supergiant stars to extend at larger eccentricities and orbital periods, in a parameter space that is unusual even for Be/XRBs.

5. Conclusions

We summarize the results of our systematic analysis in Fig. 4, making use of three characterizing quantities: two of them have been derived from the analysis of fourteen years of *INTEGRAL* observations ($DC_{18-50\ keV}$ and the average 18–50 keV luminosity, in outburst for transients), while the third one has been calculated from soft X–ray fluxes taken (or extrapolated) from the literature ($DR_{1-10\ keV}$).

We have obtained a global view of a large number of HMXBs where the different kind of sources tend to cluster mainly in different region of this 3D space, as follows:

• SgHMXBs (excluding the high luminosity RLO systems) in general show low $DR_{1-10\ keV}$ (< 40), high duty cycles ($DC_{18-50\ keV}$>10 per cent), low average 18–50 keV luminosity (~10^{36} erg s^{-1});

• SFXTs are characterized by high $DR_{1-10\ keV}$ (>100), low duty cycles ($DC_{18-50\ keV}$<5 per cent), low average 18–50 keV luminosity in outburst (~10^{36} erg s^{-1});

• Be/XRTs display a high $DR_{1-10\ keV}$ (>100), intermediate duty cycles ($DC_{18-50\ keV}$~10 per cent), high average 18–50 keV luminosity in outburst (~10^{37} erg s^{-1}).

It is worth mentioning that a number of HMXBs exist that displays intermediate properties, in particular among SgHMXB, sometimes overlapping with some region of the parameter space more typical of SFXTs, bridging together the two sub-classes. This seems to indicate that these two sub-classes have no sharp boundaries, but their phenomenology is based on continuous parameters, from persistent SgHMXBs towards the most extreme SFXT (IGR J17544-2619).

References

Hubrig S., Sidoli L., Postnov K., Schöller M., Kholtygin A. F., Järvinen S. P., & Steinbrunner P., 2018, *MNRAS*, 474, L27

in't Zand J. J. M., 2005, *A&A*, 441, L1

Kreykenbohm I., Wilms J., Kretschmar P., Torrejón J. M., Pottschmidt K., Hanke M., Santangelo A., Ferrigno C., & Staubert R., 2008, *A&A*, 492, 511

Kuehnel M., Kretschmar P., Nespoli E., Okazaki A. T., Schoenherr G., Wilson-Hodge C. A., Falkner S., Brand T., Anders F., Schwarm F.-W., Kreykenbohm I., Mueller S., Pottschmidt K., Fuerst F., Grinberg V., & Wilms J., 2015, in Proceedings of "A Synergistic View of the High Energy Sky" - 10th INTEGRAL Workshop (INTEGRAL 2014). 15-19 September 2014. Annapolis, MD, USA. Published online at http://pos.sissa.it/cgi-bin/reader/conf.cgi?confid=228 The Be X-ray Binary Outburst Zoo II. p. 78

Martínez-Núñez S., Kretschmar P., Bozzo E., Oskinova L. M., Puls J., Sidoli L., Sundqvist J. O., Blay P., *et al.* 2017, *Space Science Reviews*, 212, Issue 1-2, 59

Negueruela I., Reig P., Coe M. J., & Fabregat J., 1998, *A&A*, 336, 251

Negueruela I., & Okazaki A. T., 2001, *A&A*, 369, 108

Negueruela I., Okazaki A. T., Fabregat J., Coe M. J., Munari U., & Tomov T., 2001, *A&A*, 369, 117

Okazaki A. T., & Negueruela I., 2001, *A&A*, 377, 161

Paizis A., Mereghetti S., Götz D., Fiorini M., Gaber M., Regni Ponzeveroni R., Sidoli L., & Vercellone S., 2013, *Astronomy and Computing*, 1, 33

Paizis A., & Sidoli L., 2014, *MNRAS*, 439, 3439

Paizis A., Fiorini M., Franzetti P., Mereghetti S., Regni Ponzeveroni R., Sidoli L., & Gaber M., 2016, in Proceedings of the 11th INTEGRAL Conference Gamma-Ray Astrophysics in Multi-Wavelength Perspective. 10-14 October 2016 Amsterdam, The Netherlands (INTEGRAL2016). INTEGRAL @ INAF-IASF Milano: from Archives to Science, p. 17

Reig P., 2011, *A&Sp Sc.*, 332, 1

Romano P., Bozzo E., Mangano V., Esposito P., Israel G., Tiengo A., Campana S., Ducci L., Ferrigno C., & Kennea J. A., 2015, *A&A*, 576, L4

Shakura N., Postnov K., Kochetkova A., & Hjalmarsdotter L., 2012, *MNRAS*, 420, 216

Shakura N., Postnov K., Sidoli L., & Paizis A., 2014, *MNRAS*, 442, 2325

Sidoli L., Paizis A., Fürst F., Torrejón J. M., Kretschmar P., Bozzo E., & Pottschmidt K., 2015, *MNRAS*, 447, 1299

Sidoli L., 2017, ArXiv e-prints (arXiv:1710.03943), in Proceedings of the XII Multifrequency Behaviour of High Energy Cosmic Sources Workshop. 12-17 June, 2017 Palermo, Italy (MULTIF2017). Published online at http://pos.sissa.it/cgi-bin/reader/conf.cgi?confid=306, id.52

Sidoli L., & Paizis A., 2018, *MNRAS*, 481, 2779

Stella L., White N. E., & Rosner R., 1986, *ApJ*, 308, 669

Townsend L. J., Coe M. J., Corbet R. H. D., & Hill A. B., 2011, *MNRAS*, 416, 1556

Walter R., Lutovinov A. A., Bozzo E., & Tsygankov S. S., 2015, A&A Rev., 23, 2

Winkler C., Courvoisier T., Di Cocco G., Gehrels N., Giménez A., Grebenev S., Hermsen W., Mas-Hesse J. M., *et al.* 2003, *A&A*, 411, L1

Winkler C., Diehl R., Ubertini P., & Wilms J., 2011, *Space Science Reviews*, 161, 149

Discussion

CHATY: The distinction made with the Corbet diagram between different HMXBs (Be systems, SgHMXBs and SFXTs) seems not so valid anymore looking at all your plots (SFXTs for instance covering all parameter range).

SIDOLI: Thanks for your comment. Indeed, SFXTs sometimes cover regions typical of both Be/XRBs and SgHMXBs.

High-mass X-ray Binaries: Illuminating the Passage from Massive Binaries to Merging Compact Objects
Proceedings IAU Symposium No. 346, 2019
L. M. Oskinova, E. Bozzo, T. Bulik & D. Gies, eds.

doi:10.1017/S1743921318007615

Phase connected X-ray light curve and He II radial velocity measurements of NGC 300 X-1

S. Carpano[1], F. Haberl[1], P. Crowther[2] and A. Pollock[2]

[1]Max-Planck-Institut für extraterrestrische Physik,
Giessenbachstraße 1, 85748 Garching, Germany
emails: scarpano@mpe.mpg.de, fwh@mpe.mpg.de

[2]Department of Physics and Astronomy & Space Physics,
University of Sheffield, Sheffield S3 7RH, UK
emails: paul.crowther@sheffield.ac.uk, a.m.pollock@sheffield.ac.uk

Abstract. NGC 300 X-1 and IC 10 X-1 are currently the only two robust extragalactic candidates for being Wolf-Rayet/black hole X-ray binaries, the Galactic analogue being Cyg X-3. These systems are believed to be a late product of high-mass X-ray binary evolution and direct progenitors of black hole mergers. From the analysis of Swift data, the orbital period of NGC 300 X-1 was found to be 32.8 h. We here merge the full set of existing data of NGC 300 X-1, using XMM-Newton, Chandra and Swift observations to derive a more precise value of the orbital period of 32.7932±0.0029 h above a confidence level of 99.99%. This allows us to phase connect the X-ray light curve of the source with radial velocity measurements of He II lines performed in 2010. We show that, as for IC 10 X-1 and Cyg X-3, the X-ray eclipse corresponds to maximum of the blueshift of the He II lines, instead of the expected zero velocity. This indicates that for NGC 300 X-1 as well, the wind of the WR star is completely ionised by the black hole radiation and that the emission lines come from the region of the WR star that is in the shadow. We also present for the first time the light curve of two recent very long XMM-Newton observations of the source, performed on the 16th to 20th of December 2016.

Keywords. X-rays: individuals: NGC 300 X-1, X-rays: binaries, Stars: Wolf-Rayet, Accretion, accretion disks, Methods: data analysis

1. Introduction

Wolf-Rayet/black hole X-ray binaries are believed to be a late product of high-mass X-ray binaries and direct progenitors of black hole mergers (Bulik *et al.* 2011; Bogomazov 2014). Following the standard model, when the massive secondary star completes its main sequence evolution, it becomes a supergiant filling its Roche Lobe. The system is then embedded in a common envelope and the binary components may get closer. The envelope from the massive star is lost leaving the system composed of a Wolf-Rayet star and a relativistic object (here a black hole).

The association of the X-ray source NGC 300 X-1 with a Wolf-Rayet star was reported by Carpano *et al.* (2007a); Crowther *et al.* (2007). From the analysis of Swift data, the orbital periods for NGC 300 X-1 and IC 10 X-1, were found to be 32.8 h (Carpano *et al.* 2007b) and 34.9 h (Prestwich *et al.* 2007), respectively, and were confirmed in the optical band by radial velocity (RV) measurements of He II lines (Crowther *et al.* 2010, Silverman & Filippenko 2008).

A more precise value of 34.84306 h for the period of IC 10 X-1 has been provided by Laycock *et al.* (2015a), by combining XMM-Newton and Chandra data. This allowed to phase connect the X-ray light curve and the RV data showing that the X-ray eclipse is associated to the maximum of the blueshift of the He II line, instead of the expected zero velocity, found for example for the X-ray binary M33 X-7 (Orosz *et al.* 2007). This feature was also reported for the Galactic Wolf-Rayet/black-hole X-ray candidate Cyg X-3 (van Kerkwijk 1993, van Kerkwijk *et al.* 1996, Hanson *et al.* 2000). The model invoked to explain this 1/4 phase shift, is that the wind of the Wolf-Rayet star is completely ionised by the black hole radiation, except in the region shadowed by the star from which most of the emission lines are originating. As explained in more detailed in Laycock *et al.* (2015b) for IC10 X-1, during X-ray eclipse when the WR star lies between the black hole and the observer, the wind velocity vector lies along our line of sight producing the maximum blueshift of the emission lines. At a quarter of phase earlier or later, the wind in the shielded sector is orthogonal to the line of sight, producing zero Doppler shift.

2. Observations and data reduction

2.1. *Swift observations*

For this work we used 86 Swift XRT observations from the 5th of September 2006 to 19th June 2018 performed in photon counting mode and with total exposures varying from ∼300 s to ∼10000 s. The first long set was triggered on December 2016, after the discovery that NGC 300 X-1 was associated with a Wolf-Rayet star, and led to the discovery of the 32.8 h orbital period.

The event files were barycenter-corrected using the `ftool barycorr` from the HEASoft package. For every observation, we extracted the number of source and background counts obtained in every GTI defined in the file header. Net count rates were then obtained by subtracting the number of background counts from the source counts after scaling for the respective areas and by dividing by the GTI length.

2.2. *XMM-Newton observations*

There are currently 7 XMM-Newton observations available for NGC 300, performed between December 2000 and December 2016. The first four, published in Carpano *et al.* (2005), Carpano *et al.* (2007a), have an exposure of about ∼40 ks each, while the fifth, pointing on a supernova, was lasting only ∼20 ks. The last two observations performed on three consecutive days have a very long exposure of 140+80 ks, to cover two orbital periods. These were carried out almost simultaneously to a 163 ks long NuSTAR observation, starting on the 16th of December 2016, to study the hard (3–79 keV) component of the electromagnetic spectrum.

The data were reduced following standard procedures using the XMM-Newton SAS data software version 15.0.0. Single to quadruple events were used for MOS (Turner *et al.* 2001) cameras, with FLAG=0, while NGC 300 X-1 was partially or totally in a pn (Strueder *et al.* 2001) CCD gap for all observations. The event files were barycenter-corrected using the SAS `barycen` tool.

2.3. *Chandra observations*

There are currently 5 Chandra observations available with NGC 300 X-1 in the field-of-view, 4 of them with the Advanced CCD Imaging Spectrometer (ACIS) instrument (Garmire *et al.* 2003) and one with the High Resolution Camera (Murray *et al.* 2000), in the period between June 2006 and November 2014, with exposures from 10 to 65 ks. The data were reduced using the Chandra X-ray data analysis software, `ciao` version 4.9. The event files were barycenter-corrected using the `axbary` tool.

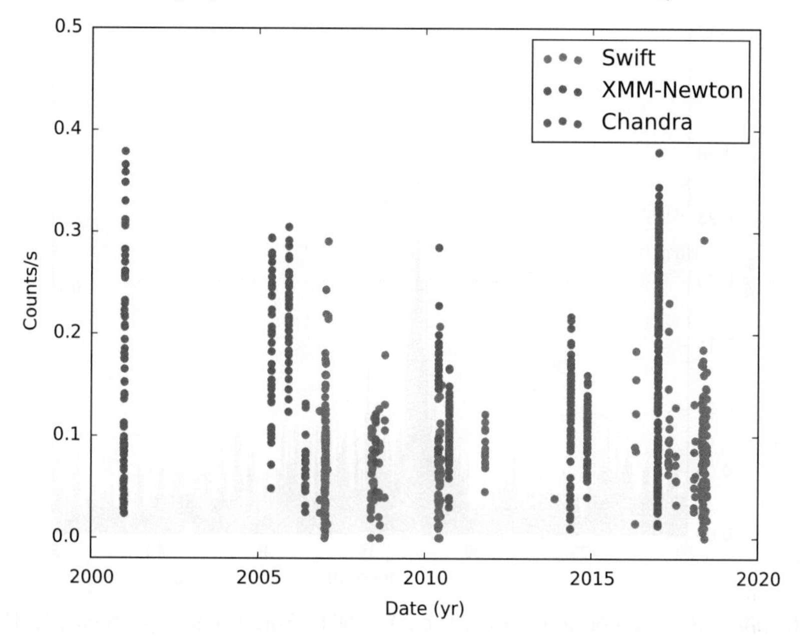

Figure 1. Long-term light curve of NGC 300 X-1, containing the Swift (green), XMM-Newton (blue) and Chandra (red) data and taking into account differences in the effective area from the various instruments and vignetting issues. The count rate values represent what is expected for MOS1+MOS2 data.

3. Light curve analysis

3.1. *Refinement of the orbital period*

We refined the value of the orbital period by combining all available observations from Swift, XMM-Newton and Chandra. The source extraction region was centered around the Chandra coordinates derived by Binder *et al.* (2011), which is $\alpha_{J2000} = 00^h55^m10\fs00$, $\delta_{J2000} = -37^\circ42'12\farcs2$. For the XMM-Newton observations, we only merged MOS1 and MOS2 data, since the source lies partially or totally in a pn CCD gap for all observations. We take into account differences in the effective area from the various instruments and vignetting issues. For this, we created simulated source spectra for every observation using the `fakeit` command of XSPEC with the relevant response files. The model spectrum is an absorbed power-law (`phabs*power`) with an N_H value of $5{\times}10^{20}\,cm^{-2}$ and Γ=2.45, similar to the model fitted to the first 4 XMM-Newton observations in Carpano *et al.* (2007a). The count rate of the simulated spectra, extracted in the 0.2–10.0 keV band, is then used to scale the observed count rates, since these represent the values expected for a constant source. The light curve segments from XMM-Newton and Chandra are rebinned to 1000 s, while for Swift, there is a data point for every valid GTI. The long-term light curve, spanning over more than 17 years, is shown in Fig. 1 with a count rate value scaled to what is expected for MOS1+MOS2.

We then determined the new value of the orbital period using the Lomb-Scargle periodogram analysis (Lomb 1976, Scargle 1982), in the full (0.2–10 keV) energy band and in the period range from 20 to 50 h, using the Python algorithm from the `gatspy` library. Fig. 2 shows the result of the period search, with a high peak at 32.7932 h above a confidence level of 99.99%, measured using Monte Carlo simulations and assuming white noise. We estimate the period error to be 0.0029 h by associating it to the 1σ width of a gaussian function fitted on the highest peak.

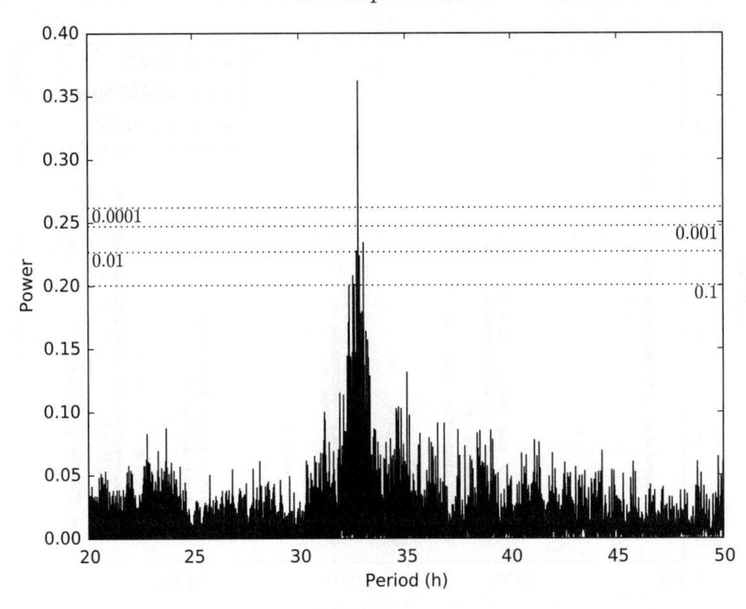

Figure 2. Lomb-Scargle periodogram in the 20 to 50 h, from the merged Swift, XMM-Newton and Chandra light curve shown in Fig. 1, with confidence levels at 90%, 99%, 99.9% and 99.99%, assuming white noise.

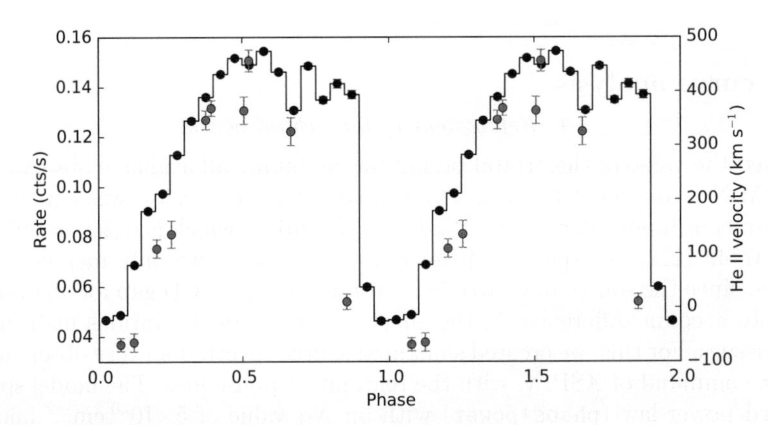

Figure 3. X-ray light curve of NGC 300 X-1 of Fig. 1 with the VLT/FORS He II radial velocity measurements (red) folded at the best period (~32.7932 h). Phase 0 corresponds to the start of the first XMM-Newton observation.

3.2. *Phase connected folded X-ray light curve and RV measurements*

In Fig. 3, we fold the long term X-ray light curve of NGC 300 X-1 of Fig. 1 with the best period derived from the periodogram. We also overlay the radial velocity measurements of the VLT/FORS He II lines (λ4686) reported in Table 1 of Crowther *et al.* (2010) folded at the same period, after converting times to the same reference. Phase 0 corresponds to the start of the first XMM-Newton observation performed on the 26th of December 2000. As for IC 10 X-1 and Cyg X-3, the X-ray eclipse corresponds to maximum of the blueshift of the He II lines, instead of the expected zero velocity.

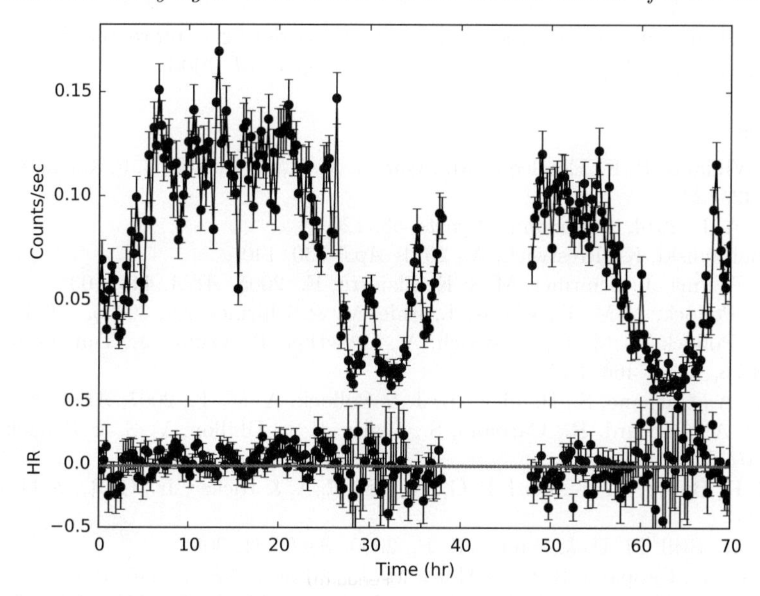

Figure 4. Lightcurve from the 17th to 20th of December 2016 XMM-Newton observation (top) and hardness ratios HR=(H-S)/(H+S), with S=0.2–1 keV and H=1–10 keV.

3.3. *Analysis of the 2016 XMM-Newton observations*

NGC 300 X-1 was observed during two long (135+80 ks) observations from 17th to 20th of December 2016, covering a bit more than two orbital cycles. The observation was performed simultaneously with NuSTAR, but the source was too soft to be detected in this harder 3–79 keV band. Fig. 4 (top) shows the corresponding light curve, expanding over about 70 h with a gap of about 9 h, and the hardness ratio (bottom) defined as HR=(H-S)/(H+S) with the soft and hard bands being 0.2–1 keV and 1–10 keV respectively.

It is the first time the full orbit is covered continuously allowing a proper visualization of the light curve structure. We can notice that there is some erratic X-ray variability in the entire light curve likely caused by X-ray scattering of the photons by the wind of the WR star. There is no significant variation in the hardness ratios between inside and outside eclipse. We also note the presence of flares during the eclipse.

4. Conclusions

We have shown in this work that one has to be cautious when measuring black hole or neutron star masses using radial velocity measurements. For all three Wolf-Rayet/compact object systems NGC 300 X-1, IC 10 X-1 and Cyg X-3, a shift of 1/4 of orbital phase was observed between the folded radial velocity measurements and the X-ray light curves, using the He II lines. This indicates that somehow the lines are not emitted isotropically and therefore do not trace the motion of the Wolf-Rayet star preventing to use them to calculate the compact object mass function. An explanation for this anisotropy could be that the Wolf-Rayet wind is fully ionised by the black hole radiation except in the part that is in the shadow.

One should therefore look for other absorption/emission lines that have the right phase in the RV curve with respect to the X-ray eclipse (zero velocity at minimum X-ray flux), in order to estimate the compact object mass. For Cyg X-3, He I absorption lines, whose RV curve had the right phase but a lower amplitude, were detected during an observation when the system was in outburst. This helped to derive a more reliable value of the compact object mass function of 0.027 $M_\odot$ (Hanson *et al.* 2000).

References

Binder, B., Williams, B. F., Eracleous, M. , Garcia, M. R., Anderson, S. F. & Gaetz, T. J., 2011, *ApJ*, 742, 128
Bogomazov, A. I., 2014, *Astronomy Reports*, 58, 126
Bulik, T., Belczynski, K., Prestwich, A., 2011, *ApJ*, 730, 140
Carpano, S., Wilms, J., Schirmer, M. & Kendziorra, E., 2005, *A&A*, 443, 103
Carpano, S., Pollock, A. M. T., Wilms, J., Ehle, M. & Schirmer, M., 2007a, *A&A*, 461, L9
Carpano, S., Pollock, A. M. T., Prestwich, A., Crowther, P., Wilms, J., Yungelson, L. & Ehle, M., 2007b, *A&A*, 466, L17
Crowther, P. A., Carpano, S., Hadfield, L. J. & Pollock, A. M. T., 2007, *A&A*, 469, L31
Crowther, P. A., Barnard, R., Carpano, S., Clark, J. S., Dhillon, V. S. & Pollock, A. M. T., 2010, *MNRAS*, 403, L41
Garmire, G. P., Bautz, M. W., Ford, P. G., Nousek, J. A. & Ricker, Jr., G. R., 2003, *Proc. SPIE*, 4851, 28
Hanson, M. M., Still, M. D. & Fender, R. P., 2000, *ApJ*, 541, 308
Laycock, S. G. T., Cappallo, R. C. & Moro, M. J., 2015a, *MNRAS*, 446, 1399
Laycock, S. G. T., Maccarone, T. J. & Christodoulou, D. M., 2015b, *MNRAS*, 452, L31
Lomb, N. R., 1976, *Ap&SS*, 39, 447
Murray, S. S.,Chappell, J. H., Kenter, A. T., Juda, M., Kraft, R. P., Zombeck, M. V., Meehan, G. R., Austin, G. K. & Gomes, J. J., 2000, *Proc. SPIE*, 4140, 144
Orosz, J. A., McClintock, J. E., Narayan, R., Bailyn, C. D., Hartman, J. D., Macri, L., Liu, J., Pietsch, W., Remillard, R. A., Shporer, A. & Mazeh, T., 2007, *Nature*, 449, 872
Prestwich, A. H., Kilgard, R., Crowther, P. A., Carpano, S., Pollock, A. M. T., Zezas, A., Saar, S. H., Roberts, T. P. & Ward, M. J., 2007, *ApJ*, 669, L21
Scargle, J. D., 1982, *ApJ*, 263, 835
Silverman, J. M. & Filippenko, A. V., 2008, *ApJ*, 678, L17
Strüder, L., Briel, U., Dennerl, K., Hartmann, R., Kendziorra, E., Meidinger, N. *et al.*, 2001, *A&A*, 365, L18
Turner, M. J. L., Abbey, A., Arnaud, M., Balasini, M., Barbera, M., Belsole, E., Bennie *et al.*, 2001, *A&A*, 365, L27
van Kerkwijk, M. H., 1993, *A&A*, 276, L9
van Kerkwijk, M. H., Geballe, T. R., King, D. L., van der Klis, M. & van Paradijs, J., 1996, *A&A*, 314, 521

Discussion

ANDREAS SANDER: Can you tell me anything about the Hydrogen content? Assuming that it is similar to Cyg X-3, which is hydrogen-free, one could assume that the moderate terminal velocity compared to single hydrogen-free WNs might be an argument for the wind being significantly affected by irradiation.

STEFANIA CARPANO: Indeed, like for Cyg X-3, there are no Hydrogen lines in the spectrum of the Wolf-Rayet companion star of NGC 300 X-1.

High-mass X-ray Binaries: Illuminating the Passage from Massive Binaries to Merging Compact Objects
Proceedings IAU Symposium No. 346, 2019
L. M. Oskinova, E. Bozzo, T. Bulik & D. Gies, eds.

doi:10.1017/S1743921318008426

On the origin of Supergiant Fast X-ray Transients

Swetlana Hubrig[1], Lara Sidoli[2], Konstantin A. Postnov[3], Markus Schöller[4], Alexander F. Kholtygin[5] and Silva P. Järvinen[1]

[1]Leibniz-Institut für Astrophysik Potsdam (AIP), An der Sternwarte 16, 14482 Potsdam, Germany
email: shubrig@aip.de

[2]INAF, Istituto di Astrofisica Spaziale e Fisica Cosmica, Via E. Bassini 15, 20133 Milano, Italy

[3]Sternberg Astronomical Institute, Moscow M.V. Lomonosov State University, 119234 Moscow, Russia

[4]European Southern Observatory, Karl-Schwarzschild-Str. 2, 85748 Garching, Germany

[5]Saint-Petersburg State University, Universitetskij pr. 28, 198504 Saint-Petersburg, Russia

Abstract. A fraction of high-mass X-ray binaries are supergiant fast X-ray transients. These systems have on average low X-ray luminosities, but display short flares during which their X-ray luminosity rises by a few orders of magnitude. The leading model for the physics governing this X-ray behaviour suggests that the winds of the donor OB supergiants are magnetized. In agreement with this model, the first spectropolarimetric observations of the SFXT IGR J11215-5952 using the FORS 2 instrument at the Very Large Telescope indicate the presence of a kG longitudinal magnetic field. Based on these results, it seems possible that the key difference between supergiant fast X-ray transients and other high-mass X-ray binaries are the properties of the supergiant's stellar wind and the physics of the wind's interaction with the neutron star magnetosphere.

Keywords. stars: magnetic fields, stars: individual (IGR J08408–4503, IGR J11215-5952), (stars:) supergiants, (stars:) binaries: general, X-rays: stars

---

1. Introduction

Among the bright X-ray sources in the sky, a significant number contain a compact object (either a neutron star or a black hole) accreting from the wind of a companion star with a mass above 10 $M_{\odot}$. Such systems are called high-mass X-ray binaries (HMXBs). They are young (several dozen million years old) and can be formed when one of the initial binary members loses a significant part of its mass through stellar wind or mass transfer before a first supernova explosion occurs (van den Heuvel & Heise 1972).

Supergiant Fast X-ray Transients (SFXTs) are a subclass of HMXBs associated with early-type supergiant companions, and characterized by sporadic, short and bright X–ray flares reaching peak luminosities of 10^{36}–10^{37} erg s^{-1} and typical energies released in bright flares of about 10^{38}–10^{40} erg – see the review by Sidoli (2017) for more details. Their X-ray spectra in outburst are very similar to accreting pulsars in HMXBs. In fact, half of them have measured neutron star spin periods similar to those observed from persistent HMXBs (Shakura *et al.* 2015; Martinez-Nunez *et al.* 2017). The physical mechanism driving their transient behavior, probably related to the accretion of matter from the supergiant wind by the compact object, has been discussed by several authors and is still a matter of debate. The leading model for the existence of SFXTs invokes their different wind properties and magnetic field strengths that lead to distinctive accretion

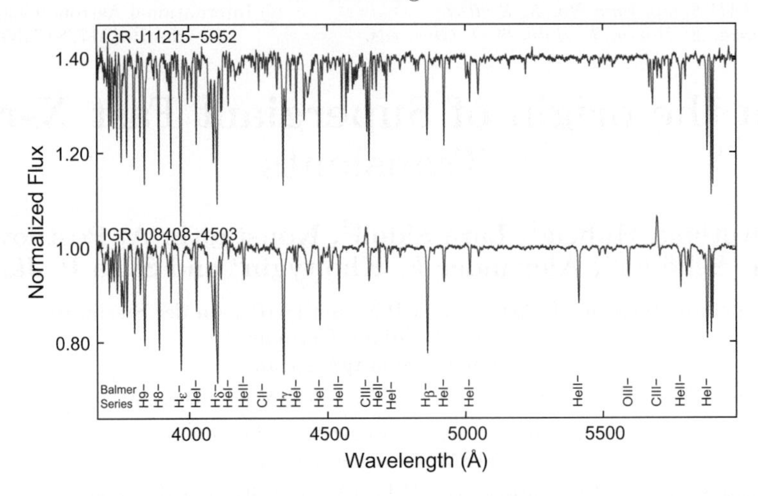

Figure 1. Normalised FORS 2 Stokes I spectra of IGR J08408−4503 and IGR J11215−5952. Well known spectral lines are indicated at the bottom. The spectrum of IGR J11215−5952 was vertically offset by 0.4 for clarity.

regimes (Shakura *et al.* 2012; Postnov *et al.* 2015; Shakura *et al.* 2015; Shakura & Postnov 2017). The SFXTs' behaviour can be explained by sporadic capture of magnetized stellar wind. The effect of the magnetic field carried by the stellar wind is twofold: first, it may trigger rapid mass entry to the magnetosphere via magnetic reconnection in the magnetopause (a phenomenon that is well known in the dayside of Earth's magnetosphere), and secondly, the magnetized parts of the wind (magnetized clumps with a tangent magnetic field) have a lower velocity than the non magnetised parts (or the parts carrying the radial field; Shakura *et al.* 2015). The model predicts that a magnetized clump of stellar wind with a magnetic field strength of a few 10 G triggers sporadic reconnection, allows accretion, and results in an X-ray flare. Typically, the neutron star orbital separation is a few $R_{*,\mathrm{RSG}}$. Thus, the expected required magnetic field on the stellar surface is of the order of 100 – 1000 G.

2. Magnetic field measurements

To investigate the magnetic nature of SFXTs, we recently observed the two optically brightest targets, IGR J08408-4503 ($P_{\mathrm{orb}} = 9.5$ d) and IGR J11215-5952 ($P_{\mathrm{orb}} = 165$ d), using the FOcal Reducer low dispersion Spectrograph (FORS 2; Appenzeller et al. 1998) mounted on the 8 m Antu telescope of the Very Large Telescope in spectropolarimetric mode. No significant magnetic field was measured in the spectra of IGR J08408-4503, with the highest value $\langle B_{\mathrm{z}}\rangle_{\mathrm{hyd}} = -184 \pm 97$ G at a significance level of 1.9σ. On the other hand, a definite magnetic field detection was achieved for IGR J11215-5952 in 2016 December with $\langle B_{\mathrm{z}}\rangle_{\mathrm{hyd}} = 416 \pm 110$ G with a significance at the 3.8σ level. The measurement obtained in 2016 May yielded $\langle B_{\mathrm{z}}\rangle_{\mathrm{hyd}} = -978 \pm 308$ G at a significance level of 3.2σ (Hubrig et al. 2018). The spectral appearance of IGR J08408−4503 and IGR J11215−5952 in the FORS 2 spectra is presented in Fig. 1.

In Fig. 2, we present Stokes V spectra of IGR J11215−5952 obtained on these two nights in the spectral region around the Hβ line. For best visibility of the Zeeman features, we overplot the Stokes V spectra of IGR J11215−5952 with the Stokes V spectra of the two well-known magnetic early B-type stars HD 96446 and ξ^1 CMa.

Regarding the significance of the magnetic field detections in massive stars at significance levels around 3σ, we note that the two clearly magnetic Of?p stars HD 148937 and CPD $-28°5104$ have been for the first time detected as magnetic in our FORS 2

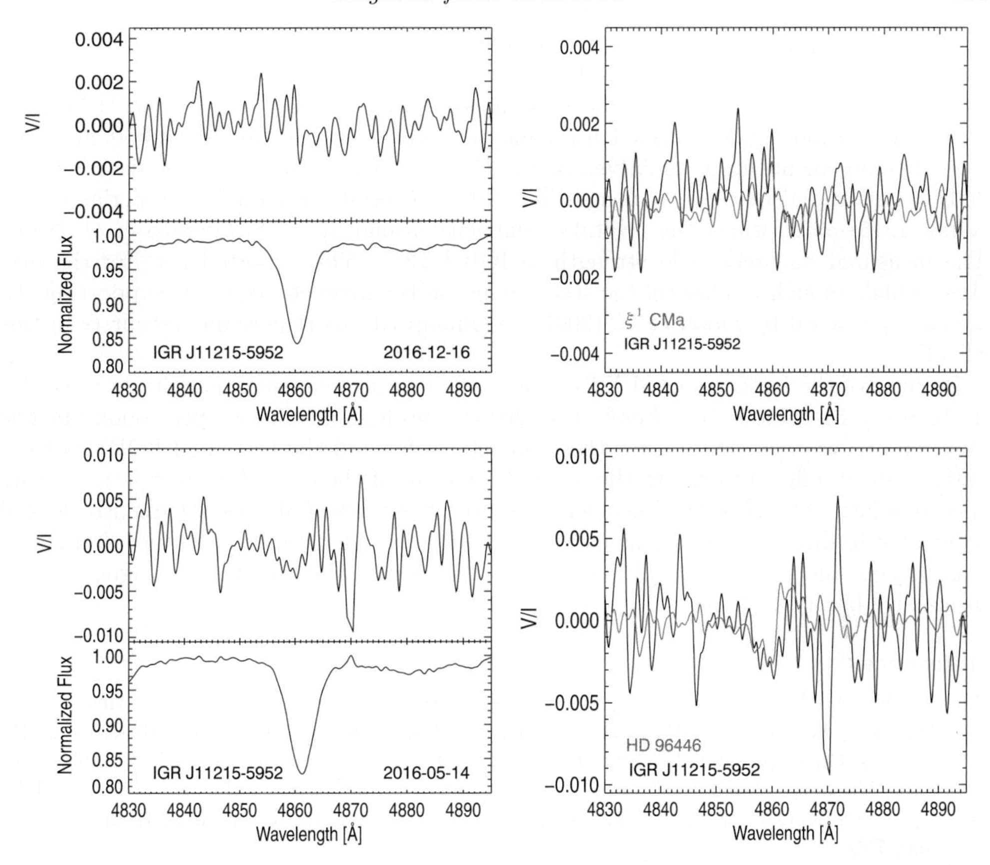

Figure 2. *Left panel*: Stokes V and Stokes I spectra of IGR J11215−5952 in the spectral region around the Hβ line at two different epochs. *Right panel*: Stokes V spectra of IGR J11215−5952 overplotted with the Stokes V spectra of the two well-known magnetic early-B type stars ξ^1 CMa ($\langle B_z \rangle_{\rm hydr} = 360 \pm 49$ G) and HD 96446 ($\langle B_z \rangle_{\rm hydr} = -1590 \pm 74$ G) for best visibility of the Zeeman features.

observations at significance levels of 3.1σ and 3.2σ, respectively (Hubrig *et al.* 2008, 2011). The detection of a magnetic field in IGR J11215−5952 at significance levels of 3.2σ and 3.8σ indicates that this target likely possesses a kG magnetic field. Although no significant magnetic field was measured in the spectra of IGR J08408-4503, due to the presence of distinct Zeeman features in its spectra it appears still valuable to obtain additional spectropolarimetric observations on other epochs corresponding to different orbital phases.

3. Discussion

Our spectropolarimetric observations of IGR J11215−5952 revealed the presence of a magnetic field on two occasions. This target is the only SFXT where strictly periodic X-ray outbursts have been observed, repeating every 164.6 d (Sidoli *et al.* 2006, 2007; Romano *et al.* 2009). To explain these short periodic outbursts, Sidoli *et al.* (2007) proposed that they are triggered by the passage of the neutron star inside an equatorial enhancement of the outflowing supergiant wind, focussed on a plane inclined with respect to the orbit. This configuration of the line-driven stellar wind might be magnetically channeled (ud-Doula & Owocki 2002). The effectiveness of the stellar magnetic field in focussing the wind is indicated by the wind magnetic confinement parameter η defined as $\eta = B_*^2 R_*^2 \, / \, (\dot{M}_{\rm w} v_\infty)$, where B$_*$ is the strength of the magnetic field at the surface of the supergiant, R$_*$ is the stellar radius (R$_*$=40 R$_\odot$), v_∞ is the wind terminal velocity ($v_\infty = 1200$ $km\ s^{-1}$) and $\dot{M}_{\rm w}$ is the wind mass loss rate ($\dot{M}_{\rm w} = 10^{-6}$ M$_\odot$ yr^{-1}; Lorenzo *et al.* 2014). Adopting B$_* \geq 0.7$ kG at the magnetic equator, we estimate $\eta \geq 500$, implying a wind confinement, up to an Alfvén radius $R_A = \eta^{1/4} R_* \geq 4.73\, R_*$ (ud-Doula & Owocki 2002). This radial distance is compatible with the orbital separation at periastron in IGR J11215−5952, where the orbital eccentricity is high ($e > 0.8$; Lorenzo *et al.* 2014). The measured magnetic field strength in IGR J11215−5952 reported here for the first time is high enough to channel the stellar wind on the magnetic equator, supporting the scenario proposed by Sidoli *et al.* (2007) to explain the short periodic outbursts in this SFXT.

Because of the faintness of SFXTs – most of them have a visual magnitude $m_V \geq 12$, up to $m_V \geq 31$ (Sidoli 2017; Persi *et al.* 2015) – no high-resolution spectropolarimetric observations were carried out for these objects so far and the presented FORS 2 observations are the first to explore the magnetic nature of the optical counterparts. Future spectropolarimetric observations of a representative sample of SFXTs are urgently needed to be able to draw solid conclusions about the role of magnetic fields in the wind accretion process and to constrain the conditions that enable the presence of magnetic fields in massive binary systems.

References

Appenzeller, I., Fricke, K., Fürtig, W., Gässler, W., Häfner, R., Harke, R., Hess, H.-J., Hummel, W., Jürgens, P., Kudritzki, R.-P., Mantel, K.-H., Meisl, W., Muschielok, B., Nicklas, H., Rupprecht, G., Seifert, W., Stahl, O., Szeifert, T., & Tarantik, K. 1998, *The ESO Messenger*, 94, 1

Hubrig, S., Schöller, M., Schnerr, R.S., González, J. F., Ignace, R., & Henrichs, H. 2008, *A&A*, 490, 793

Hubrig, S., Schöller, M., Kharchenko, N. V., Langer, N., de Wit, W. J., Ilyin, I., Kholtygin, A. F., Piskunov, A. E., Przybilla, N., & Magori Collaboration 2011, *A&A*, 528, A151

Hubrig, S., Sidoli, L., Postnov, K., Schöller, M., Kholtygin, A. F., Järvinen, S. P., & Steinbrunner, P. 2018, *MNRAS*, 474, L27

Lorenzo, J., Negueruela, I., Castro, N., Norton, A. J., Vilardell, F., & Herrero, A. 2014, *A&A*, 562, A18

Martinez-Nunez, S., Kretschmar, P., Bozzo, E., Oskinova, L. M., Puls, J., Sidoli, L., Sundqvist, J. O., Blay, P., Falanga, M., Fürst, F., Gímenez-García, A., Kreykenbohm, I., Kühnel, M., Sander, A., Torrejón, J. M., & Wilms, J. 2017, *SSRv*, 212, 59

Persi, P., Fiocchi, M., Tapia, M., Roth, M., Bazzano, A., Ubertini, P., & Parisi, P. 2015, *AJ*, 150, 21

Postnov, K. A., Gornostaev, M. I., Klochkov, D., Laplace, E., Lukin, V. V., & Shakura, N. I. 2015, *MNRAS*, 452, 1601

Romano, P., Sidoli, L., Cusumano, G., Vercellone, S., Mangano, V., & Krimm, H. A. 2009, *ApJ*, 696, 2068

Shakura, N., Postnov, K., Sidoli, L., & Paizis, A. 2014, *MNRAS*, 420, 216

Shakura, N. I., Postnov, K. A., Kochetkova, A. Yu., Hjalmarsdotter, L., Sidoli, L., & Paizis, A. 2015. *ARep*, 59, 645

Shakura, N., & Postnov, K. 2017, *Accretion Processes in Cosmic Sources, September 5-10, 2016, St-Petersburg, also:* arXiv:1702.03393

Sidoli, L., Paizis, A., & Mereghetti, S. 2006, *A&A*, 450, L9

Sidoli, L., Romano, P., Mereghetti, S., Paizis, A, Vercellone, S., Mangano, V., & Götz, D. 2007, *A&A*, 476, 1307

Sidoli, L. 2017, *Proc. of the "XII Multifrequency Behaviour of High Energy Cosmic Sources Workshop", 12-17 June, 2017 Palermo, Italy*, 52

van den Heuvel, E. P. J., & Heise, J. 1972, *Nature Physical Science*, 239, 67

ud-Doula, A., & Owocki, S. P. 2002, *ApJ*, 576, 413

High-mass X-ray Binaries: Illuminating the Passage from Massive Binaries to Merging Compact Objects
Proceedings IAU Symposium No. 346, 2019
L. M. Oskinova, E. Bozzo, T. Bulik & D. Gies, eds.

doi:10.1017/S1743921319000024

Modeling of hydrodynamic processes within high-mass X-ray binaries

Petr Kurfürst[1,2] and Jiří Krtička[1]

[1]Department of Theoretical Physics and Astrophysics, Masaryk University, Kotlářská 2, CZ-611 37 Brno, Czech Republic
emails: petrk@physics.muni.cz, krticka@physics.muni.cz

[2]Institute of Theoretical Physics, Charles University, V Holešovičkách 2, 180 00 Praha 8, Czech Republic
email: petrk@physics.muni.cz

Abstract. High-mass X-ray binaries belong to the brightest objects in the X-ray sky. They usually consist of a massive O or B star or a blue supergiant while the compact X-ray emitting component is a neutron star (NS) or a black hole. Intensive matter accretion onto the compact object can take place through different mechanisms: wind accretion, Roche-lobe overflow, or circumstellar disk. In our multi-dimensional models we perform numerical simulations of the accretion of matter onto a compact companion in case of Be/X-ray binaries. Using Bondi-Hoyle-Littleton approximation, we estimate the NS accretion rate. We determine the Be/X-ray binary disk hydrodynamic structure and compare its deviation from isolated Be stars' disk. From the rate and morphology of the accretion flow and the X-ray luminosity we improve the estimate of the disk mass-loss rate. We also study the behavior of a binary system undergoing a supernova explosion, assuming a blue supergiant progenitor with an aspherical circumstellar environment.

Keywords. massive stars, evolution, mass-loss, accretion, supernovae

1. Be/X-ray binaries

X-ray emission in Be/X-ray binaries comes from accretion of the material of the disk onto neutron star (NS) (Reig 2011). Its binary separation D provides, due to the disk truncation, the determining constraint on the outer disk radius. We employ Bondi-Hoyle-Littleton (BHL) approximation (excluding the cases of very small disk scale-height H very close to the Be star, where BHL approximation may be an excessive simplification (Okazaki & Negueruela 2001)) where the NS accretes within the radius

$$r_{\rm acc} = \frac{2GM_X}{v_{\rm rel}^2}, \tag{1.1}$$

where M_X is the mass of the NS and $v_{\rm rel}$ is the relative velocity of NS and the disk. We distinguish two limiting cases for the disk-NS systems (Krtička *et al.* 2015):

- the NS is corotating with the disk material, $v_{\rm rel} = v_R$,
- the disk is truncated far from NS; in this case $v_{\rm rel}^2 \approx v_R^2 + v_\phi^2$.

Figure 1 shows the accretion radii of the two cases up to the distance $R = 10^4\, R_\star$. In systems with low orbital eccentricity we expect the disk truncation at the $3:1$ resonance radius (Okazaki & Negueruela 2001), that is, $R_{\rm trunc}/D \approx 0.48$. If $r_{\rm acc}$ exceeds the disk scale-height H, the NS accretes all the disk material and the X-ray luminosity is

$$L_X = \frac{GM_X\dot{M}}{R_X}, \tag{1.2}$$

where $\dot{M}$ is the accretion rate and R_X is the NS radius (Krtička *et al.* 2015).

Table 1. Parameters of selected Be/X-ray binaries

Binary	Sp. Type	$T_{\rm eff}$ [kK]	$R\,[R_\odot]$	$D\,[R_\odot]$	L_X [W]
V831 Cas	B1V	24	4.5	480	2×10^{28}
IGR J16393-4643	BV	24	4.5	18.8	4×10^{28}
V615 Cas	B0Ve	26	4.9	43	5×10^{28}
HD 259440	B0Vpe	30	5.8	510	1.2×10^{26}
HD 215770	O9.7IIIe	28	12.8	260	6.5×10^{29}
CPD-632495	B2Ve	34	7.0	177	3.5×10^{27}
GRO J1008-57	B0eV	30	5.8	390	3×10^{30}

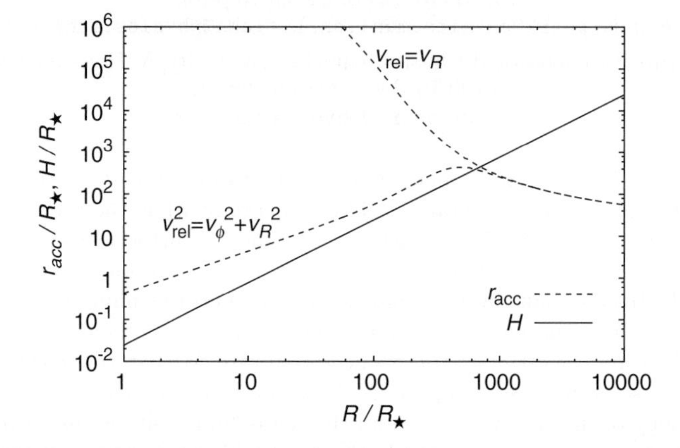

Figure 1. Comparison of $r_{\rm acc}$ and H for a corotating NS ($v_{\rm rel}=v_R$) and for the case when the disk is truncated far from the NS, where $v_{\rm rel}^2 \approx v_R^2+v_\phi^2$. Adapted from Krtička *et al.* (2015).

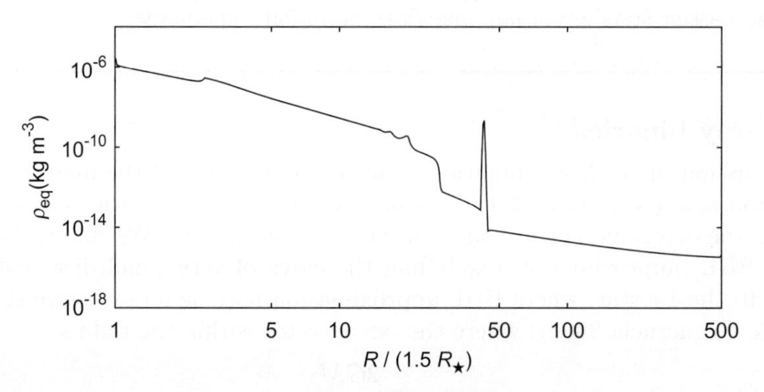

Figure 2. Disk midplane (equatorial plane) density profile $\rho_{\rm eq}$ in the direction of the corotating NS GROJ100857 (see Tab. 1), the parameters of the model are: $D\approx 67\,R_\star$, $T_{\rm eff}=32\,000$ K, $L_X = 3\times10^{28}$ J s^{-1}, $r_{\rm acc}/H \propto 10^4$. The distance is scaled to the equatorial radius $R_{\rm eq}$ of a critically rotating star, $R_{\rm eq}=1.5\,R_\star$.

We analyzed the sample of Be/X-ray binaries (see Tab. 1) for which $r_{\rm acc}>H$. With known X-ray luminosity we estimated from Eq. (1.2) $\dot M \sim 10^{-13}$ - $10^{-9}\,M_\odot\,{\rm yr}^{-1}$ (Krtička *et al.* 2015) for aligned systems. For the selected sample we calculated detailed self-consistent 2D numerical models of density and thermal structure of Be-stars' disk with included NS gravity and X-ray heating of the ambient disk gas, assuming vertical hydrostatic, thermal and radiative equilibrium (see examples for GRO J1008-57 in Figs. 2 and 3). Figure 2 shows that the disk density truncation in the disk - NS plane begins approximately at the expected resonance radius. We included the irradiative flux from the central star whose effects are calculated using method of short characteristics. We

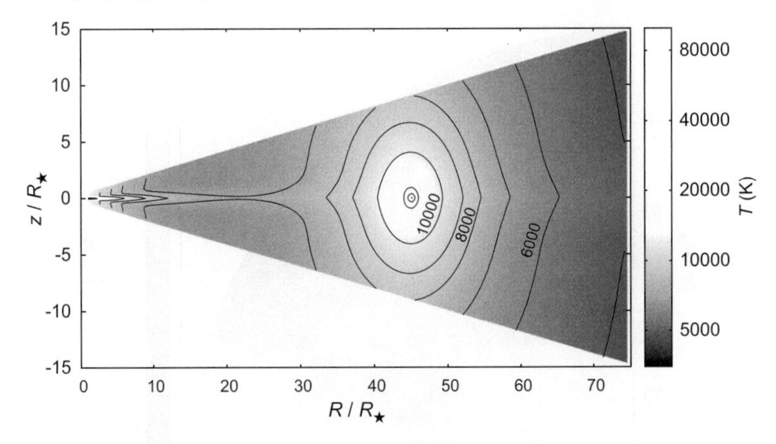

Figure 3. Thermal structure of the 2D model with the same parameters as in Fig. 2.

calculated the thermal and radiative flux near the disk midplane as a diffusion approximation. We used our own Eulerian hydrodynamic code and special cylindrical-conical coordinate system (Kurfürst *et al.* 2018).

2. Supernova explosion

We perform two separate models of interaction of an expanding supernova (SN) envelope with an asymmetric circumstellar medium (CSM) including a dense equatorial disk. We adopted the initial density and temperature profile of the progenitors from MESA model and SNEC code (Morozova *et al.* 2015). We assumed a spherically symmetric component of CSM (created, for example, by the stellar wind of the progenitor) with the initial density profile $\rho \propto r^{-2}$ while we input the disk component with a Gaussian vertical and radially decreasing density profile.

The upper panel of Fig. 4 shows the snapshot of the density structure in case of a blue supergiant (BSG) progenitor with parameters $M_\star = 40\,M_\odot$, $R_\star = 50\,R_\odot$, the explosion energy $E = 10^{44}$ J (where the total $\dot{M} = 10^{-1}\,M_\odot\,\mathrm{yr}^{-1}$), at time $t \approx 500$ hrs after the emergence of the shock. The velocities of the expansion front reach in this case values of approximately $2\,000\,\mathrm{km\,s^{-1}}$ in the polar direction z and $1\,000\,\mathrm{km\,s^{-1}}$ in the equatorial direction x. The lower panel of Fig. 4 shows the snapshot of density structure in case of a BSG progenitor with parameters $M_\star = 40\,M_\odot$, $R_\star = 50\,R_\odot$, the explosion energy $E = 10^{44}$ J (where the total $\dot{M} = 10^{-2}\,M_\odot\,\mathrm{yr}^{-1}$), at time $t \approx 180$ hrs after the emergence of the shock. The velocities of the expansion front reach in this case values of approximately $4\,500\,\mathrm{km\,s^{-1}}$ in the polar direction z and $1\,800\,\mathrm{km\,s^{-1}}$in the equatorial direction x. Due to interaction with asymmetric CSM the expanding SN envelopes are significantly aspherical toward large distances.

We investigate (see Fig. 5) the density and velocity profile in case of the interaction of the SN expanding envelope (of the BSG progenitor with the same parameters and the same CSM as in the lower panel of Fig. 4) with a binary companion with mass $M_\star = 8\,M_\odot$ (with uniform internal density for simplicity), with its own spherically symmetric CSM whose initial density profile is assumed as $\rho \propto r^{-2}$. The 1D snapshots show the values of density and expansion velocity at time $t \approx 50$ hrs after the emergence of the shock, when the shock front approximately reaches the distance of the companion, however, in later time the expansion velocity behind the companion increases up to the value of approximately $4\,500\,\mathrm{km\,s^{-1}}$, thus equating the shock front speed in the polar direction.

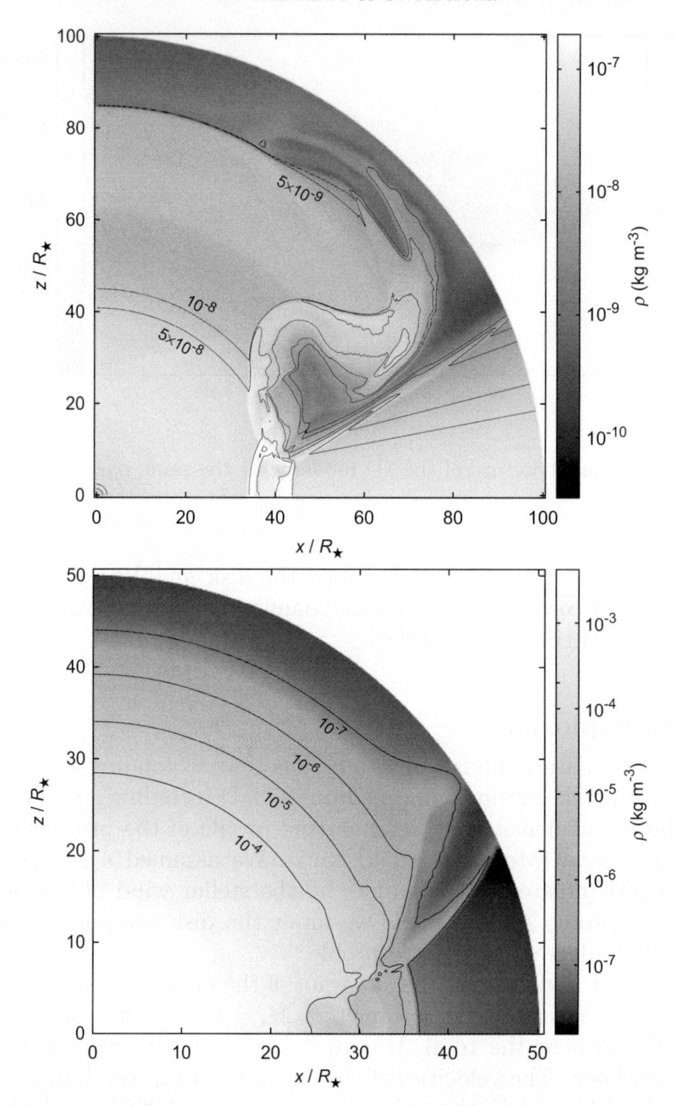

Figure 4. *Upper panel*: Density structure of the interaction of SN with asymmetric CSM that forms a dense equatorial disk (total $\dot{M} = 10^{-1}\,M_{\odot}\,\mathrm{yr}^{-1}$), at time $t \approx 500$ hrs after the emergence of the shock, the SN progenitor is a BSG star. *Lower panel*: Density structure of the BSG SN interaction with asymmetric CSM that forms a Keplerian equatorial disk (total $\dot{M} = 10^{-2}\,M_{\odot}\,\mathrm{yr}^{-1}$), at time $t \approx 180$ hrs after the emergence of the shock.

3. Conclusions

The corotating Be/X-ray binary models show that the inner disk structure is not affected significantly by presence of the NS companion and that the disk density truncation in the NS direction begins approximately at the 3:1 resonance radius (cf. Okazaki & Negueruela 2001). The disk is truncated relatively near the central star (inside the sonic point radius), therefore in case of a critically rotating star $\dot{M}$ should increase due to the angular momentum conservation. Future plans include calculations of models with eccentric and inclined (non-aligned) NS orbits and models where $r_{\mathrm{acc}} < H$.

The SN explosion models show the aspherical evolution of the density, velocity, and temperature of the ejecta that however in basic features confirm the analytic self-similarity relations developed, e.g., in Chevalier (1989). The dense equatorial disk and

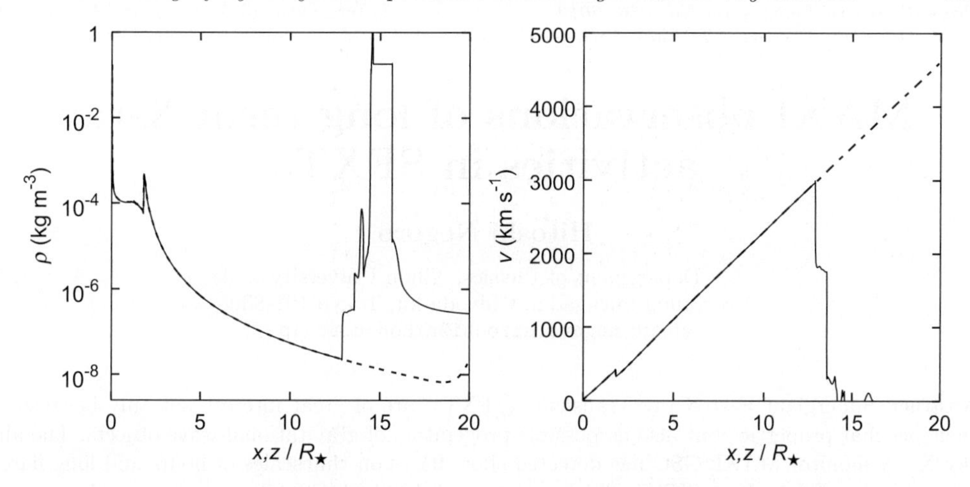

Figure 5. 1D slices of density (*left panel*) and velocity (*right panel*) in the plane of a companion star (*solid line*) and in the plane perpendicular to it (*dashed line*).

similarly the companion star may effectively block the SN expansion producing significantly over-dense and over-heated regions with developing Kelvin-Helmholtz and Rayleigh-Taylor instabilities.

Acknowledgement

The access to computing facilities owned by the National Grid Infrastructure MetaCentrum, provided under 'Projects of Large Infrastructure for Research, Development, and Innovations' (LM2010005) is appreciated. This work was supported by grant GAČR 18-05665S.

References

Chevalier, R. A. 1989, *ApJ*, 258, 790C

Krtička, J., Kurfürst, P., & Krtičková, I. 2015, *A&A*, 573, 20K

Kurfürst, P., Feldmeier, A., & Krtička, J. 2014, *A&A*, 569, 23K

Kurfürst, P., Feldmeier, A. & Krtička, J. 2018, *A&A*, 613, A75

Lee, U., Osaki, Y., & Saio, H. 1991, *MNRAS*, 250, 432L

Morozova, V., Piro, A. L., & Renzo, M. 2015, *ApJ*, 814, 63

Okazaki, A. T., & Negueruela, I. 2001, *A&A*, 377, 161

Reig, P. 2011, *Ap&SS*, 332, 1R

Rivinius, T., Carciofi, A. C., & Martayan, C. 2013, *A&ARv*, 21, 69R

High-mass X-ray Binaries: Illuminating the Passage from Massive Binaries to Merging Compact Objects
Proceedings IAU Symposium No. 346, 2019
L. M. Oskinova, E. Bozzo, T. Bulik & D. Gies, eds.

doi:10.1017/S1743921319001194

MAXI observations of long-term X-ray activities in SFXTs

Hitoshi Negoro

Department of Physics, Nihon University
1-8 Kanda-Surugadai, Chiyoda-ku, Tokyo 101-8308
email: negoro.hitoshi@nihon-u.ac.jp

Abstract. Supergiant Fast X-ray Transients (SFXTs) are of great interest not only because of their peculiar properties but also as possible progenitors of gravitational-wave objects. The all-sky X-ray monitor MAXI/GSC has detected short flares on timescales of hours and long flares on timescales of days from SFXTs. Using nine-years of MAXI/GSC data, I attempted to search periodicity of eight SFXTs of which the one-day average fluxes were below the detection limit ($\sim$10 mCrab), and confirmed the orbital periods of IGR J18483−0311 and IGR J17544−2619. This demonstrates that MAXI data are useful to find periodicities of sources even if the sources are undetectable in one day.

Keywords. X-rays: binaries, stars: early-type, neutron

1. Introduction

Monitor of All-sky X-ray Image (MAXI, Matsuoka *et al.* 2009) on the International Space Station has been observing all the X-ray sky since August 2009. Recently, MAXI plays a more important role to find an electromagnetic counterpart of a binary neutron star merger, i.e., gravitational wave (GW) objects (Kawai *et al.* 2017 for GW 150914, Serino *et al.* 2016 for GW 151226, and Sugita *et al.* 2018 for GW 170817). MAXI will be operated at least until March, 2021.

Studies of the current number of high mass X-ray binaries (HMXBs) and the binary parameters, e.g., binary separation and eccentricity, must also be important to understand the binary evolution of the HMXBs to GW objects (e.g., Tauris & van den Heuvel 2006). Thus, such studies for a relatively recently discovered subgroup of HMXBs, Supergiant Fast X-ray Transients (SFXTs, e.g., Sguera *et al.* 2005, Negueruela *et al.* 2006, Walter *et al.* 2015, and also see Martínez-Núñez *et al.* 2017), are important to cover various binary parameter spaces of HMXBs. Here, the MAXI/GSC results of the detection of the sources and their orbital periods are presented.

2. Short Flares

Due to the limited spatial resolution ($\sim$1 deg) of MAXI/GSC (Mihara *et al.* 2011; Sugizaki *et al.* 2011), about half of SFXTs can not be resolved (Sakamaki & Negoro 2017, hereafter SN17). It is also difficult to observe short and hard X-ray flares from SFXTs because of monitoring observations. Nevertheless, short X-ray enhancements probably due to activities of IGR J08408−4503, AX J1739.1−3020/XTE J1739−302, AX J1841.0−0536/IGR J18410−0535, and IGR J18483−0311 sometimes triggered the MAXI nova-alert system (Negoro *et al.* 2016; SN17). Two examples are shown in Fig. 1.

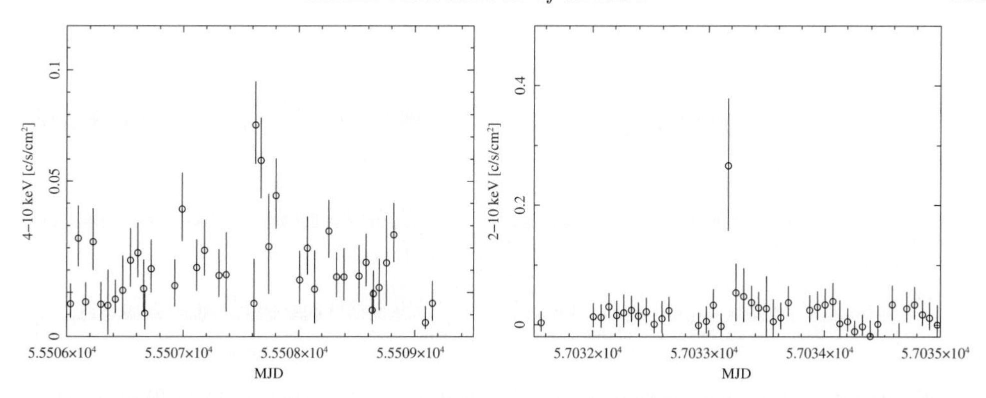

Figure 1. X-ray flares probably from AX J1841.0−0536/IGR J18410−0535 (*left*, Negoro *et al.* 2011) and AX J1739.1−3020/XTE J1739−302 (*right*, Negoro *et al.* 2015).

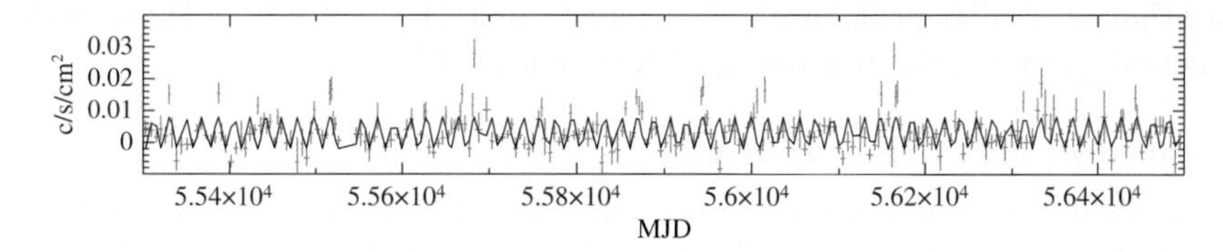

Figure 2. Longterm 4-10 keV X-ray light curve of IGR J18483−0311 observed with MAXI/GSC, and the fitted sine curve.

3. Periodicity

From IGR J18483−0311, MAXI often detected long flares lasting a few days, not short flares lasting a few hours or less. As shown in Fig. 2, bright (more than 10 mCrab, ~ 0.01 c/s/cm^2 at 4-10 keV) and periodic (18.54 days) flares can be seen. The periodicity is thought to be related with the orbital period just like a type-I outburst. To investigate such a property in a soft X-ray band, using nine-years of MAXI/GSC data I searched the periodicity of the following eight SFXTs, IGR J08408−4503, AX J1739.1−3020, AX J1841.0−0536, IGR J18483−0311, AX J1845.0−0433/IGR J18450−0435, IGR J17544−2619, XTE J1901+014, and 2XMM J185114.3−000004.

Since most SFXTs are on the galactic plan and close to other bright sources (SN17), the image fitting analysis (Morii *et al.* 2016) was performed to obtain reliable light curves taking into account the point spread functions of the cameras and the X-ray count leaks from nearby sources. Then, Lomb-Scargle periodograms (Scargle 1982) for unevenly spaced data were calculated.

IGR J18483−0311 : The periodogram in Fig. 3 (*left*) clearly shows a strong peak at 6.24×10^{-7} Hz (18.54 d). This period is consistent with previous measurements (e.g., 18.55 ± 0.03 d obtained with RXTE/ASM by Levine *et al.* (2006), also SN17). Note that other peaks are due to the ISS orbital precession (1.6×10^{-7} Hz ~ 72 d) and its harmonics.

IGR J17544−2619 : The second example of known-period detection is that of IGR J17544−2619, for which we measured a period of 4.93 days (2.35×10^{-6} Hz, Fig. 3 (*right*)). This is compatible with the period of 4.926 days reported by Clark *et al.* (2009). In addition to strong peaks (harmonics) below 2×10^{-6} Hz due to severe count leaks from nearby bright persistent sources and the ISS orbital precession, some peaks are also detected at 1.11×10^{-5} (1.04 d), 1.16×10^{-5}, and 1.18×10^{-5} Hz. However, in AX J1739.1−3020, peaks at similar frequencies can also be seen at 1.13×10^{-5} and 1.18×10^{-5} Hz. Currently, there is no consolidated explanation to interpret these peaks.

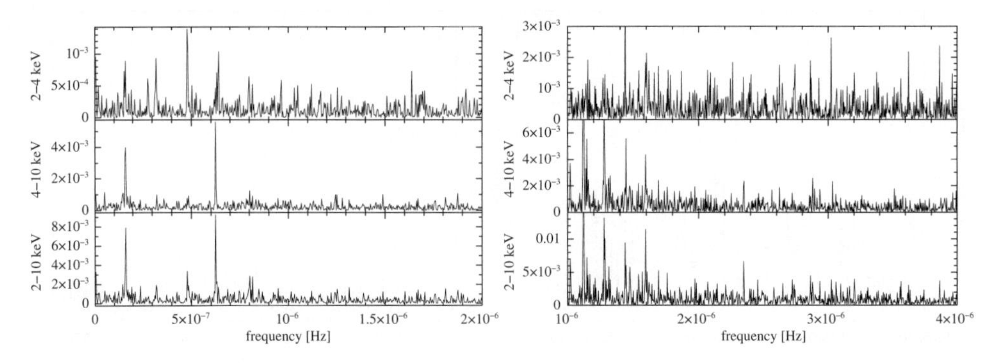

Figure 3. Preliminary periodograms obtained for IGR J18483−0311 (*left*) and IGR J17544−2619 (*right*).

We note that no similar peaks in the frequency range (1.1–1.2)$\times 10^{-5}$ Hz were deteceted in the periodograms of the remaining SFXTs analyzed here.

4. Discussion and other HMXBs

We successfully obtained the periods of the two SFXTs, IGR J18483−0311 and IGR J17544−2619, average fluxes of which were below the 1-d MAXI/GSC detection limit by more than one order of magnitude (e.g., Romano 2015). On the other hand, the periodicities of AX J1739.1−3020, AX J1841.0−0536 and AX J1845.0−0433 with similar average fluxes were not confirmed. This is not easy to be explained assuming the relations between outburst duration, frequency and rate shown by Sidoli (2013).

For bright HMXBs, long-term MAXI monitoring data are informative on periodic and aperiodic variability (e.g., see Asai *et al.* 2014 for Cir X–1, and Sugimoto *et al.* 2016 for Cyg X–1). This work demonstrates that MAXI data are also useful to find periodicities in the X-ray emission of sources that cannot be detected with 1 day-long integrations.

Acknowledgement

I thank the editor Dr. Enrico Bozzo for his careful reading of this manuscript. This work was financially supported by JSPS KAKENHI Grant Number JP17H06362.

References

Asai, K *et al.* 2014, *PASJ*, 66, 79
Clark, D. J. *et al.* 2009 *MNRAS*, 399, L113
Kawai, N. *et al.* 2017, *PASJ*, 69, 84
Levine, A. *et al.* 2006, *ApJS*, 196, 6
Martínez-Núñez, S. *et al.* 2017, *Space Sci. Rev.*, 212, 59
Matsuoka, M. *et al.* 2009, *PASJ*, 61, 999
Mihara T. *et al.* 2011, *PASJ*, 63, S623
Morii, M. *et al.* 2016, *PASJ*, 68, S00
Negoro, H. *et al.* 2011, *The Astronomer's Telegram*, 3018
Negoro, H. *et al.* 2015, *The Astronomer's Telegram*, 6900
Negoro, H. *et al.* 2016, *PASJ*, 68, S1
Negueruela, I. *et al.* 2006, *ApJ*, 638, 982
Romano, P. 2015, *Journal of High Energy Astrophysics*, 7, 126
Sakamaki, A., & Negoro, H. 2017, in: M. Serino *et al.* (eds.), *7 years of MAXI Monitoring X-ray Transients* (RIKEN), p. 167
Scargle, J. D. 1982, *ApJ*, 263, 835
Serino, M. *et al.* 2017, *PASJ*, 69, 85

Sguera, V. *et al.* 2005, *A&A*, 444, 221
Sidori, L. 2013, *Proceedings of Science, arXiv:1301.7574*
Sugimoto, J. *et al.* 2016, *PASJ*, 68, S17
Sugita, S. *et al.* 2018, *PASJ*, 70, 80
Sugizaki, M. *et al.* 2011, *PASJ*, 63, S625
Tauris, T. M. & van den Heuvel, E. P. J. 2006, in: W. Lewin and M. van der Klis (eds.), *Compact Stellar X-Ray Sources* (Cambridge University Press), p. 623
Walter, R. *et al.* 2015, *AARev*, 23, 2

High-mass X-ray Binaries: Illuminating the Passage from Massive Binaries to Merging Compact Objects
Proceedings IAU Symposium No. 346, 2019
L. M. Oskinova, E. Bozzo, T. Bulik & D. Gies, eds.

doi:10.1017/S1743921319001339

On the long-term variability of high massive X-ray binary Cyg X-1

Eugenia Karitskaya[1] **Nikolai Bochkarev**[2], **Vitalij Goranskij**[2] **and Natalia Metlova**[2]

[1]Institute of Astronomy RAS, 48 Pyatnitskaya Str., Moscow, Russia.
email: karitsk@yandex.ru

[2]Sternberg Astronomical Institute Lomonosov Moscow State University, Moscow, Russia

Abstract. We continue our study of spectral and photometric variability of Cyg X-1 on the basis of the 45-year long series of multicolor photometric observations and many-year-long series of spectral observations we have accumulated up to now. The mean level of star brightness continues to decrease since 1999 with the variations on smaller time scales superimposed. There is a connection between X-ray and optical changes. The chaotic variations of X-ray flux sometimes reaching to "hard" - "soft" state irregular changes switch on when U brightness decrease and He I λ 4713 Å absorption line depth increase. And inversely - they switch off during U brightness increasing and He I λ 4713 Å absorption line depth decreasing. This may be connected with star size variations, causing outflow gas instability. It is concluded that the fundamental parameters of the supergiant in the system of Cyg X-1 continue to vary on the time scales of years - decades.

Keywords. stars: individual (HD 226868, V1357 Cyg, Cyg X-1), stars: variables: Cyg X-1, stars: fundamental parameters, stars: O-supergiant, stars: spectral variations, stars: photometrical variations

1. Introduction

Cyg X-1 = V1357 Cyg = HDE 226868 is an X-ray binary system (with the orbital period $P = 5.6^d$) whose relativistic component is historically the first candidate to a black hole (BH). It is a prototype of high massive X-ray binaries with a black hole being the brightest one among these objects ($m_V = 9^m$). Cyg X-1 is a microquasar having a jet during its low "hard"- state. The optical component, an O9.7 Iab supergiant, is responsible for about 95% of the system optical luminosity. The X-ray satellite UHURU discovered "hard" - "soft" states of X-ray spectrum. Besides of orbital variation, so called precession variation with $P = 294^d$ (Priedhorsky *et al.* 1983, Kemp, Karitskaya, Kumsiashvili *et al.* 1987), a lot of variability types in X-ray and optical ranges on different time scales (see for example Karitskaya *et al.* 2000, Karitskaya *et al.* 2001) were revealed.

The long-time variability investigations demand long and homogeneous observations. In 1971 V. M. Lyuty started his 35-year long homogeneous photometric series of UBV observations performed at Crimean Laboratory of Sternberg Astronomical Institute (SAI). This multicolor row of observations performed on the same equipment and undergone by the same reduction has permitted to reveal the supergiant light and color variation on the time scale of tens of years (Karitskaya, Lyuty, Bochkarev *et al.* 2006). According to this study the object brightness was slowly increasing between 1985 and 1995, and then decreasing to 2003. The largest variation was in the U band. Comparison of the spectra obtained in 1997 at the Crimean Astrophysical Observatory (the 2.6-meter telescope, resolution R=35000) with the spectra from the Peak Terskol Observatory (the 2-m telescope, R=13000) and BOAO (Korea, the 1.8-m telescope, R=30000) taken in

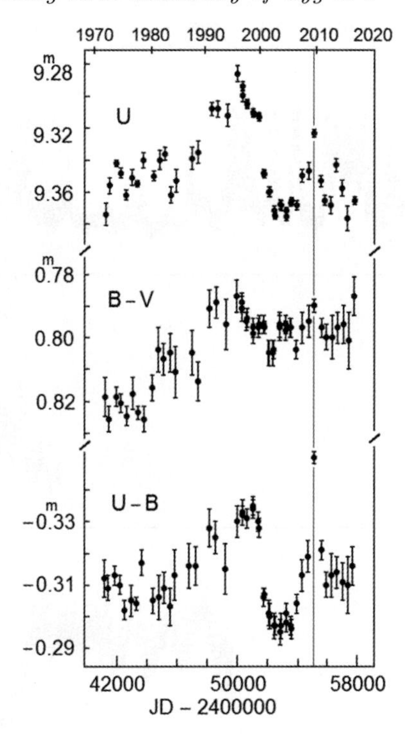

Figure 1. Long-term light curve of Cyg X-1 in U band and in $B-V$, $U-B$ color curves. Points are annualy averaged values. Vertical line marks the strong U-band flare of 2009.

2003-2004 showed that the He I λ 4713 Å line had became considerably deeper. Non-LTE simulations of this line profile and photometrical variations lead to the conclusion that the star radius has grown by about 2% from 1997 to 2004, while the temperature decreased by about 2000 K (Karitskaya, Lyuty, Bochkarev *et al.* 2006). These variations of supergiants parameter agree with the rise of X-ray activity just during the same time. The strong chaotic X-ray flux variations grow and sometimes leading to frequent changes of soft-hard states. The increasing of the degree of the Roche lobe filling leads to intensification and instability of the matter outflow towards the X-ray source. Therefore, both photometrical and spectral variations point to supergiant parameter changes.

2. Broad band photometrical variability

For continuation our study of Cyg X-1 long-term variability after the death of V. M. Lyuty, the unique homogeneous photometric UBV series was extended by N. V. Metlova up to now with the same equipment on the 60-cm telescope at the Crimean Station of the Moscow University. The long-term variability light curves for UBV see in Karitskaya, Bochkarev, Goranskij *et al.* (2017) and Karitskaya, Bochkarev, Goranskij *et al.* (2018). The largest variation is in the U band so Fig. 1 shows U light and color curves representing the object long-term variability during 45 years. The brightness drop starting from 1999 and revealed by Karitskaya, Lyuty, Bochkarev *et al.* (2006) continues up to now, although separate maxima sometimes take place. The most interesting feature is a very blue flare in 2009 which is pronounced better in the $U-B$ color curve.

3. Comparison of optical and X-ray light curves

Fig. 2 compares the long-term U-band light curve with the X-ray variations in time between 1996 and 2016. The upper panel shows the RXTE-ASM $1.5 \div 12$ keV

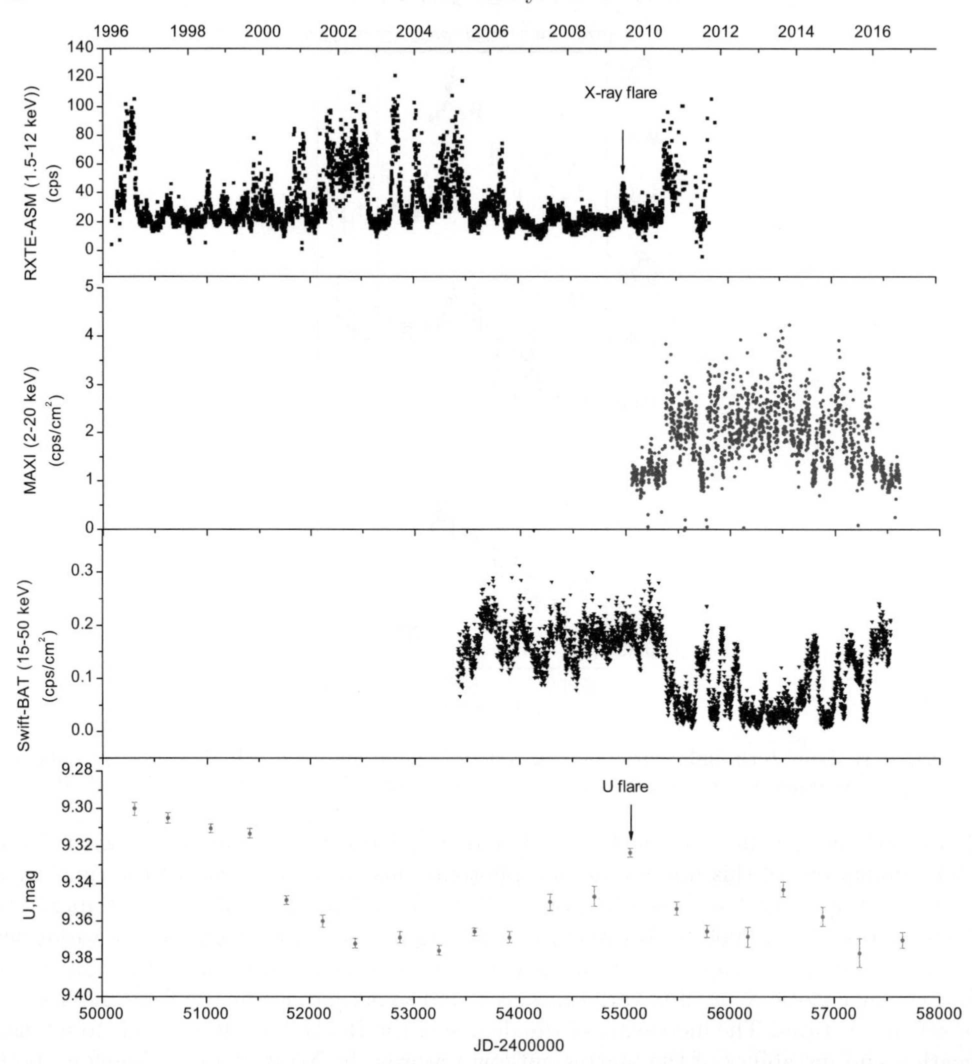

Figure 2. X-ray (RXTE-ASM, MAXI, Swift-BAT) and U-band light curves of Cyg X-1 from 1996 to 2016. The U-band data are yearly averaged. The X-ray flare coinciding with the U-band flare of 2009 is marked (see Karitskaya, Bochkarev, Goranskij *et al.* 2018).

data (http://xte.mit.edu/). RXTE-ASM data after $JD2455200$ are affected by instrumental decline before the spacecraft mission finishing. The observations in the same X-ray energy range were prolonged by Monitor of All-sky X-ray Image (MAXI energy band $2 \div 20$ keV). The second panel shows the data from this cosmic equipment (http://maxi.riken.jp/top/ index.php?cid=1&jname=J1958+352). It is very important to compare them with the hard X-ray data because of their different behavior - the light curves anti-correlate to each others. The third panel shows the light curve for $15 \div 50$ keV obtained by Swift gamma-ray burst satellite and Burst Alert Telescope (Swift-BAT) (http://swift.gsfc.nasa.gov/results/ transients/). On the lowest panel, the yearly averaged data in U-band are given.

In Fig. 2 we can see time intervals with chaotic variations of X-ray flux sometimes reaching to "hard" - "soft" state irregular changes. Earlier the transition from ordinary low "hard" state to high "soft" state occurred only once per several years. We can see at least two such events, in 2000 and 2010. They are coincide with U band brightness decreasing. For 2000 year it was noted by Karitskaya, Lyuty, Bochkarev *et al.* (2006). Inversely - when the brightness in the U band begins to grow, the X-ray activity turns off. In the time interval $JD2454000 \div 2455000$ the source was in quiet "hard" state while U-band brightness was increasing. After $JD2455000$ U-band brightness began to fall and the X-ray instability (chaotic variations of X-ray flux) appeared. This small U-band brightness growth was followed by a relative stabilization at the "soft" state, and than, the U-band brightness began to fall and X-ray instability is resumed.

The situation for 2000 year is described by Karitskaya, Lyuty, Bochkarev *et al.* (2006) being explained by supergiant parameter variation. When the temperature was decreasing, the supergiant radius was increasing approaching to the critical Roche lobe. It results to unstable material flowing between the components and resuming the X-ray activity.

In Fig. 2 we can see already mentioned very blue flare 2009. It is interesting that it coincides with an X-ray flare. This flare observed only in the U band is probably an evidence of an appearance of the very hot gas in the Cyg X-1 system.

4. He I absorption line variation

The variability of photospheric spectral line depths found in the available high-resolution spectra is also in agreement with above mentioned idea - the variation of supergiant fundamental parameters. In the papers Karitskaya, Bochkarev, Goranskij *et al.* (2017) and Karitskaya, Bochkarev, Goranskij *et al.* (2018) we have demonstrated the He I λ4713 Å line profile variations during 1997 - 2014 by using precise high resolution spectra. They where obtained on different instruments - at the Crimean 2.6 meter telescope (1997), at the observatories BOAO and Peak Terskol (2003-2004), at the 6-meter telescope of the SAO RAS (2005-2013), at 3.6 meter telescope CFHT (Hawaii) with the ESPaDOnS spectrograph (Sep 12, 2014). The first spectra from Crimean 2.6 meter telescope in 1997 were obtained in a very narrow spectral range, where He I λ4713 Å was the only one among absorption photospheric lines. Therefore we study behavior of this line. The depth variations of the line (up to 30%) are visible.

We revealed a general tendency of the line profile depth variations. During U brightness decrease the He I λ 4713 Å absorption line depth increases. And inversely - during U brightness increase the line depth is getting shellower. During the 1997 – 2004 U-band brightness decay, the He I λ 4713 Å line profile became deeper. By using model calculation in the paper Karitskaya, Lyuty, Bochkarev *et al.* (2006) it is explained by decreasing of the supergiant surface temperature: the lower is the temperature – the stronger is this photospheric absorption line. The profile depth on 12.11.2005 is approximately the same as when U brightness is in at minimum of the light curve. In 2006–2008 the line depth is decreasing while U-brightness is growthing and reaching maximum. The line profiles observed in 2013 with the 6-m BTA/NES and in 2014 with the CFHT/ESPaDOnS show the biggest depths although U-band light curve has small maximum.

Fig. 3 presents the comparison of changes in the depth of the He I λ4713 Å absorbtion line with the brightness variations in the U band in 2015. We used our low-resolution spectral observations conducted at the SAO RAS 1-m telescope with the UAGS spectrograph (the range 3850–5200 Å, R= 2000, $S/N \sim 200 - 300$). The advantage of these observations in respect to the observations with the 6-m telescope is in their frequency and regularity allowing a better comparison of the variability in line profiles with the brightness variability. The low panel of Fig. 3 presents ΔU light curve for 2015 after orbital light curve subtraction from U photometrical data. The point scattering is caused by observational errors and quick variations. Anti-correlation may be noticed, which is consistent with conclusion by Karitskaya, Lyuty, Bochkarev *et al.* (2006) on the variability of the fundamental parameters of the optical component. The moment of the

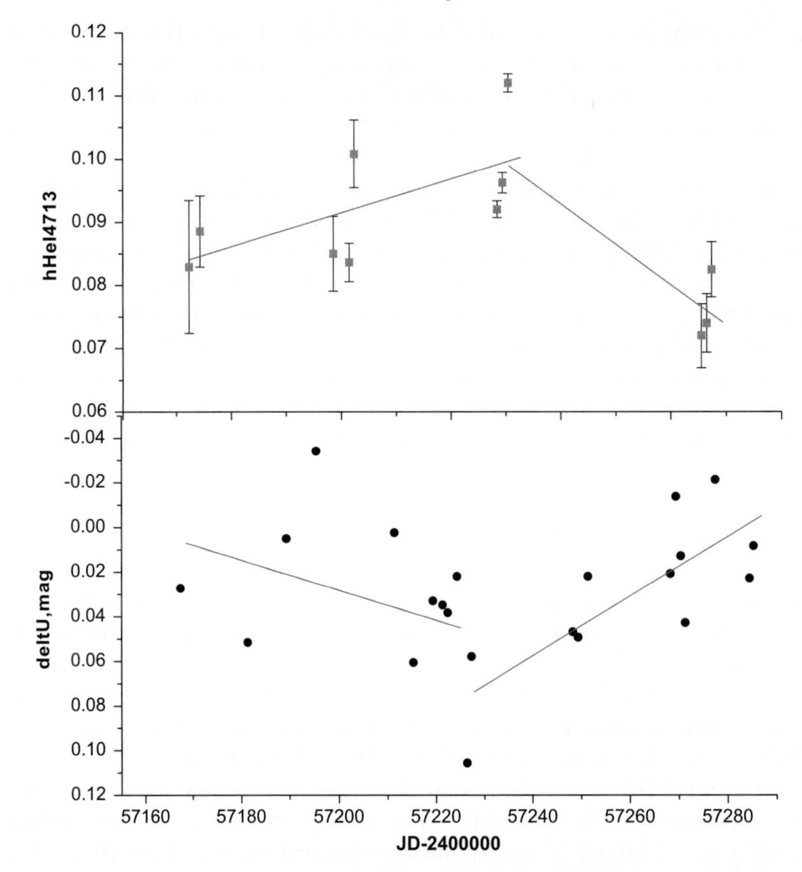

Figure 3. Comparison of the U-band brightness with the He I λ 4713 Å line profile depth variations observed at the SAO RAS 1-m telescope in 2015. U-band photometrical data are presented after averaged orbital light curve subtraction (see Karitskaya, Bochkarev, Goranskij *et al.* 2018).

U-band brightness minimum appears to be a turning point, after which the radius of the supergiant began to decrease and the temperature to rise.

5. Conclusions

As a result of our multi-year study of the photometric and spectral variability of Cyg X-1, we have established that the mean level of star brightness continues to decrease since 1999 with the variations on smaller time scales superimposed.

We revealed a very blue flare in 2009 observed only in the U band. It is probably an evidence of appearance of a very hot gas in the Cyg X-1 system. It coincides with the X-ray flare.

Our precise high resolution spectral observations and with the SAO RAS 1-meter telescopes show variations of absorption line depths up to 30%. There is a connection between X-ray and optical changes. The chaotic variations of X-ray flux sometimes reaching to "hard" - "soft" state irregular changes switch on when U brightness decrease and He I λ 4713 Å absorption line depth increase. And inversely - they switch off during U brightness increasing and He I λ 4713 Å absorption line depth decreasing.

All above mentioned facts suggest that the fundamental parameters of the supergiant in the system of Cyg X-1 continue to vary. When the temperature of the optical component (9.7Iab supergiant) was decreasing, its size was increasing approaching to the critical

Roche lobe. It was stimulating the unstable material flow and initiating the X-ray activity. Inversely - when the temperature was increasing the supergiant getting smaller, moving away from the critical Roche lobe, and the X-ray activity stops. But the conduction of research of these variations was complicated by a possible presence of a variable optically thin gas, the evidences of which were revealed.

References

Karitskaya, E. A., Goranskij, V. P., Grankin, E. N., & Melnikov, S. Yu. 2000, *Astron. Let.*, 26, 22

Karitskaya, E. A., Voloshina, I. B., & Goranskii, V. P. *et al.* 2001, *Astron. Reports*, 45, 350

Karitskaya, E. A., Lyuty, V. M., Bochkarev, N.G. *et al.* 2006, *IBVS*, No.5678, 1

Karitskaya, E. A., Bochkarev, N. G., Goranskij, V.P. *et al.* 2017, *ASP Conf. Ser.*, 510, 408

Karitskaya, E. A., Bochkarev, N. G., Goranskij, V.P. *et al.* 2018, *AApTr*, 30, 421

Kemp, J. C., Karitskaya, E. A., Kumsiashvili, M. I., *et al.* 1987, *Soviet Astron.*, 31, 170

Priedhorsky, W. C., Terrell, J., & Holt, S. S. 1983, *ApJ*, 270, 233

High-mass X-ray Binaries: Illuminating the Passage from Massive Binaries to Merging Compact Objects
Proceedings IAU Symposium No. 346, 2019
L. M. Oskinova, E. Bozzo, T. Bulik & D. Gies, eds.

doi:10.1017/S1743921319001856

Prospecting the wind structure of IGR J16320–4751 with XMM-*Newton* and *Swift*

F. García[1], F. A. Fogantini[2], S. Chaty[1] and J. A. Combi[2]

[1]AIM, CEA, CNRS, Université Paris-Saclay, Université Paris Diderot, Sorbonne Paris Cité, F-91191 Gif-sur-Yvette, France
email: federico.garcia@cea.fr

[2]Instituto Argentino de Radioastronomía (CONICET; CICPBA) & Facultad de Ciencias Astronómicas y Geofísicas, Universidad Nacional de La Plata, Argentina

Abstract. The *INTEGRAL* satellite has revealed a previously hidden population of absorbed High Mass X-ray Binaries (HMXBs) hosting supergiant (SG) stars. Among them, IGR J16320–4751 is a classical system intrinsically obscured by its environment, with a column density of $\sim 10^{23}$ cm^{-2}, more than an order of magnitude higher than the interstellar absorption along the line of sight. It is composed of a neutron star (NS) rotating with a spin period of $\sim$1300 s, accreting matter from the stellar wind of an O8I SG, with an orbital period of $\sim$9 days. We analyzed all existing archival XMM-*Newton* and *Swift*/BAT observations of the obscured HMXB IGR J16320–4751 performing a detailed temporal and spectral analysis of the source along its orbit. Using a typical model for the supergiant wind profile, we simultaneously fitted the evolution of the hard X-ray emission and intrinsic column density along the full orbit of the NS around the SG, which allowed us to constrain physical and geometrical parameters of the binary system.

Keywords. X-ray: individual objects (IGR J16320–4751), stars: supergiants, stars: neutron, X-ray: binaries

---

1. IGR J16320–4751 as an obscured sgHMXB

For more than 15 years, the *INTEGRAL* mission has discovered several obscured low luminosity X-ray sources which are difficult to observe in the soft X-rays, below 3 keV. Follow-up observations with *Chandra* or *XMM-Newton* brought arc-second level astrometric accuracy allowing for subsequent optical/IR identification of their companions (Chaty *et al.* 2008; Coleiro *et al.* 2013). Many of these sources were found to be obscured high-mass X-ray binaries (HMXB) with early-type companion stars (see Chaty (2013), Walter *et al.* (2015) and Martínez-Núñez *et al.* (2017)). In some cases, even a periodicity could be detected in the hard X-rays, attributed to their orbital motions (Corbet 1986).

IGR J16320–4751 is one these highly absorbed binary systems. The source was discovered on Feb 1, 2003, with the *INTEGRAL* observatory showing a significant variability in the 15−40 keV energy range (Tomsick *et al.* 2003). Follow-up XMM-*Newton* observations on Mar 4, 2003 showed several flares without significant hardness variations (Rodriguez *et al.* 2003) and allowed Lutovinov *et al.* (2005) to identify a pulsation period of $P = 1309 \pm 40$ sec, confirming that the system is a HMXB with an accreting neutron star (NS). Later on, using a *Swift*/BAT light curve extending from Dec 21, 2004, to Sep 17, 2005 (Corbet *et al.* 2005), an orbital period of 8.96$\pm$0.01 days was also found. An infrared counterpart was identified by Chaty *et al.* (2008) and, in a recent study, Coleiro *et al.* (2013) performed NIR spectroscopy and classified the stellar component as an BN0.5 Ia supergiant (SG).

Table 1. XMM-*Newton* PN observations used along this work.

OBSID	Name	Start date [UTC]	End date [UTC]	Phase	Exp. [ks]	GTI [ks]	ER ["]
0128531101	1285	2003-03-04 20:58	2003-03-05 03:12	–	4.78	4.47	0
0201700301	2017	2004-08-19 13:28	2004-08-20 03:20	–	38.0	33.9	10
0556140101	101	2008-08-14 22:41	2008-08-15 01:12	0.408 ± 0.006	5.33	4.64	4
0556150201	201	2008-08-16 17:38	2008-08-16 19:52	0.606 ± 0.005	1.63	1.42	0
0556140301	301	2008-08-18 13:33	2008-08-18 15:31	0.809 ± 0.005	1.23	1.07	4
0556140401	401	2008-08-20 07:34	2008-08-20 10:41	0.006 ± 0.007	11.2	9.78	4
0556140501	501	2008-08-21 07:02	2008-08-21 07:39	0.109 ± 0.001	1.34	1.16	6
0556140601	601	2008-08-22 03:54	2008-08-22 07:20	0.212 ± 0.008	12.4	10.8	10
0556140701	701	2008-08-24 18:28	2008-08-24 20:59	0.500 ± 0.006	6.03	5.21	4
0556140801	801	2008-08-26 13:33	2008-08-26 16:13	0.700 ± 0.006	9.63	8.37	8
0556141001	1001	2008-09-17 01:25	2008-09-17 03:31	0.090 ± 0.005	4.33	3.77	0

Here we report a temporal and spectral study arising from an XMM-*Newton* and *Swift*/BAT monitoring of the source. In Sect. 2 we provide details about the observations and data analysis. On Sect. 3 we describe the our main results and on Sect. 4 we analyze them using a simple model for the supergiant wind binary modulation. Finally, on Sect. 5 we summarize our conclusions.

2. XMM-*Newton* and *Swift*/BAT monitoring

XMM-*Newton* observed IGR J16320–4751 twice in Mar 2003 and Aug 2004, and nine times between Aug 14 and Sep 17, 2008 (Zurita Heras *et al.* 2009). As the PN camera effective area is larger than the MOS CCDs, and the latter have long integration times, heavily affected by pile-up, we concentrate on the analysis of the PN data set. We reduced the XMM-*Newton* data using Science Analysis System (SAS) version 16.0.0. In order to exclude high-background periods we produced background light curves for events with energies above 10 keV and Good time intervals (GTI) were obtained. We found that eight of the eleven observations were affected by pile-up and we followed the standard procedures suggested by the XMM-*Newton* calibration team to determine the excision radii (ER) necessary to mitigate the pile-up effects in each of the observations. The set of observations with their corresponding dates, exposure times, GTIs and ERs are presented on Table 1. Shortened ObsIDs are shown in the “Name” column. Phases correspond to an orbital period of 8.99 days and a central epoch corresponding to the middle of the 0556140701 exposure (phase=0.5). We also used the full *Swift*/BAT data available up to September 26, 2017 of daily and orbital light curves to obtain a refined period of 8.99±0.01 days, consistent with Corbet *et al.* (2005) for the first year of BAT data.

3. Results

We extracted background-corrected light curves using the annular regions determined by the pile-up analysis with an outer radius of 35". Considering an average count rate of 3.56 cts s^{-1}, we chose a binning time of 20 s to analyze the X-ray variability and flaring activity of the source. Based on the spectral shape, we defined two bands (*soft:* 0.5–6.0 keV and *hard:* 6.0–12.0 keV energy ranges) and we generated *soft*, *hard*, and *soft/hard* or color ratio curves that we used to search for long-term variability, short flaring, and possible spectral changes of the source.

Our visual inspection of the light curves indicated high variability on several timescales but keeping the hardness ratio almost constant between flaring and non-flaring periods. In some of them a clear modulation can be seen on a timescale of ~1300 s, as was first pointed out by Lutovinov *et al.* (2005), a period that was attributed to the spin of the NS in the system. In ObsID 0556140801, two short and bright flares were detected when the source increases its rate by a factor of ~10 for ~ 300 s, keeping a constant color ratio,

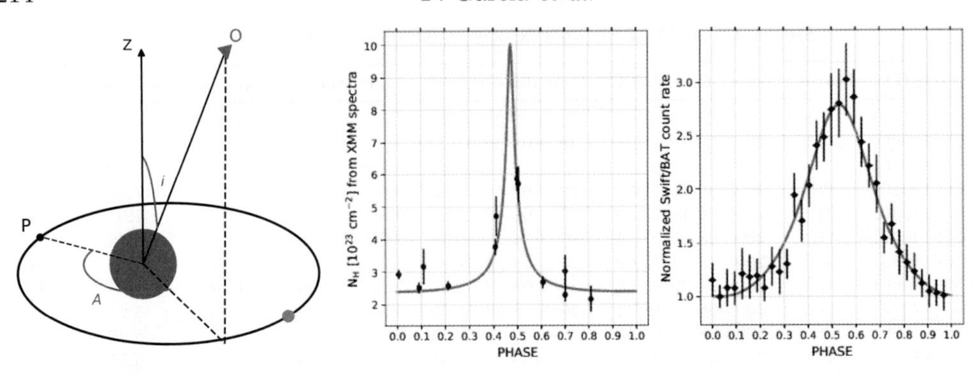

Figure 1. *Left panel:* Schematic view of our model. The NS describes an ellipse around the SG. Inclination, i, and azimuth, A, with respect to the orbital plane and the periastron, P, define the observer O. *Center and right panels:* $N_{\rm H}$ fitted to XMM-*Newton* spectra and normalized *Swift*/BAT count rate as a function of the orbital phase. Red lines represent our fitted model.

which points to a broadband flux increase and not to variations in the absorption. For subsequent spectral analysis, we separated the flaring intervals from the rest of the observation. In turn, two strong variations in hardness were detected in ObsIDs 0556140101 and 0556140701, which we thus splited into two A and B intervals for spectral analysis. In this case, the soft band is more affected and suggests a possible change in the absorption.

We extracted spectra under the same regions indicated above. We generated redistribution response matrices (RMF) and ancillary response files (ARF) using RMFGEN and ARFGEN tasks, respectively. The spectra were binned to 25 counts per bin and fitted with XSPEC v12.9.1 (Arnaud 1996) in the full 0.5–12.0 keV energy range. The spectra of IGR J16320–4751 are characterized by a highly-absorbed continuum at soft energies with a clear Fe-edge at $\sim$7 keV as well as prominent Fe-Kα line. In the best-quality spectra also fainter Fe-Kβ and Fe XXV can be detected. We fitted the spectra with a thermally Comptonized COMPTT model and three narrow Gaussian lines. We used two TBABS absorption components to model the interstellar medium (ISM) and the intrinsic absorption. We fixed the hydrogen column density of the ISM absorption model to 2.1×10^{22} cm^{-2} as in Rodriguez *et al.* (2003) and we only fitted the second $N_{\rm H}$ column.

For the majority of the spectra, we obtained $N_{\rm H} \sim 20 - 30 \times 10^{22}$ cm^{-2}, compatible with previously reported values (Rodriguez *et al.* 2003). However, in Obs 101A, 101B, 701A, and 701B, we found significantly higher values of $N_{\rm H} \sim 35 - 60 \times 10^{22}$ cm^{-2} without noticing strong changes in the continuum emission parameters, which favours a geometrical effect over a local sudden change in the accretion process. Moreover, as previously noted in Giménez-García *et al.* (2015), we recover a correlation between the Fe Kα flux and the continuum level, which is expected for fluorescence emission from a small region of matter in the very close vicinity of the accreting NS. We also confirm a Baldwin effect given by the correlation between the X-ray flux and the Fe Kα EW.

4. A simple wind model for the orbital modulation

Using the whole set of observations, we found a clear modulation of $N_{\rm H}$ with the orbital phase (Fig. 1), and correlations with the flux of the Fe Kα line, suggesting a physical link between absorbing and fluorescent matter. These evidence points towards the stellar wind being the main contributor to both continuum absorption and Fe Kα line emission.

To better understand the orbital modulations found in the $N_{\rm H}$ and the BAT hard X-ray light curve we propose a simple model consisting of a NS orbiting embedded in the dense wind of a SG companion (Fig. 1). We model the SG wind profile by means of a typical CAK model (Castor *et al.* 1975) with $\dot{M} = 3 \times 10^{-6}$ M$_\odot$ yr^{-1}, $\beta = 0.85$ and

$v_\infty = 1300$ km s^{-1}. In a close orbit, the NS accretes matter from this wind as it moves along its elliptical orbit, being able to produce a persistent X-ray emission exhibiting flux variability associated to the local density and to the integrated $N_{\rm H}$ column along the line of sight. We compute the $N_{\rm H}$, for each orbital phase (or time), integrating the wind density profile along the line of sight. We assume that the *Swift*/BAT count rate is proportional to the local wind density at the position of the accreting NS.

Under these assumptions, we simultaneously fitted the *Swift*/BAT folded light-curve and the $N_{\rm H}$ phase evolution obtaining an optimal solution given by $e = 0.20 \pm 0.01$, $A = 146.3^\circ\ ^{+3.7}_{-2.9}$ and $i = 62.1^\circ\ ^{+0.3}_{-1.5}$ (90% confidence intervals). Here, we note that e is mainly constrained by the BAT amplitude, i depends strongly on the $N_{\rm H}$ increase profile and the phase difference between the maxima of both modulations (∼0.47 vs 0.53) arises as a consequence of the observer azimuth A.

5. Conclusions

Assuming a simple model for the supergiant stellar wind we were able to explain the orbital modulation of the absorption column density and the high-energy *Swift*/BAT flux of IGR J16320–4751. The model simultaneously reproduces both the sudden change in absorption column density, measured with XMM-*Newton*, and the smooth evolution of hard X-ray *Swift*/BAT lightcurve, as well as the phase shift of their maxima. Furthermore, given the correlations found for the Fe Kα line with the $N_{\rm H}$ column density and total continuum flux, we unambiguously show that the orbital modulation of this three physical quantities is driven by intrinsic absorption of matter surrounding the NS, modulated by the stellar wind density profile, as viewed by a distant observer.

Similar spectral evolution analysis sampled along the orbit of other obscured sgHMXB sources, both in the infrared and soft/hard X-ray bands, will be of high value to better constrain the physical and geometrical properties of the sgHMXB class.

Acknowledgements

We thank F. Fortin for insightful discussions. FG and SC were partly supported by the LabEx UnivEarthS, Interface project I10, "From evolution of binaries to merging of compact objects". This work was partly supported by the Centre National d'Etudes Spatiales (CNES), and based on observations obtained with MINE: the Multi-wavelength *INTEGRAL* NEtwork. FG and JAC acknowledge support by PIP 0102 (CONICET) and PICT-2017-2865 (ANPCyT).

References

Arnaud, K. A. 1996, *ASP Conf. Ser. 101*, 101, 17
Castor, J. I., Abbott, D. C., & Klein, R. I. 1975, *ApJ*, 195, 157
Chaty, S., Rahoui, F., Foellmi, C., *et al.* 2008, *A&A*, 484,783
Chaty, S. 2013, *AdSpR*, 52, 2132
Coleiro, A., Chaty, S., Zurita Heras, *et al.* 2013, *A&A*, 560, A108
Corbet, R. H. D. 1986, *MNRAS*, 220, 1047
Corbet, R., Barbier, L., Barthelmy, S., *et al.* 2005, *ATel*, 649
Giménez-García, *et al.* 2015, *A&A*, 576, A108
Lutovinov, A., Rodriguez, J., Revnivtsev, M., *et al.* 2005, *A&A*, 433, L41
Martínez-Núñez, S., Kretschmar, P., Bozzo, E., *et al.* 2017, *Space Science Reviews*, 212, 59
Rodriguez, J., Tomsick, J. A., Foschini, L., *et al.* 2003, *A&A*, 407, L41
Tomsick, J. A., Lingenfelter, R., Walter, R., *et al.* 2003, *IAU Circular*, 8076, 1
Walter, R., Lutovinov, A. A., Bozzo, E., & Tsygankov, S. S. 2015, *Astronomy and Astrophysics Review*, 23, 2
Zurita Heras, J. A., Chaty, S., Prat, L., *et al.* 2009, *AIP*, 1126, 313

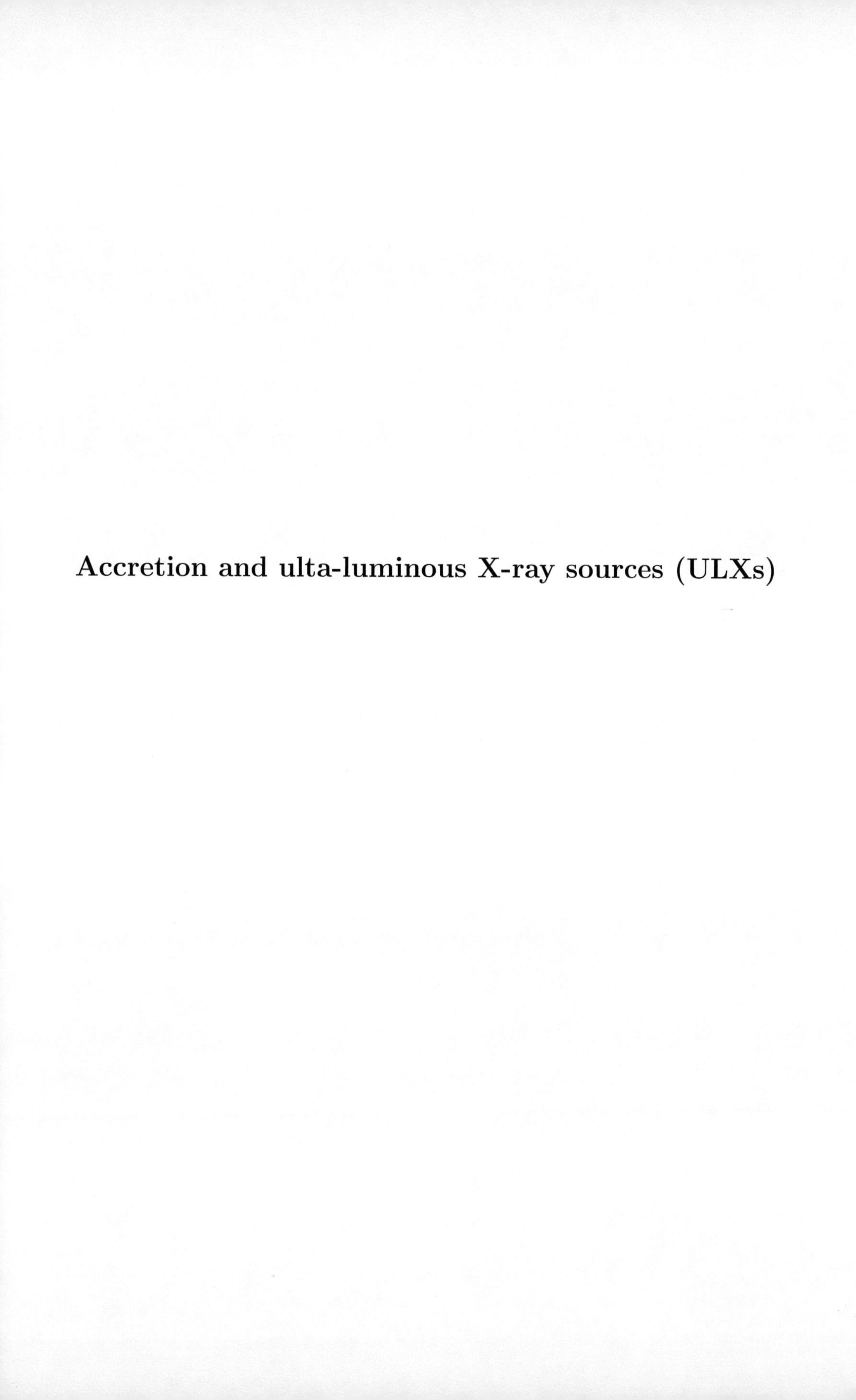

Accretion and ulta-luminous X-ray sources (ULXs)

High-mass X-ray Binaries: Illuminating the Passage from Massive Binaries to Merging Compact Objects
Proceedings IAU Symposium No. 346, 2019
L. M. Oskinova, E. Bozzo, T. Bulik & D. Gies, eds.

doi:10.1017/S174392131900125X

X-ray binaries with neutron stars at different accretion stages

Konstantin A. Postnov[1,2], **Alexander G. Kuranov**[1,3] **and Lev R. Yungelson**[4]

[1]Sternberg Astronomical Institute, Moscow M.V. Lomonosov State University
13 Universitetskij pr., 119234 Moscow, Russia
email: pk@sai.msu.ru

[2]Kazan Federal University, 18 Kremlyovskaya str., 420008 Kazan, Russia

[3]Russian Foreign Trade Academy, 4 Pudovkin str., 119285 Moscow, Russia

[4]Institute of Astronomy, RAS, 48 Pyatnitskaya str., 119017 Moscow, Russia
email: lry@inasan.ru

Abstract. Different accretion regimes onto magnetized NSs in HMXBs are considered: wind-fed supersonic (Bondi) regime at high accretion rates $\dot{M} \gtrsim 4 \times 10^{16}$ g s^{-1}, subsonic settling regime at lower $\dot{M}$ and supercritical disc accretion during Roche lobe overflow. In wind-fed stage, NSs in HMXBs reach equilibrium spin periods P^* proportional to binary orbital period P_b. At supercritical accretion stage, the system may appear as a pulsating ULX. Population synthesis of Galactic HMXBs using standard assumptions on the binary evolution and NS formation is presented. Comparison of the model $P^* - P_b$ (the Corbet diagram), $P^* - L_x$ and $P_b - L_x$ distributions with those for the observed HMXBs (including Be X-ray binaries) and pulsating ULXs suggests the importance of the reduction of P^* in non-circular orbits, explaining the location of Be X-ray binaries in the model Corbet diagram, and the universal parameters of pulsating ULXs depending only on the NS magnetic fields.

Keywords. accretion, accretion disks; binaries: close; stars: neutron

---

1. Introduction

High-mass X-ray binaries (HMXBs) are important tools to study binary evolution (see other papers in this volume). Here we focus on two particular aspects of accretion onto magnetized neutron stars (NSs) in HMXBs that have important observational manifestations: wind accretion in the settling (subsonic regime), which is relevant at moderate and low accretion rates onto NS, $\dot{M} \lesssim 4 \times 10^{16}$ g s^{-1}. For wind accretion in HMXBs with $\dot{M} > 4 \times 10^{16}$ g s^{-1}, the supersonic Bondi-Hoyle-Littleton accretion takes place. We separately consider HMXBs with supercritical disc accretion at $\dot{M} \gtrsim 10^{18}$ g s^{-1}, which is relevant to the growing class of pulsating ultraluminous X-ray sources (ULXs) (see F. Harrison and F. Fürst papers in this volume).

2. Settling wind accretion in HMXBs: an overview

The settling accretion theory is designed to describe the interaction of accreting plasma with magnetospheres of slowly rotating NSs (see Shakura *et al.* (2018) for a detailed derivation and discussion). The key feature of the model is the calculation of the plasma entry rate into the magnetosphere due to the Rayleigh-Taylor (RT) instability regulated by plasma cooling (Compton or radiative). RT instability can be suppressed by the fast rotation of the magnetosphere (Arons & Lea 1980), and settling accretion regime can be realized for NS spin periods $P^* > 27[\mathrm{s}]\dot{M}_{16}^{1/5}\mu_{30}^{33/35}(M_x/M_\odot)^{-97/70}$. At shorter NS periods, accretion regime at any X-ray luminosity will be supersonic because of efficient plasma penetration into the magnetosphere via Kelvin-Helmholtz instability (Burnard *et al.* 1983). Here and below, accretion rate $\dot{M} = \dot{M}_{16} \times 10^{16}$ g s^{-1}, NS magnetic moment $\mu = \mu_{30} \times 10^{30}$ G cm^3. Compton cooling of accreting plasma by X-ray photons generated near the NS surface enables a free-fall (Bondi-type) supersonic accretion onto the NS magnetosphere provided that the X-ray luminosity is above the critical value $L_x^\dagger \simeq 4 \times 10^{36}$ erg s^{-1} (Shakura *et al.* 2012), corresponding to the mass accretion rate onto NS $\dot{M}_x \simeq 4 \times 10^{16}$ g s^{-1}. At lower X-ray luminosities, the plasma above the magnetosphere remains hot enough to avoid an effective inflow into the magnetosphere due to RT instability, and the plasma entry rate is controlled by the cooling rate. This results in the formation of a hot, convective shell above the magnetosphere. In this shell, the plasma gravitationally captured by the neutron star from the stellar wind of the companion (basically, at the Bondi-Hoyle-Littleton rate, $\dot{M}_B$) settles towards the NS magnetosphere at a rate $\dot{M}_x = f(u)\dot{M}_B$, with the dimensionless factor $f(u) \lesssim 0.5$†, whose precise value depends on the plasma cooling regime (Compton or radiative). With a good accuracy, this factor can be written as $f(u) \approx (t_{\rm ff}/t_{\rm cool})^{1/3}$, where $t_{\rm ff}$ is the free-fall time at the magnetospheric boundary (the Alfvén radius, R_A), $t_{\rm cool}$ is the characteristic plasma cooling time. Clearly, in the case $t_{\rm cool} \gg t_{\rm ff}$, this factor can be very small, leading to an effective decrease in the mass accretion rate onto the NS surface compared to the maximum possible Bondi rate, $\dot{M}_B$.

In a circular binary system with the components separation a, Bondi-Hoyle-Littleton accretion rate can be estimated as

$$\dot{M}_B \approx \frac{1}{4}\dot{M}_{\rm o}\frac{v}{v_w}\left(\frac{R_B}{a}\right)^2, \tag{2.1}$$

where $\dot{M}_{\rm o}$ is the optical star wind mass-loss rate, gravitational Bondi radius $R_B = \delta\frac{2GM_x}{v^2}$, M_x – NS mass, $v^2 = v_{\rm orb,x}^2 + v_w^2$ – relative stellar wind velocity, $v_{\rm orb,x}$ – NS orbital velocity. The estimate (2.1) assumes a spherically-symmetric wind from the optical star and is applicable for $R_B \ll a$. Numerical factor $\delta \sim 1$ takes into account the actual location of the bow shock in the stellar wind (Hunt 1971). Modern numerical calculations (see, e.g., Liu *et al.* 2017) suggest smaller (up to an order of magnitude) mass accretion rates onto a gravitating mass from the stellar wind than given by the standard Bondi-Hoyle-Littleton formula, which can be reformulated in terms of smaller values of the parameter $\delta \simeq 0.3 - 0.5$. Other studies (e.g., de Val-Borro *et al.* 2017) claim that accretion rate can exceed the Bondi one due to focusing of the wind. Therefore, the numerical factor δ can differ from unity by a factor of a few.

The Alfvén radius is defined from the pressure balance between the accreting plasma and the magnetic field at the magnetospheric boundary and depends on the actual mass accretion rate, $\dot{M}_x$ (i.e., the observable X-ray luminosity L_x), and the neutron star magnetic moment, μ: $R_A \sim (f(u)\mu^2/(\dot{M}_x\sqrt{GM_x}))^{2/7}$. Factor $f(u)$ is

$$f(u)_{\rm Comp} \approx 0.22\zeta^{7/11}\dot{M}_{x,16}^{4/11}\mu_{30}^{-1/11} \tag{2.2}$$

and

$$f(u)_{\rm rad} \approx 0.1\zeta^{14/81}\dot{M}_{x,16}^{6/27}\mu_{30}^{2/27} \tag{2.3}$$

† In a radial accretion flow with an account of the Compton cooling, the value $f(u) \gtrsim 0.5$ also corresponds to the location of the sonic point below the magnetospheric boundary R_A (Shakura *et al.* 2012) enabling a free-fall plasma flow down to R_A and the formation of a shock above the magnetosphere (Arons & Lea 1976).

for the Compton and radiative cooling, respectively (Shakura & Postnov 2017; Shakura *et al.* 2018). Here $\zeta \lesssim 1$ is the numerical factor determining the characteristic scale of the growing RT mode (in the units of the Alfvén radius R_A).

Thus, specifying the plasma cooling mechanism, we are able to estimate the expected reduction in the mass accretion rate at the settling accretion stage, $f(u) = F(\dot{M}_x, ...) = F(f(u)\dot{M}_B, ...)$, and by solving for $f(u)$, to find the explicit expression for $\dot{M}_x$ as a function of $\dot{M}_B$ and other parameters.

Putting all things together, we are able to express the expected accretion rate onto the NS (which can be directly probed by the observed X-ray luminosity L_x) using the Bondi capture rate $\dot{M}_B$, which can be calculated from the known mass-loss rate of the optical companion $\dot{M}_o$, stellar wind velocity v_w and binary system parameters:

$$\dot{M}_{x,16}^{\rm Comp} \simeq 0.1\zeta \dot{M}_{B,16}^{11/7}\mu_{30}^{-1/7} \tag{2.4}$$

for the Compton cooling and

$$\dot{M}_{x,16}^{\rm rad} \simeq 0.05\zeta^{2/9}\dot{M}_{B,16}^{9/7}\mu_{30}^{2/21} \tag{2.5}$$

for the radiative cooling. The actual accretion rate onto NS is taken to be $\dot{M}_x = \max\{\dot{M}_x^{\rm Comp}, \dot{M}_x^{\rm rad}\}$ and determines the cooling regime. Matching Eq. (2.4) and Eq. (2.5) shows that for a given NS magnetic momentum, the Compton cooling of plasma dominates for the gravitational capture mass rate $\dot{M}_{B,16} \gtrsim 0.1\zeta^{-2/9}\mu_{30}^{5/6}$.

During the settling accretion stage, a hot convective shell formed above the NS magnetosphere mediates the angular momentum transfer to/from the rotating NS enabling long-term spin-down episodes with spin-down torques correlated with the X-ray luminosity, as observed, for example, in GX 1+4 (Chakrabarty *et al.* 1997; González-Galán *et al.* 2012). Turbulent stresses acting in the shell lead to an almost iso-momentum angular velocity radial distribution, $\omega(r) \sim 1/r^2$, suggesting the conservation of the specific angular momentum of gas captured near the Bondi radius R_B, $j_w = \eta\omega_b R_B^2$, $\eta \approx 1/4$ (Illarionov & Sunyaev 1975), where $\omega_b = 2\pi/P_b$ is the orbital angular frequency. The numerical coefficient η here can vary in a wide range due to inhomogeneities in the stellar wind and can be even negative (Ho *et al.* 1989) leading to a temporal formation of retrograde accretion discs. In our calculations, we varied this parameter in the range $\eta = [0, 0.25]$.

The condition for a quasi-spherical accretion to occur is that the specific angular momentum of a gas parcel is smaller than the Keplerian angular momentum at the Alfvén radius: $j_w \leqslant j_K(R_A) = \sqrt{GM_x R_A}$. In the opposite case, $j_w > j_K(R_A)$, the formation of an accretion disc around the magnetosphere is possible.

The equation for spin evolution of a NS at the settling accretion stage can be written in the form (Shakura *et al.* 2012, 2018)

$$\frac{dI\omega^*}{dt} = \eta Z \dot{M}_x \omega_b R_B^2 - Z(1 - z/Z)\dot{M}_x \omega^* R_A^2, \tag{2.6}$$

where I is the NS momentum of inertia, the coupling coefficient of the plasma-magnetosphere interaction is $Z = 1/f(u)(u_c/u_{ff})$, $u_c/u_{ff} \lesssim 1$ is the ratio of the convective velocity in the shell and the free-fall velocity, $z \lesssim 1$ is a numerical coefficient which takes into account the angular momentum of the falling matter at the NS surface. The account of the variable specific angular momentum of the gravitationally captured stellar wind leads to the correction of the equilibrium NS period derived in Shakura *et al.* (2012, 2018) by the factor $0.25/\eta$. In the possible case of a negative value of η, NS would rapidly spin-down and even start to temporarily rotate in the retrograde direction. However, it is difficult to imagine that such a situation holds much longer than the orbital binary period. Therefore, the possible episodes with negative η would

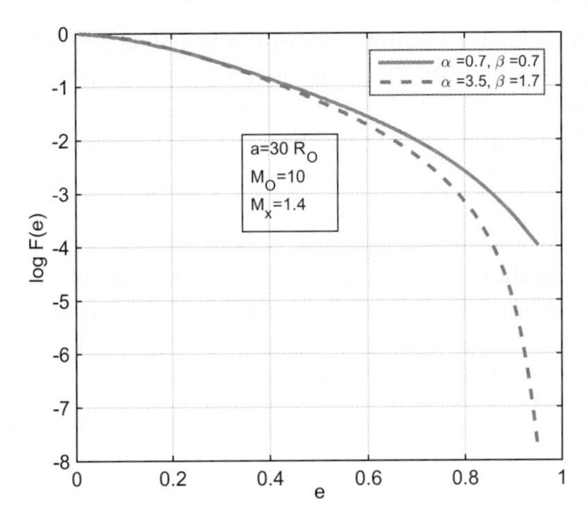

Figure 1. Reduction factor $F(e)$ for the equilibrium NSs periods at the settling accretion stage in the non-circular orbits for the range of radiative stellar wind parameters α and β.

somewhat increase the NS equilibrium period P_{eq} which we recalculate at each time step of our population synthesis calculations. Therefore, we will consider only positive values of η and restrict the NS spin rotation by the orbital binary periods, $P^* \leqslant P_b$.

Orbital eccentricity effect.

Here we describe the effect of orbital eccentricity e on the value of the equilibrium NS period at the settling accretion stage. Because of the orbital eccentricity, the spin-up and spin-down torques applied to NS change and should be averaged over the orbital period. In the standard Keplerian problem, the separation between barycenters of the binary components $r = p/(1 + e\cos\theta)$, where $p = a(1 - e^2)$ – orbital semilatus rectum, a – large semi-major axis of the orbit, e – orbital eccentricity, θ – true anomaly. Stellar wind is assumed to be spherically symmetric and centered at the optical star barycenter. The radius of the optical star is R_o, and the stellar wind velocity as a function of distance from the star is normalized to the parabolic velocity at the optical star photosphere, $v_{esc} = \sqrt{2GM_o/R_o}$. It can be written as $v_w(r) = v_{esc}f(r)$. For example, for the radiative-driven accelerating winds from early type stars $f(r) = \alpha\,(1 - R_o/r)^{\beta}$ where $\alpha = 0.7...3.8$ and $\beta = 0.7...2.0$ are obtained from recent numerical simulations (Vink 2018).

In the case of a non-circular orbit, specific angular momentum of captured wind matter is determined by the normal component of the NS orbital velocity and changes along the orbit: $j(\theta) \sim v_n(\theta) R_B(\theta)[R_B(\theta)/r(\theta)]$ (here $v_n = \sqrt{GM/p}(1 + e\cos\theta)$, $M = M_o + M_x$), while the gravitational capture Bondi radius $R_B(\theta)$ depends on the module of the sum of the orbital velocity vector $\mathbf{v_{orb}}$ and the wind velocity vector $\mathbf{v_w}$: $R_B(\theta) \sim 2GM_x/v(\theta)^2$, where $v(\theta) = \sqrt{[v_w(r(\theta)) - v_r(\theta)]^2 + v_n^2(\theta)}$ (here the radial component of the orbital velocity is $v_r(\theta) = v_n = \sqrt{GM/p}(1 + e\cos\theta)$).

Thus, by averaging the spin-up/spin-down Eq. (2.6) over orbit $r(\theta)$, we find NS equilibrium period:

$$\overline{P_{eq}} = 2\pi \frac{\langle \dot{M}_B R_A^2 \rangle}{\langle \dot{M}_B v_n R_B (R_B/r) \rangle} \,. \tag{2.7}$$

The orbit-averaged value $\overline{P_{eq}}$ can be written using the equilibrium period calculated for a circular orbit, $P_{eq} \approx P_b(R_A/R_B)^2$, as $\overline{P_{eq}} = F(e)P_{eq}$, where the factor $F(e) < 1$ is plotted in Fig. 1. A significant decrease of the equilibrium NSs periods may occur for highly

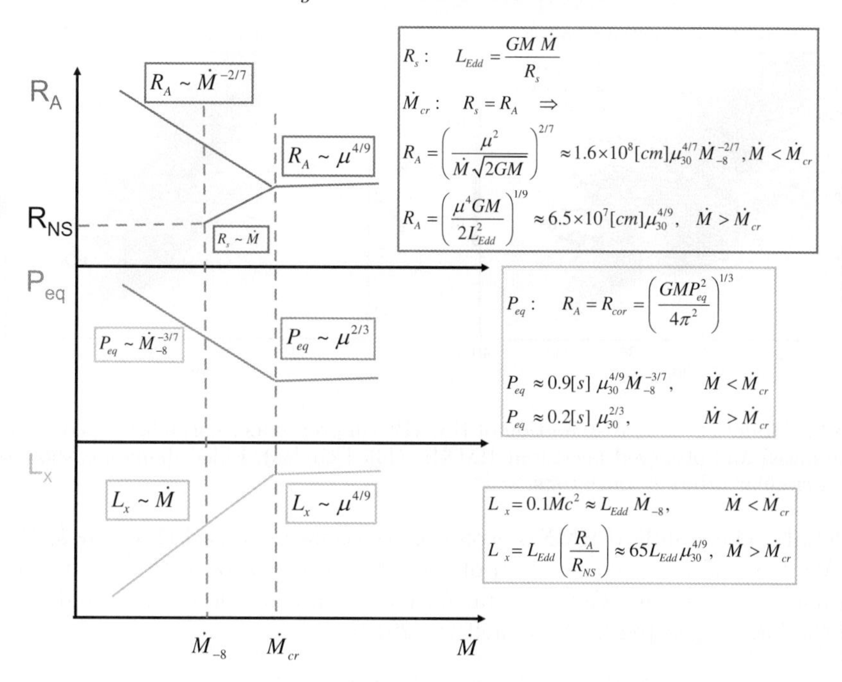

Figure 2. The Alfvén radius (top), equilibrium NS period (middle) and X-ray luminosity (bottom) for supercritical accretion discs around magnetized NSs (see the text).

eccentric binaries (e.g., Be X-ray binaries, BeXRBs). Indeed, they are located below HMXBs with almost circular orbits in the Corbet diagram. Thus, the eccentricity effect can explain the observed diversity of X-ray pulsar populations in HMXBs produced by different types of supernova noted by Knigge *et al.* (2011).

3. Supercritical accretion onto magnetized NSs

In HMXBs with Roche overflow by the optical star, a supercritical disc accretion regime must occur (Shakura & Sunyaev 1973). In this regime, a radiation-driven outflow occurs in the disc starting from the spherization radius $R_s = GM_x\dot{M}/L_{Edd}$, where $L_{Edd} \simeq 10^{38}(M_x/M_\odot)$ erg s^{-1} is the Eddington luminosity corresponding to $\dot{M}_{Edd} \simeq 10^{18}$ g s^{-1}. A feature of the supercritical accretion onto a magnetized NS is the appearance of a critical luminosity (mass accretion rate) at which the Alfvén radius R_A becomes comparable to the spherization radius R_s as shown in the scheme in Fig. 2 (see also Lipunov (1982); King *et al.* (2017)), $\dot{M}_{cr} \simeq 3\times10^{19}\mu_{30}^{4/9}$ g s^{-1}. As long as mass accretion rate in the supercritical disc increases and $\dot{M}_{Edd} < \dot{M} < \dot{M}_{cr}$, NS magnetospheric radius decreases, $R_A \sim \dot{M}^{-2/7}$, its equilibrium period decreases, $P^* \sim \dot{M}^{-5/7}$, and the X-ray luminosity generated at the NS surface increases, $L_x \sim \dot{M}$. After reaching $\dot{M}_{cr}$, NS magnetosphere stops to decrease because of the wind outflow from $R_s = R_A$ (see Fig. 2), and for $\dot{M} > \dot{M}_{cr}$ all three quantities, R_A, P_{eq} and L_x become dependent on the NS magnetic field only (the latter to within a logarithmic correction $1 + \ln(\dot{M}/\dot{M}_{Edd})$, Shakura & Sunyaev (1973)).

4. Population synthesis of HMXBs

We use a modified version of the openly available BSE code (Hurley *et al.* 2000, 2002) appended by the block for calculation of spin evolution of magnetized neutron stars (Lipunov *et al.* 2009) with account for settling accretion regime (see also Lü *et al.* 2012, for

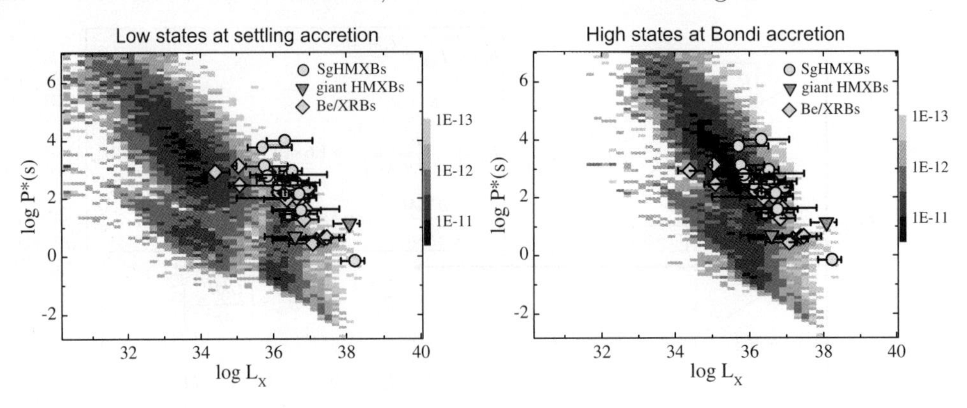

Figure 3. Model $P^* - L_x$ distribution of HMXBs (in grey scale, normalized to one $M_\odot$ of the Galactic mass) and observed persistent HMXBs (filled circles). Filled diamonds show BeXRBs with non-circular orbits in outbursts.

more detail). The standard HMXB evolutionary scenario is used (Postnov & Yungelson 2014). We have used the convolution of the HMXB formation rate after instantaneous star-formation burst with Galactic star-formation rate in the Galactic bulge and thin disc in the form suggested by Yu & Jeffery (2010):

$$\frac{\mathrm{SFR}(t)}{M_\odot\,\mathrm{yr}^{-1}} = \begin{cases} 11e^{-\frac{t-t_0}{\tau}} + 0.12(t-t_0) & t \geqslant t_0, \\ 0 & t < t_0 \end{cases} \tag{4.1}$$

with time t in Gyr, t_0=4 Gyr, $\tau = 9$ Gyr. Assumed Galactic age is 14 Gyr. This model gives total mass of Galactic bulge and thin disk $M_G = 7.2 \times 10^{10} M_\odot$, which we will use for normalization of our calculation results.

To illustrate the effect of different regimes of wind accretion in HMXBs, in Fig. 3 we show in grey scale the model Galactic distribution of equilibrium periods P^* of NSs in HMXBs vs. their X-ray luminosity L_x normalized to one $M_\odot$. The range of observed L_x of persistent wind-accreting HMXBs is taken from Sidoli & Paizis (2018). Outburst X-ray luminosity of BeXRBs with eccentric orbits are shown by filled blue diamonds. As discussed above, a wind-accreting NS can reach long spin periods at the subsonic settling accretion stage. However, when the accretion X-ray luminosity exceeds $\sim 4 \times 10^{16}$ erg s^{-1}, the supersonic Bondi accretion regime sets in, at which the NS acquires angular momentum from captured wind matter. Therefore, we plot separately the model $P^* - L_x$ distribution for the settling accretion regime (left panel in Fig. 3) and the Bondi X-ray luminosity for the same sources ($L_{x,B} = 0.1\dot{M}_B c^2$) (right panel in Fig. 3). Clearly, the observed location of persistent HMXBs and BeXRBs on this diagram better corresponds to the Bondi accretion luminosity (but see the discussion of Vela X-1 and GX 301-2 in (Shakura *et al.* 2012)), and the model left and right panels show the expected X-ray luminosity range between settling and Bondi accretion stages. The actual situation may be due to a selection effect: the brightest sources correspond to the more luminous Bondi accretion, and a bulk of possible dimmer sources with about solar X-ray luminosity are yet to be identified in more sensitive observations.

In Fig. 4 we plot the calculated distribution of sources in the X-ray luminosity – orbital period plane. Like in Fig. 3, the left and right panels correspond to settling and Bondi accretion states, respectively. The model P^*–P_b distribution (the Corbet diagram) for wind accreting HMXBs is shown in the left panel of Fig. 5. Separately, in the right panel of Fig. 5, model HMXBs with supercritical Shakura-Sunyaev accretion discs are shown (in grey scale). Observed pulsating ULXs with known orbital periods (Bachetti *et al.* 2014; Fürst *et al.* 2016; Israel *et al.* 2017a,b) are shown by red open circles. In Fig. 6 we plot

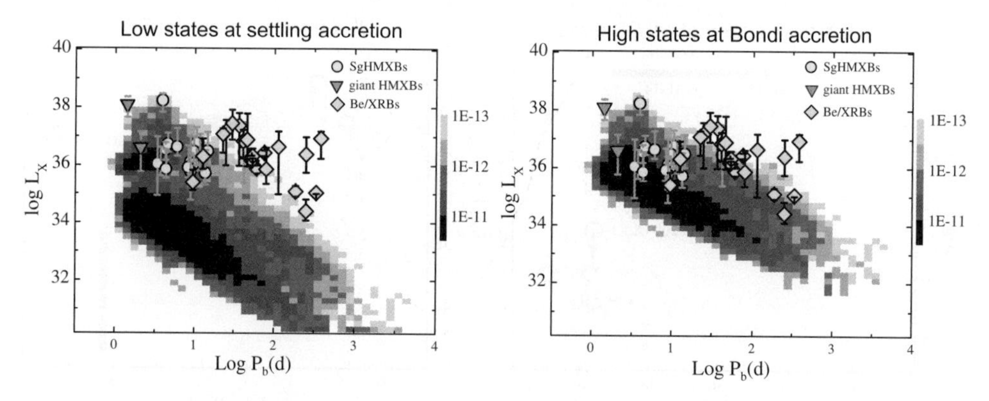

Figure 4. Model X-ray luminosity L_x – binary orbital period diagram for HMXBs (in grey scale) and observed steady sources (filled circles). Filled diamonds show BeXRBs with non-circular orbits in outbursts. Data from Sidoli & Paizis (2018).

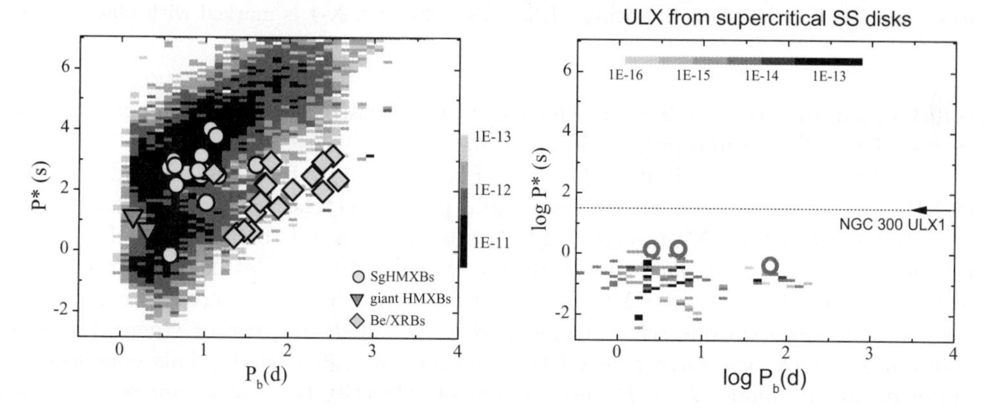

Figure 5. Left: model NS spin period P^* – binary orbital period P_b diagram (the Corbet diagram) for HMXBs (in grey scale) and observed steady sources (filled circles). Filled diamonds show BeXRBs with non-circular orbits. Data from Sidoli & Paizis (2018). Right: the Corbet diagram for NSs at supercritical accretion disc stage. Open red circles show pulsating ULXs, dashed line indicates P^* of NGC 300 ULX1.

the model distribution of HMXBs with neutron stars with supercritical accretion discs and observed pulsating ULXs (red open circles). In addition to the observed pulsating ULXs shown in the Corbet diagram (right panel of Fig. 5), we plot there also candidate ULX NGC 300 ULX1 with $P^* \simeq 31.7\,\mathrm{s}$, $L_x \simeq 4.7 \times 10^{39}$ erg s^{-1} (Carpano *et al.* 2018).

5. Discussion and conclusions

In wind accreting HMXBs with magnetized NSs, two regimes of accretion should be distinguished: the supersonic Bondi-Hoyle-Littleton accretion at high accretion rates, $\dot{M} \gtrsim 4 \times 10^{16}$ g s^{-1} (high X-ray luminosity $L_x > 4 \times 10^{36}$ erg s^{-1}) and subsonic settling accretion (lower L_x). In the regime of settling accretion, a hot shell is formed around the NS magnetosphere enabling angular momentum transfer to/from the rotating NS, and NS reaches an equilibrium spin period proportional to the orbital binary period P_b. Only spin-up of NS is possible in the Bondi-Hoyle accretion stage.

The Corbet diagram for HMXBs. We performed population synthesis calculations of HMXBs with magnetized NSs under standard assumptions on the massive binary star

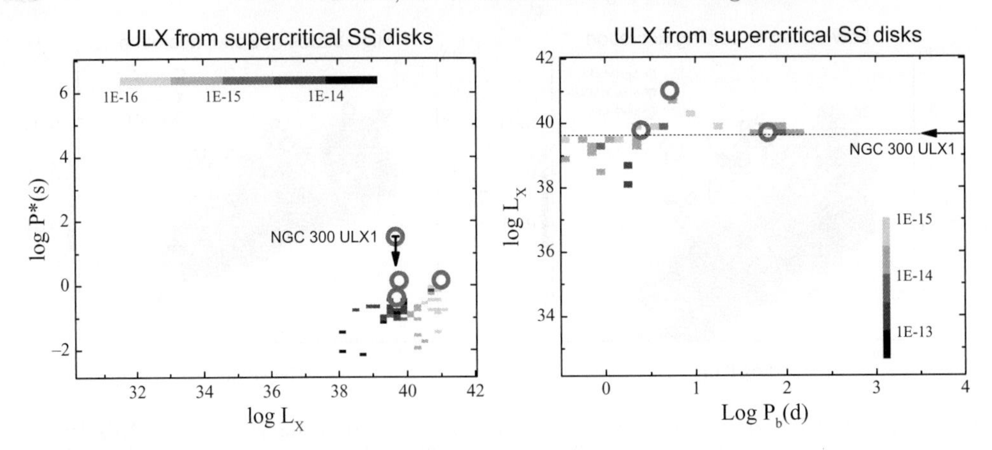

Figure 6. Model $P^* - L_x$ and $L_x - P_b$ diagrams (left and right panels, respectively) for NSs in HMXBs at supercritical accretion disc stage (in grey color, normalized to one $M_\odot$ of Galactic mass). Open red circles show pulsating ULX. NGC 300 ULX X-1 is marked with black arrow.

evolution, parameters of stellar wind for early type massive stars and magnetic fields of NSs as inferred from radio pulsars observations.

Inspection of model distribution $P^* - L_x$ (Fig. 3) suggests that NSs in wind-accreting HMXBs have reached the equilibrium spin period at the early settling accretion stage, but the observed X-ray luminosities better correspond to the Bondi-Hoyle mass accretion rate from the stellar wind (right panel of Fig. 3). The latter holds true for the model L_x-P_b distribution of HMXBs (Fig. 4): the Bondi-Hoyle accretion rates better correspond to observations (right panel of Fig. 4), suggesting earlier stage of settling accretion at which NSs have reached the equilibrium spin periods. This conclusion is supported by the model $P^* - P_b$ distribution of HMXBs (the Corbet diagram) shown in Fig. 5, left panel. Parameters of the observed quasi-steady HMXBs (filled yellow circles and triangles) are in good agreement with the model. However, the observed Be X-ray binaries shown by filled blue diamonds are located systematically lower in this diagram, most likely, due to the reduction of the equilibrium spin period due to orbital eccentricity by the factor $F(e)$ (see Eq. (2.7)) presented in Fig. 1.

The Corbet diagram for pulsating ULXs. When a massive optical star in HMXB overfills its Roche lobe, a supercritical accretion disc can be formed around a magnetized NS. Then the formulas shown in Fig. 2 should be used to calculate the NS equilibrium spin period P^* and X-ray luminosity L_x (in the case where $\dot{M}$ in the disc exceeds $\dot{M}_{cr}$ defined in Sec. 3). Population synthesis calculations of the HMXBs that during later evolution produced objects hosting supercritical accretion discs shown in Fig. 5, right panel (the Corbet diagram for ULXs) and Fig. 6 suggest that pulsating ULXs with known spin P^* and orbital periods P_b fall almost precisely into the expected distribution (shown in grey scale normalized to one $M_\odot$ of the Galactic mass). We stress that in these calculations, we have not done any special assumptions about NS properties (e.g., the initial magnetic field distribution, which is assumed to be a Gaussian centered at 10^{12} G) or binary stellar evolution, i.e. the wind-accreting HMXBs and pulsating ULXs represent single evolutionary related population. Here we stress that the equilibrium NS spin periods and luminosities at supercritical accretion stages depend only on the NS magnetic field (see formulas in Fig. 2). Therefore, the observed dispersion in spin periods of pulsating ULXs around ~ 1 sec reflects the dispersion of magnetic fields of superaccreting NSs in

HMXBs. It is clear that to explain these spin periods, no hypothesis of unusual magnetic fields of NSs is needed.

The 31.7-s NGC 300 ULX1 shown as the upper circle in the left panel of Fig. 6, is worth special mentioning. Its record-high observed spin-up rate (Carpano *et al.* 2018) $\dot{P} \sim 5 \times 10^{-7}$ s s^{-1} suggests a NS spinning-up towards equilibrium. Our model presented in Fig. 2 immediately implies that for $\dot{M} > \dot{M}_{cr}$ (which is the case for the observed X-ray luminosity $\sim 5 \times 10^{39}$ erg s^{-1}), the spin-up rate far from equilibrium is $\dot{P} = P^2/(2\pi I)\sqrt{GM_x R_A} \approx P^2/(2\pi I)\sqrt{GM_x \times 6.5 \cdot 10^7 (L_x/(65 L_{Edd}))} \simeq 5 \times 10^{-7}$ s s^{-1}, in perfect agreement with observations.

Acknowledgements

KP acknowledges grant RSF 16-12-10519 for partial support (Section 1-3). The work of AK (Section 4) is supported by RSF grant No. 14-12-00146.

References

Arons, J. & Lea, S. M. 1976, *ApJ*, 207, 914
Arons, J. & Lea, S. M. 1980, *ApJ*, 235, 1016
Bachetti, M. *et al.* 2014, *Nature*, 514, 202
Burnard, D. J., Arons, J., & Lea, S. M. 1983, *ApJ*, 266, 175
Carpano, S., Haberl, F., Maitra, C., & Vasilopoulos, G. 2018, *MNRAS*, 476, L45
Chakrabarty, D. *et al.* 1997, *ApJL*, 481, L101
de Val-Borro, M., Karovska, M., Sasselov, D. D., & Stone, J. M. 2017, *MNRAS*, 468, 3408
Fürst, F. *et al.* 2016, *ApJL*, 831, L14
González-Galán, A. *et al.* 2012, *A&A*, 537, A66
Ho, C., Taam, R. E., Fryxell, B. A., Matsuda, T., & Koide, H. 1989, *MNRAS*, 238, 1447
Hunt, R. 1971, *MNRAS*, 154, 141
Hurley, J. R., Pols, O. R., & Tout, C. A. 2000, *MNRAS*, 315, 543
Hurley, J. R., Tout, C. A., & Pols, O. R. 2002, *MNRAS*, 329, 897
Illarionov, A. F. & Sunyaev, R. A. 1975, *A&A*, 39, 185
Israel, G. L. *et al.* 2017a, *Science*, 355, 817
Israel, G. L. *et al.* 2017b, *MNRAS*, 466, L48
King, A., Lasota, J. P., & Kluźniak, W. 2017, *MNRAS*, 468, L59
Knigge, C., Coe, M. J., & Podsiadlowski, P. 2011, *Nature*, 479, 372
Lipunov, V. M. 1982, *SvA Lett.*, 26, 54
Lipunov, V. M., Postnov, K. A., Prokhorov, M. E., & Bogomazov, A. I. 2009, *Astronomy Reports*, 53, 915
Liu, Z. W., Stancliffe, R. J., Abate, C., & Matrozis, E. 2017, *ApJ*, 846, 117
Lü, G. L. *et al.* 2012, *MNRAS*, 424, 2265
Postnov, K. A. & Yungelson, L. R. 2014, *Living Reviews in Relativity*, 17, 3
Shakura, N. & Postnov, K. 2017, *ArXiv e-prints*
Shakura, N., Postnov, K., Kochetkova, A., & Hjalmarsdotter, L. 2012, *MNRAS*, 420, 216
Shakura, N., Postnov, K., Kochetkova, A., & Hjalmarsdotter, L. 2018, in: N. Shakura, ed., *Accretion Processes in Astrophysics*, chap. 8, Springer
Shakura, N. I. & Sunyaev, R. A. 1973, *A&A*, 24, 337
Sidoli, L. & Paizis, A. 2018, *MNRAS*, 481, 2779
Vink, J. S. 2018, *ArXiv e-prints*, 1808.06612
Yu, S. & Jeffery, C. S. 2010, *A&A*, 521, A85

High-mass X-ray Binaries: Illuminating the Passage from Massive Binaries to Merging Compact Objects
Proceedings IAU Symposium No. 346, 2019
L. M. Oskinova, E. Bozzo, T. Bulik & D. Gies, eds.

doi:10.1017/S1743921318007482

Monitoring of the eclipsing Wolf-Rayet ULX in the Circinus galaxy

Yanli Qiu[1] and Roberto Soria[2]

[1]National Astronomical Observatories, Chinese Academy of Sciences, 20A Datun Rd, Beijing 100101, China
email: qiuyanli824@163.com

[2]College of Astronomy and Space Sciences, University of the Chinese Academy of Sciences, Beijing 100049, China
email: rsoria@nao.cas.cn

Abstract. We studied the eclipsing ultraluminous X-ray source CG X-1 in the Circinus galaxy, re-examining two decades of *Chandra* and *XMM-Newton* observations. The short binary period (7.21 hr) and high luminosity ($L_X \approx 10^{40}$ erg s^{-1}) suggest a Wolf-Rayet donor, close to filling its Roche lobe; this is the most luminous Wolf-Rayet X-ray binary known to-date, and a potential progenitor of a gravitational-wave merger. We phase-connect all observations, and show an intriguing dipping pattern in the X-ray lightcurve, variable from orbit to orbit. We interpret the dips as partial occultation of the X-ray emitting region by fast-moving clumps of Compton-thick gas. We suggest that the occulting clouds are fragments of the dense shell swept-up by a bow shock ahead of the compact object, as it orbits in the wind of the more massive donor.

Keywords. X-rays: binaries, X-rays: individual (Circinus Galaxy X-1), stars: Wolf-Rayet, binaries: eclipsing

1. Introduction

The Circinus galaxy, located at a distance of 4.2 Mpc (Tully *et al.* 2009), contains a bright, point-like X-ray source known as CG X-1, seen at a projected distance of ≈300 pc from its starburst nucleus (Figure 1). The interpretation of this source has been the subject of debate for the past two decades (Bauer *et al.* 2001; Weisskopf *et al.* 2004; Esposito *et al.* 2015). There are at least three features that make this object interesting and unusual. The first one is its high luminosity. If it is indeed located inside the Circinus galaxy (rather than being a foreground or background object), its average X-ray luminosity would be $\approx 10^{40}$ erg s^{-1}; this would place CG X-1 near the top of the luminosity distribution of ultraluminous X-ray sources (ULXs) in the local universe: a factor of 10 times above the Eddington luminosity of Galactic stellar-mass black holes (BHs), or 100 times above the Eddington limit of a neutron star (NS). When CG X-1 was first discovered (Bauer *et al.* 2001), observational and theoretical understanding of super-Eddington X-ray binaries was still in its infancy; however, today we know such sources exist and we can quantify their population properties. Based on the star formation rate of Circinus ($\approx$3–8 $M_\odot$ yr^{-1}: For *et al.* 2012) and on the X-ray luminosity function of Mineo *et al.* (2012), we expect ≈0.2–0.6 X-ray binaries with a luminosity of 10^{40} erg s^{-1} or above, in that galaxy. Alternative interpretations, for example that of a foreground CV (Weisskopf *et al.* 2004), can be rejected based on its high X-ray/optical flux ratio (Bauer *et al.* 2001; Qiu *et al.* in preparation), as well as the low probability of finding a foreground Galactic source projected onto the starforming nucleus of Circinus.

Figure 1. Left panel: true-colour optical image of the Circinus galaxy in the g,r,i filters; data from the 2.6-m ESO-VLT Survey Telescope. North is up and east to the left. The white box is the area displayed in the *Chandra* image. Right panel: adaptively smoothed *Chandra*/ACIS image of the innermost region of the galaxy; red = 0.3–1.1 keV, green = 1.1–2.0 keV, blue = 2.0–7.0 keV. The two brightest non-nuclear X-ray sources are known as CG X-1 (the ULX subject of our investigation) and CG X-2 (also known as SN 1996cr: Bauer *et al.* 2008). A powerful outflow of hot thermal plasma from the nuclear starburst is well visibile to the west of the nucleus, but the soft diffuse emission is absorbed by thick dust to the east of the nucleus.

The second interesting property of CG X-1 is the periodicity identified in its X-ray lightcurve. Most ULXs show stochastic variability by a factor of a few; in a few cases, periodic signals of a few days (*e.g.*, ≈2.5 d in M 82 X-2: Bachetti *et al.* 2014; either ≈6 d or ≈13 d in M 51 ULX-1: Urquhart & Soria 2016a; ≈8.2 d in M 101 ULX-1: Liu *et al.* 2013) or even weeks (≈63 d for NGC 7793 P13: Motch *et al.* 2014) have been identified and interpreted as the binary periods. CG X-1 stands out with an unambiguous X-ray period of only 7.2 hr, determined from its eclipsing behaviour (Esposito *et al.* 2015). This period is too short to be consistent with a supergiant donor; it suggests instead a Wolf-Rayet donor, whose radius is small enough to fit into such a compact binary system. In any case, CG X-1 is the ULX with the most precisely known binary period.

The third intriguing feature of CG X-1 is the nature of its X-ray eclipses. Although the period is stable (X-ray lightcurves observed twenty years ago can still easily be phase connected with recent observations) and the phase-averaged lightcurve is superficially consistent with the eclipse of the compact object behind the donor star, our study of the individual cycles tells a different story. Each cycle has a different pattern of eclipse duration and dipping morphology.

In this work, we will focus on the second and third property outlined above (short period and eclipse behaviour). We will discuss the general properties of the source in and out of eclipse, and the physical origin of the eclipses. We will then briefly discuss the possible origin and future evolution of this system, and how its existence compares with the detection rate of gravitational wave events.

2. The most luminous Wolf-Rayet ULX

There is an unresolved optical counterpart detected in the only (short) *Hubble Space Telescope* observation of the field, within the *Chandra* error circle for CG X-1. We estimate an apparent brightness $m_{\rm F606W} \approx V \approx 24.3 \pm 0.1$ mag in the Vega system; the distance modulus of Circinus is ≈ 28.1 mag. Unfortunately, the optical extinction is very high, because Circinus is located behind the disk of the Milky Way; the line-of-sight extinction

Table 1. Summary of candidate Wolf-Rayet X-ray binaries, in order of increasing period.

Name	Galaxy	Distance (Mpc)	Peak $L^{a}_{0.3-10}$ (erg s^{-1})	Period hr	References
CXOU J121538.2+361921	NGC 4214	3.0	$\approx 6\times10^{38}$	3.6	1
Cygnus X-3	Milky Way	0.0074	$\approx$ a few $\times10^{38}$	3.6	2,3,4,5
CXOU J123030.3+413853	NGC 4490	7.0	$\approx 1\times10^{39}$	4.8	6
CG X-1	**Circinus**	**4.2**	$\approx \mathbf{3\times10^{40}}$	**7.2**	**7,8**
CXOU J004732.0−251722	NGC 253	3.2	$\approx 1\times10^{38}$	14.5	9
CXOU J005510.0−374212 (X-1)	NGC 300	1.9	$\approx 3\times10^{38}$	32.8	10,11,12,13
CXOU J002029.1+591651 (X-1)	IC 10	0.7	$\approx 7\times10^{37}$	34.8	14,11,15,16
[b]CXOU J140332.3+542103 (ULX-1)	M 101	6.4	$\approx 4\times10^{39}$	196.8	17,18,19

References:
1: Ghosh *et al.* (2006); 2: Hjalmarsdotter *et al.* (2009); 3: Koljonen *et al.* (2010); 4: Zdziarski *et al.* (2012); 5: McCollough *et al.* (2016); 6: Esposito *et al.* (2013); 7: Esposito *et al.* (2015); 8: Qiu *et al.*, in prep.; 9: Maccarone *et al.* (2014); 10: Carpano *et al.* (2007); 11: Barnard *et al.* (2008); 12: Crowther *et al.* (2010); 13: Binder *et al.* (2011); 14: Prestwich *et al.* (2007) 15: Silverman & Filippenko (2008) 16: Laycock *et al.* (2015) 17: Kong *et al.* (2004); 18: Liu *et al.* (2013); 19: Urquhart & Soria (2016b).
Notes:
[a]De-absorbed 0.3–10 keV luminosity in the bright phase of the orbital cycle; values taken from the references listed in this Table, but rescaled to the distance adopted here.
[b]M 101 ULX-1 differs from the other seven sources because it is an ultraluminous supersoft source, it does not show eclipses, and its binary separation is too large to permit a BH-BH merger in a Hubble time.

towards the Circinus galaxy is $A_V \approx 4$ mag (Schlafly & Finkbeiner 2011), but additional local extinction is highly likely (see also Weisskopf *et al.* 2004) and uncertain, given the location of the source near dust lines. This, coupled with the lack of observations in any other optical/IR band, make it impossible to determine the nature of this object (let alone its time variability). Thus, the main constraints to the nature of the system must come from the X-ray data.

From our study of all the archival *ROSAT*, *Chandra* and *XMM-Newton* observations between 1997 March and 2018 February (see Qiu *et al.*, in prep., for a detailed log), we derived an average binary period $P = 25,970.1 \pm 0.1$ s ≈ 7.214 hr. Using the period-density relation for binary systems (Eggleton 1983), we obtain an average density of $\rho \approx 1.2\rho_{\odot} \approx 1.7$ g cm^{-3} inside the Roche lobe of the donor star, for a mass ratio $M_2/M_1 = 2$ (where M_2 is the mass of the donor star), or $\rho \approx 0.74\rho_{\odot} \approx 1.1$ g cm^{-3}, for $M_2/M_1 = 10$. This range of values already rules out main-sequence OB stars (in fact, any main-sequence star more massive than $\approx 1\ M_{\odot}$), blue supergiants, red supergiants or red giants. The persistent nature of the X-ray source over at least 20 years, and its location in a highly star-forming region, strongly suggest a young system with a donor star more massive than the compact object. A Wolf-Rayet star is consistent with all those constraints. If CG X-1 contains a 20-$M_{\odot}$ Wolf-Rayet star and a 10-$M_{\odot}$ BH, the binary separation is $\approx 5.8\ R_{\odot}$ and the size of the Roche Lobe of the star is $\approx 2.6\ R_{\odot}$: this is large enough to contain a Wolf-Rayet but not any other type of massive star. Cygnus X-3 is the prototypical example of a high-luminosity X-ray binary with a very short binary period, fed by a Wolf-Rayet star. Very few such systems are known to-date (Table 1).

From our spectral modelling, we found (Qiu *et al.*, in prep.) that the de-absorbed X-ray luminosity of CG X-1 is $\approx 10^{40}$ erg s^{-1} during the out-of-eclipse parts of the orbital cycle, with a variability range of a factor of four above and below this value, over two decades†. Such extreme luminosity implies a mass accretion rate onto the compact object of at least $\approx 10^{-6} M_{\odot}$ yr^{-1}, for a radiative efficiency $\eta \approx 0.15$. In fact, the radiative efficiency is likely to be lower, scaling as $\eta \sim 0.1(1 + \ln \dot{m})/\dot{m}$ for super-Eddington accretion, where $\dot{m}$ is the accretion rate in Eddington units (Shakura & Sunyaev 1973; Poutanen *et al.*

† The highest out-of-eclipse luminosity ($L_{\rm X} \approx 3.8 \times 10^{40}$ erg s^{-1}) was measured from the *XMM-Newton* observation of 2001 August 6, and the lowest value ($L_{\rm X} \approx 2.3 \times 10^{39}$ erg s^{-1}) was recorded in the *Chandra* observation of 2008 October 26.

2007). Moreover, the eclipsing and dipping behaviour suggests a high viewing angle, so that we cannot invoke geometric beaming of the emission via a polar funnel. Thus, it appears that the system is really accreting at least several times $10^{-6} M_\odot$ yr^{-1}. Such high accretion rate suggests that the donor star is either filling the Roche lobe, or at least that its wind is gravitationally focused towards the compact object. Note that CG X-1 is the only system among candidate Wolf-Rayet X-ray binaries with a luminosity $\sim 10^{40}$ erg s^{-1}; all others have X-ray luminosities $\lesssim 10^{39}$ erg s^{-1}, consistent with wind accretion.

3. Periodic eclipses and stochastic dipping

Folded X-ray lightcurves (Fig. 6 in Esposito *et al.* 2015; Qiu *et al.*, in prep.) show a sharp flux drop in ingress, followed by an eclipse phase lasting for $\approx 1/5$ of the period, with faint residual emission (softer than out-of-eclipse), and then a slow return to the baseline flux. This structure is reminiscent of other high-mass X-ray binaries seen at high inclination, with a proper eclipse (apart from residual scattered photons) when the accreting object is behind the donor star, and varying absorbing column density at other phases, as the compact object moves through the wind of the donor. However, an inspection of the individual lightcurves of CG X-1 from each observed cycle tells a more complicated story (Figure 2). Any model of the system must explain the following two X-ray properties:

a) the eclipse and dipping patterns and duration of the egress phase change from orbit to orbit. Clearly, the size of the donor star and the binary separation cannot change; therefore, the eclipse and the dips must be (at least partly) caused by optically thick material (*e.g.*, clouds) in front of the X-ray source, moving on timescales shorter than the binary period.

b) the transition from full eclipse to full flux level consists of a decreasing level of partial covering by an optically thick medium. A sequence of X-ray spectra show (Qiu *et al.*, in prep.) that during the egress phase, the intrinsic spectral shape and cold absorption remain the same, and the difference is only the value of a normalization constant. In other words, the flux recovery is not caused by a gradual decrease of absorbing column density. Instead, we propose that Compton-thick clouds ($N_{\rm H} > 1.5 \times 10^{24}$ cm^{-2}) occult a variable fraction (between 0 and 100%) of the emitting region during each orbital cycle.

The occulting clouds cannot be uniformly or randomly distributed along all azimuthal angles, because the observed pattern of fast ingress, total eclipse, and dips during the slow egress is regularly repeated for two decades. In some low-mass X-ray binaries seen at high inclination, regular dipping behaviour is also observed, probably caused by the thick bulge where the accretion streams impacts the disk (White & Swank 1982; Frank *et al.* 1987) 1987). However, this scenario does not work for CG X-1, because the accretion stream always trails the compact object, and would produce most dips just before or during eclipse ingress, contrary to the observed pattern. High-mass X-ray binaries sometimes also have an asymmetric eclipse profile due to a thick accretion stream (for example, Vela X-1: Doroshenko *et al.* 2013), and in those cases, too, a slow ingress is followed by a fast egress (opposite to CG X-1).

Taking those constraints into account, we suggest that the optically thick material is located between the Wolf-Rayet and the compact object, but mostly in front of the compact object. This configuration will lead to partial occultations of the X-ray emission after the compact object has passed behind the star and is moving towards us (egress), rather than before.

We know that the ULX must have a strong, radiatively driven wind (*e.g.*, Poutanen *et al.* 2007, Ohsuga & Mineshige 2011), probably comparable in speed and mass density to that of the Wolf-Rayet. We also estimate an orbital velocity $\approx$700 km s^{-1} for a typical 10-$M_\odot$ stellar-mass BH orbiting a typical Wolf-Rayet star ($M_2 \approx 20$–$30\ M_\odot$). This implies

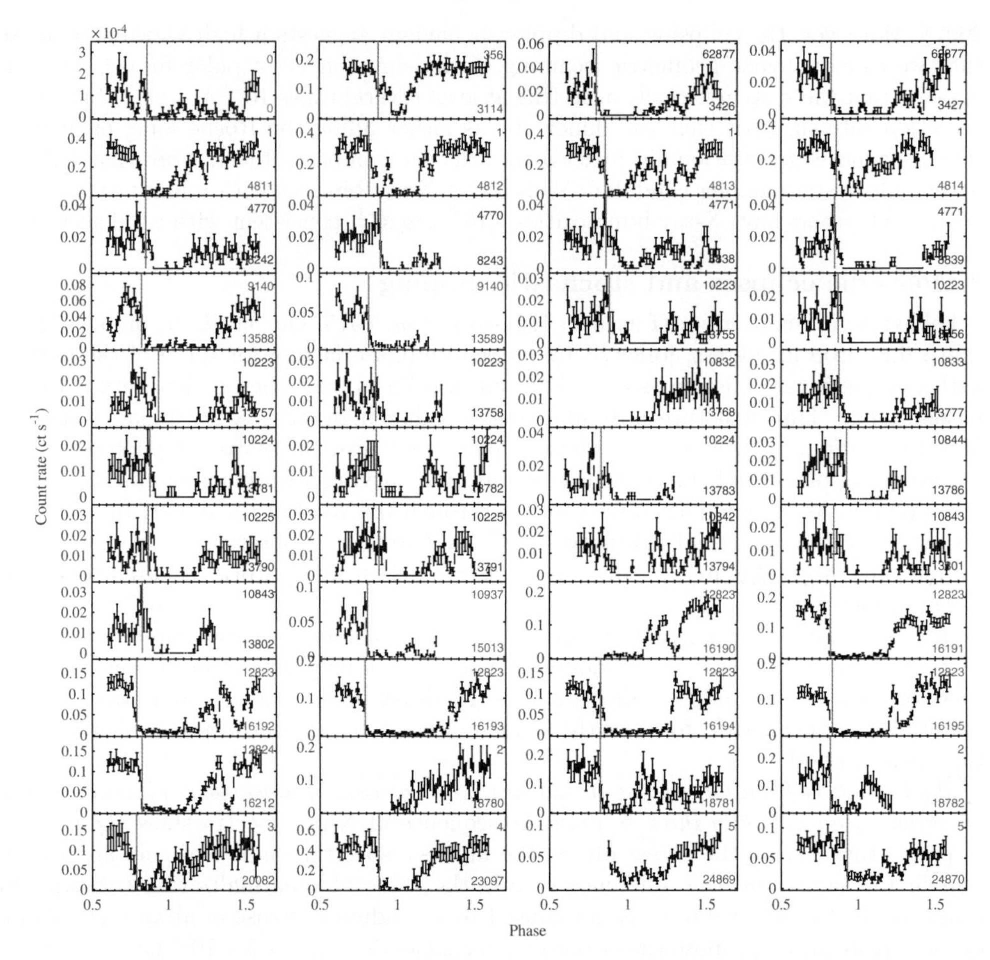

Figure 2. Twenty years of X-ray lightcurves for CG X-1, showing that each orbital cycle is different than the others, although some general underlying features remain the same. All lightcurves (apart from the one from *ROSAT*/HRI) are in the 0.3–8 keV band, and are binned to 500 s per bin. All lightcurves are folded on our best-fitting ephemeris; phase $\phi = 1$ occurs at MJD $= 50681.327156 + 0.300580(\mathrm{d}) \times N$. See Qiu *et al.*, in prep., for details of how the ingress time and the ephemeris were calculated. In each panel, the number at the bottom right corresponds to the value of N for the orbital cycle represented in that panel. The vertical magenta bar in each panel shows the estimated centre time of the eclipse ingress phase for that observation. The top left panel (labelled 0) is a *ROSAT*-HRI lightcurve from 1997; the numbers on the top right of all the other panels are short forms of the corresponding observation IDs from which those lightcurves were extracted. Red numbers (*e.g.*, 356): ObsID of a *Chandra*/ACIS-S3 observation; blue numbers (*e.g.*, 62877): ObsID of a *Chandra*/HETG observation; green numbers (*e.g.*, 1): ObsID of an *XMM-Newton*/EPIC observation, where 1, 2, 3, 4, 5 mean 0111240101, 0701981001, 0656580601, 0792382701, 0780950201, respectively. *XMM-Newton*/EPIC lightcurves are extracted from MOS1 + MOS2 for observations 1, 2, 3 and 5, and from EPIC-pn for observation 4.

that the compact object is ploughing through the thick wind of the donor star at highly supersonic speed (Mach number $\mathcal{M} \approx 20$ for a wind temperature $\approx 10^5$ K).

In general, in systems with a Wolf-Rayet star and an O star, a denser, optically-thick layer of shocked gas forms at the interface of the two winds; in CG X-1 we have the additional element that the compact object is moving at supersonic speed. We suggest that the compact object creates a bow shock along its direction of motion, and sweeps up a dense shell of shocked Wolf-Rayet wind; for a radiative shock, the density enhancement scales as $\mathcal{M}^2$, from standard bubble theory (*e.g.*, Weaver *et al.* 1977). Thus, for typical ambient densities $n_e \sim 10^{14}$ cm^{-3} (Ro & Matzner 2016) in the undisturbed wind, the density in the shocked shell can exceed 10^{16} cm^{-3} and lead to rapid cooling. By analogy with Wolf-Rayet/O-star binaries (Usov 1991), a cold dust layer may form at the contact discontinuity between the shocked Wolf-Rayet wind and the shocked accretion-disk wind. For an order-of-magnitude estimate, we can assume a radius of the shell comparable to the Roche Lobe of the accreting object ($R \sim 10^{11}$ cm) and a thickness of the shell $\sim 10^9$ cm: thus, the equivalent hydrogen column density in the swept-up shell can exceed 10^{25} cm^{-2} and cause total occultation of the X-ray emission below 10 keV.

For high Mach numbers, hydrodynamic instabilities of the swept-up, cooling shell lead to continuous fragmentation and re-formation (Park & Ricotti 2013). We argue that fragments of the swept-up shell are the optically thick structures responsible for the irregular dipping in CG X-1. If this is the case, we expect to see occultations and dips in the X-ray lightcurve mostly when the compact object moves towards us (after egress from the eclipse), rather than before ingress.

4. Conclusions

We have summarized some of the most interesting properties of the ULX CG X-1 in the Circinus galaxy. Its short orbital period (7.2 hr) makes it a strong candidate Wolf-Rayet X-ray binaries, a very rare system (less than 10 known to-date) with intriguing accretion physics. If the compact object is a stellar-mass BH and the Wolf-Rayet collapses into another BH (without disrupting the binary), the timescale for a gravitational merger is only $\approx$50 Myr (Esposito *et al.* 2015). In addition, CG X-1 is one of the most luminous ULXs in the nearby universe, reaching peak luminosities in excess of 3×10^{40} erg s^{-1} at some epochs. By contrast, all other Wolf-Rayet X-ray binaries have luminosities $\lesssim 10^{39}$ erg s^{-1}. Thirdly, the X-ray lightcurve shows a regular pattern of eclipses with fast ingress and slow egress (with a coherent phase over 20 years of observation), modified by an irregular pattern of deep dips, changing every orbital cycle. We have discussed a possible origin for the dips. We suggested that the CG X-1 differs from sub-Eddington Wolf-Rayet systems such as Cyg X-3 because both the primary and the secondary launch a massive radiatively driven outflow. In fact, the gas environment in systems such as CG X-1 may be compared to binary Wolf-Rayet systems.

In short, CG X-1 is an exceptional test case for studies of the progenitors of of gravitational wave events and their expected rate in the local universe; for studies of accretion and outflows in ULXs; and for studies of the hydrodynamics of colliding winds and shock-ionized bubbles.

In a forthcoming paper (Qiu *et al.*, in prep.), we will present a detailed X-ray spectroscopic study of the system outside eclipse, in eclipse, and during egress. We will also discuss possible origins of the system (via a common envelope phase), and whether the primary is more likely to be a NS or a BH. Finally, we will show that phasing the sharp ingress of the eclipse over 20 years of archival observations can already reveal a period derivative (more exactly, a slow increase of the period). In turns, this can constrain for example whether most of the mass lost by the Wolf-Rayet donor ends up accreted by the compact object or ejected in a wind.

Acknowledgements

We thank Alexey Bogomazov, Rosanne Di Stefano, Jifeng Liu, Michela Mapelli, Manfred Pakull, Song Wang, and Grzegorz Wicktorowicz, for useful suggestions and discussions, which greatly improved our presentation at the IAU Symposium. YQ acknowledges the Harvard-Smithsonian Center for Astrophysics, and RS thanks Curtin University and The University of Sydney, for hospitality during part of this research.

References

Bachetti, M., *et al.* 2014, *Nature*, 514, 202
Barnard, R., Clark, J. S., & Kolb, U. C. 2008, *A&A*, 488, 697
Bauer, F. E., Brandt, W. N., Sambruna, R. M., Chartas, G., Garmire, G. P., Kaspi, S., & Netzer H. 2001, *AJ*, 122, 182
Bauer, F. E., Dwarkadas, V. V., Brandt, W. N., Immler, S., Smartt, S., Bartel, N., & Bietenholz, M. F. 2008, *ApJ*, 688, 1210
Binder, B., Williams, B. F., Eracleous, M., Garcia, M. R., Anderson, S. F., & Gaetz, T. J. 2011, *ApJ*, 742, 128
Carpano, S., Pollock, A. M. T., Prestwich, A. H., Crowther, P., Wilms, J., Yungelson, L., & Ehle, M. 2007, *A&A*, 466, L17
Crowther, P. A., Barnard, R., Carpano, S., Clark, J. S., Dhillon, V. S., & Pollock, A. M. T. 2010, *MNRAS*, 403, L41
Doroshenko, V., Santangelo, A., Nakahira, S., Mihara, T., Sugizaki, M., Matsuoka, M., Nakajima, M., & Makishima, K. 2013, *A&A*, 554, 37
Eggelton, P. P. 1983, *ApJ*, 268, 368
Esposito, P., Israel, G. L., Sidoli, L., Mapelli, M., Zampieri, L., & Motta, S. E. 2013, *MNRAS*, 436, 3380
Esposito, P., Israel, G. L., Milisavljevic, D., Mapelli, M., Zampieri, L., Sidoli, L., Fabbiano, G., & Rodríguez Castillo, G. A. 2015 *MNRAS*, 452, 1112
For, B.-Q., Koribalski, B. S., & Jarrett, T. H. 2012, *MNRAS*, 425, 1934
Frank, J., King, A. R., & Lasota, J.-P. 1987, *A&A*, 178, 137
Ghosh, K. K., Rappaport, S., Tennant, A. F., Swartz, D. A., Pooley, D., & Madhusudhan, N. 2006, *ApJ*, 650, 872
Hjalmarsdotter, L., Zdziarski, A. A., Szostek, A., & Hannikainen, D. C. 2009, *MNRAS*, 392, 251
Koljonen, K. I. I., Hannikainen, D. C., McCollough, M. L., Pooley, G. G., & Trushkin, S. A. 2010, *MNRAS*, 406, 307
Laycock, S. G. T., Maccarone, T. J., & Christodoulou, D. M. 2015, *MNRAS*, 452, L31
Kong, A. K. H., Di Stefano, R., & Yuan, F. 2004, *ApJ*, 617, L49
Liu, J.-F., Bregman, J. N., Bai, Y., Justham, S., & Crowther, P. 2013, *Nature*, 503, 500
Maccarone, T. J., Lehmer, B. D., Leyder, J. C., Antoniou, V., Hornschemeier, A., Ptak, A., Wik, D., & Zezas, A. 2014, *MNRAS*, 439, 3064
McCollough, M. L., Corrales, L., & Dunham, M. M. 2016, *ApJ*, 830, L36
Mineo, S., Gilfanov, M., & Sunyaev, R. 2012, *MNRAS*, 419, 2095
Motch, C., Pakull, M. W., Soria, R., Grisé, F., & Pietrzyński, G. 2014, *Nature*, 514, 198
Ohsuga, K., & Mineshige, S. 2011, *ApJ*, 736, 2
Park, K. H., & Ricotti, M. 2013, *ApJ*, 767, 163
Poutanen, J., Lipunova, G., Fabrika, S., Butkevich, A. G., & Abolmasov, P. 2007, *MNRAS*, 377, 1187
Prestwich, A. H., *et al.* 2007, *ApJ*, 669, L21
Ro, S., & Matzner, C. D. 2016, *ApJ*, 821, 109
Schlafly, E. F., & Finkbeiner, D. P. 2011, *ApJ*, 737, 103
Shakura, N. I., & Sunyaev, R. A. 1973, *A&A*, 24, 337
Silverman, J. M., & Filippenko, A. V. 2008, *ApJ*, 678, L17
Tully, R. B., Rizzi, L., Shaya, E. J., Courtois, H. M., Makarov, D. I., & Jacobs, B. A. 2009, *AJ*, 138, 323
Urquhart, R. T., & Soria, R. 2016a, *ApJ*, 831, 56
Urquhart, R. T., & Soria, R. 2016b, *MNRAS*, 456, 1859
Usov, V. V. 1991, *MNRAS*, 252, 49
Weaver, R., McCray, R., Castor, J., Shapiro, P., & Moore, R. 1977, *ApJ*, 218, 377
Weisskopf, M. C., Wu, K., Tennant, A. F., Swartz, D. A., & Ghosh, K. K. 2004, *ApJ*, 605, 360
White, N. E., & Swank, J. H. 1982, *ApJ*, 253, L61
Zdziarski, A. A., Maitra, C., Frankowski, A., Skinner, G. K., & Misra, R. 2012, *MNRAS*, 426, 1031

High-mass X-ray Binaries: Illuminating the Passage from Massive Binaries to Merging Compact Objects
Proceedings IAU Symposium No. 346, 2019
L. M. Oskinova, E. Bozzo, T. Bulik & D. Gies, eds.

doi:10.1017/S174392131800772X

Analysis of spectrum variations in Hercules X-1

Denis Leahy

Dept. of Physics & Astronomy, University of Calgary, Calgary, Canada
email: leahy@ucalgary.ca

Abstract. Hercules X-1 (Her X-1) was observed extensively by the Rossi X-ray Timing Explorer (RXTE) over its 17 year lifetime. Here, the archival RXTE/PCA observations of Her X-1 are analyzed with emphasis on the 35-day cycle dependence. Spectral fits are carried out and the 35-day phase dependences are characterized. The regular behaviours of the changes are interpreted in terms of the precessing accretion disk. We find that the most important variation is caused by the changing illumination of the inner edge of the disk, but other variations with different causes are also seen.

Keywords. X-rays: binaries, accretion, accretion disks, stars: neutron

1. Introduction

HZ Herculis/Hercules X-1, also known as HZ Her/Her X-1, is one of the brightest low-mass X-ray binary systems, and among the most studied, e.g. Gerend & Boynton (1976), Crosa & Boynton (1980), Leahy (2002), Leahy (2003), Ji *et al.* (2009) and Klochkov *et al.* (2009). Leahy & Abdallah (2014) showed the system is located at a distance of $\simeq$6.5 kpc from Earth. It has an A7 type (which varies between late A and early B with the orbital phase) main sequence stellar companion, HZ Her, and a neutron star of approximate masses 2.2 $M_{\odot}$ and 1.5 $M_{\odot}$, respectively (Reynolds *et al.* 1997). The system is characterized by a great variety of phenomena, including 1.24 s pulsations, 1.7-day orbital period eclipses, and a 35-day X-ray intensity cycle, e.g. (Scott *et al.* 2000), (Scott & Leahy 1999) and (Boynton *et al.* 1980). Her X-1 is one of the few systems to have low interstellar absorption, which makes it feasible for observation and study at optical, UV and X-ray wavelengths.

The 35-day flux and pulse profile cycles, unique features among the known accretion-powered pulsars, are produced by the counter-precessing, tilted and twisted neutron star accretion disk. This was demonstrated in studies including those of Leahy (2004a), Leahy (2004b), Leahy (2003), Leahy (2002), Scott *et al.* (2000), Scott & Leahy (1999), Wijers & Pringle (1999), Maloney & Begelman (1997), Maloney *et al.* (1996), Larwood *et al.* (1996) and Schandl & Meyer (1994).

The 35-day cycle consists of a Main High (MH) state, a Short High (SH) state and fainter Low States (LS) which separate the MH and SH states. Scott & Leahy (1999) showed that the Main High (MH) state that covers 35-day phases 0 to 0.31, the Short High (SH) state covers phases 0.57-0.79.

The extensive RXTE/PCA archival observations of Her X-1 have been analyzed, in part, previously by our group. Leahy & Igna (2011) analyzed the full set of archival observations to produce light curves in different energy bands folded on the orbital and 35-day cycles, providing orbital and 35-day light curves superior to any produced previously. Dips were shown to occur at all orbital phases: the frequency of dips increases with increasing orbital phase and the fraction of time in dips is higher for Short High state

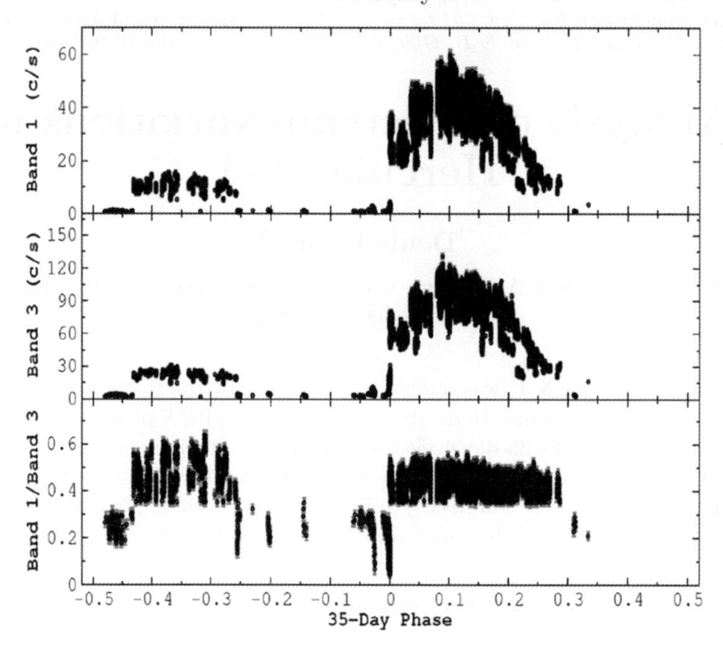

Figure 1. The 2-4 keV (top panel) and 9-20 keV (middle panel) light curves vs. 35-day phase for Hercules X-1, from all RXTE/PCA archival observations but with dips during Main High state (35-day phase 0 to 0.3) and Short High state (35-day phase 0.57 to 0.74) removed. Bottom panel: 2- 4 keV/7-20 keV softness ratio vs 35-day phase.

(average ~50%) than for Main High state (average ~15%) at all orbital phases. Igna & Leahy (2011) analyzed the extensive set of dips from Her X-1 and studied their properties, showing dips occur at preferential regions in the 35-day phase vs. orbital phase diagram. Igna & Leahy (2012) used the accretion stream and its impact with the accretion disk to explain the orbital and 35-day phase timing of the dips.

Leahy & Abdallah (2014) used RXTE/PCA timing of eclipse ingresses and egresses to determine the stellar radius of the companion star HZ Her, and combined it with archival spectrophotometry of HZ Her to determine the distance, the companion's stellar properties and its evolutionary state. Abdallah & Leahy (2015) analyzed RXTE spectra during Short High state and showed that a significant component of the flux and spectrum is best explained by X-ray reflection off the face of HZ Her.

Leahy (2015) analyzed all RXTE/PCA eclipse light curves during Main High state to detect a large scale (few times 10^{11} cm) electron-scattering corona in the binary system, and determine its density profile, assuming spherical symmetry. The corona could be a hot wind outflow from the system, or fairly static and supported by thermal pressure. The latter seems more likely as the heating rate by the X-ray flux from the neutron star is adequate to thermally support the corona. In addition, the mass-loss rate for a wind with escape velocity is much higher than the accretion rate onto the neutron star.

2. Summary of new results

The new analysis of the RXTE/PCA observations is systematic spectral fitting of Her X-1 over the 35-day cycle with a single spectral model. Fig. 1 shows the 2-4 keV light curve constructed from the RXTE/PCA data. The rise and fall of Main High and Short High states are clearly seen. To remove the effects of eclipses we omitted orbital phases from 0.93 to 0.07. Next dips were removed by selecting data with 2-4 keV/ 9-20 keV softness ratio characteristic of dips. Plots of individual Main High states (all ~15 years

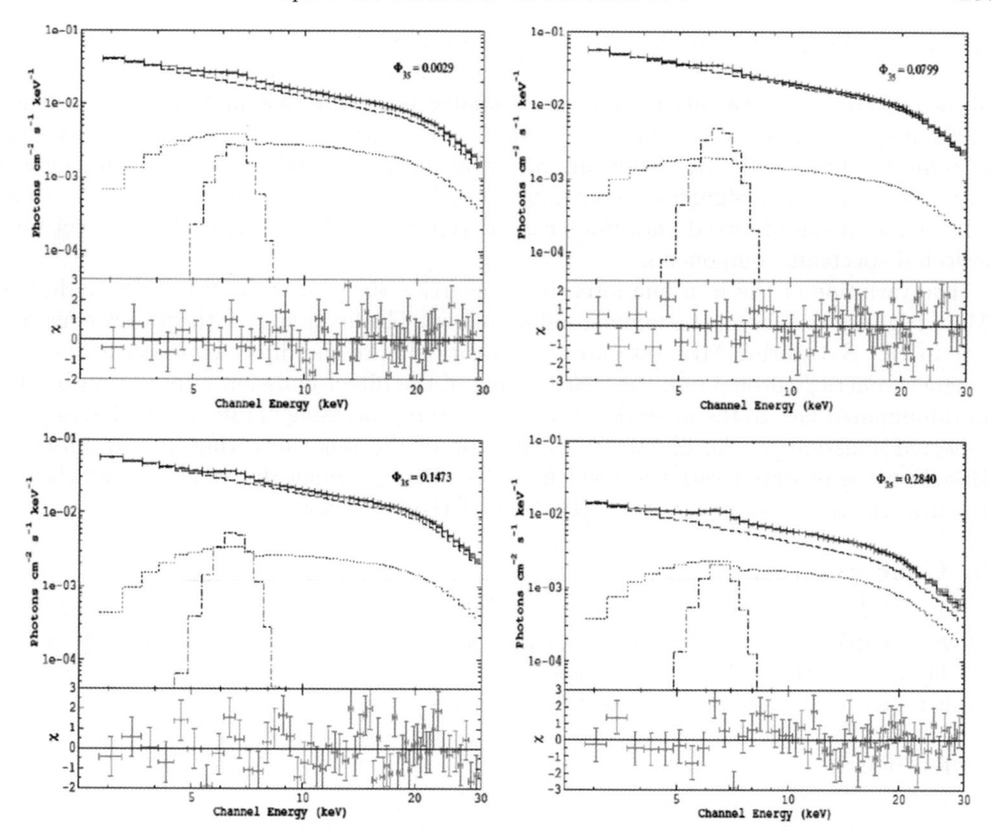

Figure 2. RXTE/PCA spectra of Hercules X-1, at the 4 different 35-day phases during Main High state. The top part of each panel shows the observed spectrum (histogram), and 3 model components (dashed line- unabsorbed continuum component, dotted line- absorbed continuum component, dot-dashed line: Fe line). The lower part of each panel shows the residuals between the model and the observed spectrum.

of RXTE/PCA data) demonstrates that the range of count rates seen in Fig. 2 is due to cycle-to-cycle variation rather than variation within a single cycle.

The spectral model that consistently describes all Main High state data (less eclipse and dip data) is the power-law with exponential cutoff model (powerlaw/highecut in XSPEC). This was multiplied by the partial covering absorption model to yield 2 components: an absorbed and an unabsorbed component. Finally, to describe the data adequately, we added a gaussian line for the well-known fluorescent Fe line. Spectra were extracted for time intervals of 256 s which had constant softness ratio, resulting in several hundred spectra covering Main High state. The spectral parameters and their errors were determined for each spectrum using XSPEC and the XSPEC error command.

Systematic variations of spectral parameters are found during the Main High state that depend on 35-day phase. The main dependencies are as follows. (i) Photon index is constant for phases 0-0.05, then decreases for phases 0.05 to 0.2, and is constant or rising for 0.2-0.3 (end of Main High). (ii) Cutoff energy is constant for phases 0-0.12, then decreases for phases 0.12 to 0.3. (iii) Iron line energy rises for phases 0-0.15, then decreases for phases 0.15 to 0.3.

The interpretation of the spectral variations is based on the disc model for Her X-1 constructed by Leahy (2002), the pulse evolution model of Scott *et al.* (2000) and the pulse beam model of Leahy (2004c). We can understand the variations of the continuum spectrum of Her X-1 during the MH-state as a superposition of three components: (i) the spectrum of the direct emission from the pulsar (fan and pencil beams); (ii) the spectrum of the reflected X-ray radiation off of the irradiated inner disc ring; and (iii) the spectrum of the magnetospheric/coronal emission which is primarily caused by electron-scattering of components (i) and (ii). Significant variations in photon index and cutoff energy are expected if there is a significant contribution of X-rays reflected off the inner disc ring. The shape of the observed variations can be reproduced by including both direct and reflected spectrum components.

The variation of the iron line intensity is nearly sinusoidal with orbital phase during Main High state. This can be understood as follows. The likely sites of the iron K emission line in Her X-1 during MH-state are: (i) reprocessing of the direct emission in the hot magnetospheric/ coronal gas; (ii) reprocessing of the direct emission in reflection by the far illuminated face of the inner disc ring; and (iii) reprocessing of direct emission and/or reflected emission by the illuminated near side of the inner disc ring in transmission. Based on the observed variation, which peaks when the inner disc ring is most visible, the iron emission is dominated by reflection off the inner disc.

3. Conclusion

The RXTE/PCA archival observations of Her X-1 are a rich record of the behaviour of this complex system. The current analysis supports the precessing disk model for the 35-day cycle of Her X-1 and the detailed picture of the twisted tilted disc that has been built up with years of studies of this fascinating system.

References

Abdallah, M. H. & Leahy, D. A. 2015, *MNRAS*, 453, 4222
Boynton, P. E., Crosa, L. M., & Deeter, J. E. 1980, *ApJ*, 237, 169
Crosa, L., & Boynton, P. E. 1980, *ApJ*, 235, 999
Gerend, D., & Boynton, P. E. 1976, *ApJ*, 209, 562
Igna, C. D. & Leahy, D. A. 2011, *MNRAS*, 418, 2283
Igna, C. D. & Leahy, D. A., 2012, *MNRAS*, 425, 8
Ji, L., Schulz, N., Nowak, M., Marshall, H. L., & Kallman, T. 2009, *ApJ*, 700, 977
Klochkov, D., Staubert, R., Postnov, K., Shakura, N., & Santangelo, A. 2009, *A&A*, 506, 1261
Larwood, J. D., Nelson, R. P., Papaloizou, J. C. B., & Terquem, C. 1996, *MNRAS*, 282, 597
Leahy, D. A. 2002, *MNRAS*, 334, 847
Leahy, D. A. 2003, *MNRAS*, 342, 446
Leahy, D. A. 2004, *MNRAS*, 348, 932
Leahy, D. A. 2004, *Astron. Nachr.*, 325, 205
Leahy, D. A. 2004, *ApJ*, 613, 517
Leahy, D. A. & Igna, C. D. 2010, *ApJ*, 713, 318
Leahy, D. A. & Igna, C. D. 2011, *ApJ*, 736, 74
Leahy, D. A. & Abdallah, M. H. 2014, *ApJ*, 793, 79
Leahy, D. A. 2015, *ApJ*, 800, 32
Maloney, P. R., Begelman, M. C., & Pringle, J. E. 1996, *ApJ*, 472, 582
Maloney, P. R., & Begelman, M. C. 1997, *ApJ*, 491, L43
Reynolds, A. P., Quaintrell, H., Still, M. D., Roche, P., Chakrabarty, D., & Levine, S. E. 1997, *MNRAS*, 288, 43
Schandl, S., & Meyer F. 1994, *A&A*, 289, 149
Scott, D. M., & Leahy, D. A. 1999, *ApJ*, 510, 974
Scott, D. M., Leahy, D. A., & Wilson, R. B. 2000, *ApJ*, 539, 392
Wijers, R. A. M. J., & Pringle, J. E. 1999, *MNRAS*, 308, 207

High-mass X-ray Binaries: Illuminating the Passage from Massive Binaries to Merging Compact Objects
Proceedings IAU Symposium No. 346, 2019
L. M. Oskinova, E. Bozzo, T. Bulik & D. Gies, eds.

doi:10.1017/S174392131800813X

Long-term optical variability of Her X-1

T. İçli, D. Koçak and K. Yakut

Department of Astronomy & Space Sciences, University of Ege,
35100, Bornova–İzmir, Turkey
email: icli.tugce@gmail.com

Abstract. Long-term and short-term multicolor photometric variations of the X-ray binary system Her X-1 (HZ Her) has been studied. We obtained new VRI observations of the system by using the 60cm Robotic telescope at the TÜBİTAK National Observatory (TUG) in 2018. Using newly obtained data, we modified the orbital period of the binary system with a neutron star component.

Keywords. stars:binaries, X-rays: binaries, stars: individual (HZ Her, Her X-1)

1. Introduction

Her X-1 (HZ Her, 4U1656+35) was discovered in 1972 from UHURU satellite observations (Tananbaum *et al.* 1972). The system is an eclipsing intermediate-mass X-ray binary (IMXB) system with a 1.7 days orbital period. The binary system contains an accreting neutron star with 1.5 $M_\odot$ and 2.2 $M_\odot$ optical component. The spin period of the compact star is about 1.24 seconds. The distance of Her X-1 was estimated by Reynolds *et al.* (1997) as 6.6 kpc which is similar to the Gaia result.

Her X-1/HZ Her observed in various wavelengths including optical, ultraviolet, radio and X-ray bands. Both short and long-term optical observations of the system have been obtained by different studies (e.g., Shakura *et al.* 1997; Cherepashchuk *et al.* 1974; Simon *et al.*, 2002; Içli 2016). The light curve of the binary system shows peculiar variability. In additional to an optical variation, the system shows a 35 days variability in X-rays (Scott *et al.*, 2000; Leahy & Abdallah 2014; Postnov *et al.* 2013). The 35-day cycle variations could be related with an accretion disc.

Observation of the long-term variability of X-ray binary system with neutron star component provides information about the physical processes with a time scale of few decades. These variations can be due to the stellar activity of the component star, as well as hot stellar winds, accretion properties, the disrupted shape of the component and the eclipsing feature of the systems. In this work, we aimed to study long-term multi-color (VRI) light variations of binary systems consisting of a neutron star and an early-type component. We present our preliminary results on Her X-1.

2. New Observations

The new optical observations of Her X-1 was obtained using the 60cm robotic telescope (T60) at the TUBITAK National Observatory (TUG) between years 2015 and 2018 in VRI filters. The exposure times were set to 60 s for all filters. The recent observations of 2018 are presented in this study. Observations of each data set were reduced separately with the standard procedure by using IRAF/PHOT. The frame reduction was performed by subtracting the bias and dark frames and finally dividing by flat-field frames. For each system, four comparison stars have been selected that were used in previous studies.

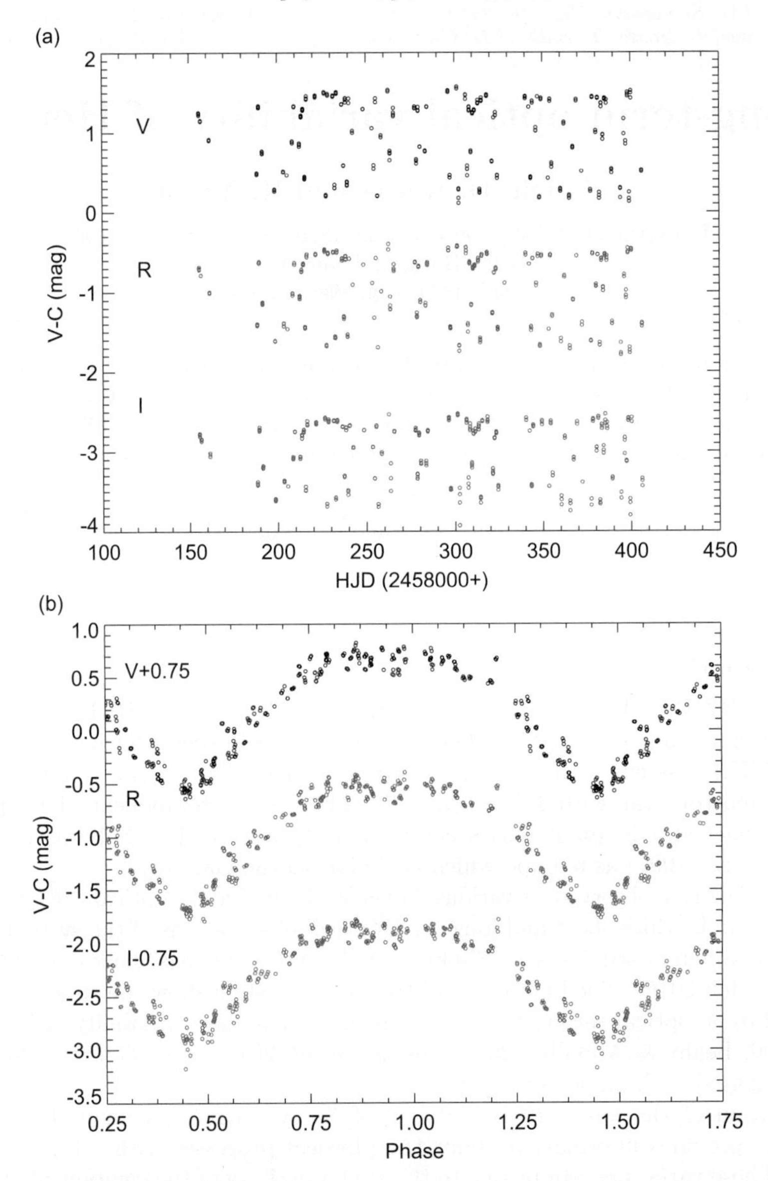

Figure 1. Long-term VRI light variation of Her X-1 (a) and light curves of the system (b) obtained in 2018.

AAVSO-135, AAVSO-132, AAVSO-150, and AAVSO-136 were chosen as a comparison and check stars, respectively.

Obtaining our new VRI observations, we decide to derive an orbital period of the system. We obtained the orbital period of the system as 1.699646(8). Figure 1a shows the long-term light variation of Her X-1 spreading over 250 nights. The complete light curves of the system, obtained in V, R and I bands are shown in Fig 1b. Long-term optical monitoring of Her X-1 is still ongoing.

Acknowledgements

This study is supported by the Turkish Scientific and Research Council-TÜBİTAK (117F188). We thank TÜBİTAK for partial support in using T60 telescope with project number respectively (15AT60-776 and 18AT160-1298). The authors would like to acknowledge the contribution of the COST (European Cooperation in Science and Technology) Action CA15117 and CA16104.

References

Fender R., 2006, *AIPC*, 856, 23

Gallo E., 2010, *HEAD*, 42, 5.05

İçli, T., 2016, MSc Thesis, Binary systems with neutron star components, University of Ege.

Leahy D. A., & Abdallah M. H., 2014, *ApJ*, 793, 79

Postnov K., Shakura N., Staubert R., Kochetkova A., Klochkov D., & Wilms J., 2013, *MNRAS*, 435, 1147

Reynolds A. P., Quaintrell H., Still M. D., Roche P., Chakrabarty D., & Levine S. E., 1997, *MNRAS*, 288, 43

Russell D. M., Fender R. P., & Jonker P. G., 2007, *MNRAS*, 379, 1108

Russell D. M., Fender R. P., Hynes R. I., Brocksopp C., Homan J., Jonker P. G., & Buxton M. M., 2006, *MNRAS*, 371, 1334

Shakura N. I., Smirnov A. V., & Ketsaris N. A., 1997, *ASPC*, 121, 379

Wang X., & Wang Z., 2014, *ApJ*, 788, 184

High-mass X-ray Binaries: Illuminating the Passage from Massive Binaries to Merging Compact Objects
Proceedings IAU Symposium No. 346, 2019
L. M. Oskinova, E. Bozzo, T. Bulik & D. Gies, eds.

doi:10.1017/S1743921318007822

NGC 300 ULX1: A new ULX pulsar in NGC 300

Chandreyee Maitra, Stefania Carpano, Frank Haberl and Georgios Vasilopoulos†

Max-Planck-Institut für extraterrestrische Physik, Giessenbachstraße 1, 85748 Garching, Germany

Abstract. NGC 300 ULX1 is the fourth to be discovered in the class of the ultra-luminous X-ray pulsars. Pulsations from NGC 300 ULX1 were discovered during simultaneous *XMM-Newton* / *NuSTAR* observations in Dec. 2016. The period decreased from 31.71 s to 31.54 s within a few days, with a spin-up rate of -5.56×10^{-7} s s^{-1}, likely one of the largest ever observed from an accreting neutron star. Archival *Swift* and *NICER* observations revealed that the period decreased exponentially from ~45 s to ~17.5 s over 2.3 years. The pulses are highly modulated with a pulsed fraction strongly increasing with energy and reaching nearly 80% at energies above 10 keV. The X-ray spectrum is described by a power-law and a disk black-body model, leading to a 0.3–30 keV unabsorbed luminosity of 4.7×10^{39} erg s^{-1}. The spectrum from an archival *XMM-Newton* observation of 2010 can be explained by the same model, however, with much higher absorption. This suggests, that the intrinsic luminosity did not change much since that epoch. NGC 300 ULX1 shares many properties with supergiant high mass X-ray binaries, however, at an extreme accretion rate.

Keywords. stars: neutron – pulsars: individual: NGC 300 ULX1 – galaxies: individual: NGC 300 – X-rays: binaries

---

1. Introduction

Ultra-luminous X-ray sources (ULXs) are point-like non-nuclear sources that emit at luminosities in excess of $\sim 10^{39}$ erg s^{-1} which is approximately the Eddington limit for a spherically symmetric accretion onto a stellar mass black hole of $10 M_{\odot}$. Although initially believed to harbour super-critically accreting stellar mass or intermediate mass black holes in order to support the exceedingly high super-Eddington luminosity, recent years have provided undisputed evidence that a substantial fraction of ULXs harbour highly magnetized accreting neutron stars (Bachetti *et al.* 2014, Fürst *et al.* 2016, Israel *et al.* 2017).

NGC 300 ULX1 is the fourth to be discovered in this class, and is the newly identified Ultra-luminous X-ray pulsar (ULXP) in NGC 300, located at a distance of 1.88 Mpc (Carpano *et al.* 2018b). The system was initially discovered in the optical wavelengths as a supernova in 2010 (Monard 2010), but was classified as a supernova impostor event due to the high X-ray flux associated with a brightening in the optical and infrared regime (Binder *et al.* 2011). NGC 300 ULX1 was later identified as a possible supergiant B[e] HMXB owing to the spectroscopic and photometric information in the UV and infrared wavelengths (Lau *et al.* 2016). NGC 300 ULX1 was observed serendipitously in December 2016 in two consecutive *XMM-Newton* observations (Obsid 0791010101 and

† Present address: Yale Department of Astronomy P.O. Box 208101, New Haven, CT 06520-8101, USA

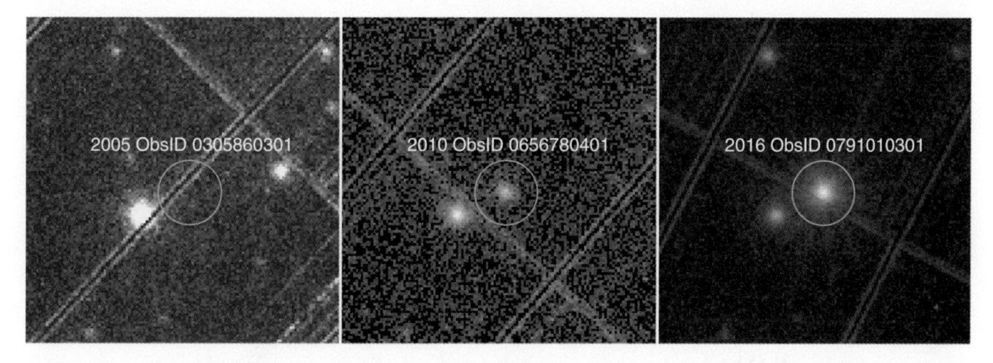

Figure 1. *XMM-Newton* EPIC-pn images of NGC 300 ULX1 at different epochs marked in the figure. The red circles denote the source extraction region used for the analysis.

0791010301) performed simultaneously with *NuSTAR* (Obsid 90401005002). The *XMM-Newton* observations were performed for a duration of 139+82 ks and the *NuSTAR* observation had an exposure of 163 ks. NGC 300 ULX1 was detected in its brightest state ever during this simultaneous observing campaign. NGC 300 ULX1 was also observed with *XMM-Newton* in 2010 when the source was reported in outburst for the first time (Obsid 0656780401), and in 4 other observations from 2000 to 2005 when the source was in the field of view but was not detected. Fig. 1 shows the *XMM-Newton* EPIC-pn image of NGC 300 ULX1 during the 2005, 2010 (Obsid 0305860301) and 2016 observations.

2. Discovery of pulsations and the extreme spin-period evolution of NGC 300 ULX1

Using the 2016 observations, Carpano *et al.* (2018a) reported the discovery of a strong periodic modulation in the X-ray flux with a pulse period of 31.6 s and a very rapid spin-up rate. A refined timing analysis was further performed using a Bayesian method (see Carpano *et al.* 2018b), to probe the spin period evolution in detail. The EPIC-pn data was split into 4 ks intervals (53 intervals from the two observations, 0.2–10 keV band), and the *NuSTAR* data was split into in 21 intervals of 15 ks in an energy band of 3–20 keV. The evolution of the spin period is shown in Fig. 2. The spin period of NGC 300 ULX1 decreased linearly from ~31.71 s at the start of the *NuSTAR* observation to ~31.54 s at the end of the *XMM-Newton*/*NuSTAR* observations. The period derivative inferred from a model with a constant and linear term fitted to the *XMM-Newton* and *NuSTAR* data is $(-5.563 \pm 0.024) \times 10^{-7}$ s s^{-1} with a spin period of 31.68262 ± 0.00036 s at the start of the of the EPIC-pn exposure (MJD 57739.39755). The pulsed fraction (0.2–10 keV), which was defined as the proportion of flux integrated over the pulse profile above minimum flux relative to the total integrated flux, increased slightly from 56.3 ± 0.3% during the first 2016 *XMM-Newton* observation to 57.4 ± 0.3% in the second. The pulsed fraction also increased strongly with energy with 72.1 ± 0.4% in the *NuSTAR* data (3–20 keV).

3. Comparison between the 2010 and 2016 *XMM-Newton* X-ray spectra

The broadband X-ray spectra were fit with a two-component model consisting of a power law with high-energy cutoff and a soft thermal component (disk black-body). The details are provided in Carpano *et al.* (2018b). The residuals after fitting this model indicated the presence of a further softer spectral component which can be attributed to the scattering and reprocessing of the X-ray photons originating in the vicinity of the neutron star, by an additional absorbing component. This was modelled using a

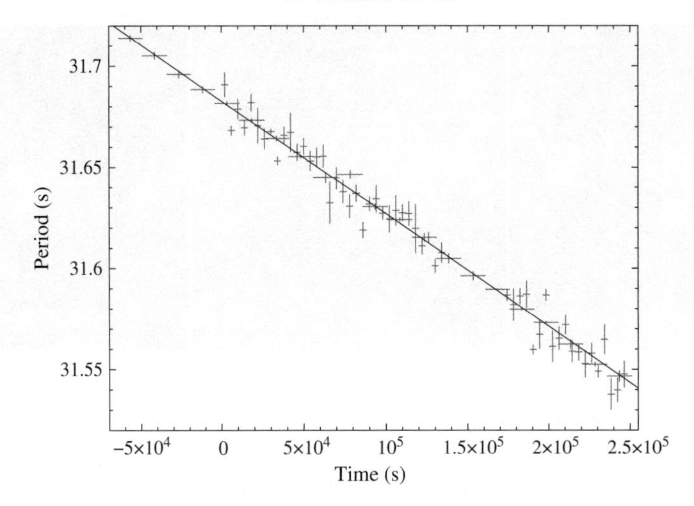

Figure 2. Spin period evolution of NGC 300 ULX1 obtained from 4 ks intervals of EPIC-pn (red crosses) and 15 ks intervals of *NuSTAR* data (blue crosses). The straight line represents the best-fit model of a linear period decrease applied to both data sets. Time zero corresponds to the start of the EPIC-pn exposure. Figure is taken from Carpano *et al.* (2018b).

partial-covering absorber component applied to the power-law and black-body components together. The model represents a physical scenario where the underlying continuum consists of a combination of power-law component (originating from the vicinity of the neutron star) plus a disk black-body component (originating from the inner accretion disk), modified by scattering and absorption by additional material (most likely located in the clumpy wind of the supergiant companion or inner part of the circum-stellar disk of a Be star). The broadband unabsorbed luminosity of the source in 2016 (*XMM-Newton* + *NuSTAR* observations) was 4.7×10^{39} erg s^{-1} in the energy range of 0.3–30 keV.

The *XMM-Newton* spectrum taken in 2010 was drastically different compared to that in 2016, with a soft component seen at energies <2 keV, an almost flat spectrum between 2–4 keV and a bump-like feature above 5 keV. This can be explained if the column density was significantly higher in 2010 and the direct component of the emission was reduced drastically. We investigated this by performing a simultaneous spectral fit of the three EPIC-pn observations, assuming the same underlying continuum spectrum as used in the broad-band spectral fit, and allowing all the absorption components to vary. In order to account for an intrinsic variation in the X-ray luminosity of the source, the power-law normalisation was also left to vary. The spectra and the best-fit model are shown in Fig. 3. The 2010 *XMM-Newton* observation can be explained by a similar intrinsic luminosity, but affected by a high partial absorption (i.e. equivalent hydrogen column density $N_{\rm H} \sim 10^{23}$ cm^{-2}), compared to the 2016 observations where the partial absorption component was lower by a factor of 100.

4. Spin and count rate evolution from the *Swift* and *NICER* observations

Given the extreme spin-up rate of NGC 300 ULX1 as derived from Carpano *et al.* (2018b), the system is a rare opportunity to probe the relation between the accretion torque and the spin-up of a neutron star at super-critical mass accretion rates. While a detailed study presenting the spin evolution of NGC 300 ULX1 and comparison with standard accretion torque models will be presented in a forthcoming paper (Vasilopoulos *et al.* 2018), we present here the spin-up history of the source using data from the *Swift*/XRT and *NICER* monitoring campaigns of NGC 300 ULX1 . The data spans

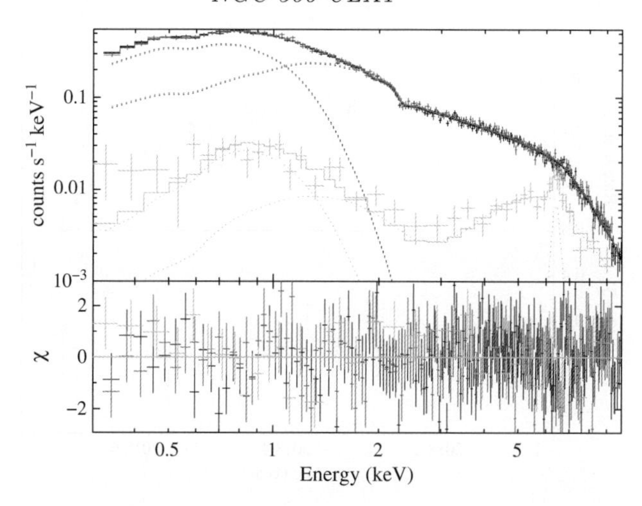

Figure 3. Simultaneous spectral fit of NGC 300 ULX1 using the EPIC-pn spectra together with the residuals as above. Observation 0791010101 is marked in black, 0791010301 in red and 0656780401 (from 2010) in green. Figure is taken from Carpano *et al.* (2018b).

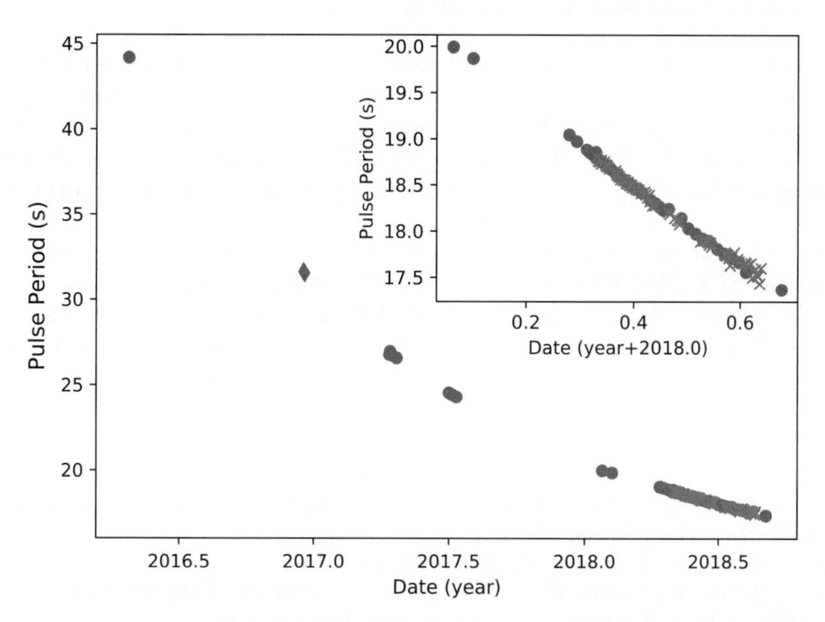

Figure 4. Long-term time evolution of the pulse period of NGC 300 ULX1 from 2016-04-25 to 2018-08-22. The *Swift* observations are marked in blue, and the *NICER* observations in magenta. The simultaneous *XMM-Newton* /*NuSTAR* observation is marked in red.

from MJD 57502.275375 to MJD 58352.427225 where the source spins up from ~45 to ~17.5 s. Fig. 4 shows the spin evolution of NGC 300 ULX1 indicating a steady spin-up of the source for a span of 2.3 years. Fig. 5 shows the recent count rate and hardness ratio evolution of the source starting from 2018. Although the count rate exhibits a gradual decline, NGC 300 ULX1 continues to spin up.

5. Summary

NGC 300 ULX1 was discovered as the fourth ULXP, exhibiting an extreme spin-up rate and a relatively constant luminosity over a long span of time. The secular spin period

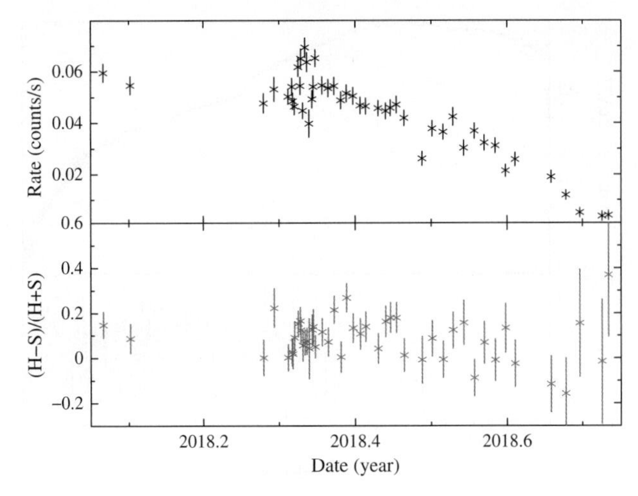

Figure 5. Evolution of the count rate (0.2–10 keV) and the hardness ratio using from the *Swift* observations corresponding to the inset in Fig. 4 (2018-01-24 to 2018-08-22). The hardness ratio is defined as shown in the second panel of the figure. The soft band (S) corresponds to 0.2–1.5 keV and the hard band (H) to 1.5–10 keV.

derivative of -5.56×10^{-7} s s^{-1} seen over three days is one of the highest ever observed from an accreting neutron star, and the strong spin evolution is further supported by archival *Swift*/XRT and *NICER* observations spanning over 2 years. The broadband X-ray spectrum derived from the 2016 observations is similar to what is observed from supergiant HMXBs, although the power law is quite steep and the high-energy cutoff starts at a relatively low energy. The archival *XMM-Newton* spectrum from 2010 can be modelled by the same continuum, however requires a much higher absorption. This indicates that NGC 300 ULX1 was also in the ULX state at similar intrinsic X-ray luminosity, albeit highly absorbed in 2010. NGC 300 ULX1 provides a rare opportunity to probe the spin evolution of an accreting neutron star at extreme accretion rates, and to understand the similarities between ULXPs and supergiant HMXBs.

References

Bachetti, M., Harrison, F. A., Walton, D. J., *et al.* 2014, *Nature*, 514, 202

Binder B., Williams B. F., Kong A. K. H., Gaetz T. J., Plucinsky P. P., Dalcanton J. J., & Weisz D. R. 2011 *ApJ*, 739, L51

Carpano, S., Haberl, F., & Maitra, C. 2018 *ATel*, 11158, 1C

Carpano, S., Haberl, F., Maitra, C., & Vasilopoulos, G. 2018 *MNRAS*, 476, L45

Fürst, F.,Walton, D. J., Harrison, F. A., *et al.* 2016 *ApJ*, 831, L14

Israel, G. L., Belfiore, A., Stella, L., *et al.* 2017 *Science*, 355, 817

Lau R. M., *et al.* 2016 *ApJ*, 830, 142

Monard L. A. G., 2010 2010 *Central Bureau Electronic Telegrams* 2289

Vasilopoulos, G., Haberl, F., Carpano, S., Maitra, C. 2018 *A&A*, submitted

High-mass X-ray Binaries: Illuminating the Passage from Massive Binaries to Merging Compact Objects
Proceedings IAU Symposium No. 346, 2019
L. M. Oskinova, E. Bozzo, T. Bulik & D. Gies, eds.

doi:10.1017/S1743921318008086

Ultraluminous X-ray source populations in the Chandra Source Catalog 2.0

Konstantinos Kovlakas[1,2], Andreas Zezas[1,2,3], Jeff J. Andrews[1,2], Antara Basu-Zych[4,5], Tassos Fragos[6,7], Ann Hornschemeier[4,8], Bret Lehmer[9] and Andrew Ptak[4,8]

[1]Department of Physics, University of Crete
Voutes University Campus, 71003 Heraklion, Crete, Greece
email: kkovlakas@physics.uoc.gr

[2]Institute of Electronic Structure & Laser, Foundation for Research & Technology - Hellas
Voutes University Campus, 71003 Heraklion, Crete, Greece

[3]Harvard-Smithsonian Center for Astrophysics
60 Garden Street, Cambridge, MA 02138, USA

[4]NASA Goddard Space Flight Center
8800 Greenbelt Rd, Greenbelt, MD 20771, USA

[5]CRESST, Department of Physics, University of Maryland Baltimore County
Baltimore, MD 21250, USA

[6]DARK, Niels Bohr Institute, University of Copenhagen
Juliane Maries Vej 30, DK-2100 Copenhagen, Denmark

[7]Geneva Observatory, University of Geneva
Chemin des Maillettes 51, 1290 Sauverny, Switzerland

[8]Department of Physics & Astronomy, Johns Hopkins University
3400 North Charles Street, Baltimore, MD 21218, USA

[9]Department of Physics, University of Arkansas
825 West Dickson Street, Fayetteville, AR 72701, USA

Abstract. The nature and evolution of ultraluminous X-ray sources (ULXs) is an open problem in astrophysics. They challenge our current understanding of stellar compact objects and accretion physics. The recent discovery of pulsar ULXs further demonstrates the importance of this intriguing and rare class of objects.

In order to overcome the difficulties of directly studying the optical associations of ULXs, we generally resort in statistical studies of the stellar properties of their host galaxies. We present the largest such study based on the combination of Chandra archival data with the most complete galaxy catalog of the Local Universe. Incorporating robust distances and stellar population parameters based on associated multi-wavelength information, and we explore the association of ULXs with galaxies in the (star formation rate, stellar mass, metallicity) space.

We confirm the known correlation with morphology, star formation rate and stellar mass, while we find an excess of ULXs in dwarf galaxies, indicating dependence on age and metallicity.

Keywords. X-rays: binaries, X-rays: galaxies, catalogs

---

1. Introduction

Ultraluminous X-ray sources are X-ray binaries exceeding the Eddington limit for a stellar black hole ($L_X \gtrsim 10^{39}$ erg s^{-1}), challenging our understanding of accretion physics and binary evolution (for a review, see Kaaret *et al.* 2017). Links with cosmological (IGM heating during the reionization era; e.g. Madau & Fragos 2017) and gravitational wave research (ULXs as progenitors of binary mergers; e.g. Marchant *et al.* 2017) further demonstrate their importance. While studies of individual objects provide significant input for understanding accretion physics in ULXs, statistical studies probe their formation and evolution via the connection with their environment, e.g. the star formation rate (SFR), stellar mass ($M_\star$) and metallicity (Z) of the host galaxy (Swartz *et al.* 2011; Wang *et al.* 2016). Such studies provide input for population synthesis models (e.g. Fragos *et al.* 2015), addressing key questions, such as the nature of the compact objects in ULXs and their evolutionary paths.

Using the new *Chandra Source Catalog 2.0* (*CSC 2.0*; Evans *et al.* 2010) we revisit the correlation of the ULX populations with global parameters of their host galaxies by creating the most up-to-date census of ULXs in the Local Universe. By associating the *CSC 2.0* with a complete catalog of galaxies, we will

(*a*) provide the most up-to-date census of ULXs and host galaxy properties
(*b*) study the correlation of the number of ULXs with stellar population parameters
(*c*) constrain the rate of Hyperluminous X-ray sources
(*d*) study the high-end of the luminosity function of High Mass X-ray Binaries

2. The Sample

The first step is to create a sample of galaxies in the Local Universe for which we have reliable information on their SFRs, $M_\star$ and Z. We select all *HyperLEDA* (Paturel *et al.* 2003; Makarov *et al.* 2014) galaxies with recession velocity less than 14000 km s^{-1} ($\lesssim$ 200 Mpc.) We collect all redshift-independent distances from *NED-D* (Steer *et al.* 2017) and derive the distance and its uncertainty for the galaxies with distance measurements. Via a Kernel Regression technique, the redshifts and distances of these galaxies are used to derive redshift-dependent distances for the the rest of the sample.

We incorporate stellar population parameters (SFR, $M_\star$, Z) or compute them using multi-wavelength information, by associating our sample with the:

- Revised Bright Galaxy Sample (Sanders *et al.* 2003)
- Revised IRAS-FSC Redshift Catalogue (Wang *et al.* 2014)
- 2MASS Extended Source Catalogue (Jarrett *et al.* 2000)
- 2MASS Large Galaxy Atlas (Jarrett *et al.* 2003)
- GALEX-SDSS-WISE Legacy Catalog (Salim *et al.* 2016)
- WISE forced photometry for SDSS galaxies (Lang *et al.* 2016)
- Firefly Stellar Population parameters (Comparat *et al.* 2017)
- MPA-JHU (Kauffmann *et al.* 2003; Brinchmann *et al.* 2004; Tremonti *et al.* 2004)

The result is the Heraklion Extragalactic CATaloguE (*HECATE*), a compilation of $\sim$163000 galaxies in the Local Universe, providing photometric/spectral data, star formation rates, stellar masses, metallicities, AGN content, etc.

By cross-matching *HECATE* with the current state of *CSC 2.0* and correcting for background contamination, we obtain the largest census of ULXs in the widest up-to-date range of environments (four orders of magnitude in SFR and $M_\star$). Out of the 13286 X-ray sources in 1678 galaxies with *Chandra* data, we find 2352 ULX candidates ($L_X > 10^{39}\,\mathrm{erg\,s^{-1}}$) in 1435 galaxies.

3. First Results

For the following analysis, we select 280 galaxies with distance less than 40 Mpc (to avoid source confusion effects) for which we have reliable estimates on the star formation rate and stellar mass. After accounting for AGN contamination, in these galaxies we estimate to have about 300 ULXs.

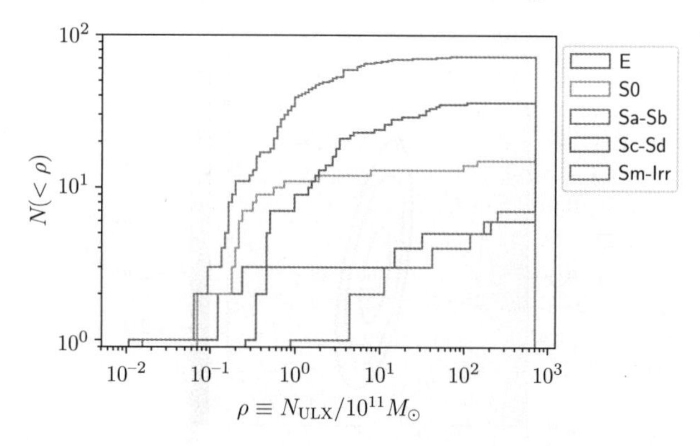

Figure 1. Cumulative number of galaxies in our sample of various morphological types (see legend), as a function of number of ULXs per 10^{11}M$_\odot$ stellar mass.

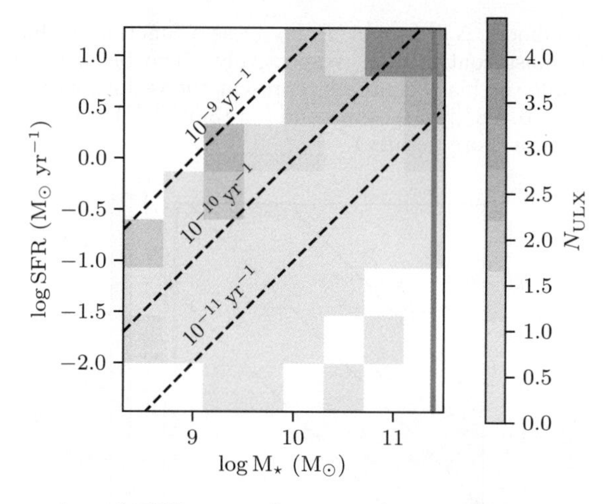

Figure 2. Average number of ULXs per galaxy as a function of SFR and stellar mass of their host galaxies. Diagonal lines correspond to equal specific SFRs. The blue vertical shows the stellar mass for which the contamination by low mass X-ray binaries (LMXB) is expected to be significant (one source above 10^{39} erg s^{-1} assuming the LXMB luminosity function from Zhang *et al.* 2012). The average number of ULXs per galaxy increases with SFR and stellar mass.

Using the stellar mass as a proxy for the galaxy size, we compute the number of ULXs per stellar mass. In Figure 1 we present the cumulative distribution of this quantity for various morphological types. We find that early-type galaxies are characterized by small number of ULXs, while late-type galaxies typically host one ULX per $10^{11}M_\odot$. Additionally, irregulars host about 10 times more ULXs than spirals of the same mass.

In Figure 2, we show the average number of ULXs per galaxy as a function of SFR and stellar mass of the host galaxy. To quantify the correlation, we employ a Maximum Likelihood Estimator where we model the observed number of ULXs ($n_{\rm obs}$) as a Poisson variate with rate depending linearly on its SFR and $M_\star$, while accounting for the contamination by background AGN, $n_{\rm bkg}$, from Kim *et al.* (2007):

$$n_{\rm obs} \sim {\rm Pois}\left(a \times {\rm SFR} + b \times {\rm M}_\star + n_{\rm bkg}\right) \tag{3.1}$$

We find $a = 0.408^{+0.040}_{-0.037}$ M$_\odot^{-1}$ yr, in agreement with with previous results (e.g. Swartz *et al.* 2011). For the stellar mass scaling factor we find $b = \left(3.98^{+1.35}_{-1.27}\right) \times 10^{-12}$ M$_\odot^{-1}$ which is consistent with the LMXB contribution that we calculate from Zhang *et al.* (2012) (one luminous LMXB per 2.5×10^{11} M$_\odot$ stellar mass; see blue line in Figures 2 and 4.) In

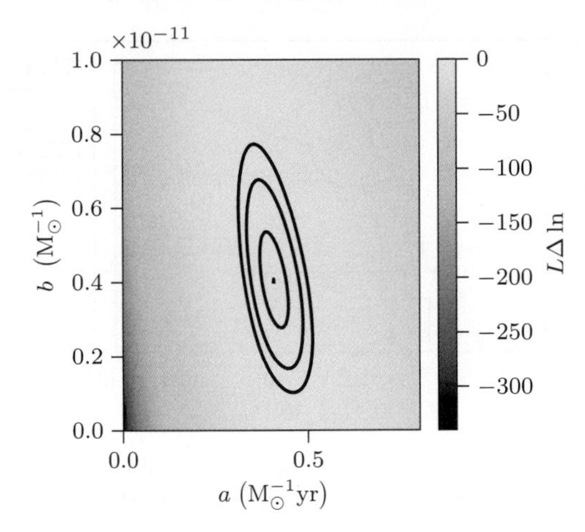

Figure 3. Relative likelihood, $\Delta \ln L = \ln (L/L_{\max})$ as a function of the scaling factors a and b (for SFR and stellar mass contribution, respectively.) The black point corresponds to the maximum likelihood, while the black contours represent the regions with confidence levels 68%, 95% and 99%. We find $a = 0.408^{+0.040}_{-0.037}$ $M_\odot^{-1}$ yr and $b = \left(3.98^{+1.35}_{-1.27}\right) \times 10^{-12}$ $M_\odot^{-1}$, consistent with previous results (see text for more details.)

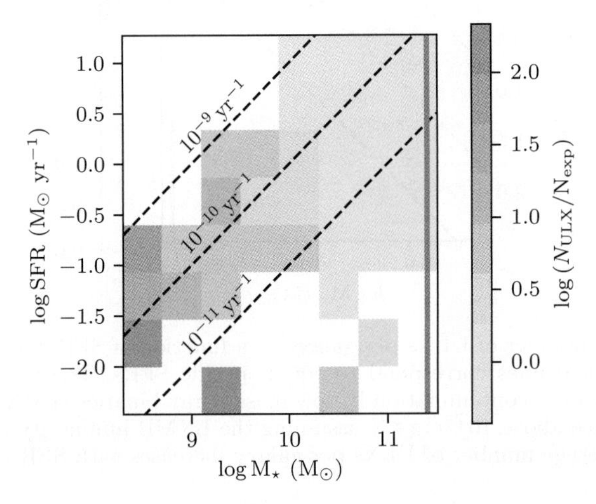

Figure 4. Same as Figure 2 but now the number of ULXs is divided by the expected number of ULXs based on the fitted model of Equation 3.1. An excess is found for galaxies with low $M_\star$, indicating higher formation efficiency of ULXs in the low-Z environment of dwarf galaxies.

Figure 3 we plot the likelihood relative to the maximum likelihood, and confidence regions on the parameter space of the fit. The SFR scaling factor is better constrained than the $M_\star$ factor, indicating that the observed ULX populations are primarily correlated with the SFR of the host galaxy. We attibute the covariance between a and b to the correlation of SFR and $M_\star$ in star forming galaxies (e.g. Maragkoudakis *et al.* 2017; Rodighiero *et al.* 2011).

Dividing the number of ULXs in different bins of the SFR-$M_\star$ plane by their expected number according to the fitted model (see Equation 3.1), we can probe the region of the parameter space that can form ULXs more efficiently. As we see in Figure 4, galaxies with low $M_\star$ or SFR are more efficient in forming ULXs, in agreement with previous results (Lehmer *et al.* 2016; Basu-Zych *et al.* 2016). As galaxies with low $M_\star$ are generally metal-poor (e.g. Sánchez *et al.* 2017), we interpret this excess of ULXs as the metallicity effect suggested by theoretical studies (e.g. Linden *et al.* 2011; Fragos *et al.* 2013) and observations (e.g. Prestwich *et al.* 2013).

Acknowledgments

The research leading to these results has received funding from the *European Research Council* under the European Union's *Seventh Framework Programme* (FP/2007-2013) / ERC Grant Agreement n. 617001. This project has received funding from the European Union's Horizon 2020 research and innovation programme under the Marie Sklodowska-Curie RISE action, grant agreement No 691164 (ASTROSTAT). We sincerely thank the reviewer for suggestions that helped to improve this manuscript.

References

Basu-Zych, A. R., Lehmer, B., Fragos, T., *et al.* 2016, *ApJ*, 818, 140
Brinchmann, J., Charlot, S., White, S. D. M., *et al.* 2004, *MNRAS*, 351, 1151
Brorby, M., Kaaret, P., & Feng, H. 2015, *MNRAS*, 448, 3374
Comparat, J., Maraston, C., Goddard, D., *et al.* 2017, arXiv:1711.06575
Evans, I. N., Primini, F. A., Glotfelty, K. J., *et al.* 2010, *ApJS*, 189, 37
Fragos, T., Lehmer, B., Tremmel, M., *et al.* 2013, *ApJ*, 764, 41
Fragos, T., Linden, T., Kalogera, V., & Sklias, P. 2015, *ApJL*, 802, L5
Grimm, H.-J., Gilfanov, M., & Sunyaev, R. 2003, *MNRAS*, 339, 793
Jarrett, T. H., Chester, T., Cutri, R., *et al.* 2000, *AJ*, 119, 2498
Jarrett, T. H., Chester, T., Cutri, R., Schneider, S. E., & Huchra, J. P. 2003, *AJ*, 125, 525
Kaaret, P., Feng, H., & Roberts, T. P. 2017, *ARAA*, 55, 303
Kauffmann, G., Heckman, T. M., White, S. D. M., *et al.* 2003, *MNRAS*, 341, 33
Kim, M., Wilkes, B. J., Kim, D.-W., *et al.* 2007, *ApJ*, 659, 29
Lang, D., Hogg, D. W., & Schlegel, D. J. 2016, *AJ*, 151, 36
Lehmer, B. D., Basu-Zych, A. R., Mineo, S., *et al.* 2016, *ApJ*, 825, 7
Linden, T., Kalogera, V., Sepinsky, J., *et al.* 2011, Evolution of Compact Binaries, 447, 121
Madau, P., & Fragos, T. 2017, *ApJ*, 840, 39
Makarov, D., Prugniel, P., Terekhova, N., Courtois, H., & Vauglin, I. 2014, *A&A*, 570, A13
Maragkoudakis, A., Zezas, A., Ashby, M. L. N., & Willner, S. P. 2017, *MNRAS*, 466, 1192
Marchant, P., Langer, N., Podsiadlowski, P., *et al.* 2017, *A&A*, 604, A55
Paturel, G., Petit, C., Prugniel, P., *et al.* 2003, *A&A*, 412, 45
Prestwich, A. H., Tsantaki, M., Zezas, A., *et al.* 2013, *ApJ*, 769, 92
Rodighiero, G., Daddi, E., Baronchelli, I., *et al.* 2011, *ApjL*, 739, L40
Salim, S., Lee, J. C., Janowiecki, S., *et al.* 2016, *ApJS*, 227, 2
Sánchez, S. F., Barrera-Ballesteros, J. K., Sánchez-Menguiano, L., *et al.* 2017, *MNRAS*, 469, 2121
Sanders, D. B., Mazzarella, J. M., Kim, D.-C., Surace, J. A., & Soifer, B. T. 2003, *AJ*, 126, 1607
Skrutskie, M. F., Cutri, R. M., Stiening, R., *et al.* 2006, *AJ*, 131, 1163
Steer, I., Madore, B. F., Mazzarella, J. M., *et al.* 2017, *AJ*, 153, 37
Swartz, D. A., Soria, R., Tennant, A. F., & Yukita, M. 2011, *ApJ*, 741, 49
Tremonti, C. A., Heckman, T. M., Kauffmann, G., *et al.* 2004, *Apj*, 613, 898
Wang, L., Rowan-Robinson, M., Norberg, P., Heinis, S., & Han, J. 2014, *MNRAS*, 442, 2739
Wang, S., Qiu, Y., Liu, J., & Bregman, J. N. 2016, *ApJ*, 829, 20
Zhang, Z., Gilfanov, M., & Bogdán, Á. 2012, *A&A*, 546, A36

High-mass X-ray Binaries: Illuminating the Passage from Massive Binaries to Merging Compact Objects
Proceedings IAU Symposium No. 346, 2019
L. M. Oskinova, E. Bozzo, T. Bulik & D. Gies, eds.

doi:10.1017/S1743921318008529

Multicolour photometry of SS 433

D. Koçak, T. İçli and K. Yakut

Department of Astronomy & Space Sciences, University of Ege,
35100, Bornova–İzmir, Turkey
email: dolunay.kocak@gmail.com

Abstract. We presented long-term optical observations of the high mass X-ray binary system SS 433 (V1343 Aql) with a black hole component. New observations have been obtained by using the 0.6m telescope at the TÜBİTAK National Observatory (TUG) in B, V, R and I filters. We aim to investigate the long-term photometric behavior of the system.

Keywords. stars:binaries, X-rays: binaries, stars: individual (SS 433, V1343 Aql)

1. Introduction

Following a core collapse/supernova, massive stars leave behind either a neutron star or a black hole at the end of their evolution. The processes may differ for a single star and for a star that is a member of a binary system. In binary systems, the massive component evolves faster to form an X-ray binary system. One of the components in an X-ray binary system is a black hole or a neutron star. X-ray binary systems can be divided into three classes as low mass X-ray binaries (LMXB), intermediate mass X-ray binaries (IMXB), and high mass X-ray binaries (HMXB); depending on their physical features. The number of known IMXB systems, with intermediate mass stellar components, is relatively low. Podsiadlowski, Rappaport, & Pfahl (2001) presented an evolutionary study of the LMXBs and IMXBs. HMXBs contain O-B type highly massive stars that lose mass via hot stellar winds. In addition to the stellar winds, active regions on the stellar surface and the oblate shape of the component can play important roles in the light variation of a system. Long-term light variations of these kinds of systems have been studied in the literature by, e.g. Goranskij (2011); Reig & Fabregat (2015), and Kocak (2016).

SS 433 (V1343 Aql, α= 19 11 49.56, δ= +04 58 57.8) is an eclipsing HMXB system ($P_{\rm orb}$=13.08223 days) containing a ~5-16 $M_\odot$ black hole and A3-7 type supergiant component (Cherepashchuk 1981, Gladyshev 1981, Margon 1984, Fabrika 2004) with a distance of 5.5 kpc. The system was the first Galactic microquasar to be identified that shows relativistic ($V_{jet} \sim 0.3c$) jets (Abell & Margon 1979; Fabian & Rees 1979, Fabrika 2004). Various researchers observed the system SS 433 at different wavelengths. The system was observed photometrically by Volkov (2012), Kurochkin (1988),Goranskii *et al.* (1998), Goranskij (2011), and Chakrabarti *et al.* (2003).

2. New Observations

New multicolor observations of SS 433 were carried out at the TÜBİTAK National Observatory with the 0.6-meter telescope (TUG-T60). The system was observed in Bessell B, V, R, and I filters in 2015, 2016, 2017, 2018 using the TUG-T60 robotic telescope. The frame reduction performed by subtracting the bias and dark frames and finally dividing by flat frames. We used 60 and 90 seconds of exposure time in the observations. We

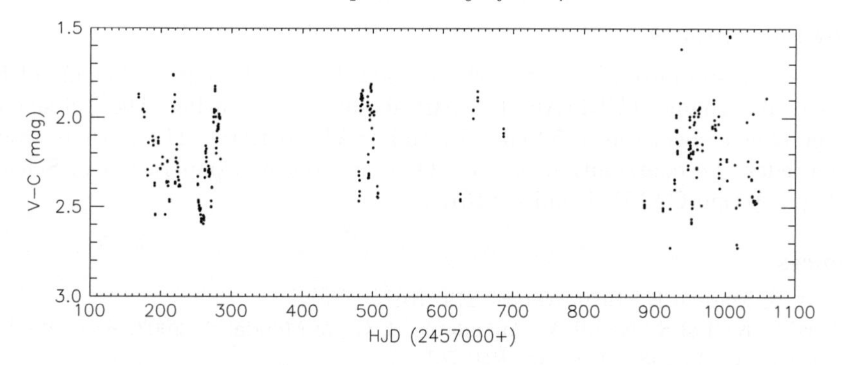

Figure 1. Long-term light variation of the system in R filter.

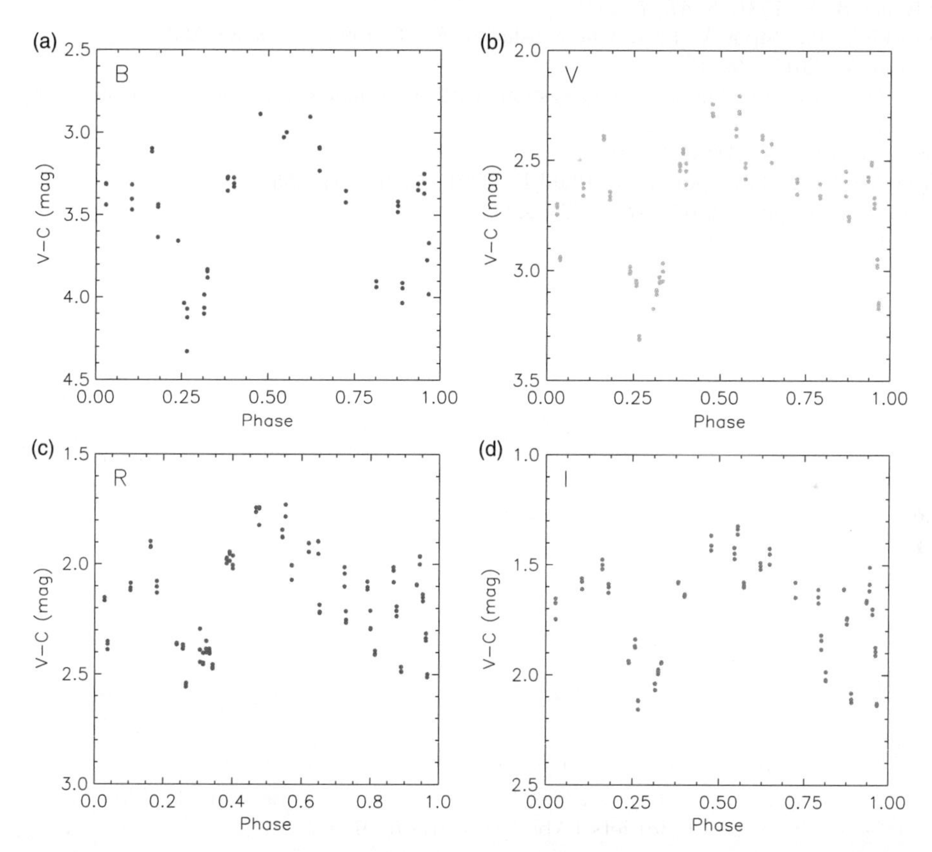

Figure 2. The light curves of the eclipsing X-ray binary V1343 Aql in B (a), V (b), R (c) and I (d) filters.

present the long-term light variation of the system in Figure 1 and multi-color (BVRI) light curves for SS 433 in Figures 2a-d. The phases were calculated using the ephemeris given by Koçak (2016).

Light curves of the system show a significant variation with the amplitude of 1.3 mag in B, 1.1 mag in V, 0.9 mag in R and 0.8 mag in I bands. All the new observation details for the selected X-ray binaries will be discussed in our next study (Koçak *et al.* 2019).

Acknowledgements

This study is supported by the Turkish Scientific and Research Council-TÜBİTAK (117F188). We thank TÜBİTAK for partial support in using T60 telescope with project number respectively (15AT60-775 and 18AT160-1298). The authors would like to acknowledge the contribution of the COST (European Cooperation in Science and Technology) Action CA15117 and CA16104.

References

Abell G. O., & Margon B., 1979, *Nature*, 279, 701
Chakrabarti S. K., Pal S., Nandi A., Anandarao B. G., & Mondal S., 2003, *ApJ*, 595, L45
Cherepashchuk A. M., 1981, *MNRAS*, 194, 761
Fabian A. C., & Rees M. J., 1979, *MNRAS*, 187, 13P
Fabrika S., 2004, *ASPRv*, 12, 1
Gladyshev S. A., 1981, *SvAL*, 7, 594
Goranskii V. P., Esipov V. F., & Cherepashchuk A. M., 1998, *ARep*, 42, 336
Goranskij V., 2011, *PZ*, 31,
Koçak, D., 2016, MSc Thesis, Binary systems with black hole component, University of Ege.
Kurochkin N. E., 1988, *PZ*, 22, 661
Margon B., 1984, *ARA&A*, 22, 507
Podsiadlowski P., Rappaport S., & Pfahl E., 2001, *ASSL*, 264, 355
Reig P., & Fabregat J., 2015, *A&A*, 574, A33
Volkov I., 2012, *IBVS*, 6022, 1

High-mass X-ray Binaries: Illuminating the Passage from Massive Binaries to Merging Compact Objects
Proceedings IAU Symposium No. 346, 2019
L. M. Oskinova, E. Bozzo, T. Bulik & D. Gies, eds.

doi:10.1017/S1743921319001169

Rapid evolution of the relativistic jet system SS 433

V. P. Goranskij[1], E. A. Barsukova[2], A. N. Burenkov[2], A. F. Valeev[2], S. A. Trushkin[2], I. M. Volkov[1,3], V. F. Esipov[1], T. R. Irsmambetova[1] and A. V. Zharova[1]

[1]Sternberg Astronomical Institute, Moscow State University,
Universitetskii prospect, 13, Moscow, 119234, Russia

[2]Special Astrophysical Observatory, Russian Academy of Sciences,
Nizhnij Arkhyz, Karachai-Cherkesia, 369167, Russia

[3]Institute of Astronomy, Russian Academy of Sciences,
Pyatnitskaya Street, 48, Moscow, 119017, Russia

Abstract. We analyzed a 40-year set of multicolor photometry and a 15-year set of synoptic monitoring of SS 433 along with fragmentary spectral and radio data. This system contains a neutron star and an A3–A7 I giant. The system is found to be either close, in contact, or it has a common envelope from time to time. The A-type giant is now in transition to the dynamical mass transfer.

Keywords. accretion, dense matter, X-rays: binaries; individual (SS 433, V1343 Aql)

SS 433 is an eclipsing system with moving emission lines in the spectrum. The moving components of Balmer and He I lines are formed by a pair of oppositely directed, highly collimated and precessing relativistic gaseous jets moving with a velocity of 0.26*c*. The orbital period of 13.082 day, the jet precession period of 162 days, and jet nodding period of 6.28 day are all represented in photometric data. Components of the system are A3–A7 I giant (Gies *et al.* 2002; Hillwig *et al.* 2004) and a neutron star (Goranskij 2011, 2013). The mass of the A type star was estimated in the range between 9.4 and 12.7 $M_\odot$ (Kubota *et al.* 2010) or between 8.3 and 11.0 $M_\odot$ (Goranskij 2011, 2013), the absolute magnitude $-5^m.9 \leq M_V \leq -5^m.0$, and reddening $E(B-V) = 2^m.65 \pm 0^m.03$. The mass of the neutron star was limited by 1.25 and 1.65 $M_\odot$ (Goranskij 2011, 2013). Based on these data, with the Stephan-Boltzmann law and Kepler's Third law, we may conclude that the system is in contact, or overcontact, it may have a common envelope, and the Roche lobe of the neutron star may be filled by matter. The system is located in the center of the radio structure W50 interpreted as a 10000-year old supernova remnant. The distance to SS 433 is well known from radio interferometry to be of 5.12 ± 0.27 kpc.

To verify the proximity of the system components, we analyzed a 40-year photometry and perform special synoptic observations with a small 25-cm telescope and electronic image tube equipped with a microchannel plate.

First, we confirmed the assumption by Barnes *et al.* (2006) that the photometric 162-day periodicity is caused mostly by precession of an expanding circumstellar disc masking the binary in the certain phases. The binary is well presented in the light curve only near precession T_3 phases, the phases of the largest divergence of moving emission lines. However, its contribution is distorted or completely invisible in other precession phases (Fig. 1). Other confirmation facts of such a disk are the following: (a) the lag of photometric precession maximum in the Rc band relative to spectroscopic T_3 phase for

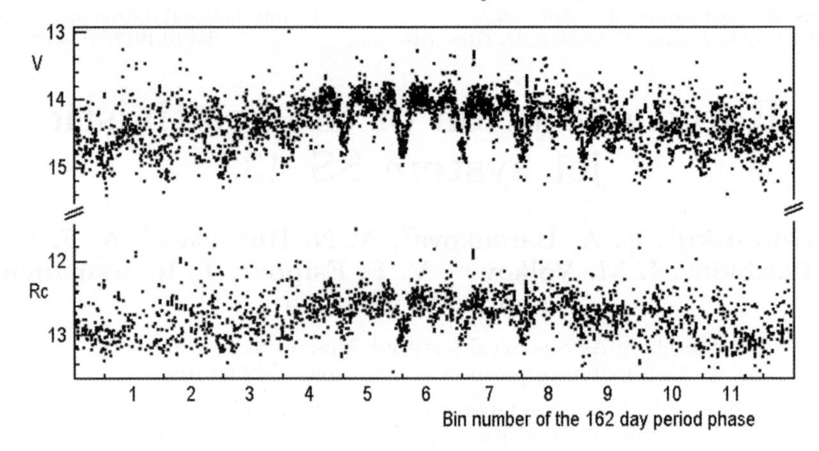

Figure 1. Evolution of the orbital light curve of SS433 depending on the phase of precession 162-day period. In this figure, 13-day light curves calculated in small phase bins are connected in order to increase precession phase. The T_3 moment corresponds to the bin No. 6.

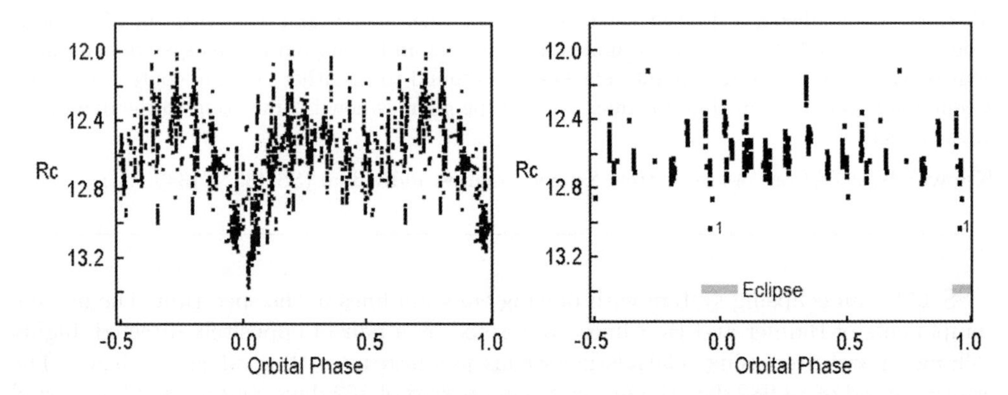

Figure 2. Light curves of SS 433 plotted in the precession phase range between −0.02 and 0.24 versus phase of the orbital elements Min I = JD 2450023.83 + 13.082454·E. Left: typical light curve. Right: observations in the common envelope episode without eclipses related to the time range between August 26 and October 8, 2016. We marked with "1" the first shallow eclipse during this time interval.

∼10 days; (b) the infrared excess in the R and I bands (Goranskij 2011) which is radiated by the external parts of this disk; (c) the equatorial outflow seen at radio interferometry (Paragi *et al.* 1999).

Second, we observed episodes of complete disappearance of the eclipses. The photometry usually shows the variability of widths and depths of the eclipses depending on the precession phase, outbursts and active states. Our synoptic monitoring revealed an episode of disappearing of eclipses near the T_3 phase in September – October 2016 which began with a shallow eclipse on September 6. The following two eclipses were absent. Fig. 2 shows the non-eclipsing orbital phase light curve compared with the typical eclipsing light curve. In this figure, neither the typical "ellipsoidal variations" out of eclipses nor other orbital periodic variations are seen.

We explain this episode as a formation of the common envelope of the A-type star with the neutron star inside. The envelope radius was larger than the radius of the A-star's Roche lobe, and therefore the envelope could not be masked by the external circumstellar disk at the wide range of precessing phases around T_3. When the neutron star is inside the common envelope with the A-star, there is nothing to eclipse. On 2016 October 8.76

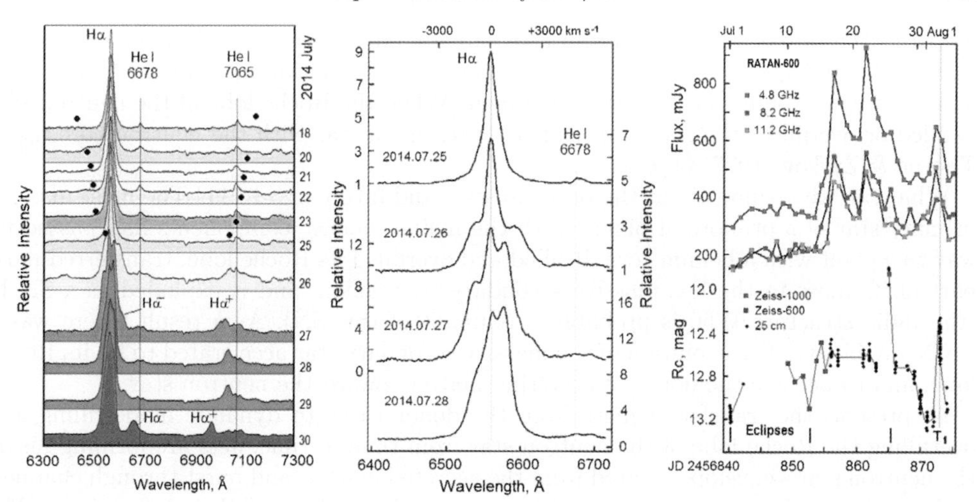

Figure 3. Left: Spectra of Hα region with the relativistic components. The forthcoming places of invisible components are marked with black points. Center: Hα-line profiles before the July 26 explosion, and during explosion (from top to bottom). Right: RATAN radio flux curves in three frequency bands (top); optical light curve in the Rc band (bottom). Note, that radio fluxes were strong and flaring before the optical outburst, when the relativisic jet components were invisible.

UT, we have taken a spectrum of SS 433 with the SAO 1-m Zeiss telescope and UAGS spectrograph. The relativistic components of Hα line were visible in the spectrum, so the jets were not blocked in this episode.

On July 26, 2014, a powerful outburst of SS 433 was observed with an unusually strong infrared excess in eclipse (Goranskij & Spiridonova 2014; Charbonnel *et al.* 2014). Fortunately, simultaneous multi-frequency monitoring observations were taken with the RATAN-600 radio telescope, and spectroscopy provided by one of us, V. F. Esipov, with the 125-cm telescope at the SAI Crimean Station. The spectral observations began 6 days before the outburst when the relativistic lines were invisible. This indicates that the jets were blocked. The forthcoming locations of these lines are marked by black circles in Fig. 3. In the outburst, the Hα profile showed a narrow central component and wide pedestal with FWZI $\sim$ 7000 km s^{-1}. Charbonell *et al.* describe this profile as a three-component one, and note its rapid variability. At the time of the outburst, the relativistic lines reappeared at the forthcoming places. Fig. 3 (left) shows the Hα component superimposed on the Hα line profile, and its motion can explain the variability of the pedestal. In a few days, the intensity of the pedestal increased gradually with decreasing its width, the central component weakened, and then the brightness of the star fell down to the normal level for its proper precession phase (Fig. 3, right). Taking into account the strong IR excess in the outburst, it becomes clear that this event was the ejection of the large mass envelope accompanied by the recovery of jets. The expanding envelope interacted with the earlier released stellar wind.

And third, we have found a secular decrease of amplitudes of all three periodic variations during 40 years of observations. The amplitudes decrease due to the increasing donor's radius and brightness. Probably, this process is irregular and spasmodic.

The A-type star is on the way to red giants, and its shell expands being truncated by the Roche lobe. Mass transfer from the A-type donor passes into a dynamical timescale (Podsiadlowski 2014). This happens with an expanding supergiant having a convective envelope and losing its mass. Its Roche lobe shrinks when the mass is transferred to a less massive star, what makes the supergiant to overfill the Roche lobe in larger amount and increases the rate of mass transfer in a dynamic mode. Overfilling of the neutron star's Roche lobe and sporadic events of common envelope formation interrupted by ejections of a large mass volume become more frequent. When the Roche lobe of the neutron star is filled or overfilled, the accretion gainer represents a star with the neutron core, i.e. a Thorne & Zytkov (1977) object.

What can we assume about the past, present, and future of SS 433? The most massive primary star of a primordial binary with the main-sequence components had passed its way to a giant with a helium core. It filled and overfilled its Roche lobe, transferred most part of its mass to the less massive secondary companion, and it exploded as a SN Ia (the radio structure W50 is probably a remnant of this SN). As a result, there was a Be/X-ray binary. The evolution of the massive secondary star accelerated to fill its Roche lobe, and to become the donor of accreting matter toward the neutron star.

At present, the transfer of mass from the donor turns to dynamic mode filling and overfilling the Roche lobe of the neutron star from time to time. Jets are forming inside the neutron star's envelope created from the accreted matter, and burst through channels and nozzles, which are seen as bright spots on the surface of the photosphere. The visibility conditions of the nozzles provide an essential contribution in shaping the source lightcurve with the periodic nutation wave having an amplitude of 0.22 mag and a period P = 6.289 day. The system loses its mass through the external Lagrangian point L_2 and the stellar wind. This is a short-time stage associated with the approach of the neutron star with the expanding photosphere of the donor.

In the future, the neutron star will be absorbed inside the common envelope with the expanding A-type giant. Jets will be blocked and disappear completely. Then the neutron star will spiral to the center of this star and form a single star, a massive Thorne-Zhytkov object. Cherepashchuk (2013) has described two scenarios of this development depending on the accretion mode on the central neutron star: (1) The lower-temperature mode when neutrinos do not bring the accretion energy out. This will be a luminous star at the Eddington limit with the lifetime of about 10^6–10^8 years. (2) The high-temperature accretion mode with the energy carried away by neutrinos. Then, if the envelope is massive enough, a single black hole will arise. In this case, the lifetime of the envelope is very short, similar to the dynamical timescale. The other scenario is (3) the explosion of the donor as a supernova before the merger of its nucleus with a neutron star. The remnant may be a binary of two neutron stars or a black hole – neutron star binary.

References

Barnes, A. D., Casares, J., Charles, P. A., *et al.* 2006, *MNRAS*, 365, 296
Gies, D. R., Huang, W., & McSwain, M. V. 2002, *ApJ*, 578, L67
Charbonnel, S., Garde, O., & Edlin, J. 2014, *Astronomer's Telegram*, 6355
Cherepashchuk, A. M. 2013, in: *Close Binary Stars, Part II* (Moscow: FizMatLit), p. 264
Goranskij, V. P. 2011, *Variable Stars*, 31, 5
Goranskij, V. P. 2013, *Central European Astrophys. Bull.*, 37, 1, 251
Goranskij, V. P., Spiridonova, O. I. 2014, *Astronomer's Telegram*, 6347
Hillwig, T. C., Gies, D. R., Huang, W., *et al.* 2004, *ApJ*, 615, 422
Kubota, K., Ueda, Y., Fabrika, S., *et al.* 2010, *ApJ*, 709, 1374
Paragi, Z., Vermeulen, R. C., Fejes, I., *et al.* 1999, *A&A*, 348, 910
Podsiadlowski, P. 2014, in: *Accretion Processes in Astrophysics.* (Cambridge Univ. Press), p. 45
Thorne, K. S. & Zytkov, A. N. 1977, *ApJ*, 212, 832

High-mass X-ray Binaries: Illuminating the Passage from Massive Binaries to Merging Compact Objects
Proceedings IAU Symposium No. 346, 2019
L. M. Oskinova, E. Bozzo, T. Bulik & D. Gies, eds.

doi:10.1017/S1743921319001121

Ultra-luminous X-ray sources as neutron stars propelling and accreting at super-critical rates in high-mass X-ray binaries

M. Hakan Erkut and K. Yavuz Ekşi

Physics Engineering Department, Faculty of Science and Letters,
Istanbul Technical University, 34469, Istanbul, Turkey
emails: mherkut@gmail.com, eksi@itu.edu.tr

Abstract. Ultra-luminous X-ray sources (ULXs) are off-nuclear point sources in nearby galaxies with luminosities well exceeding the Eddington limit for stellar-mass objects. It has been recognized after the discovery of pulsating ULXs (PULXs) that a fraction of these sources could be accreting neutron stars in high-mass X-ray binaries (HMXBs) though the majority of ULXs are lacking in coherent pulsations. The earliest stage of some HMXBs may harbor rapidly rotating neutron stars propelling out the matter transferred by the massive companion. The spin-down power transferred by the neutron-star magnetosphere to the accretion disk at this stage can well exceed the Eddington luminosities and the system appears as a non-pulsating ULX. In this picture, PULXs appear as super-critical mass-accreting descendants of non-pulsating ULXs. We present this evolutionary scenario within a self-consistent model of magnetosphere-disk interaction and discuss the implications of our results on the spin and magnetic field of the neutron star.

Keywords. X-rays: binaries, stars: neutron, stars: mass loss, accretion, accretion disks

---

1. Introduction

A significant fraction of ultra-luminous X-ray sources (ULXs) may consist of neutron stars as indicated by the recent detection of pulsations from M82 X-2 (Bachetti *et al.* 2014), ULX NGC 7793 P13 (Fürst *et al.* 2016), ULX NGC 5907 (Israel *et al.* 2017), and NGC 300 ULX1 (Carpano *et al.* 2018). In addition to pulsating ULXs (PULXs), ultra-luminous super-soft sources (ULSs) (Di Stefano & Kong 2003; Fabbiano *et al.* 2003; Kong & Di Stefano 2003) and other non-pulsating ULXs emerge as seemingly different subclasses of the ULX population. The lack of pulsations can be due to the optically thick envelope fed by the outflows of the accretion disk around the neutron star (Ekşi *et al.* 2015) and/or the propeller effect (Illarionov & Sunyaev 1975) of the neutron-star magnetosphere on the disk matter (Tsygankov *et al.* 2016).

We consider the spin and luminosity evolution of neutron stars in high-mass X-ray binaries (HMXBs). The magnetosphere of the newborn rapidly rotating neutron star with spin periods of a few milliseconds interacts with the wind-fed disk in the very early stage of the X-ray binary (Erkut *et al.* 2018). The donor, as an already evolved massive star, produces dense winds with high mass-loss rates ($\dot{M}_{\rm w} \gtrsim 10^{-6}\, M_{\odot}\, {\rm yr}^{-1}$). Such an evolutionary scheme can be illustrated by a neutron star–helium star binary that is expected to form soon after the common-envelope phase of a twin massive binary (Brown 1995; Dewi *et al.* 2006). It takes $\sim 10^6$ years for the helium-burning stage to end. During its lifetime ($\sim 10^6$ years), the massive helium star loses mass at a rate of $\dot{M}_{\rm w} \sim$

$10^{-5}\,M_\odot\,\mathrm{yr}^{-1}$. Following the fusion of elements heavier than helium within $\lesssim 10^4$ years, the helium-star core collapses to give rise to the birth of the second neutron star. Double neutron-star systems might therefore be the descendants of helium star–neutron star X-ray binaries (Dewi *et al.* 2006).

2. Model

The fraction of the mass-loss rate of a $\sim 20\,M_\odot$ wind donor to be captured by a $1.4\,M_\odot$ neutron star can be as high as $\dot{M}_0/\dot{M}_\mathrm{w} \sim 0.3$ due to photoionization of the wind matter irradiated by the X-rays emitted from the neutron star. Deceleration of the wind matter usually leads to the formation of an extensive disk around the neutron star with high mass-transfer rates (Čechura & Hadrava 2015).

In our picture, the mass transfer from the massive helium companion (Wolf-Rayet star) with mass-loss rates of $\dot{M}_\mathrm{w} \gtrsim 10^{-5}\,M_\odot\,\mathrm{yr}^{-1}$ to the neutron star occurs at super-Eddington (super-critical) rates ($\dot{M}_0 \gtrsim 10^{-6}\,M_\odot\,\mathrm{yr}^{-1}$). The innermost disk radius (magnetopause) is smaller than the spherization radius, i.e., $R_\mathrm{in} < R_\mathrm{sp}$. The system is thus in the super-critical regime (Shakura & Sunyaev 1973). The neutron star acts as a super-critical propeller for $R_\mathrm{co} < R_\mathrm{in} < R_\mathrm{sp}$. The super-critical accretion regime is realized when $R_\mathrm{in} < R_\mathrm{co} < R_\mathrm{sp}$. The corotation radius in the disk, $R_\mathrm{co} \equiv (GM/\Omega_*^2)^{1/3}$, is determined by the neutron-star spin period, $P = 2\pi/\Omega_*$. The ejector phase is realized when $R_\mathrm{L} < R_\mathrm{in}$, where $R_\mathrm{L} = c/\Omega_*$ is the light-cylinder radius.

To obtain the spin-period evolution of the neutron star of moment of inertia I, we solve the torque equation,

$$-\frac{2\pi I}{P^2}\frac{dP}{dt} = N. \tag{2.1}$$

In the ejector phase, the torque, N, acting on the neutron star is due to the magnetic dipole radiation, i.e., $N \simeq -(2/3)\mu^2\Omega_*^3/c^3$. For the super-critical propeller and accretion regimes, we write

$$N = n\,(\omega_*)\,\dot{M}_\mathrm{in}\sqrt{GMR_\mathrm{in}} \tag{2.2}$$

with the fastness parameter $\omega_* \equiv P_\mathrm{K,in}/P$ in terms of the mass inflow rate and the Keplerian period at R_in. As a function of the fastness parameter, the dimensionless torque becomes $n < 0$ and $n > 0$ for the propeller and accretion regimes, respectively. We use $n \simeq 1$ for the accretion torque and $n \simeq 1 - \omega_*$ for the propeller torque (Erkut *et al.* 2018).

The total luminosity of the disk around a propeller with subcritical mass-inflow rates can be written as $L_\mathrm{tot} = GM\dot{M}/R_\mathrm{in} - I\Omega_*\dot{\Omega}_* - \dot{M}_\mathrm{out}v_\mathrm{out}^2/2$ (Ekşi *et al.* 2005). In the super-critical propeller regime, however, each term contributing to the total luminosity of the neutron star–disk system must be treated accordingly by taking into account the regulation of the accretion flow inside the spherization radius. Noting that the mass-inflow rate, throughout the disk, is $\dot{M} = \dot{M}_0(R/R_\mathrm{sp})$ for $R < R_\mathrm{sp}$ and $\dot{M} = \dot{M}_0$ for $R > R_\mathrm{sp}$ (Shakura & Sunyaev 1973), we calculate the total luminosity of the neutron-star-disk system in the super-critical accretion regime using $L_\mathrm{tot} = L_\mathrm{acc} + L_\mathrm{out} + L_\mathrm{G}$ and in the super-critical ejector and propeller regimes using $L_\mathrm{tot} = L_\mathrm{sd} + L_\mathrm{out} + L_\mathrm{G}$. As a result of super-critical mass inflow, $L_\mathrm{out} < 0$ represents the energy-loss rate due to outflows from the disk. The rate of gravitational energy release throughout the disk is given by L_G. The spin-down power and accretion luminosity can be written as $L_\mathrm{sd} = -2\pi N/P$ and $L_\mathrm{acc} = GM\dot{M}_\mathrm{in}/R_*$ with $\dot{M}_\mathrm{in} = \dot{M}_0(R_\mathrm{in}/R_\mathrm{sp})$, respectively, for a neutron star of mass M and radius R_* (Erkut *et al.* 2018).

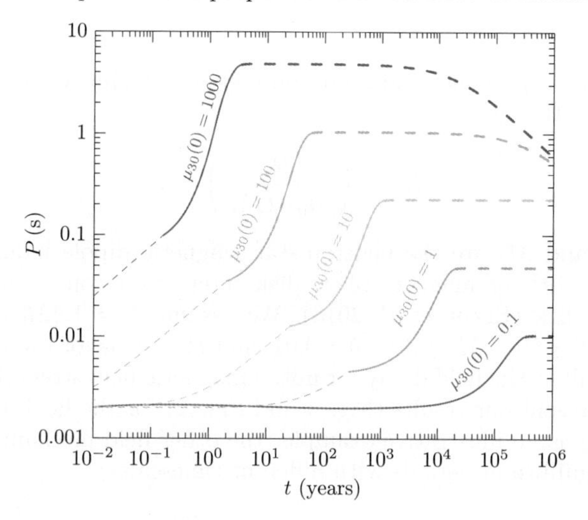

Figure 1. Spin-period evolution with super-critical ejector (thin dashed), propeller (solid), and accretor (thick dashed) phases for a set of initial magnetic moments. We assume that the dipole magnetic fields stronger than $B = 10^9$ T ($\mu_{30} = 10$ at $t = 0$) decay according to the scenario B (Erkut *et al.* 2018).

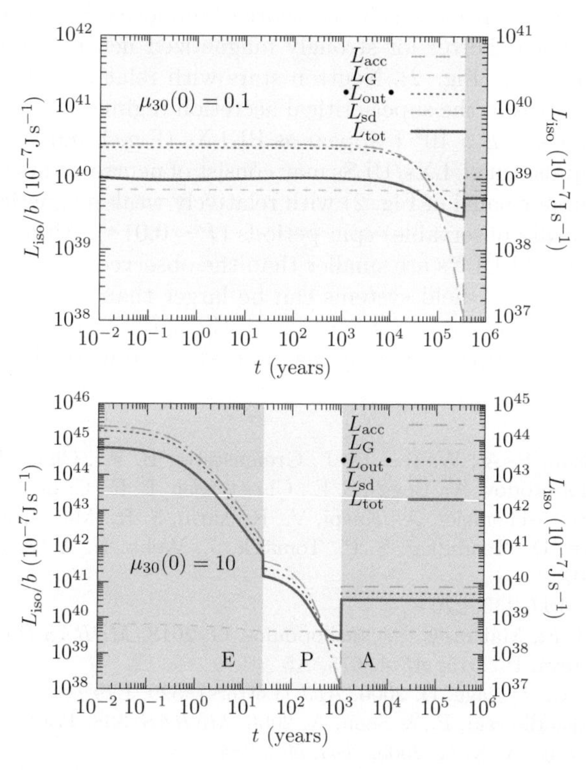

Figure 2. Luminosity evolution for two different initial magnetic moments. The luminosity for isotropic emission is on the right vertical axis. The left vertical axis represents the observed luminosity if the geometrical beaming is $b = 0.1$. Shaded regions correspond to phases such as ejector (E), propeller (P), and accretor (A) (Erkut *et al.* 2018).

3. Results

In the super-critical propeller and accretion regimes, the inner disk radius depends on several parameters, that is,

$$R_{\rm in} \simeq \left(\frac{\mu^2 R_{\rm sp} \delta}{\dot{M}_0 \sqrt{GM}} \right)^{2/9}, \tag{3.1}$$

where μ, $R_{\rm sp}$, δ, and $\dot{M}_0$ are the neutron-star magnetic dipole moment, the spherization radius, the width of magnetosphere-disk interaction zone, and the mass inflow rate in the outer disk (Erkut *et al.* 2018). We assume $M = 1.4 M_{\odot}$, $R = 10\,{\rm km}$, $\dot{M}_0 = 4 \times 10^{17}\,{\rm kg\,s^{-1}} (\sim 6 \times 10^{-6}\,M_{\odot}\,{\rm yr^{-1}})$, $\delta = 0.01$, and $P_0 = 2\,{\rm ms}$ for the initial period of the neutron star. We allow the field decay for initial magnetic field strengths in the magnetar range. Here, we present our results (Fig. 1 and Fig. 2) for the field-decay mechanism B (Colpi *et al.* 2000) as an illustrative example (the other field mechanisms such as A and C yield similar equilibrium periods with different timescales).

4. Conclusions

As shown in Figure 1, the observed spin periods of PULXs ($P \sim 1$ s) can be realized for sufficiently strong initial magnetic fields in the $B \sim 10^9 - 10^{11}$ T ($10^{13} - 10^{15}$ G) range. In the very early (ejector) stage of the luminosity evolution, neutron stars with such strong initial fields can even appear as supernova impostors (lower panel of Fig. 2). The super-critical propeller stage, during which the source luminosity becomes comparable with those of ULXs, is much shorter for strongly magnetized neutron stars than for weakly magnetized neutron stars (Fig. 2). Neutron stars with relatively strong magnetic fields spend most of their lives in the super-critical accretion regime. It is therefore more likely that the neutron stars of $B > 10^9$ T appear as PULXs (Fig. 1 and lower panel of Fig. 2).

Most of the non-pulsating ULXs/ULSs may consist of neutron stars in the super-critical propeller regime (upper panel of Fig. 2) with relatively weak magnetic fields ($B \sim 10^7$ T) and shorter (but hardly observable) spin periods ($P \sim 0.01$ s). Although the equilibrium periods of the weak-field ULXs are smaller than the observed typical periods of PULXs, the population of the weak-field systems can be larger than the population of PULXs. Yet, it would relatively be more difficult, due to the smaller size of the magnetosphere, to observe pulsations from these weakly magnetized neutron-star ULXs/ULSs.

References

Bachetti, M., Harrison, F. A., Walton, D. J., Grefenstette, B. W., Chakrabarty, D., Fürst, F., Barret, D., Beloborodov, A., Boggs, S. E., Christensen, F. E., Craig, W. W., Fabian, A. C., Hailey, C. J., Hornschemeier, A., Kaspi, V., Kulkarni, S. R., Maccarone, T., Miller, J. M., Rana, V., Stern, D., Tendulkar, S. P., Tomsick, J., Webb, N. A., & Zhang, W. W. 2014, *Nature*, 514, 202

Brown, G. E. 1995, *ApJ*, 440, 270

Carpano, S., Haberl, F., Maitra, C., & Vasilopoulos, G. 2018, *MNRAS* (Letters), 476, L45

Čechura, J., & Hadrava, P. 2015, *A&A*, 575, A5

Colpi, M., Geppert, U., & Page, D. 2000, *ApJ* (Letters), 529, L29

Dewi, J. D. M., Podsiadlowski, P., & Sena, A. 2006, *MNRAS*, 368, 1742

Di Stefano, R., & Kong, A. K. H. 2003, *ApJ*, 592, 884

Ekşi, K. Y., Hernquist, L., & Narayan, R. 2005, *ApJ* (Letters), 623, L41

Ekşi, K. Y., Andaç, İ. C., Çıkıntoğlu, S., Gençali, A .A., Güngör, C., & Öztekin, F. 2015, *MNRAS* (Letters), 448, L40

Erkut, M. H., Ekşi, K. Y., & Alpar, M. A. 2018, *ApJ* (Submitted)

Fabbiano, G., King, A. R., Zezas, A., Ponman, T. J., Rots, A., & Schweizer, F. 2003, *ApJ*, 591, 843

Fürst, F., Walton, D. J., Harrison, F. A., Stern, D., Barret, D., Brightman, M., Fabian, A. C., Grefenstette, B., Madsen, K. K., Middleton, M. J., Miller, J. M., Pottschmidt, K., Ptak, A., Rana, V., & Webb, N. 2016, *ApJ* (Letters), 831, L14

Illarionov, A. F., & Sunyaev, R. A. 1975, *A&A*, 39, 185

Israel, G. L., Belfiore, A., Stella, L., Esposito, P., Casella, P., De Luca, A., Marelli, M., Papitto, A., Perri, M., Puccetti, S., Castillo, G. A. R., Salvetti, D., Tiengo, A., Zampieri, L., D'Agostino, D., Greiner, J., Haberl, F., Novara, G., Salvaterra, R., Turolla, R., Watson, M., Wilms, J., & Wolter, A. 2017, *Science*, 355, 817

Kong, A. K. H., & Di Stefano, R. 2003, *ApJ* (Letters), 590, L13

Shakura, N. I., & Sunyaev, R. A. 1973, *A&A*, 24, 337

Tsygankov, S. S., Mushtukov, A. A., Suleimanov, V. F., & Poutanen, J. 2016, *MNRAS*, 457, 1101

High-mass X-ray Binaries: Illuminating the Passage from Massive Binaries to Merging Compact Objects
Proceedings IAU Symposium No. 346, 2019
L. M. Oskinova, E. Bozzo, T. Bulik & D. Gies, eds.

doi:10.1017/S1743921319001133

Analytical solution for magnetized thin accretion disk in comparison with numerical simulations

Miljenko Čemeljić, Varadarajan Parthasarathy and Włodek Kluźniak

Nicolaus Copernicus Astronomical Center, Bartycka 18, 00-716 Warsaw, Poland
email: `miki,varada,wlodek@camk.edu.pl`

Abstract. We obtained equations for a thin magnetic accretion disk, using the method of asymptotic approximation. They cannot be solved analytically-without solutions for a magnetic field in the magnetosphere between the star and the disk, only a set of general conditions on the solutions can be derived. To compare the analytical results with numerical solutions, we find expressions for physical quantities in the disk, using our results from resistive and viscous star-disk magnetospheric interaction simulations.

Keywords. stars: formation, stars: magnetic fields

1. Introduction

Gravitational infall of matter onto a rotating central object naturally forms a rotating accretion disk. Matter from the disk is then fed inwards through an accretion column. Examples of single objects with a disk around them are protostars and young stellar objects, and in close binary systems a disk can form when matter from donor star falls onto a white dwarf or a neutron star.

Analytical hydro-dynamical model of a thin accretion disk, with viscosity parameterized by Shakura & Sunyaev α-prescription, has been given in Kluźniak & Kita (2000). We extend this model, obtaining the equations for a magnetic thin disk. Analytical solution in the magnetic case cannot be given without knowing the solutions in a star-disk magnetosphere, only general conditions on solutions could be derived.

We perform numerical simulations of star-disk magnetospheric interaction, adding a stellar rotating surface and a magnetic field to the analytical hydro-dynamical disk solution used as an initial condition in simulations. A quasi-stationary solution is obtained, from which we find simple matching expressions for physical quantities in the disk. Those expressions can be compared with requirements from the analytically obtained equations for the magnetic disk and with the hydro-dynamic analytical and numerical solutions.

2. Numerical setup

We use the publicly available PLUTO code (v.4.1) (Mignone *et al.* 2007, 2012), with logarithmically stretched grid in radial direction in spherical coordinates, and uniformly spaced latitudinal grid. Resolution is $R \times \theta$=[217×100] grid cells, stretching the domain to 30 stellar radii. We perform axisymmetric 2D star-disk simulations in resistive and viscous magneto-hydrodynamics, following Zanni & Ferreira (2009) - see also Čemeljić *et al.* (2017).

Table 1. Coefficients k in the expressions from results in our simulation with $B_{\star}$=0.05 T.

k_1	k_2	k_3	k_4	k_5	k_6	k_7	k_8	k_9
0.88	-0.09	3.8×10^{-5}	0.255	-0.4	-0.15	-1.11	0.006	0.01

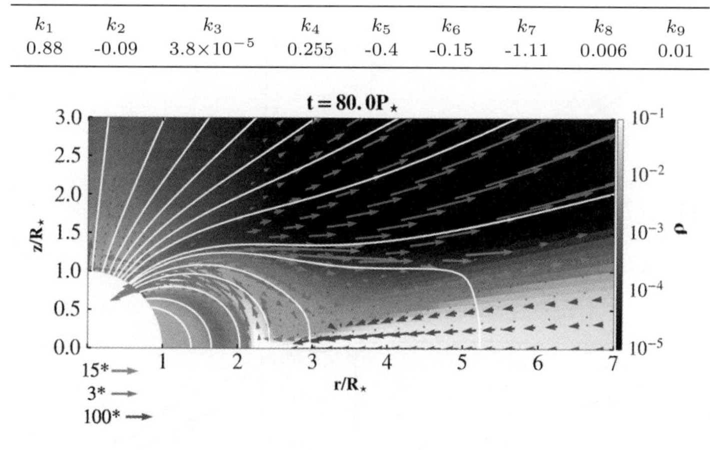

Figure 1. A zoom into our simulation result after T=80 rotations of the underlying star. Shown is the density in logarithmic color grading in code units, with a sample of magnetic field lines, depicted with white solid lines. Velocities in the disk, column and stellar wind are shown with vectors, depicted in different normalizations with respect to the Keplerian velocity at the stellar surface.

The equations solved by the PLUTO code are, in the cgs units:

$$\frac{\partial \rho}{\partial t} + \nabla \cdot (\rho \mathbf{v}) = 0$$

$$\frac{\partial \rho \mathbf{v}}{\partial t} + \nabla \cdot \left[\rho \mathbf{v}\mathbf{v} + \left(P + \frac{\mathbf{BB}}{8\pi} \right) \mathbf{I} - \frac{\mathbf{BB}}{4\pi} - \tau \right] = \rho \mathbf{g}$$

$$\frac{\partial E}{\partial t} + \nabla \cdot \left[\left(E + P + \frac{\mathbf{BB}}{8\pi} \right) \mathbf{v} + \underbrace{\eta_{\mathrm{m}} \mathbf{J} \times \mathbf{B}/4\pi - \mathbf{v} \cdot \tau}_{\text{heating terms}} \right] = \rho \mathbf{g} \cdot \mathbf{v} - \underbrace{\Lambda}_{\text{cooling}}$$

$$\frac{\partial \mathbf{B}}{\partial t} + \nabla \times (\mathbf{B} \times \mathbf{v} + \eta_{\mathrm{m}} \mathbf{J}) = 0.$$

Symbols have their usual meaning. The underbraced Ohmic and viscous heating terms and the cooling term are removed in our computations, to prevent the thermal thickening of the accretion disk. This equals the assumption that all the heating is radiated away from the disk. The solutions are still in a non-ideal magneto-hydrodynamics regime, because of the viscous term in the momentum equation, and the resistive term in the induction equation.

3. Analytical solutions ver. numerical solutions

We extended the asymptotic approximation from a hydro-dynamic (Kluźniak & Kita 2000) thin accretion disk solution to a magnetic case. Without knowing the solution for a magnetic field in a star-disk magnetosphere, which is connected with the solution in the disk, obtained equations cannot be solved analytically. In addition to the solutions which remain the same as in the hydro-dynamical case, only general constraints on the magnetic solution could be derived.

We compare the obtained analytical solutions and constraints in the magnetic case with the results from simulations. To do this, we write the solutions from simulations in terms of matching expressions-which are not formal fits, but simple matching functions chosen to represent the result within 10% of the value from the simulation, when there is

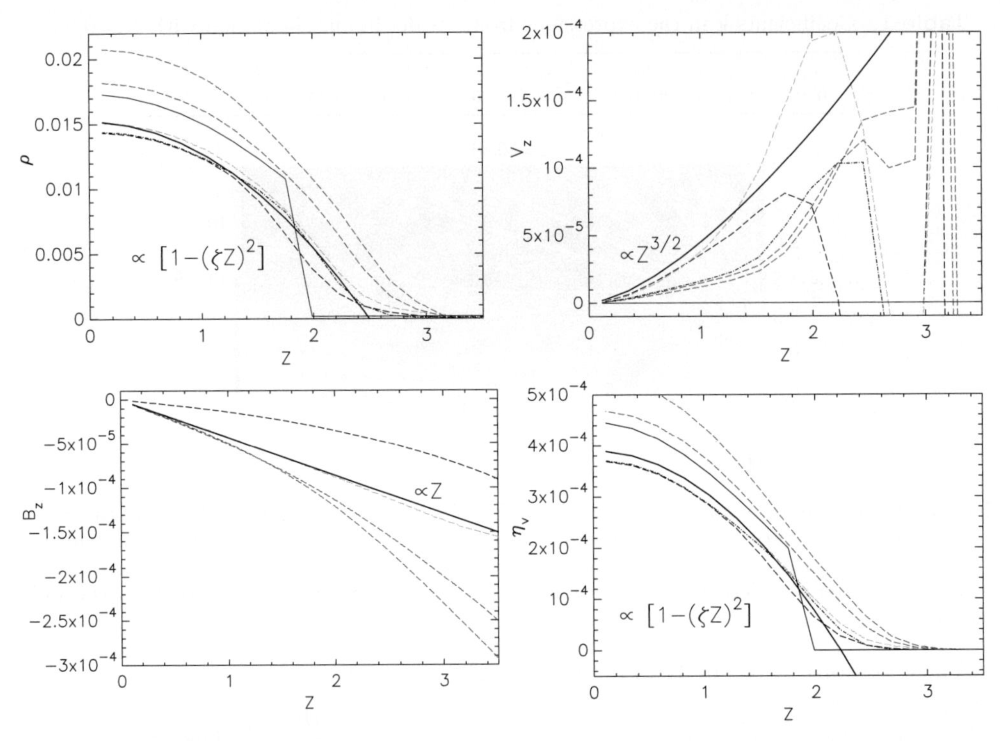

Figure 2. Trends with the increasing stellar magnetic strength in the disk density, velocity, magnetic field and viscosity. The values are taken at $R=15R_\star$ in the vertical direction, with Z expressed in the units of $R_\star$. With the thin solid line is shown the KK00 solution, with the dash-dotted line is shown the solution without the magnetic field, and in black, green, blue and green dashed lines are shown the solutions with $B_\star$=0.025, 0.05, 0.075 and 0.1 T, respectively. With the thick solid line is shown a match to the $B_\star$=0.05 T case. Expressions for the match in $B_\star$=0.05 T case are written in each panel, with ζ in the density equal to 0.4, and in the viscosity to 0.45. The density, velocity, dynamic viscosity and magnetic field are expressed in the code units $\rho_{d0}=\dot{M}_0/(R_\star^2 v_{K\star})$, $v_0=v_{K\star}$, $\eta_{v0}=\rho_{d0}R_\star v_{K\star}$, and $B_0=v_{K\star}\sqrt{\rho_{d0}}$, where $\dot{M}_0$ is the free parameter in the simulations, denoting the disk accretion rate, and $v_{K\star}$ is the Keplerian velocity at the stellar surface.

no oscillations. In all the cases, the match is chosen to the solution in the middle of the disk, further away from the star than the corotation radius, where the disk co-rotates with the star, which is at $R_{\rm cor}=2.9R_\star$.

We confirm that in the middle part of the disk, at $R=15R_\star$, the numerical solution in the magnetic case does not differ much from the hydro-dynamical analytical and numerical solutions. The difference is only in the proportionality coefficients k in the expressions for physical quantities and in the corresponding power laws.

The expressions we obtain are:

$$\rho(r,z)=\frac{k_1}{r^{3/2}}[1-(0.4z)^2],$$

$$v_r(r,z)=\frac{k_2}{r^{3/2}}[1+(0.5z)^2],\ v_z(r,z)=\frac{k_3}{r}z^{3/2},\ v_\varphi(r,z)=\frac{k_4}{\sqrt{r}},$$

$$B_r(r,z)=\frac{k_5}{r^3}z,\ B_z(r,z)=\frac{k_6}{r^3}z,\ B_\varphi(r,z)=\frac{k_7}{r^3}z,$$

$$\eta(r,z)=\frac{k_8}{r}[1-(0.45z)^2],\ \eta_{\rm m}(r,z)=k_9\sqrt{r}[1-(0.3z)^2].$$

We tabulate the coefficients k in the case of a Young Stellar Object rotating with $\Omega_\star = 0.2$ of the breakup angular velocity, with the stellar field $B_\star$=0.05 T, anomalous viscosity and resistivity coefficients α_v=1 and $\alpha_m = 1$, in Table 1.

In Fig. 2 we show the trends in the density, viscosity and vertical velocity and magnetic field components in the disk, in the cases with increasing stellar magnetic field strength.

4. Conclusions

We present our results in numerical simulations of a star-disk system with magnetospheric interaction in the case of Young Stellar Object with the stellar field of 0.05 T, rotating with 20% of the breakup velocity. Quasi-stationary solutions in the disk are obtained, and we find simple expressions to match the physical quantities.

We find that the expressions in the magnetic cases differ from the results in the hydrodynamical simulation and in the analytical solution only in proportionality coefficients.

In future work we will compare the trends in numerical solutions in the cases with different stellar magnetic field strength, rotation rate, viscosity and resistivity.

Acknowledgements

MČ developed the setup for star-disk simulations while in CEA, Saclay, under the ANR Toupies grant. His work in NCAC Warsaw is funded by a Polish NCN grant no. 2013/08/A/ST9/00795 and a collaboration with Croatian STARDUST project through HRZZ grant IP-2014-09-8656 is acknowledged. VP work is partly funded by a Polish NCN grant 2015/18/E/ST9/00580. We thank IDRIS (Turing cluster) in Orsay, France, ASIAA/TIARA (PL and XL clusters) in Taipei, Taiwan and NCAC (PSK cluster) in Warsaw, Poland, for access to Linux computer clusters used for the high-performance computations. We thank the PLUTO team for the possibility to use the code.

References

Čemeljić, M., Parthasarathy, V. & Kluźniak, W. 2017, *JPhCS*, 932, 012028

Kluźniak, W., & Kita, D. 2000, *arXiv*, astro-ph/0006266

Mignone, A., Bodo, G., Massaglia, S., Matsakos T., Tesileanu O., Zanni C., Ferrari A. 2007, *ApJS*, 170, 228

Mignone, A., Zanni, C., Tzeferacos, P., van Straalen, B., Colella, P., and Bodo, G. 2012, *ApJS*, 198, 7

Zanni, C., & Ferreira, J., 2009, *A&A*, 512, 1117

High-mass X-ray Binaries: Illuminating the Passage from Massive Binaries to Merging Compact Objects
Proceedings IAU Symposium No. 346, 2019
L. M. Oskinova, E. Bozzo, T. Bulik & D. Gies, eds.

doi:10.1017/S1743921319001200

Optical and X-ray study of *V*404 *Cyg* during its activity in the Summer 2015

Evgeniya A. Nikolaeva[1,2], Ilfan F. Bikmaev[1,2], Maxim V. Glushkov[1], Eldar N. Irtuganov[1,2] and Irek M. Khamitov[3]

[1]Kazan Federal University, 420008, Kremlyovskaya St. 18, Kazan, Russian Federation
email: evgeny.nikolaeva@gmail.com

[2]Academy of Sciences of Tatarstan, 420111, Bauman St. 20, Kazan, Russian Federation

[3]TUBITAK National Observatory, 07058, Akdeniz University Campus, Antalya, Turkey

Abstract. The black hole X-ray binary *V*404 *Cyg* was studied during of the 2015 outburst. Optical photometry and spectroscopy were performed by using 1.5-meter Russian-Turkish telescope (RTT-150) facilities at the TUBITAK National Observatory (Antalya, Turkey). From June 22 to June 28, 2015, shell expansion velocity decreased from 650 to 400 km s^{-1} as measured by Hα and Hβ lines and from 450 to 330 km s^{-1} as measured by HeI and HeII lines. Thus, the shell expansion occurred with deceleration, where the hydrogen and helium line formation regions are at different radial distances from the center of the star. The correlation of flow variability in the optical and X-ray ranges is caused by fluctuations in the rate of accretion near a compact source where X-ray photons are generated.

Keywords. *V*404 *Cyg*, low-mass X-ray binary

1. Introduction

We have observed optical outburst of the black hole X-ray binary *V*404 *Cyg* during its X-ray activity (Barthelmy *et al.* 2015, Negoro *et al.* 2015) in Summer 2015 by using 1.5-meter Russian-Turkish optical telescope (RTT-150) equipped with TFOSC instrument. We performed photometry, polarimetry and middle resolution (5 Å) spectroscopy during the period June 20-28, 2015. Optical spectra dominated by strong broad emission signatures from HI, HeI and HeII. It associated to a nova-like nebula formed by the cooling remnant of strong accretion disc winds (Rahoui *et al.* 2017).

2. Overview

The velocity of shell expansion was measured based on the FWHM of the hydrogen and helium lines. From June 22 to June 28, 2015, shell expansion velocity decreased from 650 to 400 km s^{-1} as measured by Hα and Hβ lines, from 450 to 330 km s^{-1} as measured by lines of HeI 6678 Å, 7065 Å and HeII 4685 Åline Table 1. Thus, we suppose that the shell expansion occurred with deceleration in the indicated period. We found also the asymmetry of the profiles of the hydrogen lines (Fig. 1), which may be associated with an uneven decrease in the rate of expansion of the shell in the different directions.

We performed polarimetric observations in June 23-24, 2015, and found that V-band emission of the source is polarized with $P \sim 8 \pm 0.5\%$. Our polarization degree value is in agreement with the published values (Panapoulou *et al.* 2015, Blau *et al.* 2015, Itoh *et al.* 2015). This suggests that the polarization might be of interstellar origin or occurred by dust scattering on the rings and diffuse structure (found by SWIFT XRT observations,

Table 1. Line profile's parameters.

date	HJD	λ_{peak}, Å	W, Å	FWHM/2, km s^{-1}
		Hα		
22.06.2015	2457196.56524933	6564.8	-70.8	667
24.06.2015	2457198.49295458	6563.8	-82.0	545
26.06.2015	2457200.57518956	6563.3	-233.8	533
27.06.2015	2457201.55512845	6563.9	-671.2	443
28.06.2015	2457202.55643330	6562.2	-22.3	401
		Hβ		
22.06.2015	2457196.56524933	4863.0	-16.8	557
24.06.2015	2457198.49295458	4861.6	-12.7	384
26.06.2015	2457200.57518956	4861.7	-22.7	425
27.06.2015	2457201.55512845	4862.1	-15.4	490
		HeI (7065 Å)		
22.06.2015	2457196.56524933	7067.9	-6.5	487
24.06.2015	2457198.49295458	7065.3	-6.9	319
26.06.2015	2457200.57518956	7065.5	-14.2	343
		HeI (6678 Å)		
22.06.2015	2457196.56524933	6680.0	-6.3	446
24.06.2015	2457198.49295458	6678.1	-7.4	290
26.06.2015	2457200.57518956	6678.3	-14.0	321
		HeII (4686 Å)		
22.06.2015	2457196.56524933	4685.8	-7.9	468
24.06.2015	2457198.49295458	4684.9	-12.5	297
26.06.2015	2457200.57518956	4685.1	-7.3	328

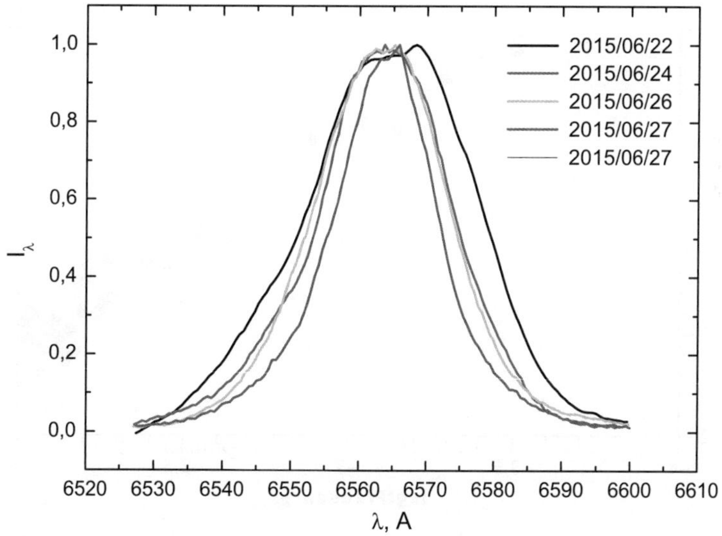

Figure 1. Changing of Hα Line Profiles.

Beardmore *et al.* 2015). These rings could be associated with previous flaring episodes from the central binary system.

The X-ray light curve of *V*404 *Cyg* was analysed during its flare activity in June and December 2015 using the open public XRT data of the SWIFT orbital observatory. RTT-150 optical observations made during the period from June 20 to June 28, 2015 in V, B, R and I filters (Fig. 2).

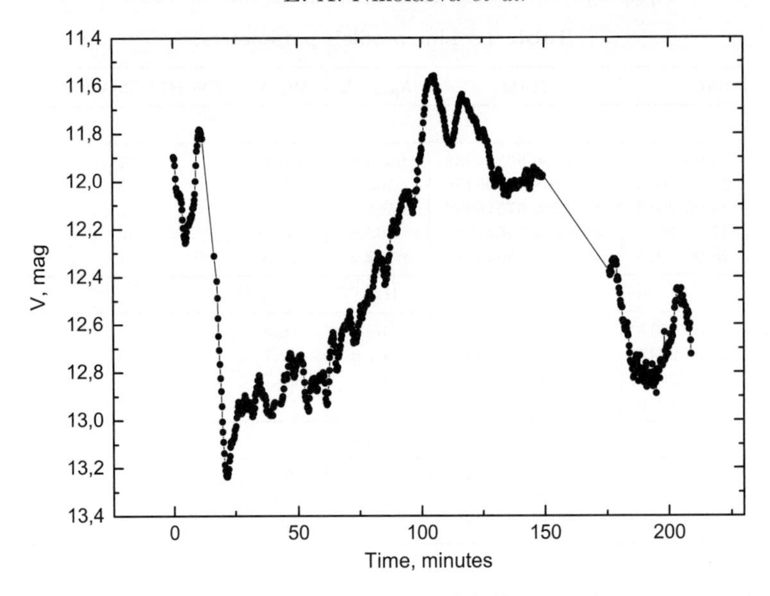

Figure 2. V-band light curve in June 20, 2015.

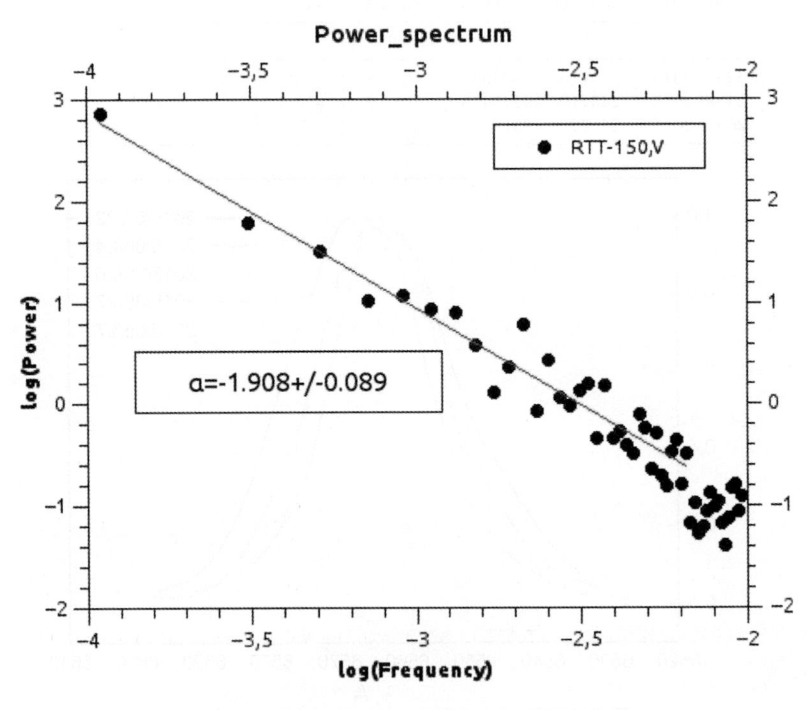

Figure 3. X-ray Power spectrum of $V404\,Cyg$.

LS-spectra (Least-Square) were constructed for the analysis of X-ray and optical data by the Lomb-Scargle method (Fig. 3, 4). Those power spectra show the same slope within the error limits: -1.9 ± 0.09 in the optical range and -1.85 ± 0.06 in the X-ray range. Within the frequency range from 0.002 Hz and below the X-ray power spectrum shows a "flat area". The X-ray power spectrum exhibits possible QPO at the frequency of 0.0137 Hz, which nature remains unknown (Fig. 5).

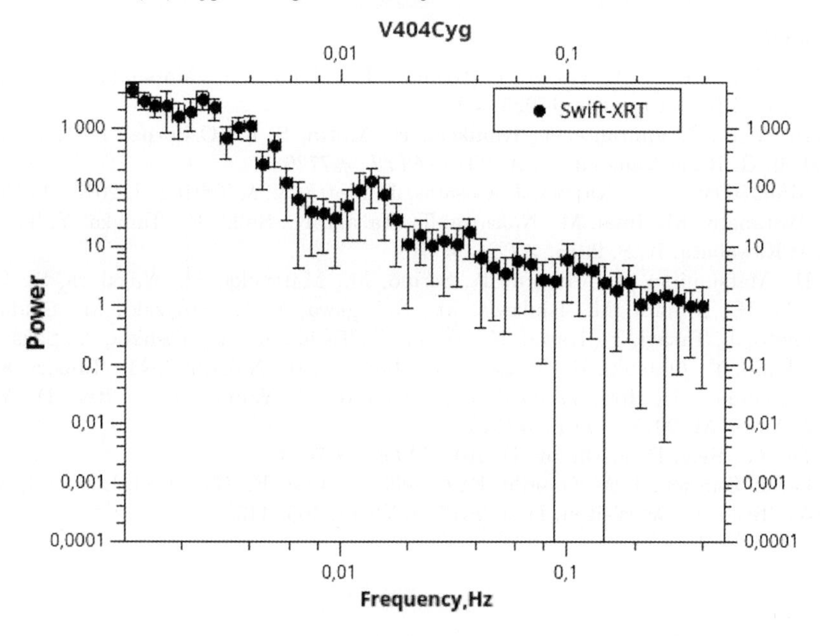

Figure 4. Optical Power spectrum of *V*404 *Cyg*.

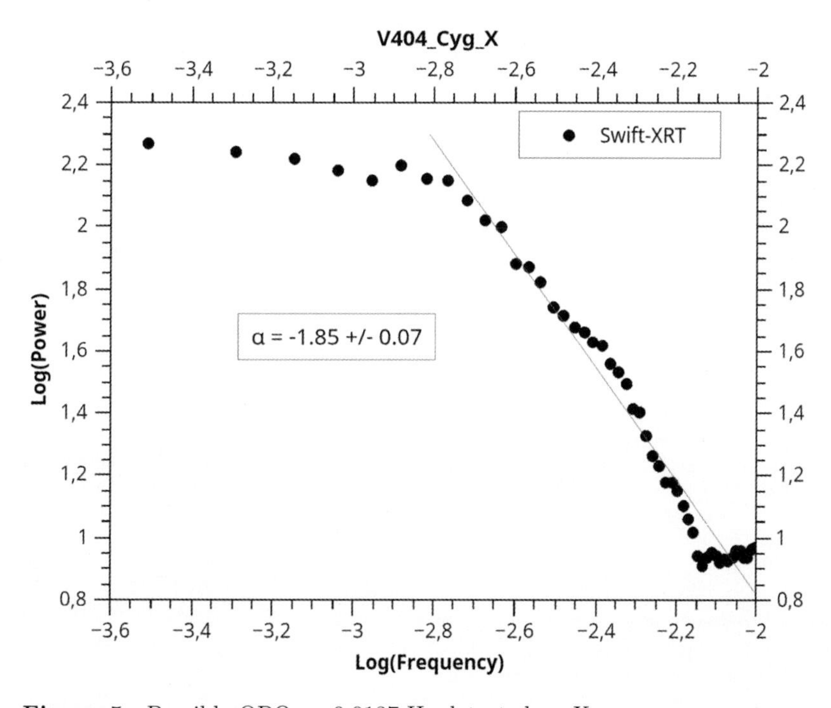

Figure 5. Possible QPO on 0.0137 Hz detected on X-ray power spectrum.

3. Acknowledgments

Authors thank TUBITAK, KFU and AST, for partial support in using RTT-150 (the Russian-Turkish 1.5-m telescope in Antalya). This work was funded by the subsidy 3.9780.2017/8.9 allocated to Kazan Federal University for the state assignment in the sphere of scientific activities.

References

Barthelmy, S.D., D'Ai, A., D'Avanzo, P., Krimm, H.A., Lien, A.Y., Marshall, F.E., Maselli, A., & Siegel, M.H. 2015, *GCN*, 17929, #1

Beardmore, A. P., Altamirano, D., Kuulkers, E., Motta, S. E., Osborne, J. P., Page, K. L., Sivakoff, G. R., & Vaughan, S. A. 2015, *ATEL*, #7736

Blay, P., Munoz-Darias, T., Kajava, J., Casares, J., Motta, S., & Telting, J. 2015, *ATEL*, #7678

Itoh, R., Watanabe, M., Imai, M., Nakaoka, T., Takaki, K., Shiki, K., Tanaka, Y. T., Uemura, M., & Kawabata, K. S. 2015, *ATEL*, #7709

Negoro, H., Matsumitsu, T., Mihara, T., Serino, M., Matsuoka, M., Nakahira, S., Ueno, S., Tomida, H., Kimura, M., Ishikawa, M., Nakagawa, Y. E., Sugizaki, M., Shidatsu, M., Sugimoto, J., Takagi, T., Kawai, N., Yoshii, T., Tachibana, Y., Yoshida, A., Sakamoto, T., Kawakubo, Y., Ohtsuki, H., Tsunemi, H., Imatani, R., Nakajima, M., Tanaka, K., Ueda, Y., Kawamuro, T., Hori, T., Tsuboi, Y., Kanetou, S., Yamauchi, M., Itoh, D., Yamaoka, K., & Morii, M. 2015, *ATEL*, #7646

Panopoulou, G., Reig, P., & Blinov, D. 2015, *ATEL*, #7674

Rahoui, Farid, Tomsick, J. A., Gandhi, P., Casella, P., Frst, F., Natalucci, L., Rossi, A., Shaw, A. W., Testa, V., & Walton, D. J. 2017, *MNRAS*, 465, 4468

High-mass X-ray Binaries: Illuminating the Passage from Massive Binaries to Merging Compact Objects
Proceedings IAU Symposium No. 346, 2019
L. M. Oskinova, E. Bozzo, T. Bulik & D. Gies, eds.

doi:10.1017/S1743921319001224

Statistical study of magnetic reconnection in accretion disks systems around HMXBs

Luís H.S. Kadowaki[1], Elisabete M. de Gouveia Dal Pino[1] and James M. Stone[2]

[1]Universidade de São Paulo, Instituto de Astronomia, Geofísica e Ciências Atmosféricas
1226 Matão Street São Paulo, 05508-090, Brasil
email: luis.kadowaki@iag.usp.br

[2]Department of Astrophysical Sciences, Peyton Hall, Princeton University
Princeton, NJ 08544, USA

Abstract. Highly magnetized accretion disks are present in high-mass X-ray binaries (HMXBs). A potential mechanism to explain the transition between the High/Soft and Low/Hard states observed in HMXBs can be attributed to fast magnetic reconnection induced in the turbulent corona. In this work, we present results of global general relativistic MHD (GRMHD) simulations of accretion disks around black holes that show that fast reconnection events can naturally arise in the coronal region of these systems in presence of turbulence triggered by MHD instabilities, indicating that such events can be a potential mechanism to explain the transient non-thermal emission in HMXBs. To find the zones of fast reconnection, we have employed an algorithm to identify the presence of current sheets in the turbulent regions and computed statistically the magnetic reconnection rates in these locations obtaining average reconnection rates consistent with the predictions of the theory of turbulence-induced fast reconnection.

Keywords. accretion disks, magnetohydrodynamics (MHD), instabilities, turbulence, magnetic reconnection

1. Introduction

Accretion disks systems are believed to be very common structures in the Universe (for reviews, see, e.g., Pringle 1981; Balbus & Hawley 1998; Abramowicz & Fragile 2013). These systems are associated with Black Hole Binaries (BHBs), Active Galactic Nuclei (AGNs) and Young Stellar Objects (YSOs). In particular, high-energy (HE) and very-high-energy (VHE) emissions are frequently observed in BHBs and AGNs. For instance, the X-ray transitions observed in BHBs (see, e.g., Fender *et al.* 2004; Belloni *et al.* 2005; Remillard & McClintock 2006; Kylafis & Belloni 2015) are characterized by a high/soft state attributed to the thermal emission of a geometrically thin, optically thick accretion disk (Shakura & Sunyaev 1973), and a low/hard state attributed to inverse Compton of soft X-ray photons by relativistic particles in a geometrically thick, optically thin accretion flow (see Esin *et al.* 1997, 1998, 2001; Narayan & McClintock 2008). Besides, a fast transient state (of the order of a few days; see Remillard & McClintock 2006) is identified between these two states. VHE emission (gamma-rays in GeV and TeV band) has also been observed in BHBs, such as Cgy-X1 (Albert *et al.* 2007) and Cgy-X3 (Aleksić *et al.* 2010). In particular, the origin of the latter is uncertain due to the poor resolution of current gamma-ray detectors. Kadowaki *et al.* (2015) and Singh *et al.* (2015) found that turbulent fast magnetic reconnection (Lazarian & Vishiniac 1999) operating at the coronal region of accretion disks can explain the gamma-ray emission as coming from the core region of BHBs. According to this model, reconnection events between the magnetic field lines lifting from the accretion disk corona and those anchored into the horizon of black hole could accelerate relativistic particles in a first-order Fermi process (see, de Gouveia Dal Pino & Lazarian 2005; de Gouveia Dal Pino *et al.* 2010a; Kowal *et al.* 2012; del Valle *et al.* 2016). These particles, interacting with the density, magnetic and radiation fields are able to produce gamma-ray emission. Recently, Khiali *et al.* (2015) considered the power released by turbulent fast magnetic reconnection events to develop an analytical, single zone scenario to produce leptons and hadrons to obtain the SEDs of Cgy-X1 and Cgy-X3. The comparison with the observed SEDs shows that this core model assuming magnetic reconnection as a source of acceleration of the particles explains very well the VHE emission of the BHBs (see also Ramirez-Rodriguez, de Gouveia Dal Pino & Alves-Batista, in these proceedings).

Despite these studies, numerical simulations are still required to probe the viability of turbulent fast magnetic reconnection events in BHBs and AGNs core regions. In this work, we have performed global general relativistic MHD (GRMHD) simulations of accretion disks around black holes and evaluated the presence of turbulent fast reconnection driven by MHD instabilities, such as the magnetorotational instability (MRI; Chandrasekhar 1960; Balbus & Hawley 1991; Hawley *et al.* 1995). To find the zones of fast reconnection, we have employed an algorithm to identify the presence of current sheets in the turbulent regions (see, Zhdankin *et al.* 2013; Kadowaki *et al.* 2018) and computed statistically the magnetic reconnection rates in these locations (see, Kadowaki *et al.* 2018; for an application in shearing-box simulations).

2. Global GRMHD simulations

We have used the ATHENA++ code (White *et al.* 2016) to perform global GRMHD simulations of a torus (thick disk, see, Fishbone & Moncrief 1976) around a rotating black hole in a two-dimensional domain with 512 cells in the radial direction and 512 in the polar direction (in Kerr-Schild coordinates). We have assumed a black hole with mass $M = 1$, spin $a/M = 0.95$, and adiabatic index $\Gamma = 13/9$. The grid was set between 0.98 times the outer horizon radius $M + \sqrt{M^2 - a^2}$ and $r = 20$ in the radial direction, and between $\theta = 0$ and $\theta = \pi$ in the polar direction. We have imposed outflow conditions in the radial boundaries and reflecting conditions in the polar boundaries. An LLF (local LaxFriedrichs) Riemann solver was used.

We have adapted the algorithm used in Zhdankin *et al.* (2013) and Kadowaki *et al.* (2018) to measure the magnetic reconnection rate (see, Kowal *et al.* 2009) in a General Relativistic approach (see also Ball *et al.* 2018). Figure 1 shows the magnetic reconnection rate ($V_{rec} = V_{in}/V_A$)† measured by an observer in the coordinate frame (top diagram) and the profiles of the magnetic field intensity and the current density (bottom diagram). This model corresponds to a torus with an initial weak poloidal magnetic field (represented by closed loops inside the torus) with the ratio of maximum gas pressure to maximum magnetic pressure equal 100 (see, White *et al.* 2016). The bottom diagram of Figure 1 shows the formation of turbulent structures due to the MRI, allowing the accretion process and the development of magnetic reconnection sites (filled circle symbols). The black symbols correspond to the local maxima current density identified by the algorithm, and the white symbols correspond to the confirmed magnetic reconnection sites (see more details in Kadowaki *et al.* 2018). The top diagram of Figure 1 shows the time evolution of the averaged values of V_{rec} evaluated in the confirmed magnetic reconnection sites (white symbols in the bottom diagram). We have obtained averaged values between 0.01 and

† We have evaluated the magnetic reconnection (V_{rec}) rate in a similar way to the method used by Kowal *et al.* (2009), where we averaged the ratio between the inflow velocity (V_{in}) of the opposite magnetic fluxes and the Alfvén speed (V_A) at the reconnection site.

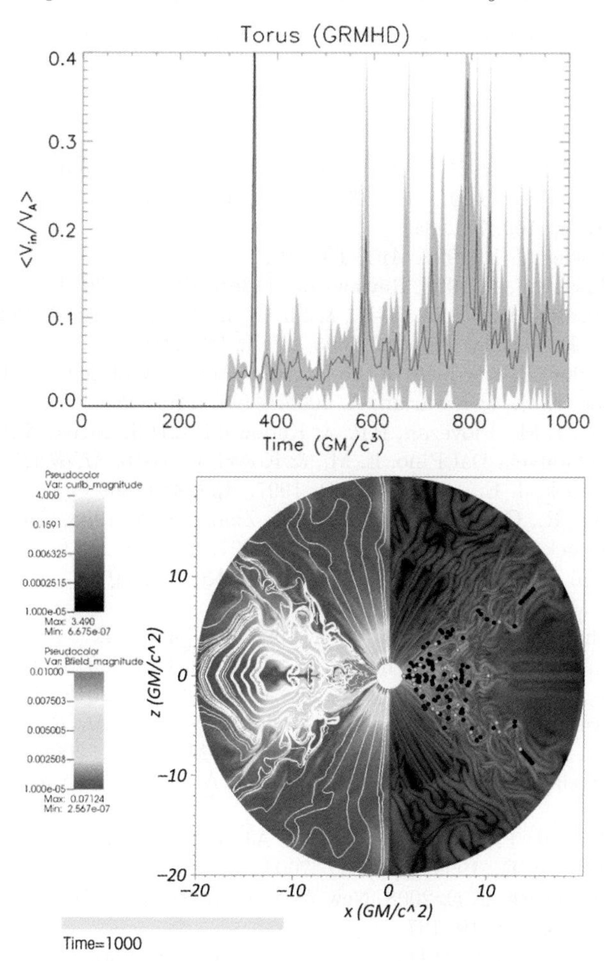

Figure 1. The top diagram shows the time evolution of the magnetic reconnection measured by an observer in the coordinate frame. The bottom diagram shows the system at $t = 1000$ (in units of GM/c^3). As time goes by the MRI sets in, allowing the accretion process and the formation of a turbulent environment. The black symbols correspond to the local maxima identified by the algorithm and the white symbols correspond to the confirmed magnetic reconnection sites.

0.7 consistent with the predictions of the theory of turbulence-induced fast reconnection (Lazarian & Vishiniac 1999).

3. Conclusions

In this work, we have employed an algorithm to identify the presence of current sheets in the turbulent regions of a torus around a black hole and computed statistically the magnetic reconnection rates in these locations. Preliminary results of our GRMHD simulations have revealed the development of turbulence due to the MRI and we have detected the presence of fast reconnection events, obtaining average reconnection velocities in Alfvén speed units of the order of 0.01 and 0.7, as predicted by the theory of turbulence-induced fast reconnection (Lazarian & Vishiniac 1999). This result strengthens the scenario where turbulent fast magnetic reconnection can take place in the core region of BHBs (de Gouveia Dal Pino & Lazarian 2005; de Gouveia Dal Pino *et al.* 2010a; Kadowaki *et al.* 2015; Singh *et al.* 2015; see also Ramirez-Rodriguez et al., in these proceedings), where the magnetic energy released by these events can accelerate relativistic particles by a first-order Fermi process and produce HE and VHE emissions observed in these sources.

References

Abramowicz, M. A., & Fragile, P. C. 2013, Living Reviews in Relativity, 16, 1
Albert, J., Aliu, E., Anderhub, H., *et al.* 2007, *ApJL*, 665, L51
Aleksić, J., *et al.*2010b, *ApJ*, 721, 84
Balbus, S. A., & Hawley, J. F. 1991, *ApJ*, 376, 214
Balbus, S. A., & Hawley, J. F. 1998, Reviews of Modern Physics, 70, 1
Ball, D., Özel, F., Psaltis, D., Chan, C.-K., & Sironi, L. 2018, *ApJ*, 853, 184
Belloni, T., Homan, J., Casella, P., *et al.* 2005, *A&A*, 440, 207
Chandrasekhar, S. 1960, Proceedings of the National Academy of Science, 46, 253
de Gouveia Dal Pino, E.M., & Lazarian, A. 2005, *A&A*, 441, 845
de Gouveia Dal Pino, E.M., Piovezan, P.P., & Kadowaki, L.H.S. 2010a, *A&A*, 518, A5
del Valle, M. V., de Gouveia Dal Pino, E. M., & Kowal, G. 2016, *MNRAS*, 463, 4331
Esin, A. A., McClintock, J. E., & Narayan, R. 1997, *ApJ*, 489, 865
Esin, A. A., Narayan, R., Cui, W., Grove, J. E., & Zhang, S.-N. 1998, *ApJ*, 505, 854
Esin, A. A., McClintock, J. E., Drake, J. J., *et al.* 2001, Apj, 555, 483
Fender, R. P., Belloni, T. M., & Gallo, E. 2004, *MNRAS*, 355, 1105
Fishbone, L. G., & Moncrief, V. 1976, *ApJ*, 207, 962
Hawley, J. F., Gammie, C. F., & Balbus, S. A. 1995, *ApJ*, 440, 742
Kadowaki, L. H. S., de Gouveia Dal Pino, E. M., & Singh, C. B. 2015, *ApJ*, 802, 113
Kadowaki, L. H. S., de Gouveia Dal Pino, E. M., & Stone, J. M. 2018, *ApJ*, 864, 52
Khiali, B., de Gouveia Dal Pino, E. M., & del Valle, M. V. 2015, *MNRAS*, 449, 34
Kowal, G., Lazarian, A., Vishniac, E. T., Otmianowska-Mazur, K., 2009, *ApJ*, 700, 63
Kowal, G., de Gouveia Dal Pino, E.M., & Lazarian, A. 2012, Physical Review Letters, 108, 241102
Kylafis, N. D., & Belloni, T. M. 2015, *A&A*, 574, A133
Lazarian, A., & Vishniac, E., 1999, *ApJ*, 517, 700
Narayan, R., & McClintock, J. E. 2008, New Astronomy Reviews, 51, 733
Pringle, J. E. 1981, *ARA&A*, 19, 137
Ramirez-Rodriguez, J., de Gouveia Dal Pino, E.M ., & Alves-Batista, R. 2018, these Proceedings
Remillard, R. A., & McClintock, J. E. 2006, *ARA&A*, 44, 49
Shakura, N. I., & Sunyaev, R. A. 1973, *A&A*, 24, 337
Singh, C. B., de Gouveia Dal Pino, E. M., & Kadowaki, L. H. S. 2015, *ApJL*, 799, L20
White, C. J., Stone, J. M., & Gammie, C. F. 2016, *ApJS*, 225, 22
Zhdankin, V., Uzdensky, D. A., Perez, J. C., & Boldyrev, S. 2013, *ApJ*, 771, 124.

High-mass X-ray Binaries: Illuminating the Passage from Massive Binaries to Merging Compact Objects
Proceedings IAU Symposium No. 346, 2019
L. M. Oskinova, E. Bozzo, T. Bulik & D. Gies, eds.

doi:10.1017/S1743921319001832

The possible origin of high frequency quasi-periodic oscillations in low mass X-ray binaries

ChangSheng Shi[1,2], ShuangNan Zhang[3,5,6] and XiangDong Li[2,4]

[1]College of Material Science and Chemical Engineering, Hainan University, Hainan 570228, China
email: shics@hainu.edu.cn

[2]Key Laboratory of Modern Astronomy and Astrophysics (Nanjing University), Ministry of Education, Nanjing 210046, China

[3]Key Laboratory of Particle Astrophysics, Institute of High Energy Physics, Chinese Academy of Sciences, Beijing 100049, China

[4]Department of Astronomy, Nanjing University, Nanjing 210046, China;

[5]National Astronomical Observatories, Chinese Academy of Sciences, Beijing 100012, China

[6]Physics Department, University of Alabama in Huntsville, Huntsville, AL 35899, USA

Abstract. We summarize our model that high frequency quasi-periodic oscillations (QPOs) both in the neutron star low mass X-ray binaries (NS-LMXBs) and black hole LMXBs may originate from magnetohydrodynamic (MHD) waves. Based on the MHD model in NS-LMXBs, the explanation of the parallel tracks is presented. The slowly varying effective surface magnetic field of a NS leads to the shift of parallel tracks of QPOs in NS-LMXBs. In the study of kilohertz (kHz) QPOs in NS-LMXBs, we obtain a simple power-law relation between the kHz QPO frequencies and the combined parameter of accretion rate and the effective surface magnetic field. Based on the MHD model in BH-LMXBs, we suggest that two stable modes of the Alfv́en waves in the accretion disks with a toroidal magnetic field may lead to the double high frequency QPOs. This model, in which the effect of the general relativity in BH-LMXBs is considered, naturally accounts for the 3:2 relation for the upper and lower frequencies of the QPOs and the relation between the BH mass and QPO frequency.

Keywords. accretion, accretion disks, MHD, X-rays: binaries.

---

1. Introduction

Quasi-periodic oscillations are a key phenomenon of variability in many X-ray binaries (XBs), such as neutron star low-mass X-ray binaries (NS-LMXBs), black hole low-mass X-ray binaries (BH-LMXBs) (see Strohmayer *et al.* 1996; van der Klis 2006), also in high mass X-ray binaries (James *et al.* 2010). Accretion from the companion star is a general phenomenon in those XBs and there is a certain relation between frequencies of QPOs and accretion rate. However, there does not exist a one-to-one relation between the kHz QPOs and the X-ray intensity in systems such as in 4U 1636–53 and 4U 1608–52. This interesting phenomenon is called "parallel tracks".

As a promising phenomenon to explore the general relativity effect, QPOs were widely studied in many models (e.g. Miller, Lamb & Psaltis 1998; Stella & Vietri 1999; Osherovich & Titarchuk 1999; Abramowicz & Kluźniak 2001, 2003; Erkut, Psaltis & Al-par 2008; Shi & Li 2009, 2010; Shi, Zhang & Li 2014, 2018). Both in NS-LMXBs and BH-LMXBs, some disturbance frequently emerges in an accretion disk because of

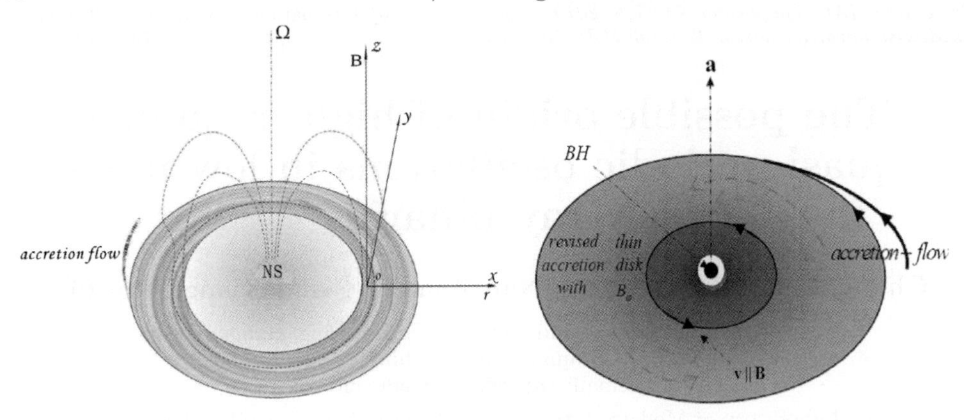

Figure 1. The sketch about an accretion process (Left: an accreting NS with a dipolar magnetic field; Right: an accreting BH with a toroidal magnetic field).

some instability and thus magnetohydrodynamic (MHD) waves are produced easily. We have solved fundamental dispersion equations and compared these frequencies with the frequencies of the QPOs in NS-LMXBs and BH-LMXBs. Then it is found that MHD waves are a promising origin of HFQPOs in LMXBs (Shi & Li 2009, 2010; Shi, Zhang & Li 2014, 2018).

2. NS-LMXBs and BH-LMXBs

As known to us, NSs are very different from BHs in many characteristics. (1) A NS has a rigid surface, which is effective to the release of gravitational energy of the accreted material, but a BH has not. (2) A NS has a dipolar magnetic field but a BH does not due to that any matter (including magnetic fields) can flow only into the event horizon. However, some weak toroidal magnetic field may emerge in the accretion disk that is rotating around the BH. Therefore, periodic pulses can be produced in a NS-XB but can not be found in a BH-XB and it is considered as a criterion to differentiate between a NS-XB and a BH-XB. Therefore the accretion matter can be funneled to the two poles by the magnetic field in NS-LMXBs but it does not happen in BH-LMXBs.

3. KHz QPOs in NS-LMXBs

Shi, Zhang & Li (2014, 2018) considered that MHD waves are generated at the magnetosphere radius and obtained the dispersion equations by solving a group of disturbing MHD equations in the accretion mode of NS-LMXBs. Then the modes of MHD waves from these disturbance are calculated and suggested to be the origin of the kilohertz quasi-periodic oscillations (kHz QPOs) in NS-LMXBs.

When the compressed magnetosphere is considered, the magnetic field is not a dipolar magnetic field (Shi, Zhang & Li, 2018). The nondipolar magnetic field can not be expressed as $B_{\rm NS}R^3$ but as $B_*R^3 = cB_{\rm NS}R^3$, where $B_{\rm NS}$ is the surface magnetic field for the non-deformation magnetosphere, B_* the effective surface magnetic field of a NS, c is a simple form factor, and R is the radius of a NS.

After calculating the frequencies of the MHD waves with different effective magnetic fields, the parallel tracks can be reproduced (Shi, Zhang & Li, 2018). When $\dot{M}/B_*^2$ as an integrated parameter is considered, the one-to-one relation between the frequencies and $\dot{M}/B_*^2$ is shown. Also, the similar relation was shown in Shi, Zhang & Li (2014). Namely the tracks for different B_* from all the data of kHz QPOs converge into a group of curves. Then we can test if the model and the frequencies of the twin kHz QPOs are the best parameters to describe these observational results. The variable instantaneous accretion rate as a important factor leads to the changing of kHz QPO frequency, and the shift of one track in "parallel tracks" originates from the slowly varying effective magnetic field.

4. HFQPOs in BH-LMXBs

Different from the model of kHz QPOs, we considered an inner advection-dominated accretion flow (ADAF) surrounded by an outer thin disk in the very high state of a BH-LMXBs (Shi & Li 2010). After the general relativistic magnetohydrodynamic equations of the perturbed plasma at the boundary between ADAF and the thin accretion disk (r_{tr}) have been solved, two stable modes of the Alfvén waves in the accretion disks with toroidal magnetic fields are obtained. The toroidal magnetic fields are widely believed to be generated by the dynamo mechanism. These two modes produced in the transition region (r_{tr}) are suggested to lead to the double high frequency QPOs. Then the velocities of the twin Alfvén waves in general relativity can be obtained and they can be converted as the same form with the expression of De Villiers & Hawley (2003) in the special relativity.

The ratio of the upper and lower frequencies between the calculated twin modes is very close to 3:2. Therefore the MHD waves with the ratio of the frequencies may be the promising origin of the HFQPO pairs. When the structure of the accretion disk with toroidal magnetic field (Begelman & Pringle 2007) is considered, the relation $\nu \propto M^{-1}$ can be obtained, where ν is the frequency of a HFQPO in BH-LMXBs and M is the mass of the BH. It is consistent with the observation of the QPO frequencies in the three BHs (GRO J1655−40, GRS 1915+105 and XTE J1550−56).

5. Discussion and conclusion

In many models, the kinematic frequency (e.g. the Kepler frequency) of orbital motion of hot spots and clumps is considered as one of the twin HFQPOs. Barret *et al.* (2005) measured the quality factor (Q) for the HFQPOs in 4U 1608−52, and found that Q $\sim$ 200. They believed that such high coherency is not possible to achieve from kinematic effects in orbital motion. However, in the axisymmetric, ideal magnetohydrodynamic simulation of Parthasarathy *et al.* (2017) for an oscillating cusp-filling tori orbiting a non-rotating neutron star, they considered different eigenmodes lead to the difference of the quality factor of the twin kHz QPOs. In our model, two different modes correspond to the twin HFQPOs and they may also lead to the difference of the quality factor as Parthasarathy *et al.* (2017) obtained, which is an important field to be explored next in the future.

As described above, the kHz QPOs in NS-LMXBs and the HFQPOs in BH-LMXBs may both be produced from MHD waves in the certain accretion process. However, the different magnetic fields and accretion processes lead to different results. In a word, the MHD waves (including Alfven waves) are therefore a promising origin.

References

Abramowicz, M. A., Bulik, T., Bursa, M., & Kluźniak, W. 2003, *A&A*, 404, L21
Abramowicz, M. A., & Kluźniak, W. 2001, *A&A*, 374, L19
Barret, D., Kluźniak, W., Olive, J. F., Paltani, S., & Skinner, G. K. 2005, MNRAS, 357, 1288
Begelman, M. C., & Pringle, J. E. 2007, MNRAS, 375, 1070
De Villiers, J. P., & Hawley, J. F. 2003, *ApJ*, 589, 458
Erkut M. H., Psaltis D., Alpar M. A. 2008, *ApJ*, 687, 1220
James, M., Paul, B., Devasia, J., & Indulekha, K. 2010, MNRAS, 407, 285
Li, X.-D., & Zhang, C.-M. 2005, *ApJL*, 635, L57
Miller, M. C., Lamb, F. K., & Psaltis, D. 1998, *ApJ*, 508, 791
Osherovich, V., & Titarchuk, L. 1999, *ApJL*, 522, L113
Parthasarathy V., Kluźniak W., Čemeljić M. 2017, MNRAS, 470, L34

Shi, C.-S. 2011, *Research in Astronomy and Astrophysics*, 11, 1327
Shi, C.-S., & Li, X.-D. 2010, *ApJ*, 714, 1227
Shi, C.-S., Zhang, S.-N., & Li, X.-D. 2014, *ApJ*, 791, 16
Shi, C. 2010, *Science China Physics, Mechanics, and Astronomy*, 53, 247
Shi, C., & Li, X.-D. 2009, *MARAS*, 392, 264
Stella, L., & Vietri, M. 1999, *Physical Review Letters*, 82, 17
Strohmayer, T. E., Zhang, W., Swank, J. H., *et al.* 1996, *ApJL*, 469, L9
van der Klis, M. 2006, in: W. Lewin & M. van der Klis (eds.), *Compact stellar X-ray sources*, Cambridge Astrophysics Series No. 39 (Cambridge, UK: Cambridge University Press), p. 39
Zhang, C. 2004, *A&A*, 423, 401

High-mass X-ray Binaries: Illuminating the Passage from Massive Binaries to Merging Compact Objects
Proceedings IAU Symposium No. 346, 2019
L. M. Oskinova, E. Bozzo, T. Bulik & D. Gies, eds.

doi:10.1017/S1743921319002151

On the nature of the 35-day cycle in the X-ray binary Her X-1/HZ Her

N. Shakura[1,2], D. Kolesnikov[1], K. Postnov[1,2], I. Volkov[1,3], I. Bikmaev[2], T. Irsmambetova[1], R. Staubert[4], J. Wilms[5], E. Irtuganov[2], P. Shurygin[2], P. Golysheva[1], S. Shugarov[1,6], I. Nikolenko[3,7], E. Trunkovsky[1], G. Schonherr[8], A. Schwope[8] and D. Klochkov[4]

[1]Sternberg Astronomical Institute, Moscow State University, 119234, Moscow, Russia
email: nikolai.shakura@gmail.com

[2]Kazan Federal University, Kazan, Russia

[3]Institute of Astronomy RAS, Moscow, Russia

[4]Institute for Astronomy and Astrophysics, Tübingen, Germany

[5]Astronomical Institute of the University of Erlangen-Nuremberg, Bamberg, Germany

[6]Astronomical Institute of the Slovak Academy of Scienses, Tatranska Lomnica, Slovakia

[7]Crimean Astrophysical Observatory, Nauchny, Russia

[8]Leibniz Institute for Astrophysics, Potsdam, Germany

Abstract. The X-ray binary Her X-1 consists of an accreting neutron star and the optical companion HZ Her. The 35-day X-ray variability of this system is known since its discovery in 1972 by the UHURU satellite and is believed to be caused by forced precession of the warped accretion disk tilted to the orbital plane. We argue that the observed features of the optical variability of HZ Her can be explained by free precession of the neutron star with a period close to that of the forced disk precession. The model parameters include a) the intensity (power) of the stream of matter flowing out of the optical star; b) the X-ray luminosity of the neutron star; c) the optical flux of the accretion disk; d) the X-ray irradiation pattern on the donor star; e) the tilt of the inner and outer edge of the accretion disk. A possible synchronization mechanism based on the coupling between the neutron star free precession and the dynamical action of non-stationary gas streams is discussed shortly.

Keywords. X-rays: binaries, stars: neutron, stars: binaries: close

---

1. Introduction

HZ Her / Her X-1 is an intermediate mass X-ray binary consisting of a $1.8 - 2.0\ M_\odot$ evolved sub-giant star and an $1.0 - 1.5\ M_\odot$ neutron star observed as X-ray pulsar (Tananbaum *et al.* 1972). The orbital period is 1.7 days, the X-ray pulsar spin period is 1.24 seconds. The optical star fills its Roche lobe and an accretion disk is formed around the neutron star. Due to X-ray irradiation, the optical flux from HZ Her is strongly modulated with the orbital period, as was first found by the inspection of archive photoplates (Cherepashchuk *et al.* 1972). Note that before X-ray observations, HZ Her was classified as an irregular variable.

The X-ray light curve of Her X-1 is modulated with a 35 day period. Most of the 35-day cycles last 20.0, 20.5 or 21.0 orbital periods (see, e.g. Shakura *et al.* (1998), Klochkov *et al.* (2006)). The 35-day X-ray cycle consists of a 7-day "main-on" state and a 5-day

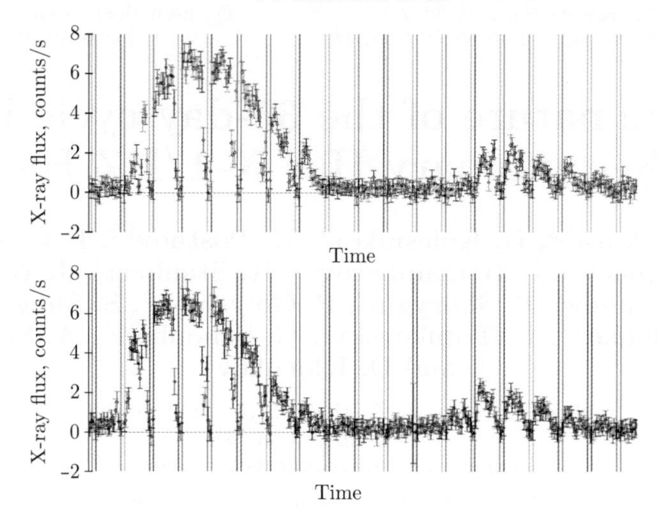

Figure 1. RXTE/ASM light curves of the 35-day X-ray cycle (Shakura *et al.* (1998), Klochkov *et al.* (2006)). Vertical lines show eclipses of the X-ray source by the donor star. Top: the "turn on" near the orbital phase 0.7. Bottom: the "turn on" near the orbital phase 0.2.

"short-on" state of lower intensity, separated by two 4-day "off" states during which X-ray radiation switches off completely (Fig. 1). The X-ray observations of the flux are well explained by the precession of the accretion disk.

2. 35-day cycle

The 35-day cycle turn-ons most frequently occur at the orbital phases ~ 0.2 or ~ 0.7, which is due to the tidal nutation of the outer parts of the disk with the double orbital frequency when the angle between the line of sight and the outer parts of the disk changes most rapidly (Katz (1973), Levine & Jernigan (1982), Boynton (1987)). The 35-day cycle of Her X-1 is explained by the accretion disk precession in the direction opposite to the orbital motion (Gerend & Boynton (1976), Shakura *et al.* (1999)). Soon after the discovery of the X-ray pulsar, the NS free precession was suggested to explain the observed 35-day modulation (Brecher 1972). Later on, the EXOSAT observations of the evolution of X-ray pulse profiles of Her X-1 were also interpreted by the NS free precession (Truemper *et al.* 1986). Extensive studies of Her X-1 suggested a warped and tilted accretion disk around the NS. Its retrograde precession results in consecutive opening and eclipses of the central X-ray source (Boynton 1987).The X-ray light curve is asymmetrical between the eclipses due to the scattering of the X-ray radiation in a hot rarefied corona above the disk. Indeed, the X-ray "turn-on" at the beginning of the "main-on" state is accompanied by a significant decrease in the soft X-ray flux because of strong absorption. There is no essential spectral change during the X-ray flux decreases, suggesting the photon scattering on free electrons of the hot corona near the disk inner edge (Becker *et al.* (1977), Davison & Fabian (1977), Parmar *et al.* (1980), Kuster *et al.* (2005)). The X-ray pulse profiles are observed to vary with the 35-day phase (Truemper *et al.* (1986), Deeter *et al.* (1998), Scott *et al.* (2000), Staubert *et al.* (2013)) differing significantly at the main turn-on and at the short-on. Such changes of the pulses are difficult to explain using the precessing disk only.

The X-ray RXTE/PCA pulse evolution with 35-day phase can be explained (Postnov *et al.* 2013) by the NS free precession with a complex magnetic field structure on the NS surface. In this model, in addition to the canonical poles (a dipole magnetic field), arc-like

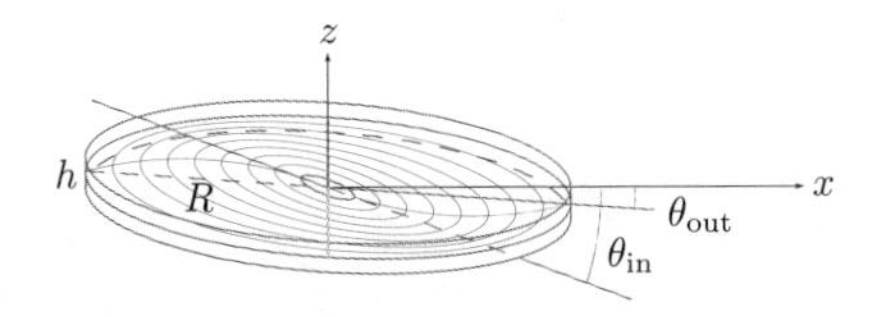

Figure 2. Model of the disk. Inner edge colored in red, outer edge colored in blue. Width h of the outer edge and radius R also showed. Twist angle on the picture equals 0 (nodal lines are coloured in green). Tilt angle of the inner edge is $\theta_{\rm in}$, tilt angle of the outer edge is $\theta_{\rm out}$. y-axis directed along the nodal lines off the reader.

magnetic regions around the magnetic poles are included, which is a consequence of a likely non-dipole magnetic field (Shakura *et al.* (1991), Panchenko & Postnov (1994)).

3. Modeling of the optical light curves of HZ Her

Here we perform a modellng of long-term B-light curves. The photometrical light curve was constructed using the following observations: 1972 – 1998 data compiled from Petro & Hiltner (1973), Davidson *et al.* (1972), Davidson *et al.* (1972), Boynton *et al.* (1973), Lyutyj (1973), Grandi *et al.* (1974), Lyutyj (1973), Cherepashhuk *et al.* (1974), Voloshina *et al.* (1990), Lyutyj & Voloshina (1989), Kilyachkov & Shevchenko (1978), Kilyachkov & Shevchenko (1980), Kilyachkov & Shevchenko (1988), Kilyachkov (1994), Kippenhahn *et al.* (1980), Gladyshev (1985), Mironov *et al.* (1986), and Goransky & Karitskaya (1986) ($\approx$ 5800 points); 2010 – 2018 data were obtained by the present authors ($\approx$ 7600 points).

The model includes two basic components:

a) an inclined, warped, forced precessing accretion disk;

b) a freely precessing neutron star.

The shape of the model optical light curve strongly depends on the X-ray shadow on the optical star produced by the warped accretion disk and on the X-ray irradiation pattern. The shadow is calculated as follows. The disk is splitted along the radius in a finite number of rings and the solid angle between each i-th and $i+1$-th ring is calculated, giving the i-th element of the shadow. As the disk is warped, the i-th and $i+1$-th rings lie in different planes. The full shadow is produced by all elements.

Geometrical parameters of the disk are given by the tilt to the orbital plane and the phase disk angle which are different for the outer and inner disk edges (See Fig. 2). The disk phase is counted opposite to the orbital motion. It is set to $0.00 - 0.05$ at the moment of the X-ray "turn on". The tilt and phase angles of the i-th ring change linearly from the outer edge to the inner edge. The difference between the inner and outer edge is called the twist angle. The twist angle and the difference between the tilt angle of the outer and inner disk edge determine the shadow size. If the twist angle is zero and the tilt of the outer and inner edge is the same, the disk shadow is determined only by the width of the outer disk.

To calculate the X-ray irradiation of the optical star we have used the model by (Postnov *et al.* 2013). This model has been modified to limit the precessional motion of the North magnetic pole to $\beta_{cr} = \arccos(\sqrt{3}/3) \approx 54^\circ\, 44'$. If the magnetic dipole axis is inclined by β_{cr} to the NS rotation axis, the magnetic torque on the inner edge of the disk vanishes (Lipunov & Shakura (1976), Lipunov *et al.* (1981), Lipunov (1987)).

To the north and to the south of this angle, the magnetic torque is non-zero; the sign of the twist angle is expected to change when crossing this critical angle. However, we found that the model with the twist angle changing sign gives a less good fit to the observation than the model with a constant sign. Therefore, we set the angle between the NS precession and axes to 80°, and the angle between the magnetic dipole and precession

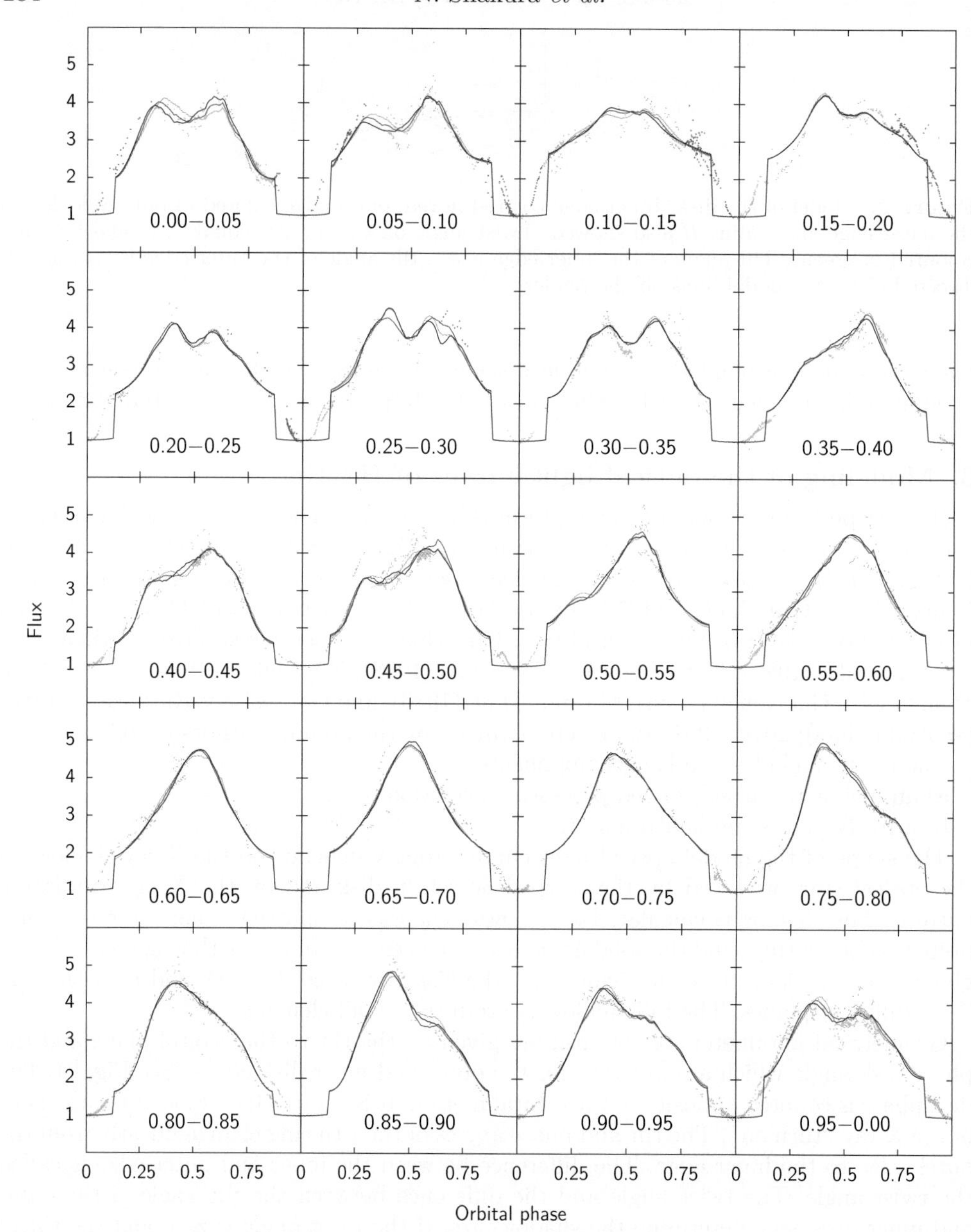

Figure 3. Synthetic light curves and observed data (points). The 35-day period is divided in 20 phase bins each. Phase 0.00 – 0.05 correspond to the X-ray "turn on". The data are colored: in green — the most robust points; in grey — less robust points; in red — points correspond to the gas stream acting on the disk (bright spot). At each phase three synthetic light curves are shown: in orange, blue and brown corresponding to a twist angle of −50, −60 and −70, respectively.

axes to 20°. In this case the angle between the dipole and NS spin axes varies within the range 60° – 100° and the magnetic torque does not change sign during the free precession period.

At the 35-day phase 0.25 – 0.30 from the "turn on", the magnetic dipole reaches a maximal angle of $80 + 20 = 100°$ to the rotation axis (this is the phase 0 of the NS free

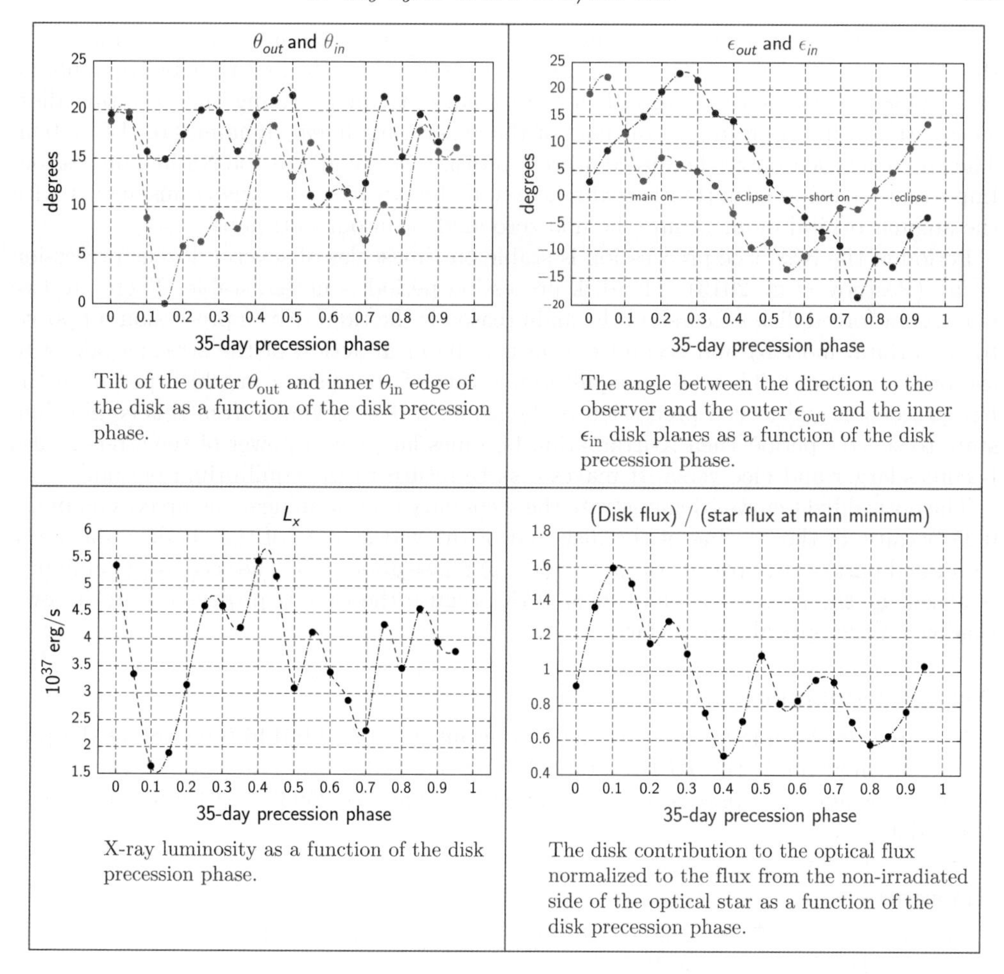

Tilt of the outer θ_{out} and inner θ_{in} edge of the disk as a function of the disk precession phase.

The angle between the direction to the observer and the outer ϵ_{out} and the inner ϵ_{in} disk planes as a function of the disk precession phase.

X-ray luminosity as a function of the disk precession phase.

The disk contribution to the optical flux normalized to the flux from the non-irradiated side of the optical star as a function of the disk precession phase.

Figure 4. Figure 4: Best-fit parameters of the model

precession). At the phase $0.75 - 0.80$, this angle is $80 - 20 = 60^\circ$ (this is phase 0.5 of the NS free precession). This difference $(0.25 - 0 = 0.75 - 0.5 = 0.25)$ between NS free precession phase and phase counted from X-ray "turn on" is the best-fit value.

Orientation of the NS in the picture plane does not affect the shape of the X-ray pulses, but has a strong effect on the optical light curve. This makes it is possible to determine the orientation of the NS spin axis with respect to orbital plane. The angle between the NS spin axis and the projection of the normal to the orbital plane on the sky is called κ.

The best-fit over all precession phases is obtained for $\kappa = 5^\circ$. The tilt and phase of the outer and inner disk, the X-ray luminosity and contribution to the optical flux from the disk have been optimized at each precessional phase (see Fig. 3). The best-fit twist angle of the disk is -60°. The minus sign means that during the precession motion, the outer disk lags behind the inner disk. Figure 4 shows the best-fit parameters of the model. Figure 3 shows the synthetic light curves with three different twist angles: -50, -60, -70, and observed B light curves of HZ Her.

The modeling was not performed for the orbital phase intervals $0.0 - 0.13$ and $0.87 - 1.0$. At these phases, the disk is eclipsed by the optical star. The brightness distribution over the disk is complex, and we leave its study for future work.

Several peak-like features at the first five precessional phase intervals shown in Fig. 3 should be noted (red dots). These features were observed for the first time by Kippenhahn *et al.* (1980). This is the result of non-stationary streams striking the accretion disk. These streams form an important part of the general nonlinear dynamics of the system. The streams synchronize the forced accretion disk precession and the NS free precession. Importantly, such a situation is realized only if the NS rotation axis is misaligned with the angular orbital momentum (the non-zero angle κ in our notation).

Period of the NS's free preceession is stable on timescales of several dozens precession cycles (Postnov *et al.* 2013). The disk precession period is not so stable. There are two reasons for forced disk precession: the main reason is the tidal-driven precession (opposite to the orbital motion) and second reason is dynamical action of the streams (along to the orbital motion). There is an equilibrium state of the system in which period of the free precession and forced precession is the same. If disk deviates from this equilibrium state to shorter period then X-irradiation becomes larger and power of the streams also becomes larger and vice versa. It makes disk to return to the equilibrium period.

The optical light curves demonstrate the secondary minimum near the precession phase 0.25 because of the passing of the disk and of the widest part of the shadow above the irradiated part of the optical star at the orbital phase about 0.5. The secondary minimum is absent at the precession phase near 0.75 because the disk is projected onto its own shadow on the optical star surface.

Acknowledgements

Research has been supported by the RSF grant No. 14-02-00146 (creation of the program and modeling), the RFBR grant No. 18-502-12025 (carrying out of the optical observations and data processing) and the DFG grant No. 259364563 (processing of the X-ray data).

References

Becker, R. H., Boldt, E. A., Holt, S. S., Pravdo, S. H., Rothschild, R. E., Serlemitsos, P. J., Smith, B. W. & Swank, J. H. 1977, *ApJ*, 214, 879

Boynton P. E. 1987, in: Giacconi R. & Ruffini R. (eds.), *Physics and Astrophysics of neutron stars and Black Holes* (Amsterdam: North-Holland publ.), p. 227

Boynton P. E. *et al.* 1973, *preprint of Seattle University Group*

Brecher K. 1972, *Nature*, 239, 325

Cherepashchuk A. M., Efremov Yu. N., Kurochkin N. E., Shakura N. I. & Sunyaev R. A. 1972, *Commision of the IAU Information bulletin on variable stars*, N720

Cherepashhuk A. M., Kovalenko V. M., Kovalenko O. N. & Mironov A. V. 1974, *Peremennye zvezdy* (in Russian), 19, 305

Davidsen, A, Henry, J. P, Middleditch, J & Smith, H. E. 1972, *ApJ*, 177, L97

Davison, P. J. N. & Fabian, A. C. 1977, *MNRAS*, 178, 1P

Deeter John E., Scott D. Matthew, Boynton Paul E., Miyamoto Sigenori, Kitamoto Shunji, Takahama Shin'ichiro & Nagase Fumiaki 2005, *ApJ*, 502, 802

Gerend D. & Boynton P. E. 1976, *ApJ*, 209, 562

Gladyshev S. A. 1985, *PhD thesis* (in Russian)

Goransky V. P. & Karitskaya E. A. 1986, *unpublished* (in Russian)

Grandi S. A., Hintzen P. M. N. O., Jensen E. B., Rydgren A. E., Scott J. S., Stickney P. M., Whelan J. A. J. & Worden S. P. 1974, *ApJ*, 190, 365

Katz J. I. 1973, *Nature Physical Science*, 246, 87

Kilyachkov N. N. & Shevchenko V. S. 1978, *Pis'ma v Astronomicheskii Zhurnal* (in Russian), 4, 356

Kilyachkov N. N. & Shevchenko V. S. 1980, *Pis'ma v Astronomicheskii Zhurnal* (in Russian), 6, 717

Kilyachkov N. N. & Shevchenko V. S. 1988, *Pis'ma v Astronomicheskii Zhurnal* (in Russian), 14, 438
Kilyachkov N. N. 1994, *Pis'ma v Astronomicheskii Zhurnal* (in Russian), 20, 664
Kippenhahn R., Schmidt H. U. & Thomas H. C. 1980, *A&A*, 90, 54
Klochkov D. K., Shakura N. I., Postnov K. A., Staubert R., Wilms J. & Ketsaris N. A. 2006, *Astronomy Letters*, 32, 804
Kuster M., Wilms J., Staubert R., Heindl W. A., Rothschild R. E., Shakura N. I. & Postnov K. A. 2005, *A&A*, 443, 753
Levine A. M. & Jernigan J. G. 1982, *ApJ*, 262, 294
Lipunov V. M. & Shakura N. I. 1976, *Pis'ma v AZH* (in Russian), 2, 343
Lipunov V. M., Semyonov E. S. & Shakura N. I. 1981, *Astronomicheskij Zhurnal* (in Russian), 58, 765
Lipunov V. M. 1987, *Astrofizika nejtronnykh zvezd* (Moscow: Nauka publ.)
Lyutyj V. M. 1973, *Peremennye zvezdy* (in Russian), 18, 41
Lyutyj V. M. 1973, *Astronomicheskij Zhurnal* (in Russian), 50, 3
Lyutyj V.M. & Voloshina I.B. 1989, *Pis'ma v Astronomicheskii Zhurnal* (in Russian), 15, 806
Mironov A. V., Moshkalev V. G., Trunkovskij E. M. & Cherepashhuk A. M. 1986, *Astronomicheskii Zhurnal* (in Russian), 63, 113
Panchenko I. E. & Postnov K. A. 1994, *A&A*, 286, 497
Parmar A. N., Sanford P. W. & Fabian A. C. 1980, *MNRAS*, 192, 311
Petro L. & Hiltner W. A. 1973, *ApJ*, 181, L93
Postnov K., Shakura N., Staubert R., Kochetkova A., Klochkov D. & Wilms J. 2013, *MNRAS*, 435, 1147
Scott D. Matthew, Leahy Denis A. & Wilson Robert B. 2000, *ApJ*, 539, 392
Shakura N. I., Ketsaris N. A., Prokhorov M. E. & Postnov K. A. 1998, *MNRAS*, 300, 992
Shakura N. I., Prokhorov M. E., Postnov K. A. & Ketsaris N. A. 1999, *A&A*, 348, 917
Shakura N. I., Postnov K. A. & Prokhorov M. E. 1991, *Soviet Astronomy Letters*, 17, 339
Staubert R., Klochkov D., Vasco D., Postnov K., Shakura N., Wilms J. & Rothschild R. E. 2013, *A&A*, 550, 9
Tananbaum H., Gursky H., Kellogg E. M., Levinson R., Schreier E. & Giacconi R. 1972, *ApJ*, 48, 143
Truemper J., Kahabka P., Oegelman H., Pietsch W. & Voges W. 1986, *ApJ*, 300, L63
Voloshina I. B, Luytyi V. M & Sheffer E. K 1990, *Pis'ma v Astronomicheskii Zhurnal* (in Russian), 16, 625

High-mass X-ray Binaries: Illuminating the Passage from Massive Binaries to Merging Compact Objects
Proceedings IAU Symposium No. 346, 2019
L. M. Oskinova, E. Bozzo, T. Bulik & D. Gies, eds.

doi:10.1017/S1743921319002485

V1187 Herculis: A Red Novae progenitor, and the most extreme mass ratio binary known

Ronald G. Samec[1], Heather Chamberlain[1], Daniel Caton[2], Russell Robb[3] and Danny R. Faulkner[4]

[1]Pisgah Astronomical Research Institute
email: ronald.samec@gmail.com

[2]Dark Sky Observatory, Appalachian State University

[3]University of Victoria

[4]Johnson Observatory

Abstract. Complete BVR_CI_C light curves of V1187 Her were obtained in May 2017 at the Dark Sky Observatory in North Carolina with the 0.81-m reflector of Appalachian State University. Earlier, spectra were taken at the Dominion Astrophysical Observatory with the 1.8-m telescope. The spectral type was found to be F8 ± 1 V (6250 K), so the binary is of solar-type. V1187 Her was previously identified as a low amplitude ($V < 0.2$ mag), short period, overcontact eclipsing binary (EW) with a period of 0.310726 d. Strikingly, despite its low amplitude, the early light curves show total eclipses (eclipse duration ≈ 31.5 minutes), which is a characteristic of an extreme mass ratio binary. A period study covering 11 years reveals a continuous period decrease $dP/dt = -4.7 \times 10^{-9}$ d yr^{-1}. The multi-band Wilson-Devinney light curve solution gives a fill-out of 79% and a mass ratio of only 0.0440 ± 0.0001. There is a cool spot region on the secondary component, which is 400 K hotter than the primary. The inclination is only 66.85 ± 0.05 despite the system's total eclipses.

Keywords. Stars:binaries: eclipsing, stars: individual: V1187 Her

---

1. Introduction

Many solar type binaries have been found to undergo continuously decreasing orbital periods, presumably due to magnetic braking. A binary continually undergoing such a process will slowly coalesce over time as it loses angular momentum. This is due to ion winds streaming radially outward on stiff magnetic field lines rotating with the binary. Recently, binaries with decaying periods have been found to undergo a catastrophic merger (a Red Novae, hereafter RN). This leads to the formation of a single, fast rotating, spectroscopically, earlier-type star. As a part of the evolution of the RN progenitor, the binary's mass ratio becomes more extreme and the Roche-lobe fill-out increases. It is believed that there is a limiting mass ratio (Li *et al.* 2016) leading to an instability and the occurrence of the RN. V1187 Her is apparently such a binary, and a progenitor of an RN.

V1187 Her was discovered By ROTSE-1 (see Fig. 1; Akerlof *et al.* 2000) and is listed as an EW type, with a ROTSE-1 mag of 11.740 ± 0.008, a period of 0.31076 ± 0.00005 d, and an amplitude 0.205 mag. The binary was described in Blaettler & Diethelm (2007), along with a finding chart (Fig. 2) and an ephemeris:

$$\mathrm{HJD(MinI)} = 2453877.4694 + 0.310726 \times E. \tag{1.1}$$

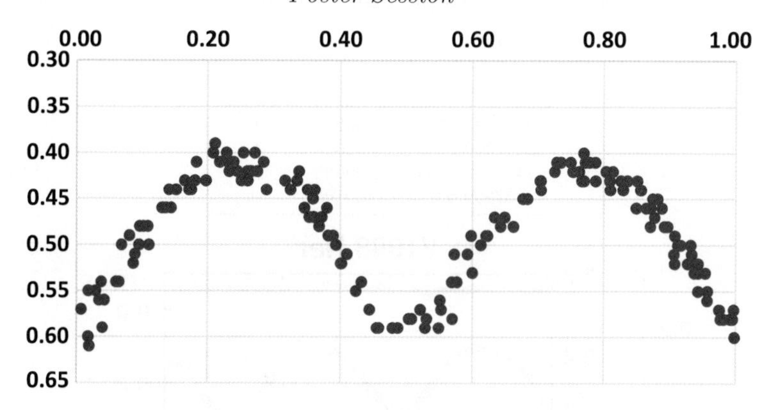

Figure 1. ROTSE-1 light curve of V1187 Her.

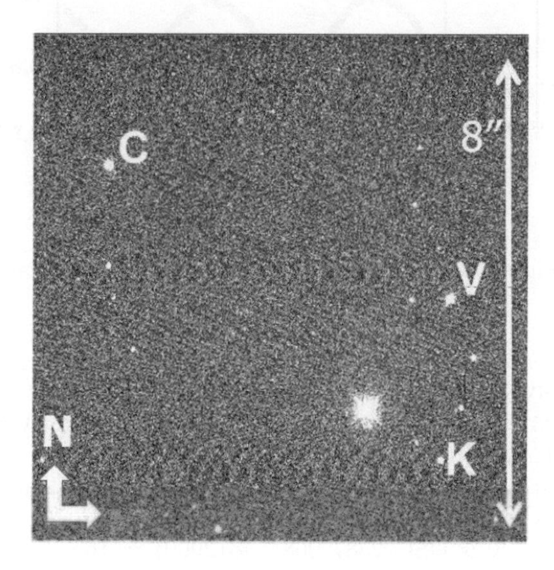

Figure 2. V1187 Her finding chart.

Eight times of minimum light from ROTSE-I observations were given in Diethelm (2007). The system also appeared in the 80th Name-List of Variable Stars (Kazarovets *et al.* 2013). Two timings of minimum light are given in Diethelm (2010a).

V1187 Her was observed as part of a student/professional collaborative study of interacting binaries program with data obtained through SARA South observations. The $BVRI$ light curves were taken with the Dark Sky Observatory (DSO) 0.81-m reflector at Philips Gap, North Carolina, on 20 and 27 May, 2017. The telescope has a thermoelectrically cooled ($-40\,^{\circ}$C) 2K × 2K Apogee Alta CCD chip. The observers were D. Caton, R. Samec, D. Faulkner, B. Hill, and D. Gentry.

On May 20, 2017, we obtained 93 observations in B, 147 in V, 156 in R_C, and 158 in I_C. The nightly C-K values stayed constant throughout the observing run with a precision of better than 1%. Exposure times were 150 s in B, 30 s in V and 20 s in R_C and I_C. The R_C and I_C light curves of May 20, 2017 are shown in Fig. 3. The May 27 observations included185 in B, 187 in V, 162 in R_C and 187 in I_C. The nightly C-K values stayed constant throughout the observing run with a precision of 1%. Exposure times varied from 250 to 275 s in B, 80 to 90 s in V and 30 to 50 s in R_C and I_C.

Table 1. New eclipse timings

Timing (HJD)	Type
2457893.94583 ± 0.0007	I
2457900.7845 ± 0.00106	I
2457893.79034 ± 0.00035	II
2457900.62773 ± 0.00105	II

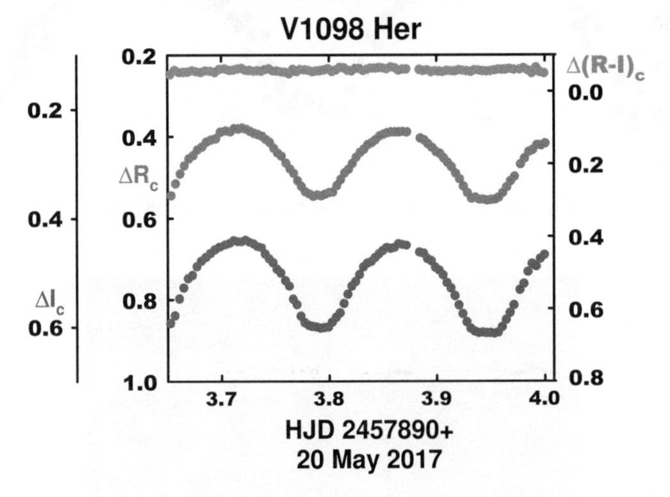

Figure 3. The May 20, 2017 R_C and I_C light curves.

2. Targets

The variable, indicated by the letter V in Fig. 2, is V1187 Her. Alternative names are GSC 2587 1888, NSVS 7913634, ROTSE1 J162919.83+353959.2 [$\alpha(2000)$ = 16h 29m 19.890s, $\delta(2000) = 35^\circ\ 40'\ 2.90''$ ICRS], $J - K = 0.341$, F8V (2MASS), TYC 2587 1888, V=15.10, $B - V = 0.489 \pm 0.134$.

The comparison star C is GSC 2587 0918, [$\alpha(2000)$ = 16h 28m 48.2441s, $\delta(2000) = 35^\circ\ 42'\ 29.300''$], 3UC252-115600, $B - V = 0.489$; $J - K = 0.44$ (2MASS), G7V, $V = 11.16$. The check star K is GSC 2587 0684, [$\alpha(2000)$ = 16h 29m 19.0219s, $\delta(2000) = 35^\circ\ 37'\ 5.340''$], 3UC252-115633, $V = 13.84$, $J - K = 0.36$, G0V, (2MASS).

3. Period Study

Four times of minimum light were calculated from our present observations, two primary and two secondary eclipse timings. They are listed in Table 1. The following linear and quadratic ephemerides were determined from all available times of minimum light:

$$\mathrm{HJD(MinI)} = 2457893.9484 \pm 0.0019 + 0.31076465 \pm 0.00000019 \times E, \quad (3.1)$$

and

$$\mathrm{HJD(MinI)} = 2457893.94565 \pm 0.0016 + 0.31076278 \pm 0.00000050 \times E + 1.42 \pm 0.36 \times E^2. \quad (3.2)$$

The O-C study covers some 11 years and about 13,000 orbits (Fig. 4). The period is decreasing as one might expect for magnetic breaking. According to the light curve solution the more massive component has a mass of 22.7 times that of the less massive one. Equation 3.2 yields a dP/dt of 2.3×10^{-7} d yr^{-1}, which corresponds to a mass flow rate $dM_2/dt = -8.9 \times 10^{-9}\ M_\odot$ d^{-1}, assuming a main sequence mass value for the primary component. It is thought that the more massive component steadily absorbs the secondary during normal evolution, so this period change follows the expected course.

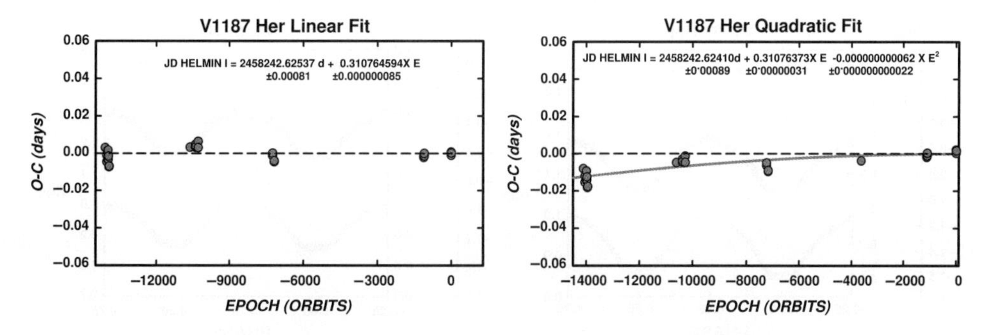

Figure 4. *L*eft: $O-C$ residuals for the linear ephemeris (Eqn. 3.1). *R*ight: $O-C$ residuals for the quadratic ephemeris (Eqn. 3.2).

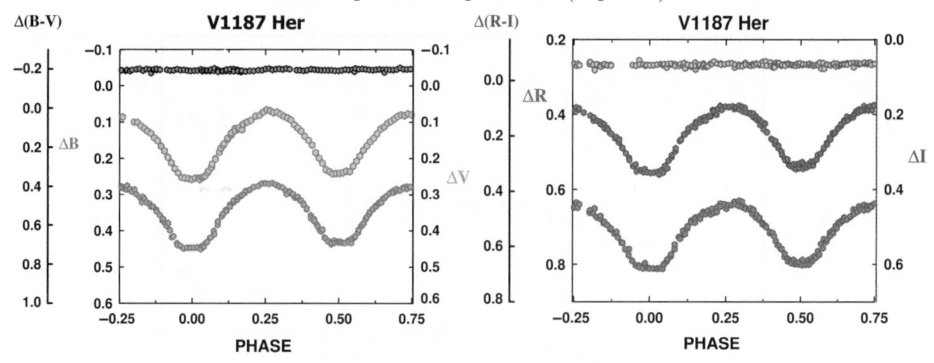

Figure 5. V1187 Her observed light curves.

The minimum mass ratio of W UMa binaries is thought to be $q=0.071$ to 0.078. Since the mass ratio of V1187 Her is 0.044, far smaller than this value, the binary is very close to coalescing.

4. Light Curve Solution

The amplitudes of the light curves are only ≈ 0.17-0.19 mag in all bands, which is very small considering the eclipses are total (Fig. 5). The O'Connell effect (difference in the maxima), classically thought of as an indicator of spot activity, is only 0.003 to 0.11 mag in all bands, and a minor spot is needed to solve the light curves. The minima show a difference of ≈ 0.015 mag in all curves, indicating that the temperature difference between the components does exist despite the tiny amplitudes. The B ,V, R_C and I_C curves were carefully pre-modeled with Binary Maker 3.0 (Bradstreet & Steelman 2002) fits in all bands. The parameters were then averaged and input into a 4-color simultaneous light curve solution using the Wilson-Devinney program (Wilson & Devinney 1971; Wilson 1990). The solution (Fig. 6) was computed in Mode 3, the over-contact mode. Convective parameters $g=0.32$, $A=0.5$ were used. Since the eclipses were total, no q-search was performed. Due to the shallow curves one important element is the possibility of third light. So third light was a part of the elements iterated throughout the process. It remained viable throughout the calculation. A second, non-third light solution was calculated showing a smaller residual.

5. Discussion

V1187 Her is found to be a short period, extreme mass ratio W UMa eclipsing binary. For the third light solution, it was found to make up only $\approx 0.36\%$ relative light in all filters. The primary component emits $\approx 92\%$ of the total system light in the final solution.

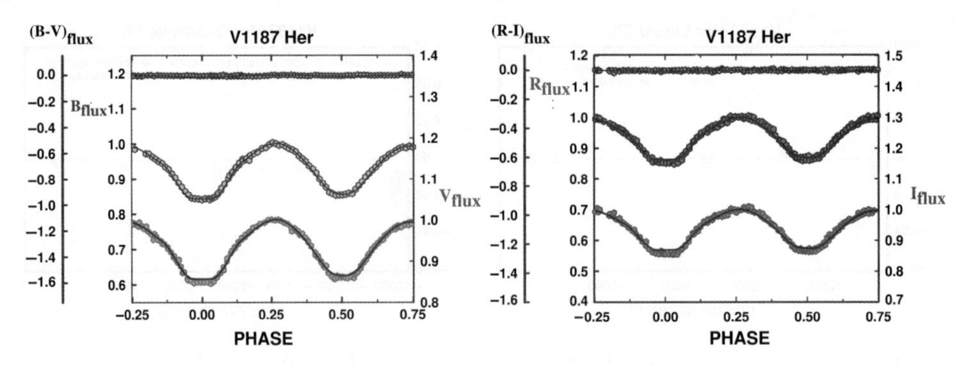

Figure 6. V1187 Her observed and computed light curves.

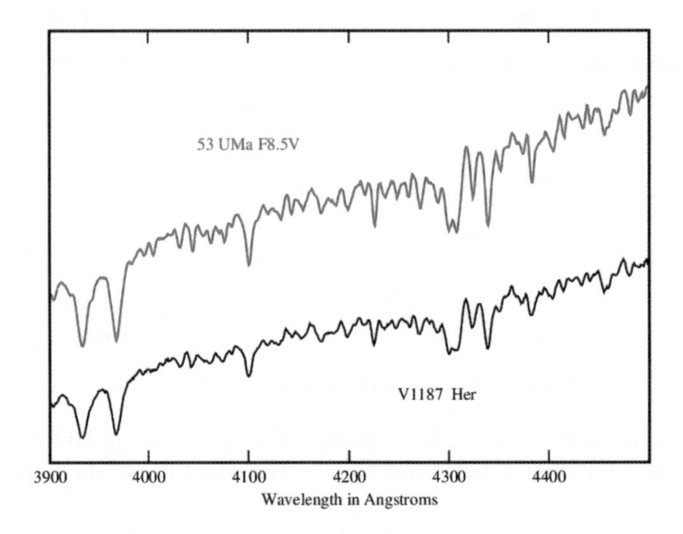

Figure 7. V1187 Her DAO spectra.

So the 0.044 mass ratio determination and the inclination (66.85 ± 0.05) is the main cause of the very shallow amplitude (< 0.2 mag). Strengthening this determination is the total eclipse which would occur at a minimum inclination of $\approx 63°$. Recently, a paper featuring a R and V light curve analysis and Hα line study of ASAS J083241+2332.4 found its mass ratio to be ≈ 0.06 (Sriram *et al.* 2016). Such extremely low mass ratio binary systems are rare and only three systems have been reported so far with mass ratios under ≈ 0.075: V53 ($q = 0.060$; Kaluzny *et al.* 2013), V857 Her ($q = 0.065$; Qian *et al.* 2005), and SX Crv ($q = 0.072$; Zola *et al.* 2004). V1187 Her is probably the most extreme of this rare class.

6. Conclusion

V1187 Her has the most extreme mass ratio of the solar type binaries. The 11 year orbital study (12,000 orbits) reveals an negative quadratic ephemeris. This decrease may be due to magnetic braking as a plasma wind is discharged along stiff rotating magnetic field lines. DAO spectra yields a F8.5V spectral type (Fig. 7). The Wilson-Devinney program solution gives a mass ratio of 0.043983 ± 0.00008 and the Roche lobe fill-out is also extreme, about 79% for this over-contact binary. Its component temperature difference is ≈ 38 K. The extreme mass ratio condition (despite its 68° inclination) allows an eclipse duration of ≈ 31.5 minutes. Despite its deep contact state, this W UMa binary is of W-type (the less massive component is hotter). Its magnetic character is attested by the cool spot modeled on its primary component. Radial velocity curves would allow determination of the binary's absolute elements.

References

Akerlof, C., Amrose, S., Balsano, R., *et al.* 2000, *AJ*, 119, 1901
Blaettler, E., & Diethelm, R. 2007, *IBVS*, 5799
Bradsteet, D. H., & Steelman, D. P. 2002, *BAAS*, 34, 1224
Diethelm, R. 2007, *IBVS*, 5781
Diethelm, R. 2010a, *IBVS*, 5920
Diethelm, R. 2010b, *IBVS*, 5922
Kazarovets, E. V., Samus, N. N., Durlevich, O. V., *et al.* 2013, *IBVS*, 6052
Kaluzny, J., Rozyczka, M., Pych, W., *et al.* 2013, Acta Astronomica, 63, 309
Li, H.-L., Wei, J.-Y., Yang, Y.-G., & Dai, H.-F. 2016, *RAA*, 16, 2
Nelson, R. H. 2010, *IBVS*, 5929
Qian, S.-B., Zhu, L.-Y., Soonthornthum, B., *et al.* 2005, *AJ*, 130, 1206
Sriram, K., Malu, S., Choi, C. S., & Vivekananda Rao, P. 2016, *AJ*, 151, 69
Wilson, R. E.1990, *ApJ*, 356, 613
Wilson, R. E., & Devinney, E. J.1971, *ApJ*, 166, 605
Zola, S., Rucinski, S. M., Baran, A., *et al.* 2004, *Acta Astronomica*, 54, 299

Population in Galaxies and X-ray Luminosity Function

High-mass X-ray Binaries: Illuminating the Passage from Massive Binaries to Merging Compact Objects
Proceedings IAU Symposium No. 346, 2019
L. M. Oskinova, E. Bozzo, T. Bulik & D. Gies, eds.

doi:10.1017/S1743921319001157

The Cartwheel galaxy as a stepping stone for binaries formation

Anna Wolter[1], Guido Consolandi[1,2] Marcella Longhetti[1], Marco Landoni[1] and Andrea Bianco[1]

[1]INAF-Osservatorio Astronomico di Brera, Via Brera 28, I-20121 Milano, Italy
email: anna.wolter@inaf.it

[2]Universitá degli Studi di Milano Bicocca, Piazza dell'Ateneo Nuovo, 1, I-20126 Milano, Italy
email: guido.consolandi@inaf.it

Abstract. Ultraluminous X-ray sources (ULXs) are end points of stellar evolution. They are mostly interpreted as binary systems with a massive donor. They are also the most probable progenitors for BH-BH, and even more, for BH-NS coalescence. Parameters of ULXs are not know and need to be better determined, in particular the link with the metallicity of the environment which has been invoked frequently but not proven strongly. We have tackled this problem by using a MUSE DEEP mosaic of the Cartwheel galaxy and applying a Monte Carlo code that jointly fits spectroscopy and photometry. We measure the metallicity of the emitting gas in the ring and at the positions of X-ray sources by constructing spatially resolved emission line ratio maps and BPT diagnostic maps. The Carthweel is the archetypal ring galaxy and the location and formation time of new stellar populations is easier to reconstruct than in more normal galaxies. It has the largest population of ULXs ever observed in a single galaxy (16 sources have been classified as ULXs in Chandra and XMM-Newton data). The Cartwheel galaxy is therefore the ideal laboratory to study the relation between Star Formation (SF Rates and SF History) and number of ULXs and also their final fate. We find that the age of the stellar population in the outer ring is consistent with being produced in the impact ($\leqslant 300 Myr$) and that the metallicity is mostly sub-solar, even if solutions can be found with a solar metallicity that account for most observed properties. The findings for the Cartwheel will be a testbed for further modelisation of binary formation and evolution paths.

Keywords. stars: binaries, galaxies: individual (Cartwheel), galaxies: peculiar, galaxies: interaction, galaxies: evolution

---

1. Introduction

Why did we choose the Cartwheel galaxy? The Cartwheel is the epitome of the Ring Galaxies (RiGs), in which many ultra luminous X-ray sources (ULXs) are produced as end points of stellar evolution. We have shown (see later) that ULXs are in general the high luminosity tail of high mass X-ray binaries (HMXB; see also contributions by Roberts or Kovlakas, this volume). They are the testimony of a recent event: the gravitational encounter of two galaxies. In RiGs, the location and formation time of new stellar populations are easier to reconstruct than in normal spirals due to the simpler geometry and dynamics. Brighter ULXs seem to be preferentially found in low metallicity environments. Two possibilities have been put forward: a) the largest black holes (BH) are constructed from direct collapse of low metallicity stars (e.g. Mapelli *et al.* 2009) or b) metal-poor X-ray binaries are more luminous than their metal-rich peers (Linden *et al.* 2010, which could be the case for both BH or neutron stars (NS) counterparts (see also contributions by Artale, this volume).

The Cartwheel belongs to a compact group of 4 members (Iovino 2002) at a distance of D = 122 Mpc. The shock wave of the encounter with one of the group galaxies has launched a shock wave that has triggered star formation and enhanced emission in all bands. The consequence of this star formation is the large number of ULXs found, especially in the ring - see Wolter *et al.* (2015).

The recent Gravitational Waves (GW) detections have increased the interest of ULXs as possible sites of coalescence. Many authors have used the ULXs as progenitors to compute the expectancy for GW detection of different kinds of merging, in particular the NS-BH event which has not been detected yet. Many recent estimates of expected rates (e.g. Inoue *et al.* 2016) do take into consideration the ULX properties (Luminosity Function, active time, frequency by galaxy mass or SFR). However, many of these parameters are very uncertain or not known (see also contributions by Giacobbo *et al.*, by Erkut *et al.* and by Fabrika *et al.* this volume).

We use an operative definition of ULX as an extra-galactic, point-like, non-nuclear, X-ray source with 10^{39} erg/s $< L_X < 10^{42}$ erg/s (Fabbiano 1989). This definition is bound to create a mixed class of sources and includes interlopers, like background AGNs, and at least a subclass of Supernovae, possibly numbering about 25% of all ULXs (Swartz *et al.* 2011). The majority of ULXs nevertheless are thought to be binary systems, with a degenerate object and a large (in most cases) companion. The engine of the system might be: a) an Intermediate Mass BH (10^{2-5} $M_\odot$ - this was the initial guess but now runs a bit out of fashion; b) a BH with a heavy stellar mass (30-100 $M_\odot$) or c) a Neutron Star - this has been proven for at least 5 cases (Bachetti *et al.* 2014, Israel *et al.* 2017a,b, Fuerst *et al.* 2016, Carpano *et al.* 2018) out of a few hundred ULXs.

2. The X-ray Luminosity Function of ULXs

First we use the X-ray luminosity function (XLF) to show that ULXs are related to HMXB, which are the subject of this meeting, and therefore deserve attention here. A number of authors, including us, has produced the XLF for individual galaxies with a large number of ULXs (Cartwheel: Wolter & Trinchieri 2004; Antennae: Zezas *et al.* 2007; NGC337: Somers *et al.* 2013; NGC 2276: Wolter *et al.* 2015) and compared it with the so-called universal luminosity function of HMXB by Grimm *et al.* (2003). The normalization of the XLF for the Star Formation Rate (SFR), measured typically via the Hα luminosity, is consistent with the idea that ULXs are linked to recent star formation bursts. The smooth connection with the lower luminosity sources and the slope of the XLF are consistent with our hypothesis.

This is also shown by the XLF of both a large number of ULXs collected in the rings of seven collisional RiGs (Wolter *et al.* 2018) and those in the collection of nearby galaxies (Swartz *et al.* 2011) observed by Chandra, which yield a similar number of ULXs and consistent results between each other, even if the RiGs have a possible excess at high luminosity. The combination of many galaxies relies on the assumption that we are witnessing a single burst of star formation in each galaxy, with a reasonable short age (possibly less than 100 Myr as per the simulations of Renaud *et al.* 2018), close to the formation time of HMXB, and that the spread in metallicity is not large. In any case, the effect of metallicity is not predominant with respect to the SFR (see e.g. Mapelli *et al.* 2010).

The Cartwheel, with its 15 out 65 sources in the collection of ring galaxies, is the best testbed in which to search for new information about the environment of formation and evolution of ULXs and HMXB in general. A measure of the metallicity of the Cartwheel is available in the literature (Fosbury *et al.* 1977), measured in the brightest HII regions. The

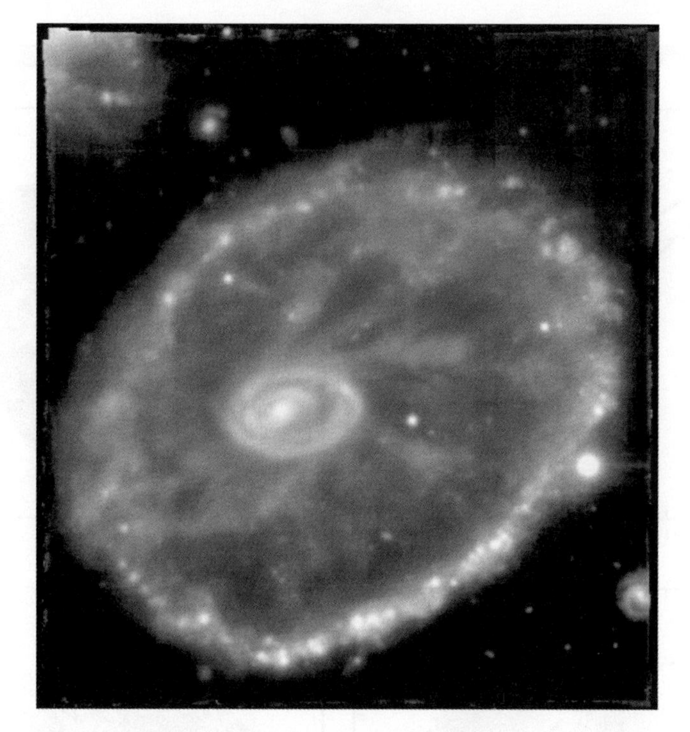

Figure 1. Three color image derived from Muse data - using filters matching as close as possible the g, i and r Gunn filters.

age of the star burst is not known, even if a few estimates of the epoch of the encounter are available in the literature: Higdon (Higdon 1996) estimates 300 Myr from the impact from the HI velocity field; Amram (Amram *et al.* 1998) estimate the age of the ring at > 200 Myr from Hα kinematics (13-30 km/s expansion); while Renaud (Renaud *et al.* 2018) run detailed simulations that imply an age < 100 Myr for the persistence of ring.

We intend therefore to study the Cartwheel to address these issues.

3. MUSE data

The Cartwheel was observed by MUSE in Aug 2014 for calibration purposes with a mosaic of four 4 ksec pointings to include the entire ring (1.4×1.5 arcmin). The data have been downloaded from the ESO archive. The MUSE 3D science data cubes have their instrumental signature removed, and are astrometrically calibrated, sky-subtracted, wavelength and flux calibrated, using the MUSE pipeline, version muse-1.4 and higher. We show in Figure 1 a color image, obtained by filtering on the Gunn filters g, i, and r the band pass of MUSE (details in Wolter *et al.* in prep).

To robustly estimate emission line fluxes, we accounted for the stellar absorptions underlying the Balmer emission lines that fall in the MUSE spectral window: Hα and Hβ. For this, we used the code GANDALF (Sarzi *et al.* 2006) complemented by the penalize pixel-fitting code (Cappellari & Emsellem 2004) to simultaneously model the stellar continuum and the emission lines in individual spaxels with S /N > 5. The stellar continuum was modeled with the superposition of stellar templates from the MILES library (Vazdekis *et al.* 2010) convolved by the stellar line-of-sight velocity distribution, whereas the emission lines and kinematics were modeled assuming a Gaussian profile. In each spaxel, the modeled stellar continuum spectrum was subtracted from the observed spectrum to obtain a final datacube of pure emission lines that is free of stellar absorption.

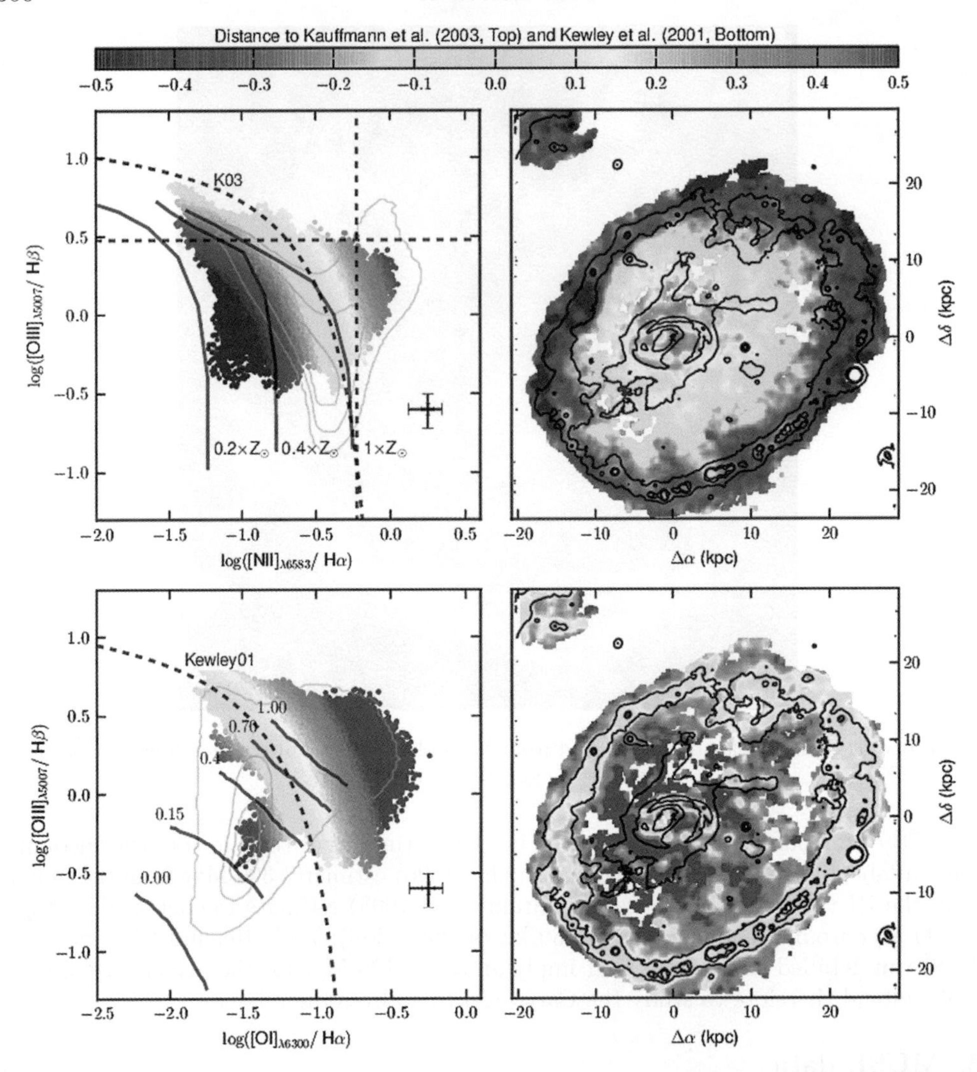

Figure 2. BTP diagrams: see text for details. *Top-left*: [OIII]λ5007/Hβ vs. [NII]λ6583/Hα. *Top-right*: map of the spaxel contributing to the BPT, color-coded as in the top-left panel. *Bottom-left*: [OIII]λ5007/Hβ vs. [OI]λ6300/Hα *Bottom-right*: map of the spaxels contributing to the BPT, color-coded as in the bottom-left panel.

3.1. *Metallicity*

We have extracted "line" images (for Hα, Hβ, [OIII], [OI] and [NII]). We can then use those images to compute standard classification diagrams resolved in spaxels, as the two BTP diagnostic maps shown in Figure 2. In the top-left panel the [OIII]λ5007/Hβ vs. [NII]λ6583/Hα is plotted. The dashed curve separates AGN from HII regions and is adopted from Kauffmann *et al.* 2003 (Ka03). Data (in the top-right panel) are color-coded according to their minimum distance to the Ka03 curve. The black crosses indicate the typical error of the ratio of lines with a S/N $\sim$ 15/5. Thick solid lines show the three different photoionization models at different metallicities (0.2 $Z_\odot$, 0.4 $Z_\odot$ and $Z_\odot$) by Kewley *et al.* (2001).

In the bottom-left panel the [OIII]λ5007/Hβ vs. [OI]λ6300/Hα is plotted. The dashed curve separates AGN from HII regions and is adopted from Kewley *et al.* 2001 (Ke01).

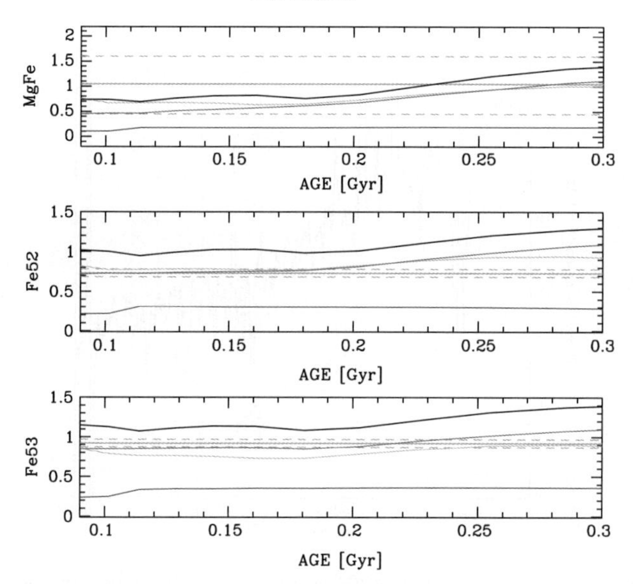

Figure 3. Comparison between BC03 models expectations and MgFe (*top panel*), Fe5270 (*middle panel*) and Fe5335 (*lower panel*) indices measures of the Cartwheel outer ring spectrum. The green lines represent our measures (*solid line*) and their errors (*short dashed line*). In each panel, expected indices values as a function of the age of a SSP are reported for solar metallicity (*black line*), Z = 0.4 $Z_\odot$ (*red line*), Z = 0.2 $Z_\odot$ (*cyan line*) and Z=0.02 $Z_\odot$ (*blue line*).

Data (in the bottom-right panel) are color-coded according to their minimum distance to the Ke01 curve. The black crosses indicate the typical error of the ratio of lines with a S/N $\sim$ 15/5. Thick solid lines show the five different shock models by Rich *et al.* 2011, indicating five different fractions (from 0 to 1) of Hα flux contributed by shocks.

A general assessment of the metallicity in the galaxy shows that the metallicity of the ring is consistently lower than in the rest of the galaxy. The small portion of G1 visible in the upper left corner shows about the same metallicity of the ring. The central region appears supersolar. However, as shown in the bottom panels, where [OI] is used instead of [NII], we cannot exclude the presence of shocks that contribute to the line ratios.

3.2. *Spectra and inferences*

It is notorious that metallicity of stars and age of the galaxy give similar effects on the optical spectra. Nevertheless we try to measure them separately by looking at some Lick narrowband spectral indices like those related to the Mg and Fe absorption features in the range 5300-5500 Å (i.e., Mgb, Fe5270, F5335, MgFe) We compare our measures with the expectations of Bruzual & Charlot 2003 (BC03) models (Simple Stellar Population SSP, Chabrier IMF) for different metallicities (i.e., Z = $Z_\odot$; Z = 0.4 $Z_\odot$; Z = 0.2 $Z_\odot$; Z = 0.02 $Z_\odot$), as shown in Fig. 3.

The star metallicity is consistent with being 20-40% solar, in particular, expecially from the iron indices Fe5335 and Fe5270 (for which our measures result to have a better accuracy), it is Z = 0.4 $\times$ solar if the age of the star population is 200 Myr, while it is 0.2 $\times$ solar if the age is 300 Myr. This would confirm the picture in which the age is small (less than 400 Myr) and, even if the correlation between age and metallicity is still present, we can constrain the interval between 20-40% solar, consistent with the majority

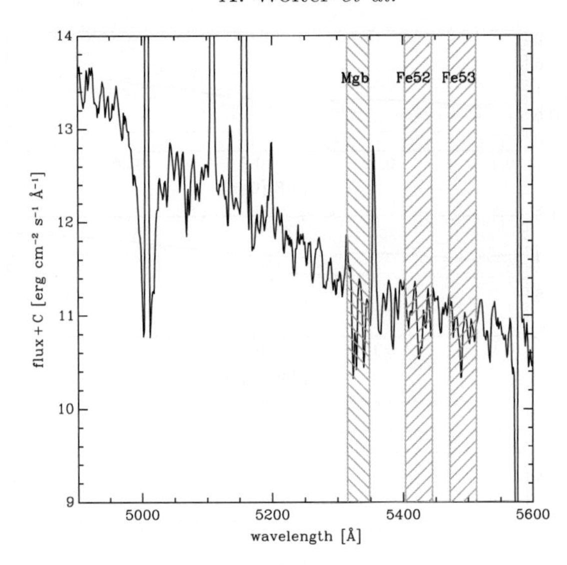

Figure 4. Cartwheel MUSE outer ring spectrum. The green dashed bands mark the spectral regions involved in the metallicity Lick indices which have been estimated in this work.

of the star having been produced in the impact, with a less than solar gas, possibly mixed between the original gas of the Cartwheel and the surrounding environment.

4. Spectro-photometric fitting with MultiNest

Is the timing right? Can we measure the age of the population? Given the age and metallicity correlation, can we disentangle them?

These are amongst the questions we would like to address with the MUSE data, in combination with all other multi-wavelength measures. To this aim, we have adapted the code by Fossati & Mendel (see Fossati *et al.* 2018 on the VESTIGE survey) which exploits a *MULTINEST* fit. It applies a Monte-Carlo Spectro-Photometric Fitter which derives the star formation history (SFH) by fitting high resolution (Bruzual&Charlot) population models. A linear interpolation of the stellar models (where Q_{age} is the look-back time of the event and T_q is the characteristic time scale) is performed, and the result is scaled in luminosity as part of the fitting procedure. The procedure includes nebular emission lines (ionizing radiation absorbed by gas) and dust attenuation.

We collect all the available photometric information in the UV-O-IR band. We select three different spatial regions: the outer ring, the middle zone and the inner (and nuclear) ring. We focus here on the outer ring, in which the majority of the stars should be produced in the impact, and which has a higher brightness, which helps in reducing degrees of freedom to the fitted models. The inner ring has a mix of old and new population which is quite more complicated to simulate with a simple star formation history. The middle zone is fainter a therefore more uncertain, even if very interesting from the point of view of confirming simulations.

We derive photometric points for the outer ring, and extract a MUSE spectrum from the same spatial region, as input to the spectrophotometric fitter. We plot the results for the solar value and for 20% solar - which are the extremes we considered - in Figure 5, where in the bottom left panel: blue indicates the young component, with stellar emission and nebular lines, while red represents the old component ($>$ 10 Myr). The shape of the SFH is a single burst, exponentially quenched. The fitted parameters are the age of the burst (Q_{age} = 300/132 Myr) and the quenching scale (T_q = 69/19 Myr) for metallicity $Z = Z_\odot$ (top) and 0.2 $Z_\odot$ (bottom) respectively.

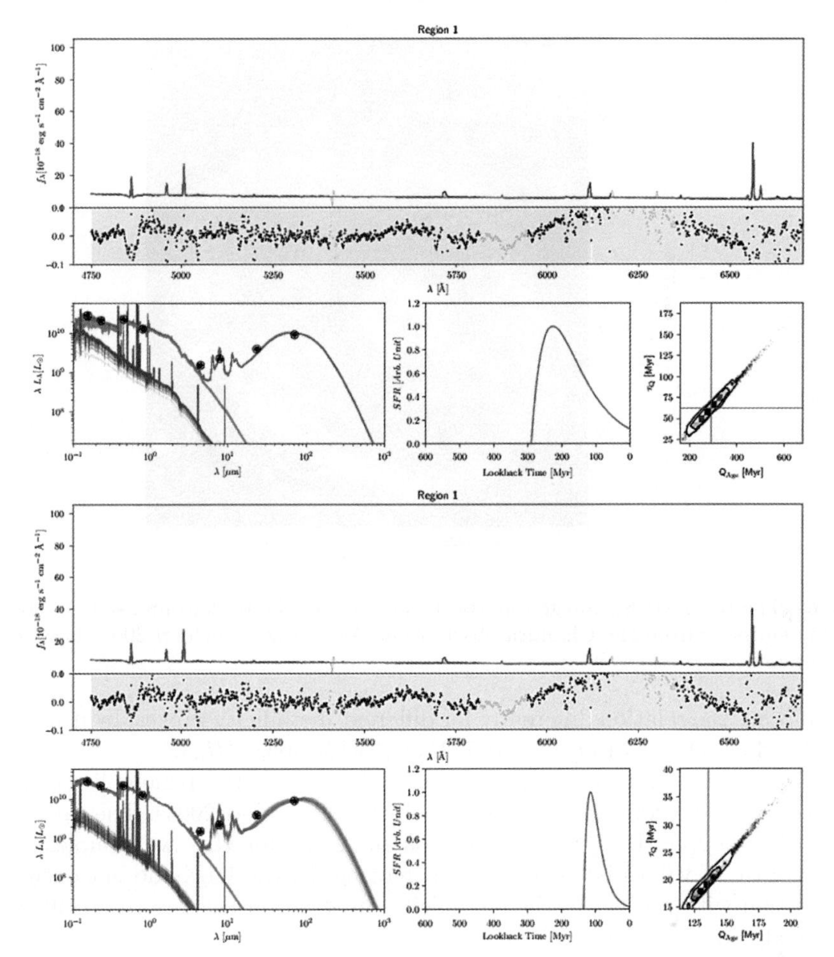

Figure 5. a) *Top figure*: solar metallicity; b) *Bottom figure*: 20% solar metallicity. Both figures are thus structured: *Top*: Results of the MC-SPF fitting for the outer ring. Upper panel: MUSE spectrum (black) and best fit model (dark red). Regions where the spectrum is plotted in grey are not used in the fit. The fit residuals ($Data - Model$) are shown below the spectrum and the grey shaded area shows the 1σ uncertainties. *Lower left panel*: photometric data points in black. The blue line represents the stellar emission with nebular lines from the young component (Age $<$ 10 Myr), the red lines are from the old component (Age $>$ 10 Myr), while the dark red lines are the total model including the dust emission. *Lower middle panel*: reconstructed SFH from the fitting procedure. *Lower right panel*: marginalised likelihood maps for the Q_{age} and T_q fit parameters. The red lines show the median value for each parameter, while the black contours show the 13 σ confidence intervals

5. X-ray band information

One of our aims is to compare the environment properties with the production of ULXs to gather information on the formation mechanism. The first step is to compare the X-ray emission observed by Chandra to the optical one measured by MUSE, as shown in Figure 6. In the same figure, red circles indicate the background objects, easily traceable in the MUSE data cube. No particularly X-ray bright interloper is found in the ring region.

First we investigate if the metallicity is different at the ULXs positions from the rest of the ring. We compute the metallicity by following the prescription of Curti *et al.* 2017,

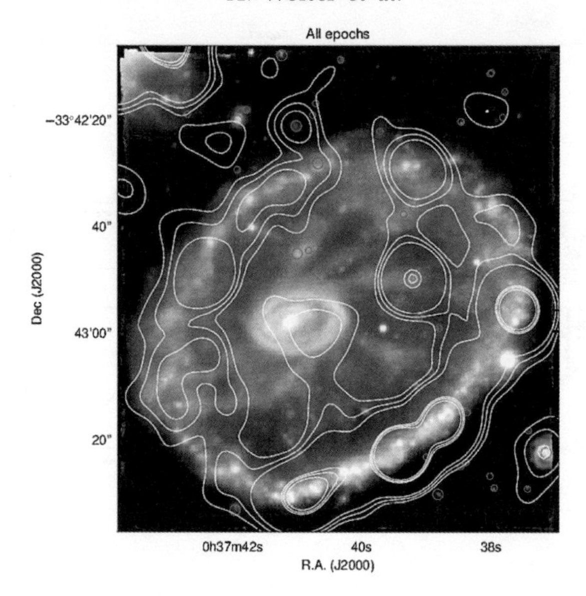

Figure 6. The RGB MUSE image reports the position of the background sources (red circles). The X-ray emission from the Chandra observation (Wolter & Trinchieri 2004) is superposed in white contours.

which calibrate the relation differently for different metallicity ranges, by using the O_3N_2 index defined as: $O_3N_2 = ([OIII]\lambda5007/H_\beta)/([NII])\lambda6584/H_\alpha)$.

We extract the metallicity value at the position of the HII regions listed in (Higdon 1995), and at the ULXs positions from (Wolter & Trinchieri 2004). The range of metallicities - 12 + log(O/H) = 8.16 – 8.47 - is the same for the two samples, indicating that a) a range of values is found even in the ring; b) the ULXs do not occupy a privileged position with respect to the metallicity distribution (see also the contribution by Kouroumpatzakis *et al.*, this volume).

6. Conclusion

ULXs are in the majority binaries, whether NS or BH, produced by a recent episode of Star Formation. The recent construction of XLF for RiGs suggests the presence of a larger number of high luminosity sources in collisional environments. This results may be applicable to constrain estimates for GW events. Galaxies like the Cartwheel represent sizable samples of ULXs and are therefore the most interesting to study. We have exploited the MUSE dataset that allows us to derive spatially resolved information to compare with multi-wavelength data. We find that metallicity is 20-40% solar in the external ring, consistently in both stars and gas, but we cannot exclude yet from our spectrophotometric fitting procedure that the metallicity is solar. From the above results it follows that the age of the stellar population produced in the encounter is $\leqslant 300$ Myr and possibily smaller, consistent with the timing of HMXB formation, which reinforces the interpretation of ULX system as the high luminosity (higher mass transfer? higher magnetic field?) tail of HMXB. ULXs are not found in special regions with respect to the rest of the galaxy, or at least of the ring, for what concerns metallicity. While this point needs to be further studied, it might lessen the importance of the environment for the formation of ULXs and strengthen the importance of the timing of the event.

In the future we plan to perform a detailed comparison with simulations (Renaud *et al.* 2018, Mapelli & Mayer 2012) for different regions of the galaxy; we will exploit available long-slit spectra for determining the ionization parameter and even better calibrate the metallicity relations. These results will be applied then to constrain formation models of ULXs (e.g. Wiktorowicz *et al.* 2017).

References

Amram, P., Mendes de Oliveira, C., Boulesteix, J. & Balkowski, C., 1998, *A&A*, 330, 881
Bachetti, M. *et al.* 2014, *Nature*, 514, 202
Bruzual, G., & Charlot, S., 2003, *MNRAS*, 344, 1000
Cappellari, M., & Emsellem, E. 2004, *PASP*, 116, 138
Carpano, S., Haberl, F., Maitra, C. & Vasilopoulos, G., 2018, *MNRAS*, 476L, 45
Curti, M., Cresci, G., Mannucci, F., Marconi, A., Maiolino, R., & Esposito, S., *MNRAS*, 465, 1384
Fabbiano, G., 1989, *ARA&A*, 27, 87
Fosbury, R. A. E. & Hawarden, T. G., 1977, *MNRAS*, 178, 473
Fossati, M., *et al.*, *A&A*, 2018, 614, 57
Fürst, F., Walton, D. J., Stern, D., Bachetti, M., Barret, D., Brightman, M., Harrison, F. A. & Rana, V., 2016, *ApJ*, 834, 77
Grimm, H.-J., Gilfanov, M. & Sunyaev, R., 2003, *MNRAS*, 339, 793
Higdon, J. L., 1995, *ApJ*, 455, 524
Higdon, J. L., 1996, *ApJ*, 467, 241
Inoue, T., Tanaka, Y.T., & Isobe, N., 2016, *MNRAS*, 461, 4329
Iovino, A., 2002, *AJ*, 124, 2471
Israel, G. L. *et al.*, 2017a, *MNRAS*, 466L, 48
Israel, G. L. *et al.*, 2017b, *Science*, 355, 817
Kauffmann, G., Heckman, T.M., Tremonti, C., *et al.* 2003, *MNRAS*, 346, 1055
Kewley, L.J., Dopita, M.A., Sutherland, R.S., Heisler, C.A., & Trevena, J. 2001, *ApJ*, 556, 121
Linden, T., Kalogera, V., Sepinsky, J. F., Prestwich, A., Zezas, A. & Gallagher, J. S., 2010, *ApJ*, 725, 1984
Mapelli, M., Colpi, M. & Zampieri, L., 2009, *MNRAS*, 395L, 71
Mapelli, M., Ripamonti, E., Zampieri, L., Colpi, M. & Bressan, A., 2010,*MNRAS*, 408, 234
Mapelli, M. & Mayer, L., *MNRAS*, 420, 1158
Renaud, F. *et al.*, 2018, *MNRAS*, 473, 585
Rich, J.A., Kewley, L.J., & Dopita, M.A. 2011, *ApJ*, 734, 87
Sarzi, M., Falcón-Barroso, J., Davies, R. L., *et al.* 2006, *MNRAS*, 366, 1151
Somers, G., Mathur, S., Martini, P., Watson, L., Grier, C.J. & Ferrarese, L., 2013, *ApJ*, 777, 7
Swartz, D. A., Soria, R., Tennant, A. F. & Yukita, M. *ApJ*, 741, 49
Vazdekis, A., Sánchez-Blázquez, P., Falcón-Barroso, J., *et al.* 2010, *MNRAS*, 404, 1639
Wiktorowicz, G., Sobolewska, M. , Lasota, J.-P. & Belczynski, K., 2017, *ApJ*, 846, 17
Wolter, A. & Trinchieri, G, *A&A*, 426, 787
Wolter, A., Esposito, P., Mapelli, M., Pizzolato, F. & Ripamonti, E., 2015, *MNRAS*, 448, 781
Wolter, A., Fruscione, A. & Mapelli, M., 2018 *ApJ*, 863, 43
Wolter, A., Consolandi, G., Longhetti, M., Landoni, M. & Bianco, A., in preparation
Zezas, A., Fabbiano, G., Baldi, A., Schweizer, F., King, A. R., Rots, A. H. & Ponman, T. J., *ApJ*, 661, 135

Discussion

INDULEKHA KAVILA: Your see misalignment between ULXs and the HII regions. Since there is a time gap between the star formation event and the appearance of the ULX, wouldn't it be more apt to look for alignment with signatures of B stars - which are the brightest objects in the "post star formation event" phase?

WOLTER: We have checked for offsets between the location of the peak of the optical emission and the ULX number density, with no particular selection of emission signatures.

Claus Leitherer: One additional step would be to apply population synthesis models with binary star evolution (BPASS; Eldridge *et al.* 2017). This would allow you to model the optical and X-ray data simultaneously and check for consistency.

Wolter: We thank you for the suggestion and we will try to implement this for the next publication.

High-mass X-ray Binaries: Illuminating the Passage from Massive Binaries to Merging Compact Objects
Proceedings IAU Symposium No. 346, 2019
L. M. Oskinova, E. Bozzo, T. Bulik & D. Gies, eds.

doi:10.1017/S1743921319001303

Spectroscopy of complete populations of Wolf-Rayet binaries in the Magellanic Clouds

Tomer Shenar[1], R. Hainich[2], W.-R. Hamann[2], A. F. J. Moffat[3], H. Todt[2], A. Sander[4], L. M. Oskinova[2], H. Sana[1], O. Schnurr[5] and N. St-Louis[3]

[1]Institute of Astrophysics, KU Leuven, Celestijnenlaan 200 D, 3001, Leuven, Belgium
email: tomer.shenar@kuleuven.be

[2]Institut für Physik und Astronomie, Universität Potsdam, Karl-Liebknecht-Str. 24/25, D-14476 Potsdam, Germany

[3]Département de physique and Centre de Recherche en Astrophysique du Québec (CRAQ), Université de Montréal, C.P. 6128, Succ. Centre-Ville, Montréal, Québec, H3C 3J7, Canada

[4]Armagh Observatory and Planetarium, College Hill, Armagh, BT61 9DG, UK

[5]Leibniz-Institut für Astrophysik Potsdam, An der Sternwarte 16, 14482 Potsdam, Germany

Abstract. Classical Wolf-Rayet stars are evolved, hydrogen depleted massive stars that exhibit strong mass-loss. In theory, these stars can form either by intrinsic mass loss (stellar winds or eruptions), or via mass-removal in binaries. The Wolf-Rayet stars in the Magellanic Clouds are often thought to have originated through binary interaction due to the low ambient metallicity and, correspondingly, reduced wind mass-loss. We performed a complete spectral analysis of all known WR binaries of the nitrogen sequence in the Small and Large Magellanic Clouds, as well as additional orbital analyses, and constrained the evolutionary histories of these stars. We find that the bulk of Wolf-Rayet stars are luminous enough to be explained by single-star evolution. In contrast to prediction, we do not find clear evidence for a large population of low-luminosity Wolf-Rayet stars that could only form via binary interaction, suggesting a discrepancy between predictions and observations.

Keywords. stars: Wolf-Rayet, binaries: spectroscopic, Magellanic Clouds, stars: evolution

---

1. Introduction

Wolf-Rayet (WR) stars define a spectral class of stars with emission-dominated spectra (e.g., Smith *et al.* 1996, Crowther *et al.* 2000) associated with strong, radiatively driven winds. Classical WR stars are understood to represent an evolved and relatively short-lived phase ($\approx$0.5 Myr) of massive stars that have stripped a substantial part of their hydrogen-rich outer layers. However, very massive stars may already exhibit WR-like spectra at birth.

Studying WR stars is essential for understanding the evolution of massive stars, the energy budget of galaxies, and the upper-mass limit of stars. WR stars easily dominate entire populations of massive stars in terms of their mechanical and radiative energy input (e.g., Doran *et al.* 2013, Ramachandran *et al.* 2018). Furthermore, as WR stars are the expected immediate progenitors of stellar-mass black holes (BHs), their attributes (e.g., mass-loss rates, masses) prior to core-collapse largely determine the properties of the BH remnant. Therefore, uncertainties with respect to WR stars directly translate to inaccurate predictions of rates and properties of gravitational wave (GW) events (e.g., de Mink *et al.* 2014, Eldridge *et al.* 2016).

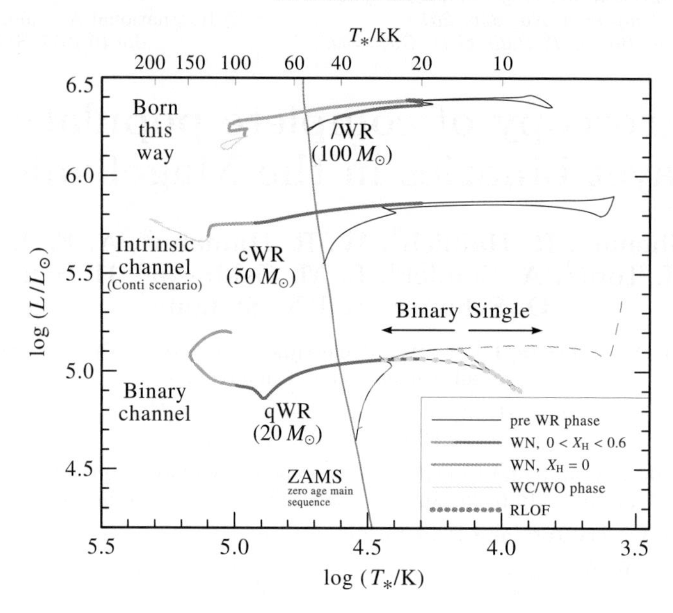

Figure 1. Evolution tracks calculated with the BPASS code for single stars with initial masses $M_i = 100, 50$, and $20\,M_\odot$, as well as a binary evolution track for the primary star with initial mass, period, and mass ratio of $M_i = 20\,M_\odot$, $P_i = 25\,d$, and $q_i = 0.3$, respectively (coloured part of lower track). The tracks illustrate the three formation channels for WR stars: the binary channel (lower tracks), the intrinsic channel (middle track), and the "born this way" channel (upper track).

In principle, WR stars may form via three channels:

(*a*) **Classical WR stars (cWR)** form through *intrinsic* mass-loss, either via stellar winds or eruptions (Conti 1976, Smith *et al.* 2006). They are evolved, hydrogen depleted (or hydrogen free) massive stars. Only stars that are sufficiently massive will reach the cWR phase. The minimum mass necessary is a strong function of the metallicity Z, and is estimated to be $\approx 20\,M_\odot$ at solar metallicity and $\approx 45 - 60\,M_\odot$ at $\approx 1/5\,Z_\odot$ (Crowther 2006, Hainich *et al.* 2014), keeping in mind that the latter values suffer from large uncertainties.

(*b*) **Binary product WR stars (stripped stars, quasi-WR, qWR)** are hydrogen depleted stars that lie close to the Eddington limit after having lost their hydrogen-rich envelope through *binary interaction*, either via Roche lobe Overflow (RLOF) or via common-envelope (CE) evolution (Paczynski 1973, Vanbeveren *et al.* 1998).

(*c*) **"Born this way" stars (/WR, WNh)** are born with such high masses ($\gtrsim 60\,M_\odot$ at solar metallicity) that they possess strong, WR-like stellar winds already at birth. Typically, spectra of such stars show P-Cygni like Hβ profiles, and they are therefore often associated with "slash stars" (Crowther 2011) or WNh stars, although very young and massive stars may also exhibit pure Hβ emission and may thus be classified as WR stars. Because they are not evolved, they are not necessarily hydrogen depleted.

These three WR-types are illustrated in Fig. 1 using evolution tracks calculated with the BPASS† (Binary Population and Spectral Synthesis) code (Eldridge *et al.* 2008, Eldridge *et al.* 2017). While the spectroscopic definition of WR stars is fairly unambiguous (e.g., Smith *et al.* 1996, Crowther 2011), it is usually not straightforward to identify a WR star's evolutionary channel from its spectrum.

One of the central problems in this context is to correctly estimate the frequency of stripped stars, or qWR stars, in a host galaxy as a function of Z. Several studies give

† bpass.auckland.ac.nz

direct evidence that the vast majority of massive stars will interact with a companion star during their lifetime (Sana *et al.* 2012). Among the WR stars, about 40 % are observed to be binaries (e.g., van der Hucht 2001).

Considering the strong tendency of the initial mass function (IMF) to form lower-mass stars and the frequency of interacting binaries, stripped WR stars (qWR stars) should be abundant in our Universe - much more abundant than classical WR stars - bearing large implications on the energy budget of galaxies (e.g., Götberg *et al.* 2018). However, to-date, only one star, HD 45166, is considered a good candidate for a stripped WR star (Oliveira *et al.* 2003, Groh *et al.* 2008). Several low-mass ($\approx 1\,M_{\odot}$) O-type subdwarfs (sdO) believed to originate from binary mass-transfer have been discovered near B-type stars (the putative mass-accretors), but their masses are too low to support a strong stellar wind and a corresponding WR-star appearance (e.g., Wang *et al.* 2018). While other peculiar WR stars have been suggested to originate from binary interactions (Schootemeijer & Langer 2017, Neugent *et al.* 2017), there is an obvious dissonance between the predicted abundance of qWR stars and their observed number. In fact, it is not even certain that a stripped star would portray a wind that is strong enough to impart on the star the appearance of a WR star: While theoretical predictions for the mass-loss rates of qWR stars exist (Vink *et al.* 2017), empirical $\dot{M}$ estimates for qWR stars are still lacking.

It is by now empirically (Mokiem *et al.* 2007, Nugis *et al.* 2007, Hainich *et al.* 2015) as well as theoretically (Kudritzki *et al.* 1987, Vink *et al.* 2001) established that the intrinsic mass-loss rates of massive stars decrease with decreasing surface metallicity, $\dot{M} \propto Z^{\alpha}$, with $0.5 \lesssim \alpha \lesssim 1$. This immediately implies that it is harder for stars at low metallicity to intrinsically peel off their outer layers and become cWR stars. In other words, the intrinsic formation channel - if dominated by mass-loss by continous stellar winds - becomes increasingly inefficient with decreasing metallicity. In contrast, the binary formation channel is, at least to first order, independent of the metallicity†. One therefore concludes that the binary channel should become increasingly dominant in the formation of observed WR populations at low metallicity.

Motivated by such predictions, Bartzakos *et al.* (2001), Foellmi *et al.* (2003a), Foellmi *et al.* (2003b), and Schnurr *et al.* (2008), conducted a large spectroscopic survey in the Small and Large Magellanic Cloud (SMC and LMC, respectively) with the goal of measuring the binary fraction in their WR populations. The LMC and SMC are both known to have a subsolar metallicity of a factor $\sim 1/3$ and $\sim 1/5$ solar, respectively (Dufour *et al.* 1982, Larsen *et al.* 2000). Following the reasoning of the previous paragraph, it is expected that the fraction of WR stars formed via the binary channel will be relatively large in the LMC, and even larger in the SMC. Relying on models by Maeder *et al.* (1994), Bartzakos *et al.* (2001) argued that virtually *all* WR stars in the SMC are expected to have been formed in binaries. This prediction remains valid even with the most recent generation of stellar evolution codes (e.g., Georgy *et al.* 2015). It was therefore surprising that the measured WR binary fraction in the SMC is only ≈40% (Foellmi *et al.* 2003a), comparable to the Galactic fraction. A similar, slightly lower binary fraction is obtained for the LMC (Foellmi *et al.* 2003b). This revealed a clear discrepancy between theory and observation which must be explained.

To explore the formation channels of the WR stars in the Magellanic Clouds, we performed spectral analyses and, when possible, orbital analyses of all known WR stars and binaries of the nitrogen sequence (WN) in the Magellanic Clouds. The results for

† At low metallicity, the radiation pressure is lower, and thus so are the stellar radii for a given initial mass and age, which in turn reduces the likelihood of binary interaction, primarily for case A mass-transfer (i.e. mass-transfer during the main sequence). However, this effect is negligible compared to the sensitivity of $\dot{M}$ to Z.

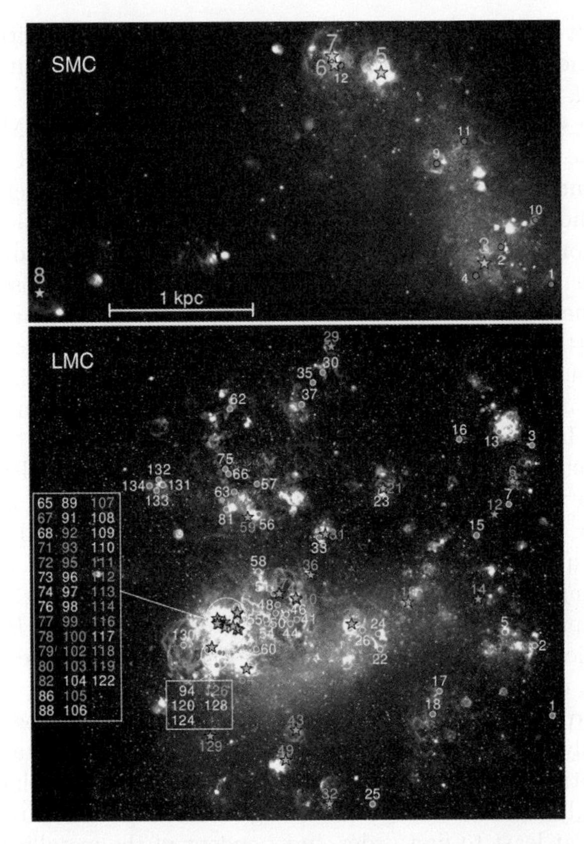

Figure 2. Upper panel: A narrow band O [III] nebular emission image of the SMC Smith *et al.* (2005) with all known WN stars and binaries marked. Nomenclature follows the catalogues Azzopardi & Breysacher (1979) and Breysacher *et al.* (1999) for the SMC and LMC, respectively. Yellow stars correspond to confirmed binary systems. Lower panel: same as the upper panel, but for the LMC in the Hα band.

the single WR stars were published by Hainich *et al.* (2014, 2015), while the binary samples were analyzed by Shenar *et al.* (2016) for the SMC and Shenar *et al.* (in prep.) for the LMC. The very massive WR+WR system BAT99 119 (R 145) in the LMC was analyzed by Shenar *et al.* (2017), while the WR quadruple/quintuple system SMC AB 6 was analyzed in Shenar *et al.* (2018). In these proceedings, we focus on the WR binaries.

2. Target selection and observations

A census of the known WR stars in the Magellanic Clouds is given by Massey *et al.* (2014) and Neugent *et al.* (2018). The positions of of all known single and binary WN stars and binaries in the SMC and LMC are displayed in Fig. 2.

The SMC sample comprises the five confirmed WR binaries SMC AB 3, 5, 6, 7, and 8, where SMC AB 5 was recently found to be a quadruple or quintuple system (Shenar *et al.* 2018). Among the 109 WR stars listed in the fourth catalog of WN stars in the LMC (Breysacher *et al.* 1999, BAT99 hereafter), Hainich *et al.* (2014) identified 43 that are either known binary/multiple systems or binary candidates. These 43 objects constitute our sample. 19 of these objects were confirmed via periodic RV variation to be binary/multiple systems, and six are considered binary candidates on the basis of RV variations for which no period could be found. Other binary candidates were identified based on their X-ray properties. A comprehensive list can be found in Hainich *et al.* (2014).

The observational dataset largely relied on the dataset used in the studies by Foellmi *et al.* (2003a), Foellmi *et al.* (2003b), and Schnurr *et al.* (2008). In some cases, archival ESO data could be retrieved. UV and far-UV data from the International Ultraviolet Explorer (IUE) and Hubble Space Telescope (HST) could also be retrieved for some of our targets. For a detailed account of the observing material, we refer to the aforementioned studies, as well as to Hainich *et al.* (2014, 2015) and Shenar *et al.* (2016, 2017, 2018).

3. Methods

The spectral analysis is performed with the non-LTE Potsdam Wolf-Rayet (PoWR) model atmosphere code, especially suitable for hot stars with expanding atmospheres†. It iteratively solves the co-moving frame radiative transfer and the statistical balance equations in spherical symmetry under the constraint of energy conservation. A more detailed description of the assumptions and methods used in the code is given by Gräfener *et al.* (2002), Hamann *et al.* (2004), and Sander *et al.* (2015). By comparing synthetic spectra generated by PoWR to observations, the stellar parameters can be derived.

By analyzing the spectra with the PoWR tool, we can derive the effective temperatures T_*, luminosities $\log L$, radii R_*, mass-loss rates $\dot{M}$, and other parameters of interest. Quantities marked with an asterisk refer to a Rosseland optical depth of $\tau_{\mathrm{Ross}}=20$, which is defined to be the inner boundary of our models. When possible, spectral disentanglement was utilized for binaries to separate the spectra into their constituents. Otherwise, composite spectra were analyzed by adding up model spectra corresponding to the various components of the systems. For a complete account of the methodology of binary analysis, we refer the reader to Shenar *et al.* (2016, 2017, 2018). In Fig. 3, we show an example for a binary analysis of the system BAT99 6.

4. Results

Figure 4 summarizes the positions of all WR binary components in the Magellanic Clouds on a Hertzsprung Russell diagram (HRD). The surface hydrogen mass fractions of the WR stars are color-coded. The same Figure also includes evolution tracks for non-rotating single stars calculated for LMC and SMC metallicity with the BPASS code.

In Shenar *et al.* (2016), we provide a detailed comparison to binary evolution tracks for each of the WR binaries in the SMC. This was done by comparing the full set of observables (e.g., periods, masses, temperatures, luminosities, rotation...) to evolution tracks calculated with the BPASS code. Through this, we derived initial masses and periods, as well as ages, for each of the binaries.

While we generally find that binary interaction can better explain the properties of the observed WR binary population in the SMC, all WR components are found to have very large initial masses ($M_{\mathrm{i}} \gtrsim 60\,M_\odot$). With such initial masses, it seems likely that the WR components would have entered the WR phase regardless of binary interaction. This stands in strong contrast to the prediction that all WR stars in the SMC must have formed via binary interaction.

Moreover, there is an apparent lack of WR components with luminosities lower than $\approx$5.8 dex. Such WR stars are expected to be abundant - much more abundant than their massive counterparts - given the tendency of the IMF to form lower-mass stars and the frequency of interacting binaries. However, none can be found in the SMC. One may think that stars at such luminosities at SMC metallicity do not retrain the appearance of a WR star. However, single WR stars in the SMC (e.g. SMC AB 2), with luminosities of $\approx$5.5 dex, do have the appearance of a WR star.

† PoWR models of Wolf-Rayet stars can be downloaded at http://www.astro.physik.uni-potsdam.de/PoWR

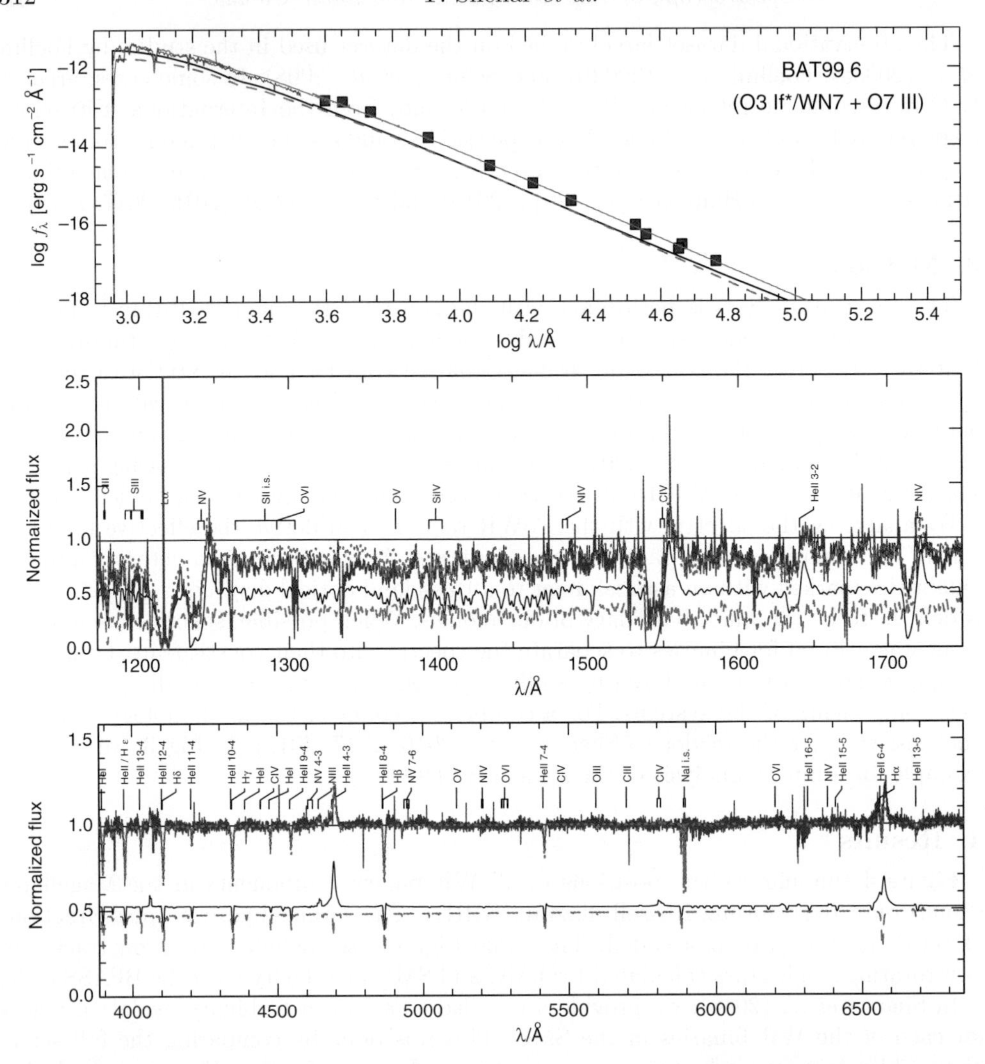

Figure 3. A spectral analysis of the system BAT99 6. The observed photometry and spectra (archival IUE, FEROS) of BAT99 6 are shown in blue. The composite synthetic spectrum (red dotted line) is the sum of the WR (black solid line) and O (green dashed line) models. The relative offsets of the model continua correspond to the light ratio between the two stars.

A similar result is obtained for the LMC sample, this time offering much better statistics. While the detailed evolutionary analyses of each binary are still underway (Shenar *et al.* in prep.), it is evident from the right panel of Fig. 4 that the majority of observed WR stars in the LMC can be explained with single-star evolution. While this does not mean that binary interaction did not take place, it again raises the question as to the lack of low-luminosity WR stars that are supposedly the product of binary interaction. This time, the LMC sample does offer a few interesting candidates (most interestingly BAT99 72), but this population again appears much smaller than predicted.

It has been suggested that many of these stars may remain very difficult to detect since they are expected to be the companions of massive, visually-brighter OB-type stars (e.g., Paczynski 1973, Schootemeijer *et al.* 2018). However, whether this can truly explain the apparent lack of qWR stars, compared to their anticipated abundance, needs to be further tested.

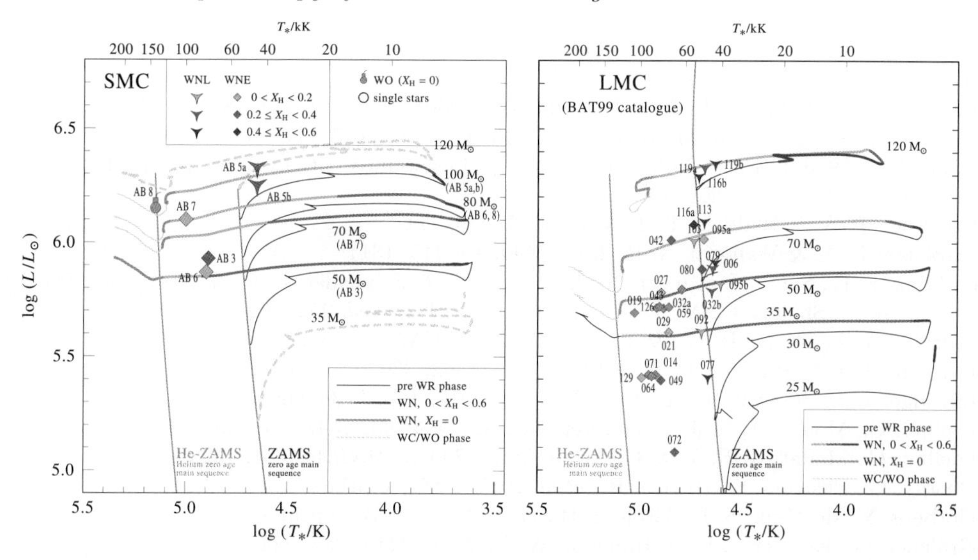

Figure 4. Comparison between observed HRD positions of our the WR components in WR binaries in the SMC (left panel) and LMC (right panel), adopted from Shenar *et al.* (2016, 2017, 2018 in prep.). Also shown are evolution tracks calculated with the BPASS code for non-rotating single stars at SMC (left panel) and LMC (right panel) metallicity for various initial masses. It is evident that the majority of WR components are luminous enough to be explained with single-star evolution.

5. Summary

We performed a spectral analysis of all known single and binary WR stars in the SMC and LMC. Our results were published in Hainich *et al.* (2014, 2015) and Shenar *et al.* (2016, 2017, 2018), with the LMC binary sample soon to be published (Shenar *et al.* in prep.). These studies provide all stellar parameters that could be derived for the WR stars in the Magellanic Clouds and, if present, for their companions. When possible, orbital analyses were performed to derive further constraints on the orbits and masses of the systems.

We generally find that the vast majority of WR components in binaries in the Magellanic Clouds are sufficiently luminous to be explained by single-star evolution. There is a very obvious lack of WR stars at low luminosities ($\log L \lesssim 5.7$ $[L_\odot]$). Such stars are expected to be very abundant given the bias of the IMF towards lower masses and the prevalence of interacting binaries. However, only few such stars are actually observed in the LMC and none in the SMC.

Observational biases offer a possible way out of this contradiction: stripped qWR stars are expected to reside near visually-brighter and more luminous mass accretors, which may render the detection of their stripped companions difficult (Paczynski 1973). However, since we do observe many WR binaries with bright OB-type stars, it is not obvious that there should be an observational cut-off precisely at luminosities these stars are expected to possess. For now, we do not find clear indications that binary interaction dominates the formation of WR stars at the low metallicity environment of the Magellanic Clouds.

Acknowledgement

This project has received funding from the European Research Council (ERC) under the European Union's DLV-772225-MULTIPLES Horizon 2020 research and innovation programme.

References

Azzopardi, M. & Breysacher, J. 1979, *A&A*, 75, 120
Bartzakos, P., Moffat, A. F. J., & Niemela, V. S. 2001, *MNRAS*, 324, 18
Breysacher, J., Azzopardi, M., & Testor, G. 1999, *A&AS*, 137, 117
Conti, P. S. 1976, in Proc. 20th Colloq. Int. Ap. Liége, university of Liége, p. 132, 193–212
Crowther, P. A. 2000, *A&A*, 356, 191
Crowther, P. A. & Hadfield, L. J. 2006, *A&A*, 449, 711
Crowther, P. A. & Walborn, N. R. 2011, *MNRAS*, 416, 1311
de Mink, S. E., Sana, H., Langer, N., Izzard, R. G., & Schneider, F. R. N. 2014, *ApJ*, 782, 7
Dufour, R. J., Shields, G. A., & Talbot, Jr., R. J. 1982, *ApJ*, 252, 461
Eldridge, J. J., Izzard, R. G., & Tout, C. A. 2008, *MNRAS*, 384, 1109
Eldridge, J. J. & Stanway, E. R. 2016, *MNRAS*, 462, 3302
Eldridge, J. J., Stanway, E. R., Xiao, L., *et al.* 2017, *PASA*, 34, e058
Foellmi, C., Moffat, A. F. J., & Guerrero, M. A. 2003a, *MNRAS*, 338, 360
Foellmi, C., Moffat, A. F. J., & Guerrero, M. A. 2003b, *MNRAS*, 338, 1025
Georgy, C., Ekström, S., Hirschi, R., *et al.* 2015, ArXiv e-prints
Götberg, Y., de Mink, S. E., Groh, J. H., *et al.* 2018, ArXiv e-prints
Gräfener, G., Koesterke, L., & Hamann, W.-R. 2002, *A&A*, 387, 244
Groh, J. H., Oliveira, A. S., & Steiner, J. E. 2008, *A&A*, 485, 245
Hainich, R., Pasemann, D., Todt, H., *et al.* 2015, *A&A*, 581, A21
Hainich, R., Rühling, U., Todt, H., *et al.* 2014, *A&A*, 565, A27
Hamann, W.-R. & Gräfener, G. 2004, *A&A*, 427, 697
Kudritzki, R. P., Pauldrach, A., & Puls, J. 1987, *A&A*, 173, 293
Larsen, S. S., Clausen, J. V., & Storm, J. 2000, *A&A*, 364, 455
Maeder, A. & Meynet, G. 1994, A&A, 287, 803
Massey, P., Neugent, K. F., Morrell, N., & Hillier, D. J. 2014, *ApJ*, 788, 83
Neugent, K. F., Massey, P., Hillier, D. J., & Morrell, N. 2017, *ApJ*, 841, 20
Neugent, K. F., Massey, P., & Morrell, N. 2018, *ApJ*, 863, 181
Mokiem, M. R., de Koter, A., Vink, J. S., *et al.* 2007, *A&A*, 473, 603
Nugis, T., Annuk, K., & Hirv, A. 2007, Baltic Astronomy, 16, 227
Oliveira, A. S., Steiner, J. E., & Cieslinski, D. 2003, *MNRAS*, 346, 963
Paczynski, B. 1973, in IAU Symposium, Vol. 49, Wolf-Rayet and High-Temperature Stars, ed. M. K. V. Bappu & J. Sahade, 143
Ramachandran, V., Hainich, R., Hamann, W.-R., *et al.*. 2018, *A&A*, 609, A7
Doran, E. I., Crowther, P. A., de Koter, A., *et al.* 2013, *A&A*, 558, A134
Sana, H., de Mink, S. E., de Koter, A., *et al.* 2012, Science, 337, 444
Sander, A., Shenar, T., Hainich, R., *et al.* 2015, *A&A*, 577, A13
Schnurr, O., Moffat, A. F. J., St-Louis, N., Morrell, N. I., & Guerrero, M. A. 2008, *MNRAS*, 389, 806
Schootemeijer, A. & Langer, N. 2017, ArXiv e-prints
Schootemeijer, A., Götberg, Y., Mink, S. E. d., Gies, D., & Zapartas, E. 2018, *A&A*, 615, A30
Shenar, T., Hainich, R., Todt, H., *et al.* 2016, *A&A*, 591, A22
Shenar, T., Richardson, N. D., Sablowski, D. P., *et al.* 2017, *A&A*, 598, A85
Shenar, T., Hainich, R., Todt, H., *et al.* 2018, *A&A*, 616, A103
Smith, L. F., Shara, M. M., & Moffat, A. F. J. 1996, *MNRAS*, 281, 163
Smith, N. & Owocki, S. P. 2006, *ApJl*, 645, L45
Smith, R. C., Points, S., Chu, Y.-H., *et al.* 2005, in Bulletin of the American Astronomical Society, Vol. 37, American Astronomical Society Meeting Abstracts, 145.01
van der Hucht, K. A. 2001, New A Rev., 45, 135
Vanbeveren, D., De Donder, E., Van Bever, J., Van Rensbergen, W., & De Loore, C. 1998, New A, 3, 443
Vink, J. S., de Koter, A., & Lamers, H. J. G. L. M. 2001, *A&A*, 369, 574
Vink, J. S. 2017, *A&A*, 607, L8
Vink, J. S., de Koter, A., & Lamers, H. J. G. L. M. 2000, *A&A*, 362, 295
Wang, L., Gies, D. R., & Peters, G. J. 2018, *ApJ*, 853, 156

Discussion

KAWAI: Do you observe rotations in WR stars? It is important as WR stars could be progenitors of γ-Ray bursts.

SHENAR: From the upper limits that could be derived, no significant rotation (above 200 km/s) could be derived. There are a few peculiar WR stars with very round emission lines (Shenar *et al.* 2014, A&A, 562, 118), but whether this is the result of rotation or not is still not clear.

MARCHANT: How do you find the best-fitting evolutionary status? When you compare your results to evolution tracks, you should use Bayesian statistics to account for the likelihood of your targets to be in the inferred positions on the tracks.

SHENAR: I use a simple χ^2-fitting with a grid of models. I agree that the inferred HRD-positions may be affected by Bayesian statistics. However, the most important result - the initial mass - is virtually independent of the fitting method.

High-mass X-ray Binaries: Illuminating the Passage from Massive Binaries to Merging Compact Objects
Proceedings IAU Symposium No. 346, 2019
L. M. Oskinova, E. Bozzo, T. Bulik & D. Gies, eds.

doi:10.1017/S1743921319001327

Different generations of HMXBs: clues about their formation efficiency from Magellanic Clouds studies

Vallia Antoniou[1,2], Andreas Zezas[3,4,1], Jeremy J. Drake[1], Carles Badenes[5], Frank Haberl[6], Jaesub Hong[1,7], Paul P. Plucinsky[1] and the SMC XVP Collaboration Team

[1]Harvard-Smithsonian Center for Astrophysics, Cambridge, MA, USA
email: vantoniou@cfa.harvard.edu

[2]Texas Tech University, Department of Physics & Astronomy, Lubbock, TX, USA

[3]University of Crete, Department of Physics, Heraklion, Greece

[4]IESL, Foundation for Research and Technology-Hellas, Heraklion, Greece

[5]University of Pittsburgh, Department of Physics & Astronomy, Pittsburgh, PA, USA

[6]Max-Planck-Institut für extraterrestrische Physik, Garching, Germany

[7]Harvard University, Department of Astronomy, Cambridge, MA, USA

Abstract. Nearby star-forming galaxies offer a unique environment to study the populations of young (<100 Myr) accreting binaries. These systems are tracers of past populations of massive stars that heavily affect their immediate environment and parent galaxies. Using a Chandra X-ray Visionary program, we investigate the young neutron-star binary population in the low metallicity of the Small Magellanic Cloud (SMC) by reaching quiescent X-ray luminosity levels (~few times 10^{32} erg/s). We present the first measurement of the formation efficiency of high-mass X-ray binaries (HMXBs) as a function of the age of their parent stellar populations by using 3 indicators: the number ratio of HMXBs to OB stars, to the SFR, and to the stellar mass produced during the specific star-formation burst they are associated with. In all cases, we find that the HMXB formation efficiency increases as a function of time up to ~40–60 Myr, and then gradually decreases.

Keywords. X-rays: binaries, (galaxies:) Magellanic Clouds, (stars:) pulsars: general, stars: neutron, stars: emission-line, Be, stars: early-type, stars: formation

1. Introduction

Based on shallow *Chandra* and *XMM-Newton* surveys of the Magellanic Clouds, we have found that the HMXBs, the Be/X-ray binaries (Be-XRBs; HMXBs with Oe- or Be-type companions), and the X-ray pulsars are observed in regions with star-formation rate (SFR) bursts ~25–60 Myr (Antoniou *et al.* 2010) and ~6–25 Myr (Antoniou & Zezas 2016) ago in the SMC and the LMC, respectively. In Fig. 1 we present the average star-formation history of regions in the Magellanic Clouds with and without young X-ray binaries showing their association with stellar populations of different ages.

In order to study the young accreting binary population of the SMC in greater detail, we have been awarded a *Chandra* X-ray Visionary Program (XVP; PI A. Zezas) totaling 1.1 Ms of exposure time to observe 11 fields in this low-metallicity, star-forming galaxy. The fields were selected based on the age of their stellar populations, while on this dataset we have added 3 *Chandra* archival observations reaching the same 100 ks depth.

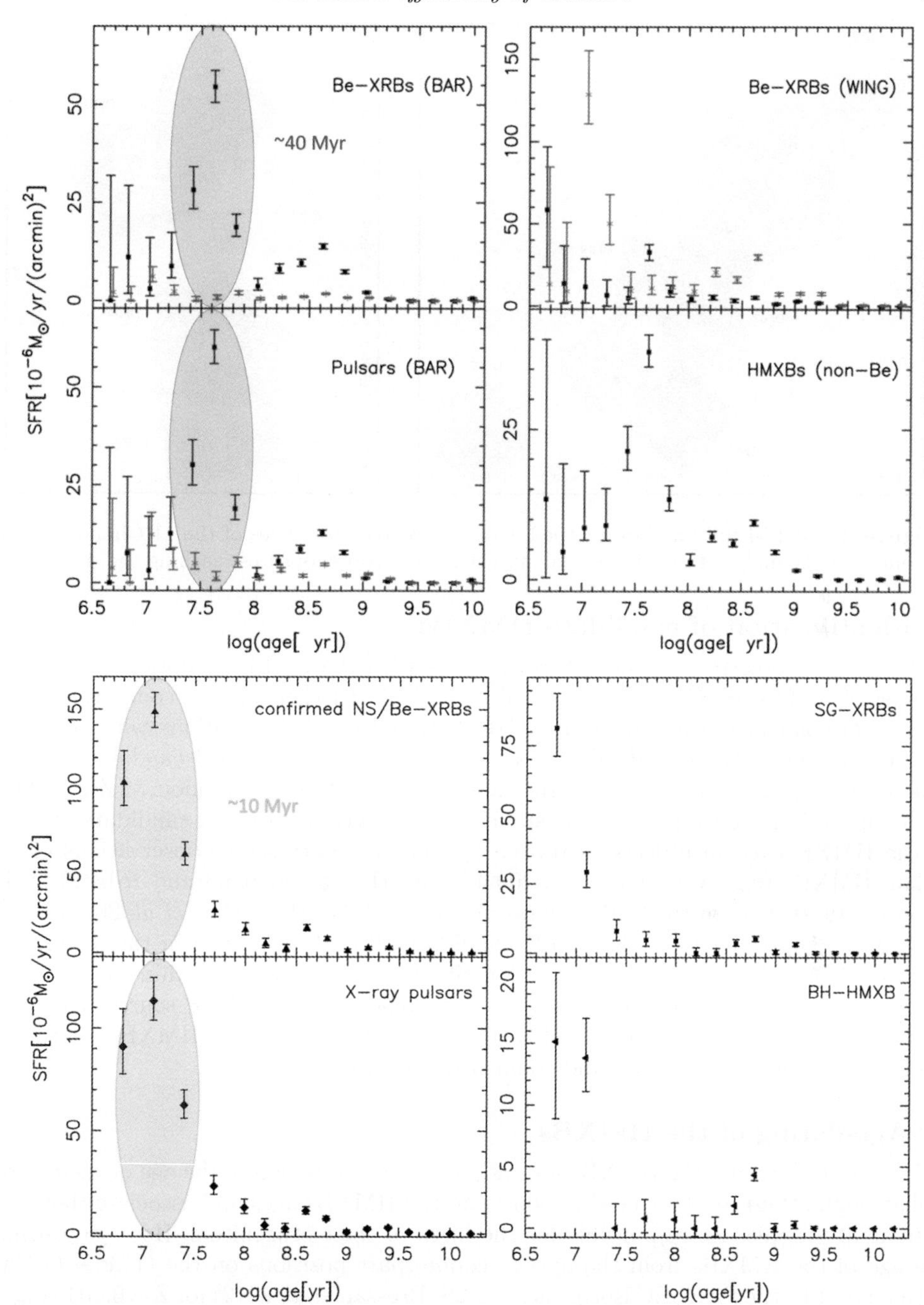

Figure 1. Average star-formation history of *(top)* SMC (Antoniou *et al.* 2010) and *(bottom)* LMC (Antoniou & Zezas 2016) regions with (black) and without (grey) young X-ray binaries showing their association with stellar populations of different ages. A clear peak in the SF history of these regions appears at ∼40 Myr and ∼10 Myr, respectively.

The details of the analysis will be presented in Antoniou et al. (2019a, in prep.), while the timing analysis and its results have been already presented in Hong *et al.* (2016) and Hong *et al.* (2017). In total, we have detected 2,393 sources down to a limiting flux of 2.6×10^{-16} erg cm^{-2} s^{-1} in the full (0.5 – 8.0 keV) band (Fig. 2).

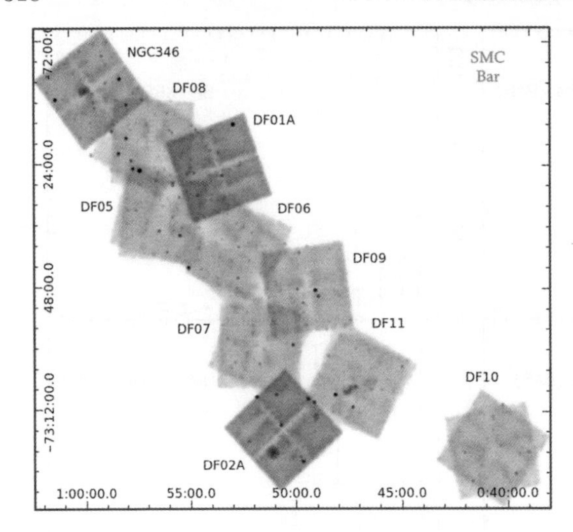

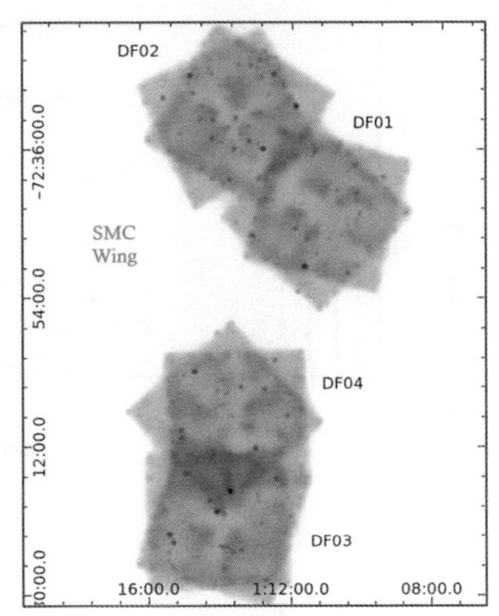

Figure 2. ACIS-I full band csmoothed exposure corrected images of the 11 Chandra X-ray Visionary fields analyzed in this work along with 3 archival exposures reaching the same depth.

2. Identification of candidate HMXBs

We then cross-correlated these X-ray sources with the OGLE-III optical photometric catalog of SMC stars (Udalski *et al.* 2008), and placed their optical counterparts on a $(V, V-I)$ color-magnitude diagram (CMD). Based on their positions with respect to the spectroscopically defined OB stars locus (Fig. 3), we identified candidate HMXBs (following Antoniou *et al.* 2009a, Antoniou *et al.* 2010, and Antoniou & Zezas 2016). This approach has been particularly successful in identifying correct candidate members of the HMXB class of objects as proven by optical spectroscopic observations of candidate HMXBs (e.g. Antoniou *et al.* 2009b), and Hα narrow-band and R-band optical photometric surveys of the SMC (Maravelias *et al.* 2017; Maravelias *et al.* 2018).

Next, we supplement the list of 127 candidate HMXBs identified in this work by 14 additional HMXBs identified by Haberl & Sturm (2016) that fall within the XVP area but have not been detected in our survey or matched with any XVP source that has at least one OB counterpart in the OGLE-III catalog. The final list of HMXBs used in this work consists of 141 candidate and confirmed HMXBs.

3. Age-dating of the HMXBs

In order to measure the HMXBs formation rate as a function of the age of their parent stellar populations, we first need to constrain the HMXBs ages and associate them with individual SF episodes responsible for the birth of their progenitors. Here, we estimate the age of the HMXBs from the optical counterpart positions on the $(V, V-I)$ CMD with respect to the PARSEC isochrones (v1.2S; Bressan *et al.* 2012) for Z = 0.004 (Fig. 3).

Following that we associate each optical counterpart with a SF episode that overlaps with the age-range of the isochrones that are consistent with its location on the CMD. For example, out of the 17 HMXBs identified in *Chandra* field DF11 (Fig. 3), only one has an optical counterpart photometrically consistent with the SF peak at 7 Myr, while the remaining 16 have ages consistent the second prominent SF peak at 42 Myr.

4. Formation efficiency of young accreting X-ray binaries

We investigate three different indicators of the formation efficiency of HMXBs as a function of age: the number of HMXBs, N(HMXBs), in different ages with respect to:

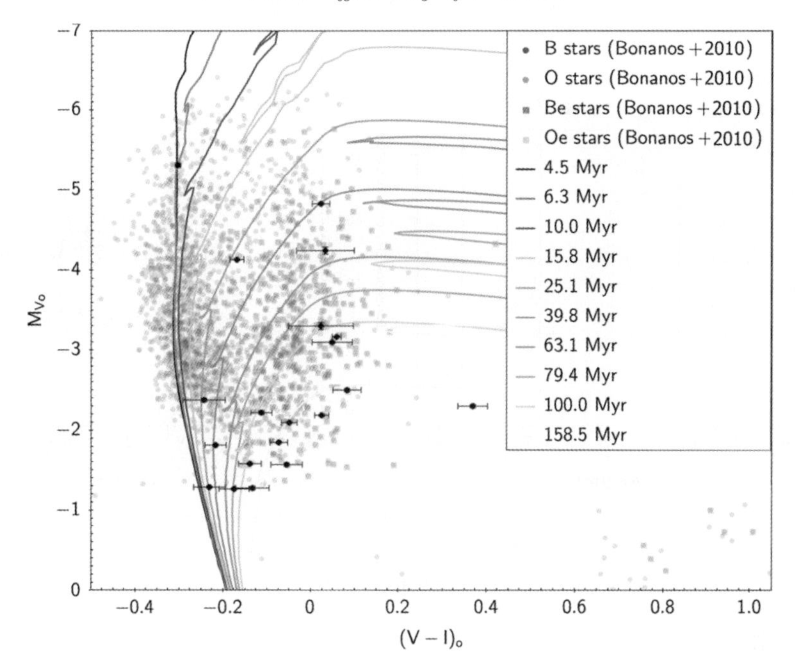

Figure 3. An example of the age determination of the HMXBs identified in *Chandra* field DF11 (black circles). V and $V-I$ have been corrected for extinction as $M_{V_o} = (m-M)_V - A_V = V - 18.96 - 0.25$, and $(V-I)_o = (V-I) - E(V-I) = (V-I) - 0.13$.

(i) the number of OB stars, N(OBs), in their respective *Chandra* field; *(ii)* the SFR of their parent stellar population; and *(iii)* the total stellar mass (M⋆) formed during the SF episode they are associated with. The OB stars are from the OGLE-III catalog (Udalski *et al.* 2008), while the SFRs as a function of age are from Harris & Zaritsky (2004). In Fig. 4, we present *for the first time* these three indicators of the formation efficiency of HMXBs as a function of age. A more detailed analysis of the production rate of these systems is presented in Antoniou et al. (2019b, subm.).

The N(HMXBs)/N(OBs) ratio is an indicator that can be calculated directly for any nearby galaxy with resolved stellar populations, without the need to derive their SF history. Therefore, it serves as a useful proxy of the relative formation rate of HMXBs that can be applied to large samples of galaxies. It also takes into account the present-day numbers of OB stars. On the other hand, N(HMXBs)/SFR takes into account the SF event that created the binaries we observe today, but not the duration of the SF burst. Finally, the ratio N(HMXBs)/M⋆ takes into account the SF burst duration (M⋆ is the integral of the SFR as a function of time), and is the fundamental relation that we were aiming to derive from this *Chandra* XVP program. We also note that the delay function of the HMXBs (Antoniou et al. 2019b, subm.) resembles best the ratio N(HMXBs)/M⋆.

In all three cases, we find an increase in the formation rate of the HMXBs for ages $\gtrsim$10–20 Myr and up to 40–60 Myr followed by a decline at older ages. This result is in agreement with studies of the formation efficiency of massive Oe/Be stars in the Magellanic Clouds (e.g. Bonanos *et al.* 2009, Bonanos *et al.* 2010), and the Milky Way (McSwain & Gies 2005) that show a peak at ages of ~20–50 Myr, matching the age of maximum production of HMXBs in the SMC.

VA acknowledges financial support from NASA/Chandra grants GO3-14051X, AR4-15003X, NNX15AR30G, NASA/ADAP grant NNX10AH47G and the Texas Tech President's Office. AZ acknowledges financial support from NASA/ADAP grant

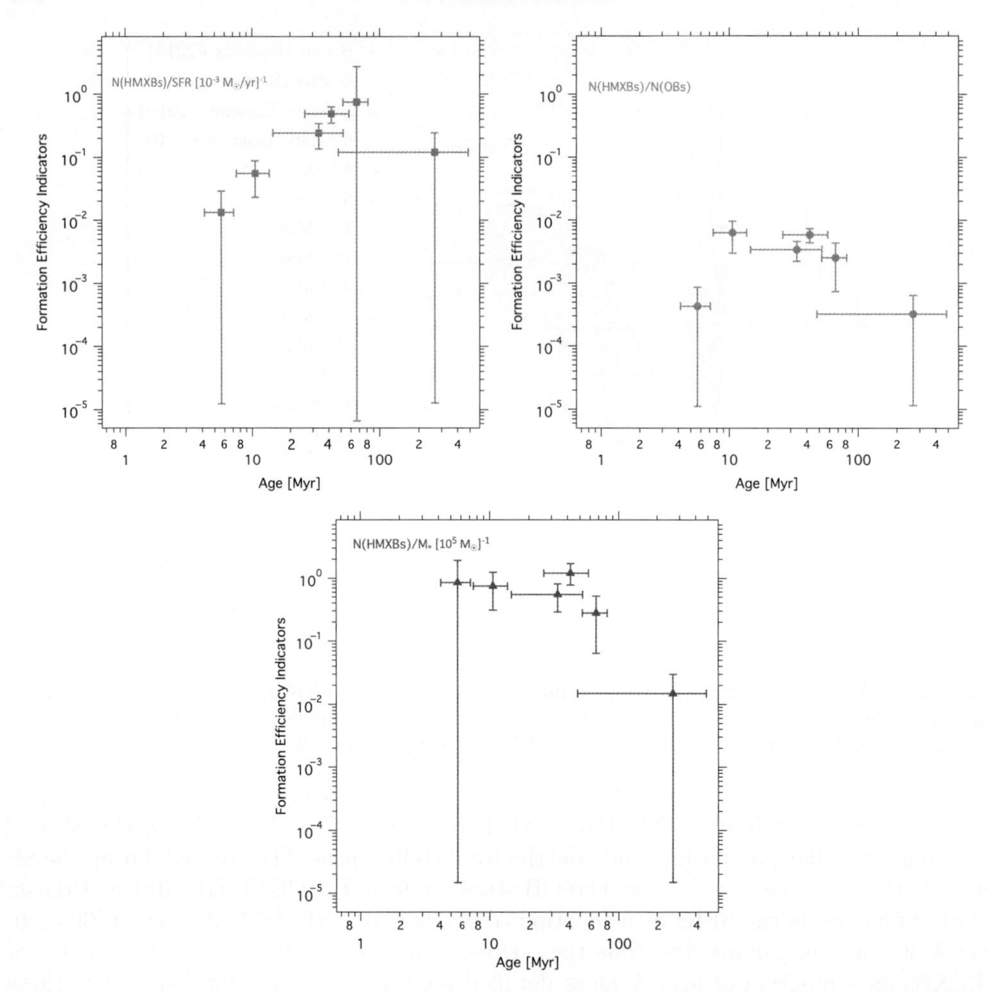

Figure 4. Different metrics of the HMXBs formation efficiency: number of HMXBs, N(HMXBs), with respect to the *(top left; green)* SFR of their parent stellar population; *(top right; blue)* number of OB stars, N(OBs); and *(bottom; red)* total stellar mass (M⋆) formed during the SF episode they are associated with. All these indicators show an increase in the HMXBs formation rate for ages $\gtrsim$10–20 Myr and up to 40–60 Myr followed by a decline at older ages.

NNX12AN05G and funding from the European Research Council under the European Union's Seventh Framework Programme (FP/2007-2013)/ERC Grant Agreement n. 617001. This project has also received funding from the European Union's Horizon 2020 research and innovation programme under the Marie Sklodowska-Curie RISE action, grant agreement No 691164 (ASTROSTAT). JJD and PPP were funded by NASA contract NAS8-03060 to the *Chandra X-ray Center.*

References

Antoniou, V., Hatzidimitriou, D., Zezas, A., & Reig, P. 2009b, *ApJ*, 707, 1080
Antoniou, V., & Zezas, A. 2016, *MNRAS*, 459, 528
Antoniou, V., Zezas, A., Hatzidimitriou, D., & Kalogera, V. 2010, *ApJ Letters*, 716, 140
Antoniou, V., Zezas, A., Hatzidimitriou, D., & McDowell, J. C. 2009a, *ApJ*, 697, 1695
Bonanos, A. Z., Lennon, D. J., Köhlinger, F., *et al.* 2010, *AJ*, 140, 416
Bonanos, A. Z., Massa, D. L., Sewilo, M., *et al.* 2009, *AJ*, 138, 1003

Bressan, A., Marigo, P., Girardi, L., *et al.* 2012, *MNRAS*, 427, 127
Haberl, F., & Sturm, R. 2016, *A&A*, 586, A81
Harris J., & Zaritsky D., 2004, *AJ*, 127, 1531
Hong, J., Antoniou, V., Zezas, A., *et al.* 2016, *ApJ*, 826, 4
Hong, J., Antoniou, V., Zezas, A., *et al.* 2017, *ApJ*, 847, 26
Maravelias, G., Zezas, A., Antoniou, V., Hatzidimitriou, D., & Haberl, F. 2017, The Lives and Death-Throes of Massive Stars, 329, 373
Maravelias, G., Zezas, A., Antoniou, V., Hatzidimitriou, D., & Haberl, F. 2018, arXiv:1811.10933
McSwain, M. V., & Gies, D. R. 2005, *ApJS*, 161, 118
Udalski, A., Soszyński, I., Szymański, M. K., *et al.* 2008, *AcA*, 58, 329

High-mass X-ray Binaries: Illuminating the Passage from Massive Binaries to Merging Compact Objects
Proceedings IAU Symposium No. 346, 2019
L. M. Oskinova, E. Bozzo, T. Bulik & D. Gies, eds.

doi:10.1017/S1743921318007470

Using High-Mass X-ray binaries to probe massive binary evolution: The age distribution of High-Mass X-ray binaries in M33

Kristen Garofali[1,2] and Benjamin F. Williams[1]

[1]Department of Astronomy, University of Washington
Box 351580, U.W., Seattle, WA, USA
emails: garofali@uark.edu, ben@astro.washington.edu

[2]Department of Physics, University of Arkansas,
825 West Dickson St, Fayetteville, AR, USA
email: garofali@uark.edu

Abstract. High-mass X-ray binaries (HMXBs) provide an exciting window into the underlying processes of both binary as well as massive star evolution. Because HMXBs are systems containing a compact object accreting from a high-mass star at close orbital separations they are also likely progenitors of gamma-ray bursts and gravitational wave sources. We present classification and age measurements for HMXBs in M33 using a combination of deep Chandra X-ray imaging, and archival *Hubble Space Telescope* data. We constrain the ages of the HMXB candidates by fitting the color-magnitude diagrams of the surrounding stars, which yield the star formation histories of the surrounding region. Unlike the age distributions measured for HMXB populations in the Magellanic Clouds, the age distribution for the HMXB population in M33 contains a number of extremely young (<5 Myr) sources. We discuss these results the context of the effect of host galaxy properties on the observed HMXB population.

Keywords. X-rays: binaries, Local Group, stars: imaging

1. Introduction

The majority of massive stars are in close binaries that will interact (e.g. via mass transfer, tides, and common envelope evolution), often dramatically altering the evolution of the component stars (Sana *et al.* 2012; Duchêne & Kraus 2013; Sana *et al.* 2014; Moe & Di Stefano 2017). The evolutionary paths taken depend strongly on often-unknown initial orbital parameters, and suffer from uncertainties in prescriptions for mass loss rates, rotation, metallicity effects, mass transfer efficiency, the common envelope phase, natal kicks, and more (Postnov *et al.* 2006; Dray 2006; Dominik *et al.* 2012; Ivanova *et al.* 2013; Smith 2014; Mandel 2016).

The uncertainties in massive binary evolution can be mitigated by observing and modeling sources at various evolutionary stages. High-mass X-ray binaries (HMXBs), systems containing a compact object (black hole [BH] or neutron star [NS]) and a high-mass ($\gtrsim 10 M_{\odot}$) star with close enough orbital separation such that accretion onto the compact object results in an X-ray-bright source, are an intermediate stage of evolution for close binaries with two massive components, and are potential precursors to close compact object binaries, which are sources of gravitational waves and short-duration gamma bursts if the systems merge within a Hubble time (Berger 2014; Belczynski *et al.* 2018).

Thus, HMXBs provide a unique snapshot in time in which to probe massive star and binary evolution, particularly as these relate to close compact object binary formation.

The Local Volume provides an ideal laboratory for studying HMXBs, as they are numerous in star-forming regions and resolvable with *Chandra* at these distances (Mineo *et al.* 2012). Surveys of resolved HMXB populations have probed key questions related to natal kick velocities, X-ray binary (XRB) formation efficiency in different environments, correlations with host galaxy star formation rate (SFR), and the evolutionary status of donor stars (Grimm *et al.* 2003; Coe *et al.* 2005; Antoniou *et al.* 2010; Mineo *et al.* 2012).

Resolved HMXB studies have also revealed subclasses of these binaries, which are typically defined by the spectral type of the donor star (nominally, the secondary in the binary). The most observationally abundant subclass of HMXBs are Be/X-ray binaries (Be-XRBs) (Liu *et al.* 2005; Liu *et al.* 2006), systems that consist of a compact object (most often a NS) accreting material as it passes through the circumstellar disk of an early B star. By contrast, there are fewer known HMXBs with supergiant companions (SG-XRBs), which are typically wind-fed systems, though may also be fed via Roche lobe overflow (RLOF). Still more rarely observed are the subclass of HMXBs that contain a compact object (usually a BH) accreting material from a Wolf-Rayet (WR) star, with only five such systems observed in the Local Volume to date (van Kerkwijk *et al.* 1996; Prestwich *et al.* 2007; Crowther *et al.* 2010; Liu *et al.* 2013; Maccarone *et al.* 2014).

Notably absent from recent HMXB population studies is M33, which is an ideal target as it is close to face-on, and does not suffer from the distance uncertainties and extinction that hamper Galactic samples, and furthermore, as a large star-forming spiral galaxy, probes an environment different from those in the well-studied, but much smaller, Magellanic Clouds (MCs). In addition, M33 has deep coverage from dedicated surveys with *Chandra*, *XMM-Newton* (Tüllmann *et al.* 2011; Williams *et al.* 2015), and the *Hubble Space Telescope* (*HST*), capable of resolving individual HMXBs and their donor stars. Furthermore, recent analysis of X-ray luminosity function (XLF) in M33 reveals a slope consistent with the universal slope expected for a population of HMXBs (Mineo *et al.* 2012), suggestive of a substantial population of HMXBs in the galaxy (Williams *et al.* 2015), of which very few ($\lesssim 5$) have been robustly characterized to date (Long *et al.* 2002; Pietsch *et al.* 2004; Pietsch *et al.* 2006; Orosz *et al.* 2007; Pietsch *et al.* 2009; Trudolyubov 2013).

Herein, we present measurement of the age distribution for new candidate HMXBs in M33. In Section 2 we briefly present the data used as part of this study, along with the source identification and SFH recovery techniques. In Section 3 we present the age distribution for all candidate HMXBs in the sample. Finally, in Section 4 and Section 5 we provide an analysis and discussion of these results in the context of binary evolution, and compact object binaries.

2. Data & Analysis Techniques

In this section, we introduce the catalogs used to find candidate HMXBs in M33, namely *Chandra* imaging from the *Chandra* ACIS Survey of M33 (ChASeM33, Tüllmann *et al.* 2011; hereafter T11) for localizing hard X-ray point sources, and archival *HST* data with at least two broad-band filters for selecting optical counterparts to X-ray sources. In addition, we discuss the color magnitude diagram (CMD) fitting technique used to produce spatially resolved SFHs, and thus ages, in the vicinities of candidate HMXBs.

2.1. *Chandra Catalog*

The X-ray sources mined for candidate HMXBs come from the ChASeM33 survey (T11), which consists of 662 X-ray sources aligned to the Two Micron All-Sky Survey

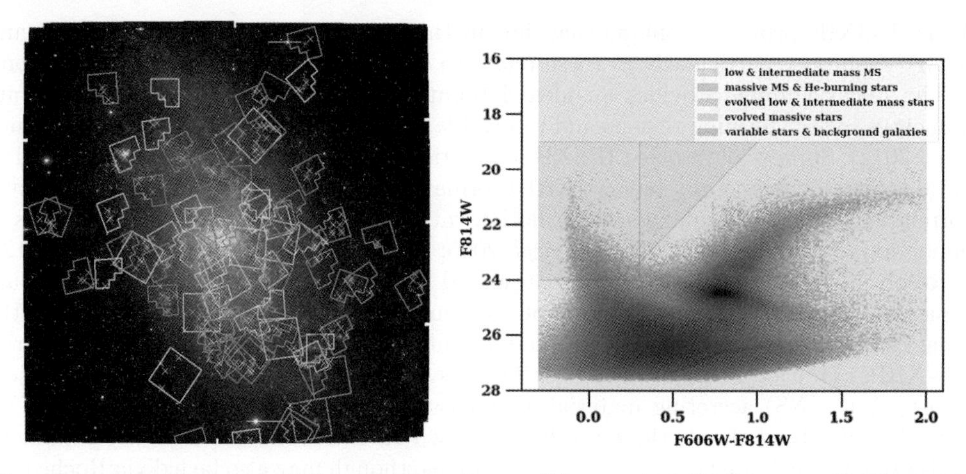

Figure 1. *Left*: R band Local Group Galaxy Survey (Massey *et al.* 2006) image of M33 overlaid with all 89 archival *HST* fields (grey and green), and all 36 fields containing candidate HMXBs (green only). Cyan crosses are X-ray sources from T11 that fall within the current archival *HST* coverage. There are 270 X-ray sources covered by 89 *HST* fields. *Right*: A Hess diagram constructed from all stars in one *HST* field, binned by 0.05 magnitudes in color and magnitude space. The shaded regions delineate sources in different mass ranges and stages of stellar evolution: green: low and intermediate mass main sequence stars; blue: massive main sequence stars (e.g. OB stars) and He-burning stars; red: evolved low and intermediate mass stars (e.g. red giant branch, red clump, and asymptotic giant branch stars); magenta: evolved massive stars (e.g. supergiants); grey: variable stars and background galaxies. Optical counterparts for HMXBs would most likely appear in the blue and magenta shaded regions.

(2MASS) (Cutri *et al.* 2003), and covers $\sim$ 70% of the D_{25} isophote of M33 down to a limiting 0.35-8.0 keV luminosity of 2.4×10^{34} erg s^{-1} with a total exposure time of 1.4 Ms.

For the purposes of selecting candidate HMXBs we used the X-ray source positions from T11 for pinpointing source locations, in addition to hardness ratios (HRs) and any prior classifications from T11 to narrow down the list of candidate HMXBs. In particular, HRs in the soft (0.35-1.1 keV), medium (1.1-2.6 keV), and hard (2.6-8.0 keV) bands were used for separating soft sources, foreground stars and supernova remnants (SNRs), from harder sources like background active galactic nuclei (AGN) and XRBs, both HMXBs and low-mass X-ray binaries (LMXBs).

2.2. *HST Photometry & Image Alignment*

Optical counterparts to the X-ray sources in T11 were selected using all available archival *HST* photometry in M33 with at least two broad-band filters, totaling 89 unique *HST* fields covering $\sim$ 40% of the T11 catalog (270 X-ray sources). The archival *HST* coverage of M33, and X-ray point sources from T11 within this coverage are shown in the left panel of Figure 1. Details of the *HST* data reduction, extraction of photometry, and specific fields used in this analysis are presented in Garofali *et al.* (2018).

Identifying *HST* counterparts to X-ray sources in T11 requires precise astrometric alignment to a common frame between data sets. As the T11 catalog is already aligned to 2MASS, we chose to align all archival *HST* fields to 2MASS as well. The alignment procedure coupled with position error from T11 results in X-ray error circles typically $\lesssim$ 0.7" in radius for sources considered here.

With *HST* fields aligned to a common frame as the X-ray data, we can search for optical counterparts to the X-ray sources from T11 with relatively low probabilities of chance coincidences, and thus spurious sources. We inspected all 89 archival *HST* fields for optical counterparts to X-ray sources, classifying optical counterparts and thus X-ray source types using color and magnitude cuts similar to those depicted in the right panel of Figure 1. We consider a strong candidate HMXB to be a hard X-ray source with an optical counterpart in its error circle that falls in the blue or magenta regions of the Hess diagram in the right panel of Figure 1. Using this criteria, we identify 55 candidate HMXBs, of which we estimate 11 ± 3 may be chance coincidences of OB stars with the X-ray error circle, and thus spurious sources.

2.3. *Color-Magnitude Diagram Fitting*

We measure ages, or formation timescales for the sample of candidate HMXBs in M33 using the stellar population surrounding each source. Below we briefly detail this process, which first involves selecting the photometric sample in the vicinity of each candidate, and then fitting the stellar photometry using models of stellar evolution.

For young sources ($\lesssim$ 50 Myr) such as HMXBs it is possible to determine the age, or formation timescale of the source based on the surrounding stellar population, as most stars are known to form in clusters, and thus in relatively co-eval populations that stay associated in regions of ~50 pc on timescales <60 Myr (Lada & Lada 2003; Gogarten *et al.* 2009; Eldridge *et al.* 2011). Thus, HMXB ages can be measured by performing fits to the CMDs of the surrounding stellar population to produce spatially resolved SFHs.

Because the onset of the HMXB phase may not immediately follow formation of the compact object, there is the possibility that an HMXB which received a strong natal kick when the primary formed a compact object could have moved away from its associated natal stellar population either into isolation, or into a region with a stellar population of unassociated age. However, both theory and observations suggest that very large natal kicks may not be common for HMXBs (Pfahl *et al.* 2002; Coe *et al.* 2005; Sepinsky *et al.* 2005; Linden *et al.* 2009; Coleiro & Chaty 2013). We therefore test stellar population region sizes of 50 pc and 100 pc for measuring SFHs, and find the results to be consistent in both cases. We discuss the results using only the 50 pc regions in Section 3.

To measure the spatially-resolved SFHs in the vicinity of each candidate HMXB, and thus recover ages for each source, we fit the CMDs of the stellar population within 50 pc of the candidate HMXB using the software `MATCH` (Dolphin 2002), with details of the fitting process and error analysis described in Garofali *et al.* (2018). The resultant SFHs return the SFR in each time bin, which can be used to derive the cumulative distribution of stellar mass formed in the timespan of interest by calculating the total amount of stellar mass formed, and then the associated fraction of the total stellar mass in each time bin. We report both the SFH as well as the cumulative distribution of stellar mass, as the latter can be interpreted as a probability distribution function for the age of the source, with the most likely age represented by the time bin containing the highest fraction of stellar mass formed.

An example of the SFH and associated cumulative stellar mass distribution out to 80 Myr for a candidate HMXB (source 013341.47+303815.9 from T11) is presented in the right and center panels of Figure 2. The most likely age (dotted line) is 7.1 Myr for this source, with the narrowest 68th percentile confidence interval in red. We also display the CMD of the stars within 50 pc of this source in the left panel of Figure 2, with the most likely donor star companion to the HMXB denoted by the cyan star, and Padova group stellar isochrones, consistent with the young age determined from the SFH, overlaid for reference.

3. Results

We identified 55 candidate HMXBs in M33 as hard X-ray point sources associated with potential massive donor star counterparts, the majority (~ 80%) of which had no classification prior to this work (Grimm *et al.* 2005; T11). The characteristics of each

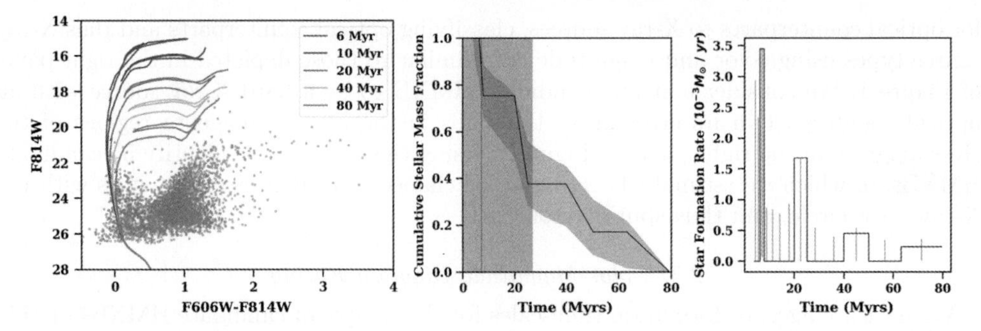

Figure 2. *Left*: The CMD representing the stellar population (grey points) within 50 pc of the HMXB candidate 013341.47+303815.9, with the candidate donor star denoted by the cyan star. Padova group isochrones shifted to the distance and extinction of M33 are overlaid for reference. *Center*: The cumulative stellar mass fraction (black line) formed within 80 Myr post-starburst. The Monte Carlo derived errors on this distribution are in grey, and the most likely age is marked with a dotted line. The narrowest 68th percentile confidence interval on the most likely age is marked in red. *Right*: The corresponding SFH for this candidate HMXB.

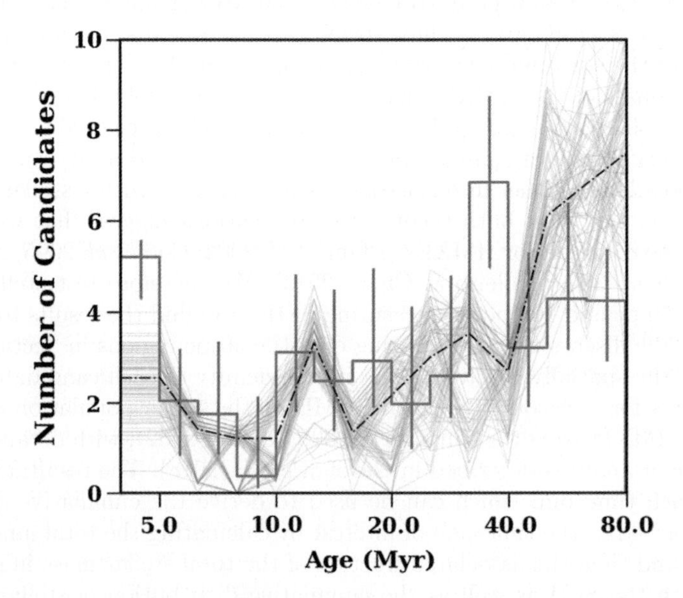

Figure 3. The age distribution calculated using 50 pc regions for all 41 HMXBs in M33 that are in fields sensitive to the MSTO at 80 Myr plotted as a solid blue line. The solid blue histogram represents the number of systems expected in each age bin given the SFHs for all sources. The light grey lines are the control sample: the equivalent age distribution for 100 random draws of 41 non-HMXB X-ray sources in M33, with the median of these 100 draws as a dash-dot line in black.

candidate HMXB are presented in Garofali *et al.* (2018). Here, we describe the age distribution for this population, as measured from the combination of the SFHs for all candidates in the archival *HST* coverage where the imaging was deep enough to be sensitive to the main sequence turn-off at 80 Myr (41 candidate HMXBs).

We built up the age distribution for all candidate HMXBs in fields that passed our depth cuts as the sum of the stellar-mass weighted star formation in each bin, normalized to a total stellar mass formed in 80 Myr for each candidate HMXB of one. This summation yields a distribution that reflects the number of HMXBs expected to form in each time bin. This analysis accounts for the entire SFH for each source, and in this way does not rely on calculating a specific age individual candidate HMXBs.

The HMXB age distribution, built up from the stacked SFHs, is shown in Figure 3 as a solid blue histogram. The light grey lines represent 100 draws of the age distribution for a "control" population, constructed from SFHs in the regions around 41 X-ray sources that are not HMXBs. The median of the 100 realizations is plotted as a dash-dot black line. Compared to the control sample, the age distribution for the candidate HMXBs contains two notable peaks: a young population present at < 5 Myr post-starburst, and a delayed population appearing at 40 Myr post-starburst. In Section 4 and Section 5 we discuss these two populations in the context of binary evolution, and compact object binaries.

4. Discussion

In this section we discuss how the host galaxy environment affects HMXB formation efficiency, and how the age distribution for HMXBs in M33 relates to different HMXB subpopulations.

4.1. *Age Distribution and Environment*

The age distribution for HMXBs has been measured from spatially-resolved SFHs in a number of other nearby galaxies, finding similar preference for HMXB formation ~ 40 Myr post-starburst in the Small Magellanic Cloud (SMC), NGC 300 and NGC 2403 (Antoniou *et al.* 2010; Williams *et al.* 2013), and 6-25 Myr post-starburst in the Large Magellanic Cloud (LMC) (Antoniou & Zezas 2016). These differing timescales for formation can stem both from differences in the host environment, as well as point to distinct HMXB subpopulations. We discuss the former here in terms of the effects of metallicity, and the latter in the following section in terms of binary evolution.

Metallicity may play a role in HMXB formation efficiency via its effect on the line-driven winds of massive stars, which may affect HMXB evolution via mass loss and orbital evolution through loss of angular momentum (Dray 2006; Linden *et al.* 2010). Because M33 has a measured metallicity gradient, from solar in the interior 3 kpc to LMC-like beyond 3 kpc (Magrini *et al.* 2007), we tested for any metallicity dependence in the HMXB subpopulations by constructing the age distribution separately for sources with galactocentric radii < 3 kpc, and those with radii > 3 kpc. We find that the small number of sources outside 3 kpc prohibits any definitive statement on the differences between HMXB populations in M33 at solar versus LMC metallicities. However, we do note that the very young sources (< 5 Myr) are found throughout the galaxy, and therefore don't appear to be a strongly metallicity dependent population, at least above the threshold metallicity that is roughly LMC-like.

4.2. *Age Distribution and Binary Evolution*

The age distribution for HMXBs in M33 has distinct peaks at very young ages (< 5 Myr), and intermediate ages (~ 40 Myr). Below, we discuss these features in the context of HMXB subpopulations and binary evolution.

4.2.1. Early Onset HMXBs: SG-XRBs with Massive Compact Objects?

The formation and appearance of candidate HMXBs < 5 Myr post-starburst in M33 implies a population of massive progenitors ($\gtrsim 40\ M_{\odot}$) with the potential for forming massive BHs (Fryer *et al.* 2012). Indeed binary population synthesis simulations predict formation of a population of BHs accreting material from the winds of their supergiant companions < 10 Myr after a burst of star formation (Linden *et al.* 2009). However mass thresholds for BH formation often rely on models of single star evolution. Binary evolution processes, such as mass transfer and envelope stripping, allow for formation of NS relatively rapidly post-starburst ($\sim$ 5 Myr), and therefore from very massive progenitors, at least some small fraction of the time (Belczynski & Taam 2008). Thus very young ages, as measured here for a subsample of candidate HMXBs in M33, do not definitively point to a population of BHs. The only source in this young subpopulation for which the compact object has been independently confirmed to be a BH is M33 X-7 (Orosz *et al.* 2007).

We can also infer HMXB characteristics by analyzing the candidate donor star photometry. In particular, if models suggest that young systems are likely SG-XRBs hosting BHs then we may expect the observed young candidate HMXBs in M33 to have the brightest donor star counterparts in the sample. In fact, for most of these young sources the *HST* colors and magnitudes for the likely donor stars are not bright enough to be indicative of supergiants. The relatively faint photometry of the optical counterparts for these young candidate HMXBs coupled with rapid formation timescales measured from SFHs suggests that these systems either have B type donor stars coupled with initially very massive primaries, that the unexpected secondary star photometry is a signature of binary evolution, or that these systems were actually formed at later times. We expound upon each of these possibilities below.

Current models of binary evolution predict that Be-BH XRBs should be rare, as the high mass ratio between the very massive primary needed to form a BH and the B star secondary will often result in a common envelope merger (Belczynski & Ziolkowski 2009). As there is only one known Be-BH binary in the Milky Way (Casares *et al.* 2014; Ribó *et al.* 2017), it seems unlikely there would be multiple such sources in M33, i.e. B star secondaries coupled with much more massive primaries. Instead of originating from systems with steep mass ratios, these young candidate HMXBs may have donor stars that due to binary evolution processes only appear under-massive. This is difficult to quantify, however, as the expected photometry of the counterpart given past history of binary interactions (e.g. mass transfer, common envelope evolution, spin-up, etc.) is difficult to constrain without knowledge of the initial binary parameters and detailed modeling. Finally, it is instead possible that these candidate HMXBs are not as young as the *most likely* age from the SFH suggests. In particular, for sources lying in regions with multiple bursts of star formation we can only quantify the probability that a source formed at a particular time, and in fact the young candidate HMXBs discussed here all have non-zero probability of forming at later times post-starburst, consistent with fainter donor stars.

The uncertainty regarding the relationship between system age and system components discussed here highlights the need for follow-up, both spectroscopically to determine donor star type, and also via novel methods for discerning compact object type and mass. In the absence of dynamical mass measurements for a large number of sources, *NuSTAR* observations of bright candidate HMXBs in M33 may help determine compact object type (Lazzarini *et al.* 2018).

4.2.2. Delayed Onset HMXBs: Be-XRBs

The peak in HMXB formation efficiency around 40 Myr has previously been ascribed to Be-XRBs (Linden *et al.* 2009; Antoniou *et al.* 2010; Williams *et al.* 2013), as these systems should form efficiently on these timescales given considerations of the IMF and binary evolution.

Under the assumption of a Kroupa IMF and a minimum mass threshold for compact object formation, there will be more NS that form from systems right at the minimum mass threshold than NS forming from progenitors more massive than this. Theory and observation suggest this minimum mass for NS formation is $\sim 8\ M_{\odot}$, which corresponds to $\sim$ 40 Myr stellar lifetime, and thus the formation of the largest number of NS on this timescale. In addition, the Be phenomenon, i.e. B star mass loss activity, has been shown observationally to peak between $\sim$ 25-80 Myr, implying that the Be phenomenon itself may be the product of binary evolution (McSwain & Gies 2005), namely B star secondaries spun up by mass transfer in a binary on these timescales.

The donor star photometry for the candidate HMXBs contributing to the 40 Myr peak in the age distribution in M33 is compatible with this picture, with colors and magnitudes consistent with B stars. Spectroscopic follow-up is still needed to determine which systems actually contain Be stars, which will also cull the sample for X-ray follow-up to determine which systems host NS compact objects.

5. Implications

The primary features of the candidate HMXB age distribution in M33, namely peaks in HMXB formation efficiency at < 5 Myr and $\sim$ 40 Myr post-starburst, are indicative of HMXB subpopulations, possibly SG-XRBs hosting BHs and Be-XRBs hosting NS, respectively. As HMXBs are ostensibly products of binary evolution that have remained bound following the death of the primary, studying the past and future evolution of these systems can be an important tool for understanding the binary evolution pathways that lead to compact object binaries, and possibly compact object mergers. Below, we discuss the observed subpopulations of HMXBs in M33 and their implications for compact object binaries and gravitational wave sources.

Young HMXBs are more likely to host BHs as compact objects, but understanding their future evolution and ultimate fate in the context of gravitational wave sources requires a detailed characterization of the system components. The only young HMXB in M33 for which this kind of characterization has been done is M33 X-7, which, with a $\sim$ 15.65 $M_\odot$ BH and a $\sim$ 70 $M_\odot$ donor star secondary, has the possibility to form a BH-BH binary if it survives common envelope evolution and avoids a merger. Even so, systems like this are clearly incapable of being precursors to some of the more massive BH-BH mergers observed by LIGO, which are more likely to originate from systems like WR-XRBs. However, young SG-XRBs with the right orbital configuration may be capable of forming BH-NS binary mergers, for which the merger rate densities are not yet constrained by LIGO.

The prevalence of Be-XRBs on $\sim$ 40 Myr timescales in M33 and other Local Group populations may be informative regarding preferred evolution channels for efficiently forming the precursors to NS-NS binaries. The appearance of Be-XRBs on these timescales requires that the binary survives the death throes of the primary, which may preferentially occur if the primary undergoes an electron capture supernova that imparts a smaller kick to the system (Linden *et al.* 2009). The efficiency with which this supernova mechanism operates to produce bound binaries, and the particular timescale on which it functions may be a function of metallicity (Podsiadlowski *et al.* 2004), though observational studies in the Local Group demonstrate that Be-XRBs form efficiently between $\sim$ 25-60 Myr across a range of metallicities from sub-solar to solar. If such Be-XRBs survive the future evolution of the Be star secondary, this implies an efficient pathway for forming close NS-NS binaries, and therefore possible NS-NS mergers in a range of environments (Tauris *et al.* 2017).

Leveraging HMXBs as probes of binary evolution, and using them to understand the pathways that lead to compact object mergers requires close feedback between theory and observations. In particular, full characterization of an HMXB, from primary and secondary star properties to orbital characteristics, provides constraints for models of binary evolution by requiring the model prescriptions for common envelope evolution, mass transfer, natal kicks, and more to effectively reproduce the observed system parameters. Likewise, well-tested models of binary evolution can be used to evolve HMXBs forward in time to better understand their connection to close compact object binary populations. The sample of candidate HMXBs in M33 and their measured age distribution discussed here provides a population ripe for follow-up to continue illuminating the connections between HMXBs and compact object mergers.

References

Antoniou, V., & Zezas, A. 2016, *MNRAS*, 459, 528
Antoniou, V., Zezas, A., Hatzidimitriou, D., & Kalogera, V. 2010, *ApJ*, 716, L140
Belczynski, K., & Taam, R. E. 2008, *ApJ*, 685, 400
Belczynski, K., & Ziolkowski, J. 2009, *ApJ*, 707, 870
Belczynski, K., Askar, A., Arca-Sedda, M., *et al.* 2018, *A&A*, 615, A91
Berger, E. 2014, *ARA&A*, 52, 43
Berghea, C. T., Dudik, R. P., Tincher, J., & Winter, L. M. 2013, *ApJ*, 776, 100
Binder, B., Williams, B. F., Kong, A. K. H., *et al.*. 2016, *MNRAS*, 457, 1636
Casares, J., Negueruela, I., Ribó, M., *et al.* 2014, *Nature*, 505, 378
Coe, M. J., Edge, W. R. T., Galache, J. L., & McBride, V. A. 2005, *MNRAS*, 356, 502
Coleiro, A., & Chaty, S. 2013, *ApJ*, 764, 185
Crowther, P. A., Barnard, R., Carpano, S., *et al.* 2010, *MNRAS*, 403, L41
Cutri, R. M., Skrutskie, M. F., van Dyk, S., *et al.* 2003, 2MASS All Sky Catalog of point sources.
—. 2002, *MNRAS*, 332, 91
Dominik, M., Belczynski, K., Fryer, C., *et al.* 2012, *ApJ*, 759, 52
Dray, L. M. 2006, *MNRAS*, 370, 2079
Duchêne, G., & Kraus, A. 2013, *ARA&A*, 51, 269
Eldridge, J. J., Langer, N., & Tout, C. A. 2011, *MNRAS*, 414, 3501
Fryer, C. L., Belczynski, K., Wiktorowicz, G., *et al.* 2012, *ApJ*, 749, 91
Garofali, K., Williams, B. F., Hillis, T., *et al.*. 2018, *MNRAS*, 479, 3526
Gogarten, S. M., Dalcanton, J. J., Williams, B. F., *et al.*. 2009, *ApJ*, 691, 115
Grimm, H.-J., Gilfanov, M., & Sunyaev, R. 2003, *MNRAS*, 339, 793
Grimm, H.-J., McDowell, J., Zezas, A., Kim, D.-W., & Fabbiano, G. 2005, *ApJS*, 161, 271
Ivanova, N., Justham, S., Chen, X., *et al.* 2013, *A&A Rev.*, 21, 59
Jennings, Z. G., Williams, B. F., Murphy, J. W., *et al.* 2012, *ApJ*, 761, 26
—. 2014, *ApJ*, 795, 170
Lada, C. J., & Lada, E. A. 2003, *ARA&A*, 41, 57
Lazzarini, M., Hornschemeier, A. E., Williams, B. F., *et al.* 2018, *ApJ*, 862, 28
Linden, T., Kalogera, V., Sepinsky, J. F., *et al.* 2010, *ApJ*, 725, 1984
Linden, T., Sepinsky, J. F., Kalogera, V., & Belczynski, K. 2009, *ApJ*, 699, 1573
Liu, J.-F., Bregman, J. N., Bai, Y., Justham, S., & Crowther, P. 2013, *Nature*, 503, 500
Liu, Q. Z., van Paradijs, J., & van den Heuvel, E. P. J. 2005, *A&A*, 442, 1135
—. 2006, *A&A*, 455, 1165
Long, K. S., Charles, P. A., & Dubus, G. 2002, *ApJ*, 569, 204
Maccarone, T. J., Lehmer, B. D., Leyder, J. C., *et al.* 2014, *MNRAS*, 439, 3064
Magrini, L., Vílchez, J. M., Mampaso, A., Corradi, R. L. M., & Leisy, P. 2007, *A&A*, 470, 865
Mandel, I. 2016, *MNRAS*, 456, 578
Massey, P., Olsen, K. A. G., Hodge, P. W., *et al.* 2006, *AJ*, 131, 2478
McSwain, M. V., & Gies, D. R. 2005, *ApJS*, 161, 118
Mineo, S., Gilfanov, M., & Sunyaev, R. 2012, *MNRAS*, 419, 2095
Moe, M., & Di Stefano, R. 2017, *ApJS*, 230, 15
Orosz, J. A., McClintock, J. E., Narayan, R., *et al.* 2007, *Nature*, 449, 872
Pfahl, E., Rappaport, S., Podsiadlowski, P., & Spruit, H. 2002, *ApJ*, 574, 364
Pietsch, W., Haberl, F., Sasaki, M., *et al.* 2006, *ApJ*, 646, 420
Pietsch, W., Mochejska, B. J., Misanovic, Z., *et al.* 2004, *A&A*, 413, 879
Pietsch, W., Haberl, F., Gaetz, T. J., *et al.* 2009, *ApJ*, 694, 449

Podsiadlowski, P., Langer, N., Poelarends, A. J. T., *et al.* 2004, *ApJ*, 612, 1044
Postnov, K. A., & Yungelson, L. R. 2006, Living Reviews in Relativity, 9, 6
Poutanen, J., Fabrika, S., Valeev, A. F., Sholukhova, O., & Greiner, J. 2013, *MNRAS*, 432, 506
Prestwich, A. H., Kilgard, R., Crowther, P. A., *et al.* 2007, *ApJ*, 669, L21
Ribó, M., Munar-Adrover, P., Paredes, J. M., *et al.* 2017, *ApJ*, 835, L33
Rizzi, L., Tully, R. B., Makarov, D., *et al.* 2007, *ApJ*, 661, 815
Sana, H., de Mink, S. E., de Koter, A., *et al.* 2012, Science, 337, 444
Sana, H., Le Bouquin, J.-B., Lacour, S., *et al.* 2014, *ApJS*, 215, 15
Sepinsky, J., Kalogera, V., & Belczynski, K. 2005, *ApJ*, 621, L37
Smith, N. 2014, *ARA&A*, 52, 487
Tauris, T. M., Kramer, M., Freire, P. C. C., *et al.* 2017, *ApJ*, 846, 170
Trudolyubov, S. P. 2013, *MNRAS*, 435, 3326
Tüllmann, R., Gaetz, T. J., Plucinsky, P. P., *et al.* 2011, *ApJS*, 193, 31
van Kerkwijk, M. H., Geballe, T. R., King, D. L., van der Klis, M., & van Paradijs, J. 1996, *A&A*, 314, 521
Williams, B. F., Binder, B. A., Dalcanton, J. J., Eracleous, M., & Dolphin, A. 2013, *ApJ*, 772, 12
Williams, B. F., Hatzidimitriou, D., Green, J., *et al.* 2014, *MNRAS*, 443, 2499
Williams, B. F., Wold, B., Haberl, F., *et al.* 2015, *ApJS*, 218, 9
Williams, B. F., Lazzarini, M., Plucinsky, P., *et al.* 2018, ArXiv e-prints, arXiv:1808.10487

Discussion

FRAGOS: Orbital periods are important constraints for determining the evolutionary history of an XRB. What is the prospect of measuring orbital periods for some of the HMXBs with identified companions?

GAROFALI: Many of the donor star companions identified in M33 are faint (22nd-24th magnitude), so a 10-meter class telescope such as Keck would be needed to first confirm companions spectroscopically, and then follow-up with dedicated radial velocity monitoring. For the smaller number of sources with companions brighter than 22nd magnitude, spectroscopic monitoring to determine donor star spectral class and orbital period would be possible with an 8-meter telescope like Gemini. The first step should be spectroscopic follow-up for the strongest candidate HMXBs to further cull the list down to the best sources for more time-intensive radial velocity monitoring to measure orbital periods.

High-mass X-ray Binaries: Illuminating the Passage from Massive Binaries to Merging Compact Objects
Proceedings IAU Symposium No. 346, 2019
L. M. Oskinova, E. Bozzo, T. Bulik & D. Gies, eds.

doi:10.1017/S1743921318007627

The High Mass X-ray binaries in star-forming galaxies

M. Celeste Artale[1], **Nicola Giacobbo**[2,3,4], **Michela Mapelli**[1,2,3] **and Paolo Esposito**[5]

[1]Institut für Astro- und Teilchenphysik, Universität Innsbruck, Technikerstrasse 25/8, 6020 Innsbruck, Austria

[2]INAF, Osservatorio Astronomico di Padova, vicolo dell'Osservatorio 5, I–35122 Padova, Italy

[3]INFN, Milano Bicocca, Piazza della Scienza 3, I–20126, Milano, Italy

[4]Dipartimento di Fisica e Astronomia "G. Galilei", Università di Padova, vicolo dell'Osservatorio 3, I-35122, Italy

[5]INAF–Istituto di Astrofisica Spaziale e Fisica Cosmica di Milano, via E. Bassini 15, 20133 Milano, Italy
email: mcartale@gmail.com

Abstract. The high mass X-ray binaries (HMXBs) provide an exciting framework to investigate the evolution of massive stars and the processes behind binary evolution. HMXBs have shown to be good tracers of recent star formation in galaxies and might be important feedback sources at early stages of the Universe. Furthermore, HMXBs are likely the progenitors of gravitational wave sources (BH–BH or BH–NS binaries that may merge producing gravitational waves). In this work, we investigate the nature and properties of HMXB population in star-forming galaxies. We combine the results from the population synthesis model MOBSE (Giacobbo & Mapelli 2018a) together with galaxy catalogs from EAGLE simulation (Schaye *et al.* 2015). Therefore, this method describes the HMXBs within their host galaxies in a self-consistent way. We compute the X-ray luminosity function (XLF) of HMXBs in star-forming galaxies, showing that this methodology matches the main features of the observed XLF.

Keywords. X-rays: binaries, galaxies: stellar content, galaxies: evolution

---

1. Introduction

High Mass X-ray Binaries (HMXBs) are systems composed of a compact object (neutron star NS, or black hole BH) and a massive companion star. Observational results have shown that HMXBs are good tracers of the star formation rate (SFR) within their host galaxies (Grimm *et al.* 2003, Mineo *et al.* 2012), and might be important heating and ionizing sources in the early Universe (e.g., Justham & Schawinski 2012, Artale *et al.* 2015, Douna *et al.* 2018, Garratt-Simthson *et al.* 2018). From a theoretical point of view, studying the population of HMXBs within galaxies is essential to understand their role in the aforementioned processes and the binary evolution.

In particular, the X-ray luminosity function (XLF) is an excellent tracer describing the global population of HMXBs in galaxies. It can also help to investigate the nature of ultraluminous X-ray sources (Mapelli *et al.* 2010, Kaaret *et al.* 2017). Several observational results show that the XLF of HMXBs is described by a power law with a slope of ~ 1.6, and normalization proportional to the SFR (Grimm *et al.* 2003, Mineo *et al.* 2012).

Population synthesis models have proved to be useful to describe the HMXB population of individual galaxies (e.g., Belczynski *et al.* 2004), and to predict the XLF of star-forming galaxies (e.g., Zuo *et al.* 2014). However, they cannot describe the diversity of stellar ages and metallicities within a galaxy in a self-consistent way.

In order to properly model star formation and metallicity evolution in galaxies, population synthesis simulations must be coupled with galaxy catalogs from galaxy formation models. Such galaxy catalogs can be obtained either from semianalytic models (Fragos *et al.* 2013), or from hydrodynamical cosmological simulations (Mapelli *et al.* 2017, 2018a, 2018b, Artale *et al.* in preparation).

In this work, we study the XLF of star-forming galaxies combining the galaxy catalogs of the hydrodynamical cosmological simulation EAGLE (Schaye *et al.* 2015) with the results from the population synthesis model MOBSE (Giacobbo & Mapelli 2018a). In Section 2 we present the methodology. We discuss our findings in Section 3.

2. Simulations and methodology

MOBSE (Giacobbo & Mapelli 2018a) is an upgraded version of BSE code (Hurley *et al.* 2002).The code includes new stellar winds prescription (Vink *et al.* 2001, 2005, Chen *et al.* 2015), electron-capture SNe (Giacobbo & Mapelli 2018b), core-collapse SNe (Fryer *et al.* 2012), pulsational pair-instability and pair-instability SNe (Spera & Mapelli 2017). MOBSE reproduces successfully the masses and merger rates of compact objects (Giacobbo & Mapelli 2018b,c) inferred by the LIGO-Virgo collaboration. In this work we adopt the simulation set referred to as $\alpha 1$ in Giacobbo & Mapelli (2018b). In this model, the common envelope parameter is set to $\alpha = 1$. This set is composed of 12 subsets at different metallicities Z = 0.0002, 0.0004, 0.0008, 0.0012, 0.0016, 0.002, 0.004, 0.006, 0.008, 0.012, 0.016 and 0.02. Each sub-set contains 10^6 binaries, hence the total number of binaries is 1.2×10^7. From the population synthesis model, we identify the HMXB sources undergoing stable mass transfer via Roche lobe overflow (RLO–HMXB), and those accreting the wind from the companion star (SW–HMXB). The X-ray luminosity of each HMXB in the catalog is computed as $L_X = \eta \frac{G\dot{M}_{acc} M_{co}}{R_{co}}$, where $\dot{M}_{acc}$, M_{co}, and R_{co} are the accretion rate, the mass, and the radius of the compact object, respectively. The parameter η is the efficiency in converting gravitational binding energy to radiation associated with accretion. We adopt that $\eta = 0.1$ for BH and NS for simplicity since BH–HMXB sources are the dominant population (see ahead in the text). On the left panel of Figure 1, we show the cumulative distribution of HMXBs normalized by the total number of sources for each metallicity sub-set. We also split the contribution of SW and RLO systems in each subsample.

The EAGLE simulation suite is a set of cosmological hydrodynamical simulations with different resolution levels and box sizes, run using an updated version of GADGET-3 code. It adopts the ΛCDM cosmology with cosmological parameters $\Omega_m = 0.2588$, $\Omega_\Lambda = 0.693$, $\Omega_b = 0.0482$, and $H_0 = 100\, h$ km s^{-1}Mpc^{-1} with $h = 0.6777$ (Planck Collaboration 2014). The simulation includes subgrid models accounting for star formation, UV/X-ray ionizing background, radiative cooling and heating, stellar evolution, chemical enrichment, AGB stars and SNe feedback, and supermassive black hole feedback. In this work we use the simulated box named as L0100N1504. This run represents a periodic box of 100 Mpc side, which initially contains 1504^3 gas and dark matter particles with masses of $m_{gas} = 1.23 \times 10^6 h^{-1} M_\odot$ and $m_{dm} = 6.57 \times 10^6 h^{-1} M_\odot$. Since this work is focused on the analysis of star forming galaxies in the local universe, we use the galaxy catalog at $z = 0$. We select a subsample of galaxies with specific star formation rate of $sSFR > 10^{-10} M_\odot yr^{-1}$ and stellar masses in the range of $M_* = 10^8 - 5 \times 10^{10}\, M_\odot$. The number of galaxies fulfilling this condition is 2596.

For each simulated galaxy in the subsample, we identify the youngest stellar particles with ages below to 100 Myr. Since the progenitors of HMXBs are systems composed

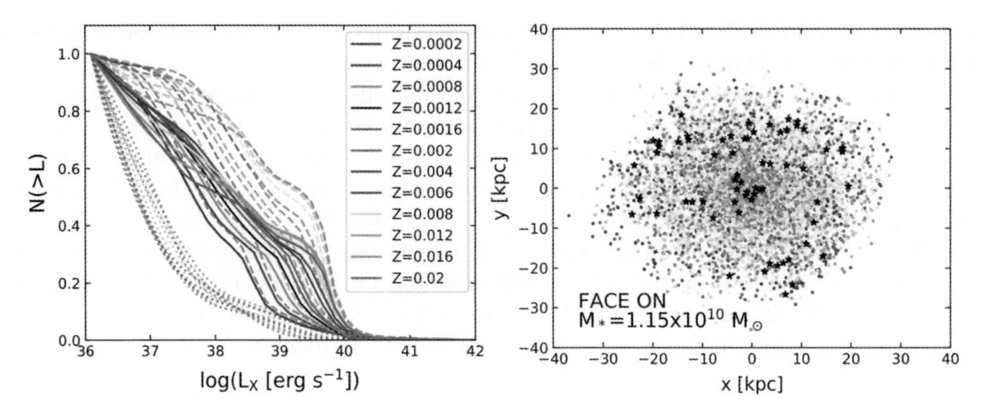

Figure 1. *Left panel:* Cumulative number of HMXBs normalized by the total number of sources, obtained with MOBSE for the 12 adopted metallicities (filled lines). We also show the contribution from SW-HMXBs (dotted lines) and RLO-HMXBs systems (dashed lines). *Right panel:* Spatial distribution of the stellar particles within one of the galaxies in the EAGLE catalog, showed face on. The color code represents the ages of the stellar particles (the oldest and the youngest populations are marked in red and in blue, respectively). Stellar particles that were formed in the last 100 Myr are shown by black stars (these particles are populated with HMXBs, see the details in Section 2).

of two massive stars, these sources are directly connected with the star-forming regions within the galaxies. The assumption to select the youngest stellar particles (< 100 Myr) accounts for this fact. In Figure 1 right panel, we show the spatial distribution of the stellar particles in one of the galaxies of the sub-sample. We find that the stellar particles with age < 100 Myr (indicated by black stars in Figure 1) are mainly in the outskirts of this galaxy, and a few of them are located close to the central region.

Hence, for each galaxy in the subsample, we identify the stellar particles that fulfil the age condition, and according to the metallicity of each particle Z_*, we compute the number of HMXBs as Mapelli *et al.* (2010),

$$N_{\rm HMXB} = \frac{N^{\rm MOBSE}_{\rm HMXB}(Z_*)}{m^{\rm MOBSE}(Z_*)} m_*^{\rm EAGLE} f_{\rm corr} f_{\rm bin}, \tag{2.1}$$

where $m_*^{\rm EAGLE}$ is the mass of the stellar particle, $\mathrm{N}^{\rm MOBSE}_{\rm HMXB}(Z_*)$ is the total number of HMXBs in the MOBSE catalog with the metallicity closer to Z_*, and $m^{\rm MOBSE}(Z_*)$ is the total mass of the MOBSE catalog at the selected metallicity. The parameter $f_{\rm corr} = 0.285$ accounts for the fact that we simulate only massive stars ($\geqslant 5$ M$_\odot$), while $f_{\rm bin} = 0.5$ is the assumed binary fraction. Hence, we randomly select a number $N_{\rm HMXB}$ of HMXBs from the sub-set of MOBSE with the metallicity closer to Z_* and we assign them to that star particle.

Our model also accounts for transient and persistent sources and includes prescriptions for Be–HMXB systems. Following Zuo *et al.* (2014), we assume that Be–HMXBs are wind-fed systems composed of an NS and a massive companion star, with orbital periods in the range of 10-300 d. We assume that only 25% of these systems are Be–HMXBs. We identify transient sources through the thermal disk instability model, where binaries with accretion rates below a critical value $\dot{M}_{\rm crit}$ are considered transient sources (see Frank, King & Raine 2002, eq. 5.105 and 5.106, p. 133). We note here that the assumptions made for transient sources are based on models for low-mass X-ray binary sources. Transient sources are in a quiescent state most of the time. Hence, we assume that its duty cycle is 1%. We also adopt a bolometric correction following Fragos *et al.* (2013).

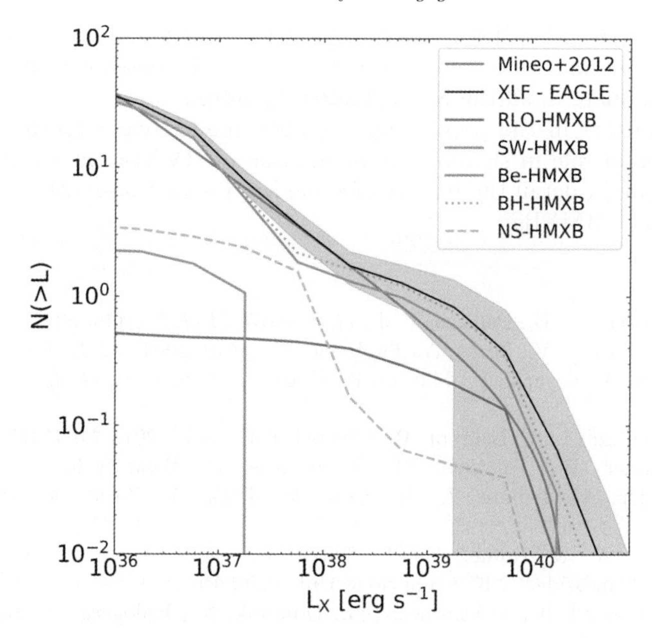

Figure 2. Mean X-ray luminosity function (XLF) normalized by the galaxy star formation rate. Black line: Mean XLF obtained by stacking together the XLFs of the star-forming galaxies in the EAGLE simulation at $z = 0$. Shaded grey area: Poissonian uncertainty on the mean XLF from the EAGLE galaxies. Solid blue (magenta) line: contribution to the simulated XLF by RLO–HMXBs (SW–HMXBs). Dotted red (dashed green) line: BH–HMXBs (NS–HMXBs). Solid red line: Be-HMXBs. Grey line: observed XLF by Mineo *et al.* (2012).

We compute the error bars assuming a Poisson distribution for the X-ray luminosities within the galaxies. We split the mean XLF of the star-forming galaxies according to the compact object (BH–HMXB and NS–HMXB) and the accretion process (RLO–HMXB, SW–HMXB, Be–HMXB) of the sources.

Using this method, for each simulated galaxy we obtain a population of HMXBs that accounts for the variability of the sources and the different emission mechanisms. In this work, we focus on studying the XLF of star-forming galaxies and compare our findings with the observational results of Mineo *et al.* (2012).

3. Results and future work

Figure 2 shows the mean XLF obtained by stacking together the XLFs of the star-forming galaxies. Our results show that the mean XLF of the simulated galaxies is in fair agreement with the observed XLF by Mineo *et al.* (2012, grey line).

In our model, BH–HMXBs are more numerous and generally brighter than NS–HMXBs. BH systems are expected to be more luminous than NS–HMXBs (Kaaret *et al.* 2017). However, observational results indicate that persistent NS–HMXBs are more numerous than persistent BH–HMXBs in the Milky Way (e.g., Lutovinov *et al.* 2013).

Nonetheless, Vulic *et al.* (2018) show that galaxies with sSFR $> 10^{-10}$ yr^{-1} have a higher number of BH-HMXBs than NS-HMXBs due to recent star formation episodes. Moreover, the fraction of BH-HMXBs and NS-HMXBs from the population synthesis model output shows that BH–HMXBs are more numerous in the simulated set. This is explained since strong interaction in binary systems tends to form more BH than NS due to mass transfer.

When we compare the HMXB according to the accretion process, we find that the RLO–HMXBs contribute only with high X-ray luminosity sources, while while SW-HMXBs dominate in all the X-ray luminosity range.

Several parameters in the population synthesis model (e.g. supernova kicks) might play a fundamental role in shaping the population of HMXBs. In a forthcoming work, we will investigate in detail the impact of some key population-synthesis parameters on the demography of HMXBs.

References

Artale, M. C., Tissera, P. B., Pellizza, L. J., *et al.* 2015, *MNRAS*, 448, 3071

Belczynski, K., Kalogera, V., Zezas, A., Fabbiano, G., *et al.* 2004, *APJL*, 601, 147

Chen, Y., Bressan, A., Girardi, L., Marigo, P., Kong, X., Lanza, A., *et al.* 2015, *MNRAS*, 452, 1068

Douna, V. M., Pellizza, L. J., Laurent, P., Mirabel, I. F., *et al.* 2018 *MNRAS*, 474, 3488

Fragos, T., Lehmer, B., Tremmel, M., Tzanavaris, P., Basu-Zych, A., Belczynski, K., Hornschemeier, A., Jenkins, L., Kalogera, V., Ptak, A., Zezas, A., *et al.* 2013, *ApJ*, 764, 41

Frank, J., King, A. & Raine, D. J. 2002, *Accretion Power in Astrophysics* ISBN 0521620538. Cambridge, UK. Cambridge University Press

Fryer, C. L., Belczynski, K., Wiktorowicz, G., Dominik, M., Kalogera, V., Holz, D. E., *et al.* 2012, *ApJ*, 749, 91

Garratt-Smithson, L., Wynn, G. A., Power, C., Nixon, C. J., *et al.* 2018, *MNRAS*, 480, 2985

Giacobbo, N., Mapelli, M. & Spera, M. 2018a, *MNRAS*, 474, 2959

Giacobbo, N. & Mapelli M. 2018b, preprint (arXiv:1805.11100)

Giacobbo, N. & Mapelli, M. 2018c, *MNRAS*, 480, 2011

Grimm, H.-J., Gilfanov, M., Sunyaev, R., *et al.* 2003, *MNRAS*, 339, 793

Hurley J. R., Tout C. A., Pols O. R., *et al.* 2002, *MNRAS* 329, 897

Justham, S. & Schawinski, K. 2012, *MNRAS*, 423, 1641

Kaaret, P., Feng, H., Roberts, T. P., *et al.* 2017, *ARAA*, 55, 303

Lutovinov, A. A., Revnivtsev, M. G., Tsygankov, S. S., Krivonos, R. A., *et al.* 2013, *MNRAS*, 431, 327

Mapelli, M., Ripamonti, E., Zampieri, L., Colpi, M., Bressan, A., *et al.* 2010, *MNRAS*, 408, 234

Mapelli, M., Giacobbo, N., Ripamonti, E., Spera, M., *et al.* 2017, *MNRAS*, 472, 2422

Mapelli, M., & Giacobbo, N. 2018a, *MNRAS*, 479, 4391

Mapelli, M., Giacobbo, N., Toffano, M., *et al.* 2018b, arXiv:1809.03521

Mineo, S., Gilfanov, M., Sunyaev, R., *et al.* 2012, *MNRAS*, 419, 2095

Planck Collaboration 2014, *A&A*, 571, 16

Spera, M. & Mapelli, M. 2017, *MNRAS*, 470, 4739

Schaye, J., Crain, R. A., Bower, R. G., Furlong, M., Schaller, M., Theuns, T., Dalla Vecchia, C., Frenk, C. S., McCarthy, I. G., Helly, J. C., Jenkins, A., Rosas-Guevara, Y. M., *et al.* 2015, *MNRAS*, 446, 521

Vink, J. S., de Koter, A., Lamers, H. J. G. L. M., *et al.* 2001, *A&A*, 369, 574

Vink, J. S., & de Koter, A. 2005, *A&A*, 442, 587

Vulic, N., Hornschemeier, A. E., Wik, D. R., Yukita, M., Zezas, A., Ptak, A. F., Lehmer, B. D., *et al.* 2018, *ApJ*, 864, 150

Zuo, Z.-Y., Li, X.-D., Gu, Q.-S., *et al.* 2014, *MNRAS*, 437, 1187

High-mass X-ray Binaries: Illuminating the Passage from Massive Binaries to Merging Compact Objects
Proceedings IAU Symposium No. 346, 2019
L. M. Oskinova, E. Bozzo, T. Bulik & D. Gies, eds.

doi:10.1017/S1743921318007548

Constraints from luminosity-displacement correlation of high-mass X-ray binaries

Zhao-yu Zuo

School of Science, Xi'an Jiaotong University, Xi'an 710049, China
email: zuozyu@xjtu.edu.cn

Abstract. We have modeled the luminosity-displacement correlation of high-mass X-ray binaries (HMXBs) with an evolutionary population synthesis (EPS) code. Detailed properties including offsets of simulated HMXBs are presented under both common envelope prescriptions usually adopted (i.e., the $\alpha_{\rm CE}$ formalism and the γ algorithm), and several theoretical models describing the natal kicks. We suggest that the distinct observational properties may be used as potential evidence to discriminate between models.

Keywords. stellar evolution, compact stars, X-ray binaries.

1. Introduction

Using data from *Chandra* and NICMOS on board *Hubble* Space Telescope (HST) of X-ray sources and star clusters, respectively, Kaaret *et al.* (2004) found that: the X-ray sources (number: 66) in three starburst galaxies (M 82, NGC 1569, and NGC 5253) are in general located near the star clusters, and brighter sources are most likely closer to the clusters. Moreover, there is an absence of luminous sources ($L_{\rm X} >\sim 10^{38}{\rm ergs\,s^{-1}}$) at relatively large displacements (>200 pc) from the clusters (i.e., $L_{\rm X}$ versus R correlation). In our work, the spatial offsets of HMXBs are modeled for a range of theoretical models describing the natal kicks and the common envelope (CE) evolution.

2. Model

We used the EPS code developed by Hurley, Pols & Tout (2000) and Hurley, Tout & Pols (2002), and updated by Zuo *et al.* (2008) to calculate the number and X-ray luminosity ($L_{\rm X}$) of X-ray binaries (XRBs) and their offsets (R) from the clusters where they were born. We calculated the X-ray luminosity for different kinds of HMXBs as in Zuo, Li & Gu (2014). We adopted the same procedure to compute the binary motion in the cluster potential (i.e., the offset R) as in Zuo & Li (2010).

We constructed seven models to investigate how the spatial offsets of HMXBs are influenced by the natal kicks we adopted. In our basic model (i.e., model BAC), the dispersion of kick velocity $\sigma_{\rm kick}$ is chosen as $150\,{\rm km\,s^{-1}}$, and we adopted a fallback prescription describing as a modified factor $\eta_{\rm m} = 1 - f_{\rm b}$, where $f_{\rm b}$ the fraction of the stellar envelope that falls back, for BH kicks. We also changed $\sigma_{\rm kick}$ (in units of ${\rm km\,s^{-1}}$) to be 50 (M1), 100 (M2) and 265 (M3) and then fixed the value of $\sigma_{\rm kick}$ at $150\,{\rm km\,s^{-1}}$ and applied three more forms of BH natal kicks (i.e., no natal kicks in model M4, high natal kicks in model M5 and momentum-conserving kicks in model M6, see Table 1), respectively.

We considered several different choices of the CE parameter $\alpha_{\rm CE}$ (see Table 2) in order to constrain its value. They are denoted as A02-A10, respectively, where the last two digits correspond to the value of $\alpha_{\rm CE}$. The maximum value of $\alpha_{\rm CE}$ should be within

Table 1. Summary of natal kick models. Here $\sigma_{\rm kick}$ is the dispersion of kick velocity, $\eta_{\rm m}$ is the modified factor, and $f_{\rm b}$ is the fraction of the stellar envelope that falls back.

Model	$\sigma_{\rm kick}$ [km s^{-1}]	$\eta_{\rm m}$	Description for NS	Description for BH
BAC	150	$1 - f_{\rm b}$	Basic[a]	Fallback; Basic
M1	50	$1 - f_{\rm b}$	Very low kicks	Fallback; Very low kicks
M2	100	$1 - f_{\rm b}$	Low kicks	Fallback; Low kicks
M3	265	$1 - f_{\rm b}$	High kicks	Fallback; High kicks
M4	150	0	Basic	No natal kicks
M5	150	1	Basic	High natal kicks
M6	150	$M_{\rm NS}/M_{\rm BH}$	Basic	Momentum-conserving kicks

[a] Basic means $\sigma_{\rm kick} = 150\,{\rm km\,s^{-1}}$.

Table 2. Different models on the treatment of the CE parameter $\alpha_{\rm CE}$.

Model	A02	A03	A04	A05	A06	A07	A08	A09	A10
$\alpha_{\rm CE}$	0.2	0.3	0.4	0.5	0.6	0.7	0.8	0.9	1.0

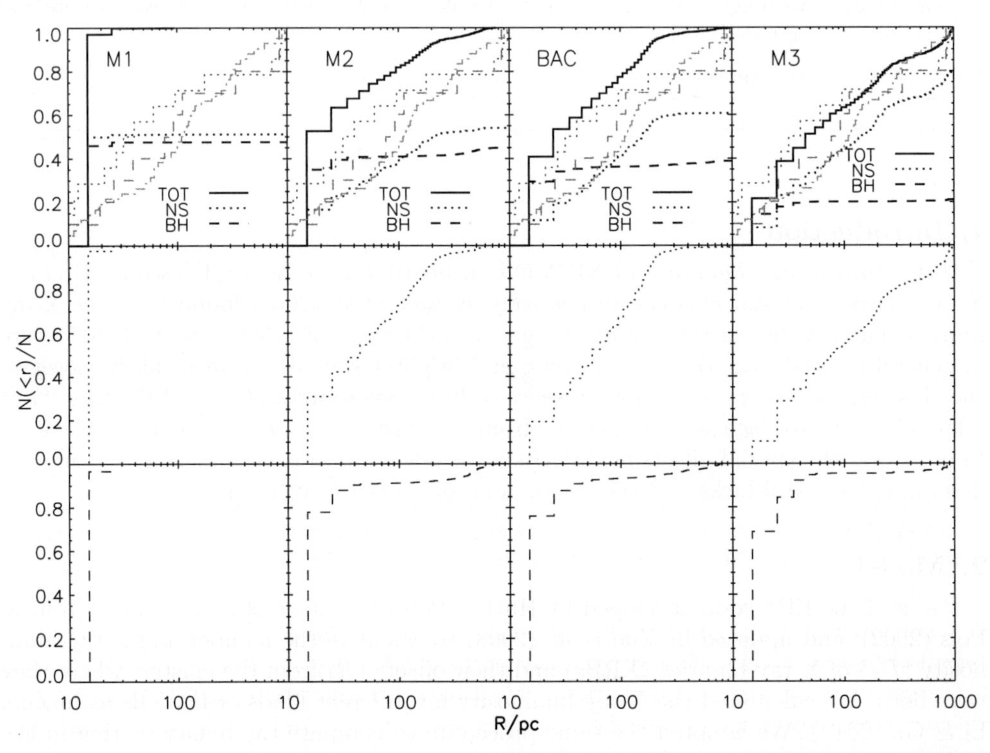

Figure 1. The normalized cumulative distribution for the numbers of ALL-XRBs (top), NS-XRBs (middle), and BH-XRBs (bottom), respectively. From left to right are models M1, M2, BAC, and M3.

unity as the potential internal energies have been considered in the quantity of the binding energy of the envelope (see Xu & Li 2010 and Loveridge *et al.* 2011 for the details).

3. Results

Fig. 1 shows the normalized cumulative distribution of HMXB offsets (top: Total, along with the relative contributions of NS- and BH- HMXBs; middle: NS-HMXBs; bottom: BH-HMXBs) in the luminosity range $10^{36} < L_{\rm X} < 10^{38}$ erg s^{-1} under different choices of $\sigma_{\rm kick}$ (see Table 1). Only sources with offsets 10 pc $<$ R $<$ 1 kpc are selected according to

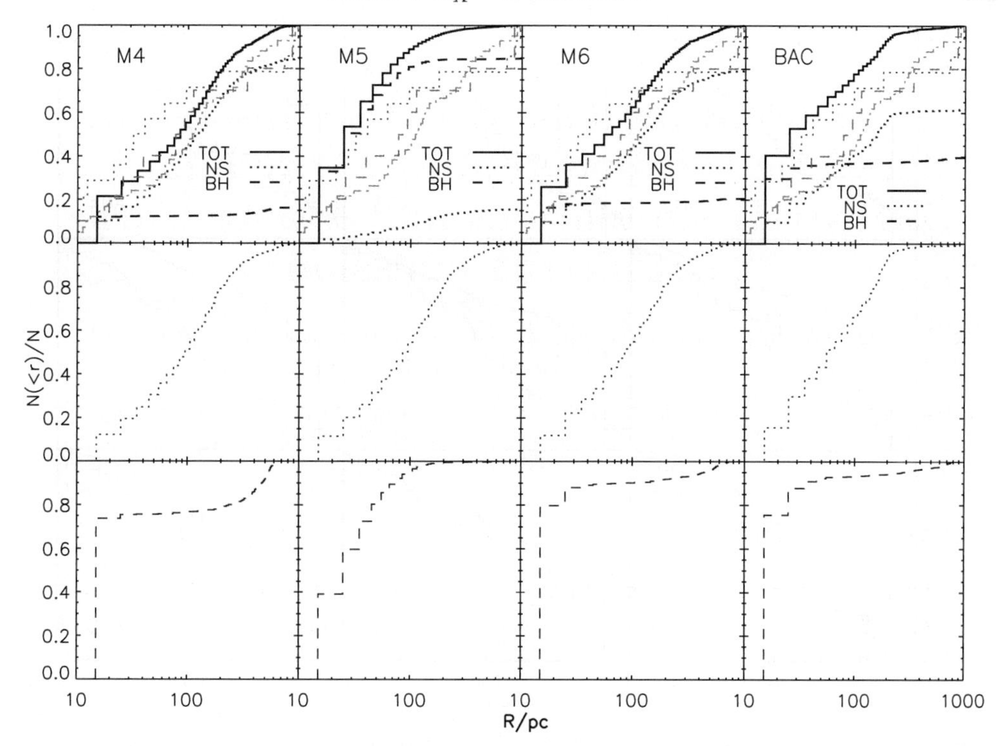

Figure 2. Same as in Fig. 1, but for sources in models M4-M6, and BAC from left to right.

the Kaaret *et al.* (2004) sample. The thin-solid, dotted, and dashed lines represent the observed cumulative distributions of source displacements in galaxies M82, NGC 1569, and NGC 5253 (see Fig. 2 in Kaaret *et al.* 2004, similarly hereinafter), respectively. The differences are remarkable. When natal kicks are too weak (i.e., on the order of several tens of $\mathrm{km\,s^{-1}}$, see model M1), no sufficient sources can be kicked to large displacements from star clusters. In order to be compatible with observations, the value of $\sigma_{\rm kick}$ should be greater than $\sim 100\,\mathrm{km\,s^{-1}}$ (see models BAC, M2, and M3).

Fig. 2 is the same as in Fig. 1, but for different treatments on the BH natal kicks (models M4-M6 versus model BAC). It is clear that when full natal kicks are applied to BHs (i.e., model M5), BH HMXBs dominate the whole population in our interested offset region, greatly different from other models and the observational data. These sources are predicted to be mainly long-period BH sources powered by stellar winds from massive ($\sim 30 - 75 M_\odot$) MS companions (see Fig. 4 in Zuo, Li & Gu 2014 for a typical evolutionary sequence), which may serve as a potential clue to further discriminate between models of BH natal kicks.

Fig. 3 is the same as in Fig. 1, but under different choices of $\alpha_{\rm CE}$ (see Table 2). Clearly models with $\alpha_{\rm CE}$ between $\sim 0.8 - 1.0$ (i.e., models A08-A10) can match the observation quite closely, while models with $\alpha_{\rm CE} <\sim 0.4$ clearly fail. In models A02-A04, very few sources can move to 100 pc away from the star clusters, which is in marked contrast with the observations. We note that this result is also consistent with the one obtained through HMXB X-ray luminosity function (XLF) simulations presented by Zuo & Li (2014).

4. Discussion and Conclusions

Our work shows that, using the apparent L_X vs. R distribution of HMXBs, it is possible to investigate the natal kick problem and constrain the value of CE parameter $\alpha_{\rm CE}$.

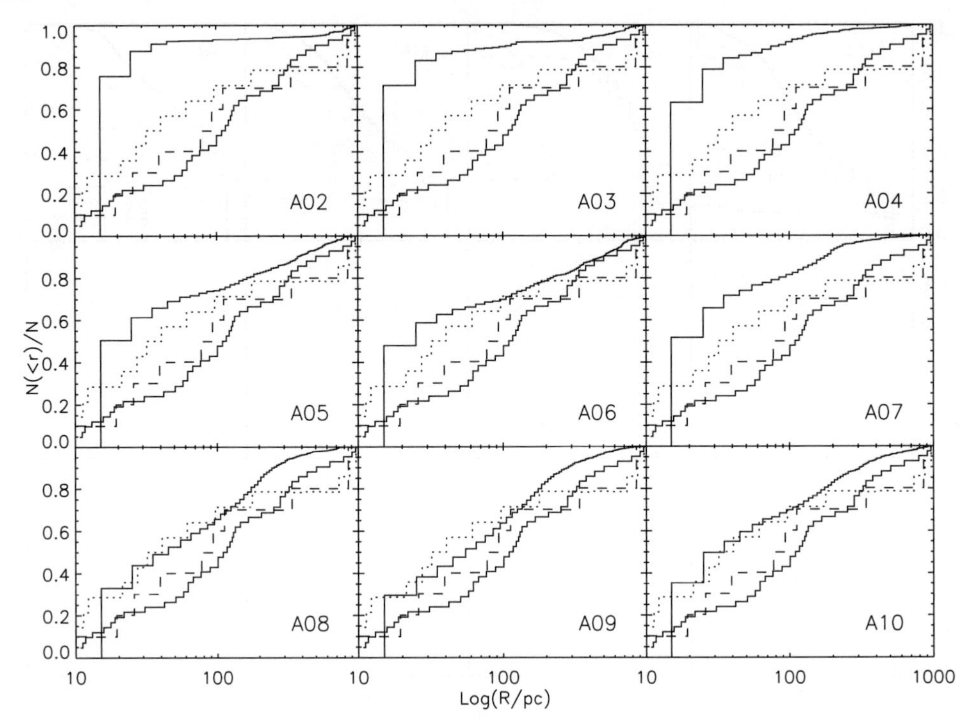

Figure 3. The normalized cumulative distribution (thick-solid line) for models A02-A10, respectively (see Table 2).

The magnitude of $\sigma_{\rm kick}$ is constrained to be larger than $\sim 100\,{\rm km\,s^{-1}}$, while $\sigma_{\rm kick}$ on the order of several tens ${\rm km\,s^{-1}}$ may be excluded. More importantly, full BH natal kicks (i.e., similar to those that NSs may receive) are likely to be ruled out though it has just been suggested recently by Repetto, Davies & Sigurdsson (2012) to explain the observed distribution of low-mass X-ray binaries with BHs. We also made constraints on the CE efficiency $\alpha_{\rm CE}$. Within the framework of the standard energy formula for CE and core definition at mass $X = 10$ per cent, a high value of $\alpha_{\rm CE}$, i.e. around $0.8-1.0$, is more preferable, while $\alpha_{\rm CE} <\sim 0.4$ likely can not reconstruct the observed $L_{\rm X}$ vs. R distribution.

The results are also subject to some uncertainties and simplified treatments. For example, we did not consider dynamical interactions in our calculation though it may play a non-negligible role in binary formation and kinematics (Mapelli *et al.* 2013). And when constraining the natal kicks, we only selected one typical value of $\alpha_{\rm CE}$ for simplicity. However CE efficiency is also an important parameter, which may not only affect the system velocity, and hence the spatial offset of the system, but also the whole population of resultant XRBs as well. In addition, the range of $\alpha_{\rm CE}$ we obtained here itself suffers from great uncertainties. It may vary greatly with different assumption describing the CE, such as different core-envelope boundaries, different energy sources used to unbind the envelope, etc.

Acknowledgements

This work was supported by the National Natural Science Foundation of China (grant 11573021), the Natural Science Basic Research Program of Shaanxi Province – Youth Talent Project (No. 2016JQ1016) and the Fundamental Research Funds for the Central Universities.

References

Hurley, J. R., Pols, O. R., & Tout, C. A. 2000, *MNRAS*, 315, 543
Hurley, J. R., Tout, C. A., & Pols, O. R. 2002, *MNRAS*, 329, 897
Kaaret P., Alonso-Herrero A., Gallagher J. S., Fabbiano G., Zezas A. & Rieke M. J. 2004, *MNRAS*, 348, L28
Loveridge, A. J., van der Sluys, M. V., & Kalogera V. 2011, *ApJ*, 743, 49
Mapelli, M., Zampieri, L., Ripamonti, E. & Bressan, A. 2013, *MNRAS*, 429, 2298
Repetto, S., Davies, M. B. & Sigurdsson, S. 2012, *MNRAS*, 425, 2799
Xu, X. J., & Li X. D. 2010, *ApJ*, 716, 114
Zuo, Z. Y., Li, X. D., & Liu, X. W. 2008, *MNRAS*, 387, 121
Zuo, Z. Y., & Li, X. D. 2010, *MNRAS*, 405, 2768
Zuo, Z. Y., Li, X. D., & Gu, Q. S. 2014, *MNRAS*, 437, 1187 (Erratum: 443, 1889)
Zuo, Z. Y., & Li, X. D. 2014, *ApJ*, 797, 45

High-mass X-ray Binaries: Illuminating the Passage from Massive Binaries to Merging Compact Objects
Proceedings IAU Symposium No. 346, 2019
L. M. Oskinova, E. Bozzo, T. Bulik & D. Gies, eds.

doi:10.1017/S1743921318008062

Emission-line diagnostics of core-collapse supernova host HII regions including interacting binary population

Lin Xiao[1], J. J. Eldridge[2], L. Galbany[3] and E. Stanway[4]

[1]CAS Key Laboratory for Research in Galaxies and Cosmology,
Department of Astronomy, University of Science and Technology of China, Hefei, China
email: lxiao33@ustc.edu.cn

[2]Department of Physics, University of Auckland, Private Bag 92019, Auckland, New Zealand
email: j.eldridge@auckland.ac.nz

[3]PITT PACC, Department of Physics and Astronomy, University of Pittsburgh,
Pittsburgh, PA 15260, USA
email: llgalbany@pitt.edu

[4]Department of Physics, University of Warwick, Gibbet Hill Road, Coventry, CV4 7AL, UK
email: e.r.stanway@warwick.ac.uk

Abstract. Considering as many as 70% of massive stars interact with a binary companion (Sana *et al.* 2012, 2014), we created a new model of the optical nebular emission of HII regions by combining the results of the Binary Population and Spectral Synthesis (BPASS, Eldridge, Stanway *et al.* 2017) code with the photoionization code (CLOUDY). This is discussed more in detail in Xiao *et al.* 2018a. Then we use this model to explore a variety of emission-line diagnostics of CCSN host HII regions from the PMAS/PPAK Integral-field Supernova hosts COmpilation (PISCO, Galbany *et al.* 2018). We determine the age, metallicity and gas parameters for H II regions associated with CCSNe, contrasting the above variables to distribution type II and type Ibc SNe. We find their nebular emission and CCSN progenitor types are largely determined by past and ongoing binary interactions, for example mass loss, mass gain and stellar mergers. However we note these two types SNe have little preference in their host environment metallicity measured by oxygen abundance or in progenitor initial mass, except that at lower metallicities supernovae are more likely to be of type II. The BPASS and nebular emission models are available from bpass.auckland.ac.nz and warwick.ac.uk/bpass.

Keywords. binaries:general, supernovae:general, HII regions.

References

Eldridge, J. J., Fraser, M., Smartt, S. J., Maund, J. R., & Crockett, R. M. 2013, *MNRAS*, 436, 774

Eldridge, J. J., Stanway, E., Xiao, L., L. A. S. McClelland, G. Taylor, M. Ng, S. M. L. Greis, J. C. Bray *et al.* 2017, *PASA*, 34, e058

Galbany, L., Anderson, J. P., Sánchez, S. F., *et al.* 2018, *ApJ*, 855, 107

Sana, H., de Mink, S. E., de Koter, A., Langer, N., Evans, C. J., Gieles, M., Gosset, E., Izzard, R. G., Le Bouquin, J.-B., & Schneider, F. R. N. 2012, *Science*, 337:444.

Sana, H., Le Bouquin, J.-B., Lacour, S., Berger, J.-P., Duvert, G., Gauchet, L., Norris, B., Olofsson, J., Pickel, D., Zins, G., Absil, O., de Koter, A., Kratter, K., Schnurr, O., Zinnecker, H., *et al.* 2014, *ApJS*, 215, 15

Smartt, S. J. 2015, pasa, 32, e016

Xiao, L., Stanway, E. R., & Eldridge, J. J. 2018a, *MNRAS*, 477, 904

Xiao, L., Galbany, L., Eldridge, J. J., & Stanway, E. R. 2018b, arXiv:1805.01213

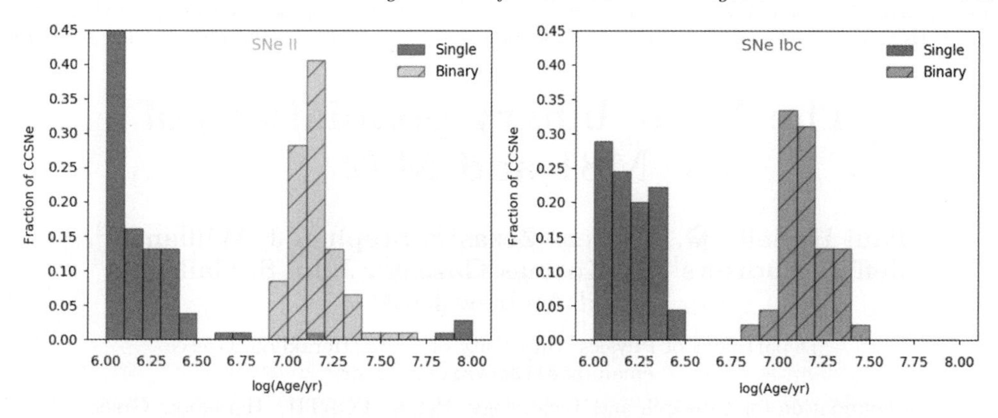

Figure 1. The fraction of CCSN number as a function of age derived from best-fitting models accounting for ionizing photon leakage. The left panel show the fraction distribution for type II SNe and the right panel for type Ibc SNe, with the black bars are from single-star models and the coloured bars from binary-star models. The binary-star models, that allow for ionizing photon loss, provide a more realistic age compared to the constraints of detected SN progenitors (Smartt 2015; Eldridge et al. 2013). We also find that type II and type Ibc SNe arise from progenitor stars of similar age, mostly from 7 to 45 Myr, which corresponds to stars with masses $< 20M_{\odot}$. (The figure taken from Xiao *et al.* 2018b)

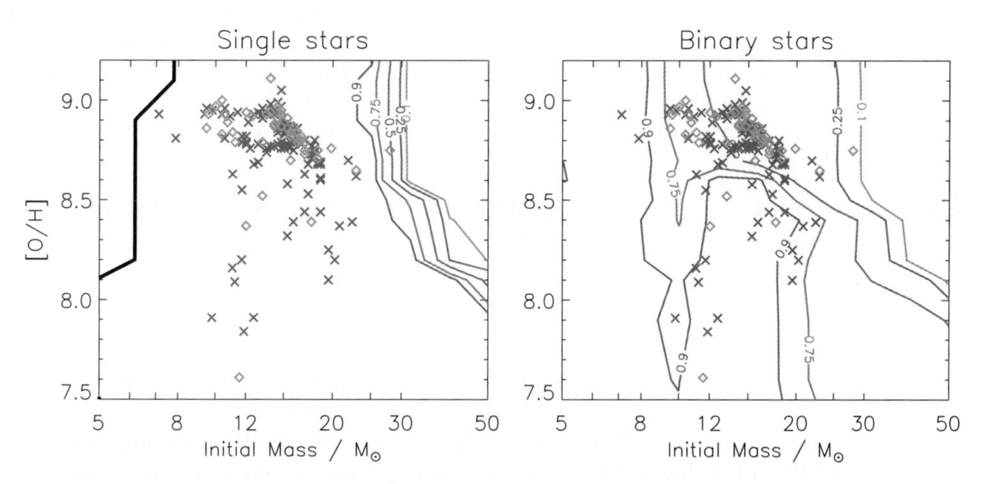

Figure 2. The predicted number count ratio of SN type as a function of initial mass and metallicity, compared to observational data (type II in red crosses, type Ibc in blue diamonds) from the best-fitting models accounting for ionizing photon loss. The contours are annotated with the fraction of supernovae that are type II at that contour. The thick black line represents the minimum initial mass for SNe to occur. The binary-star models provide a much improved prediction of the type II to Ibc ratio as a function of progenitor mass and metallicity than single-star models. (The figure taken from Xiao *et al.* 2018b)

High-mass X-ray Binaries: Illuminating the Passage from Massive Binaries to Merging Compact Objects
Proceedings IAU Symposium No. 346, 2019
L. M. Oskinova, E. Bozzo, T. Bulik & D. Gies, eds.

doi:10.1017/S1743921318008190

The X-ray binary populations of M81 and M82

Paul H. Sell[1,2], Andreas Zezas[1,2], Stephen J. Williams[1,2], Jeff J. Andrews[1,2], Kosmas Gazeas[3], John S. Gallagher[4] and Andrew Ptak[5]

[1]Department of Physics, University of Crete, Heraklion, Greece
email: psell@physics.uoc.gr

[2] Foundation for Research and Technology Hellas (FORTH), Heraklion, Greece

[3] Department of Astrophysics, University of Athens, Zografos, Athens, Greece

[4] Department of Astronomy, University of Wisconsin-Madison, Madison, WI, USA

[5] NASA Goddard Space Flight Center, Code 662, Greenbelt, MD 20771, USA

Abstract. We use deep Chandra and HST data to uniquely classify the X-ray binary (XRB) populations in M81 on the basis of their donor stars and local stellar populations (into early-type main sequence, yellow giant, supergiant, low-mass, and globular cluster). First, we find that more massive, redder, and denser globular clusters are more likely to be associated with XRBs. Second, we find that the high-mass XRBs (HMXBs) overall have a steeper X-ray luminosity function (XLF) than the canonical star-forming galaxy XLF, though there is some evidence of variations in the slopes of the sub-populations. On the other hand, the XLF of the prototypical starburst M82 is described by the canonical powerlaw ($\alpha_{\rm cum} \sim 0.6$) down to $L_X \sim 10^{36}$ erg s^{-1}. We attribute variations in XLF slopes to different mass transfer modes (Roche-lobe overflow versus wind-fed systems).

Keywords. accretion, stars: neutron, galaxies: star clusters, galaxies: individual (M81, M82), galaxies: spiral, galaxies: starburst, X-rays: binaries

---

1. Introduction

X-ray observations of star-forming galaxies reveal abundant point sources that dominate the galaxies' hard X-ray emission. Most of these sources are XRBs with a neutron star or black hole accreting matter from a high- or low-mass stellar companion through Roche lobe overflow or stellar winds. XRBs are an important evolutionary stage in binary stellar systems, occurring after one star has undergone a supernova, formed a compact object, and then brought it into contact with the other star (e.g., Tauris & van den Heuvel 2010).

XRBs have important links to various topics in astrophysics: they are responsible for feedback (winds, jets, and ionization) on multiple scales (e.g., Soria *et al.* 2010; Justham & Schawinski 2012) including preheating the intergalactic medium prior to and during reionization (Fragos *et al.* 2013; Das *et al.* 2017; Douna *et al.* 2018) and they are prime candidates for progenitors of gravitational wave events (Abbott *et al.* 2016).

Observations of elliptical galaxies and bulges of spirals in nearby galaxies enable us to isolate the low-mass XRB (LMXB) populations of galaxies (e.g., Kong *et al.* 2003; Kim & Fabbiano 2010) and study trends with stellar mass across many LMXB populations ($M_\star$; Boroson *et al.* 2011; Zhang *et al.* 2011). Similar work has been done with nearby star-forming galaxies to locate HMXBs (e.g., Zezas *et al.* 2002; Pannuti *et al.* 2011) and study their links to the recent star formation rate (SFR; Mineo *et al.* 2012; Mineo *et al.* 2014).

Regions containing HMXBs are usually mixed with other populations of sources: notably LMXBs but also supernova remnants (SNRs; Leonidaki *et al.* 2010) and hyper-accreting white dwarfs frequently seen as super-soft sources (SSSs; e.g., Di Stefano & Kong 2004). Only in rare cases with very high specific SFR (sSFR=SFR/$M_\star$) or where individual counterparts are identified can we examine these populations with very low contamination: the Milky Way (Grimm *et al.* 2002) though heavily biased (Arur & Maccarone 2018), SMC/LMC (Antoniou *et al.* 2009; Antoniou & Zezas 2016), M33 (Tüllmann *et al.* 2011; Garofali *et al.* 2018), and M31 (Lazzarini *et al.* 2018). We extend this work to two more nearby interacting galaxies: M81 and M82.

2. Source Classification in M81

The primary goal of our work on M81 is to carefully identify and study the LMXB and HMXB source populations. We briefly outline our approach using the best optical and X-ray observations of the galaxy.

For all Chandra sources, we construct 3-color stamps of the local HST stellar fields and their corresponding color-magnitude diagrams (CMDs; where necessary). We first identify interlopers (class "A" in Fig. 1): foreground stars (identified on the basis of their brightness on the HST images and their soft X-ray colors), and background galaxies (clearly seen on the HST images). Another of these categories are X-ray sources associated with clusters (identified in the basis of their spatial extent on the HST images and their location on the CMD). X-ray sources associated with globular clusters are, by definition, LMXBs (class "B" in Fig. 1). Finally, using a combination of X-ray colors and optical spectroscopic line diagnostics, we have discovered a population of SNRs (Leonidaki et al. in prep.).

For all the other sources, we follow a systematic approach for X-ray source classification based on the location of their optical associations on the CMD and their corresponding chance coincidence probability (CCP). This is determined by randomly shifting the Chandra and HST catalogs and calculating the likelihood that an X-ray source will match an HST source of the corresponding location on the CMD. The CCP is simply a ratio of the number of matched HST sources of a certain type (location on the CMD) to the total number of tries and is a function of the search radius and local stellar density. Therefore, it is normalized probability, where very small CCPs (~ 0) refer to very rare sources and large CCPs (~ 1) refer to chance associations. Extremely rare sources are more likely to be matched to their true counterparts. This procedure of matching by exclusion has been very commonly used to match source catalogs (see e.g., Antoniou *et al.* 2009 for a more detailed description of the CCP).

The search radius for each X-ray source depends on its off axis angle and number of counts. We use the prescription of Hong *et al.* (2005), which is based on extensive simulations. The full position probability at an arbitrary radius is calculated as a quadrature sum between the boresight uncertainty ($\delta_{RA,Dec} \sim 0.1''$) between the catalogs and the reported 95% confidence in Hong et al. modeled as a symmetric Gaussian. We consider all HST sources within the 3σ position error circle, which ensures that we are not missing the true counterpart.

We compare all HST sources in the X-ray error circle by assigning weights based on their chance councidence probaility. A full description of the calculation of the CCP, the weights, and their application to the source matching will be presented in a follow-up publication (Sell et al. in prep.). Examples of the HMXB and field LMXB sources in each of our remaining categories are shown in Fig. 1 (classes "C–F"). Some sources are excluded, as their classifications are ambiguous (class "G"). In Fig. 2, we present on the CMD each Chandra source with a unique/clear high-mass star counterpart.

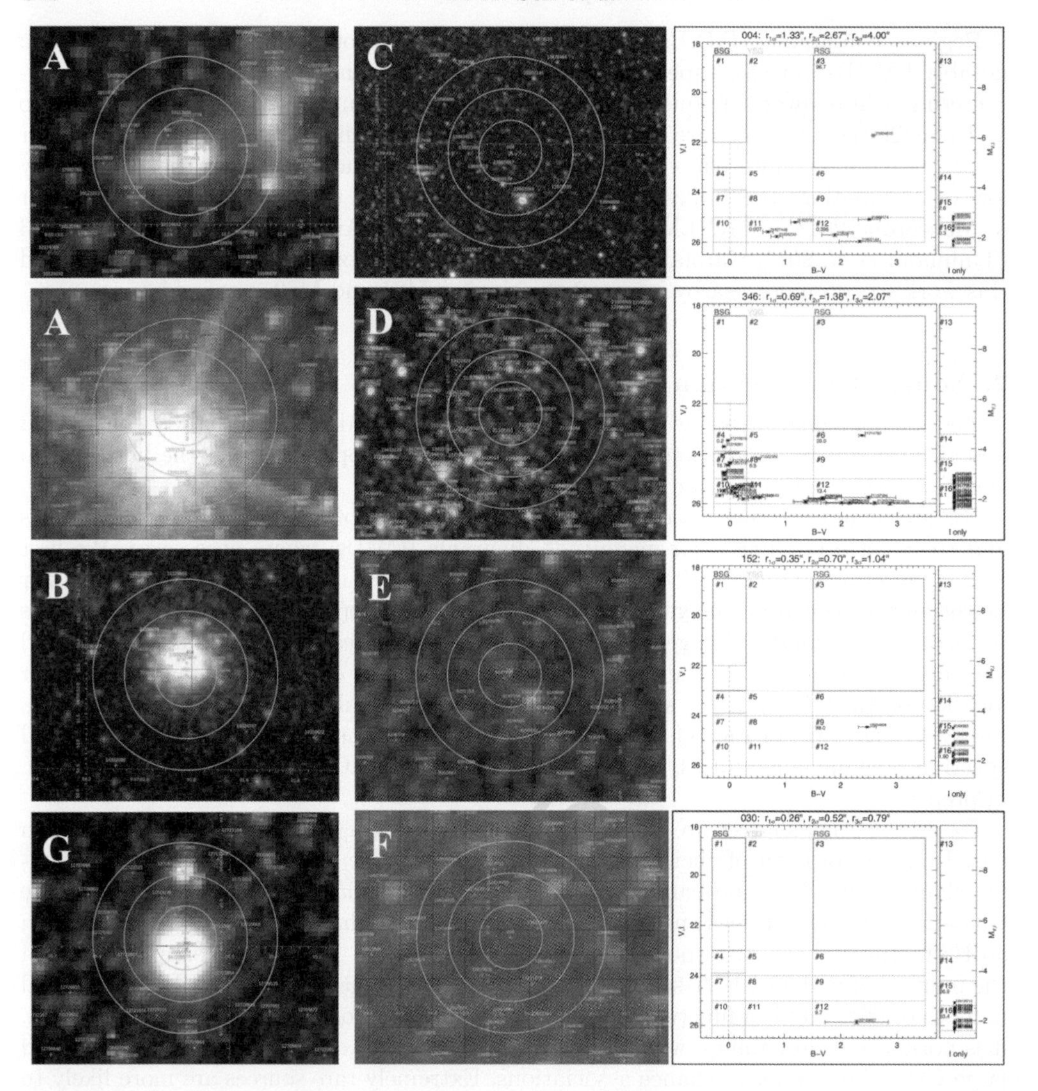

Figure 1. We show examples of various source classification categories using 3-color (B, V, and I) stamps of HST fields local to each Chandra source with observational color-magnitude diagrams when relevant: A) interlopers (background galaxies and foreground stars), B) LMXBs in globular clusters, C) uniquely classified HMXBs, D) confused HMXBs, E) uniquely classified LMXBs, F) confused LMXBs (e.g. extremely high stellar density in the bulge; only faint, low-mass stars in the field) and G) indeterminate sources, (likely either a low- or high-mass star). In the stamps, the 1σ, 2σ, and 3σ Chandra astrometric error circles are shown in cyan, the HST catalog sources with their ID numbers are in green, and the globular clusters and galaxies are in red and yellow, respectively. We construct each CMD using the observed B, V and I magnitudes of the stars within 3σ, identified by their HST source ID numbers. Each left plot includes sources with matches in all three bands and each right plot includes sources only present in the I-band. The Chandra source number and its 1σ, 2σ, and 3σ astrometric uncertainties are listed in the plot title and physically-motivated (blue, yellow and red supergiants, main-sequence stars–green) regions where we have calculated the chance coincidence are numbered. The CMD weight is below the CMD#.

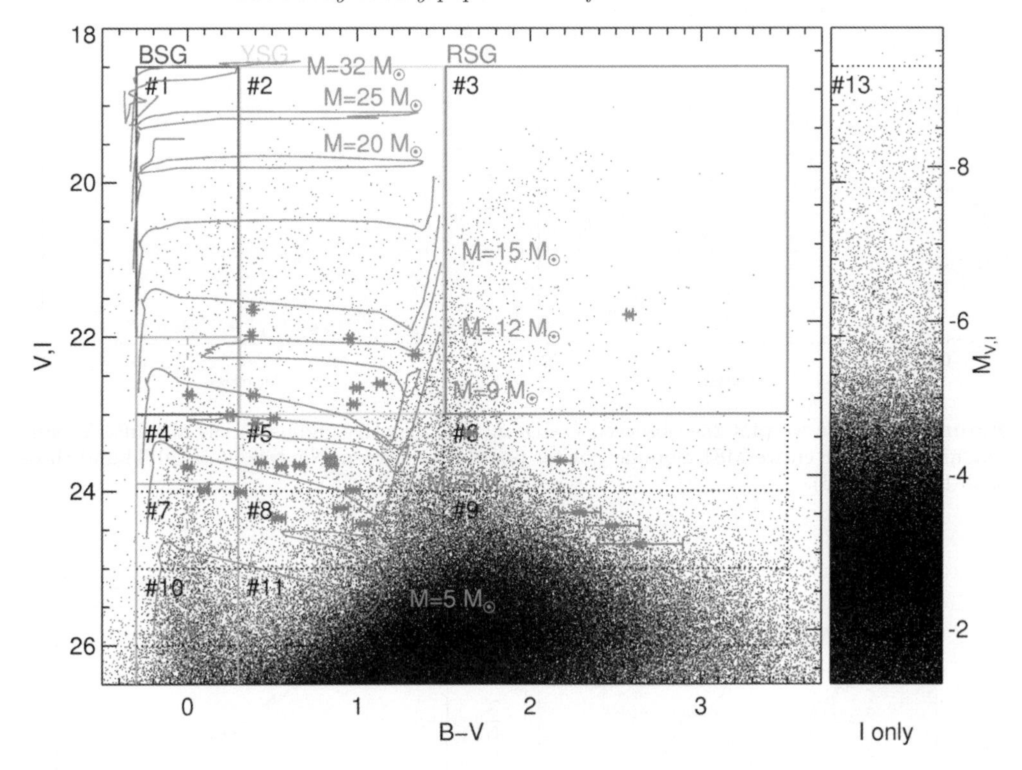

Figure 2. The CMD for all Chandra sources with unique HST counterparts (red). The black dots indicate all HST sources within 5″ of each of the sources in our X-ray catalog. Geneva single star tracks are overplotted for reference (Ekstrom *et al.* 2012).

3. Results

3.1. *Low-Mass X-ray Binaries in M81 Globular Clusters*

We examine the optical properties of the M81 globular cluster population, including those with XRBs (Fig. 3). We find that XRBs are preferentially associated with redder, more massive, and denser clusters in agreement with other work on elliptical galaxies (e.g., Sivakoff *et al.* 2007). The similarity between spiral and elliptical galaxies suggests that the same dynamical mechanisms are involved in the formation of LMXBs in GCs in both types of galaxies.

3.2. *X-ray Luminosity Functions of M81 and M82*

We examine the global properties of the X-ray point source populations of M81 and M82 through their XLFs (Fig. 4). For M81, use the X-ray luminosities from Sell *et al.* (2011). For M82, we use a very similar approach to Sell *et al.* (2011) for calculating source luminosities based on spectral fits that will be discussed in detail in a future publication (Sell et al. in prep.).

First for M81, our source-by-source classification enables us to uniquely examine XLFs of different subpopulations (Fig. 4). When there is some uncertainty in our classifications (i.e., an X-ray source has more than one optical association with comparable CCP), we draw sources into the XLFs relative to their CCPs. This is indicated by the scatter in each of the global LMXB (red) and HMXB (blue) populations in Fig. 4 (we only show one realization in the other cases for clarity). The LMXB XLF shows the typical break near $L_X \sim 10^{38}$ erg s^{-1} (e.g., Fabbiano 2006). Interestingly, we find that the HMXB XLF

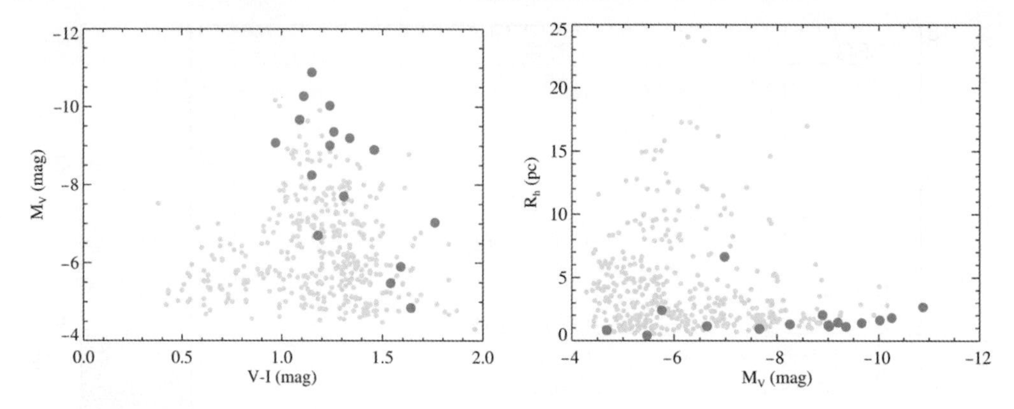

Figure 3. Above we plot the observed CMD and the effective radius versus absolute V-band magnitude for the entire M81 globular cluster population (green; Nantais *et al.* 2011) and those with LMXBs (red).

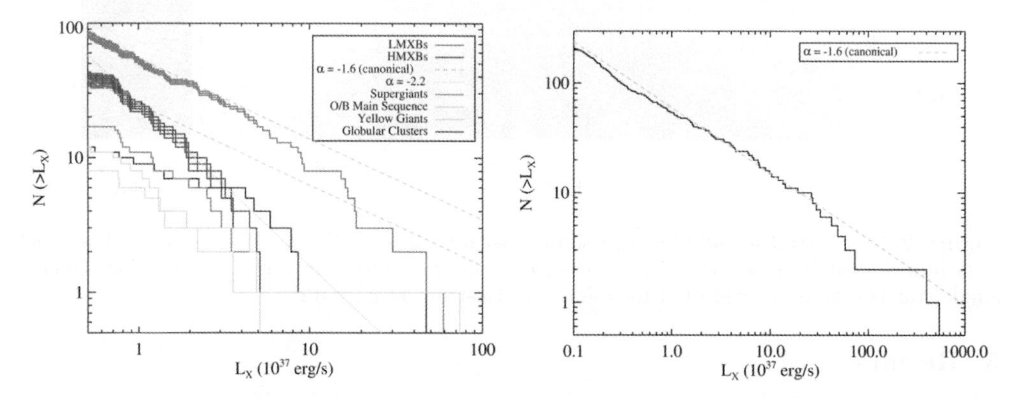

Figure 4. Left: XLFs for individually classified XRBs in M81 (incompleteness is important for $L_X \lesssim 10^{37}$ erg s^{-1}). **Right:** The XLF for M82, the deepest for a starburst galaxy. Unusually high levels of copious diffuse emission account for some incompleteness seen for $L_X \lesssim 10^{37}$ erg s^{-1}. In both plots, the canonical HMXB XLF ($dN/dL \propto L^{\alpha}$; Mineo *et al.* 2012) is overplotted in green dashed lines.

appears steeper than the canonical HMXB XLF (the green dashed line; Mineo *et al.* 2012, which may include some contamination from LMXBs). The differences in HMXB slopes could be attributed to differences in the mass transfer mode (Roche-lobe overflow vs. wind-fed systems) in the XRB populations comprising each XLF.

We compare these results with the XRB populations of the prototypical starburst galaxy, M82, for which we present one of the deepest XLFs reported for a starburst galaxy. We find that it is described by the canonical powerlaw with a cumulative slope of ~ 0.6 (Mineo *et al.* 2012) down to $L_X \sim 10^{36}$ erg s^{-1}.

Acknowledgments

The research leading to these results has received funding from the European Research Council under the European Union's Seventh Framework Programme (FP/2007-2013) / ERC Grant Agreement n. 617001. This project has received funding from the European Union's Horizon 2020 research and innovation programme under the Marie Sklodowska-Curie RISE action, grant agreement No 691164 (ASTROSTAT).

References

Abbott, B. P., *et al.* 2016, *Phys. Rev. Lett.*, 116, 1102
Antoniou, V., Zezas, A., Hatzidimitriou, D., McDowell, J. C. 2009, *ApJ*, 697, 1695
Antoniou, V. & Zezas, A. 2016, *MNRAS*, 459, 528
Arur K. & Maccarone, T. J. 2018, *MNRAS*, 474, 69
Boroson, B., Kim, D.-W., Fabbiano, G. 2011, *ApJ*, 729, 12
Das, A., Mesinger, A., Pallottini, A., Ferrara, A., & Wise, J. H. 2017, *MNRAS*, 469, 1166
Di Stefano, R. & Kong, A. K. H. 2004, *ApJ*, 609, 710
Douna, V. M., Pellizza, L. J., Laurent, P., & Mirabel, I. F. 2018, *MNRAS*, 474, 3488
Ekström, S., Georgy, C., Eggenberger, P., Meynet, G., Mowlavi, N., Wyttenbach, A., Granada, A., Decressin, T., Hirschi, R., Frischknecht, U., Charbonnel, C., & Maeder, A. 2012, *A&A*, 537, A146
Fabbiano, G. 2006, *ARAA*, 44, 323
Fragos, T., Lehmer, B. D., Naoz, S., Zezas, A., & Basu-Zych, A. 2013, *ApJ*, 776, L31
Garofali, K., Williams, B. F., Hillis, T., Gilbert, K. M., Dolphin, A. E., Eracleous, M., Binder, B. 2018, *MNRAS*, 479, 3526
Grimm, H.-J., Gilfanov, M., & Sunyaev, R. 2002, *A&A*, 391, 923
Hong, J., van den Berg, M., Schlegel, E. M., Grindlay, J. E., Koenig, X., Laycock, S., & Zhao, P. 2005, *ApJ*, 635, 907
Justham, S. & Schawinski, K. 2012, *MNRAS*, 423, 1641
Kim, D.-W., & Fabbiano, G. 2010, *ApJ*, 721, 1523
Kong, A. K. H., DiStefano, R., Garcia, M. R., & Greiner, J. 2003, *ApJ*, 585, 298
Lazzarini, M., Hornschemeier, A. E., Williams, B. F., Wik, D., Vulic, N., Yukita, M., Zezas, A., Lewis, A. R., Durbin, M., Ptak, A., Bodaghee, A., Lehmer, B. D., Antoniou, V., & Maccarone, T. 2018, *ApJ*, 862, 28
Leonidaki, I., Zezas, A., & Boumis, P. 2010, *ApJ*, 725, 842
Mineo, S., Gilfanov, M., & Sunyaev, R. 2012, *MNRAS*, 419, 25
Mineo, S., Gilfanov, M., Lehmer, B. D., Morrison, G. E., & Sunyaev, R. 2014, *MNRAS*, 437, 1698
Nantais, J. B., Huchra, J. P., Zezas, A., Gazeas, K., & Strader, J. 2011, *AJ*, 142, 183
Pannuti, T. G., Schlegel, E. M., Filipović, M. D., Payne, J. L., Petre, R., Harrus, I. M., Staggs, W. D., & Lacey, C. K. 2011, *AJ*, 142, 20
Sell, P. H., Pooley, D., Zezas, A., Heinz, S. Homan, J., & Lewin, W. H. G. 2011, *ApJ*, 735, 26
Sivakoff, G. R., Jordán, A., Sarazin, C. L., Blakeslee, J. P., Côté, P., Ferrarese, L., Juett, A. M., Mei, S., & Peng, E. W. 2007, *ApJ*, 660, 1246
Soria, R., Pakull, M. W., Broderick, J. W., Corbel, S., Motch, C. 2010, *MNRAS*, 409, 541
Tüllmann, R., Gaetz, T. J., Plucinsky, P. P., Kuntz, K. D., Williams, B. F., Pietsch, W., Haberl, F., Long, K. S., Blair, W. P., Sasaki, M., Winkler, P. F., Challis, P., Pannuti, T. G., Edgar, R. J., Helfand, D. J., Hughes, J. P., Kirshner, R. P., Mazeh, T., & Shporer, A. 2011, *ApJS*, 193, 31
Tauris, T. M. & van den Heuvel, E. P. J. 2010, *Compact Stellar X-ray Sources*, 623
Zezas, A., Fabbiano, G., Rots, A. H., & Murray, S. S. 2002, *ApJ*, 577, 710
Zhang, Z., Gilfanov, M., Voss, R., Sivakoff, G. R., Kraft, R. P., Brassington, N. J., Kundu, A., Jordán, A., Sarazin, C. 2011, *A&A*, 533, 33

High-mass X-ray Binaries: Illuminating the Passage from Massive Binaries to Merging Compact Objects
Proceedings IAU Symposium No. 346, 2019
L. M. Oskinova, E. Bozzo, T. Bulik & D. Gies, eds.

doi:10.1017/S1743921319000048

Clarifying the population of HMXBs in the Small Magellanic Cloud

Grigoris Maravelias[1,2], Andreas Zezas[1,2,3], Vallia Antoniou[3], Despina Hatzidimitriou[4,5] and Frank Haberl[6]

[1]IESL, Foundation for Research and Technology-Hellas, Heraklion, Greece
email: gmaravel@physics.uoc.gr

[2]Department of Physics, University of Crete, Heraklion, Greece

[3]Harvard-Smithsonian Center for Astrophysics, Cambridge, USA

[4]Department of Physics, National and Kapodistrian University of Athens, Zografou, Greece

[5]IAASARS, National Observatory of Athens, Athens, Greece

[6]Max-Planck-Institut für extraterrestrische Physik, Garching, Germany

Abstract. Almost all confirmed optical counterparts of HMXBs in the SMC are OB stars with equatorial decretion disks (OBe). These sources emit strongly in Balmer lines and standout when imaged through narrow-band Hα imaging. The lack of secure counterparts for a significant fraction of the HMXBs motivated us to search for more. Using the catalogs for OB/OBe stars (Maravelias *et al.* 2017) and for HMXBs (Haberl & Sturm 2016) we detect 70 optical counterparts (out of 104 covered by our survey). We provide the first identification of the optical counterpart to the source XTEJ0050-731. We verify that 17 previously uncertain optical counterparts are indeed the proper matches. Regarding 52 confirmed HMXBs (known optical counterparts with Hα emission), we detect 39 as OBe and another 13 as OB stars. This allows a direct estimation of the fraction of active OBe stars in HMXBs that show Hα emission at a given epoch to be at least $\sim 75\%$ of their total HMXB population.

Keywords. Magellanic Clouds, stars: early-type, stars: emission-line, Be, X-rays: binaries

1. Introduction

The Small Magellanic Cloud (SMC) has been a major target for X-ray surveys for a number of reasons: (i) due to the complete coverage of the galaxy, (ii) our ability to detect sources down to non-outbursting X-ray luminosities ($L_X \sim 10^{33}$ erg s^{-1}), and (iii) its impressive large number of High-Mass X-ray Binaries (HMXBs; Haberl & Sturm 2016). Thus, the SMC is a unique laboratory to examine HMXBs with a homogeneous and consistent approach. However, the X-ray properties alone cannot fully characterize the nature of each source. HMXBs consist of an early-type (OB) massive star and a compact object (neutron star or black hole), which accretes matter from the massive star either through strong stellar winds and/or Roche-lobe overflow in supergiant systems or through an equatorial decretion disk in, non-supergiant, OBe stars (Be/X-ray Binaries; BeXBs). Thus, to understand the nature of HMXBs in general we need to also study their optical counterparts. The OBe stars are massive stars that due to their decretion disks show Balmer lines in emission, of which Hα is typically the most prominent (e.g., see the review by Rivinius, Carciofi & Martayan 2013). Although the SMC is close enough to resolve its stellar population, we still lack the identification of the optical counterparts or their nature for many HMXBs. Out of the $\sim$120 HMXBs (of which almost all are actually BeXBs; Haberl & Sturm 2016), the $\sim 40\%$ is listed as candidate systems of this class of objects. By taking advantage of the fact that OBe stars display Hα in emission (i.e., easily discernible in Hα narrow-band images), we performed a wide area Hα imaging survey of the SMC to reveal prime candidates for optical counterparts to HMXBs.

2. Survey and Catalog

We used the Wide Field Imager (WFI@MPG/ESO 2.2m, La Silla, on 16/17 November, 2011) and the MOSAIC camera (@CTIO/Blanco 4m, Cerro Tololo, on 15/16 December, 2011) to observe 6 and 7 fields in the SMC, respectively, covering almost the entire main body of the galaxy. Each field was observed in the R broad-band (the continuum) and Hα narrow-band filters, in order to ensure the proper removal of the continuum and the calculation of Hα excess (for details see Maravelias 2014; Maravelias *et al.* 2017). The exposure time was selected to achieve a similar depth ($R \sim 23$ mag) in both campaigns to cover all late B-type stars at the distance of the SMC.

Using the locus of OB stars (Antoniou *et al.* 2009) we selected the best OB candidate sources, for which we calculated their (H$\alpha - R$) index. From their corresponding distribution we estimate their mean (H$\alpha - R$) and their standard deviation σ. We consider as the most reliable Hα emitting candidates the sources with: $(\mathrm{H}\alpha - R) < \langle \mathrm{H}\alpha - R \rangle - 5 \times \sigma$, and SNR$> 5$ (Maravelias *et al.* 2017). The final catalog consists of ~ 8350 Hα emitting objects out of a parent population of ~ 77000 OB stars.

3. Results

In Maravelias *et al.* (2017) we have performed an initial analysis to derive the ratios of OB stars with emission (OBe) over their total parent OB population, as well as the HMXBs over the OBe population (OBe/OB$\sim 13\%$ and HMXBs/OBe$\sim 0.002 - 0.025\%$, respectively). In this work, we focus in a more detailed treatment of the HMXB population to identify the best optical counterparts. From the most recent census of HMXBs in the SMC by Haberl & Sturm (2016) we remove all rejected and unlikely HXMBs, along with 17 sources that lie outside our Hα survey. Thus, we are left with 104 HMXBs, which we cross-correlate with our catalog to identify the counterparts of 70 HMXBs (see Fig. 1 and its caption for their representation).

We find that the optical counterpart of the source XTE J0050-731 or SXP16.6 (ID#17 in the Haberl & Sturm 2016 catalog) is an Hα emitting OB star. This is the first identification of the optical counterpart of this source (not to be confused with RX J0051.8-7310), and spectroscopy is needed to verify its nature. Moreover, we verify the, previously uncertain, optical counterparts for 17 sources to be OBe (10) or OB (7) stars. Last, we have detected 52 HMXBs, which have confirmed optical counterparts known to display Hα in emission. Out of these sources we find 39 as Hα emitters, while another 13 sources are identified only as OB stars (probably inactive Be stars due to their transient nature). Given these numbers, we estimate that at a given epoch $\sim 75\%$ of HMXBs (BeXBs) are active. Although this is in general agreement with the fraction of active Be stars identified in open Galactic clusters ranging from 50-75% (Fabregat 2003; McSwain & Gies 2005; Granada *et al.* 2018), we point out that it is actually a lower limit of the overall OBe population. This is because our selection criteria for OBe stars are conservative and this photometric approach is not sensitive to OBe stars with relatively faint Hα emission. These results might differ from those based on the census of OB stars in clusters, as we examine a much larger OBe population covering both clusters and the field, in the environment of the SMC which has a much lower metallicity than the Galaxy. This consists a direct measurement of the actual fraction of active OBe stars in HMXBs (BeXBs) that show Hα emission, i.e., an active decretion disk, at a given epoch.

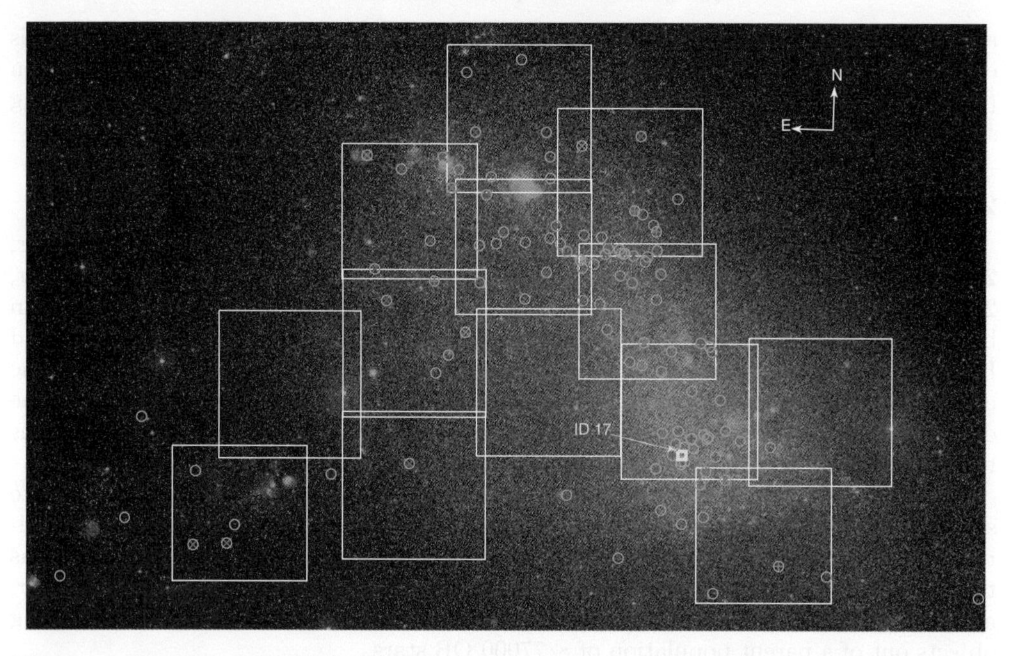

Figure 1. Our observed fields (white boxes) overplotted on a DSS image of the Small Magellanic Cloud. The green circles correspond to HMXBs from the census of Haberl & Sturm (2016), of which 104 lay within our fields and 70 have been identified. One of these is an Hα emitting OB (yellow arrow/box) coinciding with the HMXB ID#17 (Haberl & Sturm 2016; XTE J0050-731; SXP 16.6), an X-ray source without any optical identification previously. For 17 HMXBs we verify their, previously uncertain, optical counterparts to 7 OB stars (magenta crosses) and another 10 OB stars with Hα emittion (red crosses). From 52 HMXBs with confirmed Hα emitting counterparts we identify 39 sources (red X symbols) with Hα excess, and 13 sources (magenta X symbols) are identified as OB stars (inactive OBe, i.e., without Hα emission).

GM acknowledges support by an IAU travel fund. GM and AZ acknowledge funding from the European Research Council under the European Unions Seventh Framework Programme (FP/2007-2013)/ERC Grant Agreement n. 617001.

References

Antoniou, V., Zezas, A., Hatzidimitriou, D., McDowell, J. C. *et al.* 2009, *ApJ*, 697, 1695

Fabregat, J. 2003, in: Sterken, C. (ed.), *Interplay of Periodic, Cyclic and Stochastic Variability in Selected Areas of the H-R Diagram*, *ASP-CS*, Vol. 292, p. 65

Granada, A., Jones, C. E., Sigut, T. A. A., Semaan, T., Georgy, C., Meynet, G., Ekström, S. *et al.* 2018, *AJ*, 155, 50

Haberl, F. & Sturm, R. 2016, *A&A*, 586, 81

Maravelias, G. 2014, PhD Thesis, University of Crete, Heraklion, Greece

Maravelias, G., Zezas, A., Antoniou, V., Hatzidimitriou, D., Haberl, H. *et al.* 2017, in: J. J. Eldridge, J. C. Bray, L. A. S. McClelland, L. Xiao (eds.), *The Lives and Death-Throes of Massive Stars*, Proc. IAU Symposium No. 329 (CUP), p. 373

McSwain, M. V. & Gies, D. R. 2005, *ApJS*, 161, 118

Rivinius, T., Carciofi, A. C., & Martayan, C. 2013, *A&ARv*, 21, 69

High-mass X-ray Binaries: Illuminating the Passage from Massive Binaries to Merging Compact Objects
Proceedings IAU Symposium No. 346, 2019
L. M. Oskinova, E. Bozzo, T. Bulik & D. Gies, eds.

doi:10.1017/S174392131900005X

Evolution of High-mass X-ray binaries in the Small Magellanic Cloud

Jun Yang and Daniel R. Wik

Department of Physics and Astronomy, the University of Utah,
Salt Lake City, Utah 84112, USA
email: junyang@astro.utah.edu

Abstract. In order to understand the progenitor of rotation powered pulsars, we compare them with High-mass X-ray binary (HMXB) pulsars, (or X-ray pulsars), in the Small Magellanic Cloud. The plot of period period vs. period derivative shows that isolated neutron stars could be evolved from HMXBs. The pulsars with long spin period might spin up to 0.001-1 s. The mechanism is a third-body interaction that detaches the donor, leaving an isolated, small period neutron star behind.

Keywords. pulsars: general – stars: neutron – stars: evolution – X-rays: binaries – accretion, accretion disks

---

1. Introduction

When large stars end their lives in a supernova explosion, the remains of the core of these stars could form neutron stars or even black holes (Timmes *et al.* 1995). Stellar remnants with the mass between 1.35 and 2.1 $M_{\odot}$ are too heavy to exist as white dwarfs, and too light (not dense enough) to become black holes, so they will form neutron stars or possibly other strange stars or quark stars (Rosswog 2015). Kundt (2012) states that spin rates of a neutron star at birth tend to be slow, much slower than expected by conservation of angular momentum during core collapse.

A neutron star then is a type of stellar remnant that can result from the gravitational collapse of a massive star after a supernova. Neutron stars are the densest and smallest stars known to exist in the universe; with a radius of only about 12 - 13 km, they can have a mass of up to 3 $M_{\odot}$, with a surface temperature of 6×10^5 Kelvin (Kiziltan 2011; Haensel *et al.* 2007). Neutron stars have overall density of 3.7×10^{17} to 5.9×10^{17} kg/m^3 (2.6×10^{14} to 4.1×10^{14} times the density of Sun) (Shapiro & Teukolsky 2008).

Pulsars send out beams of X-ray, radio and visible light. As they rotate, the beams sweep over the earth, resulting in a periodic modulation of the received flux. Although neutron stars are very hot at birth, they do not have a source of fuel for nuclear fusion. We still detect a tremendous amount of energy from neutron stars, such as the X-ray and radio beams. There are three broad classes of pulsars depending upon their principle energy source. (1) Rotation-powered radio pulsars convert their rotational energy into radiation (Becker & Trümper 1997). These pulsars slow down very slowly at a rate attributable to magnetic dipole braking losses. (2) Accretion-powered X-ray pulsars: Most X-ray pulsars are in binary systems, and they accrete materials from their companion stars and form an accretion disc of material around them (Perna *et al.* 2006; Karino *et al.* 2008). (3) Magnetars are neutron stars with exceptionally strong magnetic fields in the range 10^{13} to 10^{16} G, compared to 10^{11} to 10^{13} G for most radio and X-ray pulsars

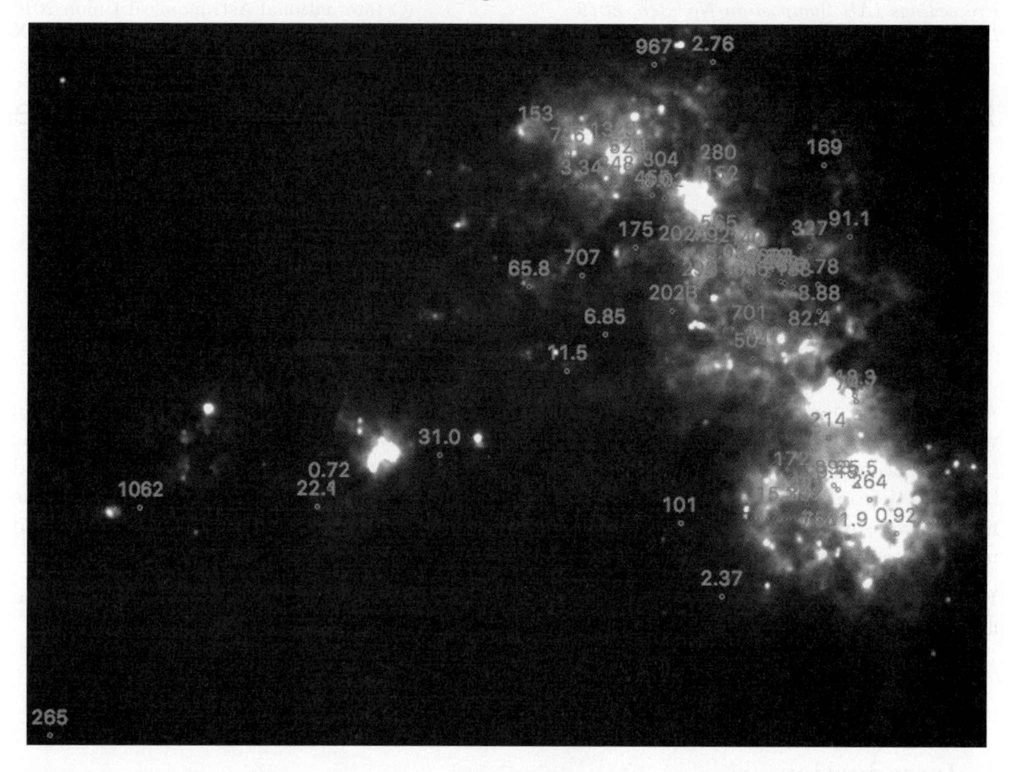

Figure 1. This figure above shows the location of the known Small Magellanic Cloud pulsars. The red numbers present their pulse periods in seconds. The background image is from the NASA/Infrared Processing and Analysis Center infrared science archive.

(Beskin *et al.* 2015). The slow decay of the magnetic field powers the radiation emission (Rees & Mészáros 2000; Heyl & Kulkarni 1998).

X-ray binaries have three distinct classes: High-mass X-ray binaries (HMXB), which have companion stars with masses $> 8\ M_\odot$, low mass (companion masses $< 1\ M_\odot$), and intermediate-mass X-ray binaries. A new class of HMXBs, Supergiant Fast X-ray Transients (SFXTs), with the unusually short transient X-ray emission and the association with blue supergiant companions, was discovered by the INTEGRAL satellite launched in October 2002 (Masetti *et al.* 2006; Negueruela *et al.* 2006; Nespoli *et al.* 2008).

2. Overview

The Small Magellanic Cloud (SMC) is a dwarf irregular galaxy near the Milky Way at a distance of about 62 kpc (Graczyk *et al.* 2013; Scowcroft *et al.* 2016). It contains a large and active population of X-ray binaries (e.g., Galache *et al.* 2008; Townsend *et al.* 2011; Klus *et al.* 2013; Coe & Kirk 2015; Christodoulou *et al.* 2016; Haberl & Sturm 2016; Yang *et al.* 2017a,b; Yang *et al.* 2018). The most numerous HMXB species in current catalogs (e.g., Yang *et al.* 2017a) are Be-HMXBs, where the companion is a Be star (Reig 2011). A Be star is a B type star (the second hottest temperature class with color temperatures in the range 20,000-40,000 K) which has at some time exhibited emission lines (signified by the letter *e*, suggesting a wind is present). The X-ray pulsars in the SMC are shown in Figure 1. The spin periods of pulsars range from seconds to thousands of seconds. The SMC provides a unique laboratory to study an important branch of stellar evolution.

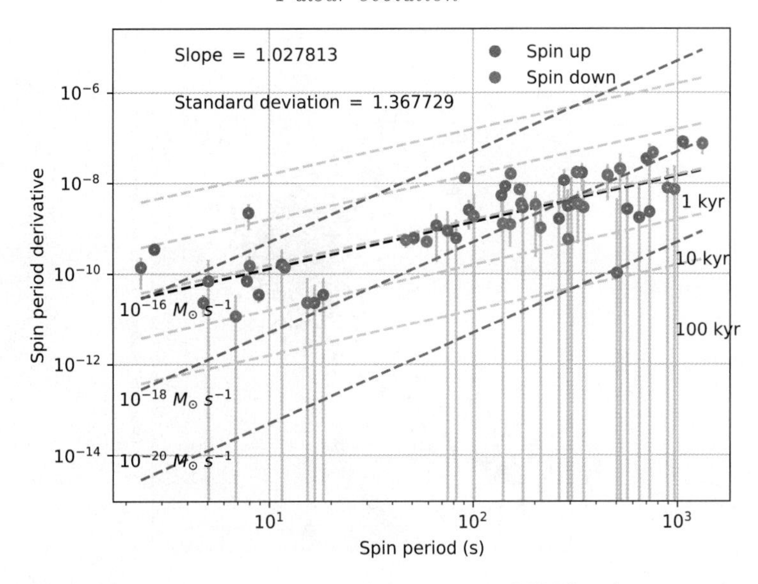

Figure 2. The filled circles show spin period derivatives of SMC pulsars as a function of spin period in seconds. Red mean this pulsar spins up and green indicates spinning down. Black dashed line is the linear fit to all pulsars in log-log space. Cyan dashed lines map the age of the pulsars. Red dashed lines are related to the mass accretion rate.

3. Implications

Bhattacharya & van den Heuvel (1991) suggest many millisecond pulsars are old, rapidly rotating neutron stars which have spun up by accreting matter from a companion star in a binary system. Massive X-ray binaries in the end might leave a single recycled neutron star which has 'evaporated' or merged with its companion star.

After accretion ceases, neutron stars could become isolated if their companions have subsequently disrupted the binary by their tidal break ups (Alpar *et al.* 1982).

Figure 2 shows the relation between spin period derivative and spin period of SMC pulsars. Red dots indicate pulsars that spin up and green symbols indicate they spin down. The error bars of the spin period derivatives are plotted in yellow. The black dashed line is the linear fit to the two variables in log-log space. Cyan dashed lines show the age map of the SMC pulsars. The majority of the sources are shown as 1 or 10 kyrs old. Interestingly, the age map is parallel to the linear fit line of spin period derivative and spin period. Within a standard deviation of 1.37, the fitting slope is consistent with 1, which implies there is no exponential relationship between spin period derivative and spin period.

Comparison of Figure 2 with the P-Pdot Diagram in Figure 3 shows that the two variables have a similar trend. In order to estimate the evolution of SMC pulsars, we consider a random source SXP 101 as and example, which spins up at the rate of 1.6×10^{-4} s/s. This pulsar is located around the 1 kyrs old line of the age map. If we assume the spin up rate is constant, after 626 kyrs later, the spin period of this pulsar will be 1s, which is located above the death line in Figure 3.

The mass accretion rate ($\dot{M}$) is plotted in red lines in Figure 2. $\dot{M}$ is estimated by assuming that the magnetic radius equals corotation radius, and the magnetic moment of the pulsar equals 10^{30} G cm^3. The deep explanation and verification about these mass accretion maps still need further study.

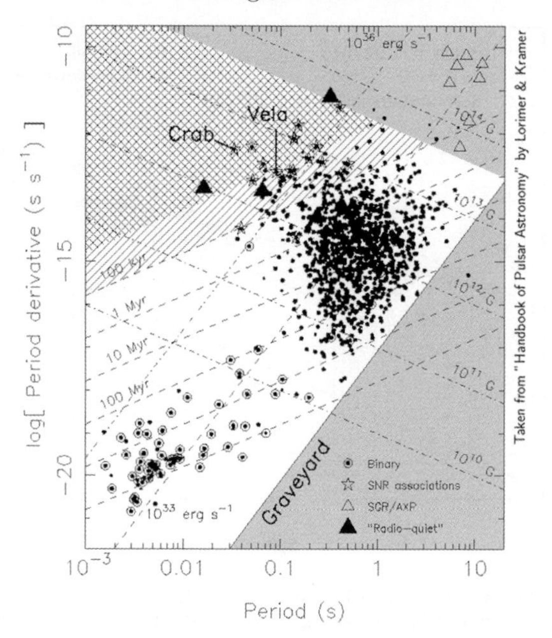

Figure 3. P-Pdot Diagram. This diagram shows the properties of the pulsars with short spin periods. Its role is similar to the Hertzsprung-Russell diagram for ordinary stars. From the spin period and spin period derivative of these pulsars, we can estimate their age, magnetic filed strength, and spin-down power (See the Handbook of Pulsar Astronomy, by Lorimer and Kramer).

References

Alpar, M., Cheng, A., Ruderman, M. & Shaham, J. (1982). A new class of radio pulsars. *Nature*, 300(5894), 728–730.

Becker, W. & Trümper, J. (1997). The x-ray luminosity of rotation-powered neutron stars. *arXiv preprint astro-ph/9708169.*

Beskin, V., Balogh, A., Falanga, M., Lyutikov, M., Mereghetti, S., Piran, T., & Treumann, R. (2015). The strongest magnetic fields in the universe. *Issues*, 1.

Bhattacharya, D. & van den Heuvel, E. P. J. (1991). Formation and evolution of binary and millisecond radio pulsars. *Physics Reports*, 203(1-2), 1–124.

Christodoulou, D. M., Laycock, S. G., Yang, J. & Fingerman, S. (2016). Tracing the lowest propeller line in magellanic high-mass x-ray binaries. *The Astrophysical Journal*, 829(1), 30.

Coe, M. & Kirk, J. (2015). Catalogue of be/x-ray binary systems in the small magellanic cloud: X-ray, optical and ir properties. *Monthly Notices of the Royal Astronomical Society*, 452(1), 969–977.

Galache, J., Corbet, R., Coe, M., Laycock, S., Schurch, M., Markwardt, C., Marshall, F. & Lochner, J. (2008). A long look at the be/x-ray binaries of the small magellanic cloud. *The Astrophysical Journal Supplement Series*, 177(1), 189.

Graczyk, D., Pietrzyński, G., Thompson, I. B., Gieren, W., Pilecki, B., Konorski, P., Udalski, A., Soszyński, I., Villanova, S., Górski, M., *et al.* (2013). The araucaria project. the distance to the small magellanic cloud from late-type eclipsing binaries. *The Astrophysical Journal*, 780(1), 59.

Haberl, F. & Sturm, R. (2016). High-mass x-ray binaries in the small magellanic cloud. *Astronomy & Astrophysics*, 586, A81.

Haensel, P., Potekhin, A. Y. & Yakovlev, D. G. (2007). *Neutron stars 1: Equation of state and structure*, volume 326. Springer Science & Business Media.

Heyl, J. S. & Kulkarni, S. (1998). How common are magnetars? the consequences of magnetic field decay. *The Astrophysical Journal Letters*, 506(1), L61.

Karino, S., Kino, M. & Miller, J. C. (2008). Funnel-Flow Accretion onto Highly Magnetized Neutron Stars and Shock Generation. *Progress of Theoretical Physics*, 119, 739–756.

Kiziltan, B. (2011). *Reassessing the fundamentals: On the evolution, ages and masses of neutron stars*. Universal-Publishers.

Klus, H., Ho, W. C., Coe, M. J., Corbet, R. H. & Townsend, L. J. (2013). Spin period change and the magnetic fields of neutron stars in be x-ray binaries in the small magellanic cloud. *Monthly Notices of the Royal Astronomical Society*, 437(4), 3863–3882.

Kundt, W. (2012). *Neutron stars and their birth events*, volume 300. Springer Science & Business Media.

Masetti, N., Morelli, L., Palazzi, E., Galaz, G., Bassani, L., Bazzano, A., Bird, A., Dean, A., Israel, G., Landi, R., *et al.* (2006). Unveiling the nature of integral objects through optical spectroscopy-v. identification and properties of 21 southern hard x-ray sources. *Astronomy & Astrophysics*, 459(1), 21–30.

Negueruela, I., Smith, D. M., Harrison, T. E. & Torrejón, J. M. (2006). The optical counterpart to the peculiar x-ray transient xte j1739–302. *The Astrophysical Journal*, 638(2), 982.

Nespoli, E., Fabregat, J. & Mennickent, R. (2008). Unveiling the nature of six hmxbs through ir spectroscopy. *Astronomy & Astrophysics*, 486(3), 911–917.

Perna, R., Bozzo, E. & Stella, L. (2006). On the Spin-up/Spin-down Transitions in Accreting X-Ray Binaries. *ApJ*, 639, 363–376.

Rees, M. J. & Mészáros, P. (2000). Fe kα emission from a decaying magnetar model of gamma-ray bursts. *The Astrophysical Journal Letters*, 545(2), L73.

Reig, P. (2011). Be/X-ray binaries., 332, 1–29.

Rosswog, S. (2015). Sph methods in the modelling of compact objects. *Living Reviews in Computational Astrophysics*, 1(1), 1–109.

Scowcroft, V., Freedman, W. L., Madore, B. F., Monson, A., Persson, S., Rich, J., Seibert, M. & Rigby, J. R. (2016). The carnegie hubble program: The distance and structure of the smc as revealed by mid-infrared observations of cepheids. *The Astrophysical Journal*, 816(2), 49.

Shapiro, S. L. & Teukolsky, S. A. (2008). *Black holes, white dwarfs and neutron stars: the physics of compact objects*. John Wiley & Sons.

Timmes, F., Woosley, S. & Weaver, T. A. (1995). The neutron star and black hole initial mass function. *arXiv preprint astro-ph/9510136*.

Townsend, L., Coe, M., Corbet, R. & Hill, A. (2011). On the orbital parameters of be/x-ray binaries in the small magellanic cloud. *Monthly Notices of the Royal Astronomical Society*, 416(2), 1556–1565.

Yang, J., Laycock, S. G. T., Christodoulou, D. M., Fingerman, S., Coe, M. J. & Drake, J. J. (2017a). A Comprehensive Library of X-Ray Pulsars in the Small Magellanic Cloud: Time Evolution of Their Luminosities and Spin Periods. *ApJ*, 839, 119.

Yang, J., Laycock, S. G. T., Drake, J. J., Coe, M. J., Fingerman, S., Hong, J., Antoniou, V. & Zezas, A. (2017b). A multi-observatory database of X-ray pulsars in the Magellanic Clouds. *Astronomische Nachrichten*, 338, 220–226.

Yang, J., Zezas, A., Coe, M. J., Drake, J. J., Hong, J., Laycock, S. G. T. & Wik, D. R. (2018). Anticorrelation between x-ray luminosity and pulsed fraction in the small magellanic cloud pulsar sxp 1323. *Monthly Notices of the Royal Astronomical Society: Letters*, 479(1), L1–L6.

High-mass X-ray Binaries: Illuminating the Passage from Massive Binaries to Merging Compact Objects
Proceedings IAU Symposium No. 346, 2019
L. M. Oskinova, E. Bozzo, T. Bulik & D. Gies, eds.

doi:10.1017/S1743921319001236

Vertical distribution of HMXBs in NGC 55: Constraining their centre of mass velocity

Babis Politakis[1], Andreas Zezas[1,2,3], Jeff J. Andrews[2] and Stephen J. Williams[2,4]

[1]University of Crete, Department of Physics, Heraklion 71003, Greece
[2]Foundation for Research and Technology - Hellas (FORTH), Heraklion 71003, Greece
[3]Harvard-Smithsonian Center for Astrophysics, Cambridge, MA 02138
[4]USNO - 3450 Massachusetts Ave, NW, Washington, DC 20392-5420

Abstract. We analyse the vertical distribution of High Mass X-ray Binaries (HMXBs) in NGC 55, the nearest edge-on galaxy to the Milky Way. Our analysis reveals significant spatial offsets of HMXBs from the star forming regions, greater than those observed in the SMC and the LMC but similar with the Milky Way. The spatial offsets can be explained by a momentum kick the X-ray binaries receive during the formation of the compact object. The difference between the scale height of the vertical distribution of HMXBs and the vertical distribution of star-forming activity is 0.48±0.04 kpc. The centre-of-mass velocity of the distribution of HMXBs in NGC 55 is moving at a velocity of 52.4±11.4 km s^{-1}, greater than the corresponding velocity of HMXBs in the SMC and LMC, but consistent with velocities of Milky Way HMXBs.

Keywords. NGC 55, X-rays, HMXB

1. Introduction

High mass X-ray binaries (HMXBs) are among the brightest sources of X-ray emission in galaxies. The observed X-ray emission is powered either by accretion of wind material or mass transfer through stable Roche-lobe overflow from the companion star to the compact object. Although HMXBs are typically associated with star-forming regions (Grimm *et al.* 2003, Fabbiano 2006), there is observational evidence for a population of HMXBs that are somewhat offset from star-forming regions (Zezas *et al.* 2002). The observed displacement may be due to kicks after an asymmetric supernova explosion during the formation of the compact object (e.g. Fryer & Kalogera 1997). If they are large enough, kicks will extend the vertical distribution of HMXBs, which can potentially be measured in nearby edge-on galaxies such as NGC 55, the edge-on galaxy nearest the Milky Way, with high angular resolution X-ray telescopes. Although there are large quantities of gas off the disk plane forming structures such as HII regions, shells, knots and filaments (Otte & Dettmar 1999), due to the distance of NGC 55 (at 1.94 Mpc) the projected density of star-forming regions is too large to allow us to identify the possible birthplace of each individual HMXB. We therefore measure the displacement of HMXBs from the mid-plane of NGC 55 which is the region with the highest star-forming activity.

2. The HMXB population in NGC 55

The population of X-ray sources in NGC 55 has been analysed as part of the Chandra Local Volume Survey (Binder *et al.* 2015). The source list consists of 154 X-ray sources down to a flux of 7×10^{-16} erg s^{-1} cm^{-2}, classified by Binder *et al.* (2015) as: one as a Ultraluminous X-ray source (ULX), 65 as Active Galactic Nuclei (AGN), 10 as

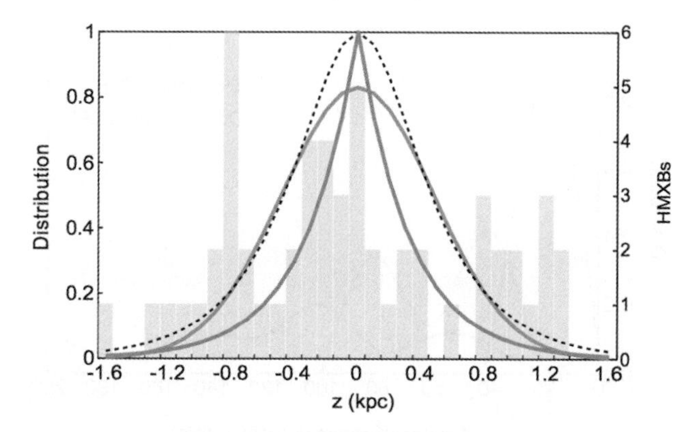

Figure 1. Simulated vertical distribution of HMXBs (dotted, black) from the mid-plane, as the convolution between a smearing of 0.48 kpc (green) on top of a star formation distribution of 0.33 kpc (red). The yellow histogram presents the observed vertical number density of HMXBs.

foreground stars, 11 as Supernova Remnants (SNRs) and 67 as X-ray binaries (XRBs) among which 24 as Low Mass X-ray binaries (LMXBs) and 43 as HMXBs. Based on our expectation of the AGN density in the Universe at our limiting X-ray sensitivity, we expect that 15 of the 67 sources classified as XRBs could be background AGN.

3. Vertical distribution of the HMXB population in NGC 55

We define the mid-plane of NGC 55 and measure the SFR density, by means of the highest resolution and least absorption-dependent SFR indicator we have available. This is the 8.0μm IRAC image that traces polycyclic aromatic hydrocarbon (PAH) emission from the star forming regions. We create a "star-formation" 8.0μm image by subtracting the 3.6μ m IRAC image from the 8.0μm image in order to remove the contribution from the non-star-forming, stellar populations. We then take slices with length equal to the major axis of NGC 55, at an angle equal to the position angle of NGC 55 and thickness equal to the FWHM of Spitzer's IRAC band 4 PSF. The slice with the highest surface brightness is selected as the mid-plane of NGC 55. We measure the vertical distribution of the SFR density calculated in bins parallel to the mid-plane. We then model the SFR density with an exponentially declining profile and fit the data to derive a star-formation scale height of $z_{\rm sfr}$=0.330±0.090 kpc (red line in Figure 1). The vertical number density of HMXBs is derived by measuring the number of HMXBs in bins of the same size as in the case of SFR (yellow histogram in Figure 1). The bins are finely spaced enough to capture the substantial displacement of HMXBs beyond the SFR density. The observed vertical displacement of HMXBs from their parent star-forming regions, may be due to kicks after an asymmetric supernova explosion during the formation of the compact object. We treat the contribution of kicks as a Gaussian smearing function of standard deviation σ that broadens the spatial distribution of HMXBs compared with their birth distribution. Based on the observed projected distances of HMXBs, we simulate the vertical distribution of HMXBs from the mid-plane as the convolution $C(\sigma, z_{\rm sfr}, z)$ between this Gaussian and the exponential distribution of SFR with a scale height of $z_{\rm sfr}$=0.330 kpc:

$$\begin{aligned} C(\sigma, z_{\rm sfr}, \mathrm{z}) = \sqrt{\frac{\pi\sigma^2}{2}} \mathbf{exp}\left(\frac{\sigma^2}{2z_{\rm sfr}^2} - \frac{\mathrm{z}}{z_{\rm sfr}}\right) \times \\ \left[\mathbf{erfc}\left[\sqrt{\frac{\sigma^2}{2}}\left(\frac{z}{\sigma^2} - \frac{1}{z_{\rm sfr}}\right)\right] - \mathbf{exp}\left(\frac{2\mathrm{z}}{z_{\rm sfr}}\right)\mathbf{erfc}\left[\sqrt{\frac{\sigma^2}{2}}\left(\frac{z}{\sigma^2} + \frac{1}{z_{\rm sfr}}\right)\right] - 2\right] \end{aligned} \tag{3.1}$$

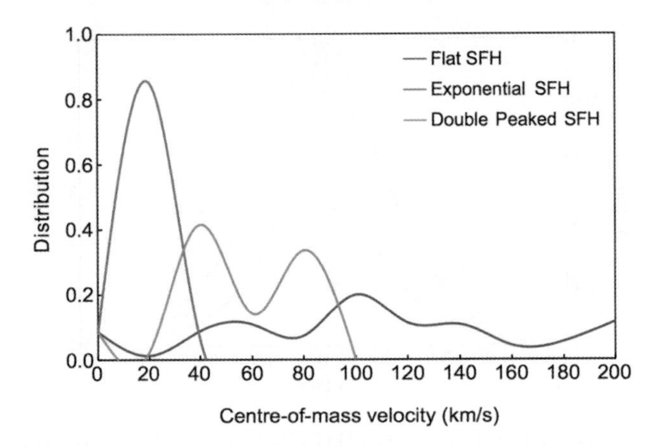

Figure 2. Distribution of the centre-of-mass transverse velocity for the HMXB population in NGC 55 for three different star-formation histories. The derived velocity depends strongly on the choice of the star-formation history adopted.

We apply the maximum likelihood method and find that the best estimate of the Gaussian smearing parameter σ is 0.48±0.04 kpc, which represents the scale height corresponding to the displacement between HMXBs and star-forming regions. Figure 1 shows the simulated vertical distribution of HMXBs (dotted, black) as a result of the convolution between a Gaussian smearing of 0.48 kpc (green) on top of a star formation distribution of 0.33 kpc (red).

We calculate the centre-of-mass transverse velocity of HMXBs by adopting HMXB travel times predicted by binary population synthesis codes for three different star formation history (SFH) models:

1. A flat SFH where stars are formed at a constant rate.
2. An exponentially declining SFH.
3. A double peaked exponentially declining SFH with bursts at 30 and 50 Myrs.

We select random offsets from the distribution of HMXBs (Figure 1) and divide them with random travel times from the distribution of SFH models. This gives the distribution of the centre-of-mass transverse velocity for the three different SFH models and is shown in Figure 2. We opt for the double peaked SFH model, in accordance with observations in star-forming galaxies that favour bursting star-formation models and estimate the centre-of-mass velocity of HMXBs in NGC 55 at 52.4±1.4 km s^{-1}. Similar measurements, show that the centre-of-mass velocities of HMXBs are lower at the MCs (12.4±7.0 km s^{-1} for the LMC, Antoniou & Zezas 2016 and 16 km s^{-1} for the SMC, Coe 2005) but similar in the Milky Way (42±14 km s^{-1}, van den Heuvel *et al.* 2000). We attribute the difference to the lack of information on the recent star formation history of NGC 55 (the value of the centre-of-mass velocity strongly depends on the choice of the star-formation history adopted), in contrast with the MCs where detailed SFH is available.

4. Vertical distribution of HMXBs in the MW

The HMXB population in our Galaxy provide an excellent benchmark for comparison with NGC55. We measure the positions of HMXBs with parallaxes available from the second data release of Gaia and fit the distribution of their vertical distances with an exponentially declining profile and find that the scale height of HMXBs in the Galaxy is h_z=145±23 kpc. The distribution of vertical distances of OB-stars from the Galaxy plane has a scale height of 103.1±3.0 (Kong & Zhu 2008). Therefore, the scale height corresponding to the displacement between HMXBs and star-forming regions in the Milky

Way is $\sim$ 0.040$\pm$0.002 kpc, considerably lower than the corresponding scale height in NGC 55. We attribute this difference to the greater gravitational potential of the Galactic disk (the stellar mass of the MW is 25 times greater than the stellar mass of NGC 55) that confines HMXBs more closely to the Galactic plane.

References

Antoniou, V. & Zezas, A. 2016, *MNRAS*, 459, 528
Binder, B., Williams, B. F., Eracleous, M., *et al.* 2015, *AJ*, 150, 94
Blaauw, A. 1961,*BAIN*, 15, 265
Coe, M. J. 2005, *MNRAS*, 358, 1379
Fabbiano, G. 2006, *ARAA*, 44, 323
Fryer, C. & Kalogera, V. 1997, *ApJ*, 489, 244
Grimm, H.-J., Gilfanov, M. & Sunyaev, R. 2003, *MNRAS*, 339, 793
Kong, D.-L. & Zhu, Z. 2008, *Chinese Astronomy and Astrophysics*, 32, 360
Kudritzki, R. P., Castro, N., Urbaneja, M. A., *et al.* 2016, *ApJ*, 829, 70
Okazaki, A. T., Hayasaki, K. & Moritani, Y. 2013, *PASJ*, 65, 41
Otte, B. & Dettmar, R.-J. 1999, *A&A*, v.343, pp. 705-712
van den Heuvel, E. P. J., Portegies Zwart, S. F., Bhattacharya, D. & Kaper, L. 2000, *AAP*, 364, 563
Zezas, A., Fabbiano, G., Rots, A. H. & Murray, S. S. 2002, *ApJ*, 577, 710

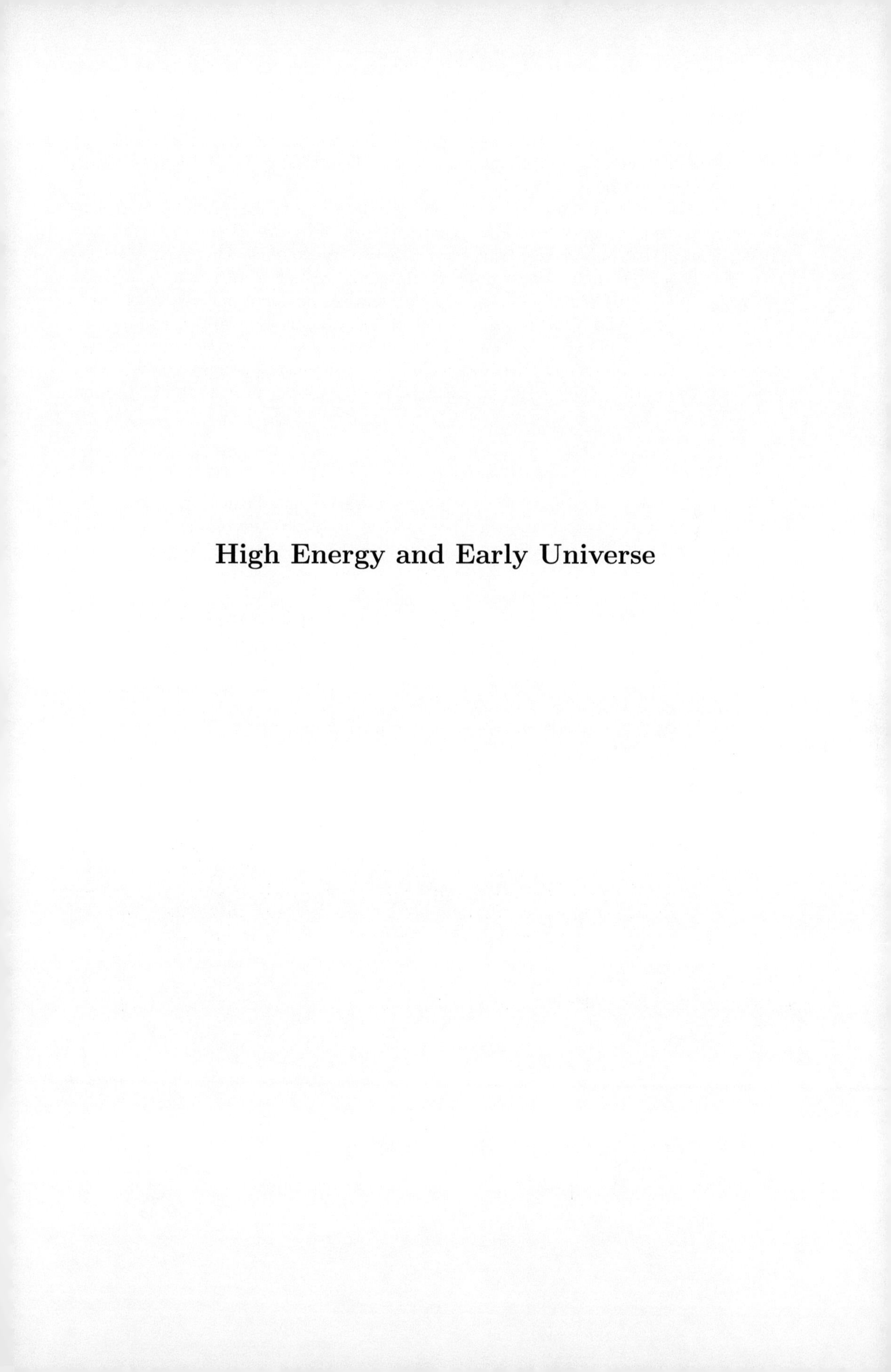

High Energy and Early Universe

High-mass X-ray Binaries: Illuminating the Passage from Massive Binaries to Merging Compact Objects
Proceedings IAU Symposium No. 346, 2019
L. M. Oskinova, E. Bozzo, T. Bulik & D. Gies, eds.

doi:10.1017/S1743921319002084

Black hole high mass X-ray binary microquasars at cosmic dawn

I. F. Mirabel[1,2]

[1]Institute of Astronomy and Space Physics. CONICET - Universidad de Buenos Aires, Ciudad Universitaria, Av. Cantilo S/N , 1428 Buenos Aires - Argentina
email: `mirabel@iafe.uba.ar`

[2]Laboratoire AIM-Paris-Saclay, CEA/DSM/Irfu/DAPCNRS, CEA-Saclay, pt courrier 131, 91191 Gif-sur-Yvette, France
email: `felix.mirabel@cea.fr`

Abstract. Theoretical models and observations suggest that primordial Stellar Black Holes (Pop-III-BHs) were prolifically formed in HMXBs, which are powerful relativistic jet sources of synchrotron radiation called Microquasars (MQs).

Large populations of BH-HMXB-MQs at cosmic dawn produce a smooth synchrotron cosmic radio background (CRB) that could account for the excess amplitude of atomic hydrogen absorption at z$\sim$17, recently reported by EDGES.

BH-HMXB-MQs at cosmic dawn precede supernovae, neutron stars and dust. BH-HMXB-MQs promptly inject into the IGM hard X-rays and relativistic jets, which overtake the slowly expanding HII regions ionized by progenitor Pop-III stars, heating and partially ionizing the IGM over larger distance scales.

BH-HMXBs are channels for the formation of Binary-Black-Holes (BBHs). The large masses of BBHs detected by gravitational waves, relative to the masses of BHs detected by X-rays, and the high rates of BBH-mergers, are consistent with high formation rates of BH-HMXBs and BBHs in the early universe.

Keywords. X-rays: binaries, microquasars, early universe, (cosmology:) black hole physics, gravitational waves

---

1. HMXBs in the heating and reionization epochs: X-rays

It is well established that between 380.000 and 1 billion years after the Big Bang the IGM underwent a phase transformation from cold and fully neutral to warm ($\sim$$10^4$ K) and ionized. Whether this phase transformation was fully driven and completed by photoionization from young hot stars is a question of topical interest in cosmology. Mirabel *et al.* (2011) and Fragos *et al.* (2013) proposed that besides the UV radiation from massive stars of populations III and II, and the soft X-rays from core-collapse SNe (Furlanetto *et al.* 2004), BH-HMXBs, the remnants of the first generations of massive binary stellar systems, likely played an important role in the process of heating, and possibly a secondary, complementary role to the reionization of the IGM that took place $\leq$$10^9$ years after the Big Bang. Because X-ray photons from HMXBs have longer mean free paths than UV photons from their massive stellar progenitors, the X-rays and possibly the relativistic jets from BH-HMXB-MQs formed from Pop III and Pop II stars would have heated the IGM across larger volumes of space during reionization. In fact, X-ray photons ionize hydrogen and helium, and the free particles deposit their kinetic energy in the IGM by secondary ionization and free-free heating.

An extensive presentation of the theoretical and observational grounds for the hypothesis of large populations of BH-HMXRBs in the early universe was published by Mirabel *et al.* (2011). From that study was concluded the following:

1. The ratio of BHs to neutron stars (NS) and the ratio of BBHs to solitary BHs should increase with redshift; that is, the rate of formation of BH-HMXBs is significantly larger in the early Universe than at present.

2. Feedback from a Galactic BH-HMXBs during a typical whole lifetime of 10^7 years is $\sim 3 \times 10^{52}$ erg, $\sim$30 times larger than the photonic and baryonic energy from a typical core collapse supernova. However, due to the low metallicities of the stellar progenitors BH-HMXBs in the early universe are likely more energetic.

3. An accreting BH in a HMXB emits a total number of ionizing photons that is comparable to its progenitor star, but one X-ray photon emitted by an accreting BH may cause the ionization of several tens of hydrogen atoms in a fully neutral medium.

4. The most important effect of BH-HMXBs in the early universe could be heating of the IGM. Soft X-rays and inverse-Compton scattering from relativistic electrons produced by BH-HMXBs could heat the low-density medium over large volumes to temperatures of $\sim 10^4$ K, which would limit the recombination rate of hydrogen keeping the IGM ionized.

5. A temperature of the IGM of $\sim 10^4$ K would limit the formation of faint galaxies at high redshifts. It constrains the total mass of dwarf galaxies to $\geq 10^9$ $M_\odot$, leaving dark matter halos of $\leq 10^9$ $M_\odot$ with no barionic mass.

6. Therefore, BH-HMXBs in the early universe could be important ingredients for reconciling the apparent disparity between the observed number of faint dwarf galaxies in the Galactic halo with the much larger number of low-mass galaxies predicted by the cold dark matter model of the universe.

7. An additional effect of metallicity in the formation of BH-HMXBs is to boost the formation of BBHs (Mirabel 2011), as the more likely first detected sources of gravitational waves than NS-NS systems (Belczynski *et al.* 2010). These predictions have been confirmed by the first observations of the LIGO-Virgo collaboration.

After the publication by Mirabel *et al.* (2011) the observational grounds for the hypothesis of a high rate of HMXB formation in the early universe has been re-enforced by new observational results (e.g. Fragos *et al.* (2013); Basu-Zych *et al.* (2013); kaaret (2014); Douna *et al.* (2015); Lehmer *et al.* (2016); Brorby *et al.* (2016)). Lehmer *et al.* (2016) found an increase in LMXB and HMXB scaling relations with redshift as being due to declining host galaxy stellar ages and metallicities, respectively, and discussed how emission from HMXRBs could provide an important source of heating to the IGM in the early universe, exceeding that of active galactic nuclei (AGN), as shown in Figure 1.

2. X-rays, Gamma-rays, and relativistic jets from HMXB-MQs

Most previous works on the impact of HMXBs in the heating and reionization epochs of the universe have considered feedback in the form of X-ray radiation. However, in recent years we have come to the realization that HMXBs are also Microquasar sources of relativistic jets and massive outflows that dissipate a large fraction of the liberated accretion power in the form of relativistic outflows (Mirabel & Rodríguez 1999). In fact, the best studied HMXBs in our Galaxy, Cygnus X-1, SS433 and Cygnus X-3, are sources of powerful jets that are equally energetic or even more energetic, than their photonic feedbacks (Mirabel & Rodríguez 1999). High energy gamma-ray emissions have also been reported from these three HMXB-MQs, and other HMXB classes (short reviews in Mirabel 2006 & Mirabel 2012)

Associated with Cygnus X-1, Gallo *et al.* (2005) found at a distance of $\sim$5 pc from the BH-HMXB a ring-like shock structure produced by the jets from the accreting BH, with a kinetic energy injection of at least that of the total X-ray luminosity. GeV emission from Cygnus X-1 most likely associated with the relativistic jets was reported by Zanin *et al.* (2016).

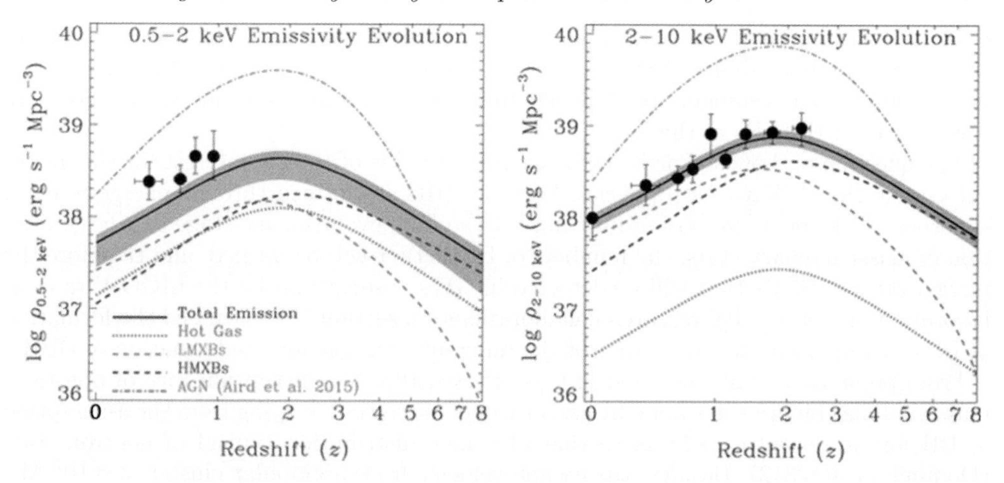

Figure 1. : These figures from Lehmer *et al.* (2016) show how the volume-averaged emission of X-rays from different X-ray sources evolves over cosmic time. The AGN curve, shown in magenta, decreases abruptly at z≥6, while the total emission from galaxies (shown by the black solid line and gray shaded region) dominates at higher z. The high mass X-ray binaries (blue dashed line) are the main contributors of X-rays within galaxies at z≥6.

In SS433 relativistic jets moving at ∼0.26c carry heavy nuclei with a mechanical energy $\geq 10^{39}$erg s^{-1}. They impact at a distance of several tens of parsecs on the interstellar medium and inflate laterally the W50 nebula that hosts the HMXB (Mirabel & Rodríguez 1999). Recently, The High Altitude Water Cherenkov (HAWC 2018) reported from SS433/W50 detection of gamma-ray emission of at least 25 TeVs, from the lobes of W50 in which the jets terminate about 40 pc from the central source.

Cygnus X-3 is at a larger distance of 7.4±1.1 kpc, the donor star shows Wolf-Rayet features and the orbital period is 4.8 hs. This HMXB-MQ is one of the most powerful X-ray transient sources of relativistic radio jets in the Milky Way, which produce synchrotron fluxes at radio waves of up to several tens of Jy. This source has been detected by the gamma-ray satellites AGILE (Tavani *et al.* 2009) and Fermi-LAT (2009). Cygnus X-3 is in a crowded region and the later secured that the high energy emission actually comes from Cygnus X-3, by the orbital period in gamma rays, as well as with the correlation of the LAT flux with radio emission from the relativistic jets.

On the other hand, in nearby dwarf galaxies are found inflated cavities by relativistic jets and massive outflows from HMXBs (Feng & Soria 2011). For instance, Pakull, Soria & Motch (2010) found in NGC 7793 a jet-inflated bubble, with a diameter of 300 pc, surrounding a HMXB-MQ. Pakull *et al.* (2010) estimated for this HMXB a mechanical kinetic energy injection of $\sim 10^4$ that of X-rays.

Heinz & SunyaevHeinz & Sunyaev (2002) studied the physics of the interaction of jets from MQs with the interstellar medium (ISM) showing that cosmic rays are produced by shocks of the relativistic jets with the ISM. The relativistic particles of the jets deposit a fraction of their kinetic energy at the interface between the jet and the ambient medium (working surface), into random, isotropic particle energy, likely contributing to the heating of the IGM. Following that work Tueros *et al.* (2014) proposed that microquasar jets contribute to the reionization of the IGM at cosmic dawn. More recently, Douna *et al.* (2018) by Monte Carlo simulations find that the contribution of microquasar jets to the heating of the IGM is of the same order of magnitude as that of cosmic rays from SNe.

3. Stellar black holes formed by implosion

Theoretical models predict that massive stars with decreasing metallicity produce weaker stellar winds, allowing larger numbers of massive stellar binaries to become

HMXBs, with more massive compact objects formed by direct collapse (BHs), and with more luminous donor stars. Therefore, it is expected that HMXBs formed in the low metallicity environments of the early universe should be more numerous and more energetic than HMXBs in the local universe.

The question on how stellar black holes are formed is of topical interest for the incipient Gravitational-Wave Astrophysics. Whether BHs are formed through energetic natal supernova kicks or by implosion may impact the final evolutionary stage of a large fraction of massive binary stars, the numbers of BBHs that will be formed, and therefore the merger rates of BBHs that will be detected in GWs observations by the LIGO-Virgo collaboration and other GW research collaborations. In section 7 is discussed the formation of BHs by implosion in the context of the different channels for the formation of BBHs.

From population synthesis models it is inferred that the disruption rate of non-tight massive stellar binaries increases by two orders of magnitude varying from the assumption of BH formation with no kicks to that of a kick distribution typical of neutron stars (Dominik *et al.* 2012). Besides, the escape velocity from a globular cluster of $\leq 10^7$ $M_\odot$ is few tens km s^{-1}, and if formed with a kick distribution typical of neutron stars most BHs would be kick out from globular clusters by SN natal kicks.

Theoretical models set progenitor masses for BH formation by implosion, but observational evidences have been elusive. The kinematics of BH-X-ray binaries can provide observations to contrast the theories on the formation of BHs. If a compact object is accompanied by a mass-donor star in an X-ray binary, it is possible to determine the distance, proper motion, and radial velocity of the center of mass of the system, from which can be derived the velocity in three dimensions of space, and in some cases, also infer the site of birth of the BH. Here are summarized the recently improved determination, in most cases by VLBI at radio wavelengths, of the kinematics of two Galactic BH-X-ray binaries formed by direct collapse. When available, short comments are given on preliminary results obtained with GAIA.

Cygnus X-1 is a X-ray binary at a distance of 1.86 $\pm$ 0.1 kpc composed by a BH of 14.8 $\pm$ 1.0 $M_\odot$ and a 09.7lab donor star of 19.2 $\pm$1.9 $M_\odot$ with an orbital period of 5.6 days and eccentricity of 0.018 $\pm$ 0.003. Cygnus X-1 appears to be at comparable distance and moving together with the association of massive stars Cygnus OB3 (Mirabel & Rodrigues 2003). Therefore, it had been proposed that the BH in Cygnus X-1 was formed in situ and did not receive an energetic trigger from a natal or nearby supernova.

More recently, a trigonometric parallax that corresponds to a distance of 1.86$\pm$0.12 kpc was obtained with the Very Long Baseline Array by Reid *et al.* (2011). These authors also measured the proper motion of Cygnus X-1 which, when coupled to the distance and Doppler shift, gives the three-dimensional space motion of the system. Reid *et al.* (2011) conclude that the binary did not experience a large "kick" at BH formation.

On the other hand, the comparison with the results from a new reduction of the Hipparcos data by Melnik & Dambis (2009) to infer the mean distance and proper motion of Cygnus OB3, reaffirmed the conjecture that Cygnus OB3 is the parent association of Cygnus X-1. From a preliminary analysis of the GAIA second data release it is found that in fact Cygnus OB3 most likely is the parent association of Cygnus X-1 (García *et al.* 2019) as proposed by Mirabel & Rodrigues (2003). However, so far there is a discrepancy between the GAIA and the VLBI radio parallaxes of Cygnux X-1, that could be attributable either to intrinsic orbital wobble, or to systematic pipeline measurement uncertainties in the GAIA data (Gandhi *et al.* 2019). The motions on the plane of the sky of Cygnus X-1 and Cygnus OB3 are shown on the left side of Figure 2.

The upper limit of the velocity in three dimensions of Cygnus X-1 relative to the mean velocity of Cygnus OB3 is 9 $\pm$ 2 km s^{-1}, which is typical of random velocities of stars in expanding associations of massive stars.

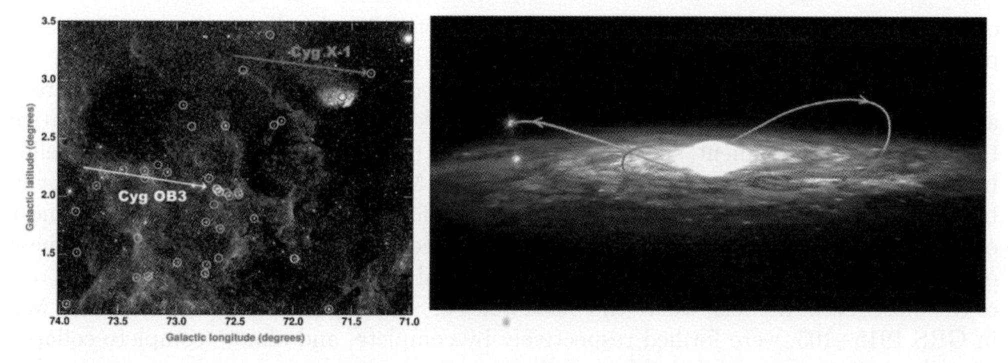

Figure 2. Left: Optical image of the sky around the BH X-ray binary Cygnus X-1 and the association of massive stars Cygnus OB3. The red and yellow arrows show the magnitudes and directions of the motion in the plane of the sky of the radio counterpart of Cygnus X-1 and the average Hipparcos motion of the massive stars of Cygnus OB3 (circled in yellow) for the past 0.5 Millions years, respectively. After the formation of the BH, Cygnus X-1 remained anchored in the parent association of massive stars Cygnus OB3. From Mirabel & Rodrigues (2003). **Right**: Schematic Galactic orbit of the Galactic Halo BH X-ray binary XTE J1118+480 (red curve) during the last ~240 Myr, which corresponds to the orbital period of the Sun around the Galactic Center. The source left the plane towards the Northern Galactic hemisphere with a galactic-centric velocity of 348±18 km s^{-1}, which after subtraction of the velocity vector due to Galactic rotation, corresponds to a peculiar space velocity of 217±18 km s^{-1} relative to the Galactic disk frame, and a component perpendicular to the plane of 126±18 km s^{-1}. The galactic orbit of XTE J1118+480 has an eccentricity of 0.54. At the present epoch XTE J1118+480 is at a distance from the Sun of only 1.9±0.4 kpc flying through the Galactic local neighborhood with a velocity of 145 km s^{-1}. From Mirabel *et al.* (2001).

From the equations for spherical mass ejection at BH formation in massive stellar binaries, it is estimated that the maximum mass that could have been suddenly ejected to accelerate the binary without binary disruption to a velocity of 9 ± 2 km s^{-1} is $\leq$1±0.3 $M_\odot$. Indeed, there are no observational evidences for a SN remnant in the radio continuum, X-rays, and atomic hydrogen surveys of the region where Cygnus X-1 was most likely formed.

Mirabel & Rodrigues (2003) estimated that the initial mass of the progenitor of the BH is ~40 ± 5 $M_\odot$ which may have lost ~25 $M_\odot$ by stellar winds during a Wolf-Rayet stage and mass exchange with the presently massive stellar donor. The observational mass lower limit of ~40 $M_\odot$ for the progenitor is the same as the mass theoretically predicted for the transition from fall-back to complete collapse for a BH progenitor of solar metallicity (Fyer *et al.* 2012), and the complete collapse of stellar helium cores in stars of solar metallicity (Sukhbold *et al.* 2016).

GRS 1915+105 is a low-mass X-ray binary containing a BH of 10.1 ± 0.6 $M_\odot$ and a donor star of spectral type K-M III of 0.5 ± 0.3 $M_\odot$, with a 34 day circular orbital period. The companion overflows its Roche lobe and the system exhibits episodic superluminal radio jets (Mirabel & Rodríguez 1994).

From a decade of astrometry with the NRAO Very Long Baseline Array it was determined a parallax distance of of 8.6 ± 1.8 kpc and proper motion, that together with the published radial velocity, is determined a modest peculiar velocity of 22 ± 24 km s^{-1} (Reid *et al.* 2014), which is consistent with the earlier proposition that the BH in GRS 1915+105 was formed without a strong natal kick, like the 14.8 ± 1.0 $M_\odot$ BH in Cygnus X-1 (Mirabel *et al.* 2001). The modest peculiar speed of 22 ± 24 km s^{-1} at the parallax distance and a donor star in the giant branch suggest that GRS 1915+105 is an old system that has orbited the Galaxy many times, acquiring a peculiar velocity component on the galactic disk of 20-30 km s^{-1}, consistent with the velocity dispersions of ~20 km s^{-1}

of old stellar systems in the thin disk, due to galactic diffusion by random gravitational perturbations from encounters with spiral arms and giant molecular clouds.

In summary, the kinematics in three dimensions of Cygnus X-1 relative to the parent association of massive stars Cygnus OB3, and the kinematics of GRS 1915+105 relative to its Galactic environment, suggest, irrespective of their origin in isolated massive stellar binaries or in typical dense clusters of less than 10^7 $M_\odot$, that the black holes in the X-ray binaries Cygnus X-1 and in GRS 1915+105, were formed in situ by complete or almost complete collapse of massive stars, with no energetic kicks from natal supernovae. These observational results are consistent with theoretical models, and in particular, with one of the most recent models, where the black holes of $\sim$15 $M_\odot$ in Cygnus X-1, and $\sim$10 $M_\odot$ in GRS 1915+105, were formed respectively by complete, and almost complete collapse of stellar helium cores (Sukhbold *et al.* 2016).

4. Black hole X-ray binaries with peculiar velocities

GRO J1655-40 is a X-ray binary with a BH of 5.3 $\pm$ 0.7 $M_\odot$ and a F6-F7 IV donor star. It had been estimated that this BH-XRB is moving with a peculiar velocity with respect to its environment of 112 $\pm$ 18 km s^{-1}, in a highly eccentric (e=0.34 $\pm$ 0.05) Galactic orbit (Mirabel *et al.* 2002).

From a preliminary analysis of the GAIA second data release it is estimated that the parallax-inversion distance of this X-ray binary corresponds to a distance of 3.66$\pm$1.01 kpc, and the peculiar velocity relative to Galactic rotation would be 150.6$\pm$13 km s^{-1} (García *et al.* 2019). The runaway linear momentum and kinetic energy of this X-ray binary are similar to those of solitary runaway neutron stars and millisecond pulsars with the most extreme runaway velocities.

XTE J1118+480 is a high-galactic-latitude (l=157°.78, b=+62°.38) X-ray binary with a BH of 7.6 $\pm$ 0.7 $M_\odot$ and a 0.18 $M_\odot$ donor star of spectral type K7V-M1V. It is moving in a highly eccentric orbit around the Galactic center region, as some ancient stars and globular clusters in the halo of the Galaxy. On the right two panels of Figure 2 are shown the top and side views of the orbital path of this X-ray binary relative to the Galactic disk (Mirabel *et al.* 2001).

V404 Cyg (GS 2023+338) is a low mass X-ray binary system composed of a BH of 9.0 $\pm$ 0.6 $M_\odot$ and a 0.75 $M_\odot$ donor of spectral type K0 IV. From astrometric VLBI observations, it was measured for this system a parallax distance of 2.39 $\pm$ 0.14 kpc (Miller-Jones *et al.* 2009a). Together with the proper motion Miller-Jones *et al.* (2009b) derived a peculiar velocity of 39.9 $\pm$ 5.5 km s^{-1}, with a component on the Galactic plane of 39.6 km s^{-1}, that is $\sim$2 times larger than the expected velocity dispersion in the Galactic plane.

From a preliminary analysis of the GAIA second data release Gandhi *et al.* (2018) estimated a parallax-inversion distance of 2.28$\pm$0.52 kpc and a peculiar velocity relative to Galactic rotation of 51.5$\pm$16.0 km s^{-1}, which -within the errors- are consistent with the radio VLBI observations.

On the interpretation of the anomalous velocities of black hole X-ray binaries. GRO J1655-40 and V404 Cyg are in the Galactic disk (b $\leq$ 3.2^o; distances from the plane z $\leq$ 0.15 kpc), where were likely formed. XTE J118+480 is in the Galactic halo (b = 62.3^o; z = 1.5 kpc) and could either have been propelled to its present position from the Galactic disk by a supernova explosion, or have been formed in a globular cluster from which it could have escaped with a mild velocity of few tens of km s^{-1}.

If these X-ray binaries were formed from binary stars in a field of relative low density, one would conclude that the three BHs with $\leq$10 $M_\odot$ were formed with significant natal kicks, whereas the BHs with $\geq$10 $M_\odot$ in Cygnus X-1 and GRS 1915+105 were formed with no energetic natal supernovae kicks.

However, if these "runaway" BH-XRBs were formed in dense stellar clusters, the anomalous velocities of the XRBs barycenter could have been caused by dynamical interactions in the stellar cluster, or by the explosion of a nearby massive star, and not necessarily by the explosion of the progenitor of the "runaway" BH. Furthermore, the possible supernova nuclear-synthetic products in the atmospheres of the donor stars in GRO J1655-40, XTE 1118+480 and V404 Cyg, could be due to contamination by the explosion of a nearby star in a high density natal environment, rather than to the explosion of the BH progenitor.

In fact, the three BH "runaway" XRBs have low mass donors, and their linear momenta are comparable to those of runaway massive stars likely ejected from multiple stellar systems by the Blaauw mechanism. Therefore, without knowing the origin of a "runaway" X-ray binary, it is not possible to constrain from its peculiar velocity alone, the strength of a putative natal supernova kick to the compact object in the XRB. It is expected that future parallax distances and proper motions of BHs and their environment, determined at radio wavelengths by VLBI and at optical wavelengths with GAIA, will provide further observational constraints on the physical mechanisms of stellar BH formation.

5. Signatures of BH-HMXB-MQs in the earlier universe

Stellar BHs in the Local Universe.

It is estimated that in the Galaxy there have been formed 10^8 to 10^9 stellar-mass BHs (e.g. Brown & Bethe 1994; Timmes *et al.* 1996). Assuming a mass of 10 $M_\odot$ for each BH, the integrated mass of those stellar BHs would be $\sim 10^3$ times the mass of the supermassive BH of $\sim 4\times 10^6$ $M_\odot$ at the center of the Galaxy. The majority of stellar BHs of that large population presently are inactive, but in earlier epochs of Galactic evolution a significant fraction may have had periods of strong feedback activity. Until now out of that putative large population of stellar BHs only a few tens of BH-XRBs have been dynamically identified, Cygnus X-1, and likely Cygnus X-3 and SS 433, being the best studied Galactic HMXB-MQs believed to harbor stellar BHs.

BH-HMXBs are sources of powerful collimated relativistic jets and massive outflows, capable of blowing large two sided cavities in the ISM (Mirabel & Rodríguez 1999; Gallo *et al.* 2005). Similar HMXB-MQs have been identified in nearby dwarf galaxies of relative low metallicity (Feng & Soria 2011). As shown in sections 1 and 2, it is expected that a large fraction of the immediate remnants of the first massive stars in the early universe are BH-HMXB-MQs, with powerful relativistic jets that generate synchrotron radiation (Mirabel & Rodríguez 1999). In this context one may ask whether that large population of BH-HMXB-MQs at cosmic dawn could have produced a diffuse cosmic radio synchrotron background, and whether it could have already been observed.

A smooth radio synchrotron background of unknown origin?

From observations with a balloon-borne instrument called ARCADE 2, Fixsen *et al.* (2011) reported an excess radio synchrotron background, consistent with the Cosmic Microwave Background (CMB) radiation at high frequencies, but significantly deviating from a blackbody spectrum at low frequencies (see Figure 3-Left). A possible radio background had been suggested before by Bridle (1967), and its possible existence was later reaffirmed with ARCADE 2 and other observations (Singal *et al.* 2018 and references therein).

The origin of this radio synchrotron background excess is a subject of debate because the observed level of surface brightness is substantially larger than expected from observed radio counts, unresolved emission from the known radio point source population, the spectrum of this excess emission being flatter than the much deeper spectrum of radio point sources, and it cannot be explained by CMB spectral distortions (Seiffert *et al.* 2011).

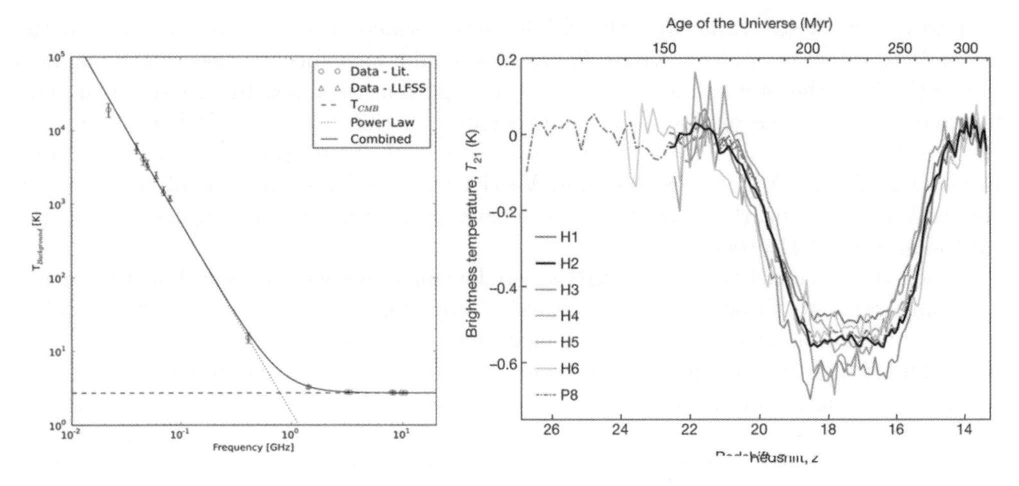

Figure 3. Left: Modeled extragalactic temperature as a function of frequency from LLFSS and other maps by Dowell & Taylor (2018). The solid line shows the best fit to the sum of a power law (dotted line) with a spectral index of 2.58 and the CMB (dashed line) at 2.722 K. **Right**: Best-fitting 21-cm absorption profiles for each hardware case of EDGES by Bowman *et al.* (2018a). Each profile for the brightness temperature is plotted against redshift z and the corresponding age of the Universe. The thick black line is the model fit for the hardware and analysis configuration with the highest signal-to-noise ratio, processed using 60 to 99 MHz and a four-term polynomial for the foreground model.

Condon *et al.* (2012) used the Very Large Array (VLA) to image at 3 GHz a primary beam at high resolution and high sensitivity. They conclude from radio source count considerations that neither AGN-driven sources or star-forming galaxies can provide the bulk of the observed level of the radio synchrotron background reported by the observations with ARCADE 2. Condon *et al.* (2012) estimate that any new discrete-source population able to produce such a bright and smooth background is far too numerous to be associated with galaxies brighter than m_{AB}=+29.

It was believed that this reported background radiation may originate from very early cosmic times because it does not seem to be associated with far-infrared thermal emission from dust. If it were produced by sources that follow the correlation between radio emission and the far-infrared radiated by dust observed in galaxies, the cosmic far-infrared background (CIB) would be overproduced above the observed levels with Planck (Ysard & Lagache 2012). Although the FIR-Radio correlation of galaxies may evolve with redshift (Ivison *et al.* 2010), Magnelli *et al.* (2015) conclude that star burst galaxies responsible for the CIB contribute $\leq 10\%$ of a putative cosmic radio background (CRB). In this context, $\geq 90\%$ of a CRB would come from sources at cosmic dawn that produce radio synchrotron emission, but do not produce radiation by dust.

Caveats on the existence of that smooth radio background have been formulated by Vernstrom *et al.* (2015). However, Dowell & Taylor (2018) studied more recently the radio background between 40 and 80 MHz using the all-sky maps from the LWA1 Low Frequency Sky Survey (LLFSS). They confirm the strong, diffuse radio background suggested by the ARCADE 2 observations in the 3 to 10 GHz range. Dowell & Taylor (2018) modeled this radio background by a power law with a spectral index of 2.58 ± 0.05 and a temperature at the rest frame 21 cm frequency of 603 ± 97 mK. But Subrahmanyan & Cowsik (2013) claim that instead modeling Galactic emission as a plane parallel slab, a more realistic modeling yields estimates for the smooth extragalactic brightness that are consistent with expectations from known extragalactic radio source populations.

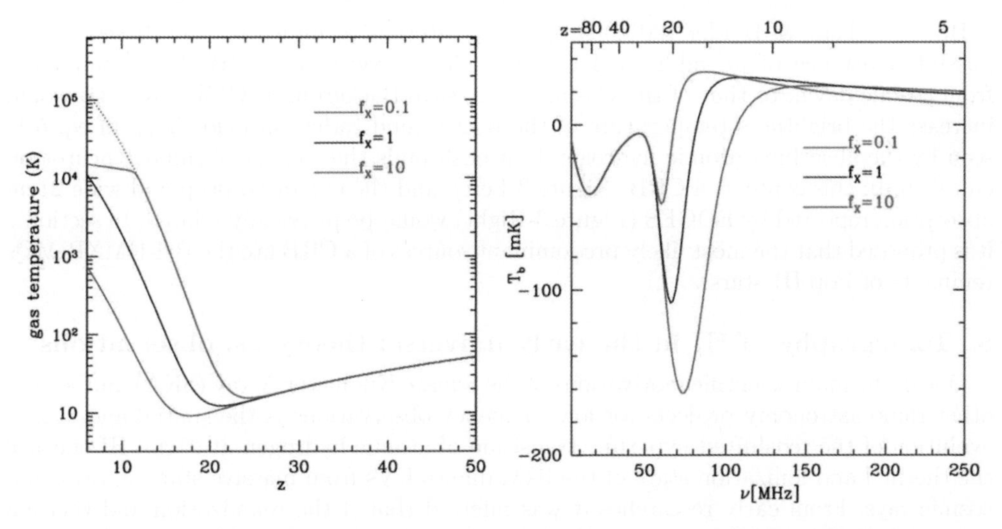

Figure 4. Left: Thermal history of the neutral intergalactic medium (IGM) including heating by BH-HMXBs. The gas temperature of the IGM is plotted as a function of redshift z for three possible values of f_X as defined by Eq. (6) in Mirabel *et al.* (2011). **Right**: Predicted brightness temperature of the hyperfine transition of the ground state of atomic hydrogen of 21 cm, averaged over the entire sky, as a function of redshift z, for the same three different values of f_X. In this model the predicted frequency of the largest absorption amplitude against the Cosmic Microwave Background (CMB) for $f_X\sim0.1$ is at $\sim$73 MHz and the gas temperature $\sim$10 K. These values are close to the central frequency of 78 MHz (z$\sim$17) of the absorption reported by EDGES (Fig. 3-Right), which would correspond to $f_X\leq0.1$, IGM gas temperatures $\leq$10 K, and hydrogen densities $N_H\geq5\times10^{23}$ cm^{-2}. Both figures reproduced from Mirabel *et al.* (2011).

A deep HI absorption at z$\sim$17.

Figure 3-Right shows a deep absorption feature of $\sim$0.5 K centered at 78 MHz (z=17) in the averaged sky of the redshifted 21 cm spectrum of atomic hydrogen found by the Experiment to Detect the Global EoR Signatures (EDGES) from Bowman *et al.* (2018a). The absorption feature is $\geq$2 times stronger than the 21-cm signal absorption expected to be observed against the Cosmic Microwave Background (CMB) by standard astrophysical models of cosmic dawn (e.g. Figure 4-Right in next section).

The search for redshifted 21 cm radiation of HI at cosmic dawn between 50 and 200 MHz require a careful and accurate subtraction of the sky foreground, since the searched signal is orders of magnitude below the foreground. Hills *et al.* (2018) expressed concerns about the modelling of the EDGES data in the reported absorption by Bowman *et al.* (2018a). These concerns were discussed in a reply by Bowman *et al.* (2018b). While waiting for an independent detection of this absorption signal, in the following it will be tentatively assumed that it is produced by neutral gas at cosmic dawn.

Madau *et al.* (1997) studied the dependence of the HI signal amplitude δT_b on the spin temperature T_s, the temperature of the Cosmic Microwave Background T_{CMB}, and an additional possible background T_{rad}. Following that study the amplitude of the HI absorption can be expressed as:

$$\delta T_b \propto \left(1 - \frac{T_{\rm CMB} + T_{\rm rad}}{T_{\rm s}}\right) \tag{5.1}$$

One interpretation of this unexpected large and frequency wide absorption was the decreased of T_s by an increased cooling of the HI gas by interaction with dark matter (e.g. Barkana 2018).

However, Feng & Holder (2018) propose an alternative interpretation, based on the possible existence of an additional cosmic radio background (CRB) field component from cosmic dawn, to that of the Cosmic Microwave Background (CMB). A CRB would increase the brightness temperature of the background radiation field (T_{rad} in eq. 5.1) seen by the absorbing atomic hydrogen that enshrouds the sources of radio synchrotron emission. In this context, a CRB (Figure 3-Left), and the enhanced deep and wide 21cm absorption reported by EDGES (Figure 3-Right) would be physically related. In section 7 it is proposed that the most likely predominant sources of a CRB are the BH-HMXB-MQs remnants of Pop III stars.

6. Tomography of HI in the early universe: theory vs. observations

One of the main scientific motivations of the Square Kilometer Array (SKA) and several other radio astronomy projects for low frequency observations is the spatial and timing evolution of the redshifted hyperfine transition of atomic hydrogen at 21cm. HI encodes the thermal and ionization state of the IGM due to UVs from massive stars, X-rays, and cosmic rays. From early researches it was inferred that if the reionization and thermal state of the IGM are fully driven and completed by photoionization from young hot stars, the tomography of HI should exhibit a morphology with prominently marked frontiers between the neutral and ionized regions, namely, the morphology should look like the morphology of Swiss cheese.

The tomography of cosmic heating and reionization inferred from the evolution of the redshifted 21cm line of atomic hydrogen for different values of the IGM temperature at redshifts $z \geq 6$ was modeled incorporating feedback in the X-rays from different types of sources by several groups of researchers. In particular, Mirabel *et al.* (2011) proposed BH-HMXB-MQs as such sources, and Haiman (2011) remarked that if hard X-rays from these sources are important, the tomography of HI should be smoother than previously thought.

Figure 4 produced by Pritchard and published in Mirabel *et al.* (2011), shows the expected HI brightness temperature as a function of frequency and redshift for different values of f_X, a variable derived from Furlanetto (2006), which denotes the fraction of the total luminosity that emerges to the IGM in the assumed relatively hard 2-10 keV band radiation of BH-HMXB-MQs.

A comparison of the model predictions by Mirabel *et al.* (2011) in Figure 4-Right, with the EDGES observed absorption in Figure 3-Right, shows that the best agreement of the model with observations would be for $f_X \leq 0.1$, concerning the redshift of the maximum amplitude of the Gaussian-shape deep absorption and the overall frequency extension of the absorption.

But there are striking differences in the amplitude and shape of the absorption profile between the expected Gaussian profile in Figure 4-Right, and the bottom-flat profile reported by EDGES in Figure 3-Right. The excess amplitude of the observed absorption may be due to the existence of a CRB. The non-Gaussian, flat bottom-shape of the EDGES signal can be accounted for by a normalized 21-cm bi-spectrum in fully-numerical simulations of the IGM heated by stellar sources and HMXBs (Watkinson *et al.* 2018).

$f_X \leq 0.1$ implies that the space density of the hydrogen that enshrouds the sources of X-rays and synchrotron radio emission should be high enough to block the X-rays, but not the radio synchrotron emission from the sources. Ewall-Wice *et al.* (2018) point out that in order to avoid heating the IGM over the EDGES trough, the sources that produce the X-rays would need to be obscured by hydrogen column depths of $N_H \sim 5 \times 10^{23}$ cm^{-2}. In fact, N_H column densities even larger than these are observed in star forming regions in the Milky Way, as well as in the central regions of extreme starburst far-infrared galaxies (Sanders & Mirabel 1996).

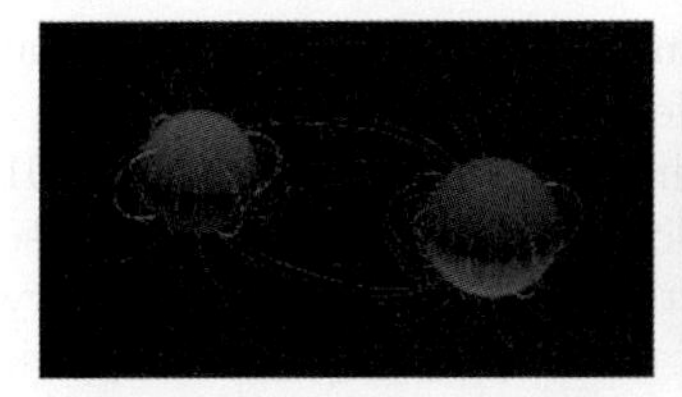
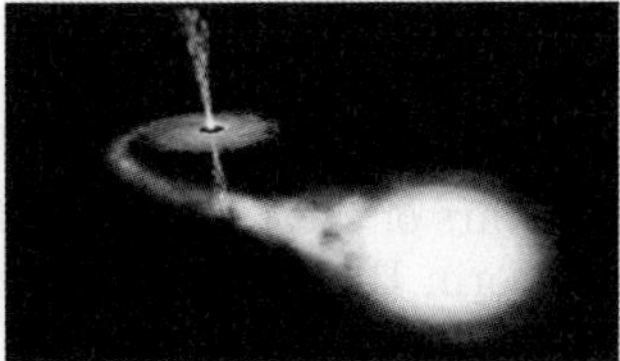
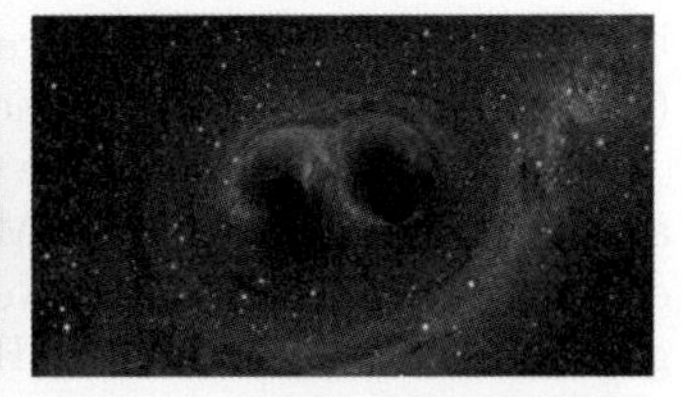

Figure 5. Binary Black Holes may be formed from relatively isolated massive stellar binaries (Belczynski *et al.* 2016) or contact massive stellar binaries (de Mink & Mandel 2016; Marchant *et al.* 2016), through an intermediate phase of BH-HMXB-MQs. Stars of $\geq$18-25 $M_\odot$ and metallicity Z$\leq$0.001 collapse directly or by failed SNe, and form BBHs that merge during Hubble time producing gravitational waves.

If the UV and X-ray photon escape fraction from stars and BH-HMXB-MQs of Pop III is $f_{esc} \leq 0.1$, the constraint imposed by the Planck measurement of Thomson scattering would not be violated (Inayoshi *et al.* 2016). Anyway, there may be scenarios in which some heating is produced while the reported large absorption is still possible.

7. Population III stars, BH-HMXB-MQs, CRB, and BBHs

UV radiation, X-rays, gamma-rays, gamma-ray bursts, and synchrotron jets can be produced by a plethora of sources: Pop III stars (e.g. Loeb 2010), gamma-ray bursts (e.g. Consumano *et al.* 2006), SN explosions (e.g. Furlanetto *et al.* 2004; Garratt-Smithson *et al.* 2018), Intermediate-Mass BHs (e.g. Madau *et al.* 2004; Ewall-Wice *et al.* 2018), Supermassive BHs (e.g. Biermann *et al.* 2014) and HMXBs (e.g. Mirabel 2011; Mirabel *et al.* 2011; Fragos *et al.* 2013). Some, if not all those sources may play some role at cosmic dawn, but a discussion of their relative roles is beyond the scope of this review.

In the local universe it is observed that BH-HMXB-MQs are sources of hard X-rays and powerful synchrotron relativistic jets, which are frequently found in low metallicity dwarf galaxies. Therefore, it is natural to expect that BH-HMXB-MQs of Pop III are likely sources of a smooth synchrotron CRB. In fact, the most recent results from theoretical simulations show that BH-HMXB-MQs are the immediate compact remnants of Pop III stars (Figure 5).

Massive stars of Pop III. Early simulations had suggested that stars of Pop III have very large masses. However, with more powerful computational ability to follow the evolution of gas physics for adequately long periods of time at sufficiently high resolution, it was later found that the primordial gas is able to fragment. This fragmentation leads to a broad top-heavy Population III, extended to low masses ($\sim$1 $M_\odot$), but with the majority of stellar mass contained within the most massive stars of tens of solar masses, structured in stellar binaries and systems of larger multiplicity (Stacy *et al.* 2016 and references therein).

Observations of stars in nearby Galactic open clusters by Sana *et al.* (2012) have shown that more than 70% of all massive stars are in binaries or larger multiple systems. The fraction of systems with orbital periods $\leq$100 days is $\sim$85% and the mass ratios of the binary components are in the range of 0.2 to 1. From those observations it was concluded that the most common end product of massive-star formation is an interacting rather close binary, where 20 to 30% of all O stars will merge.

From a following more extensive observational campaign Sana *et al.* (2014) concluded that massive stars are formed nearly exclusively in multiple systems, and if corrected for observational biases, the true multiplicity fraction might very well come close to 100%, despite different environments, sample ages and even metallicities. From a surprising dearth of short-period massive binaries in the very young ($\leq$1 Myr) massive star forming region M17, Sana *et al.* (2017) tentatively suggest that massive stars may be born with large orbital separations, and then harden or migrate to produce the typical (untruncated) power-law period distribution observed in few Myr-old OB binaries.

BH-HMXB-MQs are the next evolutionary state of primordial binaries with $Z \leq 0.001$ and primary stars of $\geq$18-25 $M_{\odot}$. Stars of those low metallicities and high masses collapse directly or through failed SNe to form BHs (e.g. Heger *et al.* 2003), leaving the binary gravitationally bound as a BH-HMXB-MQ (Figure 5).

BH-HMXB-MQs at cosmic dawn are the first steady high energy sources formed from the most massive stars of Pop III, and would naturally produce a smooth synchrotron CRB. BH-HMXB-MQs of Pop III precede SN explosions, the formation of neutron stars, and the production of dust. They promptly inject hard X-rays and relativistic jets that overtake the slowly expanding HII regions, pre-heating the IGM over larger distance scales.

A complete model of BH-HMXB-MQs formed from Pop III stars is being aimed by Sotomayor & Romero (2018), in the context of a spectral energy distribution of the radiation produced by the accretion disk and the relativistic particles in the jets, in the framework of a lepto-hadronic model. The Pop III binary system undergoes a BH-HMXB-MQ phase for $\sim$2x10^5years. During this lifetime, the donor star loses $\sim$15 $M_{\odot}$, the BH mass remains approximately constant, so almost all the accreted matter is ejected from the system in the form of jets & massive outflows.

The BH accretes matter in a super-Eddington regime. This regimen of accretion first theoretically examined by Shakura & Sunyaev (1973) in a supercritical accretion slim disk, was proposed for SS 433 by Fabrika (2004), and recently confirmed with NuSTAR by Middleton *et al.* (2018). Such steady super-Eddington accretion in SS433 may be a unique property among microquasars so far observed in the Milky Way.

BBHs with components of 20-30 $M_{\odot}$ as the sources of gravitational waves detected by LIGO-Virgo (e.g. GW150914 Abbott *et al.* 2016), can be formed by three main evolutionary channels as reviewed by Mirabel (2017): (1) BBHs formed from BH-HMXB-MQs by isolated evolution of massive stellar binaries (Belczynski *et al.* 2016), (2) BBHs formed from BH-HMXB-MQs by the evolution of tight binaries with fully mixed chemistry (de Mink & Mandel 2016; Marchant *et al.* 2016), and (3) BBHs formed by dynamical interaction in dense stellar clusters (Rodriguez *et al.* 2016 and references therein).

In model (1) the BH members of BBHs are formed by direct collapse with no BH natal kicks that would unbind the binary. Model (2) avoids the physics uncertainties in mass transfer, common envelope mass ejection events, and the still unconstrained BH kicks. However, model (2) assumes massive tight binary progenitors of BBHs and has preference for BBHs with large masses as in GW150914, but the relative low BH masses of $\leq$10 $M_{\odot}$ as seems to be the case in GW151226, are difficult to reproduce in this model. In model (3) it is tacitly assumed that the members of BBHs are also formed with no BH natal triggers that would eject the BHs from the stellar cluster before BBH formation.

BBHs formed by either direct collapse and sufficiently low natal kicks from relatively isolated massive stellar binaries (channel 1) or contact massive binaries (channel 2), will remain in situ and ultimately merge in galactic disks. BBHs formed by dynamical interactions in the cusps of dense stellar clusters, may be ejected from their birth place and ultimately merge in galactic haloes, like the sources of short gamma-ray bursts.

A gravitational wave background (GWB) from Population III BBHs may be detectable by the future O5 LIGO/Virgo (Inayoshi *et al.* 2016). These authors argue that Pop III stars may form more BBHs but inject less ionizing photons into the IGM due to strong absorption by very high density of hydrogen. In that case, the GWB would be consistent with the cosmic early reionization by Pop III stars and the recent Planck measurement of Thomson scattering of 0.055 ± 0.09. Inayoshi *et al.* (2016) conclude that the detection of a flattening of the spectral index of the GWB at frequencies as low as

30 Hz would be an unique smoking gun of a high-chirp mass and high-redshift BBH population expected from Pop III stars.

Inayoshi *et al.* (2017) studied the formation pathway of Pop III coalescing BBHs through stable mass transfer in Pop III binary stars without common envelope phases. For Pop III binaries with large and small separations they find that ∼10% of the total Pop III binaries form BBHs only through stable mass transfer, and ∼10% of these BBHs merge within the Hubble time. They conclude that the Pop III BBH formation scenario can explain the mass-weighted merger rate of the LIGOs O1 events with the maximal Pop III formation efficiency inferred from the Planck measurement, even without BBHs formed by unstable mass transfer or common envelope phases.

8. Summary

1) The theoretical and observational grounds for the hypothesis of a high formation rate of BH-HMXBs in the early universe (Mirabel 2011; Mirabel *et al.* 2011; Fragos *et al.* 2013) has been re-enforced in the last decade (e.g. Lehmer *et al.* 2016; section 1).

2) Besides sources of hard X-rays and gamma-rays, HMXBs are also sources of powerful synchrotron relativistic jets and massive outflows, as revealed by observations of the best studied HMXBs in the Galaxy: Cygnus X-1, SS433 and Cygnus X-3 (section 2).

3) Stellar black holes of $\geq$10 $M_{\odot}$ in the Milky Way (e.g. Cygnus X-1 & GRS 1915+105), are formed by direct or failed SN collapse of massive stars of $Z \geq Z_{\odot}$. Therefore, it is expected that a significant fraction of Pop III stars of $\geq$18-25 $M_{\odot}$ and $Z \leq 0.001\ Z_{\odot}$ in binaries or larger multiple systems, collapse with no energetic SN kicks, remain in situ, ending as BH-HMXB-MQs of Pop III (section 3).

4) BH-HMXB-MQs of Pop III are the most likely sources of a smooth synchrotron cosmic radio background (CRB). This CRB can provide an excess background temperature (Feng & Holder 2018) to boost the deep absorption of HI centered at z∼17 reported by EDGES (Bowman *et al.* 2018a).

5) Theoretical models of the tomography of atomic hydrogen at cosmic dawn that incorporate heating of the IGM by sources with a hard X-ray spectral distribution like BH-HMXBs (e.g. Furlanetto 2006; Mirabel *et al.* 2011, Cohen *et al.* 2017), predict the mean redshift (z∼17) of the deep and frequency wide (z∼14 to 21) HI absorption reported by EDGES, for UV and X-ray escape values $\leq$0.1. This implies column depths $N_H \geq 5\times10^{23}$ cm^{-2} (Ewall-Wice *et al.* 2018) for the cold hydrogen IGM that enshrouds the high energy sources. These column depths block the X-rays but are transparent for the synchrotron CRB (section 6).

6) Massive stars in the Milky Way are formed nearly exclusively in multiple systems, despite different environments, sample ages and even metallicities (Sana *et al.* 2014). From theoretical simulations it is concluded that population III stars are of broad top-heavy masses, extended to low masses (∼1 $M_{\odot}$), with the majority of stellar mass contained within the most massive stars of tens of solar masses, structured in stellar binaries and systems of larger multiplicity (Stacy *et al.* 2016; section 7).

7) BH-HMXB-MQs in stellar clusters of Pop III are formed before the appearance of SN explosions, neutron stars and dust, which would be consistent with an absence of a cosmic far-infrared thermal background associated with a Cosmic Radio Background. BH-HMXB-MQs promptly inject hard X-rays and relativistic jets in the cold hydrogen that enshrouds the slowly expanding HII regions ionized by the most massive progenitor stars of Pop III (section 7).

8) BH-HMXB-MQs of Pop III would be different to most known BH-MQs in the Milky Way. The steady super-Eddington accreting source SS433 may be the only observed Galactic analog of BH-HMXB-MQs of Pop III (Sotomayor & Romero 2019; Section 7).

9) The detection of a flattening of the spectral index of the Gravitational Wave Background at frequencies as low as ∼30 Hz (which corresponds to merging BHs of 30-40 $M_{\odot}$), would be the smoking gun of a high-chirp mass, high-redshift BH-HMXBs and BBHs populations, expected from Pop III stars (Inayoshi *et al.* 2016; section 7)

10) Future observations with more sensitive X-ray missions (e.g. Athena) will allow more precise quantitative determinations to test the impact of HMXBs in Cosmology. The future enhanced statistics of sources of Gravitational Waves (GWs) by GW observatories (e.g. LIGO-Virgo) will provide larger data bases to test the cosmic evolution of HMXBs and BBHs.

Acknowledgments

I thank Dale Fixsen for comments on ARCADE 2 results, Bret Lehmer, Jayce Dowel, Gregory Taylor and Judd Bowman for permission to use their published figures.

References

Abbott, B. P. *et al.* 2016, *PRL*, 116, 061102
Barkana 2018, *Nature 555, 71-74*
Basu-Zych, A. R., Lehmer, B. D., Hornschemeier, A. E., *et al.* 2013, *ApJ*, 774, 152B
Belczynski, K., Dominik, M., Bulik, T., *et al.* 2010, *ApJ*, 715, L138
Belczynski, K., Holz, D. E., Bulik, T., *et al.* 2016, *Nature*, 534, 512
Biermann, P. L., Biman, B. N., Caramete, L. I., *et al.* 2014, *MNRAS*, 441, 1147
Bowman, J. D., Rogers, A. E E., Monsalve, R. A. *et al.* 2018a, *Nature*, 555, 67
Bowman, J. D., Rogers, A. E E., Monsalve, R. A. *et al.* 2018b, *Nature*, 564, E35
Bridle, A. H. 1967, *MNRAS 136, 219*
Brorby, M., Kaaret, P., Prestwich, A., *et al.* 2016, *MNRAS*, 457, 21
Brown, G. E. & Bethe, H. A. 1994, *ApJ*, 423, 659
Cohen, A., Fialkov A., Barkana, R. *et al.* 2017, *MNRAS* 472, 1915
Condon, J., Cotton, W., Fomalont, E., *et al.* 2012, *ApJ* 758, 23
Cusumano, G, Mangano, V., Chincarini, G. *et al.* 2006, *Nature* 440, 9
de Mink, S. E. & Mandel, L. 2016, *MNRAS*, 460, 3545
Dominik, M., Belczynski, K, Fryer, C. *et al.* 2012, *ApJ*, 759, 52
Douna, V. M., Pellizza L. J., Mirabel, I. F. *et al.* 2015, *A&A* 579, A44
Douna, V. M., Pellizza L., Laurent, Ph. *et al.* 2018, *MNRAS* 474, 3488
Dowell, J. & Taylor, G. B. 2018, *ApJL*, 858, L9
Ewall-Wice, A., Chang, T.-C, Lazio, J. *et al.* 2018, *ApJ* 868, 63
Fabrika, S. 2004, *ASPRv*, 12, 1-152
Feng, C. & Holder, G. 2018 *ApJL*, 858, L17
Feng, H. & Soria, R. 2011, *New Astron. Revs. 55*, 166-183
Fermi LAT Collaboration 2009, *Science*, 326, 1512
Fialkov, A., Barkana, R., & Visbal, E. 2014, *Nature* 506, 197
Fixsen D. J., Kogut A., Levin, S. *et al.* 2011, *ApJ*, 734, 5
Fragos, T., Lehmer, B. D., Naoz, S. *et al.* 2013, *ApJ*, 776, L31
Fryer, C. L., Belczynski, K., Wiktorowicz, G. *et al.* 2012, *ApJ*, 749, 91
Furlanetto, S. R. 2006, *MNRAS* 371, 867
Furlanetto, S. R., Zaldarriaga M., Hernquist, L. 2004, *ApJ* 613, 16
Gallo E., Fender, R., Kaiser, Ch. *et al.* 2005, *Nature*, 436, 819-821
Gandhi P., Rao, A., Johnson, M. A. C. *et al.* 2018, *MNRAS*, in press
García F., Mirabel, I. F., Chaty, S. 2019, *In preparation*
Garratt-Smithson *et al.* 2018, *MNRAS*, 480, 2985
Haiman, Z. 2011, *Nature*, 472, 47
HAWC Collaboration 2018, *arXiv:1812.05682*
Heger, A., Fryer, C. L., Woosley, S. E. *et al.* 2003, *ApJ*, 591, 288
Heinz, S. & Sunyaev 2002, *A&A*, 390, 751

Hills, R., Kulkarni, G., Meerburg, D. *et al.* 2018, *Nature*, 564, E32-E34
Inayoshi, K., Kashiyama, K., Visbal, E., Haiman, Z. 2016, *MNRAS*, 461, 2722
Inayoshi *et al.* 2017, *MNRAS*, 468, 2020
Ivison *et al.* 2010, *MNRAS*, 402, 245
Kaaret, P. 2014, *MNRAS*, 440, L26
Lehmer, B. D., Basu-Zych, A. R., Mineo, S. *et al.* 2016, *ApJ* 825, 7L
Loeb, A. 2010, *Princeton University Press, 2010. ISBN: 978-1-4008-3406-8*
Madau, P., Meiksin, A., & Rees, M. J. 2004, *ApJ*, 475, 429
Madau, P., Rees, M. J., Volonteri, M. *et al.* 2004, *ApJ*, 604, 484
Magnelli, B. *et al.* 2015, *A&A*, 573, 45
Marchant, P., Langer, N., Podsiadlowski, Ph. *et al.* 2016, *A&A*, 588, A50
Mel'nik, A. M. & Dambis A. K. 2009, *MNRAS*, 400, 518
Middleton, M. J. *et al.* 2018, *arXiv:1810.10518*
Miller-Jones, J. C. A., Jonker, P. G., Dhawan, V. *et al.* 2009a, *ApJ*, 706, L230
Miller-Jones, J. C. A., Jonker, P. G., Nelemans, G. *et al.* 2009b, *MNRAS*, 394, 1440
Mirabel, I. F. & Rodríguez, L. F. 1994, *Nature*, 371, 46
Mirabel, I. F. & Rodríguez, L. F. 1998, *Nature*, 392, 673-677
Mirabel, I. F. & Rodríguez, L. F. 1999, *ARAA*, 37, 409-443
Mirabel, I. F., Dhawan, V., Mignani, R. P. *et al.* 2001, *Nature*, 413, 139
Mirabel, I. F., Mignani, R., Rodrigues, I. *et al.* 2002, *A&A*, 395, 595
Mirabel, I. F. & Rodrigues, I. 2003, *Science*, 300, 1119
Mirabel, I. F. 2006, *Science*, 312, 1759
Mirabel, I. F. 2011, *Proc. IAU Symp 275. Jets at all Scales. Cambridge University Press, pages 2-8. ed. G. E. Romero, R. A. Sunyaev, T. Belloni*, 275, 2-8
Mirabel, I. F., Dijkstra, M., Laurent, Ph., *et al.* 2011, *A&A*, 528, A149
Mirabel, I. F. 2012, *Science*, 335, 175
Mirabel, I. F. 2017, *New Astron. Revs. 78, 1*
Mirocha, J., Harker G. J. A., & Burns J. O. 2013, *ApJ*, 777, 118
Mirocha, J. Mebane, R. H., Furlanetto, S. R. *et al.* 2018, *MNRAS*, 478, 5591
Pakull M. W., Soria R., Motch C. 2010, *Nature*, 466, 209
Reid, M. J., McClintock, J. E., Narayan, R. *et al.* 2011, *ApJ*, 742, 83
Reid, M. J., McClintock, J. E., Steiner, J. F. *et al.* 2014, *ApJ*, 796, 2
Rodriguez, C. L., Haster, C-J, Chatterjee, S. *et al.* 2016, *ApJL*, 824, L8
Sana, H., de Mink, S. E., de Koter, A. *et al.* 2012, *Science*, 337, 444
Sana, H., Le Bouquin, J.-L., Lacour, S. *et al.* 2014, *ApJS*, 215, 15
Sana, H., Ramirez-Tannus, M. C., de Koter, A. *et al.* 2017, *ApJS*, 215, 15
Sanders, D. B. & Mirabel, I. F. 1996, *ARAA*, 34, 749-789
Seiffert *et al.* 2011, *A&A*, 599, L9
Shakura, N. I., & Sunyaev, R. A. 1973 *A&A*, 24, 337
Singal J. *et al.* 2018, *PASP*, 130, 036001
Sotomayor, C. P. & Romero, G. E. 2018, *A&A*, in press
Stacy, A., Bromm, V., Lee, A. T. 2016, *MNRAS*, 462, 1307
Subrahmanyan, R. & Cowsik, R. 2013, *ApJ*, 776, 42
Sukhbold, T., Ertl, T. Woosley, S. E. *et al.* 2016, *ApJ*, 821, 38
Tavani, M. etal 2009, *Nature* 462, 620
Timmes F. X., Woosley S. E., Weaver T. A. 1996, *ApJ* 457, 834
Tueros M., del Valle M. V., Romero G. E. 2014, *A&A*, 570, L3
Verstrom, T., Norris, R. P., Scott, D. *et al.* 2015, *MNRAS*, 447, 2243
Watkinson, C. A. *et al.* 2019, *MNRAS*, 482, 2653
Ysard, N. & Lagache, G. 2012, *A&A*, 547, A53
Zanin, R. *et al.* 2016, *A&A*, 596, A55

High-mass X-ray Binaries: Illuminating the Passage from Massive Binaries to Merging Compact Objects
Proceedings IAU Symposium No. 346, 2019
L. M. Oskinova, E. Bozzo, T. Bulik & D. Gies, eds.

doi:10.1017/S1743921318008025

The analogy of K-correction in the topic of gamma-ray bursts

Levente Borvák[1], **Attila Mészáros**[2] and **Jakub Řípa**[2,3,4]

[1]Department of Physics, University of Dallas, 1845 East Northgate Drive, Irving, Texas, 75062 USA
email: lborvak@udallas.edu

[2]Astronomical Institute, Faculty of Mathematics and Physics, Charles University, CZ-180 00 Prague 8, V Holešovičkách 2, Czech Republic
email: meszaros@cesnet.cz

[3]MTA-Eötvös Lóránd University, Lendület Hot Universe Research Group, Pázmány Péter sétány 1/A, Budapest, 1117, Hungary

[4]Eötvös Lóránd University, Institute of Physics, Pázmány Péter sétány 1/A, Budapest, 1117, Hungary
email: jripa@caesar.elte.hu

Abstract. It is well-known that there are two types of gamma-ray bursts (GRBs): short/hard and long/soft ones, respectively. The long GRBs are coupled to supernovae, but the short ones are associated with the so called macronovae (also known as kilonovae), which can serve as the sources of gravitational waves as well. The kilonovae can arise from the merging of two neutron-stars. The neutron stars can be substituded by more massive black holes as well. Hence, the topic of gamma-ray bursts (mainly the topic of short ones) and the topic of massive binaries, are strongly connected.

In this contribution, the redshifts of GRBs are studied. The surprising result - namely that the apparently fainter GRBs can be in average at smaller distances - is discussed again. In essence, the results of Mészáros *et al.* (2011) are studied again using newer samples of GRBs. The former result is confirmed by the newer data.

Keywords. cosmology: miscellaneous, cosmology: observations, gamma rays: bursts

---

1. Inverse behavior of the peak-flux and fluence of GRBs: The theory

In the article Mészáros *et al.* (2011) (in what follows M11) a remarkable property of the gamma-ray bursts (GRBs) was found. It can be briefly explained as follows (for details see the mentioned paper).

Given a GRB with measured peak-flux $P(z)$ (with the dimension of ph/(cm^{2}s), where "ph" means photon), if the object is at redshift z, then its isotropic peak-luminosity $\tilde{L}_{\rm iso}(z)$ (in units of ph/s) is related to the peak-flux by the expression

$$P(z) = \frac{(1+z)\tilde{L}_{\rm iso}(z)}{4\pi d_l(z)^2}, \tag{1.1}$$

where $d_l(z)$ is the luminosity distance of the object.

An instrument measures the peak-flux at an interval $E_1 < E < E_2$, where E_1 and E_2 are the limiting photon energies given by the instrument, and E is the measured energy of the photon. So the peak-luminosity must be taken from the interval $E_1(1+z)$ and $E_2(1+z)$, not simply from E_1 and E_2.

The same relation is also expected for the fluence if it has the dimension erg/cm^2.

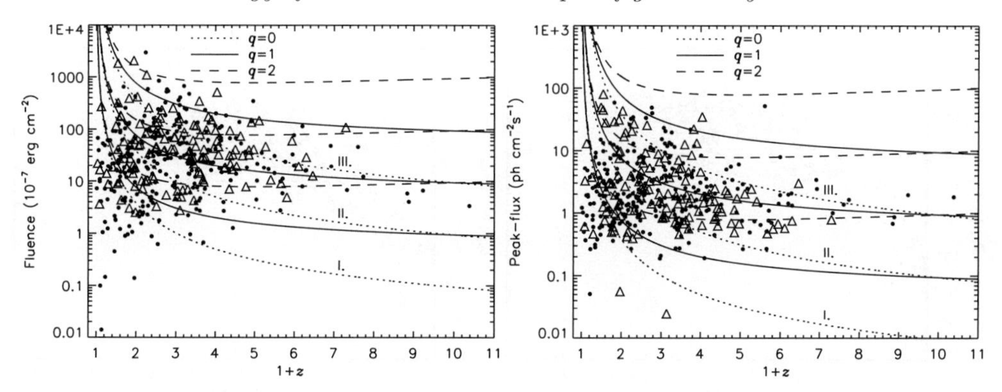

Figure 1. The 412 fluences (left panel) and the 411 peak-fluxes (right panel), respectively, versus $1 + z$ are shown in the figure. The 132 GRBs mentioned in M11 are denoted by triangles. The curves with different q illustrate the expected behavior, if either $\tilde{E}_{\text{iso}}$ or $\tilde{L}_{\text{iso}}$ are proportional to $(1+z)^q$, i.e. $\tilde{E}_{\text{iso}} = \tilde{E}_o(1+z)^q$ (constants $\tilde{E}_o$ are: I. 10^{51} erg; II. 10^{52} erg; III. 10^{53} erg) and $\tilde{L}_{\text{iso}} = \tilde{L}_o(1+z)^q$ (constants $\tilde{L}_o$ are: I. 10^{57} ph/s; II. 10^{58} ph/s; III. 10^{59} ph/s). For the inverse behavior, a value of $q > 1$ is needed for the biggest z. This trend was first mentioned by M11, and the larger sample indicate this possible behaviour, too. The figure is in essence an analogy of Fig. 3 of M11 with a larger sample.

It is standard cosmology that $d_l(z)$ increases with the redshift. For the exact formula, see Carroll *et al.* (1992). For any reasonable Ω factors it holds that $\lim_{z\to\infty}(d_l(z)/(1+z))$ is a finite value.

However, because $\tilde{L}_{\text{iso}}(z)$ depends on z, it is possible that $\tilde{L}_{\text{iso}}(z)$ increases with z. In M11 it is argued that in some cases $\tilde{L}_{\text{iso}}(z)$ can increase faster than $d_l(z)^2/(1+z)$ and hence an "inverse" behavior can occur: an apparently fainter GRB can be at a smaller redshift than a brighter one.

2. Data of the *Swift* satellite in M11

This theoretical expectation was shown to happen in M11 for *Swift*† data. For the period from 20 November 2004 to 30 April 2010, 132 GRBs were observed by *Swift* with known redshifts. They were used in M11. We show here that this behavior also holds with the currently available Swift GRB database

3. New sample from the *Swift* satellite

Below we show a comparison of the data used by M11 to the currently (until 10 May 2018) observed 416 GRBs. Five GRBs had no measured peak-fuxes, 4 GRBs had no measured fluences. Hence, 412 is the total number of GRBs with measured fluences (in the 15-350 keV band) and $\tilde{E}_{\text{iso}}$, but the total number of GRBs with measured peak-fuxes (1-s peak photon fluxes in the 15-350 keV band) and $\tilde{L}_{\text{iso}}$ is 411. The total energy emitted by the GRBs, assuming isotropic emission, is denoted as $\tilde{E}_{\text{iso}}$. The peak-luminosity, assuming isotropic emission, is denoted as $\tilde{L}_{\text{iso}}$.

In Figure 1 these fluences and peak-fluxes, respectively, are compared with $1 + z$.

For the 412 fluences and 411 peak-fluxes, respectively, $\tilde{E}_{\text{iso}}$ and $\tilde{L}_{\text{iso}}$ were calculated using Eq. 1.1. These values are shown in Figure 2.

4. Remarks

1. As a comparison for $\tilde{E}_{\text{iso}}$, remember that $M_\odot c^2 = 1.8 \times 10^{54}$ erg, where c is the velocity of light in vacuum, and $M_\odot$ is the mass of Sun.

† https://swift.gsfc.nasa.gov

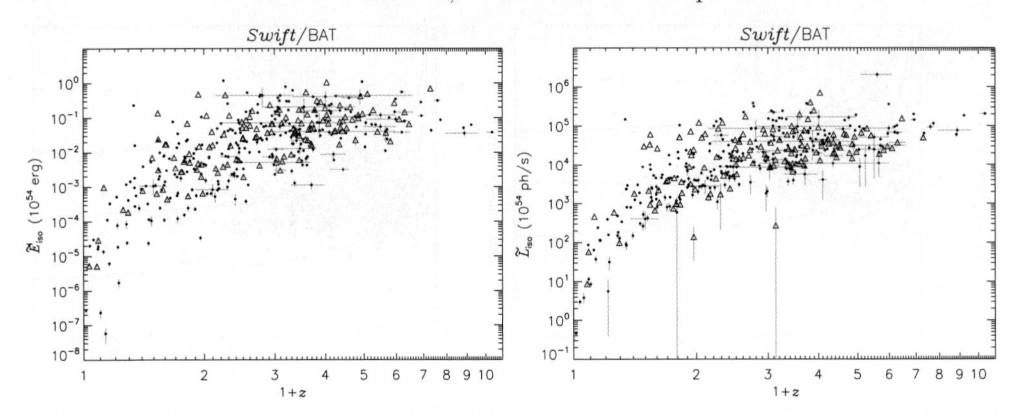

Figure 2. $\tilde{E}_{\rm iso}$ and $\tilde{L}_{\rm iso}$, respectively, versus $1+z$ are shown in the figure. The number of all GRBs is 412 (411) on the left (right) plot. The 132 GRBs, mentioned in M11, are denoted by triangles. BAT means the Burst Alert Telescope. The figure is in essence an analogy of Fig. 4 of M11 with a larger sample.

2. If the photon has energy of $100y$ keV (y is a positive unitless parameter), then (because 100 keV=1.6×10^{-7} erg), 10^{58} ph/s corresponds to $1.6y \times 10^{51}$ erg/s.

3. This behavior of $\tilde{E}_{\rm iso}$ and $\tilde{L}_{\rm iso}$ is analogous to the K-correction in optical astronomy (Hubble 1936, Humason *et al.* 1956).

5. Conclusion

The expanded dataset from the *Swift* satellite hints towards a possible inverse behavior that was first mentioned by M11.

References

Carroll, S. M., *et al.* 1992, *ARA&A*, 20, 499
Hubble, E. 1936, *ApJ*, 84, 517
Humason, M.L., *et al.* 1956, *AJ*, 61, 97
Mészáros, A., Řípa, J., & Ryde, F. 2011, *A&A*, 529, A55 (M11)

High-mass X-ray Binaries: Illuminating the Passage from Massive Binaries to Merging Compact Objects
Proceedings IAU Symposium No. 346, 2019
L. M. Oskinova, E. Bozzo, T. Bulik & D. Gies, eds.

doi:10.1017/S1743921319000036

Gamma-ray bursts: A brief survey of the diversity

Attila Mészáros[1] and **Jakub Řípa**[1,2,3]

[1]Astronomical Institute, Faculty of Mathematics and Physics, Charles University, CZ-180 00 Prague 8, V Holešovičkách 2, Czech Republic
email: meszaros@cesnet.cz

[2]MTA-Eötvös Lóránd University, Lendület Hot Universe Research Group, Pázmány Péter sétány 1/A, Budapest, 1117, Hungary

[3]Eötvös Lóránd University, Institute of Physics, Pázmány Péter sétány 1/A, Budapest, 1117, Hungary
email: jripa@caesar.elte.hu

Abstract. The separation of the gamma-ray bursts (GRBs) into short/hard and long/soft subclasses, respectively, is well supported both theoretically and observationally. The long ones are coupled to supernovae type Ib/Ic - the short ones are connected to the merging of two neutron stars, where one or even both neutron stars can be substituted by black holes. These short GRBs - as merging binaries - can also serve as the sources of gravitation waves, and are observable as the recently detected macronovae. Since 1998 there are several statistical studies suggesting the existence of more than two subgroups. There can be a subgroup having an intermediate durations; there can be a subgroup with ultra-long durations; the long/soft subgroup itself can be divided into two subclasses with respect to the luminosity of GRBs. The authors with other collaborators provided several statistical studies in this topic. This field of the GRB-diversity is briefly surveyed in this contribution.

Keywords. cosmology: miscellaneous, cosmology: observations, gamma rays: bursts

---

1. Introduction

After the discovery of the gamma-ray bursts (GRBs) (Klebesadel *et al.* 1973) it was shown from the statistical analyses of observational data that they were two types of GRBs (Mazets *et al.* 1981). Since that time the separation of GRBs into subgroups is a blistering problem. In this contribution a brief survey of this topic is provided.

2. Separation into two subgroups

Since 1981 the separation of GRBs into short/hard and long/soft subgroups was confirmed by several other studies. For the GRBs, not having directly measured redshifts, the so called hardness can be used in statistical tests together with the duration (for a survey and the relevant references see, e.g., Mészáros 2006).

Today it is clear that the short/hard bursts are given by the neutron star - neutron star mergers, where one or even two neutron stars can be substituted by black holes, and are observable as macronovae (Tanvir *et al.* 2013). The cosmologically near short/hard bursts can also serve as the source for the detectable gravitational waves (Abbott *et al.* 2017). This means that the short/hard GRBs arose at the final stage of compact binaries.

The long/soft bursts are associated with Ib/c type supernovae (Hjorth *et al.* 2003). The idea that the GRBs were coupled to supernovae was formulated in essence simultaneously with the discovery of bursts (Colgate 1968, Colgate 1974).

The separation should be done with respect to the duration at $\simeq 2$ second. But this limiting duration - as a strict one - is in doubt, because much longer GRBs were also observed, which resemble the short/hard GRBs (Gehrels *et al.* 2006). This means that it is better to speak about two basic types of GRBs (Kann *et al.* 2010).

3. Three subgroups?

In year 1998 two simultaneous papers declared the existence of a third subgroup (Mukherjee *et al.* 1998, Horváth 1998). This claim came from the statistical studies of the dataset of BATSE instrument being on the Compton Gamma Ray Burst Observatory†. Since that time several other papers declared the same result for the BATSE dataset (Horváth 2002, Hakkila *et al.* 2003, Hakkila *et al.* 2004, Horváth *et al.* 2006). This third subgroup should have an intermediate duration. It was found also at the Swift dataset‡ (Veres *et al.* 2010). For the RHESSI¶ satellite the existence of third subgroup was also declared (Řípa *et al.* 2012). On the other hand, no intermediate subgroup was found in the Fermi's‖ observations (Tarnopolski 2015, Narayana Bhat *et al.* 2016). Similarly, no third subgroup is declared to exist both in the Suzaku†† database (Ohmori *et al.* 2016), and in the Konus/WIND‡‡ catalog (Tsvetkova *et al.* 2017).

It must be added that even in the case, when the three subgroups are found by statistical tests, it is not sure that there are really three astrophysically different phenomena. Different biases, selection effects, etc... can play a role (Hakkila *et al.* 2003, Tarnopolski 2016). For example, in the Swift database the third group is found by tests, but a more detailed study shows that the third group is given by the so-called X-Ray Flashes (XRFs) - which are in essence long GRBs (Veres *et al.* 2010). But, on the other hand, in some cases it is claimed that the third subgroup cannot entirely be given by long GRBs. For example, for the BATSE and mainly for the RHESSI database the identification of the intermediate GRBs with XRFs cannot be done (Řípa & Mészáros 2016).

4. Four or even more subgroups?

There are studies claiming the existence of other subgroups - being not identical - to the intermediate one.

For example, in the BATSE database there were hints for the separation of the long GRBs themselves into the harder and softer parts (Pendleton *et al.* 1997).

The longest GRBs can also form an extra - ultra-long - subgroup (Tikhomirova & Stern 2005, Virgili *et al.* 2013, Levan *et al.* 2014). Because there are only few cases in this subgroup, from the statistical point of view this subgroup hardly can be declared as an astrophysically different phenomenon.

Recently, in the Fermi database five subgroups were found (Acuner & Ryde 2018). The long subgroup should be further separated. Theoretically, it is thought that seven different subgroups should exist (Ruffini *et al.* 2018).

5. Newest studies on the samples with known redshifts

Because the number of GRBs, which have directly measured redshifts from the afterglows observations, is increasing at the last years it is already possible to study the diversity of GRBs also from other intrinsic quantities. The intrinsic luminosity (L_{iso})

† https://heasarc.gsfc.nasa.gov/docs/cgro/index.html
‡ https://swift.gsfc.nasa.gov
¶ https://hesperia.gsfc.nasa.gov/rhessi3
‖ https://fermi.gsfc.nasa.gov
†† http://global.jaxa.jp/projects/sat/astro_e2/index.html
‡‡ http://www.ioffe.ru/LEA/kw/

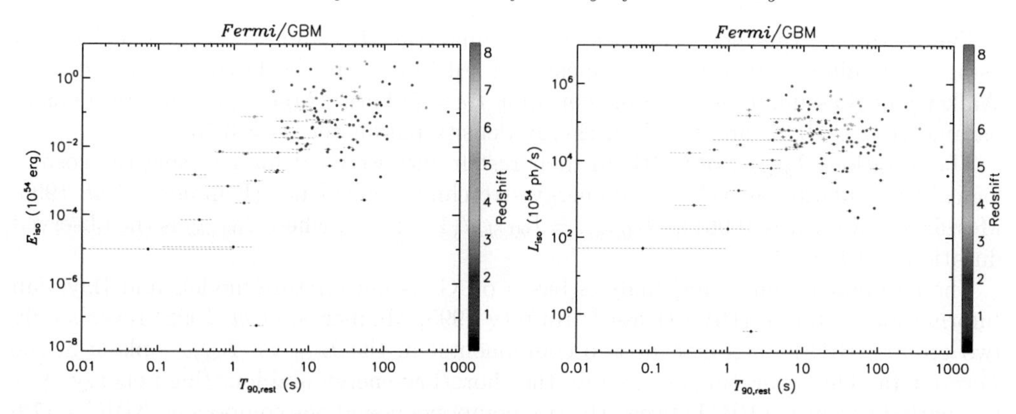

Figure 1. Plotted are 124 *Fermi*/GBM GRBs with known redshifts. Isotropic equivalent energy E_{iso} (calculated from the fluence in the 10-1000 keV band) and isotropic equivalent luminosity L_{iso} (calculated from the 1024 ms peak photon fluxes in the 10-1000 keV band), respectively, versus duration $T_{90,rest}$ in the rest frame of a GRB are shown.

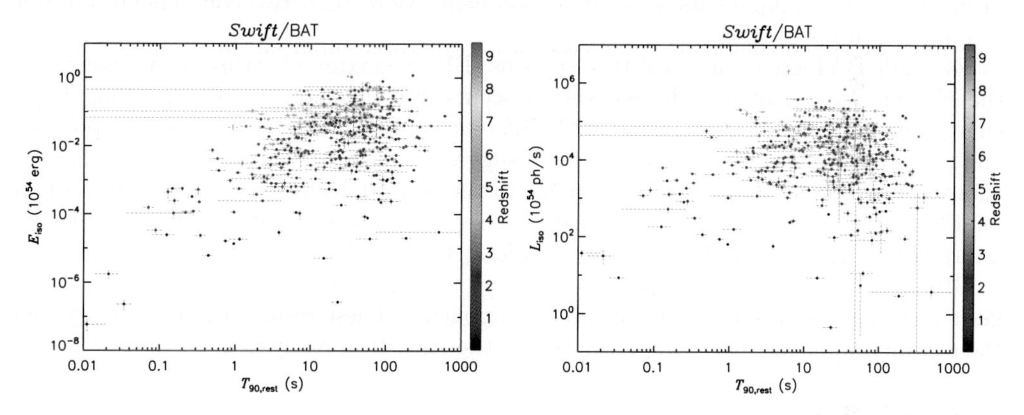

Figure 2. Plotted are 412 *Swift*/BAT GRBs with known redshifts. Isotropic equivalent energy E_{iso} (calculated from the fluence in the 15-350 keV band) and isotropic equivalent luminosity L_{iso} (calculated from the 1 s peak photon fluxes in the 15-350 keV band), respectively, versus duration $T_{90,rest}$ in the rest frame of a GRB are shown.

and the intrinsic total emitted energy (E_{iso}) can be calculated for a given GRB, if its redshift is known. Then these quantities can also be studied instead of hardness.

If such a testing is provided (Levan *et al.* 2014) on the duration vs. L_{iso} (E_{iso}, respectively) plane, one should obtain the subgroups, too. On Fig. 2 of Levan *et al.* (2014) such a study is done. Several possible subgroups are seen beyond the long and short GRBs (soft gamma repeaters, low luminosity long GRBs, ultra-long GRBs, tidal disruption events). On the other hand, there is no intermediate subgroup.

We present here the preliminary results of a similar effort. We examined data samples of GRBs with measured redshifts from *Fermi*/GBM and *Swift*/BAT instruments. The *Fermi*/GBM sample was created using the FERMIGBRST catalog†. It contains 124 GRBs with the first burst GRB 080804 and the last one GRB 180314A (see Fig. 1). The *Swift*/BAT sample was created using the *Swift* BAT Gamma-Ray Burst Catalog‡. It contains 412 GRBs with the first burst GRB 050126 and the last one GRB 180510B (see Fig. 2). The *Swift* data sample is preliminary. We are working on its completion.

† https://heasarc.gsfc.nasa.gov/W3Browse/fermi/fermigbrst.html
‡ https://swift.gsfc.nasa.gov/results/batgrbcat/

The isotropic equivalent energies and luminosities E_{iso} and L_{iso} were calculated using the luminosity distances (for measured redshifts z) in the Friedmann-Robertson-Walker model assuming the Hubble constant $H_0 = 71\,\mathrm{km\,s^{-1}Mpc^{-1}}$, the matter density parameter $\Omega_{\mathrm{M}} = 0.27$, and the dark energy density parameter $\Omega_\Lambda = 0.73$.

The durations $T_{90,rest}$ of GRBs in their rest frames were calculated using the cosmological time dilation as well as the energy-stretching dependence (Fenimore *et al.* 1995, Mészáros & Mészáros 1996) as $T_{90,rest} = T_{90,obs}/(1+z)^{1-k}$, where $T_{90,obs}$ is the observed duration and $k = 0.4$.

The preliminary clustering analysis based on Gaussian mixture models and Bayesian Information Criterion (BIC) (Kass & Raftery 1995, Mukherjee *et al.* 1998) reveals only two groups (BIC is maximal for two components) in the E_{iso} vs. $T_{90,rest}$ plane in the *Fermi* data. The separation occurs into the short/low-energy and long/high-energy clusters with difference in BIC between the two components and one component $\Delta\mathrm{BIC} = 17.6$ which suggests a very strong evidence in favour of two components. Similarly, analysis in L_{iso} vs. $T_{90,rest}$ plane in the *Fermi* data reveals only two groups separated into the low-luminosity and high-luminosity with $\Delta\mathrm{BIC} = 33.4$ suggesting a very strong evidence in favour of two components over one component. More than two components are not favoured in terms of BIC.

For *Swift*/BAT the data reveal three groups (BIC is maximal for three components) in the E_{iso} vs. $T_{90,rest}$ plane. The separation occurs into the short/low-energy, long/high-energy and long/low-energy clusters with difference in BIC between the three components and the two components $\Delta\mathrm{BIC} = 4.7$ which suggests a positive evidence in favour of three components. For the L_{iso} vs. $T_{90,rest}$ plane in the *Swift* data, three groups (BIC is maximal for three components) are revealed. The separation occurs into the low-luminosity, intermediate-duration/high-luminosity and long/high-luminosity clusters with difference in BIC between the three components and the two components $\Delta\mathrm{BIC} = 8.0$. This suggests a strong evidence in favour of three components. These results are preliminary and the completion of the data sample as well as the statistical analysis is ongoing.

6. Conclusion

There are known several statistical tests, theories, ideas, modelling, etc... about the third or even about more subgroups. A brief - never complete - survey was provided here. Summarizing these works it can be said that the existence of any astrophysically different phenomenon - beyond the two (short/hard and long/soft) types - is further in doubt. Both the intermediate subgroup and the possible subgroup of the low-luminosity long GRBs are not proven yet unambiguously. Eventual further subgroups are also in doubt.

References

Abbott, B. P., *et al.* 2017, *ApJ* (Letters), 848, L13
Acuner, Z., & Ryde, F. 2018, *MNRAS*, 475, 1708
Colgate, S. A. 1968, *Canadian Journal of Physics* (Supplement), 46, 476
Colgate, S. A. 1974, *ApJ*, 187, 333
Fenimore, E. E., *et al.* 1995, *ApJ* (Letters), 448, L101
Gehrels, N., *et al.* 2006, *Nature*, 444, 1044
Hjorth, J., *et al.* 2003, *Nature*. 423, 847
Hakkila, J., *et al.* 2003, *ApJ*, 582, 320
Hakkila, J., *et al.* 2004, *Baltic Astronomy*, 13, 2011
Horváth, I. 1998, *ApJ*, 508, 757
Horváth, I. 2002, *A&A*, 392, 791
Horváth, I., *et al.* 2006, *A&A*, 447, 23

Kann, D. A., *et al.* 2010, *ApJ*, 720, 1513
Kass, R. E., & Raftery, A. E. 1995, *Journal of the American Statistical Association*, 90, 773
Klebesadel, R. W., Strong, I. B., & Olson, R. A. 1974, *ApJ* (Letters), 182, L85
Levan, A. J., Tanvir, N. R., & Starling, R. L. C. 2014, *ApJ*, 781, 13
Mazets, E. P., *et al.* 1981, *ApSS*, 80, 3
Mészáros, A., & Mészáros, P. 1996, *ApJ*, 466, 29
Mészáros, P. 2006, *Reports on Progress in Physics*, 69, 2259
Mukherjee, S., *et al.* 1998, *ApJ*, 508, 314
Narayana Bhat, P., *et al.* 2016, *ApJ* (Suppl.), 223, id.28
Ohmori, N., *et al.* 2016, *PASJ*, 68, S30
Pendleton, G. N., *et al.* 1997, *ApJ*, 489, 175
Řípa, J., *et al.* 2012, *ApJ*, 756, id.44
Řípa, J., & Mészáros, A. 2016, *ApSS*, 361, id.370
Ruffini, R., *et al.* 2018, *ApJ*, 859, id.30
Tanvir, N. R., *et al.* 2013, *Nature*, 500, 547
Tarnopolski, M. 2015 *A&A*, 581, id.A29
Tarnopolski, M. 2016, *MNRAS*, 458, 2024
Tikhomirova, Ya. Yu., & Stern, B. E. 2005, *AstL*, 31, 291
Tsvetkova, A., *et al.* 2017, *ApJ*, 850, id.161
Veres, P., *et al.* 2010, *ApJ*, 725, 1955
Virgili, F. J., *et al.* 2015, *ApJ*, 778, id.54

High-mass X-ray Binaries: Illuminating the Passage from Massive Binaries to Merging Compact Objects
Proceedings IAU Symposium No. 346, 2019
L. M. Oskinova, E. Bozzo, T. Bulik & D. Gies, eds.

doi:10.1017/S1743921319001261

Numerical models of VHE emission by magnetic reconnection in X-ray binaries: GRMHD simulations and Monte Carlo cosmic-ray emission

J. C. Rodríguez-Ramírez, E. M. de Gouveia Dal Pino and R. Alves Batista

Instituto de Astronomia, Geofísica e Ciências Atmosféricas (IAG-USP), Universidade de São Paulo. Cidade Universitaria R. do Matão, 1226 05508-090 São Paulo, SP Brasil
email: juan.rodriguez@iag.usp.br

Abstract. Galactic microquasars have been detected at very-high-energies (VHE) (>100 GeV) and the particle acceleration mechanisms that produce this emission are not yet well-understood. Here we investigate a hadronic emission scenario where cosmic-rays (CRs) are accelerated in magnetic reconnection events by the turbulent, advected-dominated accretion flow (ADAF) believed to be present in the hard state of black hole binaries. We present Monte Carlo simulations of CR emission plus γ-γ and inverse Compton cascades, injecting CRs with a total energy consistent with the magnetic energy of the plasma. The background gas density, magnetic, and photon fields where CRs propagate and interact are modelled with general relativistic (GR), magneto-hydrodynamical simulations together with GR radiative transfer calculations. Our approach is applied to the microquasar Cygnus X-1, where we show a model configuration consistent with the VHE upper limits provided by MAGIC collaboration.

Keywords. radiation mechanisms: non-thermal, X-rays: binaries, magnetohydrodynamics, accretion disks

1. Introduction

Galactic microquasars are natural laboratories to study high energy emission processes having the advantage of being relatively close and undergoing processes in shorter time scales than high energy extragalactic sources like active galactic nuclei (AGNs) or quasars. Interestingly, there is no current unified model able to explain the very-high-energy (VHE) emission from these galactic sources: whether they have a hadronic and/or leptonic origin and if they are produced in outflows of pc scales and/or in compact regions.

The diverse nature of microquasrs' VHE emssion can be seen in the case of the sources SS 433 and Cygnus X-1. The VHE emission from SS 433 has recently been localised in outflows of pc scales launched by the binary system, with physical conditions that appear to favour a leptonic origin (Abeysekara *et al.* 2018). On the other hand, the multi-wavelength analysis on Cygnus X-1 by MAGIC Collaboration *et al.* (2017), ruled out any VHE emission correlated with the one-sided jet during the hard state (HS), suggesting that the VHE activity observed by MAGIC (Albert *et al.* 2007), might be produced inside the binary. Thus, improved observations as well as further development of theoretical models are required to clarify the origin of VHE emission from microquasars.

In this work, we investigate a hadronic emission scenario where CRs are accelerated by magnetic reconnection in a turbulent magnetised advection dominated accretion flow (ADAF) (see Singh *et al.* 2015; Kadowaki *et al.* 2015), thought to be present in the

HS of microquasars (Narayan & McClintock 2008; Yuan & Narayan 2014). Turbulent magnetic reconnection has been previously investigated with analytical and numerical approaches (de Gouveia Dal Pino & Lazarian 2005, de Gouveia Dal Pino *et al.* 2010, Kowal *et al.* 2011, 2012, Kadowaki, de Gouveia Dal Pino & Stone 2018, Kadowaki *et al.*, these proceedings) indicating that this mechanism could efficiently accelerate CRs in the surrounding of BH accretion flows (from microquasar to AGNs). This work is also motivated by previous hadronic emission models aimed to explain the VHE radiation from microquasars (Romero, Vieyro & Vila 2010; Vieyro & Romero 2012; Khiali de Gouveia Dal Pino & del Valle 2015).

The study presented here is based on Monte Carlo (MC) simulations of CRs that interact with a target environment obtained from GRMHD simulations together with GR radiative transfer calculations. With this approach we aim to provide a more self-consistent emission model with fewer free parameters in comparison with previous one-zone approaches (Khiali *et al.* 2015). In the next section we describe the numerical background model for the gas density, magnetic, and radiation field that we employ as the target for CRs interactions. In Section 3 we describe the Monte Carlo CR simulation applied to model the VHE emission of Cygnus X-1. Finally, we state our conclusions in Section 4.

2. The numerical ADAF background environment

If CRs are accelerated by magnetic reconnection in the turbulent accretion flow to the BH in the binary systems, they will mainly interact with the gas density, magnetic, and photon field of the accreting plasma, as well as with the radiation of the companion star. In the binary system Cygnus-X-1 the companion is inferred to be a O9.7 supergiant star (Orosz *et al.* 2011) and its radiation is expected to be important for VHE processes, specially around the superior conjunction of the compact object (Bosch-Ramon, Khangulyan & Aharonian 2008, Khiali *et al.* 2015). The present study is limited to consider only the radiation produced by the accretion flow; we will explore additional effects due to the stellar radiation field in a forthcoming work.

We adopt a GRMHD ADAF model and numerically obtain the gas density, magnetic, and photon fields employing the GRMHD axi-symmetric `harm` code (Gammie *et al.* 2003) together with the post-processing GR radiative transfer `grmonty` code (Dolence *et al.* 2009). In Fig. 1 we show the gas density, magnetic and photon fields of the snapshot that we set up as the environment where CRs propagate and interact.

These snapshot maps correspond to an accretion flow around a rotating BH of mass $M = 14.8$ $M_{\odot}$, with dimensionless spin parameter $a = 0.93$, at a evolution time $t = 2500 R_g/c$, being $R_g = GM/c^2 \simeq 2.2 \times 10^6$cm the gravitational radius. The plasma evolved from an initial equilibrium state where the accretion is triggered by magneto-rotational-instabiliy, similar to numerical GRMHD studies done by Mościbrodzka *et al.* (2009), O' Riordan *et al.* (2016) and others. The gas density is normalised to have a maximum value of $n = 5 \times 10^{17}$ cm^{-3} and we use an ion-to-electron temperature of $T_p/T_e = 25$ to calculate the background photon field produced by synchrotron+IC scattering. The SED associated to this photon field is the cyan histogram plotted in Fig. 2 (left), assuming a distance to the binary of 1.8 kpc (Reid *et al.* 2011). We employ these numerical profiles for the background environment to simulate the propagation of CRs and their secondaries, as we describe in the next section.

3. Monte Carlo calculations of CR emission

We simulate the propagation and interactions of CRs employing the `CRPropa3` code (Alves Batista *et al.* 2016) where we inject CRs, all of them assumed to be protons,

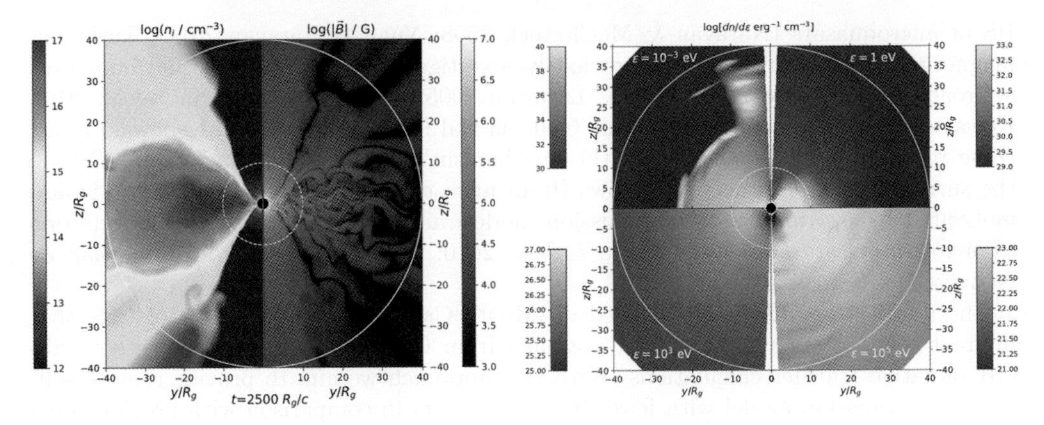

Figure 1. Left: gas density and magnetic field maps of the accretion flow environment where we simulate the propagation and interaction of CRs. This snapshot profiles are obtained with the GRMHD `harm` code. Right: photon field maps of synchrotron+IC radiation obtained by performing post-processing radiative transfer on the plasma snapshot on the left, employing the `grmonty` code. Each panel displays the photon field density for a fixed photon energy as indicated. The inner dashed circle represents the boundary within which we inject CRs assumed accelerated by magnetic reconnection. The outer solid circle represents the spherical boundary where we detect the radiation flux.

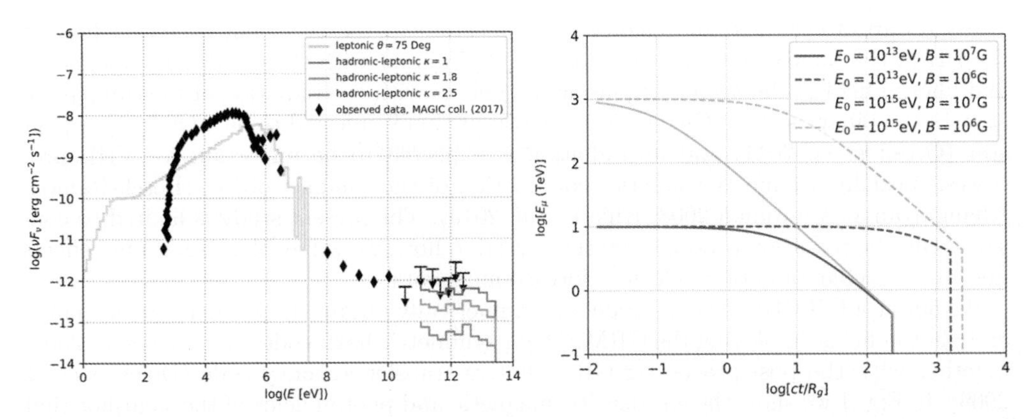

Figure 2. Left: SED of Cygnus X-1 corresponding to the hard state. The cyan histogram is the radiation flux obtained with leptonic synchrotron+IC employing the `grmonty` code (see Section 2). The histograms at energies > 100 GeV are γ-ray fluxes obtained from simulated hadronic interaction plus γ-γ and IC cascading employing the `CRPropa3` code (see Section 3). The data points and upper limits are taken from MAGIC Collaboration *et al.* (2017). Right: Estimation of muon synchrotron cooling. The curves represent energy of muons as a function of time (scale with the BH mass) in a uniform magnetic field, and with initial energy E_0 as indicated in the plot. The discontinuity in the curves indicates the time at which muons decay into neutrinos.

and accelerated by magnetic reconnection in the BH accretion flow of the binary. The 3D trajectories of charged particles (CRs and secondary leptons) are calculated by integrating the relativistic Lorentz force equation with the magnetic field of the GRMHD snapshot of Fig. 1. The gas density and photon field of the same snapshot are used to calculate the interactions rates for particles and photons along their trajectories, where these interactions are simulated with a Monte Carlo approach.

We then simulate (i) photo-pion processes, (ii) proton-proton interactions of CRs with thermal ions, (iii) γ-γ and IC cascading, and (iv) synchrotron cooling of charged particles. VHE γ-rays are produced out of photo-pion and proton-proton reactions and their attenuation is accounted by the γ-γ/IC cascades. We do not follow the photons produced by synchrotron cooling, nevertheless this process is employed to account for energy losses of charged particles (CRs and secondary leptons). We inject 10^4 isotropic CRs within a radius of $10R_g$ with energies in the range of 10^{14}-10^{16} eV, with a power-law energy distribution of index 1. Particles and photons are tracked until they complete a total trajectory of $150R_g$, or attain a minimum energy of 10^{11}eV, or else cross the spherical boundary at $40R_g$ (represented with the outer circle in Fig. 1).

The observed SED is calculated neglecting further emission and/or absorption of γ-rays beyond the detection boundary at 40 R_g from the BH (the radiation of the companion will be considered in a forthcoming work). Thus, the flux of radiation within the energy bin ϵ is given as

$$\nu F_\nu = (4\pi R_s^2)^{-1}\epsilon^2 \frac{N_\epsilon}{\Delta\epsilon\Delta t}(\text{erg s}^{-1}\text{cm}^{-2}), \tag{3.1}$$

where, $R_s = 1.8$ kpc is the distance to the binary system (Reid *et al.* 2011), Δt is the detection time interval, and N_ϵ is the number of detected physical photons within the energy bin ϵ.

The number of physical photons produced by CRs injected within the energy bin ϵ_0, which we label as $N_{\epsilon(\epsilon_0)}$, is calculated from the simulation output considering the condition $N_{\epsilon(\epsilon_0)}/N^{sim}_{\epsilon(\epsilon_0)} = N_{CR,\epsilon_0}/N^{sim}_{CR,\epsilon_0}$. The first term of this condition is the ratio of physical to simulated γ-rays within the energy bin ϵ and the second term is the ratio of physical to simulated CRs within the energy bin ϵ_0. Thus, the number of physical γ-rays as a function of the simulated γ-rays produced by CRs with initial energy ϵ_0 is calculated as

$$N_{\epsilon(\epsilon_0)} = \frac{A_\kappa \epsilon_0^{-\kappa}}{A_1 \epsilon_0^{-1}} N^{sim}_{\epsilon(\epsilon_0)} = \frac{E_{rec}}{N_{CR}} \frac{(2-\kappa)\ln(\epsilon_{max}/\epsilon_{min})}{\epsilon_{max}^{2-\kappa} - \epsilon_{min}^{2-\kappa}} \epsilon_0^{1-\kappa} N^{sim}_{\epsilon(\epsilon_0)}. \tag{3.2}$$

In the second term of the equations (3.2), A_κ is the normalisation constant of the power-law distribution of physical CRs with power-law index κ, and A_1 is the normalisation constant of the power-law distribution of simulated CRs with power-law index=1. The distribution of physical CRs is normalised with the total energy E_{rec} released by magnetic reconnection, and the distribution of simulated CRs with the total number of injected CRs, N_{CR}. In the third term of the equations (3.2), $\epsilon_{min} = 10^{14}$ eV and $\epsilon_{max} = 10^{16}$ eV are the maximum and minimum energy of injected CRs. In this manner, we weigh the simulation output, performed with a power-law index=1 of injected CRs, to obtain γ-ray fluxes corresponding to different power-law indices of CR injection.

In Fig. 2 we show the calculated emission flux that results from a simulation with $E_{rec} = 5 \times 10^{35}$ erg stored in CRs (which is consistent with magnetic reconnection energy released by an accretion plasma with $B \sim 10^7$ G within $10R_g$, according to the model by Singh, de Gouveia Dal Pino & Kadowaki 2015). The different histograms for $E > 11$ eV correspond to different power-law injection indices, as indicated.

Photo-pion and proton-proton reactions also produce neutrinos and electrons. However, due to the strong magnetic fields close to the BH of the binary, charged pions and muons are expected to cool down considerably before decaying into neutrinos and electrons. Here we do not report the neutrino fluxes associated to the simulated hadronic processes, as the propagation of pions and muons are not considered in the present calculation$*$. Nevertheless, here we estimate muon cooling by synchrotron radiation (see Fig. 2, right) from which we expect that neutrinos that originated close to the BH would be detected

$*$ Pions and muons are assumed to decay instantaneously in CRPropa3, as this code is originally aimed for cosmological propagation of CRs where the cooling of charged pions and muons is negligible.

only below 10^{13} eV energies (see also the paper by Reynoso & Romero 2009, where this effect has been investigated in detail).

4. Summary and discussion

Here we investigate the VHE emission from the accretion flow of Cygnus X-1 with a numerical approach based on:

- a GRMHD ADAF plus GR radiative transfer simulated model of the accretion flow,
- particle acceleration by magnetic reconnection, and
- MC simulations of CRs plus γ-γ/IC cascading.

Assuming that CRs are efficiently accelerated by magnetic reconnection in the turbulent accretion flow (Kadowaki *et al.* these proceedings.), we simulate a burst-like injection of CRs (appropriate to study the flare nature of the VHE emission from Cygnus-X1) and obtain a radiation flux consistent with the VHE upper limits of Cygnus X-1. This VHE flux is obtained injecting CRs with a total energy of 5×10^{35} erg (consistent with the energy release by reconnection in an ADAF with $B \simeq$ of 10^7G inside a region of $10R_g$), within the energy range of 10^{14}-10^{16}eV.

Neutrinos fluxes are not calculated from our simulation, as propagation of pions and muons is required. However, we estimate that neutrino fluxes, if produced by reconnection close to the BH of Cygnus X1, would be detected with energies below 10^{13} eV.

We applied our method to Cygnus X-1 in this work, but this model can also be applicable to other microquasars displaying VHE emission, as well as to AGNs (Rodríguez-Ramírez *et al.* 2018). To obtain more consistent predictions, we intend include the companion's radiation in our models, together with pion and muon propagation to also obtain predictions for neutrino emission. The results of these implementations will be reported in a forthcoming paper.

Acknowledgments

We acknowledge support from the Brazilian agencies FAPESP (grant 2013/10559-5) and CNPq (grant 308643/2017-8). The simulations presented in this lecture have made use of the computing facilities of the GAPAE group (IAG-USP) and the Laboratory of Astroinformatics IAG/USP, NAT/Unicsul (FAPESP grant 2009/54006-4). RAB is supported by the FAPESP grant 2017/12828-4 and JCRR by the FAPESP grant 2017/12188-5.

References

Abeysekara, A. *et al.* 2018, *Nature*, 562, 82
Albert, J. *et al.* 2007, *ApJ*, 665, L51
Alves Batista, R., Dundovic, A., Erdmann, M., Kampert, K.-H., Kuempel, D., Mller, G., Sigl, G.; van Vliet, A., Walz, D., Winchen, T. 2016, *JCAP*, 05, 038A
Bosch-Ramon V., Khangulyan D., & Aharonian F. A. 2008, *A&A*, 489, L21
de Gouveia Dal Pino, E. M. & Lazarian, A. 2005, *A&A*, 441, 845,
de Gouveia Dal Pino, E. M., Piovezan, P. P., & Kadowaki, L. H. S. 2010, *A&A*, 518, A5
Dolence, J., Gammie, C., Mościbrodzka, M., & Leung, P-A. 2009, *ApJ*, 184, 387
Gammie, C., McKinney, J., & Toth, G. 2003, *ApJ*, 598, 444
Kadowaki, L. H. S., de Gouveia Dal Pino, E. M., & Singh, C. B. 2015, *ApJ*, 802, 113
Kadowaki, L. H. S., de Gouveia Dal Pino, E. M., & Stone, J. 2018, *ApJ*, 864, 52
Khiali, B., de Gouveia Dal Pino, E. M., & del Valle, M. V. 2015, *MNRAS*, 449, 34
Kowal G., de Gouveia Dal Pino E. M., & Lazarian, A. 2011, *ApJ*, 735, 102
Kowal G., de Gouveia Dal Pino E. M., & Lazarian, A. 2012, *PRL*, 108, 1102
MAGIC Collaboration 2017, *MNRAS*, 472, 3474

Mościbrodzka, M., Gammie, C. F., Dolence, J. C., Shiokawa, H., & Leung, P. K. 2009, *ApJ*, 706, 497

Narayan, R., & McClintock, J. E. 2008, *NewAr*, 51, 733

O' Riordan, M., Pe'er, A. & McKinney, J. 2016, *ApJ*, 819, 95

Orosz J. A., McClintock J. E., Aufdenberg J. P., Remillard R. A., Reid M. J., Narayan R., & Gou L. 2011, *ApJ*, 742, 84

Reid M. J., McClintock J. E., Narayan R., Gou L., Remillard R. A., & Orosz J. A. 2011, *ApJ*, 742, 83

Reynoso, M. M., & Romero, G. E. 2009, *A&A*, 493, 1

Rodríguez-Ramírez, J. C., de Gouveia Dal Pino E. M., & Alves Batista, R. A. 2018, arXiv:1811.02812

Romero, G. E., Vieyro, F. L. & Vila, G. S. 2010, *A&A*, 519, 109

Singh C. B., de Gouveia Dal Pino E. M., & Kadowaki L. H. S. 2015, *ApJ*, 799, L20

Vieyro, F. L. & Romero G. E. 2012, *A&A*, 542, A7

Yuan, F. & Narayan, R. 2014, *ARA&A*, 52, 529

HMXB and LMXB evolution and their links with gravitational wave sources

High-mass X-ray Binaries: Illuminating the Passage from Massive Binaries to Merging Compact Objects
Proceedings IAU Symposium No. 346, 2019
A.C. Editor, B.D. Editor & C.E. Editor, eds.

doi:10.1017/S1743921319001868

Dynamical versus isolated formation channels of gravitational wave sources

Michela Mapelli[1,2,3,4]

[1]Dipartimento di Fisica e Astronomia G. Galilei, Università degli Studi di Padova, Vicolo dell'Osservatorio 3, I–35122, Padova, Italy
email: michela.mapelli@unipd.it

[2]Institut für Astro- und Teilchenphysik, Universität Innsbruck, Technikerstrasse 25/8, A–6020 Innsbruck, Austria

[3]INAF – Osservatorio Astronomico di Padova, Vicolo dell'Osservatorio 5, I–35142 Padova, Italy

[4]INFN – Sezione di Padova, Via F. Marzolo 8, I–35131 Padova, Italy

Abstract. What are the formation channels of merging black holes and neutron stars? The first two observing runs of Advanced LIGO and Virgo give us invaluable insights to address this question, but a new approach to theoretical models is required, in order to match the challenges posed by the new data. In this review, I discuss the impact of stellar winds, core-collapse and pair instability supernovae on the formation of compact remnants in both isolated and dynamically formed binaries. Finally, I show that dynamical processes, such as the runaway collision scenario and the Kozai-Lidov mechanism, leave a clear imprint on the demography of merging systems.

Keywords. black hole physics – gravitational waves – (stars:) binaries: general – stellar dynamics – stars: winds – galaxies: star clusters

1. Lesson learned from the first direct gravitational wave detections

On September 14 2015, the LIGO interferometers captured a gravitational wave (GW) signal from two merging black holes (BHs, Abbott *et al.* 2016a). This event, named GW150914, is the first direct detection of GWs, about hundred years after Einstein's prediction. To date, nine more BH mergers have been reported (GW151012, GW151226, GW170104, GW170608, GW170729, GW170809, GW170814, GW170818 and GW17082 Abbott *et al.* 2016b,c, 2017a,c,b, 2018a,b). In particular, GW170814 was the first BH merger detected jointly by three interferometers: the two LIGO detectors in the United States (Aasi *et al.* 2015) and Virgo in Italy (Acernese *et al.* 2015).

Astrophysicists have learned several revolutionary concepts about BHs from GW detections (Abbott *et al.* 2016d). First, GW150914 has confirmed the existence of double BH binaries (BHBs), i.e. binaries composed of two BHs. BHBs have been predicted a long time ago (e.g. Tutukov & Yungelson 1973; Thorne 1987; Schutz 1989; Kulkarni *et al.* 1993; Sigurdsson *et al.* 1993; Bethe & Brown 1998; Portegies Zwart & McMillan 2000; Colpi *et al.* 2003; Belczynski *et al.* 2004), but their observational confirmation was still missing. Second, GW detections show that some BHBs are able to merge within a Hubble time.

Finally, seven out of ten merging BHBs detected so far host BHs with mass in excess of 30 $M_\odot$. This result was a genuine surprise for the astrophysicists, because the only stellar BHs for which we have a dynamical mass measurement, i.e. about a dozen of BHs in X-ray binaries, have mass < 20 $M_\odot$ (see Figure 1 for a compilation of measured BH

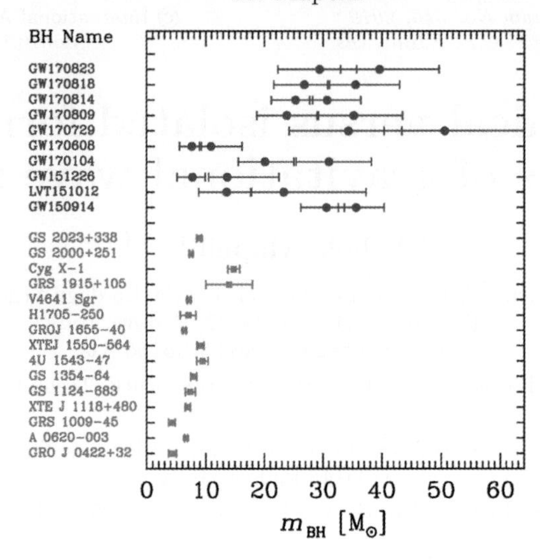

Figure 1. A compilation of BH masses $m_{\rm BH}$ from observations. Red squares: BHs with dynamical mass measurement in X-ray binaries (Orosz *et al.* 2003; Özel *et al.* 2010). This selected sample is quite conservative, because uncertain and debated results are not being shown (e.g. IC10 X-1 Laycock *et al.* 2015). Blue circles: BHs in the first published GW events (Abbott *et al.* 2016c, 2017a,b, 2018a,b).

masses). Moreover, most theoretical models did not predict the existence of BHs with mass $m_{\rm BH} > 30$ M$_\odot$ (but see Mapelli *et al.* 2009, 2010, 2013; Belczynski *et al.* 2010; Fryer *et al.* 2012; Ziosi *et al.* 2014; Spera *et al.* 2015 for several exceptions). Thus, the first GW detections have urged the astrophysical community to deeply revise the models of BH formation and evolution.

2. The formation of compact remnants from stellar evolution and supernova explosions

BHs and neutron stars (NSs) are expected to form as compact relics of massive ($\gtrsim 8$ M$_\odot$) stars. An alternative theory predicts that BHs can form also from gravitational collapse in the early Universe (the so called primordial BHs, e.g. Bird *et al.* 2016; Carr *et al.* 2016; Inomata *et al.* 2016). In this review, we will focus on BHs of stellar origin.

The mass function of BHs is highly uncertain, because it may be affected by a number of barely understood processes. In particular, stellar winds and supernova (SN) explosions both play a major role on the formation of compact remnants. Processes occurring in close binary systems (e.g. mass transfer and common envelope) are a further complication and will be discussed in the next section.

2.1. *Stellar winds and stellar evolution*

Stellar winds are outflows of gas from the atmosphere of a star. In cold stars (e.g. red giants and asymptotic giant branch stars) they are mainly induced by radiation pressure on dust, which forms in the cold outer layers (e.g. van Loon *et al.* 2005). In massive hot stars (O and B main sequence stars, luminous blue variables and Wolf-Rayet stars), stellar winds are powered by the coupling between the momentum of photons and that of metal ions present in the stellar photosphere. A large number of strong and weak resonant metal lines are responsible for this coupling.

Understanding stellar winds is tremendously important for the study of compact objects, because mass loss determines the pre-SN mass of a star (both its total mass and its core mass), which in turn affects the outcome of an SN explosion (Fryer *et al.* 1999; Fryer & Kalogera 2001; Mapelli *et al.* 2009, 2010; Belczynski *et al.* 2010).

Early work on stellar winds (e.g. Abbott 1982; Kudritzki *et al.* 1987; Leitherer *et al.* 1992) highlighted that the mass loss of O and B stars depends on metallicity as $\dot{m} \propto Z^{\alpha}$ (with $\alpha \sim 0.5 - 1.0$, depending on the model). However, such early work did not account for multiple scattering, i.e. for the possibility that a photon interacts several times before being absorbed or leaving the photosphere. Vink *et al.* (2001) accounted for multiple scatterings and found a universal metallicity dependence $\dot{m} \propto Z^{0.85}\, v_{\infty}^{p}$, where v_{∞} is the terminal velocity† and $p = -1.23$ ($p = -1.60$) for stars with effective temperature $T_{\rm eff} \gtrsim 25000$ K ($12000\,{\rm K} \lesssim T_{\rm eff} \lesssim 25000$ K).

The situation is more uncertain for post-main sequence stars. For Wolf-Rayet (WR) stars, i.e. naked Helium cores, Vink & de Koter (2005) predict a similar trend with metallicity $\dot{m} \propto Z^{0.86}$. Hainich *et al.* (2015) find an even stronger dependence on metallicity (see their Figures 10 and 11), based on a quantitative analysis of several Wolf-Rayet N stars in the Local Group performed with the Potsdam Wolf-Rayet model atmosphere code.

With a different approach (which accounts also for wind clumping), Gräfener & Hamann (2008) find a strong dependence of WR mass loss on metallicity but also on the electron-scattering Eddington factor $\Gamma_e = \kappa_e\, L\,/(4\,\pi\, c\, G\, m)$, where κ_e is the cross section for electron scattering, L is the stellar luminosity, c is the speed of light, G is the gravity constant, and m is the stellar mass. The importance of Γ_e has become increasingly clear in the last few years (Gräfener *et al.* 2011; Vink *et al.* 2011), but, unfortunately, only few stellar evolution models include this effect. For example, Chen *et al.* (2015) adopt a mass loss prescriptions $\dot{m} \propto Z^{\alpha}$, where $\alpha = 0.85$ if $\Gamma_e < 2/3$ and $\alpha = 2.45 - 2.4\,\Gamma_e$ if $2/3 \leq \Gamma_e \leq 1$. This simple formula accounts for the fact that metallicity dependence tends to vanish when the star is close to be radiation pressure dominated.

2.2. *Supernovae (SNe)*

The mechanisms triggering iron core-collapse SNe are still highly uncertain. Several mechanisms have been proposed, including rotationally-driven SNe and/or magnetically-driven SNe (see e.g. Janka 2012; Foglizzo *et al.* 2015 and references therein). The most commonly investigated mechanism is the convective SN engine (see e.g. Fryer *et al.* 2012).

Fully self-consistent simulations of core collapse with a state-of-the-art treatment of neutrino transport do not lead to explosions in spherical symmetry except for the lighter SN progenitors ($\lesssim 10$ $M_{\odot}$, Foglizzo *et al.* 2015; Ertl *et al.* 2016). Simulations which do not require the assumption of spherical symmetry (i.e. run at least in 2D) appear to produce successful explosions from first principles for a larger range of progenitor masses (see e.g. Müller & Janka *et al.* 2012a,b). However, 2D and 3D simulations are still computationally challenging and cannot be used to make a study of the mass distribution of compact remnants.

Thus, in order to study compact-object masses, SN explosions are artificially induced by injecting in the pre-SN model some amount of kinetic energy (kinetic bomb) or thermal

† The terminal velocity v_{∞} is the the velocity reached by the wind at large distance from the star, where the radiative acceleration approaches zero because of the geometrical dilution of the photospheric radiation field. Since line-driven winds are continuously accelerated by the absorption of photospheric photons in spectral lines, v_{∞} corresponds to the maximum velocity of the stellar wind. See the review by Kudritzki (2000) for more details.

energy (thermal bomb) at an arbitrary mass location. The evolution of the shock is then followed by means of 1D hydrodynamical simulations with some relatively simplified treatment for neutrinos. This allows to simulate hundreds of stellar models.

Following this approach, O'Connor & Ott (2011) propose a criterion to decide whether a SN is successful or not, based on the compactness parameter:

$$\xi_m = \frac{m/\mathrm{M}_\odot}{R(m)/1000\,\mathrm{km}}, \tag{2.1}$$

where $R(m)$ is the radius which encloses a given mass m. Usually, the compactness is defined for $m = 2.5\ \mathrm{M}_\odot$ ($\xi_{2.5}$). O'Connor & Ott (2011) measure the compactness at core bounce† in their simulations and find that the larger $\xi_{2.5}$ is, the shorter the time to form a BH (as shown in their Figure 6). This means that stars with a larger value of $\xi_{2.5}$ are more likely to collapse to a BH without SN explosion. The work by Ugliano *et al.* (2012) and Horiuchi *et al.* (2014) indicate that the best threshold between exploding and non-exploding models is $\xi_{2.5} \sim 0.2$.

The models proposed by O'Connor & Ott (2011) (see also Ertl *et al.* 2016; Sukhbold *et al.* 2014, 2016) are sometimes referred to as the "islands of explodability" scenario, because they predict a non-monotonic behaviour of SN explosions with the stellar mass. This means, for example, that while a star with a mass $m = 25\ \mathrm{M}_\odot$ and a star with a mass $m = 29\ \mathrm{M}_\odot$ might end their life with a powerful SN explosion, another star with intermediate mass between these two (e.g. with a mass $m = 27\ \mathrm{M}_\odot$) is expected to directly collapse to a BH without SN explosion. Thus, these models predict the existence of "islands of explodability", i.e. ranges of mass where a star is expected to explode, surrounded by mass intervals in which the star will end its life with a direct collapse.

Finally, it is important to recall pair-instability and pulsational pair-instability SNe (Fowler *et al.* 1964; Barkat *et al.* 1967; Rakavy *et al.* 1967; Woosley 2017). If the Helium core of a star grows above $\sim 30\ \mathrm{M}_\odot$ and the core temperature is $\gtrsim 7 \times 10^8$ K, the process of electron-positron pair production becomes effective. It removes photon pressure from the core producing a sudden collapse before the iron core is formed. For $m_{\mathrm{He}} > 135\ \mathrm{M}_\odot$, the collapse cannot be reversed and the star collapses directly in to a BH (Woosley 2017). If $135 \gtrsim m_{\mathrm{He}} \gtrsim 64\ \mathrm{M}_\odot$, the collapse triggers an explosive burning of heavier elements, which has disruptive effects. This leads to a complete disruption of the star, leaving no remnant (the so-called pair-instability SN, Heger & Woosley 2002).

For $64 \gtrsim m_{\mathrm{He}} \gtrsim 32\ \mathrm{M}_\odot$, pair production induces a series of pulsations of the core (pulsational pair instability SNe), which trigger an enhanced mass loss (Woosley 2017). At the end of this instability phase a remnant with non-zero mass is produced, significantly lighter than in case of a direct collapse.

2.3. *The mass of compact remnants*

The previous sections suggest that our knowledge of compact object mass is hampered by severe uncertainties, connected with both stellar winds and core-collapse SNe. Thus, models of the mass spectrum of compact remnants must be taken with a grain of salt. However, few robust features can be drawn.

Figure 2 is a simplified version of Figures 2 and 3 of (Heger *et al.* 2003). The final mass of a star and the mass of the compact remnant are shown as a function of the ZAMS mass. The left and the right-hand panels show the case of a solar metallicity star and of a metal-free star, respectively. In the case of the solar metallicity star, the final mass of the

† Ugliano *et al.* (2012) show that $\xi_{2.5}$ is not significantly different at core bounce or at the onset of collapse.

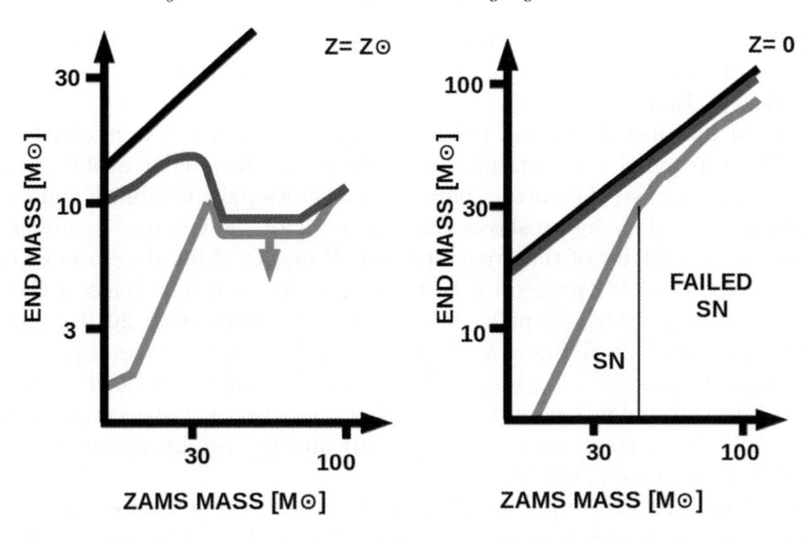

Figure 2. Final mass of a star ($m_{\rm fin}$, blue lines) and mass of the compact remnant (m_{rem}, red lines) as a function of the ZAMS mass of the star. The thick black line marks the region where $m_{\rm fin} = m_{\rm ZAMS}$. Left-hand panel: solar metallicity star. Right-hand panel: metal-free star. The red arrow on the left-hand panel is an upper limit for the remnant mass. Vertical thin black line in the right-hand panel: approximate separation between successful and failed SNe at $Z = 0$. This cartoon was inspired by Figures 2 and 3 of Heger *et al.* (2003).

star is much lower than the initial one, because stellar winds are extremely efficient. The mass of the compact object is also much lower than the final mass of the star because a core-collapse SN always takes place.

In contrast, a metal-free star (i.e. a population III star) loses a negligible fraction of its mass by stellar winds (the blue and the black line in Figure 2 are superimposed). As for the mass of the compact remnant, Figure 2 shows that there are two regimes: below a given threshold ($\approx 30 - 40$ M$_\odot$) the SN explosion succeeds even at zero metallicity and the mass of the compact remnant is relatively small. Above this threshold, the mass of the star (in terms of both core mass and envelope mass) is sufficiently large that the SN fails. Most of the final stellar mass collapses to a BH, whose mass is significantly larger than in the case of a SN explosion.

What happens at intermediate metallicity between solar and zero, i.e. in the vast majority of the Universe we know?

As a rule of thumb (see e.g. Fryer *et al.* 2012; Spera *et al.* 2015), we can draw the following considerations. If the zero-age main sequence (ZAMS) mass of a star is large ($m_{\rm ZAMS} \gtrsim 30$ M$_\odot$), then the amount of mass lost by stellar winds is the main effect which determines the mass of the compact object. At low metallicity ($\lesssim 0.1$ Z$_\odot$) and for a low Eddington factor ($\Gamma_e < 0.6$), mass loss by stellar winds is not particularly large. Thus, the final mass $m_{\rm fin}$ and the compactness $\xi_{2.5}$ of the star may be sufficiently large to avoid a core-collapse SN explosion: the star may form a massive BH ($\gtrsim 20$ M$_\odot$) by direct collapse, unless a pair-instability or a pulsational-pair instability SN occurs. At high metallicity ($\approx$ Z$_\odot$) or large Eddington factor ($\Gamma_e > 0.6$), mass loss by stellar winds is particularly efficient and may lead to a small $m_{\rm fin}$: the star is expected to undergo a core-collapse SN and to leave a relatively small compact object.

If the ZAMS mass of a star is relatively low ($7 < m_{\rm ZAMS} < 30$ M$_\odot$), then stellar winds are not important (with the exception of super asymptotic giant branch stars), regardless of the metallicity. In this case, the details of the SN explosion (e.g. energy of the explosion and amount of fallback) are crucial to determine the final mass of the compact object.

This general sketch may be affected by several factors, such as pair-instability SNe, pulsational pair-instability SNe (e.g. Woosley 2017) and an *island scenario* for core-collapse SNe (e.g. Ertl *et al.* 2016).

The effect of pair-instability and pulsational pair-instability SNe is clearly shown in Figure 3. The top panel was obtained accounting only for stellar evolution and core-collapse SNe. In contrast, the bottom panel also includes pair-instability and pulsational pair-instability SNe. This figure shows that the mass of the compact remnant strongly depends on the metallicity of the progenitor star if $m_{\rm ZAMS} \gtrsim 30\ {\rm M}_\odot$. In most cases, the lower the metallicity of the progenitor is, the larger the maximum mass of the compact remnant (Heger *et al.* 2003; Mapelli *et al.* 2009; Belczynski *et al.* 2010; Mapelli *et al.* 2010, 2013; Spera *et al.* 2015; Spera & Mapelli 2017). However, for metal-poor stars ($Z < 10^{-3}$) with ZAMS mass $230 > m_{\rm ZAMS} > 110\ {\rm M}_\odot$ pair instability SNe lead to the complete disruption of the star and no compact object is left. Only very massive ($m_{\rm ZAMS} > 230\ {\rm M}_\odot$) metal-poor ($Z < 10^{-3}$) stars can collapse to a BH directly, producing intermediate-mass BHs (i.e, BHs with mass $\gtrsim 100\ {\rm M}_\odot$).

If $Z < 10^{-3}$ and $110 > m_{\rm ZAMS} \gtrsim 60\ {\rm M}_\odot$, the star enters the pulsational pair-instability SN regime: mass loss is enhanced and the final BH mass is smaller ($m_{\rm BH} \sim 30 - 55\ {\rm M}_\odot$, bottom panel of Fig. 3) than we would have expected from direct collapse ($m_{\rm BH} \sim 50 - 100\ {\rm M}_\odot$, top panel of Fig. 3). Thus, accounting for both pair instability and pulsational pair-instability SNe leads to a *BH mass gap*† between $m_{\rm BH} \sim 60\ {\rm M}_\odot$ and $m_{\rm BH} \sim 120\ {\rm M}_\odot$.

3. Binaries of stellar black holes

Naively, one could think that if two massive stars are members of a binary system, they will eventually become a double BH binary and the mass of each BH will be the same as if its progenitor star was a single star. This is true only if the binary system is sufficiently wide (detached binary) for its entire evolution. If the binary is close enough, it will evolve through several processes which might significantly change its final fate.

The so-called binary population-synthesis codes have been used to investigate the effect of binary evolution processes on the formation of BHBs in isolated binaries (e.g. (Portegies Zwart *et al.* 1996; Hurley *et al.* 2002; Podsiadlowski *et al.* 2003; Belczynski *et al.* 2008; Mapelli *et al.* 2013; Mennekens *et al.* 2014; Eldridge *et al.* 2016; Mapelli *et al.* 2017; Stevenson *et al.* 2017; Giacobbo *et al.* 2018; Barrett *et al.* 2018; Giacobbo *et al.* 2018b; Kruckow *et al.* 2018; Eldridge *et al.* 2018; Mapelli & Giacobbo 2018; Spera *et al.* 2018; Giacobbo *et al.* 2019)). These are Monte-Carlo based codes which combine a description of stellar evolution with prescriptions for supernova explosions and with a formalism for binary evolution processes.

In the following, we mention some of the most important binary-evolution processes and we briefly discuss their treatment in the most used population-synthesis codes.

3.1. *Mass transfer*

If two stars exchange matter to each other, it means they undergo a mass transfer episode.

The Roche lobe of a star in a binary system is the maximum equipotential surface around the star within which matter is bound to the star. While the exact shape of

† The existence of a BH mass gap between ~ 50 and $\sim 100\ {\rm M}_\odot$ is currently consistent with GW detections (see e.g. Fishbach *et al.* (2017); Talbot *et al.* (2018); Wysocki *et al.* (2018); Abbott *et al.* (2018b)).

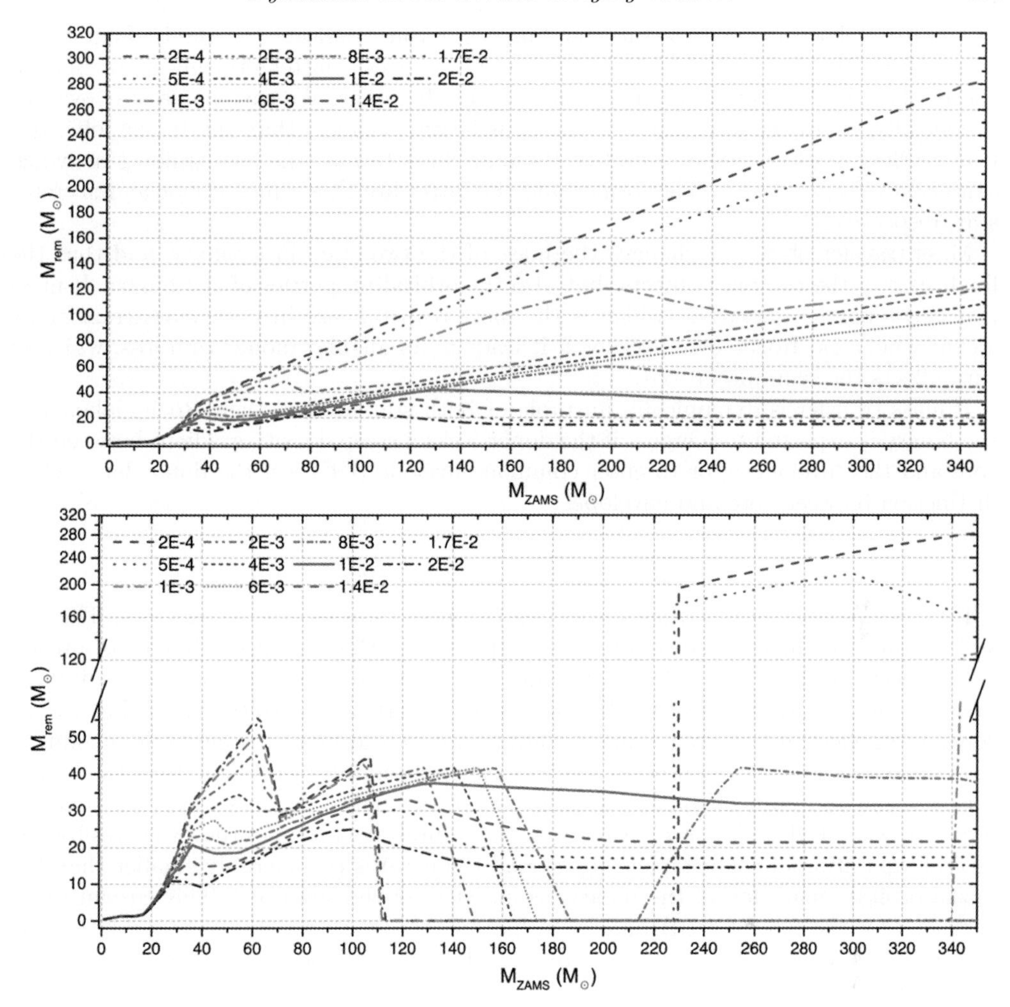

Figure 3. Mass of the compact remnant ($m_{\rm rem}$) as a function of the ZAMS mass of the star ($m_{\rm ZAMS}$). Lower (upper) panel: pulsational pair-instability and pair-instability SNe are (are not) included. In both panels: dash-dotted brown line: $Z = 2.0 \times 10^{-2}$; dotted dark orange line: $Z = 1.7 \times 10^{-2}$; dashed red line: $Z = 1.4 \times 10^{-2}$; solid red line: $Z = 1.0 \times 10^{-2}$; short dash-dotted orange line: $Z = 8.0 \times 10^{-3}$; short dotted light orange line: $Z = 6.0 \times 10^{-3}$; short dashed green line: $Z = 4.0 \times 10^{-3}$; dash-double dotted green line: $Z = 2.0 \times 10^{-3}$; dash-dotted light blue line: $Z = 1.0 \times 10^{-3}$; dotted blue line: $Z = 5.0 \times 10^{-4}$; dashed violet line: $Z = 2.0 \times 10^{-4}$. A delayed core-collapse SN mechanism has been assumed, following the prescriptions of (Fryer *et al.* 2012). This Figure was adapted from Figures 1 and 2 of Spera & Mapelli (2017).

the Roche lobe should be calculated numerically, a widely used approximate formula Eggleton (1993) is

$$r_{\rm L,1} = a\,\frac{0.49\,q^{2/3}}{0.6\,q^{2/3} + \ln\left(1 + q^{1/3}\right)}, \tag{3.1}$$

where a is the semi-major axis of the binary and $q = m_1/m_2$ (m_1 and m_2 are the masses of the two stars in the binary). This formula describes the Roche lobe of star with mass m_1, while the corresponding Roche lobe of star with mass m_2 ($r_{\rm L,2}$) is obtained by swapping the indexes.

The Roche lobes of the two stars in a binary are thus connected by the L1 Lagrangian point. Since the Roche lobes are equipotential surfaces, matter orbiting at or beyond the Roche lobe can flow freely from one star to the other. We say that a star overfills (under-fills) its Roche lobe when its radius is larger (smaller) than the Roche lobe. If a star overfills its Roche lobe, a part of its mass flows toward the companion star which can accrete (a part of) it. The former and the latter are thus called donor and accretor star, respectively.

Mass transfer obviously changes the mass of the two stars in a binary, and thus the final mass of the compact objects, but also the orbital properties of the binary. If mass transfer is non conservative (which is the most realistic case in both mass transfer by stellar winds and Roche lobe overflow), it leads to an angular momentum loss, which in turn affects the semi-major axis.

If mass transfer is dynamically unstable or both stars overfill their Roche lobe, then the binary is expected to merge – if the donor lacks a steep density gradient between the core and the envelope –, or to enter common envelope (CE) – if the donor has a clear distinction between core and envelope.

3.2. *Common envelope (CE)*

If two stars enter in CE, their envelope(s) stop co-rotating with their cores. The two stellar cores (or the compact remnant and the core of the star, if the binary is already single degenerate) are embedded in the same non-corotating envelope and start spiralling in as an effect of gas drag exerted by the envelope. Part of the orbital energy lost by the cores as an effect of this drag is likely converted into heating of the envelope, making it more loosely bound. If this process leads to the ejection of the envelope, then the binary survives, but the post-CE binary is composed of two naked stellar cores (or a compact object and a naked stellar core). Moreover, the orbital separation of the two cores (or the orbital separation of the compact object and the core) is considerably smaller than the initial orbital separation of the binary, as an effect of the spiral in†. This circumstance is crucial for the fate of a BH binary. In fact, if the binary which survives a CE phase evolves into a double BH binary, this double BH binary will have a very short semi-major axis, much shorter than the sum of the maximum radii of the progenitor stars, and may be able to merge by GW emission within a Hubble time.

In contrast, if the envelope is not ejected, the two cores (or the compact object and the core) spiral in till they eventually merge. This premature merger of a binary during a CE phase prevents the binary from evolving into a double BH binary.

The $\alpha\lambda$ formalism (Webbink 1984) is the most common formalism adopted to describe a common envelope. The basic idea of this formalism is that the energy needed to unbind the envelope comes uniquely from the loss of orbital energy of the two cores during the spiral in. This formalism relies on two free parameters, α, which describes the conversion efficiency of orbital energy into thermal energy ($\alpha = E_{\rm th}/E_{\rm b} \leq 1$, where $E_{\rm b}$ and $E_{\rm th}$ are the binding energy of the two cores and the thermal energy acquired by the envelope as an effect of the spiral-in, respectively), and λ, describing the concentration of the halo. Actually, we have known for a long time (see Ivanova *et al.* 2013 for a review) that this simple formalism is a poor description of the physics of CE, which is considerably more complicated. For example, there is a number of observed systems for which an $\alpha > 1$ is

† Note that a short-period (from few hours to few days) binary system composed of a naked Helium core and BH might be observed as an X-ray binary, typically a Wolf-Rayet X-ray binary. In the local Universe, we know a few (~ 7) Wolf-Rayet X-ray binaries, in which a compact object (BH or NS) accretes mass through the wind of the naked stellar companion (see e.g. Esposito *et al.* 2015 for more details). These rare X-ray binaries are thought to be good progenitors of merging compact-object binaries.

required, which is obviously un-physical. Moreover, λ cannot be the same for all stars. It is expected to vary wildly not only from star to star but also during different evolutionary stages of the same star. Thus, it would be extremely important to model the CE in detail, for example with numerical simulations. A lot of effort has been put on this in the last few years, but there are still many open questions.

3.3. *Alternative evolution to CE*

Massive fast rotating stars can have a chemically homogeneous evolution (CHE): they do not develop a chemical composition gradient because of the mixing induced by rotation. This is particularly true if the star is metal poor, because stellar winds are not efficient in removing angular momentum. If a binary is very close, the spins of its members are even increased during stellar life, because of tidal synchronisation. The radii of stars following CHE are usually much smaller than the radii of stars developing a chemical composition gradient (de mink & Mandel 2016; Mandel & de Mink 2016). This implies that even very close binaries (few tens of solar radii) can avoid CE.

Marchant *et al.* (2016) simulate very close binaries whose components are fast rotating massive stars. A number of their simulated binaries evolve into contact binaries where both binary components fill and even overfill their Roche volumes. If metallicity is sufficiently low and rotation sufficiently fast, these binaries may evolve as "over-contact" binaries†: the over-contact phase differs from a classical CE phase because co-rotation can, in principle, be maintained as long as material does not overflow the L2 point. This means that a spiral-in that is due to viscous drag can be avoided, resulting in a stable system evolving on a nuclear timescale.

Such over-contact binaries maintain relatively small stellar radii during their evolution (few ten solar radii) and may evolve into a double BH binary with a very short orbital period. This scenario predicts the formation of merging BHs with relatively large masses (> 20 M$_\odot$), nearly equal mass ($q = 1$), and with large aligned spins.

3.4. *Summary of the isolated binary formation channel*

In this section, we have highlighted the most important aspects and the open issues of the "isolated binary formation scenario", i.e. the model which predicts the formation of merging BHs through the evolution of isolated binaries. For isolated binaries we mean stellar binary systems which are not perturbed by other stars or compact objects.

To summarize, let us illustrate schematically the evolution of an isolated stellar binary (see e.g. Belczynski *et al.* 2016; Mapelli *et al.* 2017; Stevenson *et al.* 2017; Giacobbo *et al.* 2018) which can give birth to merging BHs like GW150914 and the other massive BHs observed by the LIGO-Virgo collaboration (Abbott *et al.* 2016a,c, 2017a,b, 2018a). In the following discussion, several details of stellar evolution have been simplified or skipped to facilitate the reading for non specialists.

The left-hand panel of Figure 4 shows the evolution of an isolated binary system composed of two massive stars. These stars are gravitationally bound since their birth. Initially, the two stars are both on the main sequence (MS). When the most massive one leaves the MS (i.e. when Hydrogen burning in the core is over, which happens usually on a time-scale of few Myr for massive stars with ZAMS mass $m_{\rm ZAMS} \gtrsim 30$ M$_\odot$), its radius starts inflating and can grow by a factor of a hundreds. The most massive star becomes a giant star with a Helium core and a large Hydrogen envelope. If its radius equals the Roche lobe (equation 3.1), the system starts a stable mass-transfer episode. Some mass is lost by the system, some is transferred to the companion. After several additional

† It is interesting to note that Hainich *et al.* (2018) actually show that the parameters required for over-compact binaries exist in observed stellar populations.

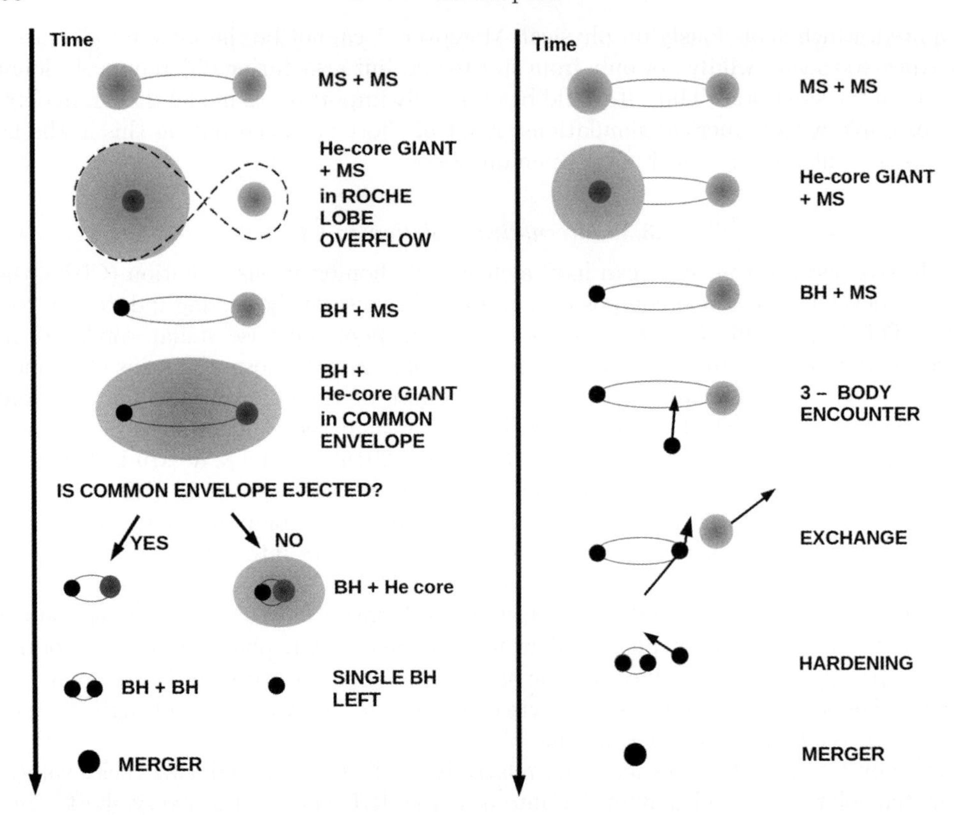

Figure 4. Left: Schematic evolution of an isolated binary which can give birth to a merging BH (see e.g. Belczynski *et al.* 2016; Mapelli *et al.* 2017; Stevenson *et al.* 2017; Giacobbo *et al.* 2018). Right: Schematic evolution of a merging BHB formed by dynamical exchange (see e.g. Downing *et al.* 2010; Ziosi *et al.* 2014; Mapelli 2016; Rodriguez *et al.* 2015, 2016; Askar *et al.* 2017).

evolutionary stages, the primary collapses to a BH (a direct collapse is preferred with respect to a SN explosion if we want the BH to be rather massive). At this stage the system is still quite large (hundreds to thousands of solar radii).

When also the secondary leaves the MS, growing in radius, the system enters a CE phase: the BH and the Helium core spiral in. If the orbital energy is not sufficient to unbind the envelope, then the BH merges with the Helium core leaving a single BH. In contrast, if the envelope is ejected, we are left with a new binary, composed of the BH and of a stripped naked Helium star. The new binary has a much smaller orbital separation (tens of solar radii) than the pre-CE binary, because of the spiral-in. If this new binary remains bound after the naked Helium star undergoes a SN explosion or if the naked Helium star is sufficiently massive to directly collapse to a BH, the system evolves into a close BHB, which might merge within a Hubble time.

The most critical quantities in this scenario are the masses of the two stars and also their initial separation (with respect to the stellar radii): a BHB can merge within a Hubble time only if its initial orbital separation is tremendously small (few tens of solar radii, unless the eccentricity is rather extreme); but a massive star (> 20 $M_\odot$) can reach a radius of several thousand solar radii during its evolution. Thus, if the initial orbital separation of the stellar binary is tens of solar radii, the binary merges before it can become a BHB. On the other hand, if the initial orbital separation is too large, the two BHs will never merge. In this scenario, the two BHs can merge only if the initial orbital separation of the progenitor stellar binary is in the range which allows the binary to enter CE and then to leave a short period BHB. This range of initial orbital separations dramatically depends on CE efficiency and on the details of stellar mass and radius evolution.

4. The dynamics of black hole binaries

In the previous sections of this review, we discussed the formation of BH binaries as isolated binaries. There is an alternative channel for BH binary formation: the dynamical evolution scenario.

4.1. *Dynamically active environments*

Collisional dynamics is important for the evolution of binaries only if they are in a dense environment ($\gtrsim 10^3$ stars pc^{-3}), such as a star cluster. On the other hand, astrophysicists believe that the vast majority of massive stars (which are BH progenitors) form in star clusters (Lada & Lada 2003; Weidner *et al.* 2006, 2010; Portegies Zwart *et al.* 2010).

Most studies of dynamical formation of BH binaries focus on globular clusters (e.g. Sigurdsson *et al.* 1993; Portegies Zwart & McMillan 2000; Mapelli *et al.* 2005; Downing *et al.* 2010; Benacquista 2013; Samsing *et al.* 2014; Rodriguez *et al.* 2015, 2016; Askar *et al.* 2017; Samsing *et al.* 2017). *Globular clusters* are old stellar systems (~ 12 Gyr), mostly very massive ($> 10^4$ M$_\odot$) and dense ($> 10^4$ M$_\odot$ pc^{-3}). They are sites of intense dynamical processes (such as the gravothermal catastrophe). However, globular clusters represent a tiny fraction of the baryonic mass in the local Universe ($\lesssim 1$ per cent, Harris *et al.* 2013).

In contrast, only few studies of BH binaries (e.g Ziosi *et al.* 2014; Mapelli 2016; Kimpson *et al.* 2016; Banerjee 2017) focus on *young star clusters*. These young ($\lesssim 100$ Myr), relatively dense ($> 10^3$ M$_\odot$ pc^{-3}) stellar systems are thought to be the most common birthplace of massive stars. When they evaporate (by gas loss) or are disrupted by the tidal field of their host galaxy, their stellar content is released into the field. Thus, it is reasonable to expect that a large fraction of BH binaries which are now in the field may have formed in young star clusters, where they participated in the dynamics of the cluster. The reason why young star clusters have been neglected in the past is exquisitely numerical: the dynamics of young star clusters needs to be studied with direct N-body simulations, which are rather expensive (they scale as N^2), combined with population-synthesis simulations. Moreover, their dynamical evolution may be significantly affected by the presence of gas. Including gas would require a challenging interface between direct N-body simulations and hydrodynamical simulations, which has been done in very few cases (Moeckel & Bate 2010; Parker & Dale 2013; Fuji *et al.* 2015; Mapelli 2017) and has been never used to study BH binaries. Finally, a fraction of young star clusters might survive gas evaporation and tidal disruption and evolve into older *open clusters*, like M67.

Another flavour of star cluster where BH binaries might form and evolve dynamically are *nuclear star clusters*, i.e. star clusters which lie in the nuclei of galaxies. Nuclear star clusters are rather common in galaxies (e.g. Böker *et al.* 2002; Ferrarese *et al.* 2006; Graham *et al.* 2009), are usually more massive and denser than globular clusters, and may co-exist with super-massive BHs (SMBHs). Stellar-mass BHs formed in the innermost regions of a galaxy could even be "trapped" in the accretion disc of the central SMBH, triggering their merger (see e.g. Stone *et al.* 2016; Bartos *et al.* 2017; McKernan *et al.* 2017). These features make nuclear star clusters unique among star clusters, for the effects that we will describe in the next sections.

4.2. *Three-body encounters*

We now review what are the main dynamical effects which can affect a BH binary, starting from three-body encounters. Binaries have a energy reservoir, their internal energy:

$$E_{\rm int} = \frac{1}{2}\,\mu\,v^2 - \frac{G\,m_1\,m_2}{r}, \tag{4.1}$$

where $\mu = m_1\,m_2/(m_1 + m_2)$ is the reduced mass of the binary (whose components have mass m_1 and m_2), v is the relative velocity between the two members of the binary, and r is the distance between the two members of the binary. As shown by Kepler's laws, $E_{\rm int} = -E_{\rm b} = -G\,m_1\,m_2/(2\,a)$, where $E_{\rm b}$ is the binding energy of the binary (a being the semi-major axis of the binary).

The internal energy of a binary can be exchanged with other stars only if the binary undergoes a close encounter with a star, so that its orbital parameters are perturbed by the intruder. This happens only if a single star approaches the binary by few times its orbital separation. We define this close encounter between a binary and a single star as a *three-body encounter*. For this to happen with a non-negligible frequency, the binary must be in a dense environment, because the rate of three-body encounters scales with the local density of stars. Three-body encounters have crucial effects on BH binaries, such as *exchanges*, *hardening*, and *ejections*.

4.3. *Exchanges*

Dynamical exchanges are three-body encounters during which one of the former members of the binary is replaced by the intruder.

Exchanges may lead to the formation of new double BH binaries. If a binary composed of a BH and a low-mass star undergoes an exchange with a single BH, this leads to the formation of a new double BH binary. This is a very important difference between BHs in the field and in star clusters: a BH which forms as a single object in the field has negligible chances to become member of a binary system, while a single BH in the core of a star cluster has good chances of becoming member of a binary by exchanges.

Exchanges are expected to lead to the formation of many more double BH binaries than they can destroy, because the probability for an intruder to replace one of the members of a binary is ≈ 0 if the intruder is less massive than both binary members, while it suddenly jumps to ~ 1 if the intruder is more massive than one of the members of the binary (Hills & Fullerton 1980). Since BHs are among the most massive bodies in a star cluster (after their massive progenitors transform into them), they are very efficient in acquiring companions through dynamical exchanges.

Thus, exchanges are a crucial mechanism to form BH binaries dynamically. By means of direct N-body simulations, Ziosi *et al.* (2014) show that > 90 per cent of double BH binaries in young star clusters form by dynamical exchange.

Moreover, BH binaries formed via dynamical exchange will have some distinctive features with respect to field BH binaries:

- double BH binaries formed by exchanges will be (on average) more massive than isolated double BH binaries, because more massive intruders have higher chances to acquire companions;
- exchanges trigger the formation of highly eccentric double BH binaries (eccentricity is then significantly reduced by circularisation induced by GW emission, if the binary enters the regime where GW emission is effective);
- double BH binaries born by exchange will likely have misaligned spins, because exchanges tend to randomize the spins.

Spin misalignments are another possible feature to discriminate between field binaries and star cluster binaries (e.g. Farr *et al.* 2017a,b). Unfortunately, there is no robust theory to predict the magnitude of the spin of a BH given the spin of its parent star (Miller & Miller 2015). However, we can reasonably state that the orientation of the spin of a BH matches the orientation of the spin of its progenitor star, if the latter evolved in isolation and directly collapsed to a BH.

Thus, we expect that an isolated binary in which the secondary becomes a BH by direct collapse results in a double BH binary with aligned spins (i.e. the spins of the two BHs have the same orientation, which is approximately the same as the orbital angular momentum direction of the binary), because tidal evolution and mass transfer in a binary tend to synchronise the spins (Hurley *et al.* 2002). On the other hand, if the secondary undergoes a SN explosion, the natal kick may reshuffle spins.

For dynamically formed BH binaries (through exchange) we expect misaligned, or even nearly isotropic spins, because any original spin alignment is completely reset by three-body encounters.

4.4. *Hardening*

If a double BH binary undergoes a number of three-body encounters during its life, we expect that its semi-major axis will shrink as an effect of the encounters. This process is called dynamical *hardening.*

Following Heggie (1975), we call hard binaries (soft binaries) those binaries with binding energy larger (smaller) than the average kinetic energy of a star in the star cluster. According to Heggie's law (1975), hard binaries tend to harden (i.e. to become more and more bound) via three-body encounters. In other words, a fraction of the internal energy of a hard binary can be transferred into kinetic energy of the intruders and of the centre-of-mass of the binary during three body encounters. This means that the binary loses internal energy and its semi-major axis shrinks.

Most double BH binaries are expected to be hard binaries, because BHs are among the most massive bodies in star clusters. Thus, double BH binaries are expected to harden as a consequence of three-body encounters. The hardening process may be sufficiently effective to shrink a BH binary till it enters the regime where GW emission is efficient: a BH binary which is initially too loose to merge may then become a GW source thanks to dynamical hardening.

It is even possible to make a simple analytic estimate of the evolution of the semi-major axis of a double BH binary which is affected by three-body encounters and by GW emission (equation 9 of Colpi *et al.* 2003):

$$\frac{da}{dt} = -2\,\pi\,\zeta\,\frac{G\,\rho}{\sigma}\,a^2 - \frac{64}{5}\,\frac{G^3\,m_1\,m_2\,(m_1+m_2)}{c^5\,(1-e^2)^{7/2}}\,a^{-3}, \tag{4.2}$$

where $\zeta \sim 0.2 - 1$ is a dimensionless hardening parameter (which has been estimated through numerical experiments, Hills 1983), ρ is the local mass density of stars, σ is the local velocity dispersion, c is the light speed, e is the eccentricity. The first part of the right-hand term of equation 4.2 accounts for the effect of three-body hardening on the semi-major axis. It scales as $da/dt \propto -a^2$, indicating that the wider the binary is, the more effective the hardening. This can be easily understood considering that the geometric cross section for three body interactions with a binary scales as a^2.

The second part of the right-hand term of equation 4.2 accounts for energy loss by GW emission. It is the first-order approximation of the calculation by (Peters 1964). It scales as $da/dt \propto -a^{-3}$ indicating that GW emission becomes efficient only when the two BHs are very close to each other.

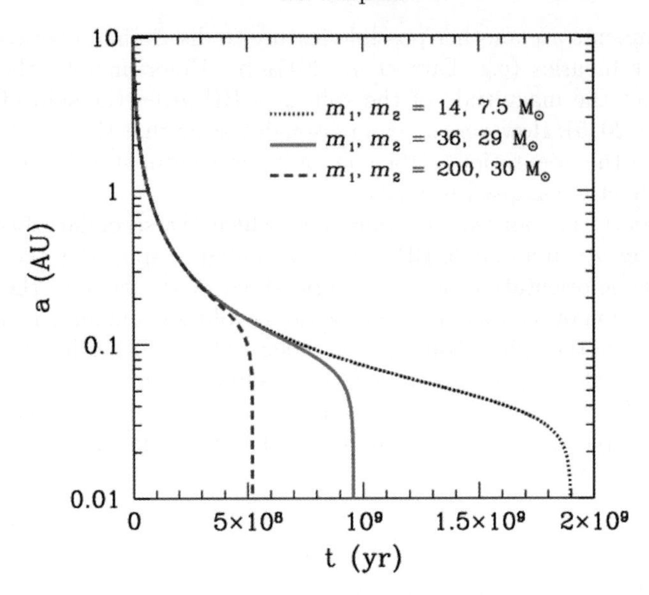

Figure 5. Time evolution of the semi-major axis of three BH binaries estimated from equation 4.2. Blue dashed line: BH binary with masses $m_1 = 200$ M$_\odot$, $m_2 = 30$ M$_\odot$; red solid line: $m_1 = 36$ M$_\odot$, $m_2 = 29$ M$_\odot$; black dotted line: $m_1 = 14$ M$_\odot$, $m_2 = 7.5$ M$_\odot$. For all BH binaries: $\xi = 1$, $\rho = 10^5$ M$_\odot$ pc^{-3}, $\sigma = 10$ km s^{-1}, $e = 0$ (here we assume that ρ, σ and e do not change during the evolution), initial semi-major axis of the BH binary $a_{\rm i} = 10$ AU.

In Figure 5 we solve equation 4.2 numerically for three double BH binaries with different mass. All binaries evolve through (i) a first phase in which hardening by three body encounters dominates the evolution of the binary, (ii) a second phase in which the semi-major axis stalls because three-body encounters become less efficient as the semi-major axis shrink, but the binary is still too wide for GW emission to become efficient, and (iii) a third phase in which the semi-major axis drops because the binary enters the regime where GW emission is efficient.

4.5. *Formation of intermediate-mass black holes by runaway collisions*

In Section 2.3, we have mentioned that intermediate-mass black holes (IMBHs, i.e. BHs with mass $100 \lesssim m_{\rm BH} \lesssim 10^4$ M$_\odot$) might form from the direct collapse of metal-poor extremely massive stars (Spera & Mapelli 2017). Other formation channels have been proposed for IMBHs and most of them involve dynamics of star clusters. The formation of massive BHs by runaway collisions has been originally proposed about half a century ago (Colgate 1967; Sanders 1970) and was then elaborated by several authors (e.g. Portegies Zwart *et al.* 1999; Portegies Zwart & McMillan 2002; Portegies Zwart *et al.* 2004, Gürkan *et al.* 2006; Freitag *et al.* 2006; Mapelli *et al.* 2006, 2008; Giersz *et al.* 2015; Mapelli 2016).

The basic idea is the following (as summarized by the cartoon in Figure 6). In a dense star cluster, dynamical friction (Chandrasekhar 1943) makes massive stars to decelerate because of the drag exerted by lighter bodies, on a timescale which can be expressed as

$$t_{\rm DF}(M) = \frac{\langle m \rangle}{M}\, t_{\rm rlx}, \qquad (4.3)$$

where $\langle m \rangle$ is the average star mass in a star cluster (for young star clusters $\langle m \rangle \sim 1$ M$_\odot$), M is the mass of the massive star we consider ($M > \langle m \rangle$) and $t_{\rm rlx}$ is the central two body relaxation timescale of the star cluster (i.e. the timescale needed for a star to

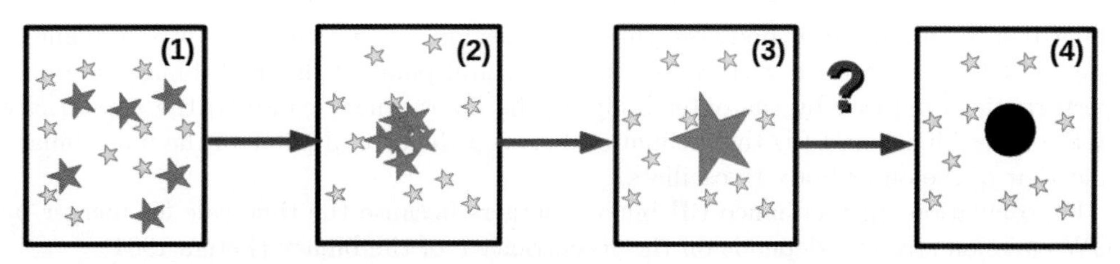

Figure 6. Cartoon of the runaway collision scenario in dense young star clusters (see e.g. Portegies Zwart & McMillan 2002). From left to right: (1) the massive stars (red big stars) and the low-mass stars (blue small stars) follow the same initial spatial distribution; (2) dynamical friction leads the massive stars to sink to the core of the cluster, where they start colliding between each other; (3) a very massive star ($\gg 100$ M$_\odot$) forms as a consequence of the runaway collisions; (4) this massive star might be able to directly collapse into a BH.

completely lose memory of its initial velocity as an effect of two-body encounters, Spitzer *et al.* 1971). For dense young star clusters

$$t_{\rm rlx} \simeq 20\,{\rm Myr}\,\left(\frac{M_{\rm cl}}{10^4{\rm M}_\odot}\right)^{1/2}\,\left(\frac{R_{\rm cl}}{1\,{\rm pc}}\right)^{3/2}\,\left(\frac{{\rm M}_\odot}{\langle m\rangle}\right), \tag{4.4}$$

where $M_{\rm cl}$ is the total star cluster mass and $R_{\rm cl}$ is the virial radius of the star cluster (Portegies Zwart *et al.* 2010).

Thus, for a star with mass $M \geq 25$ M$_\odot$, we estimate $t_{\rm DF} \leq 1$ Myr: dynamical friction is very effective in dense massive young star clusters. Because of dynamical friction, massive stars segregate to the core of the cluster, which is the centre of the cluster potential well.

If the most massive stars in a dense young star cluster sink to the centre of the cluster by dynamical friction on a time shorter than their lifetime (i.e. before core-collapse SNe take place, removing a large fraction of their mass), then the density of massive stars in the cluster core becomes extremely high. This makes collisions between massive stars extremely likely. Actually, direct N-body simulations show that collisions between massive stars proceed in a runaway sense, leading to the formation of a very massive ($\gg 100$ M$_\odot$) star (Portegies Zwart & McMillan 2002). The main open question is: "What is the final mass of the collision product? Is the collision product going to collapse to an IMBH?".

There are essentially two critical issues: (i) how much mass is lost during the collisions? (ii) how much mass does the very-massive star lose by stellar winds?

Hydrodynamical simulations of colliding stars (Gaburov *et al.* 2008, 2010) show that massive star can lose ≈ 25 per cent of their mass during collisions. Even if we optimistically assume that no mass is lost during and immediately after the collision (when the collision product relaxes to a new equilibrium), the resulting very massive star will be strongly radiation pressure dominated and is expected to lose a significant fraction of its mass by stellar winds. Recent studies including the effect of the Eddington factor on mass loss (e.g. Mapelli 2016) show that IMBHs cannot form from runaway collisions at solar metallicity. At lower metallicity ($Z \lesssim 0.1$ Z$_\odot$) approximately $10-30$ per cent of runaway collision products in young dense star clusters can become IMBHs by direct collapse (they also avoid being disrupted by pair-instability SNe).

Other possible formation channels of IMBHs include the repeated merger scenario (Miller & Hamilton 2002; Giersz *et al.* 2015) and the formation by gas drag in galactic nuclei (Miller & Davies 2012; McKernan *et al.* 2012, 2014; Stone *et al.* 2017).

4.6. *Kozai-Lidov resonance*

Unlike the other dynamical processes discussed so far, Kozai-Lidov (KL) resonance (Kozai 1962; Lidov 1962) can occur both in the field and in star clusters. KL resonance appears whenever we have a stable hierarchical triple system (i.e. a triple composed of an inner binary and an outer body orbiting the inner binary), in which the orbital plane of the outer body is inclined with respect to the orbital plane of the inner binary. Periodic perturbations induced by the outer body on the inner binary cause (i) the eccentricity of the inner binary and (ii) the inclination between the orbital plane of the inner binary and that of the outer body to oscillate.

KL oscillations may enhance BH binary mergers because the timescale for merger by GW emission strongly depends on the eccentricity e of the binary (Peters 1964).

It might seem that hierarchical triples are rather exotic systems. This is not the case. In fact, ~ 10 per cent of low-mass stars are in triple systems (Tokovinin *et al.* 2008; Raghavan *et al.* 2010). This fraction gradually increases for more massive stars, up to ~ 50 per cent for B-type stars (Sana *et al.* 2014; Moe & Di Stefano 2016).

The main signature of the merger of a KL system is the non-zero eccentricity until very few seconds before the merger. Eccentricity might be significantly non-zero even when the system enters the LIGO-Virgo frequency range (e.g. Toonen *et al.* 2017; Antonini *et al.* 2017).

4.7. *Summary of dynamics*

In this section, we have seen that dynamics is a crucial ingredient to understand BH demography. Dynamical interactions (three and few body close encounters) can favour the coalescence of BH binaries through dynamical hardening. New BH binaries can form via dynamical exchanges. Both processes suggest a boost of the BH binary merger rate in a dynamically active environment.

Moreover, exchanges favour the formation of more massive binaries, with higher initial eccentricity and with misaligned spins. Also, KL resonances favour the coalescence of more massive binaries and with higher eccentricity, even close to the last stable orbit. On the other hand, three-body encounters might trigger the ejection of compact-object binaries from their natal environment, inducing a significant displacement between the birth place of the binary and the location of its merger. Finally, dynamics can lead to the formation of IMBHs, with mass of few hundreds solar masses.

The right-hand panel of Figure 4 summarizes one of the possible evolutionary pathways of merging BHBs which originate from dynamics (the variety of this formation channel is too large to account for all dynamical channels mentioned above in a single cartoon). As in the isolated binary case, we start from a stellar binary. In the dynamical scenario, it is not important that this binary evolves through Roche lobe or CE (although this may happen). After the primary has turned into a BH, the binary undergoes a dynamical exchange: the secondary is replaced by a massive BH and a new BHB forms. The new binary is not ejected from the star cluster and undergoes further three-body encounters. As an effect of these three-body encounters the binary hardens enough to enter the regime in which GW emission is efficient: the BHB merges by GW decay.

5. Summary and outlook

We reviewed our current understanding of the astrophysics of stellar-mass BHs. The era of gravitational wave astrophysics has just begun and has already produced two formidable results: BH binaries exist and can host BHs with mass > 30 $M_\odot$ (Abbott *et al.* 2016a,d, 2018a).

According to nowadays stellar evolution and supernova theories, such massive BHs can form only from massive relatively metal-poor stars. At low-metallicity, stellar winds are quenched and stars end their life with a larger mass than their metal-rich analogues. If its final mass and its final core mass are sufficiently large, a star can directly collapse to a BH with mass $\gtrsim 30$ $M_\odot$ (Mapelli *et al.* 2009; Belczynski *et al.* 2010). An alternative scenario predicts that $\sim 30 - 40$ $M_{\odot}$ BHs are the result of gravitational instabilities in the very early Universe (primordial BHs, e.g. Carr *et al.* 2016).

The formation channels of merging BH binaries are still an open question. All proposed scenarios have several drawbacks and uncertainties. While mass transfer and common envelope are a major issue in the isolated binary evolution scenario, even the dynamical evolution is still effected by major issues (e.g. the small statistics about BHs in young star clusters, and the major simplifications adopted in dynamical simulations).

Finally, a global picture is missing, which combines stellar and binary evolution with dynamics and cosmology, aimed at reconstructing the BH merger history across cosmic time. This is crucial for the astrophysical interpretation of LIGO-Virgo data and for meeting the challenge of third-generation ground-based GW detectors.

Acknowledgements

MM acknowledges financial support by the European Research Council for the ERC Consolidator grant DEMOBLACK, under contract no. 770017. This review is based on the material of the lecture I held for the Course 200 on "Gravitational Waves and Cosmology" of the International School of Physics "Enrico Fermi". This work benefited from support by the International Space Science Institute (ISSI), Bern, Switzerland, through its International Team programme ref. no. 393 *The Evolution of Rich Stellar Populations & BH Binaries* (2017-18).

References

Abbott, B. P., Abbott, R., Abbott, T. D., *et al.* 2016, Physical Review Letters, 116, 061102
Abbott, B. P., Abbott, R., Abbott, T. D., *et al.* 2016, Physical Review Letters, 116, 241103
Abbott, B. P., Abbott, R., Abbott, T. D., *et al.* 2017, Physical Review Letters, 118, 221101
Abbott, B. P., Abbott, R., Abbott, T. D., *et al.* 2017, Astrophysical Journal Letter, 851, L35
Abbott, B. P., Abbott, R., Abbott, T. D., *et al.* 2017, Physical Review Letters, 119, 141101
Abbott, B. P., Abbott, R., Abbott, T. D., *et al.* 2016, Physical Review X, 6, 041015
The LIGO Scientific Collaboration, & the Virgo Collaboration 2018, arXiv:1811.12907
The LIGO Scientific Collaboration, & The Virgo Collaboration 2018, arXiv:1811.12940
Aasi, J., Abadie, J., Abbott, B. P., *et al.* 2015, Classical and Quantum Gravity, 32, 115012
Acernese, F., Agathos, M., Agatsuma, K., *et al.* 2015, Classical and Quantum Gravity, 32, 024001
Abbott, B. P., Abbott, R., Abbott, T. D., *et al.* 2016, Astrophysical Journal Letter, 818, L22
Tutukov, A., & Yungelson, L. 1973, Nauchnye Informatsii, 27, 70
Thorne, K. S. 1987, Three Hundred Years of Gravitation, 330
Schutz, B. F. 1989, Classical and Quantum Gravity, 6, 1761
Kulkarni, S. R., Hut, P., & McMillan, S. 1993, Nature, 364, 421
Sigurdsson, S., & Hernquist, L. 1993, Nature, 364, 423
Bethe, H. A., & Brown, G. E. 1998, Astrophysical Journal, 506, 780
Portegies Zwart, S. F., & McMillan, S. L. W. 2000, Astrophysical Journal Letter, 528, L17
Colpi, M., Mapelli, M., & Possenti, A. 2003, Astrophysical Journal, 599, 1260
Belczynski, K., Sadowski, A., & Rasio, F. A. 2004, Astrophysical Journal, 611, 1068
Mapelli, M., Colpi, M., & Zampieri, L. 2009, MNRAS, 395, L71
Mapelli, M., Ripamonti, E., Zampieri, L., Colpi, M., & Bressan, A. 2010, MNRAS, 408, 234
Belczynski, K., Bulik, T., Fryer, C. L., *et al.* 2010, Astrophysical Journal, 714, 1217
Fryer, C. L., Belczynski, K., Wiktorowicz, G., *et al.* 2012, Astrophysical Journal, 749, 91
Mapelli, M., Zampieri, L., Ripamonti, E., & Bressan, A. 2013, MNRAS, 429, 2298
Ziosi, B. M., Mapelli, M., Branchesi, M., & Tormen, G. 2014, MNRAS, 441, 3703
Spera, M., Mapelli, M., & Bressan, A. 2015, MNRAS, 451, 4086
Bird, S., Cholis, I., Muñoz, J. B., *et al.* 2016, Physical Review Letters, 116, 201301
Carr, B., Kühnel, F., & Sandstad, M. 2016, Physical Review Documents, 94, 083504

Inomata, K., Kawasaki, M., Mukaida, K., Tada, Y., & Yanagida, T. T. 2017, Physical Review Documents, 95, 123510
van Loon, J. T., Cioni, M.-R. L., Zijlstra, A. A., & Loup, C. 2005, Astronomy & Astrophysics, 438, 273
Orosz, J. A. 2003, A Massive Star Odyssey: From Main Sequence to Supernova, 212, 365
Özel, F., Psaltis, D., Narayan, R., & McClintock, J. E. 2010, Astrophysical Journal, 725, 1918
Laycock, S. G. T., Maccarone, T. J., & Christodoulou, D. M. 2015, MNRAS, 452, L31
Fryer, C. L. 1999, Astrophysical Journal, 522, 413
Fryer, C. L., & Kalogera, V. 2001, Astrophysical Journal, 554, 548
Abbott, D. C. 1982, Astrophysical Journal, 259, 282
Kudritzki, R. P., Pauldrach, A., & Puls, J. 1987, Astronomy & Astrophysics, 173, 293
Leitherer, C., Robert, C., & Drissen, L. 1992, Astrophysical Journal, 401, 596
Vink, J. S., de Koter, A., & Lamers, H. J. G. L. M. 2001, Astronomy & Astrophysics, 369, 574
Kudritzki, R.-P., & Puls, J. 2000, Annual Review of Astronomy and Astrophysics, 38, 613
Vink, J. S., & de Koter, A. 2005, Astronomy & Astrophysics, 442, 587
Hainich, R., Pasemann, D., Todt, H., *et al.* 2015, Astronomy & Astrophysics, 581, A21
Gräfener, G., & Hamann, W.-R. 2008, Astronomy & Astrophysics, 482, 945
Gräfener, G., Vink, J. S., de Koter, A., & Langer, N. 2011, Astronomy & Astrophysics, 535, A56
Vink, J. S., Muijres, L. E., Anthonisse, B., *et al.* 2011, Astronomy & Astrophysics, 531, A132
Chen, Y., Bressan, A., Girardi, L., *et al.* 2015, MNRAS, 452, 1068
Janka, H.-T. 2012, Annual Review of Nuclear and Particle Science, 62, 407
Foglizzo, T., Kazeroni, R., Guilet, J., *et al.* 2015, Publications of the Astronomical Society of Australia, 32, e009
Ertl, T., Janka, H.-T., Woosley, S. E., Sukhbold, T., & Ugliano, M. 2016, Astrophysical Journal, 818, 124
Müller, B., Janka, H.-T., & Marek, A. 2012, Astrophysical Journal, 756, 84
Müller, B., Janka, H.-T., & Heger, A. 2012, Astrophysical Journal, 761, 72
O'Connor, E., & Ott, C. D. 2011, Astrophysical Journal, 730, 70
Ugliano, M., Janka, H.-T., Marek, A., & Arcones, A. 2012, Astrophysical Journal, 757, 69
Horiuchi, S., Nakamura, K., Takiwaki, T., Kotake, K., & Tanaka, M. 2014, MNRAS, 445, L99
Sukhbold, T., & Woosley, S. E. 2014, Astrophysical Journal, 783, 10
Sukhbold, T., Ertl, T., Woosley, S. E., Brown, J. M., & Janka, H.-T. 2016, Astrophysical Journal, 821, 38
Fowler, W. A., & Hoyle, F. 1964, Astrophysical Journal Supplement, 9, 201
Barkat, Z., Rakavy, G., & Sack, N. 1967, Physical Review Letters, 18, 379
Rakavy, G., & Shaviv, G. 1967, Astrophysical Journal, 148, 803
Woosley, S. E. 2017, Astrophysical Journal, 836, 244
Heger, A., & Woosley, S. E. 2002, Astrophysical Journal, 567, 532
Heger, A., Woosley, S. E., Fryer, C. L., & Langer, N. 2003, From Twilight to Highlight: The Physics of Supernovae, 3
Spera, M., & Mapelli, M. 2017, MNRAS, 470, 4739
Fishbach, M., & Holz, D. E. 2017, Astrophysical Journal Letter, 851, L25
Talbot, C., & Thrane, E. 2018, Astrophysical Journal, 856, 173
Wysocki, D., Lange, J., & O'Shaughnessy, R. 2018, arXiv:1805.06442
Giacobbo, N., & Mapelli, M. 2019, MNRAS, 482, 2234
Portegies Zwart, S. F., & Verbunt, F. 1996, Astronomy & Astrophysics, 309, 179
Hurley, J. R., Tout, C. A., & Pols, O. R. 2002, MNRAS, 329, 897
Podsiadlowski, P., Rappaport, S., & Han, Z. 2003, MNRAS, 341, 385
Belczynski, K., Kalogera, V., Rasio, F. A., *et al.* 2008, Astrophysical Journal Supplement, 174, 223
Mennekens, N., & Vanbeveren, D. 2014, Astronomy & Astrophysics, 564, A134
Eldridge, J. J., & Stanway, E. R. 2016, MNRAS, 462, 3302
Mapelli, M., Giacobbo, N., Ripamonti, E., & Spera, M. 2017, MNRAS, 472, 2422

Stevenson, S., Vigna-Gómez, A., Mandel, I., *et al.* 2017, Nature Communications, 8, 14906
Giacobbo, N., Mapelli, M., Ripamonti, E., Spera, M. 2018, MNRAS, 474, 2959
Barrett, J. W., Gaebel, S. M., Neijssel, C. J., *et al.* 2018, MNRAS, 477, 4685
Giacobbo, N., Mapelli, M. 2018, MNRAS, 480, 2011
Mapelli, M., Giacobbo, N. 2018, MNRAS, 479, 4391
Kruckow, M. U., Tauris, T. M., Langer, N., Kramer, M., & Izzard, R. G. 2018, arXiv:1801.05433
Eldridge, J. J., Stanway, E. R., & Tang, P. N. 2018, arXiv:1807.07659
Spera, M., Mapelli, M., Giacobbo, N., *et al.* 2018, arXiv:1809.04605
Eggleton, P. P. 1983, Astrophysical Journal, 268, 368
Esposito, P., Israel, G. L., Milisavljevic, D., *et al.* 2015, MNRAS, 452, 1112
Webbink, R. F. 1984, Astrophysical Journal, 277, 355
Ivanova, N., Justham, S., Chen, X., *et al.* 2013, The Astronomy and Astrophysics Review, 21, 59
de Mink, S. E., & Mandel, I. 2016, MNRAS, 460, 3545
Mandel, I., & de Mink, S. E. 2016, MNRAS, 458, 2634
Marchant, P., Langer, N., Podsiadlowski, P., Tauris, T. M., & Moriya, T. J. 2016, Astronomy & Astrophysics, 588, A50
Hainich, R., Oskinova, L. M., Shenar, T., *et al.* 2018, Astronomy & Astrophysics, 609, A94
Belczynski, K., Holz, D. E., Bulik, T., & O'Shaughnessy, R. 2016, Nature, 534, 512
Lada, C. J., & Lada, E. A. 2003, Annual Review of Astronomy and Astrophysics, 41, 57
Weidner, C., & Kroupa, P. 2006, MNRAS, 365, 1333
Weidner, C., Kroupa, P., & Bonnell, I. A. D. 2010, MNRAS, 401, 275
Portegies Zwart, S. F., McMillan, S. L. W., & Gieles, M. 2010, Annual Review of Astronomy and Astrophysics, 48, 431
Mapelli, M., Colpi, M., Possenti, A., & Sigurdsson, S. 2005, MNRAS, 364, 1315
Downing, J. M. B., Benacquista, M. J., Giersz, M., & Spurzem, R. 2010, MNRAS, 407, 1946
Benacquista, M. J., & Downing, J. M. B. 2013, Living Reviews in Relativity, 16, 4
Samsing, J., MacLeod, M., & Ramirez-Ruiz, E. 2014, Astrophysical Journal, 784, 71
Rodriguez, C. L., Morscher, M., Pattabiraman, B., *et al.* 2015, Physical Review Letters, 115, 051101
Rodriguez, C. L., Chatterjee, S., & Rasio, F. A. 2016, Physical Review Documents, 93, 084029
Askar, A., Szkudlarek, M., Gondek-Rosińska, D., Giersz, M., & Bulik, T. 2017, MNRAS, 464, L36
Samsing, J., Askar, A., & Giersz, M. 2018, Astrophysical Journal, 855, 124
Harris, W. E., Harris, G. L. H., & Alessi, M. 2013, Astrophysical Journal, 772, 82
Mapelli, M. 2016, MNRAS, 459, 3432
Kimpson, T. O., Spera, M., Mapelli, M., & Ziosi, B. M. 2016, MNRAS, 463, 2443
Banerjee, S. 2017, MNRAS, 467, 524
Moeckel, N., & Bate, M. R. 2010, MNRAS, 404, 721
Parker, R. J., & Dale, J. E. 2013, MNRAS, 432, 986
Fujii, M. S., & Portegies Zwart, S. 2015, MNRAS, 449, 726
Mapelli, M. 2017, MNRAS, 467, 3255
Böker, T., Laine, S., van der Marel, R. P., *et al.* 2002, Astronomical Journal, 123, 1389
Ferrarese, L., Côté, P., Dalla Bontà, E., *et al.* 2006, Astrophysical Journal Letter, 644, L21
Graham, A. W., & Spitler, L. R. 2009, MNRAS, 397, 2148
Stone, N. C., Metzger, B. D., & Haiman, Z. 2017, MNRAS, 464, 946
Bartos, I., Kocsis, B., Haiman, Z., & Márka, S. 2017, Astrophysical Journal, 835, 165
McKernan, B., Ford, K. E. S., Bellovary, J., *et al.* 2017, arXiv:1702.07818
Hills, J. G., & Fullerton, L. W. 1980, Astronomical Journal, 85, 1281
Farr, W. M., Stevenson, S., Miller, M. C., *et al.* 2017, Nature, 548, 426
Farr, B., Holz, D. E., & Farr, W. M. 2018, Astrophysical Journal Letter, 854, L9
Miller, M. C., & Miller, J. M. 2015, Physics Reports, 548, 1
Heggie, D. C. 1975, MNRAS, 173, 729
Hills, J. G. 1983, Astronomical Journal, 88, 1269

Peters, P. C. 1964, Physical Review, 136, 1224
Colgate, S. A. 1967, Astrophysical Journal, 150, 163
Sanders, R. H. 1970, Astrophysical Journal, 162, 791
Portegies Zwart, S. F., Makino, J., McMillan, S. L. W., & Hut, P. 1999, Astronomy & Astrophysics, 348, 117
Portegies Zwart, S. F., & McMillan, S. L. W. 2002, Astrophysical Journal, 576, 899
Portegies Zwart, S. F., Baumgardt, H., Hut, P., Makino, J., & McMillan, S. L. W. 2004, Nature, 428, 724
Gürkan, M. A., Fregeau, J. M., & Rasio, F. A. 2006, Astrophysical Journal Letter, 640, L39
Freitag, M., Gürkan, M. A., & Rasio, F. A. 2006, MNRAS, 368, 141
Mapelli, M., Ferrara, A., & Rea, N. 2006, MNRAS, 368, 1340
Mapelli, M., Moore, B., Giordano, L., *et al.* 2008, MNRAS, 383, 230
Giersz, M., Leigh, N., Hypki, A., Lützgendorf, N., & Askar, A. 2015, MNRAS, 454, 3150
Chandrasekhar, S. 1943, Astrophysical Journal, 97, 255
Spitzer, L., Jr., & Hart, M. H. 1971, Astrophysical Journal, 164, 399
Gaburov, E., Lombardi, J. C., & Portegies Zwart, S. 2008, MNRAS, 383, L5
Gaburov, E., Lombardi, J. C., Jr., & Portegies Zwart, S. 2010, MNRAS, 402, 105
Miller, M. C., & Hamilton, D. P. 2002, MNRAS, 330, 232
Miller, M. C., & Davies, M. B. 2012, Astrophysical Journal, 755, 81
Stone, N. C., Küpper, A. H. W., & Ostriker, J. P. 2017, MNRAS, 467, 4180
McKernan, B., Ford, K. E. S., Lyra, W., & Perets, H. B. 2012, MNRAS, 425, 460
McKernan, B., Ford, K. E. S., Kocsis, B., Lyra, W., & Winter, L. M. 2014, MNRAS, 441, 900
Kozai, Y. 1962, Astronomical Journal, 67, 591
Lidov, M. L. 1962, Planetary and Space Science, 9, 719
Tokovinin, A. 2008, MNRAS, 389, 925
Raghavan, D., McAlister, H. A., Henry, T. J., *et al.* 2010, Astrophysical Journal Supplement, 190, 1
Sana, H., Le Bouquin, J.-B., Lacour, S., *et al.* 2014, Astrophysical Journal Supplement, 215, 15
Moe, M., & Di Stefano, R. 2017, Astrophysical Journal Supplement, 230, 15
Toonen, S., Hamers, A., & Portegies Zwart, S. 2016, Computational Astrophysics and Cosmology, 3, 6
Antonini, F., Toonen, S., & Hamers, A. S. 2017, Astrophysical Journal, 841, 77

High-mass X-ray Binaries: Illuminating the Passage from Massive Binaries to Merging Compact Objects
Proceedings IAU Symposium No. 346, 2019
L. M. Oskinova, E. Bozzo, T. Bulik & D. Gies, eds.

doi:10.1017/S1743921318008037

High mass X-ray binaries as progenitors of gravitational wave sources

Jakub Klencki[1] **and Gijs Nelemans**[1,2]

[1]Department of Astrophysics/IMAPP, Radboud University,
P O Box 9010, NL-6500 GL Nijmegen, The Netherlands
emails: j.klencki@astro.ru.nl, nelemans@astro.ru.nl

[2]Institute of Astronomy, KU Leuven,
Celestijnenlaan 200D, B-3001 Leuven, Belgium

Abstract. X-ray binaries with black hole (BH) accretors and massive star donors at short orbital periods of a few days can evolve into close binary BH (BBH) systems that merge within the Hubble time. From an observational point of view, upon the Roche-lobe overflow such systems will most likely appear as ultra-luminous X-ray sources (ULXs). To study this connection, we compute the mass transfer phase in systems with BH accretors and massive star donors ($M > 15\,M_{\odot}$) at various orbital separations and metallicities. In the case of core-hydrogen and core-helium burning donors (cases A and C of mass transfer) we find the typical duration of super-Eddington mass transfer of up to 10^6 and 10^5 yr, with rates of 10^{-6} and $10^{-5}\,M_{\odot}\,\mathrm{yr}^{-1}$, respectively. Given that roughly 0.5 ULXs are found per unit of star formation rate, we estimate the rate of BBH mergers from stable mass transfer evolution to be at most 10 $\mathrm{Gpc}^{-3}\,\mathrm{yr}^{-1}$.

Keywords. X-rays: binaries, (stars:) binaries (including multiple): close, stars: evolution

1. Introduction

The first discovery of a gravitational wave signal from a binary black hole (BBH) merger by the Advanced LIGO Interferometer in September 2015 (Abbott *et al.* 2016) revived the discussion on possible formation scenarios for double compact objects. A large number of channels have been put forth, especially in the case of BBH, including but not limited to the formation from isolated binaries through a common envelope event (eg. Belczynski *et al.* 2016, Eldridge & Stanway 2016, Klencki *et al.* 2018, Mapelli & Giacobbo 2018, Kruckow *et al.* 2018), or in a chemically homogenoues evolution regime (Mandel & de Mink 2016; de Mink & Mandel 2016; Marchant *et al.* 2016), dynamical formation in globular clusters (eg. Rodriguez *et al.* 2016; Askar *et al.* 2017) in nuclear clusters (Arca-Sedda & Gualandris 2018), or in disks of active galactic nuclei (Antonini & Rasio 2016; Stone *et al.* 2017), as well as formation channels involving triple stellar systems (Antonini *et al.* 2017). Given the difficulty of distinguishing the formation channels based on the gravitational wave information alone (although there is some hope connected to the measurement of the BH-BH spin-orbit misalignments, eg. Farr *et al.* 2017, 2018), as well as a lack of electromagnetic counterpart to BBH detected so far, the contribution of various channels to the entire population of gravitational wave sources is usually estimated on theoretical grounds.

In most theoretical scenarios, the estimates of the merger rate density of BBH fall within the range of about $\sim 0.1 - 20\ \mathrm{Gpc}^{-3}\,\mathrm{yr}^{-1}$. An exception is the case of common envelope (CE) evolution channel where the merger rate could possibly be as high as $\sim 100 - 200\ \mathrm{Gpc}^{-3}\,\mathrm{yr}^{-1}$ (eg. Belczynski *et al.* 2016). For comparison, the current observational limits on the merger rate density of BBHs inferred by the LIGO/Virgo are $13 - 212\ \mathrm{Gpc}^{-3}\ \mathrm{yr}^{-1}$ (Abbott *et al.* 2017).

For the moment, the CE evolution channel is a promising candidate for the origin of at least some of the BBH detected by the LIGO/Virgo. However, our understanding of the CE phase itself is still very limited (Ivanova 2011), and a number of known issues exist. In particular, recent studies of mass transfer stability from the giant donor stars (Woods & Ivanova 2011, Pavlovskii *et al.* 2017) reveal that the mass transfer remains stable for a larger parameter space than previously thought, thus avoiding a CE evolution in the majority of cases. Additionally, very massive stars have been shown to stay rather compact throughout their evolution and potentially never reach the large radii and outer convective envelope layers that are at the heart of a CE formation channel for BBH mergers.

Recently, van den Heuvel *et al.* (2017) pointed out the possibility of forming close BBH systems via stable and inconservative Roche-love overflow (RLOF) mass transfer from radiative giants onto stellar BHs in binaries with mass ratio $q = M_{\mathrm{donor}}/M_{\mathrm{accretor}} \simeq 3.0 - 3.5$, as in the case of the SS433 system (Hillwig & Gies 2008). They show that a few of the known Galactic double spectroscopic WR-O-type binaries (van der Hucht 2001) may be progenitors of such an evolutionary path. Based on the number of such systems and the expected duration of the WR phase, van den Heuvel *et al.* (2017) estimate a galactic merger rate of close BBH systems produced by stable mass transfer as $\simeq 0.5 \times 10^{-5}\ \mathrm{yr}^{-1}$ [i.e. one merger every 2×10^5 years]. For comparison, assuming a star formation rate (SFR) of $\sim 1 \mathrm{M}_{\odot}\mathrm{yr}^{-1}$ for the Milky Way, the CE channel typically produces one BBH merger every 5×10^6 yr in the case of Solar metallicity ($Z_{\odot} = 0.02$) and every 1×10^5 yr for $Z = 0.1\, Z_{\odot}$ (Klencki *et al.* 2018). This rough estimate signifies that the channel proposed by van den Heuvel *et al.* could be a competitive source of BBH mergers, especially at $Z \sim Z_{\odot}$ metallicites, which is the dominant metallicity of massive star forming galaxies in the nearby Universe.

The phase of a stable RLOF mass transfer in binaries with BH accretors and massive-star donors will inevitably be observable as a luminous high-mass X-ray binary (HMXB) phase, and most likely even as an ultra-luminous X-ray source (ULXs, Rappaport *et al.* 2005). Here we will discuss the channel proposed by Van den Heuvel from the point of view of the connection between a population of BBH systems and the population of ULXs.

2. Forming close binary black holes through stable mass transfer

Van den Heuvel et al. scenario. In order to form close BBH systems through stable mass transfer one requires RLOF to occur in BH binaries with giant donors at orbital periods of at most ~ 10 days and relatively high mass ratios $q = M_{\mathrm{donor}}/M_{\mathrm{accretor}} \sim 3$. (van den Heuvel *et al.* 2017). Such systems are natural outcomes of a futher evolution of Wolf-Rayet-O-type star binaries, a few tens of which are known in the Milky Way and Magellanic Clouds (van der Hucht 2001). The high values of q make it possible for a stable mass transfer to reduce the binary orbital period enough (to values of the order of ~ 1 day), so that the system is going to merge due to gravitational wave emission within the Hubble time. Alternatively, if the matter that is lost from the system during the inconservative mass transfer phase has high angular momentum (i.e. higher than the usually assumed orbital angular momentum in the proximity of the accretor) then smaller mass ratios could also lead to close enough BBH systems.

Stability of mass transfer from giant donors. In the first approximation, the stability of mass transfer in a binary depends mostly on the mass ratio of the components: $q = M_{\mathrm{donor}}/M_{\mathrm{accretor}}$. If the mass ratio is lower than some critical value, i.e. $q < q_{\mathrm{crit}}$ then the mass transfer will be stable, proceeding on a thermal or nuclear timescale of the donor star. If the mass ratio is too high ($q > q_{\mathrm{crit}}$) then at some point the Roche lobe of the donor star shrinks more quickly with mass loss then the donor radius. This can cause dynamical instability, leading to a very high mass transfer rate and ultimately to a CE evolution. In general, the value of $q_{\rm crit}$ depends on the exact structure of the donor star. For example, the most immediate adiabatic response of the donor envelope to mass loss is sensitive to the entropy profile (Ge *et al.* 2010; Ivanova 2015; Pavlovskii *et al.* 2017). In a simplified picture, this can be summarized as follows: in the case of donors with radiative envelope $q_{\rm crit} \approx 3.5$ (Ge *et al.* 2010), whereas in the case of donors with deep enough layers of outer convective envelope † $q_{\rm crit} \approx 1.5 - 2.2$ (Pavlovskii & Ivanova 2015).

To investigate what types of donor envelopes expected possible in binaries with BH accretors depending on the orbital period, we computed evolutionary models of single massive ($M > 15\,M_{\odot}$) stars at different metallicities using the MESA code (Paxton *et al.* 2011, 2015) until the end of core helium burning. While further evolution may result in additional mass transfer phases, they would not be long-lasting because there is not much time left until the core collapse at this point ($\sim 10^4$ yr). We model convection by using the mixing-length theory (Böhm-Vitense 1958) with a mixing-length parameter $\alpha = 1.5$, and we adopt the Ledoux criteria for convection. We account for semi-convection following (Langer *et al.* 1983) with an efficiency parameter $\alpha_{\rm SC} = 1.0$. We also account for overshooting by applying a step overshooting formalism with an overshooting length of 0.385 pressure scale heights. Stellar winds are modeled in a way described in (Brott *et al.* 2011).

In Fig. 1 we show, as a function of the donor mass (M_d) and the initial binary orbital period ($P_{\rm init}$), different cases of a possible mass transfer in a binary with a $10\,M_{\odot}$ BH accretor: from a donor star that is still during its main sequence (MS) evolution (case A), from a donor that is crossing the Hertzprung Gap (HG) during hydrogen shell burning (case B), and from a donor that is already burning helium in the core (case C). The red lines mark critical mass ratios whereas the black hatched area indicates donors with convective outer envelopes. Note that the vast majority of cases in Fig. 1 are donors with radiative envelopes. The chosen metallicity for this plot is 20% the Solar metallicity, which is a common metallicity of many ULX-host galaxies (Mapelli *et al.* 2010). At Solar metallicity we find that the parameter space for a case C mass transfer is very small due to $15 - 30\,M_{\odot}$ stars expanding all the way until the giant branch before the core-helium ignition (see also de Mink *et al.* 2008).

Additionally, we show the estimated parameters of BH-O-star binaries that are expected to be the further evolutionary stage of a sample of double-spectroscopic WR-O binaries with well known component masses (van der Hucht 2001). To do so, we make the same assumptions about the WR lifetime, mass-loss rate, and the BH formation as (van den Heuvel *et al.* 2017), see their Sect. 3.4.

<u>*Parameter space for merging BBH.*</u> For each pair M_d, $P_{\rm init}$ that falls into case B or case C mass transfer regimes in Fig. 1 we can estimate what would be the parameters of a resulting BBH system by making a few assumptions about the further evolution (again, similar to the assumptions made by van den Heuvel *et al.* 2017). In particular, we assume that the entire envelope mass of the donor is lost during the mass transfer (producing a BH-WR system), that the mass transfer is 100% non-conservative, and that the matter is expelled from the proximity of the accreting BH having its orbital specific angular momentum. We also account for WR winds and estimate a compact object formation (BH or NS) from the 'rapid' supernova engine prescription from Fryer *et al.* (2012). All BHs were assumed to form in direct collapse (i.e. no natal kick) with 10% of the collapsing core mass lost in neutrinos. Note that in some case when M_d is too small a NS is formed instead of a secondary BH.

The assumption that each donor loses its entire envelope mass can be justified and easily implemented in the case B and case C mass transfer systems for which most of the helium core has already be formed during the entire MS evolution, and the core-envelope boundary can be defined. Case A mass loss from a star, unless the RLOF occurs at the

† eg. at least 10% deep in mass coordinate

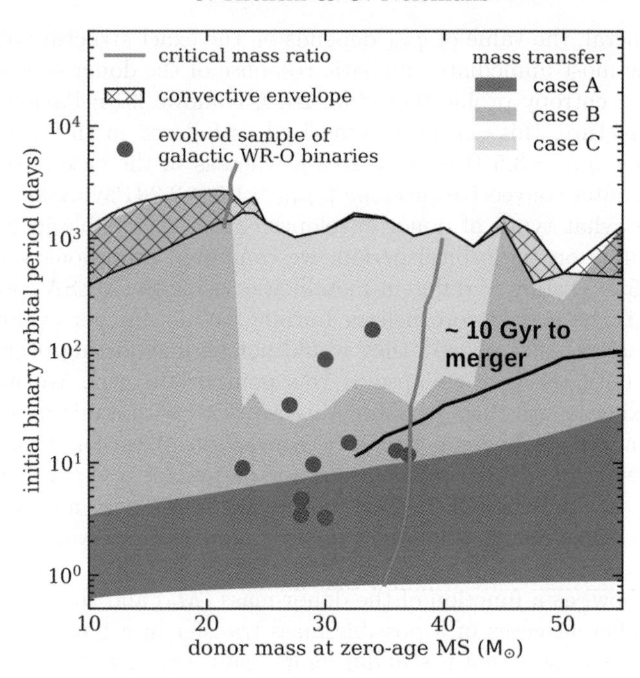

Figure 1. Different cases of mass transfer from giant donors in a binary with a $10\,M_\odot$ black hole accretor: case A, from a donor that is still a main-sequence star, case B, from a donor that is during the Hertzprung Gap evolution, and case C, from a donor that is already burning helium in the core. The mass transfer in systems to the left of the red lines is expected to be stable. Dark red points indicate estimated parameters of BH-O-star systems that are likely descendants of a sample of known spectroscopic WR-O binaries (see text for the details). The black line indicates an estimated upper initial orbital period limit on the formation of close binary black hole systems that merge in less than 10 Gyr.

very end of MS, can significantly reduce the final helium core mass of the donor with respect to a single star evolution case. Case A region in Fig. 1 thus requires detailed binary evolution modelling.

For every BBH formed from case B and case C mass transfer systems we then calculate what would be the delay time between the formation of a BBH and the merger due to gravitational wave emission. With the black line in Fig. 1 we show a threshold delay time value of ~ 10 Gyr; systems bellow the black line could merge within the Hubble time, while those above it would still be too wide after a stable mass transfer episode. The purpose of this rough estimate is to show, after (van den Heuvel *et al.* 2017), that it is possible to form BBH systems through stable mass transfer and a ULX phase with short enough orbital periods so that they will contribute to the BBH merger population. Notably, the larger the component mass ratio at RLOF the more likely it becomes that the final product will be a merging BBH. We wish to highlight at this point that the recent study of mass transfer stability of binaries with BH accretors and giant radiative donors (Pavlovskii *et al.*) suggests that the critical mass ratio in their case can be larger than the $q = 3.5$ value plotted in Fig. 1, being even as high as $\sim 6 - 8$ in some cases.

3. ULXs as progenitors of BBH mergers

From an observational point of view, binaries of compact object and massive donors that transfer mass through an accretion disc are high luminosity HMXBs. The mass transfer rates in the case of RLOF in BH binaries are most likely super-Eddington already from donors of a few Solar masses (Podsiadlowski *et al.* 2003, Rappaport *et al.* 2005), and even more so in the case of more massive donors $M > 15 - 20\,M_\odot$. Such systems are primary candidates for ultra-luminous X-ray (ULX) sources with $L_X > 10^{39}$ erg/s, both on the ground of theoretical models of accretion (Lipunova 1999; Poutanen *et al.* 2007; Lasota *et al.* 2016) and GRMHD simulations of super-critical accretion disks around BHs (Sądowski *et al.* 2014; McKinney *et al.* 2014; Sądowski & Narayan 2016). The rate of formation of BBH mergers through stable mass transfer is thus anchored to the number of ULXs systems observed. We can take advantage of this fact in order to put an upper limit on the local merger rate of BBHs from the Van den Heuvel scenario. If the number of ULXs per unit of SFR is $n_{\rm ULX}\,(M_\odot\,{\rm yr}^{-1})^{-1}$ and an average duration of the ULX phase is $t_{\rm ULX}$ then one ULX is formed per every $M_{\rm SF;ULX} = t_{\rm ULX}/n_{\rm ULX}\,(M_\odot)$ stellar mass formed. Assuming that a fraction $f_{\rm BBH}$ of ULXs will form close BBHs, and that the delay time distribution of this BBH population is $dN/dt_{\rm del}$, the local merger rate of BBHs $R_{\rm BBH;loc}\,({\rm yr}^{-1})$ that formed from ULXs can be expressed as:

$$R_{\rm BBH;loc} = \int_{t_{\rm del}=0}^{t_{\rm Hubble}} {\rm SFR}(z(t_{\rm del})) \times \frac{n_{\rm ULX}}{t_{\rm ULX}} \times f_{\rm BBH} \times \frac{dN}{dt_{\rm del}} dt_{\rm del} \tag{3.1}$$

where SFR(z) $M_\odot\,{\rm yr}^{-1}$ is the cosmic star formation rate, and $z(t_{\rm del})$ is the redshift corrsponding to a lookback time equal $t_{\rm del}$. The values of $n_{\rm ULX}$ and SFR(z) are determined observationally. We will now take a look at other terms in the above formula.

The delay time distribution of BBH mergers $dN/dt_{\rm del}$ in general depends on a particular formation scenario. However, if the distribution of the semi-major axes of the newly formed BH-BH binaries can be described by a power law $dN/da \approx a^{-\beta}$, then the distribution of the delay times is $dN/dt_{\rm del} \approx t_{\rm del}^{-\alpha}$ where $\alpha = (3+\beta)/4$ because the delay time is proportional to $t_{\rm del} \propto a^4$. One can see that even if β varies in an extreme range from 0 to 7 then α is between 0.75 and 2. Thus, for most astrophysical scenarios, α is constrained to $1 < \alpha < 2$.

It is beyond the scope of this paper to estimate the fraction of ULXs that produce merging BBH: $f_{\rm BBH}$. Hence, we treat it as a free parameter. In Fig. 1 one can see that only systems with large enough mass ratios and within the optimal orbital period range can end their evolution as close enough BBH systems. In particular, except a few promising candidates, most of the observed WR-O binaries are not expected to fall in the right parameter space. Additionally, it is known that a fraction of ULXs host NS accretors (Kaaret *et al.* 2017). BH X-ray binaries with donor masses of a few Solar masses can also contribute to the ULX population (Podsiadlowski *et al.* 2003; Rappaport *et al.* 2005), most likely the low-luminosity end of up to a few $\times 10^{39}$ erg s^{-1}. For all these reason we consider a fraction $f_{\rm BBH} = 0.1$ to be a conservative upper limit.

In X-ray binaries with stellar BH accretors, the average duration of ULX phase $t_{\rm ULX}$ is the duration of super-Eddington (or close to such) mass transfer. The Van den Heuvel scenario requires substantially large mass ratios $q \gtrsim 3$ at RLOF so that the orbital separation can shrink sufficiently. For such q values the mass transfer will be launched on the thermal timescale of the donor star (the Kelvin-Helmholz timescale, $t_{\rm KH}$). In the case of massive stars $t_{\rm KH} \approx 10^4$ yr or less. Naivly, one could take this value for the duration of the entire mass transfer episode (eg. as in discussion of Mineo *et al.* 2012).

In practice, the issue is more complicated. In addition to the component mass ratio, the mass transfer rate (and hence its duration) also depend on the response of the donor star to mass loss, which in turn depends on the donor structure in each particular case. Moreover, the component mass ratio will be changing over the course of the entire mass transfer, and even though the mass transfer may be proceeding on a thermal timescale in the beginning, it is possible that once the q becomes lower the mass transfer rate will slow down to a nuclear timescale.

In order constrain the average $t_{\rm ULX}$ more accurately, we have run a grid of binary evolution simulation models using the MESA code (with the same assumptions as described before). We have investigated cases of binaries with BH accretors ($M_{\rm BH} = 10$ or $20\,M_\odot$)

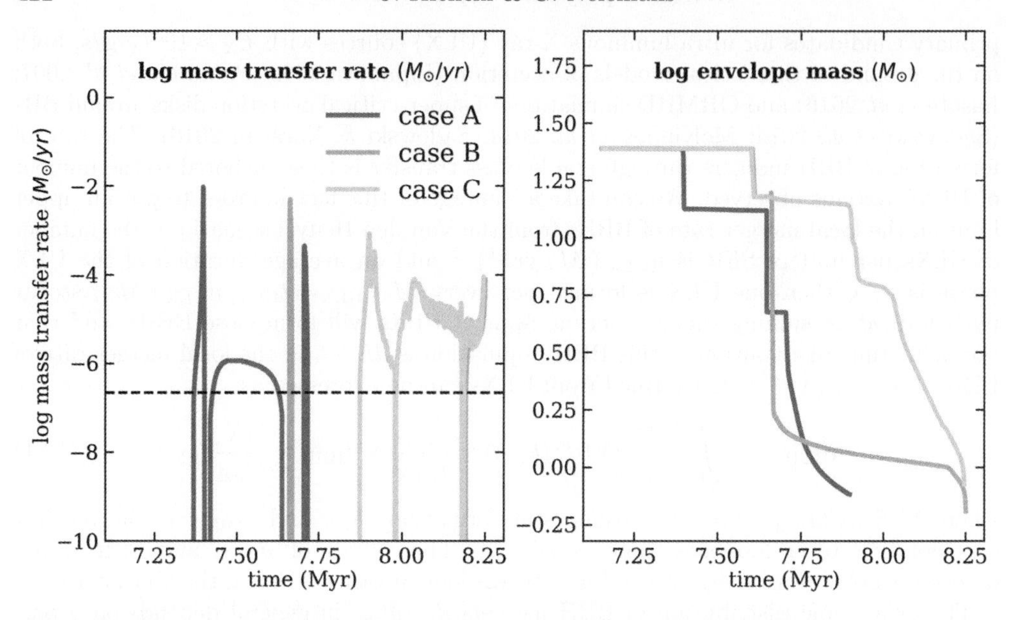

Figure 2. Examples of time-evolution of the mass transfer rate (left) and the donor envelope mass (right) computed with MESA for binaries of 10 $M_\odot$ BH accretors and 25 $M_\odot$ giant donors. Different colors indicate different initial binary separations (40, 50 and 150 $R_\odot$, corresponding to orbital periods of $\sim$ 5, 7, and 36 days, respectively), and hence different cases of mass transfer: case A, case B, and case C (see also Fig. 1). The horizontal dashed line in the left panel indicates Eddington mass transfer rate. Note that in case A and case C mass transfer the super-Eddington mass transfer continues for a duration of a few $\times 10^5$ yr, which is significantly longer than thermal timescale of the donor star.

and massive donors (M_d from 15 to 30 $M_\odot$ in 2.5 $M_\odot$ steps) for different initial binary separations (18 different values from $a = 20\,R_\odot$ to $2100\,R_\odot$).

A systematic study of the computed evolutionary tracks of BH binaries across different metallicites in underway (Klencki *et al.* in prep). Here, in Fig. 2 we present a representative example of how typically the mass transfer proceeds, depending mostly on the evolutionary stage of the donor star (i.e. the mass transfer case; see also Fig. 1). We plot three different instances of a time-evolution of the mass transfer rate and the donor envelope mass in binaries with 10 $M_\odot$ BH accretors and 25 $M_\odot$ donor stars at three different initial separations: 40, 50 and 150 $R_\odot$, which correspond to case A, case B, and case A mass transfer types respectively. Note that in all the three cases the donor has a radiative envelope at the point of mass transfer.

One can see that in all the three cases the initial mass transfer phase is very rapid, proceeding approximately on a thermal timescale of the donor star: mass transfer rates of the order of $10^{-3} \div 10^{-2}\,M_\odot\,\mathrm{yr}^{-1}$, for a duration of $0.5 \div 2 \times 10^4$ yr. This is the result of a relatively high mass ratio $q \approx 2.5$ at the moment of RLOF.

In cases A and C the mass transfer slows down when the component mass ratio drops to about $q \sim 1.4$. Afterwards, there is enough mass left in the donor envelope to power another longer phase of nuclear timescale mass transfer that is also super-Eddington. Because the nuclear timescale of helium core burning is ~ 10 times shorter than that of hydrogen core burning, the nuclear mass transfer rate in case C is about ~ 10 times higher: $10^{-5}\,M_\odot\,\mathrm{yr}^{-1}$ in case C compared to $10^{-6}\,M_\odot\,\mathrm{yr}^{-1}$ in case A. In case B mass transfer, on the other hand, this rapid phase continues for as long as the core collapses (roughly until the core helium ignition) and causes the propagation of the hydrogen burning shell through the star that is pushing the outer stellar layers even more outwards.

Once this phase is completed there is so little mass left in the envelope that no further mass transfer occurs. Only in some cases of a very late case B mass transfer is this different, and similar to a case C evolution. Notably, in case A and case C the secondary mass transfer episodes can last for a few $\times 10^5$ yr up to a few $\times 10^6$ yr if the RLOF occurs early during the MS evolution, which is much longer than the initial rapid thermal mass transfer phase. In the case of super-critical accretion discs the X-ray luminosity is a slow logarithmic function of the mass transfer rate $L_X \propto L_{\rm Edd}(1+\alpha\,\dot{m})$ where $\dot{m} = \dot{M}/M_{\rm Edd}$ and α is of the order of unity (eg. $\alpha = 3/5$ in an advection dominated disk with winds, Poutanen *et al.* 2007). For that reason, even though the mass transfer rates are significantly higher during the short-lived thermal mass transfer phases, we are more likely to observe ULXs during the longer-lasting mass transfer episodes with $\dot{M}$ of the order of $10^{-6} \div 10^{-5}\, M_\odot\, {\rm yr}^{-1}$. Especially if the radiation is beamed and the beaming is proportional to $\propto \dot{m}^2$ as suggested by King & Lasota (2016; although it is possible that the beaming is limited to a factor of a few due to advection, see Lasota *et al.* (2016) and also GRMHD simulations of Sądowski & Narayan (2016)).

4. Upper limit on BBH merger rate from stable mass transfer

Following the reasoning from the begining of Section 3, we can find an upper limit on the merger rate of BBH systems that formed through stable mass transfer from the observed number of ULXs. ULX-oriented surveys have shown that, by averaging over different galaxy types and different metallicites, there is roughly 0.5 ULX observed per unit of star formation rate ($M_\odot\, {\rm yr}^{-1}$, Mapelli *et al.* 2010; Swartz *et al.* 2011; Walton *et al.* 2011). † For an average duration of the ULX phase $T_{\rm ULX}$(yr), this implies that 1 ULX source is formed per every $0.5 \times T_{\rm ULX}(M_\odot)$ stellar mass formed. Assuming that a fraction $f_{\rm BBH}$ of ULXs are progenitors of BBH systems that are going to merge within the Hubble time, and that a typical delay time from the formation of BBH until the merger is $t_{\rm del} = 1$ Gyr, we can integrate the formation rate of ULXs over the cosmic SFR (Madau & Dickinson 2014) to calculate the local (i.e. redshift z = 0) merger rate density of BBH formed from ULXs (equation 3.1):

$$R_{\rm BBH;loc} \approx 1 \times \frac{f_{\rm BBH}}{0.1} \times \frac{10^6\,{\rm yr}}{T_{\rm ULX}}\ {\rm Gpc}^{-3}\ {\rm yr}^{-1} \tag{4.1}$$

On the basis of our simulations of mass transfer in BH binaries, we expect the average duration of the ULX phase in BH HMXBs to be of the order of $T_{\rm ULX} \approx$ a few $\times 10^5$ yr. Assuming that $f_{\rm BBH} = 0.1$ of all the ULXs form BBH systems that merge within the Hubble time, we obtain an upper limit on the local merger rate $R_{\rm BBH;loc} \lesssim 10\ {\rm Gpc}^{-3}\ {\rm yr}^{-1}$. We note that this result does not depend significantly on the choice of a particular delay time distribution.

5. Summary

We have studied the mass transfer in binaries comprised of stellar BH accretors and massive star donors ($M > 15\, M_\odot$) that overflow their Roche-lobes by computing single and binary evolution models with the MESA code. In the case of donors with radiative envelopes, which we find to be the ones that dominate the parameter space of possible systems, the mass transfer is stable up to mass ratios $q = M_{\rm donor}/M_{\rm accretor}$ of at least $q \sim 3 - 3.5$. Such systems are also the most likely descendants of a sample of known O-WR binaries (see Fig. 1). For $q \gtrsim 2$ at the point of RLOF, the mass transfer launches on a thermal timescale of the donor star. In the case of massive star donors this implies mass transfer rates of $10^{-3} \div 10^{-2}\, M_\odot\, {\rm yr}^{-1}$. It is typically assumed that during such a thermal

† In fact, the X-ray luminosity function of ULX sources appears to be a natural continuation of the X-ray luminosity function of X-ray binaries, Mineo *et al.* 2012.

mass transfer the entire donor envelope is lost, and that this entire phase lasts for only about $10^3 \div 10^4$ yr (eg. Mineo *et al.* 2012). Our simulations confirm this simple picture in the case of Solar metallicity, at which the massive stars expand significantly on a thermal timescale during the Hertzprung gap evolution before igniting helium in the core. This expansion helps to sustain a significant mass loss rates of $10^{-3} \div 10^{-2}\, M_\odot\, \mathrm{yr}^{-1}$ until only a small envelope ($\sim 1\, M_\odot$) is left. However, in the case of lower metallicity stars (eg. $Z = 0.2\, Z_\odot$), thanks to steeper density gradients and higher temperatures in the stellar cores, the helium is ignited earlier in the evolution, and the thermal expansion towards the giant branch is slowed down to a nuclear timescale at a smaller radii (eg. $\sim 100 - 200\, R_\odot$) than in the case of Solar metallicity stars (eg. $\sim 1000\, R_\odot$). This allows for a mass transfer from slowly-expanding core-helium burning giants (which we refer to as case C mass transfer) in a significant number of cases (see Fig. 1, and also deMink *et al.* 2008). In such cases, the phase of a rapid thermal mass transfer lasts only until the component mass ratio drops to about $q = 1.4$. Afterwards, there is still enough mass left in the envelope to power a secondary longer-lasting mass transfer phase on a nuclear timescale (see Fig. 2). The typical duration of that phase is of the order of a few $\times 10^5$ yr, with mass transfer rates of about $10^{-5} \div 10^{-4}\, M_\odot\, \mathrm{yr}^{-1}$. In the case of a $10\, M_\odot$ BH accretor, for which the Eddington mass transfer rate is roughly $2.2 \times 10^{-7}\, M_\odot\, \mathrm{yr}^{-1}$, all these mass transfer rates are super-Eddington. Such systems are thus expected to be observed as ultra-luminous X-ray sources (if viewed from the right angle given that the emission can be beamed). Because in the case of super-critical accretion disks the X-ray luminosity scales slowly with the mass transfer rate $\dot{M}$ as $L \propto L_{\mathrm{Edd}}[1 + ln(\dot{M}/\dot{M}_{\mathrm{Edd}})]$ (eg. Lipunova 1999, Poutanen *et al.* 2007), we argue that longer-lasting mass transfer phases with $\dot{M} \sim 10^{-5}\, M_\odot\, \mathrm{yr}^{-1}$ are more likely to observed than the shorter duration episodes of thermal mass transfer $\dot{M} \sim 10^{-3}\, M_\odot\, \mathrm{yr}^{-1}$. The case of less massive donors ($M < 15\, M_\odot$) was studied previously by Podsiadlowski *et al.* (2003) and Rappaport *et al.* (2005), who found $\dot{M}$ of up to $10^{-6}\, M_\odot\, \mathrm{yr}^{-1}$, and durations of at least several Myr. Consequently, the average duration of the ULX phase in the case of binaries with BH accretors is most likely of the order of $T_{\mathrm{ULX}} \approx 10^6$ yr. This agrees with the estimated ages of nebulae around some of the ULXs being ~ 1 Myr (Pakull & Mirioni 2002, Abolmasov & Moiseev 2008).

As an immediate application, we use the estimated value of T_{ULX} to discuss an upper limit on the merger rate of BBH systems that can form through stable mass transfer evolution, as suggested by van den Heuvel *et al.* (2017). Given that there is roughly 0.5 ULXs observed per unit of star-formation rate, and assuming that 10% of all the ULXs form merging BBH systems through the van den Heuvel scenario, we estimate the local merger rate of BBH formed this way to be about $R_{\mathrm{BBH;loc}} \approx 1 \div 10\ \mathrm{Gpc}^{-3}\ \mathrm{yr}^{-1}$.

Acknowledgements

We thank Tomasz Bulik for a helpful discussion about the delay time distribution of double compact objects. The authors acknowledge support from the Netherlands Organisation for Scientific Research (NWO).

References

Abbott, B. P., Abbott, R., Abbott, T. D., *et al.* 2016, *Physical Review Letters*, 116, 061102
Abbott, B. P., Abbott, R., Abbott, T. D., *et al.* 2017, *Physical Review Letters*, 118, 221101
Abolmasov, P. & Moiseev, A. V. 2008, *Rev. Mexicana Astron. Astrofis.*, 44, 301
Antonini, F. & Rasio, F. A. 2016, *ApJ*, 831, 187
Antonini, F., Toonen, S., & Hamers, A. S. 2017, *ApJ*, 841, 77
Arca-Sedda, M. & Gualandris, A. 2018, *MNRAS*, 477, 4423

Askar, A., Szkudlarek, M., Gondek-Rosińska, D., Giersz, M., & Bulik, T. 2017, *MNRAS*, 464, L36
Belczynski, K., Holz, D. E., Bulik, T., O'Shaughnessy, R., *et al.*, 2016, *Nature*, 534, 512
Böhm-Vitense, E. 1958, *ZAp*, 46, 108
Brott, I., de Mink, S. E., Cantiello, M., *et al.* 2011, *A&A*, 530, A115
de Mink, S. E. & Mandel, I. 2016, *MNRAS*, 460, 3545
de Mink, S. E., Pols, O. R., & Yoon, S.-C. 2008, in American Institute of Physics Conference Series, Vol. 990, First Stars III, ed. B. W. O'Shea & A. Heger, 230–232
Eldridge, J. J. & Stanway, E., 2016, *MNRAS*, 462, 3302
Farr, B., Holz, D. E., & Farr, W. M. 2018, *ApJL*, 854, L9
Farr, W. M., Stevenson, S., Miller, M. C., *et al.* 2017, *Nature*, 548, 426
Fryer, C. L., Belczynski, K., Wiktorowicz, G., *et al.*, 2012, *ApJ*, 749, 91
Ge, H., Hjellming, M. S., Webbink, R. F., Chen, X., & Han, Z. 2010, *ApJ*, 717, 724
Hillwig, T. C. & Gies, D. R. 2008, *ApJL*, 676, L37
Ivanova, N. 2015, Binary Evolution: Roche Lobe Overflow and Blue Stragglers, ed. H. M. J. Boffin, G. Carraro, & G. Beccari, 179
Ivanova, N. 2011, in Astronomical Society of the Pacific Conference Series, Vol. 447, Evolution of Compact Binaries, ed. L. Schmidtobreick, M. R. Schreiber, & C. Tappert, 91
Kaaret, P., Feng, H., & Roberts T. P. 2017, *ARAA*, 55, 303
King, A. & Lasota, J.-P. 2016, *MNRAS*, 458, L10
Klencki, J., Moe, M., Gladysz, W., *et al.* 2018, *A&A*, 619, A77
Kruckow, M. U., Tauris, Th. M., Langer, N., *et al.*, 2018, *MNRAS*, 481, 1908
Langer, N., Fricke, K. J., & Sugimoto, D. 1983, *A&A*, 126, 207
Lasota, J.-P., Vieira, R. S. S., Sadowski, A., Narayan, R., & Abramowicz, M. A. 2016, *A&A*, 587, A13
Lipunova, G. V. 1999, Astronomy Letters, 25, 508
Madau, P. & Dickinson, M. 2014, *ARAA*, 52, 415
Mandel, I. & de Mink, S. E. 2016, *MNRAS*, 458, 2634
Mapelli, M., Ripamonti, E., Zampieri, L., Colpi, M., & Bressan, A. 2010, *MNRAS*, 408, 234
Mapelli, M., & Giacobbo, N. 2018, MNRAS, 479, 4391
Marchant, P., Langer, N., Podsiadlowski, P., Tauris, T. M., & Moriya, T. J. 2016, *A&A*, 588, A50
McKinney, J. C., Tchekhovskoy, A., Sadowski, A., & Narayan, R. 2014, *MNRAS*, 441, 3177
Mineo, S., Gilfanov, M., & Sunyaev, R. 2012, *MNRAS*, 419, 2095
Pakull, M. W. & Mirioni, L. 2002, ArXiv Astrophysics e-prints [astro-ph/0202488]
Pavlovskii, K. & Ivanova, N. 2015, *MNRAS*, 449, 4415
Pavlovskii, K., Ivanova, N., Belczynski, K., & Van, K. X. 2017, *MNRAS*, 465, 2092
Paxton, B., Bildsten, L., Dotter, A., *et al.* 2011, *ApJS*, 192, 3
Paxton, B., Marchant, P., Schwab, J., *et al.* 2015, *ApJS*, 220, 15
Podsiadlowski, P., Rappaport, S., & Han, Z. 2003, *MNRAS*, 341, 385
Poutanen, J., Lipunova, G., Fabrika, S., Butkevich, A. G., & Abolmasov, P. 2007, *MNRAS*, 377, 1187
Rappaport, S. A., Podsiadlowski, P., & Pfahl, E. 2005, *MNRAS*, 356, 401
Rodriguez, C. L., Haster, C.-J., Chatterjee, S., Kalogera, V., & Rasio, F. A. 2016, *ApJL*, 824, L8
Sądowski, A. & Narayan, R. 2016, *MNRAS*, 456, 3929
Sądowski, A., Narayan, R., McKinney, J. C., & Tchekhovskoy, A. 2014, *MNRAS*, 439, 503
Stone, N. C., Metzger, B. D., & Haiman, Z. 2017, *MNRAS*, 464, 946
Swartz, D. A., Soria, R., Tennant, A. F., & Yukita M. 2011, *ApJ*, 741, 49
van den Heuvel, E. P. J., Portegies Zwart, S. F., & de Mink, S. E. 2017, *MNRAS*, 471, 4256
van der Hucht, K. A. 2001, VizieR Online Data Catalog, 3215
Walton, D. J., Roberts, T. P., Mateos, S., & Heard V. 2011, *MNRAS*, 416, 1844
Woods, T. E. & Ivanova, N. 2011, *ApJL*, 739, L48

High-mass X-ray Binaries: Illuminating the Passage from Massive Binaries to Merging Compact Objects
Proceedings IAU Symposium No. 346, 2019
L. M. Oskinova, E. Bozzo, T. Bulik & D. Gies, eds.

doi:10.1017/S1743921318007469

The black hole spin in coalescing binary black holes and high-mass X-ray binaries

Y. Qin[1,2], T. Fragos[1], G. Meynet[1], P. Marchant[2], V. Kalogera[2], J. Andrews[3,4], M. Sørensen[1] and H. F. Song[5,6]

[1]Geneva Observatory, University of Geneva, CH-1290 Sauverny, Switzerland
email: ying.qin@unige.ch

[2]Center for Interdisciplinary Exploration and Research in Astrophysics (CIERA) and Department of Physics and Astrophysics, Northwestern University, Evanston, IL 60208

[3]Foundation for Research and Technology - Hellas, IESL, Voutes, 71110 Heraklion, Greece

[4]Physics Department & Institute of Theoretical & Computational Physics, University of Crete, 71003 Heraklion, Crete, Greece

[5]College of Physics, Guizhou University, Guiyang City, Guizhou Province, 550025, P.R. China

[6]Key Laboratory for the Structure and Evolution of Celestial Objects, Chinese Academy of Sciences, Kunming 650011

Abstract. The six LIGO detections of merging black holes (BHs) allowed to infer slow spin values for the two pre-merging BHs. The three cases where the spins of the BHs can be determined in high-mass X-ray binaries (HMXBs) show that those BHs have high spin values. We discuss here scenarios explaining these differences in spin properties in these two classes of object.

Keywords. binaries: close binary stars; X-rays: binaries; black hole; gravitational waves

1. Introduction

Astrophysical BH can be fully described by its mass M and angular momentum $\vec{J}$. The dimensionless BH spin parameter a_* is defined as follows:

$$a_* = cJ/GM^2, \tag{1.1}$$

where c is the speed of light in vacuum and G is the gravitational constant. With the detection of the first gravitational wave event (Abbott *et al.* 2016a) by the Advanced Laser Interferometer Gravitational-Wave Observatory (AdLIGO) (LIGO Scientific Collaboration *et al.* 2015), the existence of the massive stellar BHs has been observationally demonstrated and a new window has been opened to directly study their properties. To date, AdLIGO has already detected six gravitational wave events and one high-significance event (Abbott *et al.* 2016a,b,c, 2017a,b,c), which are unambiguously believed to originate from the merger of binary BHs (BBHs).

Before the discovery of the GW events, X-ray binaries have been considered to be an ideal environment to indirectly measure BH's properties (McClintock 2006; McClintock, Narayan, & Steiner 2014; Reynolds 2014; Casares & Jonker 2014; Miller & Miller 2015). In low-mass X-ray binaries, the measured BH spins a_* cover the whole range (from 0 to 1) and this can be well explained via accretion from its companion after the BHs' formation (Fragos & McClintock 2015). However, for BHs in HMXBs, the currently measured spins are extremely high ($a_* > 0.8$). The donor star in a HMXB has a relatively short lifetime. Hence the BH can not accrete enough material to spin itself up. So the alternative possibility that such BHs were born with a high natal spin is preferred.

The Case-A mass transfer (MT) channel (MT is occurring when the BH progenitor is still on the MS.) was for the first time proposed by Valsecchi *et al.* (2010) to explaining the formation of M33 X-7. However, it was assumed that during the MS, solid body rotation implies necessarily that differential rotation is not considered. So this will not provide a trustable or quantitative prediction on the BH spin.

The paper is organized as follows. In §2, we introduce main methods used in the stellar and binary evolution models. The results of two BH spins in coalescing BBHs are shown in §3. We then present the main results of the BH spin in HMXBs §4. The main results are summarized in §5.

2. The main methods in the stellar and binary evolution models

To investigate BH spins in coalescing BBHs we use the Modules for Experiments in Stellar Astrophysics (MESA) code version 8118 (Paxton *et al.* 2011, 2013, 2015, 2018). The initial mass fraction of helium is given with a linearly increasing from $\mathrm{Y} = 0.2447$ (Grevesse, Noels, & Sauval 1996) at $\mathrm{Z} = 0$ to $\mathrm{Y} = 0.28$ at $\mathrm{Z} = \mathrm{Z}_{\odot}$ (0.017 is taken as the solar metallicity $\mathrm{Z}_{\odot}$ in Asplund, Grevesse, Sauval, & Scott 2009). The implementation of the stellar winds is clearly described in Marchant *et al.* (2016).

We model convection by using the standard mixing-length theory (Böhm-Vitense 1958) with a mixing-length parameter $\alpha_{ov} = 1.5$ and apply the Schwarzschild criterion to treat the boundary of the convective zones, as well as a convective core overshooting parametrized with $\alpha_{ov} = 0.1$. The angular momentum transport and chemical mixing of material are treated as a diffusion process, which includes the Eddington-Sweet circulations, the Goldreich-Schubert-Fricke instability, and secular and dynamical shear mixing with an efficiency parameter $\mathrm{f_c} = 1/30$ (Chaboyer & Zahn 1992).

Our work of the BH spin in HMXB is in preparation and the latest MESA version 10398 is used instead of version 8118. The tidal coefficient E_2 for computing the synchronization timescale of the tides is taken from Eq. 9 in Qin *et al.* (2018). Furthermore, the implementation of the tides is considered to only have an impact on the layers of outer readiative zones instead of all layers inside the star.

3. The BH spin in coalescing binary BHs

Since the first GW event was discovered, the proposed double BH formation channels can be divided into two main categories: the binary evolution channel and the dynamical formation channel. If the binary system evolves initially at a wide separation (i.e., several thousand solar radii), it will go through the "CE" phase that is still poorly understood ("CE" binary evolution channel, Phinney 1991; Tutukov & Yungelson 1993; Belczynski, Holz, Bulik, & O'Shaughnessy 2016; Tutukov & Cherepashchuk 2017; van den Heuvel, Portegies Zwart, & de Mink 2017; Inayoshi, Hirai, Kinugawa, & Hotokezaka 2017). Alternatively, if the two stars are in a close orbit and at low metallicity, both components will evolve chemically homogeneously (the CHE channel, de Mink & Mandel 2016; Marchant *et al.* 2016; Mandel & de Mink 2016; Song *et al.* 2016). In contrast, in the dynamical formation channel, the two BHs are born in different places of globular clusters and are brought together via dynamical friction.

Based on the "CE" binary evolution channel, the systematic studies on the two BHs' spins have been investigated in Qin *et al.* (2018). The progenitor of the first-born BH evolves initially like a single star, expands to a supergiant phase and then loses the hydrogen envelope via Roche-lobe overflow MT and wind mass loss. During this process, the star loses most of its angular momentum and forms a BH (first-born BH) with a negligible ($a_* \lesssim 0.1$) spin.

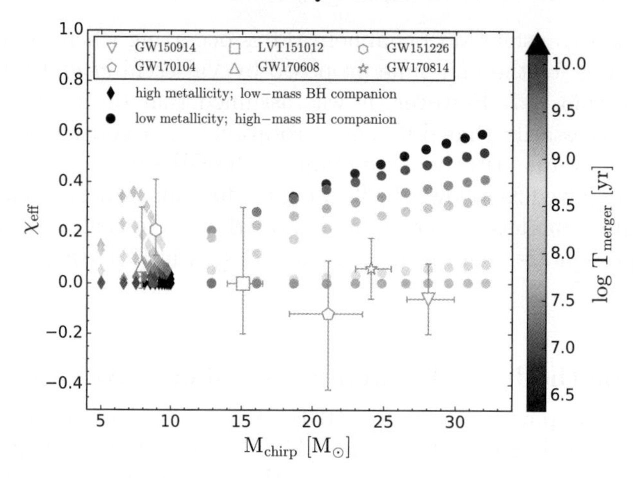

Figure 1. The time to merger (T_{merger}) as a function of the χ_{eff} and the M_{chirp}. The colored dots correspond to a low metallicity (0.01 $Z_{\odot}$, $Z_{\odot}$ is the solar metallicity) grid, and the colored diamonds refer to a high metallicity ($Z_{\odot}$). Various empty symbols represent the currently observed events with corresponding error bars.

The post-CE system consists of a helium star and a BH with an orbital period of a few days. Such binary system was systematically investigated by taking into account different initial parameters, i.e., masses of two binary components, initial orbital period, initial rotation of the helium star, and its metallicity. It was found that the dimensionless spin a_* of the second-born BH covers the whole range (form 0 to 1). After the formation of the second-born BH, the merger timescale T_{merger} due to the GW emission can be derived (Peters 1964). In Fig. 1, we see that lower χ_{eff} corresponds to a higher redshift. This is expected since to prevent the BH progenitors being accelerated by tides, the distance of the two binary components should be larger, hence a longer duration of the merger timescale T_{merger}. In order to form lower values of χ_{eff}, their corresponding T_{merger} should be longer, which means such systems must have been formed at a higher redshift. This is consistent with the current observation from AdLIGO. However, we predict that with the improvement of AdLIGO's sensitivity in the future, the events with higher χ_{eff} will be detected at a lower redshift. Furthermore, it is shown that more massive BHs (i.e. $\gtrsim$ 20 $M_{\odot}$) are not formed at a high redshit (i.e., solar metallicity). This is because the stars at a high metallicity loses more mass due to metallicity-dependent stellar winds and collapse to form less massive BHs.

Under the assumption of the direct core-collapse model, the BH progenitor directly forms a BH without any mass and angular momentum loss when it reaches the central carbon exhaustion.

4. The BH spin in high-mass X-ray binaries

A large fraction of all the massive binaries would go through the Case-A MT phase (Sana *et al.* 2012). The CHE is expected when the orbital period is shorter (shorter than about a few days) and the stars have a lower metallicity. Such a case was for the first time proposed by de Mink *et al.* (2009) in the binary evolution. In this part, we briefly introduce the main results on the study of BH HMXBs via the Case-A MT and the CHE channel. In Fig. 2, we show the detailed evolution processes of the BH progenitors' spins and orbital periods for the Case-A and the CHE. The binary system consisted of two stars (95.0 and 38.0 $M_{\odot}$ at 1/2 solar metallicity) goes through the Case-A MT channel when the initial orbital period is 3.25 days. The BH progenitor star speeds up during the MT phase, then decreases slowly and ends up with a fast-rotating BH at the end of

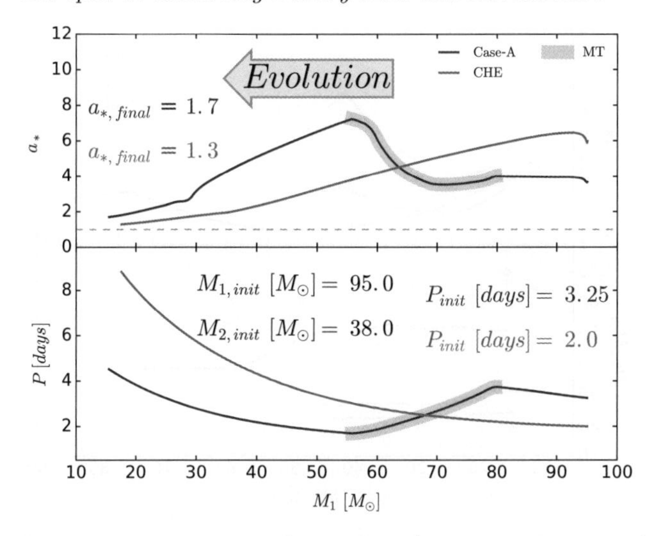

Figure 2. The spin of the resultant BH (upper panel) and orbital period (lower panel) as a function of the primary mass. Blue and red solid lines refer to the evolution of the Case-A MT and the CHE, respectively. The arrow "Evilution" represents the direction of each evolutional track and the MT phase is shown in green band. The horizontal dashed line represents the value of unity.

central carbon depletion. As shown in the lower panel of Fig. 2, the MT from the massive star (BH progenitor) onto its companion keeps the binary system tight. However, starting at the orbital period of 2.0 days, the system would go through the CHE instead of the Case-A MT channel. The BH progenitor continuously slows down and forms a BH with a_* around 1.0. Under this condition, the stellar winds mass-loss makes the orbit widening instead of shrinking. A fast-rotating BH for the two channels can be formed, while only shorter orbital period would be expected through the Case-A MT channel.

Here we highlight that the Tayler-Spruit dynamo (TS dynamo, Spruit 1999, 2002) plays a key role in forming a fast-rotating BH. For both channels, the BH progenitor keeps rotating fast during the MS. It does not evolve to a supergiant phase after the MS, but contracts after the helium surface abundance reaches a certain point. The angular momentum content of the BH progenitor will not be greatly changed during the period of this fast-shrinking phase. From this phase of fast shrinking, the evolution of the spin parameter is very different depending on the efficiency of the angular momentum transport mechanism. Only when a less efficient angular momentum transport mechanism than the one given by the TS dynamo is accounted for, a fast rotating BH is obtained. Otherwise, the BH will have a negligible spin. During that phase, tides are weak and thus have a small impact on the final spin of the BH.

We also created a big grid covering the initial parameter space of initial mass of the primary (from 20 - 100 $M_\odot$ with a step of 5 $M_\odot$), mass ratio q (from 0.25, 0.30, ..., 0.95) and orbital period (between 1 and 4 days with a step of 0.25 days, between 4 and 6 days with a step of 0.5 days). Here we only show a slice of our grid with mass ratio $q = 0.4$.

In Fig. 3, we present the best matches (blue track: M33 X-7, green track: Cygnus X-1 and red track: LMC X-1) with the current observations. Compared with the channel of the CHE, the Case-A MT channel results in a shorter orbit, which is consistent with current observations of the orbital periods. It is clearly shown in this figure, all the properties of LMC X-1 are well matched. For Cygnus X-1, the results are still acceptable and a better match will be expected with a higher resolution of the parameter space. In contrast, for M33 X-7, most the quantities are consistent with the observation, except

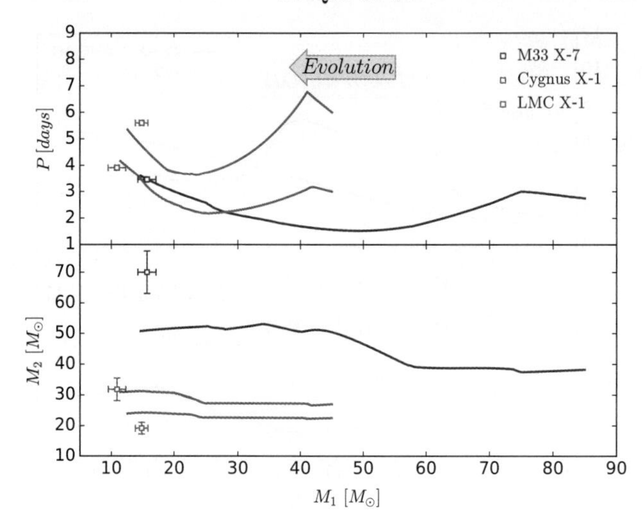

Figure 3. Orbital period (top panel) and secondary mass (bottom panel) as a function of primary mass. The properties of the observed systems are marked with blue, red and green squares for M33 X-7, Cygnus X-1 and LMC X-1, respectively. The arrow on the top panel shows the direction of the evolution.

the mass of the BH companion. This is because in our model, the companion star is being spun up due to the accreted material from the BH progenitor. When the companion star reaches its critical rotation, the MT becomes highly non-conservative. However, the MT efficiency is still uncertain. Had the MT been assumed to be conservative, as in Valsecchi *et al.* (2010), the mass of the BH companion could reach much higher values.

The nitrogen surface abundances of the BH companion stars between the Case-A MT and the CHE channel are significantly different. For the former channel, the mass is transferred from deep layers of the primary that have been reprocessed from the CNO cycle and hence this can largely enhance the nitrogen surface abundance of the secondary. In contrast, without the MT from the primary star of the CHE channel, no such high enhancement would be expected. Such results can be used for two purposes. First, the nitrogen surface abundance is a prediction that can be used to check the consistency of the models. Second, nitrogen surface abundance appears as a discriminating quantity, together with the orbital period, between the case-A MT and the CHE channel.

5. Conclusions

In this paper, we briefly describe our main results on the BH spin in two different BH binaries, namely, the coalescing BBHs and BH HMXBs. The BH in these two types of systems have very different BH spin measurements, which can be explained well by introducing different formation channels. For the "CE" binary evolution channel, the first-born BH has a negligible spin, while the second-born one covers the whole range of the spin (from 0 to 1). Besides, we expect the higher $\chi_{\rm eff}$ would be observed at lower redshits with the improvements of AdLIGO. On the other hand, with an assumption that the inefficient angular momentum transport is implemented after the MS of the BH progenitor, the currently observed BH spins in HMXBs can be well explained via the Case-A MT and the CHE channel. Compared to the CHE channel, the Case-A MT can form a HMXB in a tight orbit, which is consistent with the current observations. Hence The Case-A MT can be considered a potential channel to explain the current properties of Cygnus X-1, LMC X-1 and M33 X-7. Finally, we expect the nitrogen surface abundance of the BH companion star can be challenged from the observational point of view to distinguish the two channels.

Acknowledgements

This work was sponsored by the Chinese Scholarship Council (CSC). This project has received funding from the European Union's Horizon 2020 research and innovation programme under the Marie Sklodowska-Curie RISE action, grant agreement No 691164 (ASTROSTAT). TF is grateful for support from the SNSF Professorship grant (project number PP00P2_176868) the DNRF (Niels Bohr Professorship Program), the Carlsberg Foundation and the VILLUM FONDEN (project number 16599). The computations were performed at the University of Geneva on the Baobab computer cluster. All figures were made with the free Python module Matplotlib (Hunter 2007).

References

Abbott, B. P., Abbott, R., Abbott, T. D., *et al.* 2016, Physical Review Letters, 116, 061102
Abbott, B. P., Abbott, R., Abbott, T. D., *et al.* 2016, Physical Review Letters, 116, 241102
Abbott, B. P., Abbott, R., Abbott, T. D., *et al.* 2016, Physical Review X, 6, 041015
Abbott, B. P., Abbott, R., Abbott, T. D., *et al.* 2016, Physical Review Letters, 116, 241103
Abbott, B. P., Abbott, R., Abbott, T. D., *et al.* 2017, Physical Review Letters, 119, 161101
Abbott, B. P., Abbott, R., Abbott, T. D., *et al.* 2017, Physical Review Letters, 119, 141101
Abbott, B. P., Abbott, R., Abbott, T. D., *et al.* 2017, The Astrophysical Journal, 851, L35
Abbott, B. P., Abbott, R., Abbott, T. D., *et al.* 2017, Physical Review Letters, 118, 221101
Asplund, M., Grevesse, N., Sauval, A. J., & Scott, P. 2009, Annual Review of Astronomy and Astrophysics, 47, 481
Böhm-Vitense, E. 1958, Zeitschrift fur Astrophysik, 46, 108
Belczynski, K., Holz, D. E., Bulik, T., & O'Shaughnessy, R. 2016, Nature, 534, 512
Casares, J., & Jonker, P. G. 2014, Space Science Reviews, 183, 223
Chaboyer, B., & Zahn, J.-P. 1992, Astronomy and Astrophysics, 253, 173
de Mink, S. E., Cantiello, M., Langer, N., *et al.* 2009, Astronomy and Astrophysics, 497, 243
de Mink, S. E., & Mandel, I. 2016, Monthly Notices of the Royal Astronomical Society, 460, 3545
Fragos, T., & McClintock, J. E. 2015, The Astrophysical Journal, 800, 17
Grevesse, N., Noels, A., & Sauval, A. J. 1996, Cosmic Abundances, 99, 117
Hunter, J. D. 2007, Computing in Science and Engineering, 9, 90
Inayoshi, K., Hirai, R., Kinugawa, T., & Hotokezaka, K. 2017, Monthly Notices of the Royal Astronomical Society, 468, 5020
LIGO Scientific Collaboration, Aasi, J., Abbott, B. P., *et al.* 2015, Classical and Quantum Gravity, 32, 074001
Mandel, I., & de Mink, S. E. 2016, Monthly Notices of the Royal Astronomical Society, 458, 2634
Marchant, P., Langer, N., Podsiadlowski, P., Tauris, T. M., & Moriya, T. J. 2016, Astronomy and Astrophysics, 588, A50
McClintock, J. E. 2006, Bulletin of the American Astronomical Society, 38, 33.01
McClintock, J. E., Narayan, R., & Steiner, J. F. 2014, Space Science Reviews, 183, 295
Miller, M. C., & Miller, J. M. 2015, Physics Reports, 548, 1
Paxton, B., Bildsten, L., Dotter, A., *et al.* 2011, The Astrophysical Journal Supplement Series, 192, 3
Paxton, B., Cantiello, M., Arras, P., *et al.* 2013, The Astrophysical Journal Supplement Series, 208, 4
Paxton, B., Marchant, P., Schwab, J., *et al.* 2015, The Astrophysical Journal Supplement Series, 220, 15
Paxton, B., Schwab, J., Bauer, E. B., *et al.* 2018, The Astrophysical Journal Supplement Series, 234, 34

Peters, P. C. 1964, Physical Review, 136, 1224
Phinney, E. S. 1991, The Astrophysical Journal, 380, L17
Qin, Y., Fragos, T., Meynet, G., *et al.* 2018, Astronomy and Astrophysics, 616, A28
Remillard, R. A., & McClintock, J. E. 2006, Annual Review of Astronomy and Astrophysics, 44, 49
Reynolds, C. S. 2014, Space Science Reviews, 183, 277
Sana, H., de Mink, S. E., de Koter, A., *et al.* 2012, Science, 337, 444
Song, H. F., Meynet, G., Maeder, A., Ekström, S., & Eggenberger, P. 2016, Astronomy and Astrophysics, 585, A120
Spruit, H. C. 2002, Astronomy and Astrophysics, 381, 923
Spruit, H. C. 1999, Astronomy and Astrophysics, 349, 189
Tutukov, A. V., & Cherepashchuk, A. M. 2017, Astronomy Reports, 61, 833
Tutukov, A. V., & Yungelson, L. R. 1993, Monthly Notices of the Royal Astronomical Society, 260, 675
Valsecchi, F., Glebbeek, E., Farr, W. M., *et al.* 2010, Nature, 468, 77
van den Heuvel, E. P. J., Portegies Zwart, S. F., & de Mink, S. E. 2017, Monthly Notices of the Royal Astronomical Society, 471, 4256

High-mass X-ray Binaries: Illuminating the Passage from Massive Binaries to Merging Compact Objects
Proceedings IAU Symposium No. 346, 2019
L. M. Oskinova, E. Bozzo, T. Bulik & D. Gies, eds.

doi:10.1017/S1743921318007639

Local merger rates of double neutron stars

Martyna Chruslinska

Department of Astrophysics/IMAPP, Radboud University, P.O. Box 9010, NL-6500 GL Nijmegen, The Netherlands
email: `m.chruslinska@astro.ru.nl`

Abstract. The first detection of gravitational waves from a merging double neutron star (DNS) binary implies a much higher rate of DNS coalescences in the local Universe than typically estimated on theoretical grounds. The recent study by Chruslinska *et al.* (2018) shows that apart from being particularly sensitive to the common envelope treatment, DNS merger rates appear rather robust against variations of several factors probed in their study (e.g. conservativeness of the mass transfer, angular momentum loss, and natal kicks), unless extreme assumptions are made. Confrontation with the improving observational limits may allow to rule out some of the extreme models. To correctly compare model predictions with observational limits one has to account for the other factors that affect the rates. One of those factors relates to the assumed history of star formation and chemical evolution of the Universe and its impact on the final results needs to be better constrained.

Keywords. stars: neutron, stars: binaries, stars: evolution, gravitational waves

1. Introduction

When two stars composing a binary system complete their evolution they leave behind two compact remnants orbiting each other - so called double compact object (DCO). If the system evolves in isolation (dynamical interactions can be neglected), its subsequent orbital evolution is determined by the angular momentum loss due to gravitational wave emission. If the orbital separation is small enough, two stellar remnants merge within the Hubble time (Peters 1964). The current network of ground based gravitational wave detectors can observe the final stages of mergers of DCOs composed of neutron stars (NS) and black holes (BH) happening in the local Universe (Abbott *et al.* 2016). With the increasing number of detections and observing time those observations will allow to put progressively tighter constraints on the frequency of mergers of different types of DCOs occurring in the probed volume (local merger rate density R_{loc}). The first detection of gravitational waves from a merging double neutron star (DNS) (Abbott *et al.* 2017a) confirmed that those events are responsible for the production of at least some of the short gamma ray bursts (sGRB) and kilonovae (Abbott *et al.* 2017b). In principle, electromagnetic observations of these phenomena can be also used to set limits on R_{loc} for DNS. However, the weakly constrained sGRB jet opening angle and luminosity function (e.g. Petrillo *et al.* 2013) and small number of observations in case of kilonovae lead to substantial uncertainties and differences in the limits estimated by different groups (see Fig. 1). Moreover, a (unknown) fraction of those events may originate from NS-BH/BH-NS mergers (e.g. Berger 2014). Thus, gravitational waves provide the most direct measure of cosmological DNS merger rates.

One can estimate R_{loc} on theoretical grounds following the evolution of stars either in isolation (e.g. Tutukov & Yungelson 1993; Portegies Zwart & Yungelson 1998; Belczynski *et al.* 2002) or in dense environments (e.g. Portegies Zwart *et al.* 2004; Rodriguez *et al.* 2016; Askar *et al.* 2017), where dynamical interactions are important, and studying what

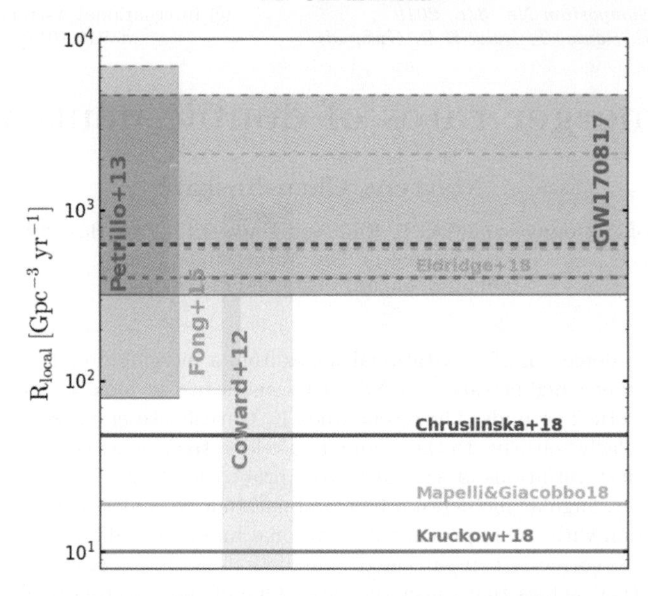

Figure 1. Shaded areas show observational limits on double neutron star (DNS) local merger rate density (R_{loc}) implied by the detection of GW170817 in gravitational waves (purple; Abbott *et al.* 2017a) and based on short gamma ray burst observations from Coward *et al.* 2012 (cyan), Petrillo *et al.* 2013 (green) and Fong *et al.* 2015 (yellow). Thick lines mark the recent population synthesis results obtained by different groups (blue - Chruslinska *et al.* 2018a, light green - Mapelli & Giacobbo 2018, gray - Kruckow *et al.* 2018, orange - Eldridge *et al.* 2018). The solid lines show results for models indicated by the authors as fiducial, while the dashed ones show the highest DNS R_{loc} found within each study.

fraction of them produces DCOs close enough to merge within the Hubble time. In this contribution we focus on the isolated evolution. The rates in this channel are usually calculated using the population synthesis method. In this approach one follows the evolution of many systems from the zero age main sequence (ZAMS) phase, when their parameters (mass of the primary star, mass ratio, separation, eccentricity) are randomly drawn from observation-based distributions (Sana *et al.* 2012; Moe & Di Stefano 2017). During the subsequent evolution the binary parameters change due to wind mass loss (e.g. Vink *et al.* 2001), interaction between the stars via mass transfer or tides, possible mass loss from the system during the formation of each of the compact objects and natal kick velocity gained by NS or BH due to asymmetries involved in the process of their formation (e.g. Gunn & Ostriker 1970; Hobbs *et al.* 2005; Fryer & Kushenko 2006; Janka 2017). Only a small fraction of the simulated binaries reach the interesting DCO stage (many systems merge during the unstable mass transfer phases or get disrupted during the formation of one of the compact objects) and have parameters allowing for their merger within the Hubble time (e.g. Chruslinska *et al.* 2018a). This fraction (and hence the merger rate) depends on the choice of a particular set of assumptions (parameters that define a model) used to describe the weakly constrained phases of binary evolution, e.g. conservativeness of the mass transfer, distribution describing the magnitude of NS and BH natal kicks (e.g. Mennekens & Vanbeveren 2014; Chruslinska *et al.* 2018a; Kruckow *et al.* 2018; Mapelli & Giacobbo 2018; Eldridge *et al.* 2018; Barrett *et al.* 2018).

Confrontation of the calculated rates with observational limits is one of the possible ways to constrain those models, hopefully allowing to rule out some of the extreme models and pointing us towards the correct understanding of the binary evolution. However, to be meaningful this comparison must take into account also other factors, besides those related directly to binary evolution, that affect the estimated R_{loc}. Those include the choice of distributions describing initial parameters of binary stars and formation and evolution of progenitor stars in the chemically evolving Universe. While the uncertainty caused by the former has been already studied by de Mink & Belczynski (2015) and Klencki *et al.* (2018), the impact of the latter still lacks discussion in the literature.

2. Binary evolution and DNS merger rates

The impact of binary evolution-related assumptions on merger rates of double neutron star systems has been studied by many groups over the years, with the use of different population synthesis codes (e.g. Tutukov & Yungelson 1993; Belczynski *et al.* 2002; Mennekens & Vanbeveren 2014; Chruslinska *et al.* 2018a; Kruckow *et al.* 2018; Mapelli & Giacobbo 2018; Eldridge *et al.* 2018; Barrett *et al.* 2018). The reported R_{loc} span more than an order of magnitude and fall between a few 10 Gpc^{-3} yr^{-1} to a few 100 Gpc^{-3} yr^{-1} when the fiducial models are considered. In this contribution we focus only on the most recent studies. In Figure 1 we show the most recent population synthesis results together with observational limits on DNS R_{loc} resulting from detection of gravitational waves from merging DNS (GW170817; Abbott *et al.* 2017a) and short gamma ray bursts observations. The short GRB limits come from Coward *et al.* 2012, Petrillo *et al.* 2013 and Fong *et al.* 2015. In case of Petrillo *et al.* 2013 the upper limit assumes the jet beaming angle of 10° and the lower limit assumes weakly collimated emission of 60° (see Figure 3 therein). The very conservative lower limit from Coward *et al.* (2012) assumes isotropic sGRB jet emission. As shown in Fig. 1, most of the population synthesis results are either below or fall on the lower side of the current observational limits, even despite the broad range of reported values. The wide range of results coming from those studies reflects our limited knowledge of the details of the evolution of massive stars in binaries. The typical evolutionary path for the formation of a merging DNS together with the major unknown factors in the evolution is summarized in Figure 2. Below we provide a short overview of the most important factors that affect the evolution towards a merging DNS. For more details see sec. 2 in Chruslinska *et al.* (2018a) and references therein.

The final DNS must end up on the very close orbit of a few solar radii to belong to the merging population (Peters 1964). Thus, its progenitor stars form on the orbit that ensures interaction via mass transfer phase(s) when one of the components over-fills its Roche lobe (RLOF; unless a forming NS receives a very favorably oriented natal kick velocity that efficiently decreases the orbital separation). The amount of material lost from the system during the mass transfer (conservativeness) and angular momentum carried away with the escaping mass are one of the unknowns in binary evolution (e.g. de Mink *et al.* 2007).

Furthermore, formation of DNS involves two core-collapse events in which NS form. These can be either iron-core collapse supernovae (CCSN), electron-capture supernovae (ECS; when the partially degenerate ONeMg core of mass $\sim 1.37 M_{\odot}$ collapses due to electron captures on Mg and Ne; e.g. Miyaji *et al.* 1980) or accretion induced collapse (AIC; e.g. Nomoto & Kondo 1991) of a massive accreting white dwarf to a NS. The exact conditions for the occurrence of different core collapse events, especially the boundary between ECS and CCSN are currently not known (e.g. Jones *et al.* 2016). Binary interaction is believed to broaden the range of masses of progenitor stars that can undergo ECS (e.g. Podsiadlowski *et al.* 2004; van den Heuvel 2007), which in case of single stars is very narrow (and hence their evolution through ECS is unlikely). Neutron stars forming due to electron-capture triggered collapse (either ECS or AIC) are believed to gain relatively small natal kick velocities ($\lesssim$50km/s; e.g. van den Heuvel 2007). Taking into account that the typical birth velocities of young single pulsars are of the order of a few 100 km/s (e.g. Hobbs *et al.* 2005), this formation path significantly increases the chance that a binary

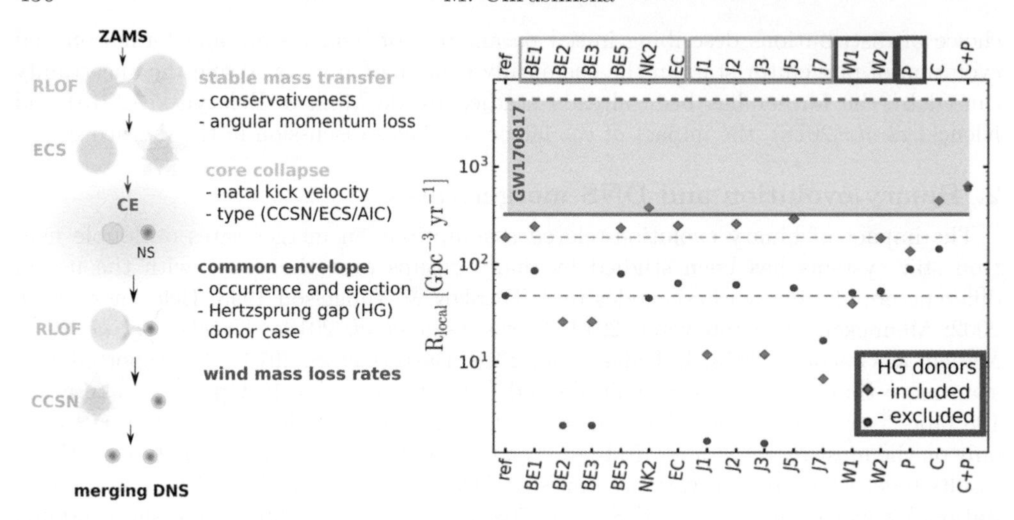

Figure 2. Left: schematic picture of the evolution leading to formation of a merging double neutron star (DNS). This path involves several stages of stable mass transfer (RLOF), two core-collapse events leading to formation of both NS (ECS-electron capture supernova, CCSN - core collapse supernova) and common envelope evolution (CE). The major evolution-related factors that affect R_{loc} calculations and need to be parametrized in population synthesis studies at each of these stages are listed in the middle.
Right: DNS R_{loc} for different population synthesis models. Colors on the top axis indicate which evolutionary phase was parametrized differently with respect to the reference (ref) model (orange - core collapse, green - mass transfer, brown - common envelope). Models C and $C+P$ involve combinations of several modifications favoring the formation of merging DNS. Each model comes in two variations depending on whether evolution through CE with HG type donors was allowed (red diamonds) or excluded, assuming to lead to merger (blue dots). The shaded purple region indicates observational limits implied by detection of GW170817. Modified figure from Chruslinska *et al.* (2018a). See the original paper for details.

will remain bound during the NS formation. It has been suggested that the magnitude of natal kick may depend on the amount of mass ejected during the supernova explosion (with smaller mass of ejecta leading to smaller birth velocity; e.g. Beniamini & Piran 2016; Bray & Eldridge 2016; Janka 2017; Bray & Eldridge 2018). Thus, NS progenitors that loose a large fraction of their envelope during mass transfer phases may potentially form with smaller natal kick velocities. Such severe envelope stripping can occur when the companion star is already a compact object, leading to ultra-stripped supernova phenomenon (e.g. Tauris *et al.* 2013; Tauris *et al.* 2015). However, the mechanism responsible for production of natal kicks is not well understood. If the role of neutrino asymmetries is dominant (e.g. (Fryer & Kushenko 2006, e.g.)), the suggested ejecta mass - natal kick correlation may turn out erroneous. The distribution of natal kicks is one of the key ingredients of population synthesis studies focusing on DNS. Different assumptions can significantly affect the size of the final population and hence the calculated merger rates. Finally, the classical formation channel of merging DNS requires at least one stage of unstable mass transfer (so-called common envelope; CE e.g. Ivanova *et al.* 2013). The radii of massive stars typically reach 100 - 1000 $R_{\odot}$, which is much larger than the required separation for the merging systems (a few solar radii for DNS binary). To explain the formation of such systems without dynamical interactions, one needs the mechanism capable of decreasing the orbital separation by even a few orders of magnitude. Common envelope is believed to provide such a mechanism. CE forms when the mass transfer rate is too high for the accretor to accrete all of the transferred material, giving rise to a short-lived phase during which both stars are immersed in a shared envelope. This causes a binary inspiral due to increased friction and, if the envelope is not ejected beforehand (e.g. at the expense of the orbital energy), may lead to its coalescence before the formation of a DNS. This evolutionary phase is particularly weakly constrained both by theory and observations (e.g. Ivanova *et al.* 2013). Neither the conditions for the onset of CE, nor for its ejection are presently known. Both are likely dependent on the structure and type of stars involved and require detailed modeling. The particularly controversial case arises when the donor star does not have a well defined core-envelope boundary, as for the stars during the Hertzsprung gap phase (referred to as HG type donors; e.g. Ivanova & Taam 2004). Such CE was believed to lead to binary merger, however recent studies by Pavlovskii & Ivanova (2015) and Pavlovskii *et al.* (2017) show that in some cases where based on earlier studies one would expect the common envelope initiated by a HG type donor to ensue, the mass transfer may in fact be stable.

2.1. *Recent population synthesis results*

Chruslinska *et al.* (2018a) revisited the topic of DNS merger rates in light of recent observational and theoretical progress in the study of the evolution of double compact objects. They conclude that apart from being particularly sensitive to the common envelope treatment and extreme assumptions about natal kicks, DNS R_{loc} appear rather robust against variations of several of the key factors probed in their study. Within 21 models calculated with the StarTrack population synthesis code (Belczynski *et al.* 2002; Belczynski *et al.* 2008) they identify only two variations leading to significant (a factor of ~10) increase in R_{loc} of NS-NS binaries with respect to the reference model. These are the models that combine several factors supporting the formation of double neutron star binaries, requiring simultaneous changes in the treatment of common envelope evolution, angular momentum loss, natal kicks and electron capture supernovae. The summary of R_{loc} for models considered in this study is shown in the right panel of Figure 2. Only three of the presented models fall above the LIGO/Virgo lower 90% confidence limit on DNS R_{loc} implied by the detection of GW170817 (Abbott *et al.* 2017a). At the same time the local BH-BH merger rate densities calculated for those three models are found to exceed the corresponding gravitational waves limits (Abbott *et al.* 2016). A possible solution to this discrepancy, as suggested by the authors, might be different CE evolution in case of the most massive stars (BH-progenitors) than in case of NS-progenitors or higher BH natal kicks than assumed in the simulations, both factors largely affecting the BH-BH merger rate calculations and currently weakly constrained. We return to this question in section 3.

A number of other groups published their results on DCO local merger rates during the last year. The highest NS-NS R_{loc} of ~600 Gpc^{-3} yr^{-1} quoted by Mapelli & Giacobbo (2018) (dashed green line in Fig. 1) was obtained assuming 5 times higher efficiency of CE ejection than in their reference model and additionally requiring that all NS form with low natal kicks (drawn from Maxwellian distribution with velocity dispersion σ=15 km/s). Note that these velocities are generally lower than in any of the models presented in Chruslinska *et al.* (2018). The highest R_{loc} found by the latter is also around 600 Gpc^{-3} yr^{-1}. Kruckow *et al.* (2018) find 'optimistic' DNS R_{loc} of ~400 Gpc^{-3} yr^{-1} also requiring low natal kicks at NS formation. R_{loc} of ~400 Gpc^{-3} yr^{-1} was also found by Eldridge *et al.* (2018) for their fiducial model. The highest NS-NS R_{loc} found by Eldridge *et al.* 2018) reaches up to ~2100Gpc^{-3} yr^{-1} and was obtained using their updated prescription for NS natal kicks (Bray & Eldridge 2018). In this prescription natal kick velocity scales with the amount of mass ejected during the supernova and mass of the stellar remnant (see discussion in Janka 2017), as opposed to the commonly used approach in which those velocities are randomly drawn from observationally motivated distribution (e.g. Hobbs *et al.* 2005). They find that for the earlier version of this natal kick model (Bray & Eldridge 2016) the estimated NSNS local rates reasonably agree with the results obtained for the corresponding model from Chruslinska *et al.* (2018a) that employed the prescription from Bray & Eldridge (2016).

All of the above values are consistent with the 90% confidence limits on DNS R_{loc} implied by the detection of GW170817 (1540^{+3200}_{-1220} Gpc^{-3} yr^{-1}). However, except for the highest value reported by Eldridge *et al.* (2018) for the model using the Bray & Eldridge (2018) natal kick prescription, they fall decidedly on the lower side of this estimate.

3. Other factors affecting calculation of the local merger rates

3.1. *Distributions of initial parameters of binaries*

Before drawing conclusions from the comparison with observations, one has take into account other factors, besides the evolution-related assumptions, that affect the population synthesis results. One of those factors relates to the initial conditions of the simulations that have the form of distributions describing parameters of binaries at their formation. de Mink & Belczynski (2015) studied how the properties of merging DCOs change when they use initial conditions based on study of massive spectroscopic binaries (Sana *et al.* 2012) instead of the previously used distributions of initial parameters. They report a slight increase (a factor of $\lesssim$2) in the DNS merger rates, mostly due to higher binary fraction (fraction of stars assumed to form in binaries). They argue that the uncertainty in the high-mass slope of the initial mass function (IMF) may affect the results by a factor of ~4, however, recent study by Klencki *et al.* (2018) shows that this effect is in fact less significant if the assumed star formation rate (SFR) and IMF are varied consistently. Klencki *et al.* (2018) implement the empirical inter-correlated distributions of initial binary parameters reported by Moe & Di Stefano (2017) based on their analysis of over twenty massive star surveys. The variations in the merger rates due to this change stays within a factor of ~2 when compared with the simulations using initial distributions from Sana *et al.* (2012). This transition generally decreases the merger rate estimates for DNS. This effect is the strongest at high metallicities.

3.2. *Chemical evolution of the Universe*

Metallicity itself is one of the crucial factors determining the evolution of stars. It affects for instance wind mass loss rates and stellar radii, also impacting the evolution of stars in binaries and the outcome of their evolution (e.g. Meader 1992; Hurley *et al.* 2000; Vink *et al.* 2001; Belczynski *et al.* 2010). In particular, the number of close double compact binaries of a certain type created per unit of mass formed in stars (DCO formation efficiency $\chi_{\rm DCO;i}$) is known to vary significantly depending on the composition of progenitor stars (e.g. Dominik *et al.* 2012; Eldridge & Stanway 2016; Stevenson *et al.* 2017; Klencki *et al.* 2018; Giacobbo *et al.* 2018).

A double compact object that merges in the local Universe (at time t_{mr}) forms at some earlier time $t_{\rm form}$=t_{mr}-t_{del}-t_{DCO}, where t_{DCO} is time needed to complete binary evolution up to the formation of a DCO and t_{del} time needed to decrease its separation due to gravitational wave emission to the point where two stellar remnants merge. Its progenitor binary forms with metallicity that is typical for its neighborhood at time t_{form} †. Since DCOs form with different parameters, they have a range of t_{del}. As a consequence, the local merger rate density is a summed contribution of merging DCOs that formed at different cosmic times and with different metallicities. Hence, R_{loc} depends

† The metallicity of the star-forming material evolves over time due to interplay between metal enrichment by evolving stars, feedback from supernovae and active galactic nuclei and possible inflows of metal-poor material.

on both the amount of star formation happening throughout the cosmic time (which sets the number of progenitor binaries formed at a given time) and on its distribution across different metallicities (SFR(t, Z); because the fraction of progenitor binaries that end up as merging DCOs of certain type depends on metallicity). Furthermore, the dependence of $\chi_{\mathrm{DCO;i}}$ on metallicity is different for different types of DCOs (e.g. Fig. 6 in Klencki *et al.* 2018; more (recent) SFR at low metallicities generally favors formation of merging BH-BH while discouraging the formation of merging DNS) and the exact form of this dependence is generally sensitive to the population synthesis assumptions (Chruslinska *et al.* 2018b). As a result, different assumptions about SFR(t, Z) affect the ratio of R_{loc} calculated for different types of DCOs and the uncertainty associated with this choice can vary between the models.

Different approaches have been taken to determine SFR(t, Z) used to calculate merger rate densities. One can extract this information from cosmological simulations (e.g. Mapelli *et al.* 2017; Schneider *et al.* 2017), or use the available observations and/or complement observational results with theoretical inferences (e.g. Dominik *et al.* 2013; Belczynski *et al.* 2016; Eldridge *et al.* 2018; Chruslinska *et al.* in prep.). All methods have their weaknesses. Observations are prone to biases and provide complete information only in limited ranges of redshifts and luminosities of the objects of interest. On the other hand, cosmological simulations do not fully account for all of the observational relations (e.g. mass - metallicity relation) and are resolution-limited. In any case, the use of incorrect SFR(t,Z) affects the resulting cosmological merger rates and may lead to erroneous conclusions. However, the importance of this choice was not evaluated in previous studies.

Following the approach described in Belczynski *et al.* (2016), Chruslinska *et al.* (2018a) used the cosmic star formation rate density from Madau & Dickinson (2014) and mean metallicity of the Universe as a function of redshift from the chemical evolution model proposed by these authors. This metallicity was increased by 0.5 dex and a Gaussian spread of σ=0.5 dex was added to the mean to construct a distribution describing contributions from different metallicities to SFR at different times. However, observations of star-forming galaxies in the local Universe suggest that the mean metallicity at which star formation takes place is likely higher than assumed by Belczynski *et al.* (2016). Massive galaxies (with stellar masses $\gtrsim 10^9 \mathrm{M}_\odot$) showing SFR even two orders of magnitude higher than in their dwarf counterparts, dominate the star formation budget in the Universe (e.g. Brinchmann *et al.* 2004; Lara-López *et al.* 2013; Boogaard *et al.* 2018). The star forming gas found in those galaxies has relatively high metal content, which is close to the solar value $\mathrm{Z}_\odot$, even if uncertainty in the absolute metallicity calibration is taken into account (e.g. Kewley & Ellison 2008). Thus, the amount of low-metallicity ($\lesssim 0.1\mathrm{Z}_\odot$) star formation assumed in Chruslinska *et al.* (2018a) may be overestimated (see also discussion in Klencki *et al.* 2018 and Chruslinska *et al.* 2018b). A more detailed study of the SFR(t,Z) resulting from currently available observations of star-forming galaxies and the associated uncertainties is underway.

To demonstrate the potential effect of the assumed SFR(t,Z) on their merger rate density calculations, we contrast the assumptions made by Belczynski *et al.* (2016) and Dominik *et al.* (2013). In Figure 3 we show the mean metallicity evolution with redshift used in both cases. In the method proposed by Dominik *et al.* (2013) the average metallicity of the star formation in the local Universe is $\sim \mathrm{Z}_\odot$ (in contrast to $\sim 0.3\mathrm{Z}_\odot$ in Belczynski *et al.* 2016) and the amount of mass formed in stars at metallicity $\lesssim 0.1\mathrm{Z}_\odot$ since redshift of 10 is ~2.6 times lower when compared with the method proposed by Belczynski *et al.* (2016). For the model with DCO formation efficiency dependence on metallicity as shown in Fig. 6 from Klencki *et al.* (2018) (identical to the reference model from Chruslinska *et al.* 2018a) this difference translates to a factor of ~ 1.5 increase of

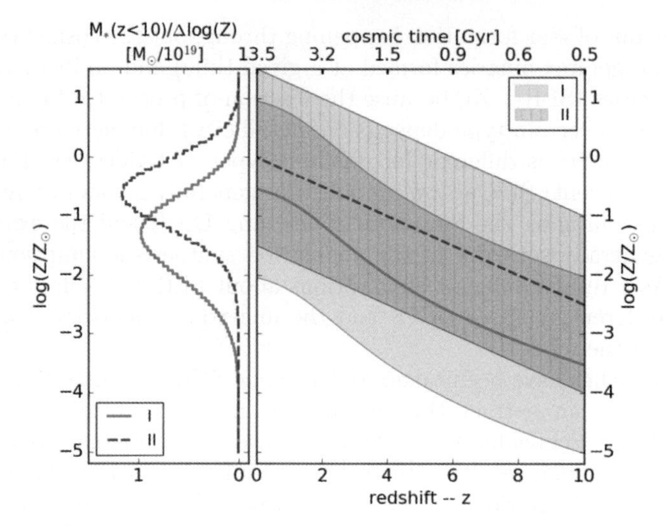

Figure 3. Right: comparison of redshift evolution of the mean metallicity as assumed by Belczynski *et al.* (2016) (I) and Dominik *et al.* (2013) (II). The shaded areas indicate 3σ spread limits around the mean.
Left: distribution of mass formed in stars since redshift z=10 at different metallicities for both versions of the metallicity evolution. The fraction of mass formed at low metallicity (< 0.1 $Z_\odot$) since redshift of 10 amounts to 71% in case I and 27% in case II. Modified figure from Chruslinska *et al.* (2018b).

R_{loc} for DNS and a factor of $\sim$4 decrease in R_{loc} for BH-BH. As shown by Chruslinska *et al.* (2018b) this difference is also sufficient to resolve the reported discrepancy between the gravitational waves limits on BH-BH R_{loc} and R_{loc} calculated for those systems in model CA from Chruslinska *et al.* (2018a) which led to DNS R_{loc} that satisfy current limits implied by the detection of GW170817. In this model several moderate modifications were introduced to their reference model, all found to favor the formation of merging DNS: Bray & Eldridge (2016) prescription for the natal kicks, reduced angular momentum loss during the mass transfer and wider limits on the helium core mass for the progenitors of stars undergoing electron-capture supernovae. Furthermore, the evolution through a common envelope phase initiated by a Hertzsprung gap donor was allowed. In this model the discrepancy can be resolved by the use of SFR(t,Z) closer to that expected based on observations of local star-forming galaxies, without the need for different evolution-related assumption in the BH- and NS-progenitors mass regime.

This highlights the importance of the choice of a particular way to distribute the cosmic star formation rate across metallicities and time and the need to better understand the uncertainties associated with that choice. Without tighter constraints on SFR(t,Z) one has to deal with another layer of degeneracy in the calculated merger rates, which makes drawing any strong conclusions from studies that aim to use cosmological rates as constraints impossible.

4. Conclusions

The first detection of gravitational waves from merging double neutron star binary implied local merger rate density (R_{loc}) of this type of systems that is much higher than most current theoretical estimates. While certain population synthesis models can produce rates consistent with the 90% confidence limits reported after this event, they fall on the low side of this estimate unless specific assumptions about natal kicks at the formation of NS are used. If the future observations support the rate close to the current most likely value reported by LIGO/Virgo, our understanding of the evolution of massive stars in binaries may need revision.

At the same time, confrontation of the theoretical results with observational limits requires knowledge of the uncertainties induced by other factors, besides those related to assumptions about the binary evolution that are accounted for in the population synthesis parameter studies.

One of those factors, necessary to calculate volumetric rates of any stellar evolution-related phenomena, relates to the assumed fraction of star formation happening at different metallicities throughout the cosmic time SFR(t,Z). Because of different dependence of the efficiency of formation of different types of merging double compact objects on metallicity the choice of SFR(t,Z) strongly affects the estimated ratio of merger rates of different types of binaries. Different assumptions about SFR(t,Z) were made in the literature, however the importance of those choices has not been discussed. The SFR(t,Z) used by Chruslinska *et al.* (2018a) likely overestimates the amount of low metallicity star formation. While a detailed study of the SFR(t,Z) resulting from observational properties of star-forming galaxies is underway, to demonstrate the possible impact of this assumption on the final results, we estimate the change in the calculated NS-NS and BH-BH rates caused by the use of SFR(t,Z) that assumes higher metallicity of the star formation in the local Universe. This change results in a factor of 1.5 increase in NS-NS R_{loc} and a factor of 4 decrease in BH-BH R_{loc} for a model with DCO formation efficiency - metallicity relation as in their reference model and is sufficient to affect some of the conclusions drawn from their study (Chruslinska *et al.* 2018b).

This highlights the importance of the choice of a particular way to distribute the cosmic star formation rate across metallicities and time and the need to better understand the uncertainties associated with that choice.

References

Abbott, B. P., Abbott, R., Abbott, T. D., Abernathy, M. R., Acernese, F., Ackley, K., Adams, C., Adams, T., Addesso, P., Adhikari, R. X. *et al.* 2016, *ApJ*, 832, L21

Abbott, B. P., Abbott, R., Abbott, T. D., Acernese, F., Ackley, K., Adams, C., Adams, T., Addesso, P., Adhikari, R. X., Adya, V. B. *et al.* 2017a, *Phys. Rev. Lett.*, 119, 161101

Abbott, B. P., Abbott, R., Abbott, T. D., Acernese, F., Ackley, K., Adams, C., Adams, T., Addesso, P., Adhikari, R. X., Adya, V. B. *et al.* 2017b, *Ap. Lett.*, 848, L12

Askar, A., Szkudlarek, M., Gondek-Rosinska, D., Giersz, M. & Bulik, T. 2017, *MNRAS*, 464, L36

Barrett, J. W., Gaebel, S. M., Neijssel, C. J., Vigna-Gmez, A., Stevenson, S., Berry, C. P. L., Farr, W. M. & Mandel, I. 2018, *MNRAS*, 477, 4685

Belczynski K., Kalogera V. & Bulik T. 2002, *ApJ* 572, 407

Belczynski K., Kalogera V., Rasio F. A., Taam R. E., Zezas A., Bulik T., Maccarone T. J. & Ivanova N. 2008, *ApJS*, 174, 223

Belczynski, K., Bulik, T., Fryer, C. L., Ruiter, A., Valsecchi, F., Vink, J. S. & Hurley, J. R. 2010, *ApJ*, 714, 1217

Belczynski, K., Holz, D. E., Bulik, T. & O'Shaughnessy, R. 2016, *Nature*, 534, 512

Beniamini P. & Piran T. 2016, *MNRAS*, 456, 4089

Berger, E. 2014, *ARA&A*, 52, 43

Bray J. C. & Eldridge J. J. 2016, *MNRAS*, 461, 3747

Bray J. C. & Eldridge J. J. 2018, *MNRAS*, 480, 5657

Brinchmann, J., Charlot, S., White, S. D. M., Tremonti, C., Kauffmann, G., Heckman, T. & Brinkmann, J. 2004, *MNRAS*, 351, 1151

Boogaard, L. A., Brinchmann, J., Bouché, N., Paalvast, M., Bacon, R., Bouwens, R. J., *et al.* 2018, *ArXiv e-prints*, arXiv:1808.04900

Chruslinska, M., Belczynski, K., Klencki, J., & Benacquista, M. 2018a, *MNRAS*, 474, 2937

Chruslinska, M., Nelemans, G. & Belczynski, K. 2018b, *ArXiv e-prints*, arXiv:1811.03565
Coward D. M., *et al.* 2012, *MNRAS*, 425, 2668
de Mink S. E., Pols O. R. & Hilditch R. W. 2007, *A&A*, 467, 1181
de Mink S. E. & Belczynski K. 2015, *ApJ*, 814, 58
Dominik M., Belczynski K., Fryer C., Holz D. E., Berti E., Bulik T., Mandel I. & OShaughnessy R. 2012, *ApJ*, 759, 52
Dominik M., Belczynski K., Fryer C., Holz D. E., Berti E., Bulik T., Mandel I. & OShaughnessy R. 2013, *ApJ*, 779, 72
Eldridge J. J. & Stanway E. R. 2016, *MNRAS*, 462, 3302
Eldridge J. J., Stanway E. R. & Tang P. N. 2018, *ArXiv e-prints*, arXiv:1807.07659
Fong W., Berger E., Margutti R. & Zauderer B. A. 2015, *ApJ*, 815, 102
Fryer C. L. & Kusenko A. 2006, *ApJS*, 163, 335
Giacobbo N., Mapelli M. & Spera M. 2018, *MNRAS*, 474, 2959
Gunn J. E. & Ostriker J. P. 1970, *ApJ*, 160, 979
Hobbs G., Lorimer D. R., Lyne A. G. & Kramer M. 2005, *MNRAS*, 360, 974
Hurley, J. R., Pols, O. R. & Tout, C. A. 2000, *MNRAS*,315, 543
Ivanova N.& Taam R. E. 2004, *ApJ*, 601, 1058
Ivanova N. *et al.* 2013, *A&AR*, 21, 59
Janka H.-T. 2017, *ApJ*, 837, 84
Jones S., Ropke F. K., Pakmor R., Seitenzahl I. R., Ohlmann S. T. & Edelmann P. V. F. 2016, *A&A*, 593, A72
Kewley, L. J. & Ellison, S. L. 2008, *ApJ*, 681, 1183
Klencki, J., Moe, M., Gladysz, W., Chruslinska, M., Holz, D. E. & Belczynski, K. 2018, *ArXiv e-prints*, arXiv:1808.07889
Kruckow, M. U., Tauris, T. M., Langer, N., Kramer, M. & Izzard, R. G. 2018, *MNRAS*, 481, 1908
Lara-López, M. A., Hopkins, A. M., López-Sánchez, A. R., *et al.* 2013, *MNRAS*, 434, 451
Madau, P. & Dickinson, M. 2014, *ARA&A*, 52, 415
Mapelli, M., Giacobbo, N., Ripamonti, E., & Spera, M. 2017, *MNRAS*, 472, 2422
Mennekens N. & Vanbeveren D. 2014, *A&A*, 564, A134
Maeder, A. 1992, *A&A*, 264, 105
Miyaji S., Nomoto K., Yokoi K. & Sugimoto D. 1980, *PASJ*, 32, 303
Moe, M. & Di Stefano, R. 2017, *ApJS*, 230, 15
Nomoto K. & Kondo Y. 1991, *ApJ*, 367, L19
Pavlovskii K. & Ivanova N. 2015, *MNRAS*, 449, 4415
Pavlovskii K., Ivanova N., Belczynski K. & Van K. X. 2017, *MNRAS*, 465, 2092
Peters, P. C. 1964, *Phys. Rev.*, 136, 1224
Petrillo C. E., Dietz A. & Cavaglia M. 2013, *ApJ*, 767, 140
Podsiadlowski P., Langer N., Poelarends A. J. T., Rappaport S., Heger A. & Pfahl E. 2004, *ApJ*, 612, 1044
Portegies Zwart, S. F. & Yungelson, L. R. 1998, *A&A*, 332, 173
Portegies Zwart, S. F., Baumgardt, H., Hut, P., Makino, J. & McMillan, S. L. W. 2004, *Nature*, 428, 724
Rodriguez, C. L., Haster, C.-J., Chatterjee, S., Kalogera, V. & Rasio, F. A. 2016, *ApJ*, 824, L8
Sana H., *et al.* 2012, *Science*, 337, 444
Schneider, R., Graziani, L., Marassi, S., Spera, M., Mapelli, M., Alparone, M., & Bennassuti, M. d. 2017, *MNRAS*, 471, L105
Stevenson S., Vigna-Gmez A., Mandel I., Barrett J. W., Neijssel C. J., Perkins D. & de Mink S. E. 2017, *Nature Communications*, 8, 14906
Tauris T. M., Langer N., Moriya T. J., Podsiadlowski P., Yoon S.-C., Blinnikov S. I. 2013, *ApJ*, 778, L23
Tauris T. M., Langer N. & Podsiadlowski P. 2015, *MNRAS*, 451, 2123

Tutukov, A. V. & Yungelson, L. R. 1993, *MNRAS*, 260, 675
van den Heuvel E. P. J. 2007, in: di Salvo T., Israel G. L., Piersant L., Burderi L., Matt G., Tornambe A. & Menna M. T. (eds), *The Multicolored Landscape of Compact Objects and Their Explosive Origins* AIP Conf. Ser. Vol. 924, (Am. Inst. Phys., New York), p. 598
Vink, J. S., de Koter, A. & Lamers, H. J. G. L. M., 2001 *A&A*, 369, 574

High-mass X-ray Binaries: Illuminating the Passage from Massive Binaries to Merging Compact Objects
Proceedings IAU Symposium No. 346, 2019
L. M. Oskinova, E. Bozzo, T. Bulik & D. Gies, eds.

doi:10.1017/S1743921318007731

Constraining the progenitor evolution of GW 150914

Jorick S. Vink

Armagh Observatory and Planetarium, BT61 9DG Armagh, College Hill, Northern Ireland
email: jorick.vink@armagh.ac.uk

Abstract. One of the largest surprises from the LIGO results regarding the first gravitational wave detection (GW 150914) was the fact the black holes (BHs) were "heavy", of order 30 - 40 $M_{\odot}$. The most promising explanation for this obesity is that the BH-BH merger occurred at low metallicity (Z): when the iron (Fe) contents is lower this is expected to result in weaker mass loss during the Wolf-Rayet (WR) phase. We therefore critically evaluate the claims for the reasons of heavy BHs as a function of Z in the literature. Furthermore, weaker stellar winds might lead to more rapid stellar rotation, allowing WR and BH progenitor evolution in a chemically homogeneous manner. However, there is as yet no empirical evidence for more rapid rotation amongst WR stars in the low Z environment of the Magellanic Clouds. Due to the intrinsic challenge of determining WR rotation rates from emission lines, the most promising avenue to constrain rotation-rate distributions amongst various WR subgroups is through the utilisation of their emission lines in polarised light. We thus provide an overview of linear spectro-polarimetry observations of both single and binary WRs in the Galaxy, as well as the Large and Small Magellanic Clouds, at 50% and 20% of solar Z, respectively. Initial results suggest that the route of chemically homogeneous evolution (CHE) through stellar rotation is challenging, whilst the alternative of a post-LBV or common envelope evolution is more likely.

Keywords. gravitational waves, polarization, stars: early-type, stars: evolution, stars: mass loss, stars: rotation, stars: winds, outflows, stars: Wolf-Rayet

---

1. Introduction

One of the main surprises regarding the first gravitational wave detections by LIGO concerning the physical merging of 2 black holes (BHs) was the fact that the masses inferred for these BHs were very heavy – of order 30 - 40 $M_{\odot}$ (Abbott *et al.* 2016). Within our own Galactic environment the maximum black hole mass is thought to be of order 10 $M_{\odot}$ (Belczynski *et al.* 2010).

For this reason there are 2 possible solutions to the problem. Either the GW150914 event took place in an environment that was low metallicity, reducing the amount of mass loss during stellar evolution, or the initial stars started off with very high masses. Even if there had been relatively little mass loss, the initial masses must have been higher than 40 $M_{\odot}$, relating to a stellar mass regime where winds are a critical ingredient for massive-star evolution (e.g. Higgins & Vink 2018). For very massive stars (VMS), defined with masses over $\simeq$100 $M_{\odot}$ (Vink *et al.* 2015), stellar winds completely dominate their evolution and fate (Woosley & Heger 2015; Hirschi 2015).

In this contribution, we constrain the physics and evolution of the progenitor of GW150914 in terms of the geometry and the amount of the mass loss from Wolf-Rayet (WR) stars, which are thought to be the direct progenitors of BHs.

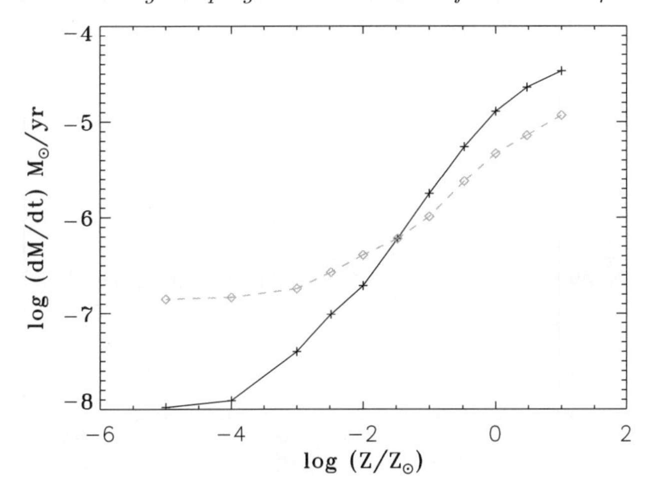

Figure 1. Predicted mass-loss rates of WR stars versus host galaxy metallicity. The black solid line indicates the steeper dependence for the nitrogen-rioch WN stars, whilst the green dashed line indicates the shallower slope for the carbon-rich WC stars. From Vink & de Koter (2005).

2. Stellar winds at low metallicity

Whilst it has been known for decades that stellar winds during the initial O star phase depend on metallicity Z (Abbott *et al.* 1982; Kudritzki *et al.* 1987; Vink *et al.* 2001), the realisation that WR stars also depend on the iron (Fe) contents of the host galaxy is rather more recent (Vink & de Koter 2005, Hainich *et al.* 2015). Until 2005, most stellar evolution modellers assumed that the host metallicity (Fe) was so low in terms of stellar abundance when compared to the self-enriched carbon during the WC phase that mass loss was assumed to be independent of host galaxy metallicity.

Using an established Monte Carlo method, Vink & de Koter (2005) showed that WR mass-loss rates depend on the host galaxy metallicity (Fe) after all. Figure 1 shows the predicted WR mass-loss versus Z dependence for both nitrogen-rich WN stars (in black solid) and the slightly shallower Z dependence for carbon-rich WC stars (in dashed green). Note that the self-enriched materials are not taken into account in the definition of Z on the x-axis. In more recent times, Vink (2017) confirmed the theoretical $Z_{\rm Fe}$ dependence for optically thin stripped helium stars using a dynamically consistent version of the Monte Carlo method (Müller & Vink 2008).

A direct consequence of including Z dependent mass-loss rates of WR stars into massive star evolution models was the prediction of more massive BHs (Eldridge & Vink 2006). Figure 2 shows the expected maximum BH mass at a given host galaxy metallicity from a single star population synthesis by Belczynski *et al.* (2010). The dashed red line indicates the older view that independent of host galaxy Z the maximum BH mass would not exceed ~10-20 $M_{\odot}$, as the WR mass-loss rates were thought to be due to self-enrichment by carbon and oxygen. Only when including the newer Vink & de Koter (2005) WR Fe dependent mass-loss rates one may expect to find heavier BHs at lower metallicities.

Exactly the same plot was also included in the astrophysical interpretation paper of GW 150914 by the LIGO consortium (Abbott *et al.* 2016) but, confusingly, the older and newer expectations of WR mass loss versus Z implementations were here referred to as 'stronger' and 'weaker' stellar winds. That is not correct, as absolute values of the mass-loss rates have not changed. Instead, it is the weaker mass loss at lower Z that is the key physics explaining the heavier BHs at low Z. At the same time, there have not been any substantial changes at solar Z, as can be seen at the intersection of the 2 curves on the left-hand side of Figure 2.

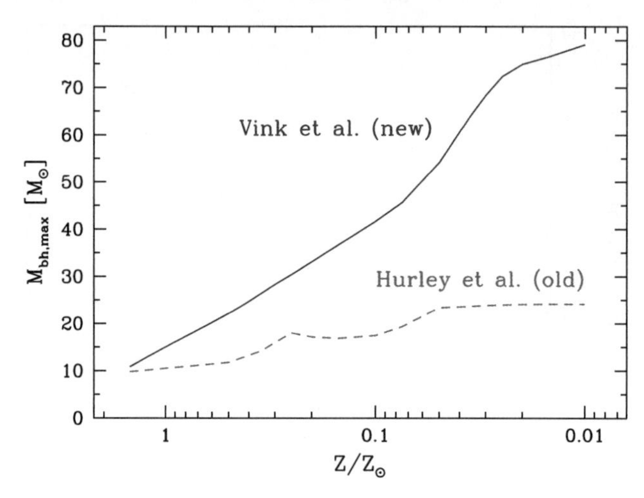

Figure 2. Predicted maximum BH mass versus metallicity. The red dashed line shows the almost Z independent maximum BH mass when including the situation around the year 2000 (Hurley *et al.* 2000), whilst including the Z dependence of not only the Vink *et al.* (2001) O-type stars, but also the WR Z dependent winds of Vink & de Koter (2005) leads to a strong sensitivity of maximum BH mass on Z, as indicated by the blue solid line. From Belczynski *et al.* (2010). The same plot with incorrect labels was included in Abbott *et al.* (2016).

Simply reading off the y-axis of Figure 2 for a 40 $M_\odot$ maximum BH mass, immediately leads to the conclusion that the chemical environment of the GW150914 progenitor should have been 1/10 $Z_\odot$ or less. Detailed single star or binary evolution does not appear to be particularly relevant for making this particular inference. Instead, given the intrinsically low Z, GW150914-type events can teach us interesting physics about winds and stellar evolution at low Z.

3. Interplay between rotation and winds

In order to not only predict compact object masses, but also their spin, it is important to consider the theory of stellar rotation and winds. The first aspect is that the relation can work in both directions: stellar winds may remove angular momentum, thus braking the star, even in the absence of a magnetic field (e.g. Langer 2012), but reversely it has been argued that stellar rotation may increase the overall mass-loss rate (e.g. Maeder & Meynet 2000). In more recent times, 2D dynamical calculations by Müller & Vink (2014) showed there are cases where the overall mass-loss rate may actually decrease with respect to their non-rotating counterparts.

The third relevant aspect is to consider the geometry of rotating winds: do we expect equatorial enhancement (which may remove angular momentum very efficiently) or polar winds? Let us briefly review the key physical ingredients. Friend & Abbott (1986) argued that as a result of a lower effective gravity from a rotating star, mass loss from the equator would be more efficient than from the pole. Later, Cranmer & Owocki (1995) found that due to the Von Zeipel gravity darkening the pole would be brighter and thus mass would preferentially be lost from the pole instead. Alternatively, taking both these competing effects into account, for certain temperatures around the bi-stability jump temperature, mass would predominately be lost from the equator after all (Pelupessy *et al.* 2000). It is clear that the situation regarding the geometry is complex, and that theory needs guidance from observations in order to make progress.

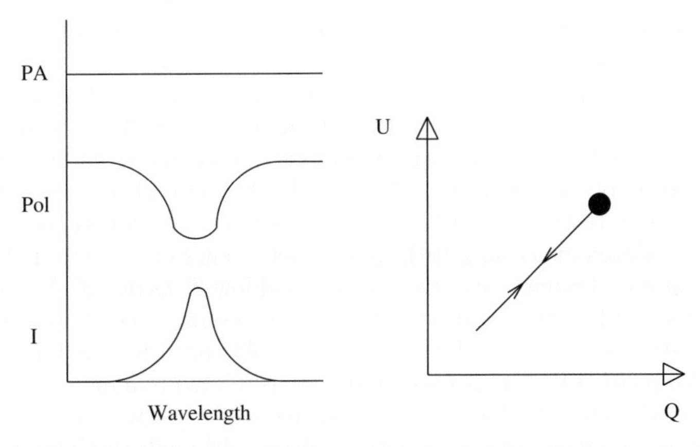

Figure 3. Polarization triplot (on the left-hand side) and the Stokes QU plane (right-hand side). For an emission line (lower panel triplot) one expects a 'depolarization' in the middle panel of the triplot if the innermost geometry is flattened. For a simple disk, the position angle (PA) of the disk remains constant, and the PA can also be read off from the QU plane (right-hand side).

4. Testing stellar rotation with polarimetry

Until astronomers are able to spatially resolve the innermost radii of hot massive stars, linear spectropolarimetry is the only tool available to dissect geometry of stellar winds. The technique was already applied in the 1970s for classical Be stars, and was later also applied to young pre-main sequence Herbig Ae/Be and T Tauri stars to understand accretion disks. One of the key strengths of *spectro*polarimetry is that there is no dependence on any dust particles between the star under consideration and the observer (due to interstellar polarization).

Figure 3 shows the expectations from a disk around a star in the polarization triplot (on the left-hand side) and the Stokes QU plane. For an emission line (lower panel triplot), one only expects to see a 'depolarization' in the middle panel of the triplot if the innermost geometry is significantly flattened, i.e. disk-like. For a simple disk, the position angle (PA) remains constant, and the PA can also be determined from the QU plane (see the right-hand side).

Given the expected lower mass-loss rates of WR stars at lower Z, one might perhaps expect disks to be more prevalent at lower Z than in our Milky Way. Indeed many B-type stars in the SMC seem to be Be stars (e.g. Castro *et al.* 2018). However, in our recent VLT-FORS polarisation study of large samples of WR stars in the low Z environments of the Magallanic Clouds (39 in the LMC; all 12 WRs in the SMC) we found the incidence of depolarisation 'line effects' (Fig. 3) to be indistinguishable from those in the Milky Way (Vink & Harries 2017).

This appears to be quite a revelation for stellar modellers attempting to explain the evolution of GW 150914 with physics related to rapid rotation, such as the rotationally-induced chemically homogeneous evolution (CHE), as we have basically no empirical evidence that WR stars at low metallicity rotate any faster than those in the Milky Way.

5. Remarks on evolutionary scenarios

The results of the VLT study of Vink & Harries (2017) suggest that WR stars in low Z environments do not rotate any faster than at high Z. What does this imply for the subsequent evolution of WR stars into BHs, and how can this information be used to constrain the evolution towards the WR phase?

Independent of any subtle binary evolution effects that will undoubtedly come into play when we wish to explain the compact object mass spectrum, we can already learn some lessons regarding the physics of either component in the merging BH binary progenitor.

First of all, the concept of CHE for the formation of WR stars at low Z due to rotationally-induced CHE (Yoon & Langer 2005) is challenging to entertain given the lack of evidence for WR rotation at low Z. This would therefore also make it harder for this evolutionary pathway to be relevant for the very specific event of GW150914 (e.g. Mandel & de Mink 2016, Marchant *et al.* 2016), unless such events are evolutionary unrelated to ordinary single star and binary evolution at high and low Z, as our VLT sample contained both single WRs and binaries, and no difference in the line-effect frequency was noted.

Vink *et al.* (2011) argued that the most likely explanation for the fact that only a 10-20% sub-population of WR stars is found to rotate is related to a more classical post-LBV like evolutionary channel. In the context of binary evolution we might translate this to common-envelope evolution (Belczynski *et al.* 2016), although the interesting discussion on CHE should certainly continue!

References

Abbott, D. C. 1982, *ApJ*, 259, 282
Abbott, B. P., Abbott, R., Abbott, T. D., *et al.* 2016, *ApJL*, 818, L22
Belczynski, K., Bulik, T., Fryer, C. L., *et al.* 2010, *ApJ*, 714, 1217
Belczynski, K., Holz, D. E., Bulik, T., & O'Shaughnessy, R. 2016, *Nature*, 534, 512
Castro, N., Oey, M. S., Fossati, L., & Langer, N. 2018, arXiv:1810.04682
Cranmer, S. R., & Owocki, S. P. 1995, *ApJ*, 440, 308
Eldridge, J. J., & Vink, J. S. 2006, *A&A*, 452, 295
Friend, D. B., & Abbott, D. C. 1986, *ApJ*, 311, 701
Hainich, R., Pasemann, D., Todt, H., *et al.* 2015, *A&A*, 581, A21
Higgins, E. R. & Vink, J. S., 2018, *A&A* submitted
Hirschi, R. 2015, *Very Massive Stars in the Local Universe*, 412, 157
Hurley, J. R., Pols, O. R., & Tout, C. A. 2000, *MNRAS*, 315, 543
Kudritzki, R. P., Pauldrach, A., & Puls, J. 1987, *A&A*, 173, 293
Langer, N. 2012, *ARA&A*, 50, 107
Maeder, A., & Meynet, G. 2000, *A&A*, 361, 159
Mandel, I., & de Mink, S. E. 2016, *MNRAS*, 458, 2634
Marchant, P., Langer, N., Podsiadlowski, P., Tauris, T. M., & Moriya, T. J. 2016, *A&A*, 588, A50
Müller, P. E., & Vink, J. S. 2008, *A&A*, 492, 493
Müller, P. E., & Vink, J. S. 2014, *A&A*, 564, A57
Pelupessy, I., Lamers, H. J. G. L. M., & Vink, J. S. 2000, *A&A*, 359, 695
Vink, J. S. 2017, *A&A*, 607, L8
Vink, J. S., & de Koter, A. 2005, *A&A*, 442, 587
Vink, J. S., & Harries, T. J. 2017, *A&A*, 603, A120
Vink, J. S., de Koter, A., & Lamers, H. J. G. L. M. 2001, *A&A*, 369, 574
Vink, J. S., Gräfener, G., & Harries, T. J. 2011, *A&A*, 536, L10
Vink, J. S., Heger, A., Krumholz, M. R., *et al.* 2015, *Highlights of Astronomy*, 16, 51
Woosley, S. E., & Heger, A. 2015, *Very Massive Stars in the Local Universe*, 412, 199
Yoon, S.-C., & Langer, N. 2005, *A&A*, 443, 643

High-mass X-ray Binaries: Illuminating the Passage from Massive Binaries to Merging Compact Objects
Proceedings IAU Symposium No. 346, 2019
L. M. Oskinova, E. Bozzo, T. Bulik & D. Gies, eds.

doi:10.1017/S1743921318007433

Common envelope evolution of massive stars

Paul M. Ricker[1], Frank X. Timmes[2], Ronald E. Taam[3] and Ronald F. Webbink[1]

[1]Department of Astronomy, University of Illinois,
1002 W. Green St., Urbana, IL 61801 USA
emails: pmricker@illinois.edu, rwebbink@illinois.edu

[2]School of Earth and Space Exploration,
Arizona State University, Tempe, AZ 85287-1404 USA
email: Francis.Timmes@asu.edu

[3]Department of Physics and Astronomy, Northwestern University,
2145 Sheridan Road, Evanston, IL 60208 USA
email: r-taam@northwestern.edu

Abstract. The discovery via gravitational waves of binary black hole systems with total masses greater than $60M_{\odot}$ has raised interesting questions for stellar evolution theory. Among the most promising formation channels for these systems is one involving a common envelope binary containing a low metallicity, core helium burning star with mass $\sim 80 - 90M_{\odot}$ and a black hole with mass $\sim 30 - 40M_{\odot}$. For this channel to be viable, the common envelope binary must eject more than half the giant star's mass and reduce its orbital separation by as much as a factor of 80. We discuss issues faced in numerically simulating the common envelope evolution of such systems and present a 3D AMR simulation of the dynamical inspiral of a low-metallicity red supergiant with a massive black hole companion.

Keywords. stars: binaries: close, stars: evolution, hydrodynamics

---

1. Introduction

Of the five statistically significant gravitational wave detections to date, three have involved binary black hole mergers with at least one component having a mass greater than $30M_{\odot}$ (Abbott *et al.* 2016, 2017a,b). While these objects were not unexpected (Lipunov, Postnov, & Prokhorov 1997), at metallicities close to solar very massive stars generally lose too much matter to winds to leave behind such massive black holes (Belczynski *et al.* 2010). There are, nevertheless, a number of different evolutionary channels through which pairs of massive binary black holes might form. The two most-studied categories of models are isolated binary channels and dynamical channels that bring together black holes formed separately (see Tutukov & Cherepashchuk 2017 for a recent review).

Here we examine the common envelope (CE) mass-transfer phase that is an ingredient of the 'classical' isolated compact binary formation channel (Tutukov & Yungelson 1973; van den Heuvel & De Loore 1973). We consider in particular the 'typical' model discussed by Belczynski *et al.* (2016). In this model, a low-metallicity binary with zero-age main sequence masses of about $96M_{\odot}$ and $60M_{\odot}$ first undergoes a mass-transfer phase that inverts the mass ratio but does not produce much orbital shrinkage. After the first black hole forms through direct collapse and the second star reaches core helium burning, the binary undergoes a CE phase that shrinks the orbit by a factor of 80 or more and ejects half of the donor star's mass. This level of orbital shrinkage, which is necessary in order to bring the final stellar remnants close enough to merge due to gravitational radiation emission within a Hubble time, requires a high envelope ejection efficiency and/or sources of energy beyond the potential energy of the binary orbit (Kruckow *et al.* 2016). Our objective is to determine if this orbital shrinkage is a realistic prediction. In the process we will examine a mass range not heretofore considered in three-dimensional CE simulations, which have for the most part focused on low-mass systems (Ivanova *et al.* 2013).

CE simulations involving very massive donor stars offer challenges beyond those required for low-mass systems. The spatial dynamic range required to simultaneously resolve the helium-burning core of a supergiant and the ejected envelope can be a factor of ten larger than that required for a low-mass AGB star. The envelope thermal timescale is comparable to or less than the dynamical timescale, requiring radiative transfer to be included. Winds may be important during CE mass transfer. Finally, the large Eddington factors in massive-star envelopes change the behavior of convection there in ways that are just beginning to be understood (e.g. Jiang *et al.* 2018). The simulations described here represent a first attempt to address the CE problem in this regime; a more complete investigation is underway and will be discussed in a forthcoming paper.

2. Numerical methods

We use the adaptive mesh refinement (AMR) code FLASH 4.5 (Fryxell *et al.* 2000; Dubey, Reid, & Fisher 2008) to simulate common envelope evolution. Simulations are carried out on an oct-tree mesh via the PARAMESH library (MacNeice *et al.* 2000) using 8^3 zones per block with the base mesh level formed from 32^3 blocks. The mesh is refined using the standard second-derivative criterion (Löhner 1987) for blocks with densities above 3×10^{-11} g cm^{-3}. The Euler equations are solved using the directionally split Piecewise Parabolic Method (PPM; Colella & Woodward 1984) with a variant of the Helmholtz equation of state that includes hydrogen and helium partial ionization via table lookup. The gravitational potential is computed using a direct multigrid solver (Ricker 2008) with isolated boundary conditions. The companion star and the core of the donor star are represented using particles. In contrast to our earlier work (Ricker & Taam 2008, 2012), instead of using particle clouds we use single particles whose gravitational accelerations are directly added to the mesh accelerations computed by differencing the mesh potential. The particles are treated as corresponding to uniform-density spheres with radius equal to three times the smallest zone spacing. We include single-group flux-limited radiation diffusion using Crank-Nicolson integration and the HYPRE linear algebra library with the Levermore & Pomraning (1981) flux limiter. Opacities are determined using the OPAL tables (Iglesias & Rogers 1996) for high temperatures and the Ferguson *et al.* (2005) tables for low temperatures, both with $Z = 0.0001$.

To construct initial conditions for FLASH, we evolve the donor star from the zero-age main sequence using MESA (Paxton *et al.* 2011, 2013, 2015, 2018). Once the star reaches maximum expansion we remove the envelope density inversion that develops by expanding the outermost layers of the star at constant entropy until hydrostatic equilibrium is reached (this procedure stands in place of a more sophisticated treatment still to be developed). We then relax the star in the binary potential together with the companion using a heavily modified version of the 3D smoothed particle hydrodynamics (SPH) code StarCrash (Rasio & Shapiro 1992; Faber & Rasio 2000). The SPH code has been modified to use a tree solver for gravitation, a variable-timestep leapfrog integrator, the same equation of state used in FLASH, a Lagrangian formulation to correctly include the effects of variable smoothing length, and the same particle cores used in FLASH. It also includes an explicit radiation diffusion solver; an implicit solver based on the method of Whitehouse, Bate, & Monaghan (2005) is under development. To initialize the SPH code from the MESA model, we employ an approach similar to that suggested

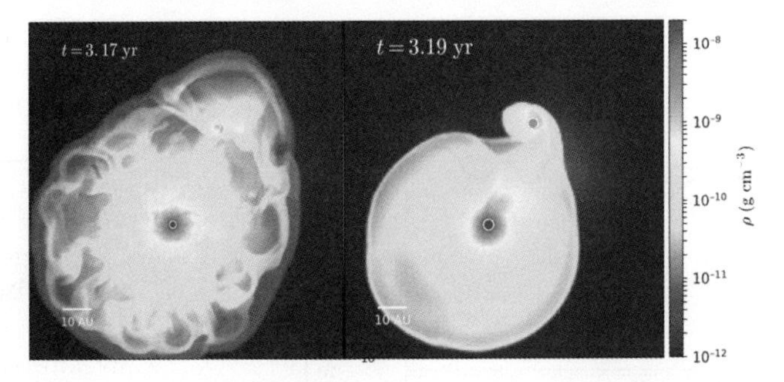

Figure 1. *Left:* Gas density in the orbital plane in the nonradiative run after 3.17 yr. Circles represent the positions and sizes of the donor core and the companion. *Right:* Same, but for the radiative run after 3.19 yr.

by Ohlmann *et al.* (2017), solving a modified Lane-Emden equation with density and entropy matched to the MESA model at the numerical core radius.

Further details of the numerical methods used and tests of the initialization procedure will be presented in a forthcoming paper.

3. Results

We conducted CE simulations of a binary system containing an $82.1M_\odot$ red supergiant (RSG) and a $35M_\odot$ black hole. The RSG had a metallicity $Z = 0.0002$ and was evolved from an $88M_\odot$ zero-age mass until it reached maximum expansion at a radius of $2891R_\odot$. After flattening the outer envelope to remove the density inversion, the star's radius increased by 11.7%. Two runs with somewhat different resolutions were conducted: a nonradiative run with box size of 375 AU, minimum zone spacing $78.6R_\odot$, and core mass $63.7M_\odot$; and a run including radiation diffusion having a box size of 274 AU, minimum zone spacing $115R_\odot$, and core mass $63.7M_\odot$. These resolutions represent a minimum physically reasonable value for this donor star, as its envelope binding energy measured relative to the surface increases sharply in magnitude inside a radius of $\sim 200R_\odot$. Removing the envelope becomes progressively more difficult at smaller radii.

A comparison of the nonradiative and radiative runs at an early stage is shown in Fig. 1. Large convective eddies seen in the nonradiative run are absent in the radiative run. This occurs because the thermal readjustment timescale of the envelope is less than or equal to the dynamical timescale. The envelope is convectively unstable, but because of the short thermal timescale the heat flux implied by the temperature gradient should be carried partly by radiative diffusion. In the nonradiative run this is not possible, so convection is far more vigorous than it should be. This behavior is not seen in CE simulations involving low-mass giants because the thermal timescales of such stars are much longer.

In Fig. 2 we show the orbital separation and gravitationally bound gas mass versus time in the radiative run. Initially the orbital separation and period are 33.1 AU and 17.6 yr, respectively. At first the stars inspiral by about a factor of two and demonstrate orbital circularization as in previous simulations in the literature. However, after about 40 yr of evolution the orbital eccentricity increases briefly, and the orbit widens again to 23.4 AU. In the nonradiative case the orbit shrinks to a separation of about 11 AU before apparently stabilizing at a constant value without passing through a phase of increasing eccentricity.

For the radiative run, the initial plunge corresponds to a reduction of about $6M_\odot$ in the bound mass, or about 1/3 of the gas initially on the grid. This unbinding appears to stall for about 20 yr before beginning again. By the end of the run (88.7 yr), $8.3M_\odot$

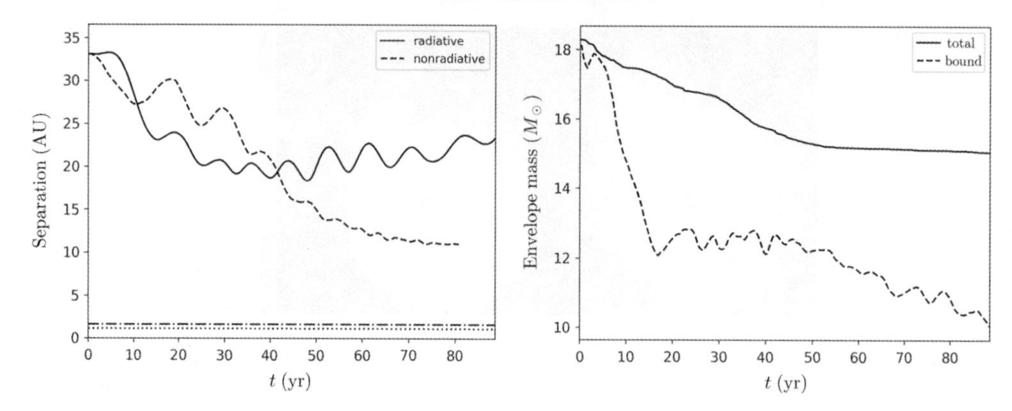

Figure 2. *Left:* Orbital separation vs. time in the nonradiative and radiative runs. Dotted and dash-dotted lines represent the core radius in the nonradiative and radiative runs, respectively. *Right:* Total and bound gas mass vs. time in the radiative run.

have been unbound, with the trend toward continued unbinding. The total mass (bound plus unbound) declines until about 55 yr before becoming nearly constant. Since the hydrodynamical method and AMR library explicitly conserve mass, this means that mass outflow from the grid has stopped at least temporarily.

4. Discussion

The lack of sufficient orbital shrinkage observed in our simulations is in line with results from low-mass CE simulations (e.g., Ricker & Taam 2008, 2012; Passy *et al.* 2012; Ohlmann *et al.* 2016; Iaconi *et al.* 2017). Energy release due to hydrogen and helium recombination has been proposed as an aid to envelope ejection (e.g., Webbink 2008) and does appear to help in double white dwarf formation (Nandez, Ivanova, & Lombardi 2015), though there is some dispute over whether this energy is simply radiated away (Soker, Grichener, & Sabach 2018; Ivanova 2018). For massive stars the available recombination energy is generally less than for low-mass stars at maximum expansion; the amount depends on the core definition but is not sensitive to metallicity (Kruckow *et al.* 2016). Since our radiative simulation includes (for the first time in 3D CE simulation work) both the effects of partial ionization on the equation of state and radiative transfer effects, we are able to address directly the question of whether recombination energy helps in envelope removal for very massive stars. The initial results presented here suggest that it does not. In a separate project we are also investigating the efficacy of recombination energy in low-mass envelopes (Zhu *et al.*, in prep.).

An alternative source of energy, which we have yet to consider, is accretion onto the compact object, leading to jets or accretion disk winds (Soker 2004, 2017; Chamandy *et al.* 2018; López-Cámara, De Colle, & Moreno Méndez 2018). However, results from 1D modeling suggest that accretion disk formation may be a transitory phase in the very massive stars we are considering here (Merguia-Berthier *et al.* 2017).

The fact that the orbit widening, outflow cessation, and resumption of unbinding all begin at roughly the same time ($\sim 40-50$ yr) suggests that they may be connected. The ratio of the kinetic plus thermal energy to the gravitational energy for the unbound material remaining on the grid is roughly constant at a value of about 1.5 until 40 yr, after which time it begins to steadily increase, reaching a value of 4 after 88.7 yr. The computational volume may not be large enough to contain all of the matter whose gravitational influence matters. An additional consideration is the fact that at the end of the radiative run the orbit appears to be re-circularizing as the bound fraction of the mass remaining on the grid continues to decrease. It is possible that longer-term evolution of the system studied here may reveal a new phase of orbital shrinkage. We will address this question in future work.

PMR acknowledges support from the National Science Foundation under grant AST 14-13367, as well as the hospitality of the Academia Sinica Institute for Astronomy and Astrophysics during a sabbatical visit. Portions of this work were completed at the Kavli Institute for Theoretical Physics, where it was supported in part by the National Science Foundation under grant NSF PHY-1125915. FLASH was developed and is maintained largely by the DOE-supported Flash Center for Computational Science at the University of Chicago. Simulations were carried out using XSEDE resources at the Texas Advanced Computing Center under allocation TG-AST040034N.

References

Abbott, B. P., *et al.* 2016, *Phys. Rev. Lett.*, 116, 061102
Abbott, B. P., *et al.* 2017a, *Phys. Rev. Lett.*, 118, 221101
Abbott, B. P., *et al.* 2017b, *Phys. Rev. Lett.*, 119, 141101
Belczynski, K., *et al.* 2010, *ApJ*, 714, 1217
Belczynski, K., *et al.* 2016, *Nature*, 534, 512
Colella, P., & Woodward, P. R. 1984, *J. Comp. Phys.*, 54, 174
Chamandy, L., *et al.* 2018, *MNRAS*, 480, 1898
Dubey, A., Reid, L. B., & Fisher, R. 2008, *Phys. Scr.*, T132, 014046
Faber, J. A., & Rasio, F. A. 2000, *Phys. Rev. D*, 62, 064012
Ferguson, J. W., *et al.* 2005, *ApJ*, 623, 585
Fryxell, B., *et al.* 2000, *ApJS*, 131, 273
Iaconi, R., *et al.* 2017, *MNRAS*, 464, 4028
Iglesias, C. A., & Rogers, F. J. 1996, *ApJ*, 464, 943
Ivanova, N., *et al.* 2013, *A&AR*, 21, 59
Ivanova, N. 2018, *ApJ*, 858, L24
Jiang, Y.-F., *et al.* 2018, *Nature*, 561, 498
Kruckow, M. U., *et al.* 2016, *A&A*, 596, 58
Levermore, C. D., & Pomraning, G. C. 1981, *ApJ*, 248, 321
Lipunov, V. M., Postnov, K. A., & Prokhorov, M. E. 1997, *MNRAS*, 288, 245
Löhner, R. 1987, *Comp. Meth. Appl. Mech. Eng.*, 61, 323
López-Cámara, D., De Colle, F., & Moreno Méndez, E. 2018, MNRAS, accepted (arXiv:1806.11115)
MacNeice, P., *et al.* 2000, *Comp. Phys. Comm.*, 126, 330
Murguia-Berthier, A., *et al.* 2017, *ApJ*, 845, 173
Nandez, J. L. A., Ivanova, N., & Lombardi, J. C. 2015, *MNRAS*, 450, L39
Ohlmann, S. T., *et al.* 2016, *ApJ*, 816, L9
Ohlmann, S. T., *et al.* 2017, *A&A*, 599, A5
Passy, J.-C., *et al.* 2012, *ApJ*, 744, 52
Paxton, B., *et al.* 2011, *ApJS*, 192, 3
Paxton, B., *et al.* 2013, *ApJS*, 208, 4
Paxton, B., *et al.* 2015, *ApJS*, 220, 15
Paxton, B., *et al.* 2018, *ApJS*, 234, 34
Rasio, F. A., & Shapiro, S. L. 1992, *ApJ*, 401, 226
Ricker, P. M. 2008, *ApJS*, 176, 293
Ricker, P. M., & Taam, R. E. 2008, *ApJ*, 672, L41
Ricker, P. M., & Taam, R. E. 2012, *ApJ*, 746, 74
Soker, N. 2004, *New Ast.*, 9, 399
Soker, N. 2017, *MNRAS*, 471, 4839
Soker, N., Grichener, A., & Sabach, E. 2018, *ApJ*, 863, L14
Tutukov, A. V., & Cherepashchuk, A. M. 2017, *Astron. Rep.*, 61, 833

Tutukov, A., & Yungelson, L. 1973, *Nauch. Inform.*, 27, 70
van den Heuvel, E. P. J., & De Loore, C. 1973, *A&A*, 25, 387
Webbink, R. F. 2008, in: E. F. Milone, D. A. Leahy, & D. W. Hobill (eds.), *Short-Period Binary Stars: Observations, Analyses, and Results* (Springer: Berlin), p. 233
Whitehouse, S. C., Bate, M. R., & Monaghan, J. J. 2005, *MNRAS*, 364, 1367

Discussion

VAN DEN HEUVEL: We heard in this meeting of the highly obscured B[e] HMXBs like Chaty's INTEGRAL source with an 80 day orbit with an enormous amount of dust around. It does not seem to be spiralling in, which may be a confirmation of your simulations.

RICKER: Possibly. But this is a low-metallicity system, so the effects of dust opacity should be much less than for the INTEGRAL source.

SANDER: The density inversion in massive stars is also something that bothers us when analyzing stars with stellar atmospheres. However, we have issues with temperatures for certain massive stars that are lower than we expect. There are also theoretical works (e.g. Gräfener *et al.*, Sanyal *et al.*) pointing towards an inflated envelope. Couldn't that be something that might eventually even help to eject the CE?

RICKER: Some inflation is already present here due to our treatment of the density inversion in the MESA model. Further expansion might help increase tidal drag, but it would also make it easier for recombination energy to escape, so it's not clear what the net effect would be.

High-mass X-ray Binaries: Illuminating the Passage from Massive Binaries to Merging Compact Objects
Proceedings IAU Symposium No. 346, 2019
L. M. Oskinova, E. Bozzo, T. Bulik & D. Gies, eds.

doi:10.1017/S1743921318007457

High-mass X-ray binaries: Evolutionary population synthesis modeling

Zhao-yu Zuo

School of Science, Xi'an Jiaotong University, Xi'an 710049, China
email: zuozyu@xjtu.edu.cn

Abstract. Using an evolutionary population synthesis code, we modeled the universal, featureless X-ray luminosity function of high-mass X-ray binaries (HMXBs) in star-forming galaxies. We put constraints on the natal kicks, super-Eddington accretion factor, as well as common envelope prescriptions usually adopted (i.e., the $\alpha_{\rm CE}$ formalism and the γ algorithm), and presented the detailed properties of HMXBs under different models, which may be investigated further by future high-resolution X-ray and optical observations.

Keywords. stellar evolution, compact stars, X-ray binaries.

---

1. Introduction

One of the most striking features of HMXB populations is that the X-ray luminosity function (XLF) takes a universal form of a single, smooth power law (slope $\sim$1.6) giving an excellent account of X-ray binaries (XRBs) containing NSs, stellar-mass BHs and probably intermediate-mass BHs over the entire X-ray luminosity range $L_{\rm X} \sim 10^{35} - 10^{40}\,{\rm ergs\,s^{-1}}$. This was first discovered by Grimm et al. (2003) and then reconfirmed by Mineo, Gilfanov & Sunyaev (2012). In this work, we applied an updated evolutionary population synthesis (EPS) technique to model the XLF of HMXBs, taking into account both the $\alpha_{\rm CE}$ formalism and the γ algorithm to describe the CE evolution. Several parameters (such as the binary fraction f, the super-Eddington factor $\eta_{\rm Edd}$, the bolometric correction factor $\eta_{\rm bol}$, the mass ratio q [index α], the initial mass function (IMF), the dispersion of the natal kick velocity $\sigma_{\rm kick}$) are also examined. The aim of the work is to constrain the model parameters, and to discriminate between models of the CE.

2. Model

We used the EPS code developed by Hurley, Pols & Tout (2000) and Hurley, Tout & Pols (2002), and modified by Zuo *et al.* (2008) to calculate the X-ray luminosity ($L_{\rm X}$) of XRBs and their numbers. We calculated the X-ray luminosity for supergiant(SG) and main-sequence(MS) HMXBs as in Zuo & Li (2010) and Be-XRBs as in Belczynski & Ziolkowski (2009). Besides the modifications made to the original code by Zuo *et al.* (2008), we also used a more physical estimate of the binding energy parameter λ (Xu & Li 2010; Loveridge *et al.* 2011) in $\alpha_{\rm CE}$ formalism to model the CE evolution.

We first manage to fit the observed XLF in the $\alpha_{\rm CE}$ formalism (Fig. 1). When the best-fit model is achieved, the parameter combination is as follows: $\alpha = 0$, $\eta_{\rm Edd,BH} = 100$, $f = 0.5$, $\sigma_{\rm kick} = 110\,{\rm km\,s^{-1}}$, $\eta_{\rm bol,BH} = 0.6$, $\eta_{\rm bol,NS} = 0.3$, Salpeter IMF **and a constant star formation (i.e., $1\,{\rm M_\odot\,yr^{-1}}$) for 50 Myr**. Then we compare between the two CE mechanisms under the same parameter combination as the above. In this case, only values of γ and $\alpha_{\rm CE}$ are changed to see their effects on the XLF. We consider different CE

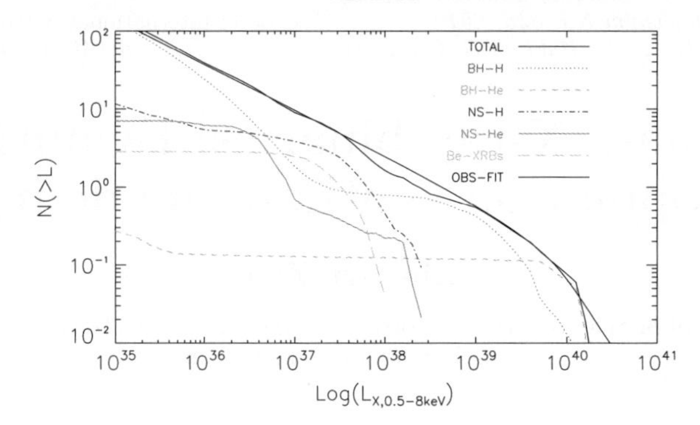

Figure 1. The detailed components of the simulated XLF in the best-fit model of $\alpha_{\rm CE}$ formalism.

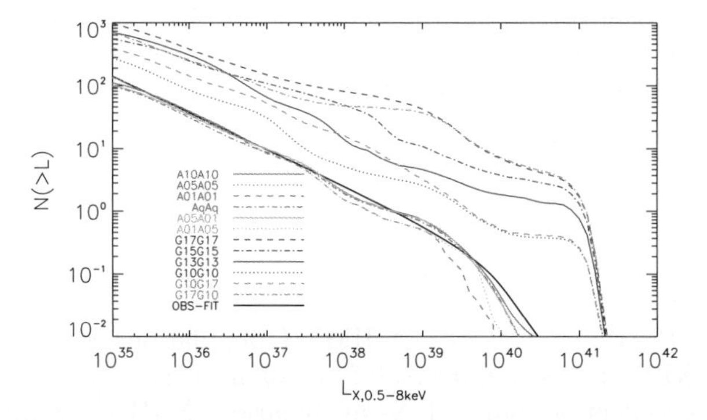

Figure 2. Simulated XLFs of different models on the treatment of the CE phase (Same parameter combination as in the best-fit model of $\alpha_{\rm CE}$ formalism).

efficiencies for the first and second CE episodes. For example, models with different values of $\alpha_{\rm CE}$ are denoted as A01A01, A05A05, A10A10, A01A05, and A05A01, respectively, where the two digits following each letter correspond to the values of $\alpha_{\rm CE}$ during the first and second CE episodes, respectively. It was done similarly for the γ algorithm as well. We also adopt the derived expression of a varied $\alpha_{\rm CE} = 0.05 \times q^{1.2}$ in De Marco *et al.* (2011), denoted as model AqAq (see Fig. 2). At last we manage to determine the best-fit model in the γ algorithm (Fig. 3) by varying all the key parameters and see their effects on the XLF.

3. Results

Fig. 1 shows the simulated XLF and its detailed components contributed by accreting NS/BH with hydrogen-rich (NS/BH-H) and helium-rich (NS/BH-He) donors, and Be-XRBs, respectively. The thick triple-dot-dashed line represents the observed average XLF (labeled as "OBS-FIT") derived by Mineo *et al.* (2012) using the data of 29 nearby star-forming galaxies (Similarly hereinafter). The high luminosity ($L_{\rm X} >\sim 10^{39}{\rm ergs\,s^{-1}}$) sources are mainly BH systems, including both wind-fed BH-XRBs with massive ($\sim 10-30M_{\odot}$) SG donors (i.e., BH-SG HMXBs), orbital period several thousand days to even hundreds of years, and RLOF-fed BH-XRBs, with less massive (typically $< 10M_{\odot}$) MS donors, and orbital period typically on the order of days. While the low luminosity sources ($L_{\rm X} <\sim 10^{37}{\rm ergs\,s^{-1}}$) are dominated by wind-fed BH systems powered by higher mass ($\sim 30-75M_{\odot}$) MS stars (i.e., BH-MS HMXBs), with orbital period from about months

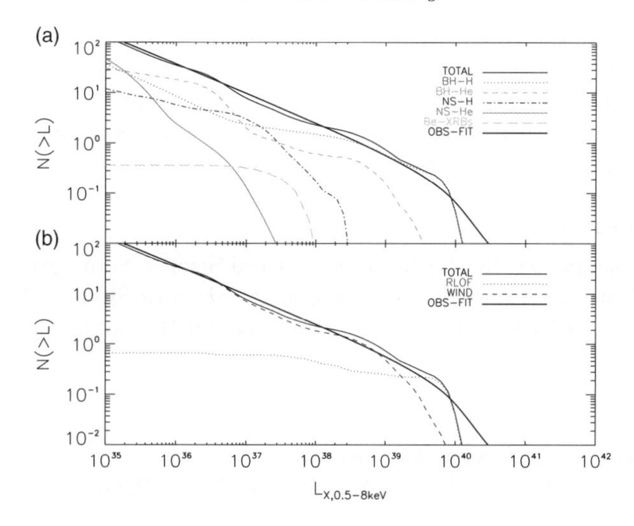

Figure 3. The detailed components of the simulated XLF in the best-fit model of γ algorithm.

to $\sim 10^3$ days. In between are dominated by wind-fed NS systems. In addition, the Be-XRBs are predicted to be very rare. We note that, quantitatively, our calculation is in general consistent with current HMXB population statistics.

Fig. 2 compares the simulated XLFs with different treatments of the CE phase. Clearly, under the same parameter combination the γ algorithm (models with initial letter "G") can produce more (up to one order of magnitude) HMXBs than in the $\alpha_{\rm CE}$ formalism (models with initial letter "A"). In the framework of $\alpha_{\rm CE}$ formalism, the XLF is not very sensitive to $\alpha_{\rm CE}$ and a high value of $\alpha_{\rm CE}$ ($\sim 0.5-1$) seems more preferable. While in the case of the γ algorithm, the number of HMXBs is rather sensitive to the value of γ, especially in the first CE phase (compare models G10G17 with G17G17 or models G10G10 with G17G10).

Shown in Fig. 3 are the detailed components of the simulated XLF (left) and the accretion modes in XRBs (right) in the best-fit model of γ algorithm. It is clear that under the γ algorithm BH-He XRBs dominate in the low luminosity range (i.e., $L_{\rm X} <\sim 10^{37}{\rm ergs\,s^{-1}}$) of the XLF while this is not the case in the $\alpha_{\rm CE}$ formalism, where BH-MS XRBs dominate instead (Zuo, Li & Gu 2014). The orbital period distribution is also distinct from that in the $\alpha_{\rm CE}$ formalism. There are much more sources with period relatively short, i.e, less than several tens of days in this case, which may provide further clues to discriminate between this two models.

4. Discussion and Conclusions

Our work suggests that in the case of HMXBs, both the $\alpha_{\rm CE}$ formalism and the γ algorithm are possible to reproduce the observed XLF. In the framework of the $\alpha_{\rm CE}$ formalism, a high value of $\alpha_{\rm CE}$ (i.e., $\sim$ 0.5-1.0) is more preferred. In addition, we also make constraints on several other parameters, such as the super-Eddington factor $\sim$ 80-100 and the dispersion of kick velocity $\sigma_{\rm kick}$ $\sim$100-150 km/s.

We also give predictions to discriminate both CE mechanisms. For low luminosity sources ($L_{\rm X} < 10^{36}{\rm ergs\,s^{-1}}$), The α-formalism gives: wind-fed BHs with massive MS companion, $M_{\rm opt} \sim 30-75 M_{\odot}$, $P_{\rm orb} \sim$ months-10^3 days, while the γ-algorithm predicts: wind-fed BHs with less massive HeMS companion, $M_{\rm opt} \sim$ several-10 $M_{\odot}$, $P_{\rm orb} \sim$ tens of days.

We concluded that the simulated HMXBs under the γ algorithm have a much larger population of short-period (less than about several tens of days) BH-He systems than in the $\alpha_{\rm CE}$ formalism, which may serve as clues to discriminate between the two kinds of models. Our work motivates further high-resolution X-ray and optical observations of HMXB populations in nearby star-forming galaxies.

Acknowledgments

This work was supported by the National Natural Science Foundation of China (grant 11573021), the Natural Science Basic Research Program of Shaanxi Province – Youth Talent Project (No. 2016JQ1016) and the Fundamental Research Funds for the Central Universities.

References

Belczynski K. & Ziolkowski J. 2009, *ApJ*, 707, 870
De Marco O., Passy J., Moe M., Herwig F., Mac Low M., & Paxton B. 2011, *MNRAS*, 411, 2277
Grimm, H.-J., Gilfanov, M., & Sunyaev, R. 2003, *MNRAS*, 339, 793
Hurley, J. R., Pols, O. R., & Tout, C. A. 2000, *MNRAS*, 315, 543
Hurley, J. R., Tout, C. A., & Pols, O. R. 2002, *MNRAS*, 329, 897
Loveridge, A. J., van der Sluys, M. V., & Kalogera V. 2011, *ApJ*, 743, 49
Mineo, S., Gilfanov, M., & Sunyaev, R. 2012, *MNRAS*, 419, 2095
Xu, X. J., & Li X. D. 2010, *ApJ*, 716, 114
Zuo, Z. Y., Li, X. D., & Liu, X. W. 2008, *MNRAS*, 387, 121
Zuo, Z. Y., & Li, X. D. 2010, *MNRAS*, 405, 2768
Zuo, Z. Y., Li, X. D., & Gu, Q. S. 2014, *MNRAS*, 437, 1187 (Erratum: 443, 1889)
Zuo, Z. Y., & Li, X. D. 2014, *ApJ*, 797, 45

High-mass X-ray Binaries: Illuminating the Passage from Massive Binaries to Merging Compact Objects
Proceedings IAU Symposium No. 346, 2019
L. M. Oskinova, E. Bozzo, T. Bulik & D. Gies, eds.

doi:10.1017/S1743921318007603

How pulses in short gamma-ray bursts constrain HMXRB evolution

Jon Hakkila[1] **and Robert D. Preece**[2]

[1]Dept. of Physics and Astronomy, College of Charleston,
66 George St., Charleston, SC, USA
email: hakkilaj@cofc.edu

[2]Dept. of Space Science, University of Alabama in Huntsville,
Huntsville, AL, USA
email: rob.preece@nasa.gov

Abstract. We demonstrate how pulse structures in Short gamma-ray bursts (SGRBs), coupled with observations of GRB/GW 170817A, constrain the geometries of dying HMXRB systems composed of merging neutron stars.

Keywords. X-rays: binaries, gamma rays: bursts, methods: data analysis, methods: statistical

1. Introduction

Binary neutron stars represent an HMXRB evolutionary end resulting in the creation of short gamma-ray bursts (SGRBs). These luminous flashes of $\gamma-$radiation occur after neutron stars merge following the decay of their orbits. The most powerful evidence linking neutron star mergers to SGRBs has been the LIGO gravitational wave 'chirp' of GW 170817 (Abbott *et al.* 2017) in which two compact objects having masses between $1.17M_{\odot}$ and $1.60M_{\odot}$ and a combined mass of $2.74M_{\odot}$ merged. SGRB 170817A was observed 1.7 seconds later by the GBM experiment on Fermi. This SGRB had a duration of roughly 2 seconds (Goldstein *et al.* 2017) and a Lorentz factor of $\Gamma > 10$ (e.g., Zou *et al.* 1995).

The dominant method of emission in GRBs is via $\gamma-$ray pulses. GRB pulse light curves are not simple smoothly-varying 'bumps.' Instead they generally exhibit structure that cannot be explained by stochastic background variations. Furthermore, this structure often has a wavelike shape that gives a GRB pulse a triple-peaked rather than a single-peaked appearance. Typical GRB pulses evolve from hard to soft but re-harden as the intensity re-brightens; this behavior is true for both SGRBs and LGRBs (long GRBs).

2. GRB Pulse Structure

In order to characterize GRB pulse structure, Hakkila *et al.* (2018a) classified GRB pulses based on their complexity as determined by a simple monotonic 'bump' overlaid by an identifiable 'wavy' structure; this simple approach is often effective. Pulses were classified as *simple* when they could be fitted by a monotonic pulse alone (using the Norris *et al.* 2005 pulse shape), *blended* when they could be fitted by a monotonic pulse with significant wavy residual structure (characterized by the Hakkila & Preece 2014 residual function), *structured* when fits were improved but not completely adequate, and *complex* when pulse light curves were too structured for a good combined fit.

Examples of SGRB pulses containing different amounts of structure are shown for BATSE SGRBs 0373 (simple), 2896 (blended), 5564 (structured), and 4955 (complex)

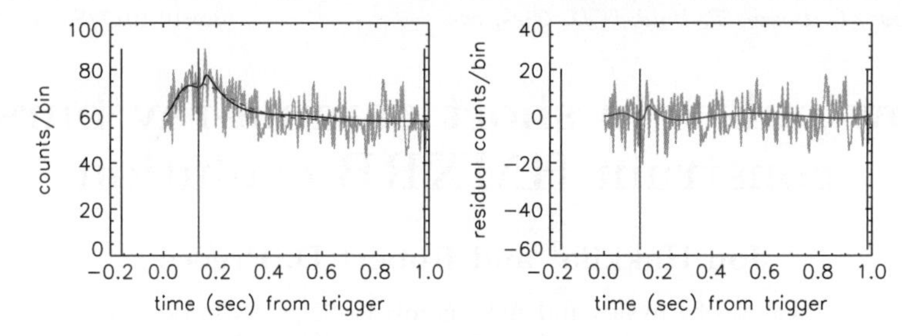

Figure 1. Simple SGRB pulse BATSE 0373. Pulse fit (left) and residual fit (right).

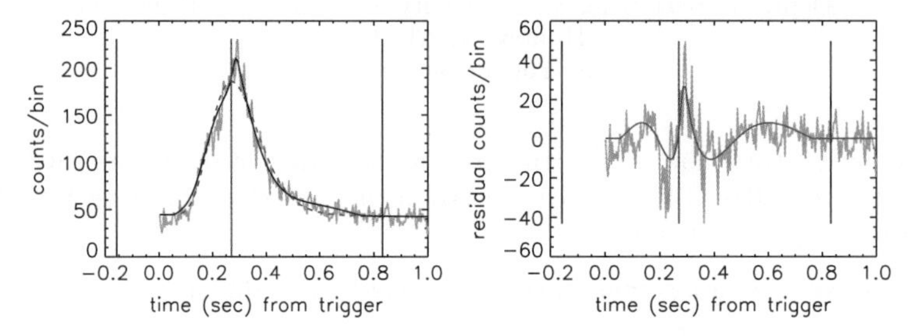

Figure 2. Blended SGRB pulse BATSE 2896. Pulse fit (left) and residual fit (right).

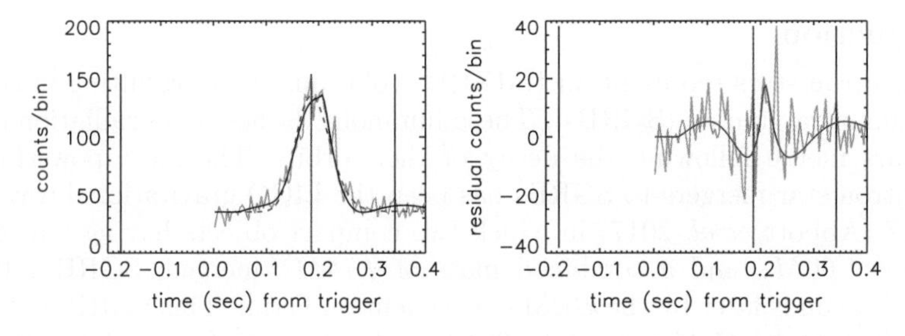

Figure 3. Structured SGRB pulse BATSE 5564. Pulse fit (left) and residual fit (right).

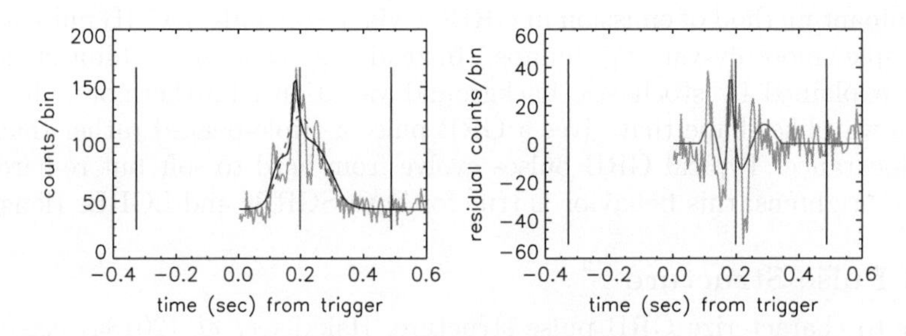

Figure 4. Complex SGRB pulse BATSE 4955. Pulse fit (left) and residual fit (right).

in Figs. 1 through 4. In each figure, the left panel demonstrates the fit obtained with the Norris *et al.* (2005) pulse shape (dashed blue line) and the Norris *et al.* (2005) pulse shape combined with the Hakkila & Preece (2014) residual structure (solid black line). The right panel indicates the residuals once the Norris *et al.* (2005) fit has been removed, overlaid by the Hakkila & Preece (2014) residual fit (solid blue line).

Detector signal-to-noise ratio (S/N) and temporal resolution are capable of smearing out intrinsically complex GRB structures and of causing GRB pulses to appear as

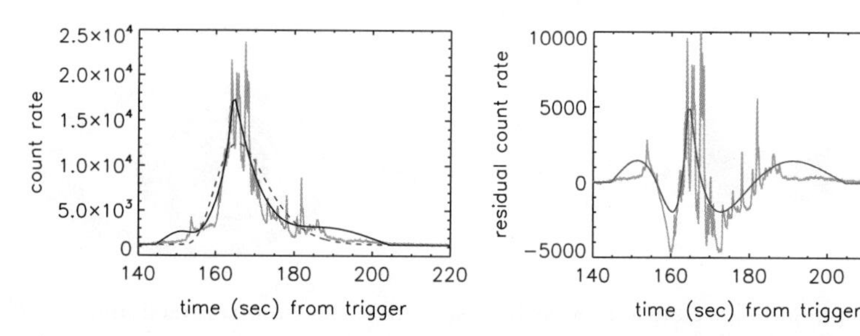

Figure 5. LGRB BATSE 7301p2, a complex, extremely bright pulse. Pulse fit (left) and residual fit (right).

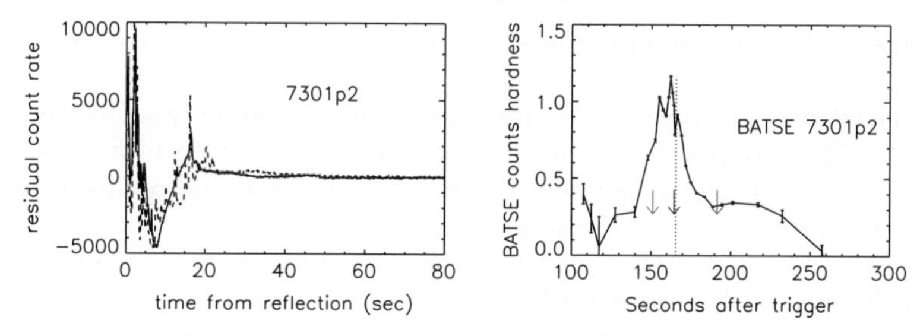

Figure 6. LGRB BATSE 7301p2. Time-reversed and stretched residuals (left) and spectral hardness evolution (right).

either monotonic shapes augmented by the triple-peaked structure or as simple monotonic shapes. Thus a GRB pulse's appearance is a combination of intrinsic structures and instrumental smearing effects. The effects of S/N on both SGRB and LGRB pulse classification have been demonstrated by Hakkila *et al.* (2018b) and Hakkila *et al.* (2018b).

3. Constraints Imposed by Time-Reversed and Stretched Residuals

Since instrumental effects can make it hard to delineate GRB pulse structure from noise, interpretation of GRB pulse physics is best understood through the study of bright GRB pulses. Hakkila *et al.* (2018b) studied six of the brightest BATSE LGRB pulses and demonstrated that the residual structure model employed previously was too simple and incomplete for describing the residuals of these pulses because the residual structure extends far beyond the temporal boundaries containing the three peaks. Furthermore, the extended wavelike structure is shown to have strange characteristics: it is both *time-reversible* and *stretched* around a *time of reflection*. In other words, the pulse residuals following the time of reflection have a memory of the residuals preceding it, but these events are repeated in reverse order after undergoing a dilation at the time of reflection.

An example of this is shown for LGRB BATSE pulse 7301 p2 in Fig. 5 and Fig. 6. The left panel of this figure shows the model fits for the pulse, both including (solid black line) and excluding (dashed blue line). The residual model is able to fit part of the residual light curve, but is inadequate in identifying and fitting all of the structure (right panel of Fig. 5). The left panel of Fig. 6 uses a new approach that recognizes the time-reversed and stretched structure of the residual model without being dependent on that model's functional form. The residuals are folded over in time and stretched until they line up with one another, so the residuals prior to the time of reflection (solid line) are shown overlaid by the time-reversed and stretched residuals preceding the time of reflection (dashed line). Further evidence that these residuals are linked together in a chain, rather than distributed randomly, is shown in the hardness evolution plot of the pulse (Fig. 6).

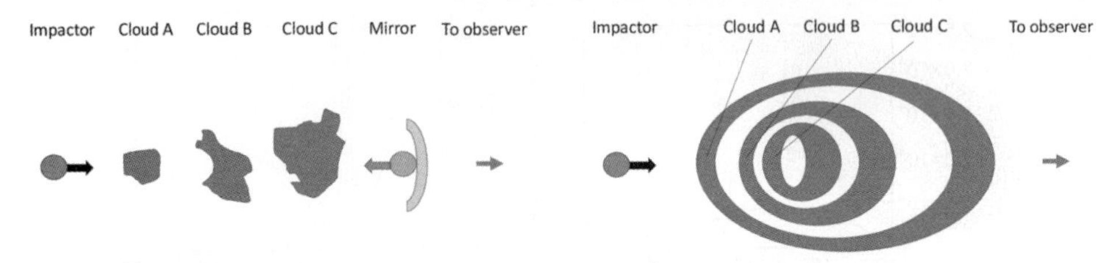

Figure 7. Two possible kinematic models for explaining the time-reversed and stretched residuals found in GRB pulse light curves: the mirror model (left panel) and bilaterally-symmetric model (right panel).

Here the pulse hardness generally evolves from hard-to-soft, with a re-hardening at each residual peak.

Time-reversed and stretched pulse residuals place remarkably strict constraints on GRB models. They couple events that happen at the beginning of a pulse with those that happen towards the end, but they further indicate that the conditions responsible for creating the pulse structure must repeat in reverse order and be time-dilated. We demonstrate two simple models in which pulse light curves with the observed characteristics might be created by jetted GRB material. We note that these models are driven solely by kinematics and geometry rather than by a specific radiation mechanism.

The first *mirror model*, shown in the left panel of Fig. 7 consists of a relativistically-moving *impactor* (shown as a red circle) ejected from the central engine. This impactor might be a soliton (shock wave) or a plasma blob that interacts with other plasma clouds in the jet (in blue) to produce radiation. The impactor slows upon striking a mirror (in yellow; presumably the jet head), which allows the clouds to catch up with it in the opposite order. The blueshifted initial motion of the impactor through the clouds produces beamed emission (A.B.C) followed by emission that is less-strongly blueshifted (C...B...A).

The second *bilaterally-symmetric model*, shown in the right panel of Fig. 7, is composed of clouds distributed in a bilaterally-symmetric fashion along the impactor's path, producing the beamed (A.B.C.C...B...A) emission pattern.

4. Constraints Imposed by the Rarity of Pulses

The standard definition of a pulse refers to a single-peaked monotonic bump. Using this definition, observers are misled into thinking that a typical GRB generally contains many pulses. The recognition that peaks are linked temporally (such that the time-reversed and stretched residuals can be used to identify all the peaks associated with a single pulse), allows the number of pulses in a GRB to be reduced dramatically.

The recent study of Hakkila *et al.* (2018b) finds that 90% of SGRBs are single-pulsed, and most of the remaining 10% are double-pulsed. Thus the mechanism producing SGRBs generally does so in the form of a single structured pulse, but this can also occur less frequently as two or maybe three pulses. Since SGRBs are produced by colliding neutron stars, it seems unlikely that each interaction is capable of producing more than a single blast wave. We thus turn to GRB geometry to explain multi-pulsed bursts.

5. An SGRB Model that Accounts for Pulse Structure and Rarity

We can use the constraints imposed by GRB pulse structure and by the rarity of multi-pulsed SGRBs, in conjunction with theoretical models of merging neutron stars, to improve physical models. Standard models of merging neutron stars suggest that they produce a thick accretion disk with a tail extending behind it as it rotates (*e.g.* Rosswog

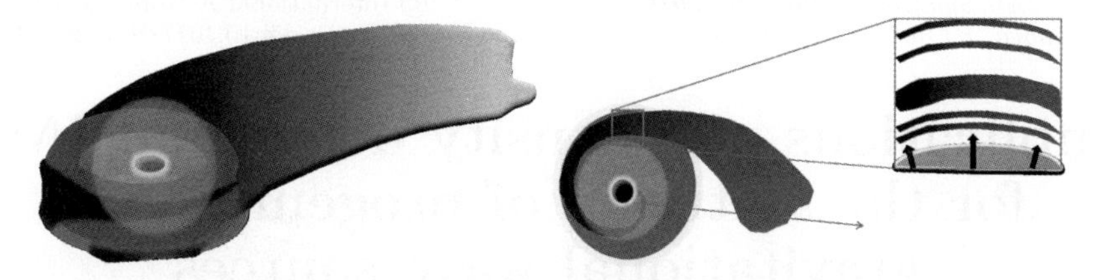

Figure 8. 3D model of ns-ns merger as seen from the side (left) and top (right) views. Most lines-of-sight produce single-pulsed SGRBs, but a line-of-sight through the accretion tail will produce two pulses. The accretion disk structure is similar to that found in the axially-symmetric model shown in Figure 7.

et al. 2014). The timescale for the existence of this disk is very short (< 20 ms). From the perspective of our model, we can consider the radial distribution of the disk to be the 'jet' and density variations in the disk itself to comprise the distribution of clouds within the jet.

In order to match our observations, the merging neutron stars likely produces a soliton at the moment of black hole formation, and this impactor expands spherically outward (denoted by the yellow sphere in Fig. 8). This model can reproduce the time-reversed and stretched residuals found in SGRB pulse light curves if the radial distribution of material is bilaterally-symmetric (seen in the enlargement of the accretion disk radial structure found on the far right side of Fig. 8). We note that double-pulsed bursts can occur if the accretion tail is pointed along the observer's line-of-sight, so that the observer sees two emitted pulses each with similar time-reversed and stretched structures. The timescale of each pulse is essentially the light travel time of the disk and of the tail, and the interpulse separation is essentially the light travel time of the gap between the disk and the tail.

This attempt to model SGRBs is among the first to incorporate constraints imposed by observations of GRB pulses. Other models are also possible, but each must be consistent with the pulse observations. We continue to study and explore these models for both SGRBs and LGRBs.

References

Abbot, B. P., *et al.* 2017, *ApJ*, 848L, 13
Goldstein, A., *et al.* 2017, *ApJ*, 848L, 14
Hakkila, J., & Preece, R. D. 2014, *ApJ*, 783, 88
Hakkila, J., *et al.* 2018, *ApJ*, 855, 110
Hakkila, J., *et al.* 2018, *ApJ*, 863, 77
Norris, J. P. *et al.* 2005, *ApJ*, 627, 324
Rosswog, S., Piran, T., & Nakar, E. 2013, *MNRAS* 430, 2585
Zou, Y-C. *et al.* 2018, *Astrophys J*, 852L, 1

High-mass X-ray Binaries: Illuminating the Passage from Massive Binaries to Merging Compact Objects
Proceedings IAU Symposium No. 346, 2019
L. M. Oskinova, E. Bozzo, T. Bulik & D. Gies, eds.

doi:10.1017/S1743921318008177

Implications of a density dependent IMF for the statistics of progenitors of gravitational wave sources

Indulekha Kavila and Megha Viswambharan

School of Pure & Applied Physics, Mahatma Gandhi University, Kottayam 686560 INDIA
emails: indulekha@mgu.ac.in, meghapv7@gmail.com

Abstract. Observations of mergers of multi-compact object systems offer insights to the formation processes of massive stars in globular clusters. Simulations of stellar clusters, may be used to understand and interpret observations. Simulations generally adopt an Initial Mass Function (IMF) with a Salpeter slope at the high mass end, for the initial distribution of stellar masses. However, observations of the nearest high mass star forming regions point to the IMF at the high mass end being flatter than Salpeter, in regions where the stellar densities are high. We explore the impact of this on the formation rate of potential GW sources, estimated from standard considerations. Globular clusters being significant contributors to the ionization history of the universe, the results have implications for the same. It impacts our ability to explore the putative mass gap, between the upper limit for neutron star masses and the lower limit for black hole masses, also.

Keywords. binaries: close, globular clusters: general, open clusters and associations: general, early universe

1. Introduction

Gravitational wave signals are clean signals. With the detection of gravitational wave (GW) signals from coalescing binary compact objects by LIGO (Abbott *et al.* 2016a, 2016c, 2017a, 2017b, 2017c, 2017d), precise determination of the masses and effective spin parameters of the coalescing objects have now become possible (Veitch *et al.* 2015; Abbott *et al.* 2016b). The masses obtained so far are of stellar order. These observations are thus probes for the nature and extent of the earliest processes of formation of Pop II stars. The length of the time interval expected, between binary formation and ultimate coalescence of the compact objects, suggests that, the progenitors of these GW sources formed very early. The oldest stars in galaxies are found in the globular clusters. Globular clusters are putative hosts for GW sources, due to the possibility of formation of binary compact objects by capture (Abbott *et al.* 2016d). Massive stars which formed in globular clusters are considered to have contributed to the reionization of the universe also (Boylan-Kolchin 2018).

The effective spin parameters of binary compact objects, which formed through dynamical interactions of evolved objects in dense stellar environments like those in globular clusters, are expected to be zero, since there will no preference for alignment between the directions of orbital and spin angular momenta (Abbott *et al.* 2016d). With a bunch of six detections of GW signals in hand, the results for all the sources are consistent with their effective spin parameters being zero. However, the results are found to be not insensitive to the prior. It has been noticed that Bayesian priors have an impact on the characterization of Binary Black Hole (BBH) mergers and that in the case of GW 151226, the odds are almost similar for a prior peaked around alignment for spin direction as for one isotropic in spin direction (Vitale *et al.* 2017). An alternative to the formation of binary compact objects by capture, is the scenario where a binary evolves through common envelope evolution, surviving the two supernova explosions (van den Heuvel E P J 1981). In this context, recent observations of non-standard results regarding the IMF in high mass star forming regions and intra-core velocity dispersion in dense, star forming cores, are analyzed to understand their impact on the formation rate of massive binaries.

2. Analysis and results

A high star density and a preponderance of massive stars have been both hypothesized as well as observed in the case of massive star clusters and starbursts, where star formation takes place in assemblies of dense gas. Examples include the Milky Way clusters Arches, Westerlund 1 and NGC 3603 (see for eg Habibi *et al.* 2013; Lim, Sung & Hur 2014; Harayama, Eisenhover & Martins 2008). The starburst region 30 Dor in the Large Magellanic Cloud is observed to have a flatter IMF and a preponderance of massive stars over and above the numbers expected from a Salpeter IMF (Schneider *et al.* 2018). Sub-virial velocity dispersions have been observed for the denser component of gas in cores (Kirk *et al.* 2010; Wilking *et al.* 2015). Both the above factors can enhance the chances for the formation of binary compact objects. Only $\sim 10\%$ of stellar clusters which form, survive bound, the rest dissolving into the field. Smaller velocity dispersion within the denser gas of cores and locally nonstandard IMFs, both enhance the probability for the formation of field BBHs with aligned spins, over and above the expectation from the standard scenario. This enhancement factor is estimated. Significant spin alignment of stars has been noticed in observations of old open clusters (Corsaro *et al.* 2017). Stellar clusters do show a hierarchical sub-clustering which is interpreted as the imprint of the density distribution of the supersonic turbulence in the gas from which they formed via fragmentation (Gouliermis *et al.* 2014). Dense substructures thus offer conditions that are favorable for the formation of binaries.

Enhancement of the probability of formation of massive binaries due to non-standard IMF: For a given mass of stars, the ratio r of the number of stars with mass greater than 8 solar mass, sampled from an IMF with slope $-\alpha$, with that sampled from an IMF with Salpeter slope, is determined (Fig. 1). The ratio r is ~ 2 for flatter IMFs with $\alpha > 2$. The Salpeter value for the slope is -2.35.

Enhancement of the probability of formation of massive binaries due to sub-virial intra-core velocity dispersion: If the expected relative speeds of protostars are only, say, a factor a times the expected virial speed, the probability for binary formation by capture, during the early phases of star formation will be higher by a factor a^{-3}. Dispersion in the intra-core line-of-sight (los) velocities of dense gas is found to be closer to thermal than virial in dense cores. The observed mean dispersion in the los velocity is $< 0.5\,\mathrm{km/s}$ where the velocity dispersion of the surrounding less dense gas is of the order of 1 - 2 km/s (Kirk *et al.* 2010; Wilking *et al.* 2015).

It may be seen from the above that, taking into account the possible non-standard nature of the IMF in dense high mass star forming regions and the low intra-core velocity dispersions, enhances the chances for the formation of BBHs by an order of magnitude.

3. Discussion

A flatter IMF for star formation taking place at high redshifts has been proposed to account for the inconsistency between the cosmic stellar mass density and the observed star formation rate up to z ~ 8 (Yu & Wang 2016). The IMF of the nearby 30 Dor

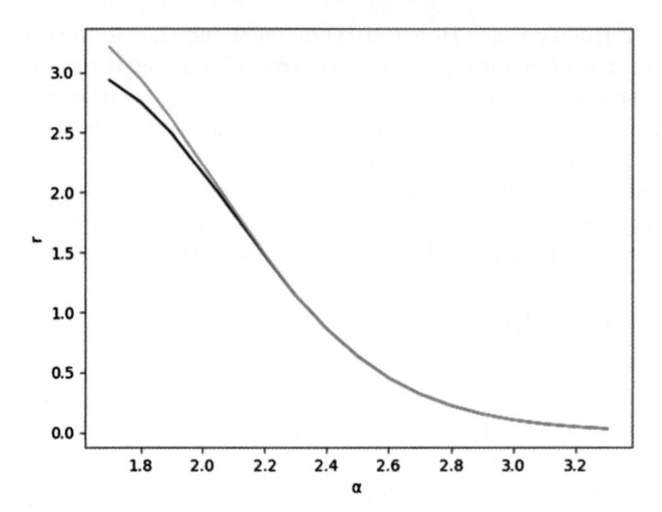

Figure 1. The ratio r of the number of stars with mass > 8 solar masses expected for sampling from an IMF with slope α with the number expected from sampling from an IMF with Salpeter slope for a cluster of given mass. The dark line is for the case where the upper limit for stellar masses is taken as 150 solar masses and the lighter one is for an upper mass limit of 100 solar masses.

starburst region, which is considered to be an analogue of distant starbursts in the early universe, has a slope ~ 1.9, for stars having masses > 15 solar masses (Schneider *et al.* 2018). It is pointed out that a flatter IMF, along with low intra star-forming-core velocities which could arise from the high dissipation rate of turbulence in the dense cores, may lead to at least an order of magnitude enhancement, of the chances for the formation of binary black holes with aligned spins, compared to the chances for the same in the standard scenario (Abbott *et al.* 2016d, 2016e; Kalogera 2000).

Globular clusters are the oldest objects accessible for observation in the nearby universe. The advent of next generation telescopes like the James Webb Space Telescope and the Thirty Meter Telescope, will allow detailed exploration of binaries, the binary fraction, and the evolution of both, in crowded and / or deeply embedded star burst fields as well as globular cluster fields. With the advent of aLIGO (Advanced LIGO) and IndIGO (Indian Initiative in Gravitational-wave Observations) it will be possible to locate GW sources in the sky, enabling multi-messenger astronomy of binary black hole mergers and identification of the hosts of the BBHs. Simulations allow us to backtrack from observations to the formation and evolution of binaries and the binary fraction of stellar clusters.

The formation of massive stars is still a mystery. The above considerations show that observations of multi-compact object mergers offer insights to massive star formation in the early universe. Considering the fact that massive stars are significant contributors to the ionization history of the early universe, the study of the details of the formation and subsequent events in the lives of massive multi-star systems attains great importance.

References

Abbott B. P. *et al.* (LIGO and Virgo Scientific Collaboration) 2016a, *PRL*, 116, 061102
Abbott B. P. *et al.* (LIGO and Virgo Scientific Collaboration) 2016b, *PRL*, 116, 241102
Abbott B. P. *et al.* (LIGO and Virgo Scientific Collaboration) 2016c, *PRL*, 116, 241103
Abbott B. P. *et al.* (LIGO and Virgo Scientific Collaboration) 2016d, *ApJL*, 818, L22
Abbott B. P. *et al.* (LIGO and Virgo Scientific Collaboration) 2016e, *ApJL*, 833, L1
Abbott B. P. *et al.* (LIGO and Virgo Scientific Collaboration) 2017a, *PRL*, 118, 221101

Abbott B. P. *et al.* (LIGO and Virgo Scientific Collaboration) 2017b, *ApJL*, 851, L35
Abbott B. P. *et al.* (LIGO and Virgo Scientific Collaboration) 2017c, *PRL*, 119, 141101
Abbott B. P. *et al.* (LIGO and Virgo Scientific Collaboration) 2017d, *PRL*, 119, 161101
Boylan-Kolchin M. 2018, *MNRAS*, 479, 332
Corsaro, E. *et al.* 2017, *NatAs*, 1E, 64
Gouliermis, D. A., Hony, S., & Klessen, R. S. 2014, *MNRAS*, 439, 3775
Habibi, M., Stolte, A., Brandner, W., Hussmann, B., & Motohara, K. 2013, *A&A*, 556, A26
Harayama. Y., Eisenhauer, F., & Martins, F. 2008, *ApJ*, 675, 1319
Kalogera V. 2000, *ApJ*, 541, 319
Kirk, H., Pineda, J., Johnstone, D., & Goodman, A. 2010, *ApJ*, 723, 457
Lim, B., Sung, H., & Hur, H. 2014, *ASPC*, 482, 225
Schneider, F. R. N., Sana, H., Evans, C. J., Bestenlehner, J. M., Castro, N., Fossati, L., Gr?fener, G., Langer, N. *et al.* 2018, *Science*, 359, 69
van den Heuvel, E. P. J. 1981, *IAUS*, 93, 155
Veitch J., Raymond V., Farr B., Farr W., Graff P., Vitale S. *et al.* 2015, *PRD*, 91, 042003
Vitale, S., Gerosa, D., Haster, C.-J., Chatziioannou, K., & Zimmerman, A. 2017, *PRL*, 119, 251103
Wilking, B. A, Vrba, F. J., & Sullivan, T. 2015, *ApJ*, 815, 2
Yu, H., & Wang, F. Y. 2016, *ApJ*, 820, 114

High-mass X-ray Binaries: Illuminating the Passage from Massive Binaries to Merging Compact Objects
Proceedings IAU Symposium No. 346, 2019
L. M. Oskinova, E. Bozzo, T. Bulik & D. Gies, eds.

doi:10.1017/S1743921318008189

Multifractal signatures of gravitational waves detected by LIGO

Daniel B. de Freitas[1], **Mackson M. F. Nepomuceno**[2] and **J. R. De Medeiros**[3]

[1]Departamento de Física, Universidade Federal do Ceará,
Caixa Postal 6030, Campus do Pici, 60455-900 Fortaleza, Ceará, Brazil
email: danielbrito@fisica.ufc.br

[2]Departamento de Ciência e Tecnologia, Universidade Federal Rural do Rio Grande do Norte-UFERSA, Campus Caraúbas, Rio Grande do Norte, Brazil

[3]Departamento de Física Teórica e Experimental,
Universidade Federal do Rio Grande do Norte-UFRN, Rio Grande do Norte, Brazil

Abstract. We analyze the data from the 6 gravitational waves signals detected by LIGO through the lens of multifractal formalism using the MFDMA method, as well as shuffled and surrogate procedures. We identified two regimes of multifractality in the strain measure of the time series by examining long memory and the presence of nonlinearities. The moment used to divide the series into two parts separates these two regimes and can be interpreted as the moment of collision between the black holes. An empirical relationship between the variation in left side diversity and the chirp mass of each event was also determined.

Keywords. gravitational waves — methods: statistical

---

1. Introduction

Since the first detection, five more signals have been confirmed as GWs: GW151226 (Abbott *et al.* 2016e), GW170104 (Abbott *et al.* 2017a), GW170608 (The LIGO Scientific Collaboration *et al.* 2017), GW170814 (Abbott *et al.* (2017b)) and GW170817 (Abbott *et al.* 2017c) (the only one coming from a system of coalescing neutron stars), and one signal remains as a suspected GW (LVT151012 Abbott *et al.* 2016e). The GW data used here are within the range of 32 seconds around the event and have a measurement frequency of 4096Hz. We will assume that these GWs, denoted by $y(t)$, are linear combinations of a deterministic signal, $d(t)$, and background noise, $n(t)$. In this context, the present analysis deals with observations that are collected over evenly spaced and discrete time intervals. In this Letter, we reports an analysis of a search for traces of multifractality in GW150914, a fact that may have strong consequences for our understanding of different characteristics of GW. A general discussion of all the GW signals detected to date (with the exception of GW170817) will be also presented. The signal of GW170817 (from coalescence of binary neutron stars) was removed from the sample since it differs in number of data from the signals produced by coalescence of black holes.

2. Multifractal analysis

In monofractal series, one exponent (the Hurst exponent, Hurst 1951) is sufficient to characterize the behavior of the series at various scales. H values of $0 < H < 0.5$ and $0.5 < H < 1$ indicate persistence and anti-persistence, respectively, while $H = 0.5$ indicates that the time series is uncorrelated. In multifractal time series, a range of values for this exponent is calculated. Thus, multifractal analysis consists of studying the scaling behavior in the time series $y(t)$. First, in accordance with the MultiFractal Detrending Moving Average (MFDMA) procedure, we calculated the mean-square function $F_\nu^2(n)$ for a ν segment of size n:

$$F_\nu^2(n) = \frac{1}{n} \sum_{i=1}^{n} [e_\nu(i)]^2, \tag{2.1}$$

where $e_\nu(i) = y(i) - \tilde{y}(i)$ is the residual series in the segment ν and $\tilde{y}(i)$ is the moving average function. However, some authors have shown that semi-sinusoidal and power-law trends in multifractal approaches, including Multifractal Detrended Fluctuation Analysis (MFDFA) and MFDMA, are not efficiently removed Eghdami *et al.* (2017). In our study, we did not encounter this problem. We then calculated the q_{th} order overall fluctuation function $F_q(n)$, which is given by

$$F_q(n) = \left\{ \frac{1}{N_n} \sum_{\nu=1}^{N_n} F_\nu^q(n) \right\}^{1/q} \quad \text{for } q \neq 0 \tag{2.2}$$

and, for $q = 0$,

$$\ln\left[F_0(n)\right] = \frac{1}{N_n} \sum_{\nu=1}^{N_n} \ln[F_\nu(n)], \tag{2.3}$$

where N_n is the number of segments non-overlaping. For larger values of n, the fluctuation function follows a power-law given by

$$F_q(n) \sim n^{h(q)}. \tag{2.4}$$

The generalized Hurst exponent $h(q)$ is related to standard multifractal analysis parameters such as the Renyi scaling exponent (τ), which is given by

$$\tau(q) = qh(q) - 1, \tag{2.5}$$

when $q = 2$, we return to using monofractal analysis, i.e., $h(2) = H$ is the Hurst exponent.

Two other important parameters are obtained using a Legendre transform, defined as

$$\alpha = \frac{d\tau(q)}{dq}, \quad \alpha \in [\alpha_{min}, \alpha_{max}] \tag{2.6}$$

and

$$f(\alpha) = q\alpha - \tau(q), \tag{2.7}$$

which are the Hölder exponent and singularity spectrum, respectively.

One way to measure the degree of multifractality $(\Delta\alpha)$ in a series is by using the width of the multifractal singularity spectrum, which Tanna & Pathak (2014) and Ashkenazy *et al.* (2003) defined as the difference between the maximum and minimum values of the Hölder exponent, i.e., $\Delta\alpha = \alpha_{max} - \alpha_{min}$.

3. Results and discussion

We analyze the data from the 6 gravitational waves signals detected by LIGO identified as GW151226, GW170104, GW170608, GW170814, GW170817 and LVT151012. All of data were extracted from LIGO. Data were analyzed using the multifractal formalism. Our aim is to study the possible sources of multifractality and to extract a set of multifractality indexes.

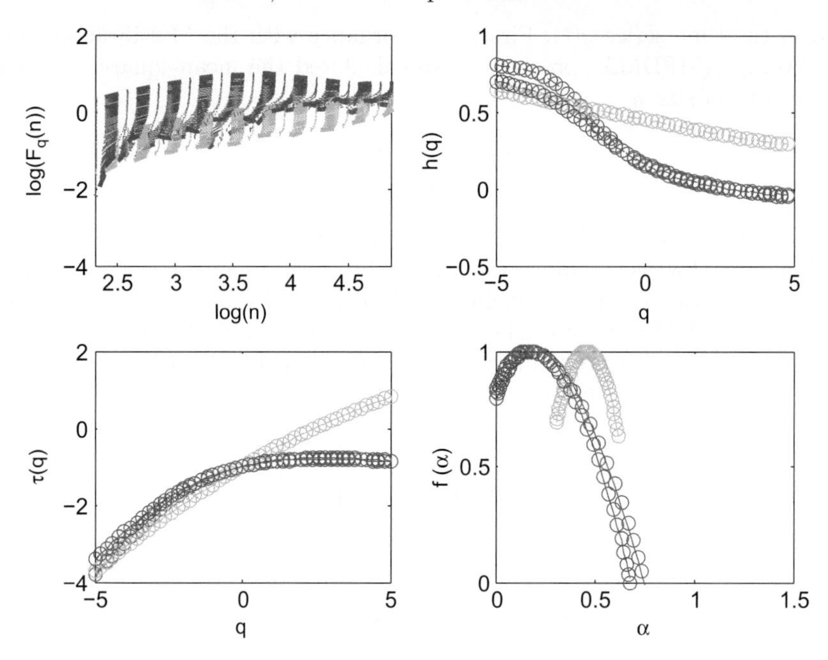

Figure 1. Multifractal analysis of the full data from GW150914 (`H1data`). The top-left panel shows the fluctuation function versus the multi-scale behavior in a log-log diagram. The original series is in red, the shuffled series is in green, and the upper and lower limits correspond to $q=5$ and $q=-5$, respectively, while the bold in the middle corresponds to $q=0$. Dependences on the q_{th} moment of the generalized Hurst exponent, $h(q)$, and the multifractal scaling exponent, $\tau(q)$, are shown in the top-right and bottom-left panels, respectively. The multifractal spectrum is shown in the bottom-right panel.

To investigate the source of multifractality, we applied the shuffled method to the original series (the green curves in Figure 1). This method destroys the memory signature, but preserves the distribution of the data with $h(q)=0.5$, if the source of multifractality in time series only presents long-range correlations (de Freitas *et al.* 2017). We realized that the multifractal behavior remains but with lowered strength. Similarly, the surrogate method (the blue curves in Figure 1) also could not eliminate the multifractality in the original series. Already, this method destroys effects of non-linearity of the original series by randomizing the Fourier phases. These results indicate that the source of the multifractality is not only related to long-range correlations but also linked to the existence of non-linear terms that produce a heavy-tailed probability density function (PDF). The same analysis described in the previous three paragraphs was applied to the other three waves and indicated similar behavior both for the Hanford and Livingston detector data.

To study the evolution of the parameters related to the multifractal singularity spectrum throughout the time series, we constructed Fig. 2, with the original time series shown in green, for the Hanford data in the top panel and Livingston in the bottom panel. The parameter values at one point in the time series data reflect the values calculated up to that point in a 50-point data window. As seen in Fig. 2, the left side diversity $(\Delta f_L(\alpha))$ of the multifractal singularity spectrum, defined as $1 - f(\alpha)^{left}_{min}$, is shown in blue and is associated with the sensitivity of the series to small-scale fluctuations with large magnitudes. In the same Figure, the right side diversity $(\Delta f_L(\alpha))$ of the multifractal singularity spectrum, denoted by $1 - f(\alpha)^{right}_{min}$, is shown in black and is linked to the sensitivity to fluctuations in the series with small magnitudes Tanna & Pathak (2014). Furthermore, the parameters $(\Delta f_L(\alpha))$ and $(\Delta f_R(\alpha))$ indicate either a left or

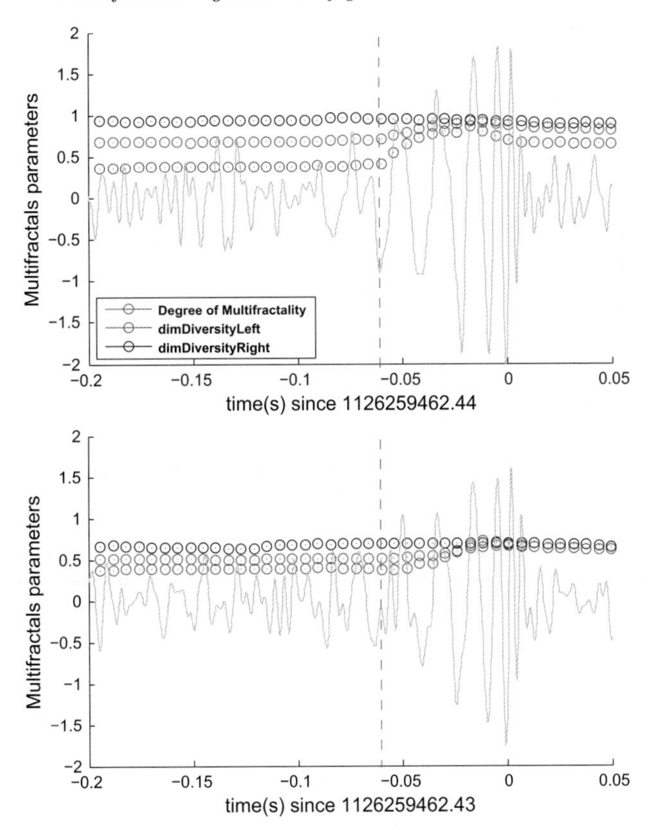

Figure 2. Point-to-point multifractal analysis for the GW150914 time series from Livingston (bottom panel) and Hanford (top panel; shifted and inverted Abbott *et al.* 2016a), illustrated in green. Red circles represent the degree of multifractality ($\Delta\alpha$) calculated in the time series up to that point; likewise, blue and black circles represent the left side diversity $f(\alpha)_{max} - f(\alpha)_{min}^{left}$ and right side diversity $f(\alpha)_{max} - f(\alpha)_{min}^{right}$, respectively. The vertical lines represent $t = -0.06s$, the time point at which the time series are divided.

right truncation of the multifractal spectrum, respectively For this analysis, the parameters were calculated for the signal in the interval between 1 second before the event (for GW150914, this time is 1126259462.44s) and 0.05s after the event. We can observe a slight increase in the left side diversity at $t = -0.06s$, which indicates the presence of a strong small-scale fluctuation. This behavior appears in the data analysis of the two advanced LIGO detectors, H1 and L1. Using this time point, we divided the original series into two parts, wherein the first is identified as `H1data1` with 3581 measurements, while the second part comprises 720 measurements and is identified as `H1data2` data.

Using the same procedure as that for the entire time series, we performed multifractal analyses on both the `H1data1` and `H1data2` data. The shuffled method has eliminated the multifractality contained in the `H1data1`, shown in the right pane, and for the shuffled series, $h(2) = 0.5429$ and $\Delta\alpha = 0.0288$. These results indicate that the multifractality present in `H1data1` is due only to long-term correlations and thus does not provide non-linear terms. These correlations can be understood as stemming from the periodic orbital motion of the black holes. As for `H1data1`, multifractality is still present for the original time series, but neither the shuffled nor surrogate methods could eliminate the multifractal behavior; i.e., the multifractal behavior is due to two possible sources, i.e., memory and non-linearity.

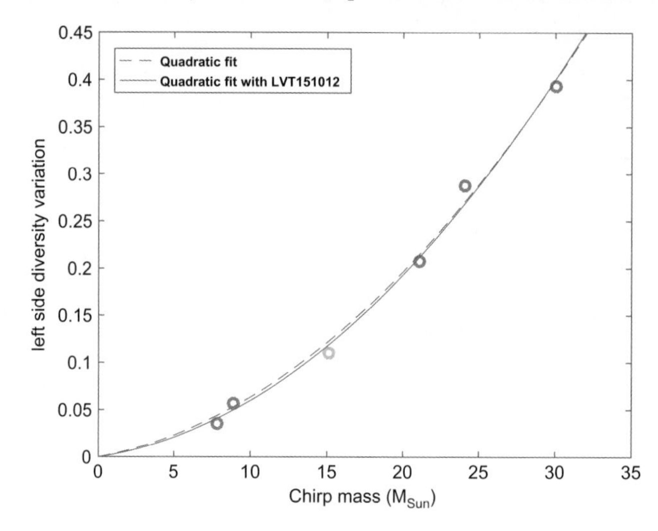

Figure 3. The correlation between left side diversity variation (ΔD_L) and chirp mass for each GW (circles in blue) and LVT151012 (circle in green). The dashed line indicates a quadratic fit adjustment without the LVT signal, and the solid line is the same fit when considering this signal.

These results have two consequences: first, the entire contribution of non-linearity in the analysis of the complete time series occurs in the second part of the series; second, as the periodic movement continues, even in the ringdown phase, the terms associated with long-term correlations continue to appear in the series. The enlargement of the PDF is because the amplitude of the strain grows somewhat in the second part of the series. Given that the strain amplitude is linked to the orbital velocity and mass that generated the gravitational wave, these nonlinear terms are caused by the collision of the black holes. In short, the contribution of long-term temporal correlation is due to the periodic motion of the orbiting black holes, and nonlinear terms occur due to the increase in the strain amplitude.

The difference, presented in the Figure 2, between the maximum and minimum value of the left side diversity in the GW amplitude region of increase can be considered as the variation in left side diversity ($\Delta f_L(\alpha)$), as indicated in Figure 2 for GW150914. We find an empirical correlation between this parameter and the chirp masses of each signal. Figure 3 illustrates these parameters in blue circles for GWs and green circles for the LVT. A quadratic fit with (solid line) and without (dashed line) the LVT151012 signal is also shown in the same Figure. The overlap of these lines indicates that the analysis of the LVT signal falls within the expected behavior according to this correlation. Since we associate the variation in left side diversity with the amplitude increase in the signal, which in turn is related to chirp mass, we are led to conclude that this is an expected correlation. The detection of new GWs can serve as a good test for the correlation found here as well as a check for the chirp mass value of the detected signal.

4. Conclusions

The statistical approach proposed in this study highlights the scenario opened by detection of the first GWs. We summarize the main results in three points: i) characterize the fractal dynamics of the signals, identifying their multifractal sources; ii) find the moment of the beginning of merger phase in black hole coalescence system, and; iii) determine the empirical relationship between the variation in left side diversity and chirp mass as an additional way for estimating this latter parameter. The methodology applied here may serve as a standard procedure for future analyses of gravitational waves. The prospect of new gravitational wave observatories, both on the ground and in space, provides more opportunities for the field of astronomy to employ the statistical tools already widely used in other areas of knowledge.

References

Aasi, J., Abbott, B. P., Abbott, R., *et al.* 2015, *Classical and Quantum Gravity*, 32, 074001
Abbott, B. P., Abbott, R., Abbott, T. D., *et al.* 2016, *Physical Review Letters*, 116, 061102
Abbott, B. P., Abbott, R., Abbott, T. D., *et al.* 2016, *Physical Review Letters*, 116, 131102
Abbott, B. P., Abbott, R., Abbott, T. D., *et al.* 2016, *Physical Review Letters*, 116, 131103
Abbott, B. P., Abbott, R., Abbott, T. D., *et al.* 2016, *Classical and Quantum Gravity*, 33, 134001
Abbott, B. P., Abbott, R., Abbott, T. D., *et al.* 2016, *Physical Review Letters*, 116, 241103
Abbott, B. P., Abbott, R., Abbott, T. D., *et al.* 2016, *ApJL*, 826, L13
Abbott, B. P., Abbott, R., Abbott, T. D., *et al.* 2017, *Physical Review Letters*, 118, 221101
Abbott, B. P., Abbott, R., Abbott, T. D., *et al.* 2017, *Physical Review Letters*, 119, 141101
Abbott, B. P., Abbott, R., Abbott, T. D., *et al.* 2017, *Physical Review Letters*, 119, 161101
The LIGO Scientific Collaboration, the Virgo Collaboration, Abbott, B. P., *et al.* 2017, arXiv:1711.05578
Alessio, E., Carbone, A., Castelli, G., & Frappietro, V. 2002, *European Physical Journal B*, 27, 197
Arneodo, A., Bacry, E., Graves, P. V., & Muzy, J. F. 1995, *Physical Review Letters*, 74, 3293
Ashkenazy, Y., Baker, D. R., Gildor, H., & Havlin, S. 2003, *Geophysics Research Letters*, 30, 2146
Blanchet, L., Damour, T., Iyer, B. R., Will, C. M., & Wiseman, A. G. 1995, *Physical Review Letters*, 74, 3515
Coyne, R., Corsi, A., & Owen, B. J. 2016, *Physical Review D*, 93, 104059
de Freitas, D. B., Nepomuceno, M. M. F., de Moraes Junior, P. R. V., *et al.* 2016, *ApJ*, 831, 87
de Freitas, D. B., Nepomuceno, M. M. F., Gomes de Souza, M., *et al.* 2017, *ApJ*, 843, 103
Eghdami, I., Panahi, H., & Movahed, S. M. S. 2017, arXiv:1704.08599
Feder, J. 2013. Fractals. (Springer Science & Business Media)
Gu, G.-F., & Zhou, W.-X. 2010, *Physical Review E*, 82, 011136
Hurst, H. E. 1951, *Transactions of the American Society of Civil Engineers*, 116, 770
Kantelhardt, J. W., Zschiegner, S. A., Koscielny-Bunde, E., *et al.* 2002, *Physica A Statistical Mechanics and its Applications*, 316, 87
Mali, P. 2016, *Journal of Statistical Mechanics: Theory and Experiment*, 1, 013201
Muzy, J. F., Bacry, E., & Arneodo, A. 1991, *Physical Review Letters*, 67, 3515
Muzy, J. F., Bacry, E., & Arneodo, A. 1994, *International Journal of Bifurcation and Chaos*, 4, 245
Norouzzadeh, P., Dullaert, W., & Rahmani, B. 2007, *Physica A Statistical Mechanics and its Applications*, 380, 333
Peng, C.-K., Buldyrev, S. V., Havlin, S., *et al.* 1994, *Physical Review E*, 49, 1685
Tanna, H. J., & Pathak, K. N. 2014, *Astrophysics and Space Science*, 350, 47

High-mass X-ray Binaries: Illuminating the Passage from Massive Binaries to Merging Compact Objects
Proceedings IAU Symposium No. 346, 2019
L. M. Oskinova, E. Bozzo, T. Bulik & D. Gies, eds.

doi:10.1017/S1743921318008323

The masses of 18 pairs of double neutron stars and implications for their origin

ChengMin Zhang[1,2,3] and YiYan Yang[1,4]

[1]CAS Key Lab of FAST, National Astronomical Observatories, Chinese Academy of Sciences, Beijing 100101, China

[2]Key Lab of Radio Astronomy, Chinese Academy of Sciences, Beijing 100101, China

[3]School of Physical Science, University of Chinese Academy of Sciences, Beijing 100049, China

[4]Astronomy Department, Beijing Normal University, Beijing 100875, China
emails: zhangcm@bao.ac.cn, yangyiyan@gznc.edu.cn

Abstract. For the observed 18 pairs of double neutron star (DNS) systems, we find that DNS mass distribution is very narrow and its mean value (about 1.34 solar mass) is less than the mean of all measured pulsars of about 1.4 solar mass. To interpret the special DNS mass characteristics, we analyze the DNS formation process, via the phases of HMXBs, by investigating the evolution of massive binary stars. Moreover, in DNSs, two classes of NSs are taken into account, formed by supernova (SN) and electron capture (EC), respectively, and generally the NS mass by SN is bigger than that by EC. Quantitatively, with various initial conditions of binary stars, the observed special DNS distribution can be satisfactorily explained.

Keywords. pulsar, double neutron star, binary system, mass

---

1. Introduction

The DNS masses are important parameters to constrain the gravitational wave by the NS merge, which has been detected by LIGO (e.g., Abbott *et al.* 2017), so the statistics of DNS masses not only infers the DNS formation process and the DNS birth rate but also implies the message of gravitational wave background in the universe. Until now, more than 80 NS masses have been measured in binary pulsar systems. However, the masses of DNSs (mean of 1.34 solar mass) are systematically lower than the mean (1.4 solar mass) of all pulsars, and show a very narrow distribution with a deviation of **0.3** solar mass (see Fig. 1 and Fig. 2) (e.g., Zhang *et al.* 2011; Miller & Miller 2015; Özel & Freire 2016; Tauris *et al.* 2017). So, this special DNS mass distribution should be involved in their formation process.

In this short paper, we propose a schematic phenomenon model to interpret the DNS mass by considering the massive binary star evolution process. It is usually believed that the first massive star in binary system will experience a SN to form a NS, then the second massive star has two choices to form a NS, experiencing a SN explosion if its mass over 10 solar masses or an electron capture if its mass ranges 8-10 solar masses (e.g., Podsiadlowski *et al.* 2004). Generally, a NS by SN should has a big kick, arising a big eccentricity system (e.g. Hulse-Taylor pulsar system: PSR B1913+16 (e.g., Hulse & Taylor 1975; Weisberg & Huang 2016)), while a NS by EC (e.g., Nomoto 1984) has a null kick, arising a circular orbit (van den Heuvel 2004; Yang *et al.* 2017) (e.g. double pulsar system: PSR J0737-3039 (e.g., Lyne *et al.* 2004; Kramer *et al.* 2006)).

2. DNS mass formation

For simplicity, we discuss a presentive case of massive binary star system, where the masses of both progenitor stars are much more massive than 10 solar masses and they

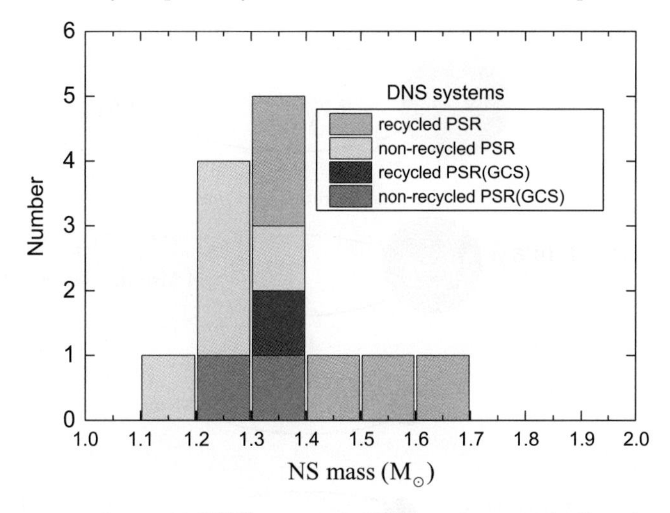

Figure 1. Histogram of measured NS masses in DNS systems, including the recycled pulsars and non-recycled ones, where GCS represents the globular cluster system.

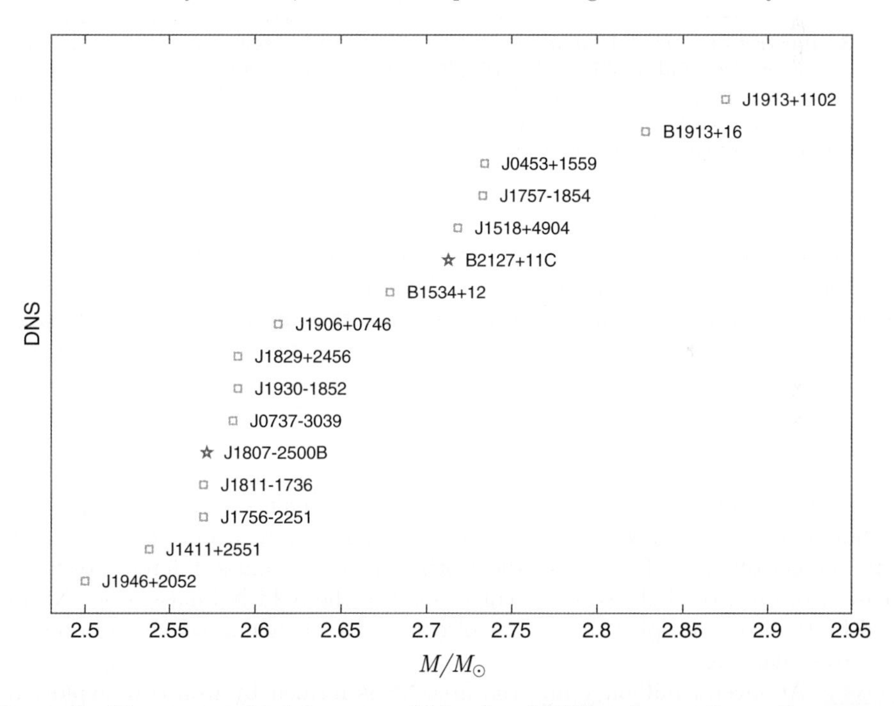

Figure 2. The measured total masses of 16 pairs of DNS systems, **where the squares (pentagrams) represent Galactic (GCS) DNSs.**

separate a distance much over the radii of stars that corresponds the orbital period of about 10 days. In such an orbital span, both stars, primary one and secondary companion, have the mass of 20 and 10 solar masses, respectively, and they will interact each other and the mass exchange happens at the initial stage of binary evolution, the illustration of which is shown in Fig. 3 (e.g., Bhattacharya & van den Heuvel 1991). On the formation process of DNS, we divide it into different evolution phases, where we estimate and discuss the mass status of both components, based on the knowledge of star and NS evolution.

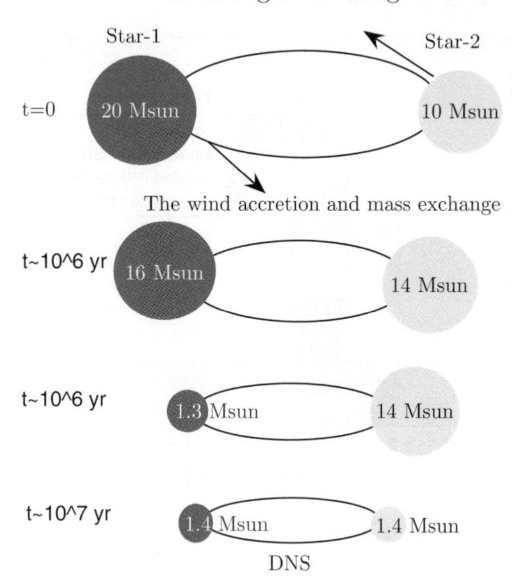

Figure 3. Illustration of DNS formation process. The orbital parameters evolution of DNS have been studied (e.g., Podsiadlowski *et al.* 2005). Tauris showed an illustration of DNS formation, which contains a pair of massive primordial binaries and goes through a supernova explosion to form DNS, and eventually merges into a black hole (e.g., Tauris *et al.* 2017). Here, we focus on the mass exchange during the DNS formation.

Phase 1. As shown in Fig. 3, at the initial phase, both stars have the mass exchange by the stellar winds before the primary enters the post-main-sequence phase, after onset of the Roche-lobe overflow a lot of mass will release from star-1 to star-2 until the SN explosion happens in star-1, while the masses of both stars are 16 and 14 solar masses respectively. This means that the mass distribution of both stars are varied, and the bigger star-1 becomes small while the less massive star-2 becomes bigger than before. The mass exchange will last several million years until star-1 explodes. The shorter the orbit scale (orbit period), more homogeneous of both star masses. If the orbit period is over 100 days, both stars should less interact, then the kick velocity of NS in such a system should generally exceed the orbital velocity and makes both DNSs separate away. This should be the reason why there is no DNS system of long orbital period of 100 days. Say, the mass exchange of binary of short orbit makes the mass difference of both stars decrease. However, the light star will correspond to the light NS mass, since NS mass is formed by the iron core plus the matter of convective elements, which is proportionally related to stellar mass.

Phase 2. At several million years, the first NS is formed by iron core explosion, and some crust matter of NS is lately formed by the fall-back matter of SN materials. At moment the NS mass is expressed by the gravitational energy coefficient 0.85 times Chandrasekha mass $M_{ch} = 1.44\ \mathrm{M}_\odot$ and fall-back mass M_f, shown as $\mathrm{M} = 0.85\ M_{ch} + M_f$. If there is no second star or it is far away, $M_f \sim$ 0.1-0.5 $\mathrm{M}_\odot$ should be all fallen onto NS, which results in a NS mass of about 1.4 $\mathrm{M}_\odot$. At the star explosion stage, outer iron core may be expelled, so the collapsed core mass can be less than M_{ch}, which can explain why some NS masses are close to 1.2 solar mass. However, in binary system, the fall-back mass is almost absorbed by the star-2, which makes $M_f \sim 0.1\mathrm{M}_\odot$. Thus, the mass of DNS1 is now about $1.3\mathrm{M}_\odot$.

Phase 3. The secondary explodes will follow the similar procedure to star-1, but with less mass. Its fall-back matter will split into two similar parts, and then almost equally shared by two NSs. For $M_{f2} = 0.2$ M$_\odot$, $0.5M_{f2} \sim 0.1$ M$_\odot$ will add both NSs, which makes NS1 bigger than 1.4 M$_\odot$ and NS2 less than 1.4 M$_\odot$, like PSR B1913+16. This picture of DNS mass formation can explain why NS1 is almost bigger than NS2. If the star-2 has more convective matter, then NS2 can be bigger than NS1. If the star-2 is a light one with 8 solar masses, the EC will happen and produce a NS of 1.25 M$_\odot$. For the EC, there is little fall-back matter, so both masses are light, 1.3 and 1.25 M$_\odot$ respectively, which can explain the case of double pulsar PSR J0737-3039.

3. Summary

The special properties of DNS masses, which has the systematical low value and is narrowly distributed, are investigated, based on the evolution of binary star. A simple and approximated formula that describes the DNS mass (M_1 and M_2 for recycled and non-recycled NS) is proposed as below,

$$M_1 = 0.85M_{ch} + 0.1M_{f1} + M_{ac} + 0.5M_{f2} \tag{3.1}$$

$$M_2 = 0.85M_{ch} + 0.5M_{f2} \tag{3.2}$$

where $M_{acc} \sim 0.05$M$_\odot$ is the accretion mass of recycled NS. For the fall-back mass of SN-type NS, $M_{f1} \sim M_{f2} \sim 0.2M_\odot$, whereas for the EC-type NS, $M_{f2} \sim 0.1$M$_\odot$, so the above formula can explain the DNS mass distribution properly. The physical detail of the formula will be studied in the subsequent work.

Acknowledgements

This research has been supported by the National Program on Key Research and Development Project (Grant No. 2016YFA0400801), the Strategic Priority Research Program of the Chinese Academy of Sciences, (Grant No. XDB23000000) and CAS Interdisciplinary Innovation Team, the National Natural Science Foundation of China NSFC (U1731238), the fundamental research funds for the central university. the Innovation Talent Team (Grant No. (2015)4015) and the High Level Creative Talents program (Grant No. (2016)-4008) of Guizhou provincial depart of Science and Technology.

References

Abbott B. P., Abbott R., Abbott T. D., *et al.* 2017, *Physical Review Letters*, 118, 221101
Bhattacharya, D., & van den Heuvel, E. P. J., 1991, *Physics Reports*, 203, 1
Hulse, R. A., Taylor, J. H. 1975, *ApJL*, 195, 51
Kramer, M., Stairs, I. H., Manchester, R. N., *et al.* 2006, *Science*, 314, 97
Lyne, A. G., Burgay, M., Kramer, M., *et al.* 2004, *Science*, 303, 1153
Miller, M. C., & Miller, J. M. 2015, *Physics Reports*, 548, 1
Nomoto, K., 1984, *ApJ*, 277, 791
Özel, F., & Freire, P., 2016, *ARAA*, 54, 401
Podsiadlowski, P., Langer, N., Poelarends, A. J. T., *et al.* 2004, *ApJ*, 612, 1044
Podsiadlowski, P., Dewi, J. D. M., & Lesaffre, P. 2005, *MNRAS*, 361, 1243
Tauris T. M., Kramer M., Freire P. C. C. *et al.* 2017, *ApJ*, 846, 170
van den Heuvel, E. P. J., 2004, *Science*, 303, 1143
Hulse, R. A., Taylor, J. H. 1975, *ApJL*, 195, 51
Weisberg, J. M., & Huang, Y. 2016, *ApJ*, 829, 55
Yang, Y. Y., Zhang, C. M., Li, D., *et al.* 2017, *ApJ*, 835, 185
Zhang, C. M., Wang, J., Zhao, Y. H., *et al.* 2011, *A & A*, 527, 83

High-mass X-ray Binaries: Illuminating the Passage from Massive Binaries to Merging Compact Objects
Proceedings IAU Symposium No. 346, 2019
L. M. Oskinova, E. Bozzo, T. Bulik & D. Gies, eds.

doi:10.1017/S1743921319000012

ONe WD+He star systems as the progenitors of IMBPs

Bo Wang[1,2] and Dongdong Liu[1,2]

[1]Yunnan Observatories, Chinese Academy of Sciences, Kunming 650216, China
email: wangbo@ynao.ac.cn

[2]Key Laboratory for the Structure and Evolution of Celestial Objects,
Chinese Academy of Sciences, Kunming 650216, China

Abstract. Previous theoretical studies can only explain part of the observed intermediate-mass binary pulsars (IMBPs) with short orbital periods. Note that an ONe white dwarf (WD) accreting mass from a He star may experience the accretion-induced collapse process and eventually form IMBPs, known as the ONe WD+He star scenario. By investigating the evolution of a large number of ONe WD+He star binaries, we found that the ONe WD+He star scenario can form IMBPs including pulsars with $5 - 340$ ms spin periods, and the orbital periods range from 0.04 to 900 d. Compared with the observed IMBPs, this scenario can cover almost all of the IMBPs with short orbital periods. Thus, we suggest that the ONe WD+He star channel is responsible for the formation of IMBPs with short orbital periods.

Keywords. binaries: close – stars: evolution – white dwarfs – stars: neutron

1. Introduction

In the observations, intermediate-mass binary pulsars (IMBPs) are composed of a pulsar with a spin period of about 10–200 ms and a heavy ($M_{\rm WD} > 0.4\,{\rm M}_\odot$) carbon-oxygen (CO) or oxygen-neon (ONe WD) (Camilo *et al.* 1996, 2001; Edwards & Bailes 2001). However, the progenitors of these IMBPs are still not understood. Generally, it has been suggested that most IMBPs originate from the intermediate-mass X-ray binaries (IMXBs) including a neutron star (NS) and a 2.0–10.0 $\rm M_\odot$ H-rich donor star (e.g. van den heuvel 1975). However, this channel cannot produce IMBPs with orbital periods shorter than 3 d (see Tauris, van den Heuvel & Savonije 2000). Chen & Liu (2013) suggested that the NS+He star channel can explain the formation of 4 IMBPs with short orbital periods.

However, these models can only explain part of IMBPs with orbital periods shorter than 3 d (e.g. Chen & Liu 2013). Another formation channel for IMBPs is the ONe WD+He star scenario, in which an ONe WD accreting mass from a H star will experience the accretion-induced collapse process when the WD mass approaches the Chandrasekhar mass limit, and can eventually form an IMBP when the He star turns to be a WD (see Tauris *et al.* 2013). However, Tauris *et al.* (2013) only considered the case with the initial mass of the primary ONe WDs $M^{\rm i}_{\rm WD} = 1.2\,{\rm M}_\odot$. In this work, we evolve a large number of ONe WD+He star systems for the formation of IMBPs with $M^{\rm i}_{\rm WD} = 1.0, 1.1, 1.2$ and $1.3\,{\rm M}_\odot$, and found that almost all of the observed IMBPs with short orbital periods can be explained by the ONe WD+He star scenario.

2. Model and Results

In order to obtain the parameter space of IMBPs from the ONe WD+He star scenario, we employed the Eggelton stellar evolution code (Eggleton 1973), and carried

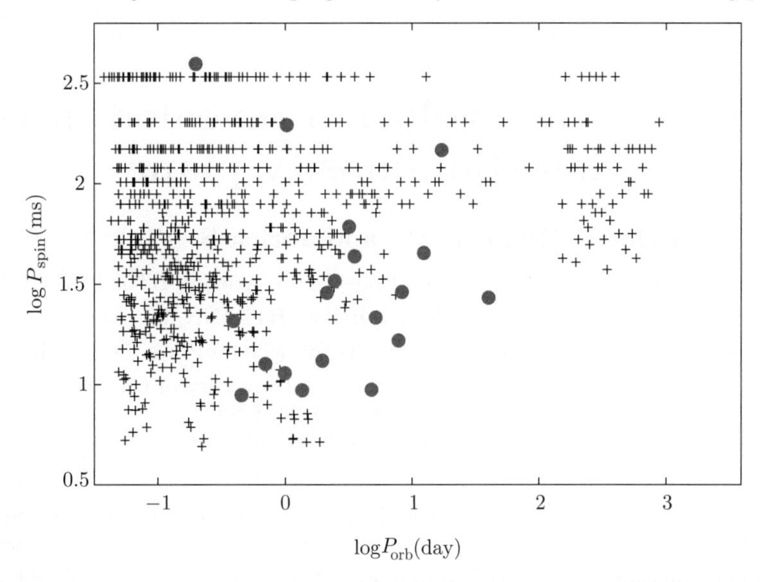

Figure 1. Parameter space of the formed IMBPs in the orbital period−spin period ($\log P_{\rm orb}-P_{\rm spin}$) plane. The orbital periods are measured at the formation moment of IMBPs. The plus symbols denote the predicted IMBPs by our simulation, and the filled circles represent the observed IMBPs (see Manchester *et al.* 2005). Source: Liu *et al.* (2018).

out a number of detailed binary evolution calculations for ONe WD+He star systems that would experience the accretion-induced collapse process and eventually produce IMBPs. We then provided the parameter space for the formed IMBPs in the orbital period−spin period plane, and compared with the observed IMBPs taken from the ATNF Pulsar Catalogue in 2017 October (Manchester *et al.* 2005; see http://www.atnf.csiro.au/research/pulsar/psrcat).

In Fig. 1, we present the parameter space of the predicted IMBPs in the orbital period−spin period plane (i.e. the Corbet diagram; Corbet 1984). In this figure, the spin periods of recycled NSs are in the range of $\sim 5-340$ ms, and the orbital periods are distributed in the range of $\sim 0.04-900$ d. In this figure, we also plot the 19 observed IMBPs by filled circles. We found that the orbital periods and spin periods of 13 observed IMBPs can be covered by the ONe WD+He star scenario, and almost all the observed IMBPs with orbital periods shorter than 3 d can be covered.

References

Camilo, F., Nice, D. J., Shrauner, J. A., & Taylor, J. H., 1996, *ApJ*, 469, 819

Camilo, F. *et al.* 2001, *ApJ*, 548, L187

Chen, W. C., & Liu, W. M. 2013, *MNRAS*, 432, L75

Corbet, R. H. D. 1984, *A&A*, 141, 91

Edwards, R. T., & Bailes, M. 2001, *ApJ*, 547, L37

Eggleton, P. P. 1973, *MNRAS*, 163, 279

Liu, D., Wang, B., Chen, W., Zuo, Z., & Han, Z. 2018, *MNRAS*, 477, 384

Manchester, R. N., Hobbs, G. B., Teoh, A., & Hobbs, M. 2005, *AJ*, 129, 1993

Tauris, T. M., Sanyal, D., Yoon, S. C., & Langer, N. 2013, *A&A*, 558, A39

Tauris, T. M., van den Heuvel, E. P. J., & Savonije, G. J. 2000, *ApJ*, 530, L93

van den Heuvel, E. P. J. 1975, *ApJ*, 198, L109

High-mass X-ray Binaries: Illuminating the Passage from Massive Binaries to Merging Compact Objects
Proceedings IAU Symposium No. 346, 2019
L. M. Oskinova, E. Bozzo, T. Bulik & D. Gies, eds.

doi:10.1017/S1743921319001170

Massive star evolution revealed in the Mass-Luminosity plane

Erin R. Higgins[1,2,3,4] **and Jorick S. Vink**[1,4]

[1]Armagh Observatory and Planetarium, College Hill, Armagh BT61 9DG, N. Ireland
[2]Queen's University of Belfast, Belfast BT7 1NN, N. Ireland
[3]Dublin Institute for Advanced Studies, 31 Fitzwilliam Place, Dublin, Ireland
[4]Kavli Institute for Theoretical Physics, University of California, Santa Barbara, CA 93106, USA
emails: eh@arm.ac.uk, jsv@arm.ac.uk

Abstract. Massive star evolution is dominated by key physical processes such as mass loss, convection and rotation, yet these effects are poorly constrained, even on the main sequence. We utilise a detached, eclipsing binary HD166734 as a testbed for single star evolution to calibrate new MESA stellar evolution grids. We introduce a novel method of comparing theoretical models with observations in the 'Mass-Luminosity Plane', as an equivalent to the HRD (see Higgins & Vink 2018). We reproduce stellar parameters and abundances of HD166734 with enhanced overshooting (α_{ov}=0.5), mass loss and rotational mixing. When comparing the constraints of our testbed to the systematic grid of models we find that a higher value of α_{ov}= 0.5 (rather than α_{ov}= 0.1) results in a solution which is more likely to evolve to a neutron star than a black hole, due to a lower value of the compactness parameter.

Keywords. stars: mass loss, stars: evolution, stars: rotation, convection

1. Introduction

The lives of massive stars ($M_{init} \geqslant 8\,M_\odot$) are influenced by the key processes acting on their structure, including stellar winds, convection and rotation. The degree by which these processes affect evolutionary paths is dictated by the initial mass, metallicity and multiplicity of the star. In the mass range of ~8-30 $M_\odot$ rotation is the dominant effect, whereas above ~30 $M_\odot$ the evolution is heavily dominated by mass loss via stellar winds, e.g Maeder & Meynet (2000). The main-sequence (MS) lifetime is dependent on the extension of the convective core by overshooting due to extra mixing of hydrogen (H) in the core. Although this process remains largely unresolved it is key in determining the final phases of evolution (e.g. final masses which form neutron stars or black holes) since the MS accounts for 90% of O star lifetimes, (Claret & Torres 2017). Martins & Palacios (2013) present an overview of current evolutionary codes e.g. Ekström *et al.* (2012), Brott *et al.* (2011b) as well as their implementations of various physical processes. The differing treatment of rotation, mass loss and convective overshooting has led to a diversity of evolutionary masses and MS-lifetimes. Weidner & Vink (2010) discuss the effects this may have on the mass determination of O stars and further discrepancies when compared to spectroscopic masses (i.e. the 'mass discrepancy' problem). In this paper we aim to compare new and existing prescriptions with one code, utilising the highly flexible MESA code e.g. Paxton *et al.* 2011. We investigate the evolution of a high mass, detached, eclipsing binary HD166734, modelled as a testbed for single star evolution with constraints on α_{ov} and $\dot{M}$. Dynamical masses of 39.5 $M_\odot$ and 33.5 $M_\odot$, for the primary

Table 1. Fundamental properties of HD 166734, from Mahy *et al.* (2017).

	Primary	Secondary
$T_{\rm eff}$ [K]	32000 ±1000	30500 ±1000
log(L/ $L_\odot$)	5.840 ± 0.092	5.732 ± 0.104
$M_{\rm dyn}$ [$M_\odot$]	39.5 ± 5.4	33.5 ± 4.6
$M_{\rm spec}$ [$M_\odot$]	37.7 ± 29.2	31.8 ± 26.6
v sin i [km s^{-1}]	95 ± 10	98 ± 10
[N/H]	8.785	8.255

and secondary respectively, have allowed exploration of the dominant processes for the entire O star mass range, since in this mass regime the effects of rotation, mass loss and overshooting all play as role as they interact and overlap. We develop a novel method of analysing the evolution of our models alongside observations of HD166734 in the Mass-Luminosity plane, and present a calibrated grid of rotating and non-rotating models with a sample of Galactic O stars.

2. Methodology

The stellar evolution code MESA has been utilised in this study in developing a set of theoretical models for calibration of physical processes such as convective overshooting, mass loss and rotation. The convective core boundary is defined by the Ledoux criterion with step-overshooting applied as the extension of the core by a factor $\alpha_{\rm ov}$ of the pressure scale height $H_{\rm p}$. Mass-loss rates are employed by the Vink *et al.* (2001) prescription, accounting for the bi-stability jump at 21kK and metallicity dependencies. We adopt the default metallicity of Z=0.02 in MESA to provide comparisons with a galactic observations. Rotation is applied in a fully diffusive approach with Eddington-Sweet circulation and dynamical and secular shear instabilities accounted for.

Mahy *et al.* (2017) provide a rare opportunity in the analysis of the non-interacting binary HD166734, since the dynamical masses of both stars are in excellent agreement with their spectroscopic masses, thus allowing calibration of evolutionary masses through our models. Table 1 highlights the main stellar parameters of HD166734, including exact positions in the HRD and surface nitrogen abundances. We apply an equal-age assumption for the binary such that they have evolved from the same initial stage, allowing for constraints of the MS-width and rotation rates.

3. Mixing and Mass loss

In order to minimise interacting processes, we firstly calculated non-rotating models which solely employ mass loss and overshooting as mixing processes, with the aim of reproducing the 3:1 ratio of surface nitrogen abundances between the primary and secondary, and HRD positions simultaneously. We explore a range of factors of the Vink *et al.* (2001) mass-loss prescription (0.1-10 times), alongside a variety of overshooting $\alpha_{\rm ov}$= 0.1-0.8, and find that all variations of this parameter space result in either negligible surface nitrogen enrichment, or a factor of 10 enrichment, representative of CN-equilibrium. We conclude that a large value of $\alpha_{\rm ov}$ coupled with an increased mass-loss rate results in stripping of the stellar envelope exposing CNO material at the core boundary. Thus another mechanism of mixing is required to reproduce any intermediate surface enrichment, we hence explore the addition of rotation to our models. Figure 1 demonstrates the surface nitrogen abundances of both non-rotating (red) and rotating (blue) models for a range of $\alpha_{\rm ov}$ and $\dot{M}$, we note that intermediate enrichment only occurs for rotating models. However, when analysing these models we were unable to simultaneously reproduce the stellar parameters in a HRD for both primary and secondary due to the dependence of rotation on mass loss. We found that only by removing the default implementation

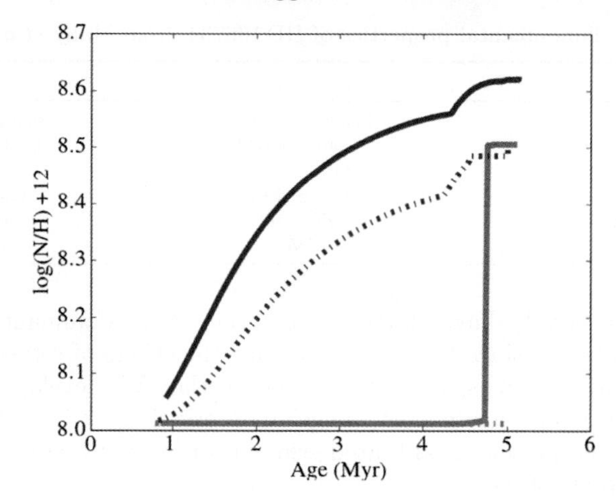

Figure 1. Surface nitrogen abundances as a function of stellar age for varying α_{ov}. The blue lines represent rotating, 40 $M_\odot$ models with an initial rotation rate of 200 km s^{-1}, α_{ov}=0.1 (dash-dotted), and α_{ov}=0.5 (solid). Red lines show the corresponding non-rotating models for the same mass and values of α_{ov} respectively.

of rotationally-induced mass loss could we disentangle each physical process to reach a solution in the HRD.

4. The Mass-Luminosity Plane

Observations of HD166734 suggest high initial masses representative of the observed luminosities, yet present challenges when comparing this to the much lower dynamical masses. We find that an extreme factor of mass-loss rate would be required to reproduce both the observed luminosities and dynamical masses, though in this case factors 2 and above result in a significant drop in luminosity such that the observed luminosity cannot be reached.

We hence developed a method of simultaneously reproducing the mass and luminosity of our observations via a novel tool termed here as the 'mass-luminosity plane' (M-L plane). Figure 2 illustrates an evolutionary model beginning the ZAMS at the red dot and evolving along the vector with time, or temperature as in the HRD. We find that at a given evolutionary stage or observed temperature, we may calculate the length of this vector with respect to initial mass and luminosity setting the initial position, and an observed temperature setting the final position or current evolutionary stage. We thus can use this vector length to constrain physical processes in our theoretical models to reproduce observations provided by the testbed HD166734. We observe that the length (from initial - final positions) can only be *further* extended by rotation or overshooting, while the gradient of the vector relies on the multiplication factor of the mass-loss rate.

We exclude extreme factors of the Vink *et al.* (2001) mass-loss rate since a factor of 0.1 results in an almost vertical vector which cannot reproduce dynamical masses, as well as factors beyond 1.5 times the Vink *et al.* (2001) prescription since this results in a much too shallow gradient to reproduce observed luminosities, leaving an accepted parameter range of 0.5-1.5 times the Vink *et al.* (2001) prescription. When constraining our models in the M-L plane with observations of HD166734 we found that enhanced overshooting was required to reproduce stellar parameters simultaneously (α_{ov}=0.3 $\pm$0.1 and 0.5 $\pm$ 0.1 for the primary and secondary respectively), since initial rotation rates of 250 km s^{-1} and 120 km s^{-1} for the primary and secondary respectively were also calibrated via surface nitrogen enrichments, and extra mixing was still required.

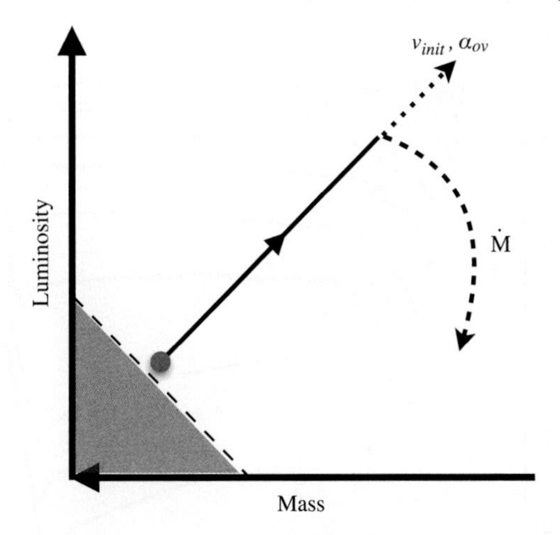

Figure 2. Illustration of the Mass-Luminosity plane with a typical evolutionary track entering the ZAMS at the red dot, evolving along the black arrow. The dotted vector suggest how increased rotation and/or convective overshooting may extend the M-L vector. The curved dashed line represents the gradient at which mass-loss rates effect this M-L vector. The red solid region represents the boundary set by the mass-luminosity relationship, and as such is forbidden.

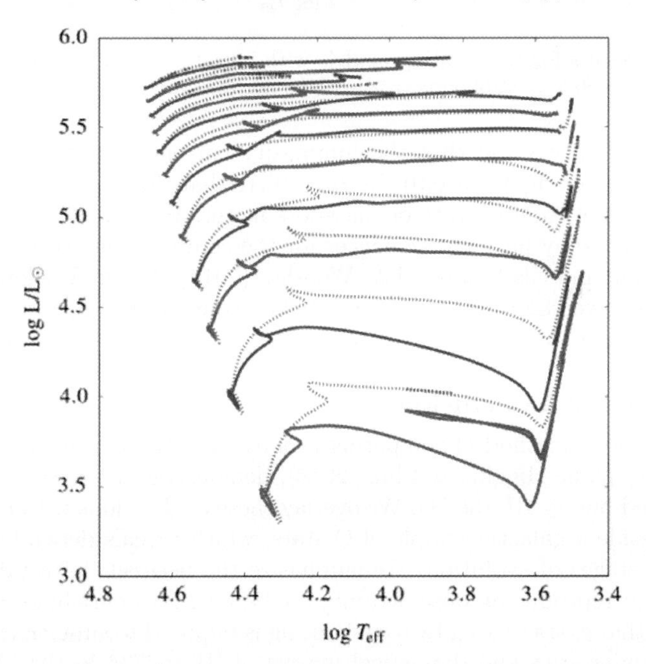

Figure 3. Grid results for the non-rotating models of mass range 8 - 60 M$_\odot$employing $\alpha_{\rm ov}$ = 0.1 (solid lines) and $\alpha_{\rm ov}$ = 0.5 (dotted lines).

5. Grid analysis

We consolidate our results for HD166734 by comparing our constraints to a galactic sample of 30 O stars from Markova *et al.* (2018). We calculated a systematic grid of models for initial masses 8-60 M$_\odot$, initial rotation rates of 0-500 km s^{-1}and for two extreme values of $\alpha_{\rm ov}$ = 0.1 and 0.5, see figures 3 and 4 for example. We identify the main calibrations from this study as extended overshooting to $\alpha_{\rm ov}$ = 0.5, and reduced mass loss with rotation by excluding rotationally-induced mass loss. We note that comparing models

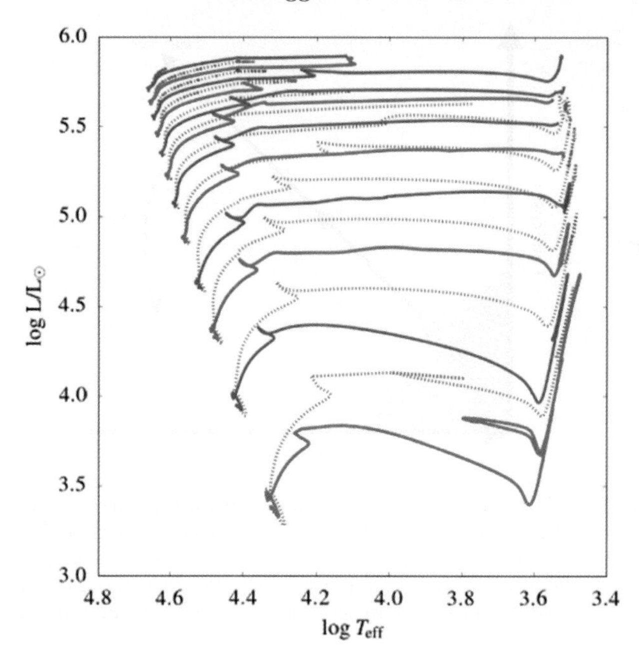

Figure 4. Grid results for the rotating models with initial rotation rates of 300 km s^{-1}for the mass range 8 - 60 M$_\odot$ employing $\alpha_{\rm ov}$= 0.1 (solid lines) and $\alpha_{\rm ov}$= 0.5 (dotted lines).

of $\alpha_{\rm ov}$ = 0.1 and 0.5, we find that the luminosity cut-off for red supergiant evolution is much higher ($\sim$ log L/ L$_\odot$ = 6.0) for $\alpha_{\rm ov}$ = 0.1, than for $\alpha_{\rm ov}$ = 0.5 ($\sim$ log L/ L$_\odot$ = 5.5-5.8), regardless of rotation rate or mass-loss factor. We also note that enlarging the overshooting region may have consequences for the compactness parameter ($\zeta_{2.5}$) in the final phases of our models Sukhbold & Woosley (2014). We find a lower value of $\zeta_{2.5}$ with an enhanced overshooting of $\alpha_{\rm ov}$ = 0.5 (when compared to $\alpha_{\rm ov}$ = 0.1), thus making it easier to create neutron stars and more challenging to evolve to black holes.

6. Discussion and Conclusion

We present a novel method of comparing evolutionary tracks with observations in the Mass-Luminosity plane Higgins & Vink (2018), demonstrated by our constraints of the testbed, detached binary HD166734. We overlay these calibrations with a systematic grid of models alongside a galactic sample of O stars, which reveals dependencies of $\alpha_{\rm ov}$ and $\dot{M}$ on the final stages of evolution. We emphasize the necessity of rotational mixing in stellar models for reproducing observed intermediate surface enrichments.

We conclude that extra mixing by overshooting is required to simultaneously reproduce the observed luminosities and dynamical masses of HD166734 in the M-L plane, while we exclude extreme factors of the mass-loss rate that lie beyond 0.5-1.5 times the Vink *et al.* (2001) rates. Finally, we omit the application of rotationally-induced mass loss since stellar parameters of HD166734 cannot be simultaneously reproduced with this theory.

References

Brott, I., de Mink, S. E., Cantiello, M., Langer, N., de Koter, A., Evans, C. J., Hunter, I., Trundle, C., & Vink, J. S. 2011, *Astronomy & Astrophysics*, 530, A115

Claret, A., & Torres, G. 2017, *Astrophysical Journal*, 849(1), 18

Ekström, S., Georgy, C., Eggenberger, P., Meynet, G., Mowlavi, N., Wyttenbach, A., Granada, A., Decressin, T., Hirschi, R., Frischknecht, U. & Charbonnel, C. 2012, *Astronomy & Astrophysics*, 537, A146

Higgins, E. R., & Vink, J. S. 2018, *Astronomy & Astrophysics*
Maeder, A., & Meynet, G. 2000, *Annual Review of Astronomy & Astrophysics*, 38(1), 143-190
Mahy, L., Damerdji, Y., Gosset, E., Nitschelm, C., Eenens, P., Sana, H. & Klotz, A. 2017, *Astronomy & Astrophysics*, 607, A96
Markova, N., Puls, J., & Langer, N. 2018, *Astronomy & Astrophysics*, 613, A12
Martins, F., & Palacios, A. 2013, *Astronomy & Astrophysics*, 560, A16
Paxton, B., Bildsten, L., Dotter, A., Herwig, F., Lesaffre, P. & Timmes, F. 2011, *Astrophysical Journal Supplement Series*, 192(1), 3
Sukhbold, T. & Woosley, S. E. 2014, *Astrophysical Journal*, 783, 10
Vink, J. S., de Koter, A. & Lamers, H. J. G. L .M. 2001, *Astronomy & Astrophysics*, 369(2), 574-588
Weidner, C., & Vink, J. S. 2010, *Astronomy & Astrophysics*, 524, A98

Summary

High-mass X-ray Binaries: Illuminating the Passage from Massive Binaries to Merging Compact Objects
Proceedings IAU Symposium No. 346, 2019
L. M. Oskinova, E. Bozzo, T. Bulik & D. Gies, eds.

doi:10.1017/S1743921319002369

High mass X-ray binaries: Beacons in a stormy universe

Douglas R. Gies

Center for High Angular Resolution Astronomy, Department of Physics & Astronomy, Georgia State University, P.O. Box 5060, Atlanta, GA 30302-5060, USA
email: gies@chara.gsu.edu

Abstract. The discovery of gravity waves from the mergers of black hole binaries has focused the astronomical community on the high mass X-ray binaries (HMXBs) as the potential progenitors of close pairs of compact stars. This symposium gathered experts in observational and theoretical work for a very timely review of our understanding of the processes that drive the X-ray luminosity of the diverse kinds of binaries and what evolutionary stages are revealed in the observed cases. Here I offer a condensed summary of some of the results about massive star properties, the observational categories of HMXBs, their accretion processes, their numbers in the Milky Way and other galaxies, and how they may be related to the compact binaries that merge in a burst of gravity waves.

Keywords. stars: early-type, stars: emission-line Be, stars: evolution, X-rays: binaries

---

1. HMXBs and Gravitational Waves from Merging Compact Objects

The LIGO detection in 2015 September of gravity waves from the merger of two black holes was a seminal moment in modern astrophysics that marked the first direct measurement of gravity waves and proved the existence of binary black holes. At the time of writing (2018 December), the number of detected black hole mergers has risen to ten, and some trends are already emerging (for example, pre-merger black hole masses do not exceed $45M_{\odot}$ and both components tend to have similar mass; The LIGO Scientific Collaboration & The Virgo Collaboration 2018). The origins of compact merging objects are now a subject of intense study, and the logical starting point is assessing anew the evolutionary stages that lead to the known binaries with black hole components, the high mass X-ray binaries (HMXBs).

This IAU Symposium offered participants a critical appraisal of our understanding of the properties and processes that define the HMXBs and how they may be related to binary black hole mergers. In particular, the presence of a luminous component in HMXB systems means that observational studies across the electromagnetic can help us explore processes that are otherwise hidden from us, and thus HMXBs are beacons in our journey of discovering how black hole mergers may occur. This meeting brought together experts from many communities including observational optical and X-ray astronomy and binary star theory, and we enjoyed a diverse and vibrant exchange that included 50 talks and 109 poster presentations. What follows are my subjective impressions of the highlights from the work presented. In general, I will refer to results by the name of the presenter only (see index), while full citations are given for work presented elsewhere. This summary includes notes on massive star evolution (§2), the observed diversity of HMXBs (§3), accretion processes (§4), numbers of HMXBs (§5), evolutionary paths to black hole mergers (§6), and a few thoughts about future directions (§7). Readers interested in the origins of the field will enjoy reading a brief review by Trimble & Thorne (2018).

2. Massive Star Evolution and Donor Stars

The evolution of massive stars towards core collapse will inevitably create neutron star (NS) and black hole (BH) remnants depending on their intial mass, metallicity, and spin (Heger). However, the numerical modeling of processes leading to a supernova is extraordinarily complex due to the high neutrino flux, convection below the shock front, and gas fall-back, and small changes in the initial conditions can decide whether or not a supernova occurs and the kind of remnant created (Müller *et al.* 2016). There are several lines of evidence that suggest that stars more massive than $20M_{\odot}$ may collapse without any supernova explosion (Smartt 2015; Adams *et al.* 2017) as predicted in some models (Heger *et al.* 2003). Theoretical models suggest that most massive stars $< 10M_{\odot}$ will form NS remnants while those with masses $> 20M_{\odot}$ will make BHs except in cases with high mass loss rates (particularly for high metallicity stars) or very high mass progenitors that are completely disrupted by pair instability supernovae. Thus, we expect that a large fraction of massive stars are destined to create neutron star and black hole remnants with the latter generally favored at higher initial mass and lower metallicity.

Investigations of the donor stars in HMXBs are particularly important because they are the source of gas that powers accretion-driven X-rays and their physical properties help us understand the evolutionary stage of the binary system. There are now computational tools available that model both the atmosphere and winds (such as the PoWR code: Sander, Hainich), and these create synthetic spectra that can be compared to observations to determine effective temperature $T_{\rm eff}$, gravity, abundances, projected rotational velocity, and mass loss rate. The dynamical state of the atmosphere will influence mass loss processes, for example through the action of sub-photospheric convective motions that create structure in the winds (Cantiello *et al.* 2009) and the constructive interference of nonradial pulsation modes that lifts gas out into the circumstellar disks of Be stars (Baade *et al.* 2018). The winds of luminous stars are very dynamic entities that are subject to both large scale (co-rotating interaction regions) and small scale (clumping) instabilities that control temporal variations in the accretion processes (§4). The wind mass loss rates are functions of luminosity, metallicity, rotation, and temperature (Vink), and the accretion properties reflect the diversity of the mass donor winds.

Massive stars such as the progenitors of the HMXBs are often born in dense, small number groups where gravitational encounters may occur and lead to the ejection of stars (Allen, Mapelli). HMXBs may attain a runaway velocity through the instantaneous mass loss of a supernova explosion, as first suggested by Blaauw (1961). However, kinematical studies indicate that most HMXBs have modest peculiar space velocities, and only those with lower mass progenitors have runaway speeds (Fragos, Gvaramadze, Mirabel). This may reinforce the idea that more massive stars collapse without exploding as a supernova (thus yielding a more massive BH). Identifying the specific progenitors of HMXBs is still speculative, but a clue is probably the presence of a He-star companion that was stripped of its hydrogen envelope through binary interaction. A number of massive WR+O binaries in the SMC are probably destined to make massive compact remnants (Shenar), and a growing number of Be stars are found with He-star companions that may be the progenitors of the Be X-ray binaries (Wang *et al.* 2018).

3. HMXB Zoo

The HMXBs form a diverse "zoo" (Reig 2011) that can be classified based upon kind of remnant, evolutionary stage, or observational properties. I will mainly focus on the categories related to the characteristics of the mass donor star, with the warning that

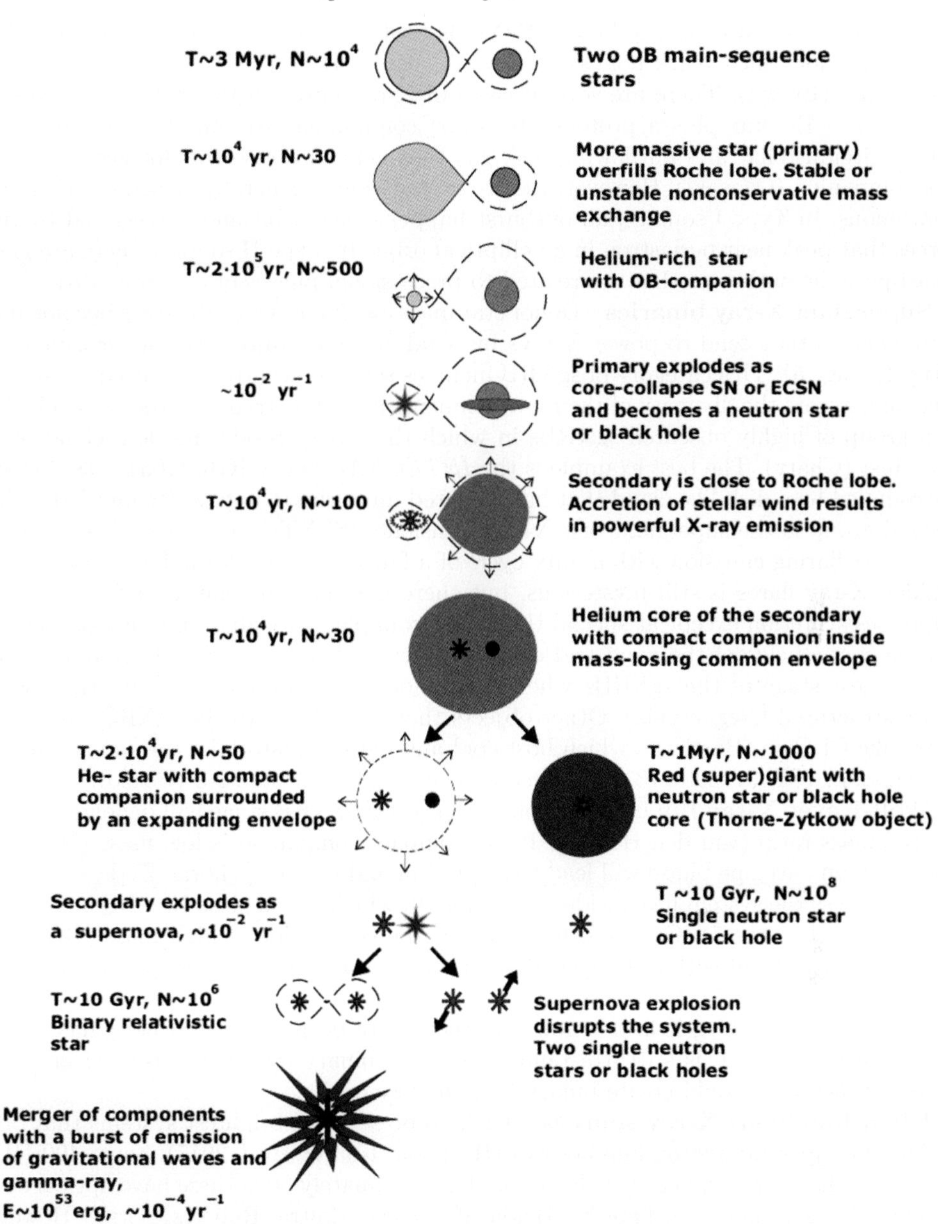

Figure 1. Evolutionary scenario for the formation of NSs or BHs in massive binaries (from Postnov & Yungelson 2014). T is the typical time scale and N is the estimated number of objects in a given evolutionary stage.

these groups are not entirely independent and that intermediate cases are known (Sidoli & Paizis 2018). The order below follows the evolutionary sequence shown in Figure 1†.

Be X-ray binaries – Be stars are rapid rotators that are shedding angular momentum to create transient disks (Rivinius), and a significant fraction of these were probably spun up through past mass transfer in a binary (Pols *et al.* 1991). If the companion is now a compact remnant, then mass transfer and accretion may power a BeXRB system

† Republished with permission of *Living reviews in relativity*, from Postnov & Yungelson (2014); permission conveyed through Copyright Clearance Center, Inc.

(Fig. 1, stage 4). Most of the known BeXRBs host neutron star companions, but there is now one identified black hole system MWC 656 = HD 215227 (Ribo) and probably more exist (Brown). There are several cases of X-ray novae in the SMC that probably consist of a Be star plus a proto-white dwarf companion (Kawai). Be stars eject gas into a circumstellar and outflowing disk gas that acts as a reservoir for accretion onto the companion (Okazaki). However, the X-ray emission is generally episodic rather than continuous. In Type I sources, an outburst happens each orbit and is triggered by tidal forces that peak near periastron in an elliptical orbit. In Type II sources, there are giant, quasi-periodic outbursts that are related to precessional phases of a warped disk.

Supergiant X-ray binaries – Donor star mass loss increases as the stars become more luminous, so they tend to power X-rays by wind accretion during the supergiant phase (Fig. 1, stage 5). This category of sgXRB includes several recently recognized sub-groups that underscore the diversity of their environments and mass transfer processes. The first is a group of highly obscured sgXRBs in which the binary resides inside a cloud of gas and dust (Chaty). The best example is the *INTEGRAL* source IGR J16318–4848 that is surrounded by a disk-like cloud that has a heated inner rim (Chaty & Rahoui 2012). The second group is the supergiant fast X-ray transients (SFXTs), and these display bright and short flaring emission with a duty cycle of a few percent (Sidoli). The cause of these sudden X-ray flares is still mysterious, but there may be some gating mechanism that supresses accretion except at special times, for example, through an interaction between the magnetic fields of the wind and compact star (Hubrig). The SFXTs may represent the earliest stage of the sgXRBs when the donors are more compact and their winds more structured (Negueruela). Other objects that are related to the sgXRBs are sgB[e] stars like CI Cam (Bartlett), which host cool and dense equatorial zones, and symbiotic systems like IGR J17329-2731, which consist of a cool giant and NS (Bozzo).

Wolf-Rayet X-ray binaries – Evolution past the sgXRB stage will depend critically on the mass ratio (van den Heuvel). If the compact companion is low mass (NS), then the common envelope phase will lead to a spiral in and merger (Thorne-Zytkow object = TZO), unless the system has a wide separation, in which case a close NS+NS binary may result (Fig. 1, stage 7). On the other hand, if the companion is high mass (BH), then the spiral-in will end with a stripped He-star and companion in a short period orbit (van den Heuvel *et al.* 2017). Until recently, the only known example of such a He-star binary (or WR-XRB) was Cyg X-3, but now six others have been found in other galaxies, and one of these, CG X-1, is an Ultra-Luminous X-ray binary (Soria; Esposito *et al.* 2015). These WR-XRBs should create binary BHs in close orbits.

Ultra-Luminous X-ray sources – ULXs represent the brightest systems that often radiate at super-Eddington luminosities (Harrison; Kaaret *et al.* 2017). Some 400 ULXs have been discovered in nearby galaxies, and approximately 50 of these have optical counterparts (Anastasopoulou, Fabrika, Heida, Kovlakas, Maitra, Roberts, Soria). However, one relatively nearby ULX was found in outburst in our galaxy, Swift J0243.6+6124, and this is probably a Be star plus NS binary (Wilson-Hodge *et al.* 2018). This system joins five others that have known pulsar companions (Harrison). There is a wide diversity among ULXs in the kinds of donor stars (hot/cool) and compact components (both neutron stars and black holes; Fürst, Carpano, Soria, Heida). ULXs may be the outcome of stable mass transfer at an advanced mass-transfer stage, such as we find in SS 433 (Pavlovskii *et al.* 2017; van den Heuvel *et al.* 2017). One particularly striking environment is the Cartwheel Galaxy, a ring galaxy that experienced a burst in the star formation rate about 100 Myr ago and now hosts some 15 ULXs (Wolter).

Gamma-ray binaries – A number of HMXBs are also emit γ-rays with an orbital-phase modulated amplitude. The emission probably originates through up-scattering of the donor star's photons by relativistic particles or through the interaction of pulsar winds or jets with the winds of the donors (Mirabel 2012). There are perhaps about 100 γ-ray binaries in the Galaxy (Dubus *et al.* 2017). However, the *Fermi* LAT instrument has detected over a thousand additional γ-ray sources that include many more binaries. The known counterparts consist of a diverse assortment of binaries including HMXBs plus pulsars, microquasars (jet sources), novae, colliding wind systems, and Low Mass X-ray Binaries (LMXBs) with a pulsar (Wilson-Hodge, Zhang).

4. Accretion Processes

The processes that control the flow of gas from the donor to the region of X-ray formation near the gainer are complex, and developing a full picture requires modeling physical processes on vastly different spatial scales (Wilms; Negueruela 2010). On scales comparable to the binary separation, gas accretion occurs primarily through Roche lobe overflow (RLOF; dominant among the LMXBs), wind capture or Bondi-Hoyle-Lyttleton accretion (dominant among the the luminous sgXRBs with high mass loss rates), and episodic tidal gas capture (in eccentric orbit BeXRBs). These different accretion regimes are recognized in the Corbet (1986) diagram of ($P_{\rm orbit}$, $P_{\rm spin}$) relating the binary orbital and pulsar spin periods. Those systems with short orbital periods and small spin periods probably experience RLOF leading to a persistent disk around the neutron star. The BeXRBs form a near-linear sequence in the diagram that probably reflects the balance between magnetic spin-down and transient accretion spin-up of the pulsar. The sgXRBs tend to occupy the mid-range orbital period and long spin period part of the diagram in which wind accretion may create only a transient disk close to the neutron star and hence allow limited angular momentum transfer to spin up the pulsar.

All of the mass transfer processes are influenced by the characteristics of the X-ray source. The stellar winds of sgXRBs become ionized in the vicinity of X-ray source, and Doppler-shifted parts of the wind lines will disappear when the over-ionized region is seen in the foreground (the Hatchett-McCray effect). Furthermore, this X-ray ionization will remove the ion-specific absorbers of the stellar flux that drive the wind outwards, so that the wind acceleration ceases. Depending upon the detailed circumstances, the lower than expected wind flow near the compact object will often power increased X-ray luminosity (Krtička). The ionization boundaries may create large-scale photoionization wakes that trail the compact object. Furthermore, the stellar winds are dynamic entities that develop large scale co-rotation interaction regions between outflows of differing speed and that form wind clumps on smaller spatial scales (Martínez-Núñez *et al.* 2017). The clumping in particular will affect the intervening column density to the X-ray source and impart an intrinsic time-variability to the accretion process (Martínez-Núñez, El Mellah, Hainich, Chaty, Grinberg). Calculating the wind mass transfer rate requires both atmospheric and hydrodynamical models of wind flows and radiative transfer codes (El Mellah, Kurfurst, Sander), but sophisticated three-dimensional models now exist that deal with the flows on scales from the orbital, through the accretion zone, and into the vicinity of the compact component. El Mellah *et al.* (2019) show how such models predict that the wind-accreted gas in sgXRBs has sufficient angular momentum to create an accretion disk around the neutron star or black hole.

The net accretion rates onto the neutron star or black hole mass gainer depend critically on the gas flows in their immediate vicinity (Postnov; Shakura 2018). The magnetic fields of neutron stars tend to direct the gas onto the polar regions where they create accretion columns with an anisotropic X-ray flux that causes the observed variations with the spin period (Harrison, Wilms; Lai 2014; Revnivtsev & Mereghetti 2014). The interaction between the neutron star magnetosphere and the surrounding disk will set the accretion rate and X-ray flux that may range from super-Eddington in the case of ULXs (Walton *et al.* 2018) to shutting off accretion by the "magnetic propeller effect" when the magnetic field is very large and/or the accretion rate is low (Torrejón).

Black hole binaries experience X-ray state changes that make a loop in the (hardness, intensity) diagram as they vary between a hard state with emission from a hot corona (when jets appear) and a soft state with emission from an optically thick accretion disk (when the jets disappear; Fender 2016). These states probably correspond to low and high net accretion rates, respectively, into the central regions. Liska *et al.* (2018) present magnetohydrodynamic simulations of thick accretion disks around rapidly spinning black holes, and they show how magnetic dynamos in the disk can launch very energetic jets. They also find that the disk-jet systems undergo precession, and the precessional periods may be related to the observed super-orbital periods (Corbet, Townsend; Larwood 1998).

5. Census of HMXBs

Our position in the disk of the Galaxy imposes limits on our ability to make a complete census of HMXBs in the Milky Way, but there are about 200 known systems at present (Haberl; Liu *et al.* 2006; Walter *et al.* 2015) and most are found close to the star forming complexes in the spiral arms (Coleiro & Chaty 2013). The situation is somewhat better for nearby galaxies: for example, some 150 HMXBs are known in the SMC (Haberl, Zezas, Sell, Fornasini), which experienced several star formation bursts 25 to 60 Myr ago. An important census was recently completed for the nearby spiral galaxy M33 (Garofali). Garofali *et al.* (2018) used surveys from *HST* and Chandra to identify optical counterparts of 55 HMXBs in M33. They examined the colors and magnitudes of the stars in the immediate vicinity of each target to make a color – magnitude diagram and estimate the probable age of the system. They find a double-peaked distribution with peaks at ages < 5 Myr (sgXRBs) and ≈ 40 Myr (BeXRB). Other key surveys of HMXBs are now available for the spiral galaxies M31 (Zezas) and M51 (Lehmer).

These surveys of nearby galaxies are key to the calibration of the relationships between the net X-ray luminosity from LMXBs and HMXBs and a galaxy's star formation rate and mass (Gilfanov). Large scale surveys of distant galaxies in the Chandra Deep Field South (Lehmer *et al.* 2016) and in the Chandra COSMOS-Legacy sample (Fornasini *et al.* 2018) show how these relations may have differed in the past (among high z galaxies). The ratio of X-ray heating by X-ray binaries to that from AGN appears to increase at high redshift, so that X-ray binaries may have played an important role in heating the intergalactic medium early in the history of the Universe (Lehmer, Mirabel).

6. Evolution and Gravitational Wave Sources

The great challenge is to understand what processes dominate in creating the kinds of BH+BH and NS+NS binaries that will lead to mergers like those discovered through gravitational waves. The time scale for orbital shrinkage by gravitational wave emission is proportional to separation to the fourth power, so in order for pairs to merge in a Hubble time, they must be brought into close proximity by other means. Thus, the fundamental problem is determining what processes lead to very close orbits of collapsed remnants (with periods less than 1 day). Three main scenarios offer promising explanations (Mandel): (1) common envelope or other shrinkage during the course of massive binary star evolution, (2) tidally forced rapid rotation in close binaries that leads to mixing and homogeneous evolution, and (3) dynamical encounters in dense stellar environments. All these channels were discussed vigorously at the meeting to explore the specific physical parameters and evolutionary stages that will lead to mergers.

There are many evolutionary paths that lead from an isolated massive binary system to a merger product (NS+NS, BH+NS, BH+BH; Belczynski; Dominik *et al.* 2012). There are two leading scenarios for creating close BH+BH systems. The first (Belczynski *et al.* 2016) begins with a very massive pair of stars in large orbit. The initially more massive star grows to fill its Roche lobe and commences RLOF with non-conservative mass loss from the binary and relatively little change in the orbital dimension. This star will subsequently collapse to form a BH without a supernova explosion or other significant mass loss. Later, the companion evolves to larger size and initiates a common envelope (CE) stage in which the black hole begins to spiral in through the envelope of the companion. In some circumstances, this will conclude with ejection of the envelope and a now much more close binary composed of a stripped He star and BH. Finally, the He star will collapse (again without a SN), yielding a close BH+BH pair. Lower mass stars follow a similar path to create a NS+NS pair, but each collapse is accompanied by a supernova explosion that has the potential to break up the binary through asymmetrical kicks (Chruslinska *et al.* 2018). A second potential scenario (Klencki, van den Heuvel; van den Heuvel *et al.* 2017) starts with a HMXB (BH+OB) with a period of order a week to several months. As long as the mass ratio is not too extreme and the donor star still has a radiative envelope, then RLOF occurs with most of the mass ejected by jets or other outflows (for example, as found in SS 433). This process will lead to a gentle spiral-in that avoids the CE stage, and it will result in a compact binary consisting of a stripped He star and BH (like Cyg X-3). Then, as in the first scenario, the He star collapses to form a BH+BH binary that is close enough to merge over a Hubble time.

The production rates associated with such binary star evolutionary channels depend upon many details of physical processes that are only partially understood. For example, the mass of the remnant depends critically upon mass loss suffered through wind loss (metallicity dependent), binary interactions (systemic mass loss), and supernova explosions (if they occur). The properties of the stars during mass transfer may change on relatively fast timescales, so stellar models using the MESA code (Paxton *et al.* 2015) are now being incorporated in binary evolutionary simulations (Klencki, Marchant). Finally, the energetic processes involved with the common envelope phase are poorly constrained (Fragos, Marchant, Ricker). These processes involve many temporal and spatial scales, and detailed hydrodynamical simulations are required to determine the extent of mass loss and the final outcome of the CE stage, i.e., a very close binary or a stellar merger. Some results indicate a low binary survival rate with a large production of merged Thorne-Zytkow objects (Bulik, Ricker). The CE episode will probably be marked by a short flux outburst (Bulik), and some of these might be observed as transient sources (Kochanek *et al.* 2014; Metzger & Pejcha 2017).

The second means of creating BH+BH binaries is through non-interacting pairs of very close binary stars (Mandel & de Mink 2016). Members of tight binaries will experience tidal interactions that can force rapid rotation that is synchronous with the binary period. Rapid rotation in turn promotes interior mixing that can lead to chemically homogeneous evolution. Instead of building up a He core, the He is mixed throughout until all the H is exhausted and a He star is formed. The star shrinks in the process and avoids binary mass transfer, so that the final BH remnants retain much of the original mass. This process assumes more importance at low metallicity (high z) because at larger metallicity massive stars have strong winds that will carry away angular momentum and cause the stars to spin down.

The third way to make close BH+BH binaries is through dynamical encounters in dense star clusters and other environments (Mapelli; Ziosi *et al.* 2014). Gravitational interactions between single and binary stars often act to eject the lowest mass component, so a close passage will often result in retention in the binary of the most massive component, which may be a BH in a cluster rich with massive stars. Subsequent interactions of a binary with cluster field stars tend to make the binary more compact (Heggie's

Law: gravitational encounters tend to make hard binaries harder and soft binaries softer, where hard and soft refer to the gravitational binding energy relative to the field star kinetic energy). We can see the results of these kinds of processes in the Galactic Center region where we find about a dozen quiescent BH binaries within 1 pc of Sgr A* (Hailey *et al.* 2018). These probably formed by tidal capture of companions (becoming orbitally bound by transforming kinetic energy into stellar oscillations) through encounters of field stars with a large pool of BHs (10^4) that have accumulated near Galactic Center (Generozov *et al.* 2018).

Presumably all these processes occur in the Universe, and in order to determine their relative significance we need to perform large scale population synthesis models that calculate the numbers of merger systems as a function of star formation rate and metallicity over the history of the cosmos (Belczynski, Chruslinska, Mapelli). These are ambitious and complex codes that must make numerous assumptions about the details of binary star properties and evolution, stellar dynamics, star formation, and galaxy evolution. They include simulations such as BPASS (Eldridge *et al.* 2017), COMBINE (Kruckow *et al.* 2018), COMPAS (Vigna-Gómez *et al.* 2018), MOBSE (Mapelli & Giacobbo 2018), and STARTRACK (Chruslinska *et al.* 2018), among others. The details of the simulations are important, because the vast majority of stellar systems never make it to become gravitational wave sources (Belczynski, Bulik), so we are studying the results of a restricted set of merger channels.

The predicted merger rate of neutron star pairs is especially interesting after the seminal discovery of gravity waves from the NS+NS merger of GW170817. This merger was also observed as a short γ-ray burster and a kilonova, verifying that such bursters are the result of NS mergers (Wilson-Hodge, Hakkila, Meszaros). With only a single NS+NS merger observation thus far, it is too early to compare predicted and observed merger rates, but the model predictions appear to be consistent with known population of radio-detected, double neutron stars (Tauris *et al.* 2017; Chruslinska *et al.* 2018; Vigna-Gómez *et al.* 2018).

There are many avenues available to create compact BH+BH binaries, and at present it is difficult gauge which processes dominate. Most models that rely on binary evolution and orbital shrinkage through a CE phase appear to produce BH+BH merger rates that are consistent with the initial LIGO estimates (Belczynski; Kruckow *et al.* 2018; Mapelli & Giacobbo 2018) even without contributions from the homogeneous evolution and dynamical capture processes. However, this is still a very young field with large uncertainties on both the observational and theoretical sides, so it is premature to assume which if any of these three processes is the prevalent one. However, future results from the gravity wave detectors on BH masses, spins, merger rates, and their metallicity and redshift dependence should provide the means to begin to discriminate between the relative contributions of the different processes (Belczynski, Fragos, Qin; Arca Sedda & Benacquista 2019).

7. Future promise

Since their discovery some 50 years ago, the study of HMXBs has grown in scope and depth in amazing ways as demonstrated by the work presented at this meeting. This growth will accelerate in the future through new opportunities in observational work and the expansion of computational facilities. The current gravitational wave experiments (LIGO, VIRGO) and those under construction (KAGRA, IndIGO, TianQin) will measure the mergers of several hundred compact objects, and this will provide the statistical basis to test theories of the origins of BH+BH, NS+NS, and also BH+NS systems (the latter may be observed as kilonovae; Gompertz *et al.* 2018). The legacy X-ray missions (XMM Newton, Integral, Chandra, Fermi, NICER, NuSTAR) and those ahead (Insight-HXMT

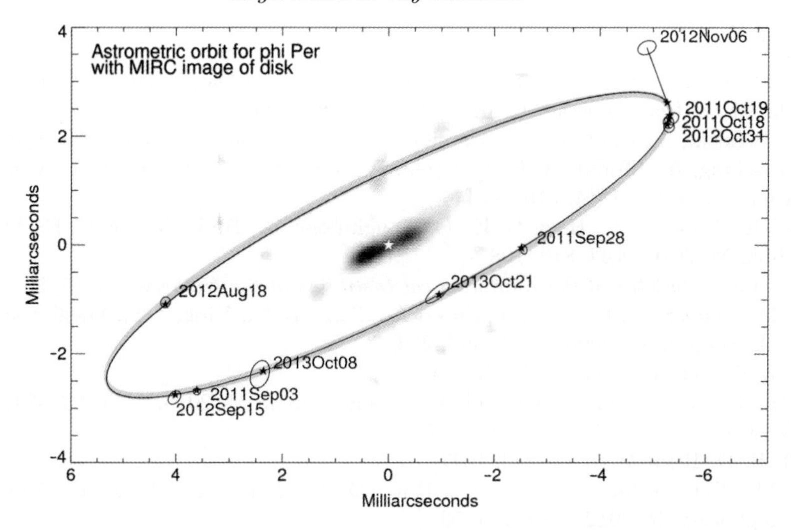

Figure 2. The Be star ϕ Per with the orbit of its He star companion.

[Zhang], eRosita, Athena, Lynx, XRISM) hold the promise of completing our picture of the populations of HMXBs in our galaxy and others. Likewise, the next generation of ground-based giant telescopes (ELT, GMT, TMT) and survey telescopes (LSST) will help characterize the mass donor stars of HMXBs.

Advances in high angular resolution work through optical and radio long baseline interferometry will be particularly striking in the near future to help us probe these binaries and their mass accretion and ejection processes. We learned at this meeting how optical long baseline interferometry with VLTI/Gravity has provided the means to explore the inner structure of the microquasars SS 433 and BP Cru (Waisberg), and the power of radio interferometry was demonstrated for a pulsar orbit measured with the Australian Long Baseline Array (Miller-Jones). One other remarkable example was the detection of the hot He star companion orbiting the Be star ϕ Persei (Fig. 2; made with the Georgia State University CHARA Array; Mourard *et al.* 2015). Objects like ϕ Per may be the precursors of BeXRBs and NS+NS mergers, so their study offers us an important opportunity to learn about this early stage of evolution. The gas disk of ϕ Per orbits in the same sense as the He star's orbit, consistent with the idea that the fast spin of the Be star was caused by a past mass transfer stage, and the age and luminosity of the He star support the idea that it has advanced into a bright, He-shell burning stage (Schootemeijer *et al.* 2018). Such work at the limits of high angular resolution will reveal the processes that forge the evolution of HMXBs.

Acknowledgments

I am deeply grateful to the Chair of IAU Symposium 346, Dr. Lidia Oskinova, and to the members of the SOC for their efforts that made this meet so successful. I thank the participants at the meeting for sharing their insights, and I extend our thanks to members of the volunteer team headed by Dr. Ines Brott (Christine Ackerl, Karan Dsilva, Huanchen Hu, and Evgeniya Nikolaeva).

The Georgia State University Center for High Angular Resolution Astronomy Array is supported by the U.S. National Science Foundation under Grant No. AST-1636624 and AST-1715788.

References

Adams, S. M., Kochanek, C. S., Gerke, J. R., Stanek, K. Z., & Dai, X. 2017, *MNRAS*, 468, 4968
Arca Sedda, M., & Benacquista, M. 2019, *MNRAS*, 482, 2991
Baade, D., Pigulski, A., Rivinius, Th., Carciofi, A. C., Panoglou, D., Ghoreyshi, M. R., Handler, G., Kuschnig, R., Moffat, A. F. J., Pablo, H., Popowicz, A., Wade, G. A., Weiss, W. W., & Zwintz, K. 2018, *A&A*, 610, A70
Belczynski, K., Repetto, S., Holz, D. E., O'Shaughnessy, R., Bulik, T., Berti, E., Fryer, C., & Dominik, M. 2016, *ApJ*, 819, 108
Blaauw, A. 1961, *Bulletin of the Astronomical Institutes of the Netherlands*, 15, 265
Cantiello, M., Langer, N., Brott, I., de Koter, A., Shore, S. N., Vink, J. S., Voegler, A., Lennon, D. J., & Yoon, S.-C. 2009, *A&A*, 499, 279
Chaty, S., & Rahoui, F. 2012, *ApJ*, 751, 150
Chruslinska, M., Belczynski, K., Klencki, J., & Benacquista, M. 2018, *MNRAS*, 474, 2937
Coleiro, A., & Chaty, S. 2013, *ApJ*, 764, 185
Corbet, R. H. D. 1986, *MNRAS*, 220, 1047
Dominik, M., Belczynski, K., Fryer, C., Holz, D. E., Berti, E., Bulik, T., Mandel, I., & O'Shaughnessy, R. 2012, *ApJ*, 759, 52
Dubus, G., Guillard, N., Petrucci, P.-O., & Martin, P. 2017 *A&A*, 608, A59
El Mellah, I., Sander, A. A. C., Sundqvist, J. O., & Keppens, R. 2019, *A&A*, 622, A189
Eldridge, J. J., Stanway, E. R., Xiao, L., McClelland, L. A. S., Taylor, G., Ng, M., Greis, S. M. L., & Bray, J. C. 2017, *PASA*, 34, 58
Esposito, P., Israel, G. L., Milisavljevic, D., Mapelli, M., Zampieri, L., Sidoli, L., Fabbiano, G., & Rodríguez Castillo, G. A. 2015, *MNRAS*, 452, 1112
Fender, R. 2016, *AN*, 337, 381
Fornasini, F. M., Civano, F., Fabbiano, G., Elvis, M., Marchesi, S., Miyaji, T., & Zezas, A. 2018, *ApJ*, 865, 43
Garofali, K., Williams, B. F., Hillis, T., Gilbert, K. M., Dolphin, A. E., Eracleous, M., & Binder, B. 2018, *MNRAS*, 479, 3526
Generozov, A., Stone, N. C., Metzger, B. D., & Ostriker, J. P. 2018, *MNRAS*, 478, 4030
Gompertz, B. P., Levan, A. J., Tanvir, N. R., Hjorth, J., Covino, S., Evans, P. A., Fruchter, A. S., González-Fernández, C., Jin, Z. P., Lyman, J. D., Oates, S. R., O'Brien, P. T., & Wiersema, K. 2018, *ApJ*, 860, 62
Hailey, C. J., Mori, K., Bauer, F. E., Berkowitz, M. E., Hong, J., & Hord, B. J. 2018, *Nature*, 556, 70
Heger, A., Fryer, C. L., Woosley, S. E., Langer, N., & Hartmann, D. H. 2003, *ApJ*, 591, 288
Kaaret, P., Feng, H., & Roberts, T. P. 2017, *ARA&A*, 55, 303
Kochanek, C. S., Adams, S. M., & Belczynski, K. 2014. *MNRAS*, 443, 1319
Kruckow, M. U., Tauris, T. M., Langer, N., Kramer, M., & Izzard, R. G. 2018, *MNRAS*, 481, 1908
Lai, D. 2014, in: E. Bozzo, P. Kretschmar, M. Audard, M. Falanga, & C. Ferrigno (eds.), *Physics at the Magnetospheric Boundary*, EPJ Web of Conferences, 64, 01001
Larwood, J. 1998, *MNRAS*, 299, L32
Lehmer, B. D., Basu-Zych, A. R., Mineo, S., Brandt, W. N., Eufrasio, R. T., Fragos, T., Hornschemeier, A. E., Luo, B., Xue, Y. Q., Bauer, F. E., Gilfanov, M., Ranalli, P., Schneider, D. P., Shemmer, O., Tozzi, P., Trump, J. R., Vignali, C., Wang, J.-X., Yukita, M., & Zezas, A. 2016, *ApJ*, 825, 7
The LIGO Scientific Collaboration & The Virgo Collaboration 2018, submitted (arXiv:1811.12940)
Liska, M., Hesp, C., Tchekhovskoy, A., Ingram, A., van der Klis, M., & Markoff, S. 2018, *MNRAS*, 474, L81
Liu, Q. Z., van Paradijs, J., & van den Heuvel, E. P. J. 2006, *A&AS*, 455, 1165
Mandel, I., & de Mink, S. E. 2016, *MNRAS*, 458, 2634
Mapelli, M., & Giacobbo, N. 2018, *MNRAS*, 479, 4391

Martínez-Núñez, S., Kretschmar, P., Bozzo, E., Oskinova, L. M., Puls, J., Sidoli, L., Sundqvist, J. O., Blay, P., Falanga, M., Fürst, F., Gímenez-García, A., Kreykenbohm, I., Kühnel, M., Sander, A., Torrejón, J. M., & Wilms, J. 2017, *Space Sci. Revs*, 212, 59

Metzger, B. D., & Pejcha, O. 2017, *MNRAS*, 471, 3200

Mirabel, I. F. 2012, *Science*, 335, 175

Mourard, D., Monnier, J. D., Meilland, A., Gies, D., Millour, F., Benisty, M., Che, X., Grundstrom, E. D., Ligi, R., Schaefer, G., Baron, F., Kraus, S., Zhao, M., Pedretti, E., Berio, P., Clausse, J. M., Nardetto, N., Perraut, K., Spang, A., Stee, P., Tallon-Bosc, I., McAlister, H., ten Brummelaar, T., Ridgway, S. T., Sturmann, J., Sturmann, L., Turner, N., & Farrington, C. 2015, *A&A*, 577, A51

Müller, B., Heger, A., Liptai, D., & Cameron, J. B. 2016, *MNRAS*, 460, 742

Negueruela, I. 2010, in: J. Martí, P. L. Luque-Escamilla, & J. A. Combi (eds.), *High Energy Phenomena in Massive Stars*, ASP Conf. Vol. 422 (San Francisco: ASP), p. 57

Pavlovskii, K., Ivanova, N., Belczynski, K., & Van, K. X. 2017, *MNRAS*, 465, 2092

Paxton, B., Marchant, P., Schwab, J., Bauer, E. B., Bildsten, L., Cantiello, M., Dessart, L., Farmer, R., Hu, H., Langer, N., Townsend, R. H. D., Townsley, D. M., & Timmes, F. X. 2015, *ApJS*, 220, 15

Pols, O. R., Cote, J., Waters, L. B. F. M., & Heise, J. 1991, *A&A*, 241, 419

Postnov, K. A., & Yungelson, L. R. 2014, *Living Rev. Relativ.*, 17, 3

Reig, P. 2011, *Ap&SS*, 332, 1

Revnivtsev, M., & Mereghetti, S. 2015, *Space Science Reviews*, 191, 293

Schootemeijer, A., Götberg, Y., de Mink, S. E., Gies, D., & Zapartas, E. 2018, *A&A*, 615, A30

Shakura, N. (ed.) 2018, *Accretion Flows in Astrophysics*, ASSL 454 (Heidelberg: Springer International Publishing)

Sidoli, L., & Paizis, A. 2018, *MNRAS*, 481, 2779

Smartt, S. J. 2015, *PASA*, 32, 16

Tauris, T. M., Kramer, M., Freire, P. C. C., Wex, N., Janka, H.-T., Langer, N., Podsiadlowski, Ph., Bozzo, E., Chaty, S., Kruckow, M. U., van den Heuvel, E. P. J., Antoniadis, J., Breton, R. P., & Champion, D. J. 2017, *ApJ*, 846, 170

Trimble, V., & Thorne, K. S. 2018, submitted (arXiv:1811.04310)

van den Heuvel, E. P. J., Portegies Zwart, S. F., & de Mink, S. E. 2017, *MNRAS*, 471, 4256

Vigna-Gómez, A., Neijssel, C. J., Stevenson, S., Barrett, J. W., Belczynski, K., Justham, S., de Mink, S. E., Müller, B., Podsiadlowski, P., Renzo, M., Szécsi, D., & Mandel, I. 2018, *MNRAS*, 481, 4009

Walter, R., Lutovinov, A. A., Bozzo, E., & Tsygankov, S. S. 2015, *A&AR*, 23, 2

Walton, D. J., Fürst, F., Heida, M., Harrison, F. A., Barret, D., Stern, D., Bachetti, M., Brightman, M., Fabian, A. C., & Middleton, M. J. 2018, *ApJ*, 856, 128

Wang, L., Gies, D. R., & Peters, G. J. 2018, *ApJ*, 853, 156

Wilson-Hodge, C. A., Malacaria, C., Jenke, P. A., Jaisawal, G. K., Kerr, M., Wolff, M. T., Arzoumanian, Z., Chakrabarty, D., Doty, J. P., Gendreau, K. C., Guillot, S., Ho, W. C. G., LaMarr, B., Markwardt, C. B., Özel, F., Prigozhin, G. Y., Ray, P. S., Ramos-Lerate, M., Remillard, R. A., Strohmayer, T. E., Vezie, M. L., Wood, K. S., & NICER Science Team 2018, *ApJ*, 863, 9

Ziosi, B. M., Mapelli, M., Branchesi, M., & Tormen, G. 2014, *MNRAS*, 441, 3703

Author index